Ice-Structure Interaction

International Union of Theoretical
and Applied Mechanics

International Association
for Hydraulic Research

S. J. Jones, R. F. McKenna, J. Tillotson,
I. J. Jordaan (Eds.)

Ice-Structure Interaction

IUTAM/IAHR Symposium
St. John's, Newfoundland Canada 1989

Springer-Verlag
Berlin Heidelberg New York
London Paris Tokyo
Hong Kong Barcelona Budapest

Dr. Stephen Jones

Joy Tillotson

National Research Council
Institute for Marine Dynamics
P. O. Box 12093
Station A, St. John's
NF A1B 3T5
Canada

Dr. Richard F. McKenna

Dr. Ian J. Jordaan

Ocean Engineering Research Centre
Faculty of Engineering and Applied Science
Memorial University of Newfoundland
St. John's
NF A1B 3X5
Canada

ISBN-13:978-3-642-84102-6 e-ISBN-13:978-3-642-84100-2
DOI: 10.1007/978-3-642-84100-2

Softcover reprint of the hardcover 1st edition 1991

Typesetting: Camera ready by authors.

61/3020 543210 – Printed on acid-free paper

GENERAL INFORMATION

Scientific Committee:

D.B. Muggeridge, Canada (Chairman)
I.J. Jordaan, Canada (Co-Chairman)
T. Carstens, Norway
F.T. Christensen, Denmark
K.R. Croasdale, Canada
D. Drucker, USA (ex-officio)
P. Duval, France
G. Frankenstein, USA
R. Frederking, Canada
Y. Fujita, Japan
S.S. Grigoryan, USSR
M. Määttänen, Finland
O. Mahrenholtz, FRG
B. Michel, Canada
R.L. Nordgren, USA
G.W. Timco, Canada

Sponsor:

International Union of Theoretical and Applied Mechanics (IUTAM)

Co-sponsors:

International Association for Hydraulic Research (IAHR)
Memorial University of Newfoundland, Faculty of Engineering and Applied Science
Centre for Cold Ocean Resources Engineering (C-CORE)
National Research Council Canada, Institute for Marine Dynamics
Mobil Oil Canada Properties

Local Organizing Committee:

I.J. Jordaan, Memorial University, Chairman
D.B. Muggeridge, Memorial University, Co-Chairman
D. Duthinh, C-CORE/Memorial University
S.J. Jones, National Research Council of Canada
R.F. McKenna, Memorial University
A.S.J. Swamidas, Memorial University
J.A. Whittick, C-CORE/Memorial University

Financial Support:

Financial support of the Symposium was generously provided by the International Union of Theoretical and Applied Mechanics (IUTAM), the National Research Council of Canada (NRC), the Natural Sciences and Engineering Research Council (NSERC) and Mobil Oil Canada Properties.

PREFACE

IUTAM-IAHR Symposium on Ice-Structure Interaction

Professor Bez Tabarrok, Chairman of the Canadian National Committee (CNC) of the International Union of Theoretical and Applied Mechanics (IUTAM) invited Professor Derek Muggeridge to organize a symposium on ice structure interaction. Dr. Muggeridge readily agreed and prepared a proposal that was endorsed by the CNC and presented to the General Assembly Meeting of IUTAM for their consideration. This Assembly gave its approval and provided the local organizing committee with the names of individuals who were willing to serve on the Scientific Committee. Dr. Muggeridge became chairman of this committee and Dr. Ian Jordaan became co-chairman of this committee as well as chairman of the local organizing committee.

The symposium followed the very successful previous meeting, chaired by Professor P. Tryde in Copenhagen, by ten years. Both symposia utilized Springer-Verlag to publish their proceedings. The Faculty of Engineering and Applied Science at Memorial University of Newfoundland were particularly pleased to host this prestigious symposium as it marked the twentieth anniversary of its Ocean Engineering Research Centre.

A co-sponsor was found for the symposium, in the form of the International Association for Hydraulics Research, and we are grateful for their help and advice. Later, other organizations were approached:
- National Research Council of Canada
- Mobil Oil Canada Properties
- Centre for Cold Oceans Resources Engineering (C-CORE)

and we were delighted to receive support from them, both financial and technical. Thus was born the "IUTAM-IAHR Symposium on Ice-structure Interaction" held at Memorial University of Newfoundland, St. John's, Newfoundland, Canada, from 14-17 August, 1989.

In accordance with the rules of the IUTAM, all the speakers and participants were invited. The programme consisted of 33 papers which are included in these proceedings.

The timing of the symposium was very appropriate. The subject matter had previously been of interest to offshore engineers for many years, but within months of this symposium two monographs on the subject were published. These books, together with these proceedings, clearly represent the state of the subject in 1990.

Many people worked hard to make the symposium the success that it was. In addition to the people on the committees listed, many students and staff from Memorial University, and from the Institute for Marine Dynamics, contributed time and effort. We have gone to some lengths to prepare a volume of good quality, and attractive appearance, and this has involved considerable retyping. Mary Brown and Deborah Quinlan have borne the brunt of this and we are very grateful to them, to Barbara Rodden for proofreading the final manuscripts, and to Barry Stone for assembling the final version. We also appreciate greatly Hanny M. Muggeridge's artistic skills in designing our logo. Finally, we thank the publishers for their co-operation and patience in this endeavour.

Stephen J. Jones

St. John's, Newfoundland
November 1990

CONTENTS

General Information v

Preface vii

SESSION I: CREEP AND VISCOELASTICITY

Creep and Rupture in Relation to Field Experiments 3
A.R.S. Ponter and A.C.F. Cocks

The Use of Tertiary Creep Rates in Ice at High Strains in Compression and Shear 21
T.H. Jacka and W.F. Budd

Creep and Settlement of Ice Structures - A Case History 37
D.M. Masterson and T. Smith

Creep Constitutive Equations for Polycrystalline Ice and Effect of Microcracking 55
P. Duval, P. Kalifa and J. Meyssonnier

Kinetics of Microcracking and Dilatation in Polycrystalline Ice 69
N.K. Sinha

Chairman's Remarks 89

SESSION II: FRACTURE

Fracture Mechanics Models of Ice-Structure Interaction 93
A.C. Palmer

The Fracture Toughness of Ice 109
J.P. Dempsey

The Study of Fracture Toughness of Bohai Sea Ice 147
W. Shen, S.Z. Lin, P. Gu, X.A. Zhou

The Tensile and Compressive Fracture of Ice 165
E.M. Schulson

SESSION III: DAMAGE AND COMPRESSIVE FRACTURE

Models for the Deformation Behavior of Viscoelastic Media with Distributed Damage and Their Applicability to Ice 191
R.A. Schapery

Microfracture and The Compressive Failure of Polycrystalline Ice 231
D.M. Cole

Some Questions on Ice and Ice Cover Fracture in Compression 251
R.V. Goldstein and N.M. Osipenko

Seasonal Compressive Strength of Beaufort Sea Ice Sheets 267
G.W. Timco and R.M.W. Frederking

Processes of Deformation and Fracture of Ice in Compression 283
I.J. Jordaan and R.F. McKenna

A Plasticity Model for the Crushing of Ice 311
J.F. Dorris

Strength and Failure Modes of Pure Ice and Multi-Year Sea Ice Under Triaxial Loading 339
S.A.F. Murrell, P.R. Sammonds and M.A. Rist

Sea Ice Under Complex Stress States: Constitutive Modelling and Testing 363
F.U. Häusler, H.G. Matthies and C.S. Moore

SESSION IV: FRICTION AND BROKEN ICE MASSES

Mechanisms of Ice Friction 385
J. Mølgaard

Crushing Friction Experiments on Freshwater Ice 405
R.E. Gagnon and J. Mølgaard

Do Cracks Reduce Thermal Ice Stresses? 423
L.Å. Fransson

SESSION V: DESIGN METHODOLOGIES

Statistical Analysis of Ice Forces 439
T.J.O. Sanderson

Probabilistic Determination of Iceberg Collision Design Loads for Floating Production Vessels 459
M.K. Fuglem, D. Duthinh, J.H. Lever and I.J. Jordaan

The Selection of Design Ice Loads and Design Ice Features For Fixed Offshore Structures in the Beaufort Sea 483
M. Nessim and I.J. Jordaan

Structural Risk Model of Arctic Shipping 507
C.G. Daley, C. Ferregut and R. Brown

Probabilistic Ice Forces on Offshore Structures 541
D.E. Nevel

Chairman's Remarks 559

SESSION VI: ICE-STRUCTURE INTERACTION

Ice Interaction with Structures 563
M. P. Määttänen

Adjusting Local Stiffness for Improving Dynamic Ice-Structure Interaction of Bohai Jackets 581
X. Jizu and W. Lingyu

Theoretical Modelling of Ice-Structure Interaction 595
K. Riska

Ice-Structure Interaction During Indentation Tests 619
D.S. Sodhi

Structure Geometry and Ice Interaction 641
K.R. Croasdale and M. Metge

Dynamics of Continuous Icebreaking by an Icebreaker Hull 669
R. Ettema and F. Stern

Impact Ice Pressure: More Questions than Answers 693
A.L. Tunik

A Method of Scaling for Ship-Ice Model Tests 715
D.B. Colbourne and J.H. Lever

Chairman's Remarks 731

Contribution to General Discussions 735

SESSION I:

Creep and Viscoelasticity

Creep and Rupture in Relation to Field Experiments

A.R.S. Ponter and A.C.F. Cocks

Department of Engineering, University of Leicester,
Leicester, U.K. LE1 7RH

Abstract The paper discusses the micromechanical deformation and fracture processes in ice under sustained loading and their relationship with observed macromechanical processes which define the forces which moving ice can exert on a stationary structure. Using reference stress concepts, originally developed for metallic structures operating at high temperatures, it is possible to directly relate the behaviour of an ice sheet or ice mass with uniaxial compression tests. The theory is necessarily approximate, but allows a simplified overview of the behaviour in terms of standard plasticity solutions.

1. Introduction

If an ice mass or ice sheet is driven at a known velocity against a relatively rigid structure, what forces are developed between them? Knowing the micromechanical processes which occur in ice, how is it possible to predict these forces using classical plasticity calculations? Attempts to answer these two questions have formed a significant part of research in ice mechanics, and it seems appropriate, in the first paper of the symposium, to provide an overview in terms of the authors' background in creep mechanics and creep design. We find that the concept of reference stress originally derived by Marriott and Leckie (1964) and Anderson et al. (1963) to overcome the problems of material variability in high temperature metallic creep, has relevance to ice mechanics when it forms part of an overall view of the mechanisms of deformation and fracture in ice.

The main purpose of the paper is to demonstrate ways in which uniaxial ice data may be related to the force induced by an indentor in the regime where maximum loads occur, i.e. the transition from pure creep behaviour to circumstances where general microcracking can locally reduce the creep compliance of the ice. Our approach may not be appropri-

S. Jones · R. F. McKenna · J. Tillotson · I. Jordaan (Eds.)
Ice-Structure Interaction
IUTAM-IAHR Symposium St. John's, Newfoundland, Canada

ate for situations where the strain rates are very low ($< 10^{-7}\text{sec}^{-1}$) where recrystallisation can become a dominant mechanism for tertiary creep. Neither are we concerned with very high strain rates ($> 10^{-2}\text{sec}^{-1}$) when propagation of single or multiple macrocracks defines the force level. Possible interactions between micro- and macrocracking have been discussed by Cocks (1989a) and Bhat (1988).

2. General Considerations

Ice crystals have a hexagonal close-packed structure. Like other crystalline materials ice creeps freely under stress close to its melting point. The elastic response of ice single crystals is nearly isotropic, but its creep properties are highly anisotropic, with the resistance to slip on the non-basal slip planes being at least 60 times greater than that on the basal plane (Ashby and Duval, 1985). Steady-state creep of polycrystalline and S2 ice requires slip to occur at matching rates on the basal and non-basal planes. This is achieved through the development of a residual stress field during primary creep as stress is transferred from the basal to the non-basal planes. In S2 ice, where the c-axes of the crystals all lie in the plane of the ice sheet, creep along the axis of the grains can only result from slip on the non-basal planes. As a result S2 ice is more creep resistant through the sheet thickness, requiring about 15 times the stress to produce the same creep rate as in the plane of the sheet (see, for example, Sanderson, 1989).

At the microlevel the residual stress field that develops during primary creep results from the formation of dislocation pile-ups where the easy slip planes meet grain boundaries, producing a significant back stress along the basal planes. If the applied stress is removed during a deformation process, significant strain recovery occurs due to the relaxation of these dislocation pile-ups. The energy stored within a pile-up can provide the driving force for the nucleation of a crack in an adjacent grain or along a grain boundary. Since this energy scales as the length of the pile-ups, which is generally of the order of the grain-size, the length of the nucleated crack also scales as the grain-size. Also the pile-ups form and develop in response to the deformation of the component. As a result nucleation of microcracks is deformation controlled.

If the macro-stress is tensile then a crack can propagate unstably once the stress has reached a critical value. For fine grained ice the stress generally needs to be increased beyond that at which cracks first form to produce failure. Above a critical grain size, which is a function of strain-rate and temperature, spontaneous propagation of a crack occurs immediately upon nucleation (Hallam, 1986). Ice has a very low fracture toughness of about 100 $kNm^{-3/2}$ and this leads to a low tensile strength, in the region of 1MPa.

Under an applied compressive stress the microcracking is much more extensive. The cracks do not propagate further after nucleating and remain open, assisted by grain-boundary sliding and basal slip (see Figure 3 of Sinha, 1989) providing a coherent structure with a reduced creep resistance. Once steady state creep is reached there is no further change in the residual stress field and, as a result, no additional microcracks are formed.

Any deformation field will, potentially, contain features of all these processes. For simplicity, we may subdivide a deformation field into four separate regions, each depending upon the local stress state:

1. Pure creep - the stresses are insufficient to cause nucleation of microcracks, but the ice creeps.

2. Stable fracturing - nucleation of microcracks can occur as the material creeps, but the stress state is such that unstable crack growth does not take place.

3. Fully fractured - either the density of microcracking is such that the region can carry little or no stress or a crack has propagated.

4. Low stress - a fully fractured region has separated this region from the boundary loading.

In uniaxial compression tests the behaviour is described by 1) and 2) in the creep and transitional regions and by 3) in the brittle region as shown in Figure 1.

As an ice mass deforms under indentation loading all four regions will normally develop, although their exact sequence of occurrence may vary between nominally identical events, as fracture is inherently sensitive to the precise details of geometry and loading. High stress concentrations will give rise to regions of types 3), tensile stress will cause crack propagation until, finally, a residual volume consisting of regions 1) and 2) will remain as, essentially, a reduced geometry, which resists the movement of the ice sheet and yields the largest forces on the structure. The

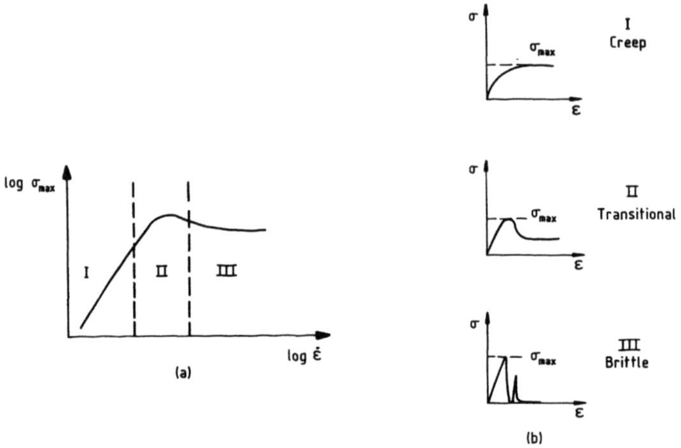

Figure 1. (a) Typical maximum stress/strain-rate plot for ice loaded in compression showing the creep, transitional and brittle regimes, which are typified by the plots of (b).

relationship between load P and indentation velocity v, Figure 2, has a similar form to Figure 1, but the boundaries between the regions are less well defined. The aim of this paper is to define, for a given geometry, the relationship between Figures 1 and 2.

In our analyses of this process we first consider an ice mass consisting entirely of region 1). We then discuss the influence of microcracking.

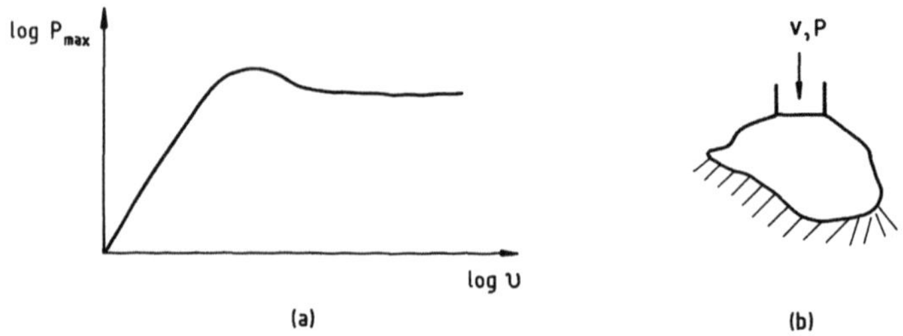

Figure 2. Variation of maximum load with indentation velocity when an ice mass is indented by a rigid structure.

3. Creep Properties of Ice

There have been many studies of the creep properties of ice under uniaxial tension and compression. Complete constitutive equations have been proposed by Sinha (1983), Ashby and Duval (1985), Cocks (1989b) and Jordaan and McKenna (1988). For the prediction of the maximum load it is sufficient to model the steady state behaviour. The uniaxial strain-rate $\dot{\varepsilon}$ is then given by Glen's law:

$$\frac{\dot{\varepsilon}}{\dot{\varepsilon}_0} = \left(\frac{\sigma}{\sigma_0}\right)^n , \tag{1}$$

where $\dot{\varepsilon}_0$ is the strain-rate at a stress σ_0 and the creep index $n = 3$ for ice. For a triaxial state of stress the relationship may be generalised to

$$\dot{\varepsilon}_{ij} = \frac{\partial}{\partial \sigma_{ij}} \left(\frac{\dot{D}^c}{n+1}\right) , \tag{2}$$

where $\dot{D}^c(\sigma_{ij})$ is the rate of energy dissipation

$$\dot{D}^c = \sigma_{ij}\dot{\varepsilon}_{ij} = \sigma_0\dot{\varepsilon}_0 \left(\frac{\bar{\sigma}}{\sigma_0}\right)^{n+1} . \tag{3}$$

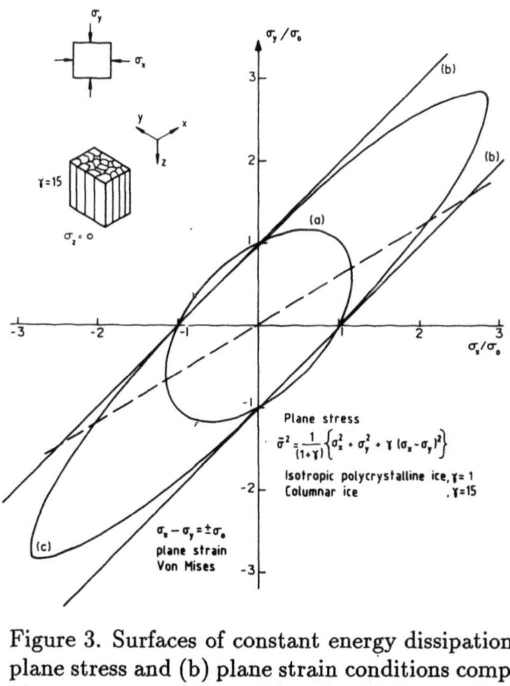

Figure 3. Surfaces of constant energy dissipation rate for an isotropic material under (a) plane stress and (b) plane strain conditions compared with (c) the surface for S2 ice under plane stress conditions. The dashed line represents the level of triaxiality under a plane strain punch.

Hence surfaces of constant $\dot{D}^c$ in stress space are represented by surfaces of constant effective stress $\bar{\sigma}$. Such surfaces are shown in Figure 3 for isotropic polycrystalline ice (the von Mises effective stress) and columnar S2 ice (Hill's generalisation) for conditions of plane stress, and for polycrystalline ice, under plane strain loading.

4. Prediction of Creep Forces

Consider the general problem of an ice mass subject to indentation by a body which is penetrating the mass at a velocity v over part of its surface, resulting in a force P, predicted by (2). From purely dimensional consideration, the relationship between P and v must be of the form

$$\frac{v}{\phi\psi\dot{\varepsilon}_R d} = \left(\frac{P}{\sigma_R d t \phi}\right), \tag{4}$$

where ϕ and ψ are functions of the creep index n, $\dot{\varepsilon}_R$ is the strain rate corresponding to a reference stress σ_R, d is the width of the indentor and t is the thickness of the ice sheet. If we choose σ_R of the form

$$\sigma_R = \frac{P}{\phi d t}, \tag{5}$$

then the creep rate $\dot{\varepsilon}_R$ is related to v by

$$\dot{\varepsilon}_R = \frac{v}{\phi\psi d}. \tag{6}$$

This form of the solution is the reference stress form where, with ϕ and ψ known, the relationship between v and P may be calculated directly from uniaxial creep data. A bound on the total energy dissipation rate is given by Ponter (1980):

$$Pv \leq \sigma_R \dot{\varepsilon}_R, \tag{7}$$

or

$$\phi^{uB} = \frac{P_L}{\sigma_y d t}, \tag{8}$$

where $\sigma_R = \frac{P}{P_L}\sigma_y$ (P_L is the limit load for a perfectly plastic material with yield surface $\bar{\sigma} = \sigma_y$ and ϕ^{uB} represents an upper bound to ϕ). In the general case V denotes the volume of the body, but for the indentation problem V means, effectively, the volume where intense deformation takes place.

From the analysis of a large number of problems where the deformation field is confined, it is found that constant values of ψ and ϕ can be obtained so that equations (5) and (6) are a close approximation to the exact solution over a range of values of n. These values may be best computed by ensuring the result is exact for two values of n, say $n = 3$ and $n = 7$. Generally the value of σ_R given by the resulting equation differs from the exact solution by only 1 or 2% over this range of n, assuming the expression for $\dot{\varepsilon}_R$ is exact. A further experienced observation is that the resulting equation, although derived for a n power law, gives good results for a wide range of nonlinear relationships between $\dot{\varepsilon}_R$ and σ_R. The bound (7) can be generalised to an arbitrary relationship with some restrictions (Ponter, 1980).

For the constitutive relationship of equation (2) the degree to which the material response is orthotropic is taken into account through the introduction of the parameter γ, which is defined in Figure 3. The response of isotropic polycrystalline ice is represented by $\gamma = 1$ and for S2 columnar ice γ is typically 15 (Sanderson, 1989). Figures 5 and 6 show the results of the reference stress calculations for those two values of γ for the wedge of ice under plane stress loading shown in Figure 4 where the exact solutions are finite element solutions using the creep equations (2) and (3) with the appropriate expression for $\bar{\sigma}$ for $\gamma = 1$ and $\gamma =$

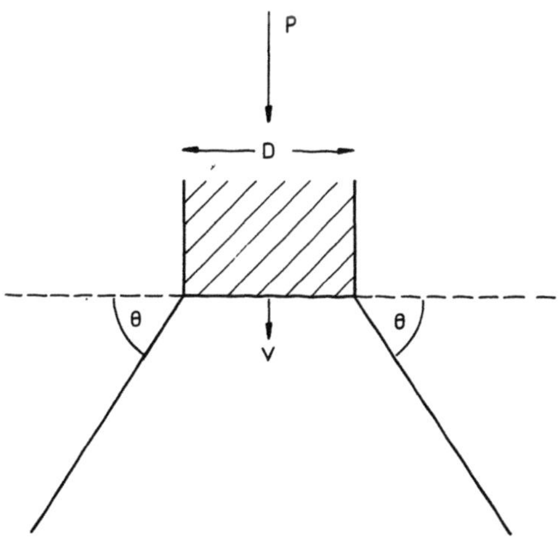

Figure 4. Indentor geometry.

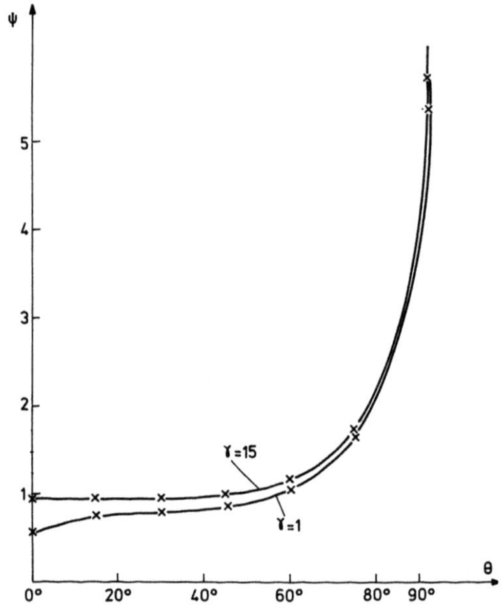

Figure 5. Variation of reference stress parameter Ψ with Θ for $\Upsilon = 1$ and $\Upsilon = 15$.

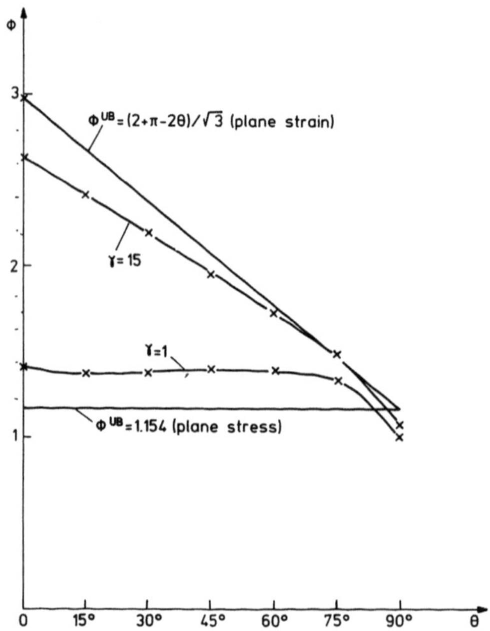

Figure 6. Variation of reference stress parameter Φ with Θ for $\Upsilon = 1$ and $\Upsilon = 15$.

15 (Ponter and Brown, 1985). The upper bound for ϕ in equation (8) requires knowledge of the perfectly plastic limit load for the appropriate expression for $\bar{\sigma}$. For the case $\gamma = 1$, this solution is known (see, for example Johnson and Mellor, 1973), but the limit load for $\bar{\sigma} = \sigma_y$ for $\gamma = 15$ is not known. However, it can be observed in Figure 3 that $\bar{\sigma}$ for $\gamma = 15$ (line c) remains close to $\bar{\sigma}$ = constant for $\gamma = 1$ for plane strain conditions over the range of stress ratios found in the plane strain plastic solution. As a result we might expect that the value of ϕ^{uB} for $\gamma = 1$ and plane strain would give a reasonable approximation to the value of ϕ for $\gamma = 15$. The comparison is given in Figure 6.

It is worth noting that $\psi \approx 1$ for a wide range of θ values for both $\gamma = 1$ and $\gamma = 15$. The rapid increase in ψ as θ approaches 90° reflects the fact that the deformation region is confined to a volume of the order of d^2t for small θ, but at $\theta = 90^o$ the deformation is uniformly distributed through the entire volume, and this change is reflected in the variation of ψ. The particular value of $\psi(90^o)$ depends upon the size of the region modelled in the finite element solution.

Such solutions are clearly able to yield good approximations to the steady state maximum load for low velocities, where creep deformation dominates. Two sets of experimental data help to put these particular solutions in context. During the measurement of strain on the ice sheet surrounding Tarsiut Island in 1983 by British Petroleum (Sanderson, 1983) one particular event occurred which resulted in the highest strain-rate during the two month period of the test programme (March 1 to April 30 1983). A wedge shaped piece of ice abutted the island with a low field velocity of v = 0.1 m/hour. The relationship between v and strain measurements have been shown to be predicted very well by a steady state creep analysis assuming $\gamma = 15$ (columnar ice) (Ponter and Brown, 1989) and, perhaps of greater interest, the stress predicted by the analysis was everywhere compressive, explaining the continuation of the high strain rates for a period of hours. The total load on the island was calculated as 288MN. If the wedge were approximated as a perfect wedge with an angle of 57°, the reference stress method yields a value of total load of 324MN, about 12% higher. This accords with the observation in creep and plasticity solutions that the boundary load is not particularly sensitive to details of the geometry, so that standard solutions can have a wide range of application. In the case of Tarsiut Island high strain-rates continued for a period of about 5 hours which roughly corresponds to the rupture life under compression at the average indentation stresses.

The significance of the wedge geometry is shown particularly well by the classic experiments of Michel and Toussaint (1977) on pure fresh water ice with a columnar structure. Square plates of ice 2.5 to 10 cm thick were indented at a constant velocity at the centre of an edge. With increasing velocity v the load response changed from ductile to brittle behaviour, the transition occurring at a velocity $v = 2.5\ 10^{-4}$ cm/s. For a particular test at this transitional velocity, the load displacement plot is shown in Figure 7. Using the reference stress method with $\theta = 0$ and $\gamma = 15$ the maximum load under creep is given by $P = 1040$kg, slightly greater than the observed peak value. The authors comment on the rapid growth of fracture regions developing from the edges of the indentor at an angle of θ=60° at higher velocities. If we assume initial failure occurs on such planes the sheet would then act as a wedge, and, as the stresses are now compressive, the creep load would then be sustained. For $\theta = 60^o$ the creep load given by the reference stress method is 678kg, about 10% higher than the observed steady state load. Hence, through the development of concentrated regions of microcracking, the ice sheet reduces itself to a geometry which may only fail by generalised microcracking.

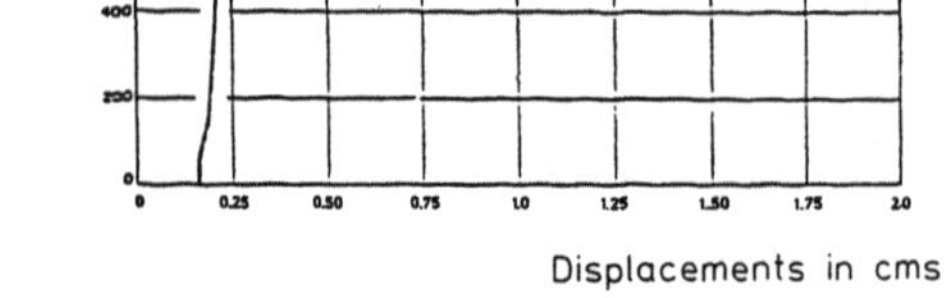

Figure 7. Loading curve for indentation experiment, D=5.09cms, t=2.54cms and v = $2.54 \cdot 10^{-4}$cm/s, from Michel and Toussaint (1977). The reference stress values of P assuming $\Theta = 0$ and $\Theta = 60^o$ closely correspond to the peak and steady state load.

The question then raised is, how can the maximum load be calculated in circumstances where a confined region, with no tensile stresses is subjected to indentation loading?

Let us suppose that for uniaxial compression tests at constant compressive strain rate the history of stress may be represented by

$$\sigma(t) = g(\omega)f(\dot{\varepsilon}) = g(\omega)\sigma^m, \tag{9}$$

where $g(\omega)$ is a function of an internal variable ω, such that $g(\omega) \leq 1$ with equality when $\sigma(\mathrm{t}) = \sigma^m$ the maximum stress. The quantity ω can be thought of as a parameter that represents the internal dislocation structure that develops during primary creep and the extent of microcracking in the material. The relationship of equation (9) could be expanded further, with g being a function of a number of internal variables, each describing a particular feature of the developing internal structure. In the analysis presented below, however, there is no need to determine the number of internal variables, the dependence of $g(\omega)$ on these variables, nor their evolution laws. When generalising equation (9) to multiaxial stress states we assume that $\dot{\varepsilon}$ and σ can be replaced by their effective values $\bar{\dot{\varepsilon}}$ and $\bar{\sigma}$. As $g(\omega)$ is assumed scalar there is an implicit assumption that the dislocation structure and damage that forms has an equal effect on each component of stress. Also restricting $g(\omega)$ to be less than or equal to 1, amounts to assuming that the material response is not sensitive to changes in the hydrostatic component of stress and the evolution of the internal state variables is only a function of the effective stress $\bar{\sigma}$. Neither of these assumptions are strictly true but for near biaxial stress states they are probably adequate.

The purpose of the following analysis is to obtain a reference stress for situations when microcracking takes place. We anticipate this result and define a fictitious power-law creeping material by taking the tangent to the curve of maximum stress in Figure 1 at the reference stress. For this fictitious material we may write

$$\dot{\varepsilon}_{ij} = \frac{\partial}{\partial \sigma_{ij}} \left\{ \frac{\dot{D}^f}{n^f + 1} \right\}, \tag{10}$$

where

$$\dot{D}^f = \sigma_{ij}\dot{\varepsilon}_{ij} = \sigma_o^f \left(\frac{\bar{\sigma}}{\sigma_o^f} \right)^{n^f+1}.$$

This expression is the same as the steady state expression of equation (2) except that the creep constants and n^f differ from the steady state values at high stresses, with n^f increasing with increasing stress up to the peak of Figure 1.

If $\dot{\varepsilon}_{ij}$ is the strain-rate at a point of the indented mass of ice at a given time t, we may write

$$\sigma_{ij}(t)\dot{\varepsilon}_{ij} \leq \sigma_{ij}^{m}\dot{\varepsilon}_{ij} \leq \sigma_{ij}^{f}\dot{\varepsilon}_{ij}, \tag{11}$$

where $\sigma_{ij}(t)$ is the actual stress and σ_{ij}^{f} is the stress predicted by equation (10) at the same strain-rate.

The convexity condition for the material of equation (10) is (Martin, 1966)

$$\frac{n^f}{n^f+1}\sigma_{ij}^1\dot{\varepsilon}_{ij}^1 - \sigma_{ij}^2\dot{\varepsilon}_{ij}^1 + \frac{\sigma_{ij}^2\dot{\varepsilon}_{ij}^2}{n^f+1} \geq 0, \tag{12}$$

where σ_{ij}^1 and σ_{ij}^2 represent two arbitrary stress states and $\dot{\varepsilon}_{ij}^1$ and $\dot{\varepsilon}_{ij}^2$ are the resulting strain rates predicted by equation (10). If we identify $\dot{\varepsilon}_{ij}^1$ with the strain-rate at a point of the indented body for the real material and $\dot{\varepsilon}_{ij}^2$ is the strain-rate predicted by equation (10) for the same indentor velocity v, integrating (12) over the volume, noting (11) and applying the principle of virtual work gives

$$P(t)v \leq P^f v, \tag{13}$$

where P(t) is the actual load at a given instant in time and P^f is that predicted by the model creeping material. The reference stress method described earlier can be used to determine P^f for a known velocity v. The maximum load, P^m, in a given situation can then be bounded by combining equation (5) with (13)

$$P^m \leq \sigma_R \phi dt, \tag{14}$$

where now σ_R is the stress on the maximum stress curve of Figure 1 which corresponds to the reference strain-rate of equation (6).

This argument is essentially rather approximate, but, in the absence of well tried constitutive equations for descriptions of internal stable fracture of ice it is not possible to be more precise. Essentially the result says that if each element of the ice goes through a maximum in the value of $\bar{\sigma}$(t), that an upper bound on P^m can be found if we assume all elements go through the maximum simultaneously, instead of at a sequence of separate instances. The result, however, is rather useful, it implies that the reference stress method can be used well into the transitional region, in accordance with observations made by Ponter et al., (1983). Figure

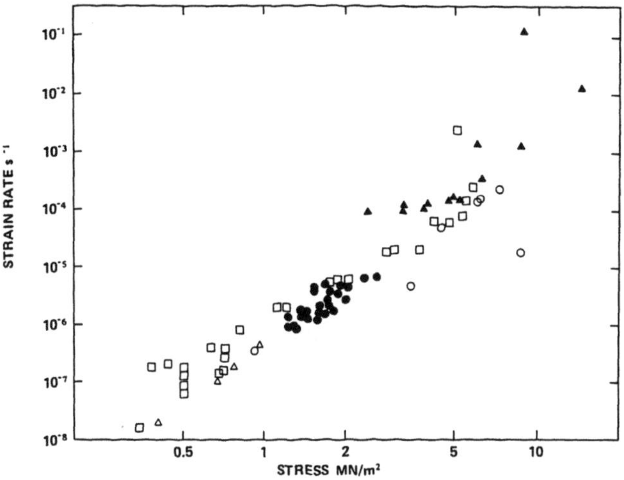

Figure 8. Comparison between indentation tests and uniaxial tests. Solid symbols are from indentation tests and open symbols are from uniaxial tests. See Ponter et al. (1983) for source of original data.

8 shows a comparison between indentation tests and uniaxial test data, taken from that paper. For high stresses the uniaxial data consistently overestimates these indentation tests in accordance with this result.

The experiment of Michel and Toussaint (1977) in Figure 7 provides additional evidence for the result as the upper bound for $\theta = 0$ is just greater than the maximum load, which is consistent with the observation that microcracking preceeded the development of highly damaged regions orientated at 60° to the surface emanating from the edges of the wedge.

5. Conclusion

The paper has attempted to demonstrate that the occasions when high loads occur in creep indentation problems correspond to circumstances where a body of ice develops after initial fracture or intense localised microcracking, where stresses are compressive and failure can only occur by widespread microcracking over a substantial volume of the body. For ice sheets this body seems to consist of a truncated wedge pointing in the direction of the ice velocity with a subtended angle of about 60°. This value must be associated, in some way, with the stress dependence of ice properties. In these circumstances the forces may be predicted using reference stress methods well into the transitional region. We are, however, discussing a rare event. Its rarity is indicated by the Tarsiut Island measurements where one such event occurred for a period of about 5 hours in two months, i.e. about 0.3% of the time. For this particular event, the ice velocity was relatively low, resulting in stress levels of the order of 1MPa, but we cannot exclude the possibility that higher ice velocities could, in principle, have produced much higher stresses. The events at Molikpaq (Jefferies and Wright, 1988) indicate that yet another mechanism can operate. Under high confining stresses ice can reduce to a powder form when sheared. The phenomenon is not well understood, presumably because laboratory experiments which simulate these particular conditions have not been carried out. At Molikpaq the vibrational response of the structure clearly had an influence on the phenomenon.

The BP collection of field data (Sanderson, 1986) tends to imply that, for wide structures, a size effect substantially reduces the average stress. The BP relationship

$$\text{Indentation pressure } \alpha \text{ (contact area)}^{-1/2}$$

can be quite well described by assuming that the effective contact area in all cases is of the order of 1m^2 (i.e. a region whose width is of the order of the ice sheet depth) which, perhaps, is no surprise. The real problem consists of predicting the probability of the rare high load event over the lifetime of a structure, a feature which is incorporated in Sanderson's analysis. The parallel problem of ship design, the size of the 10 year and the 100 year wave etc., ultimately relied upon the extensive measurement and collection of data on ship losses carried out by Lloyds of London over a considerable period of time. A passive approach to offshore structural design in ice conditions, resulting from accepting the scale effect based on a relatively small sample of field data, has inherent problems, as the probability of the rare high load event of the type discussed in this paper is not fully understood.

Acknowledgement

The authors would like to acknowledge the support of British Petroleum for the work described by Ponter and Brown (1985, 1989) and for permission to publish the results at this time. Particular thanks are due to Tim Sanderson and Sheila Hallam.

References

Anderson, R.G., Gardner, L.R.T. and Hodgkins, W.R. 1963. Deformation of uniformly loaded beams obeying complex creep laws. J. Mech. Eng. Sci., 5, pp.238-244.

Ashby, M.F., and Duval, P. 1985. The creep of polycrystalline ice. Cold Regions Sci. Tech., 11, pp.285-300.

Bhat, S.U. 1988. Analysis for splitting of ice flows during summer impact. Cold Regions Science and Technology, 15, pp.53-63.

Cocks, A.C.F. 1989a. The growth of radial cracks in ice sheets. Proc. 8th International Conference on Offshore Mechanics and Arctic Engineering (OMAE), The Hague, The Netherlands, V.4, pp.7-12.

Cocks, A.C.F. 1989b. A micromechanics based model for the creep of ice including the effect of general microcracking. In Mechanics of Creep Brittle Materials - 1, Cocks, A.C.F. and Ponter, A.R.S. (eds), Elsevier Applied Science, pp.213-229.

Hallam, S.D. 1986. The role of fracture in limiting ice forces. In Proc. 8th IAHR Int. Symp. on Ice, Iowa City, Iowa, Vol.2, pp.287-320.

Jefferies, M.G. and Wright, W.H. 1988. Dynamic response of Molikpaq to ice-structure interactions. Proc. 7th International Conference on Offshore Mechanics and Arctic Engineering (OMAE), Houston, Texas, V.4, pp.201-220.

Johnson, W. and Mellor, P.B. 1973. Engineering Plasticity. Van Nostrand Reinhold.

Jordaan, I.J. and McKenna, R.F. 1988. Constitutive relations for creep of ice. In Proc. 9th IAHR Ice Symposium, Sapporo, Japan, Vol.3, pp.47-58.

Marriott, D.L. and Leckie, F.A. 1964. Some observations on the deflection of structures during creep. Proc. Inst. Mech. Eng., Vol.178 (part 3L) pp.4-5 - 4-15.

Martin, J.B. 1966. Extended displacement bound theorem for work/hardening continua subjected to dynamic loading. Int. J. Solids Structures, 2, 9p.

Michel, B. and Toussaint, N. 1977. Mechanisms and theory of indentation of ice plates. J. of Glaciology, Vol.19, pp.285-299.

Ponter, A.R.S. 1980. On the creep modified shakedown limit. In Creep in Structures, Ponter, A.R.S. and Hayhurst, D.R. (eds), Springer-Verlag, pp.264-278.

Ponter, A.R.S. and Brown, P.R. 1985. Creep indentation solutions, variation of reference stress parameters with geometry. University of Leicester, Department of Engineering Report 85-9.

Ponter, A.R.S. and Brown, P.R. 1989. Ice forces on wide structures: field measurements at Tarsiut Island. In Mechanics of Creep Brittle Materials-1, Cocks, A.C.F. and Ponter, A.R.S. (eds), Elsevier Applied Science, pp.168-187.

Ponter, A.R.S., Palmer, A.C., Goodman, D.J., Ashby, M.F., Evans, A.G. and Hutchinson, J.W. 1983. The forces exerted by a moving ice sheet on an offshore structure, Part 1: the creep model. Cold Regions Science and Technology, Vol.8, pp.109-118.

Sanderson, T.J.O. 1983. Tarsiut Island research programme; the interesting results. BP Petroleum Development Limited, March 1983.

Sanderson, T.J.O. 1986. A pressure-area curve for ice. In Proc. 8th IAHR Int. Symp. on Ice, Iowa City, Iowa, Vol.2, pp.361-384.

Sanderson, T.J.O. 1989. Ice mechanics: risks to offshore structures. Graham and Trotman.

Sinha, N.K. 1983. Creep model of ice for monotonically increasing stress. Cold Regions Science and Technology, Vol.8, pp.25-33.

Sinha, N.K. 1989. Ice and steel, a comparison of creep and failure. In Mechanics of Creep Brittle Materials-1, Cocks, A.C.F. and Ponter, A.R.S. (eds), Elsevier Applied Science, pp.201-212.

Discussion

Q: T.J.O. Sanderson, BP International

To comment on the process of microcracking: depending on temperature and strain-rate, microcracking may continue and lead to progressive damage and eventual gross failure; alternatively, dynamic recrystallisation processes may allow a "stable" state of tertiary creep to be reached.

A: A.R.S. Ponter

Of course, Dr. Sanderson is perfectly correct that these two phenomena occur in ice. Our main concern in the paper is to describe methods of assessing the total load on an indentor for displacement rates which cover the transitional range from pure creep behaviour to brittle fracture. In this range the nucleation of microcracking and their effect upon subsequent creep behaviour dominates the behaviour. Dynamic recrystallization does not seem to be a significant mechanism in these circumstances but, of course, becomes particularly important for much lower strain rates. Similarly, microcracking does continue during steady state creep but the damage rate is much less than the rate which occurs during the initial transient phase.

Q: Oskar Mahrenholtz

I would like to know about the stabilizing phenomena following peak stress in constant strain rate experiments.

A: A.R.S. Ponter

In this paper we have considered the material response under high rates of loading. The decrease of stress after peak load in a constant strain-rate test results from the formation of microcracks in the material which form in response to the developing residual stress field as described in the paper. A steady state response is achieved when there is no further change in the microscopic residual stress field and no additional microcracks form in the material.

Q: S. Shyam Sunder, Massachusetts Institute of Technology

In the case of microcracking-induced damage, you generalize the reference stress concept through a stress versus strain-rate relation in terms of a separable, scalar damage function $g(\omega)$. Experimental evidence indicates that damage is highly anisotropic, pressure sensitive and history-dependent. Can you comment on the importance of these phenomena vis-a-vis ice force prediction and how they can be incorporated in the analysis.

A: A.R.S. Ponter

For the type of problem considered in this paper, where the loading is proportional and there is a high degree of kinematic determinacy, it is generally found that use of a scalar damage function to describe the material and component response is adequate. The most questionable assumption used in the paper is that $g(\omega) \leq 1$. For plane stress loading of isotropic polycrystalline ice sheets this assumption is probably correct. Under plane strain conditions, however, the pressure dependence of the microcracking process needs to be taken into account. One possible approach would be to use the plot of maximum effective stress against strain-rate obtained from triaxial compressive tests, at a pressure corresponding to the mean pressure under the indentor, instead of the uniaxial relationship. This approach would require an evaluation of how the pressure might change as microcracking develops under the indentor. The development of microcracks in S2 columnar ice under plane stress conditions will be a function of the in-plane mean stress. We are not aware of any biaxial tests on S2 ice that would permit the strength of this dependence, and therefore the suitability of the assumption that $g(\omega) \leq 1$, to be evaluated.

Despite these problems, the analysis presented here retains the essential features of the material and component (ice sheet) phenomena. It also provides a means of comparing uniaxial and indentation tests and a way of presenting the results of more detailed calculations when they become available.

The Use of Tertiary Creep Rates in Ice at High Strains in Compression and Shear

T.H. Jacka

Australian Antarctic Division,
Kingston, 7050, Australia

W.F. Budd

Meteorology Department, The University of Melbourne
Parkville, 3052, Australia

Abstract The steady state tertiary creep of ice which is attained at total strains in excess of $\sim 20\%$ is examined. By the term "creep" we imply that we shall examine ice deforming at $<\sim 10^{-6}\ s^{-1}$ and with no visible evidence of cracking or failure, thus strains $> 20\%$ take long times to establish in the laboratory. At these large strains, crystal anisotropy plays a most important role in the determination of creep rates.

For initially isotropic ice, the concept of a steady state secondary creep condition is invalid because of the development of crystal anisotropies, i.e. because of the transitory nature of secondary creep.

For initially isotropic or anisotropic ice, at high strains a new anisotropy develops, compatible with the applied stress configuration. If the initial and new anisotropies are similar, the strain required for the new anisotropy to develop will be small. If they are very different, the strain required will be large, but ultimately, a steady state constant creep rate condition with anisotropy compatible with stress configuration will eventuate.

Tertiary steady state creep rates are higher than the secondary isotropic creep rates which have commonly been used in the study of ice creep. For compression, the compatible crystal orientation fabric pattern is a small circle girdle, and the steady state creep rate is a factor of ~ 3 greater than the minimum secondary isotropic creep rate. For simple shear, the compatible fabric is a single maximum pattern and the creep rate is ~ 8 times the secondary creep rate. Further tests are required in other stress configurations including combinations of compression and shear.

Laboratory studies indicate that crystal size has little effect on ice creep rate. Rather, a condition is reached with the establishment of steady state creep, in which deformation and rotation are balanced by crystal growth and change and an equilibrium crystal size is attained. Crystal size then, can be regarded more as a result of the deformation to high strains, rather than a cause of it.

S. Jones · R. F. McKenna · J. Tillotson · I. Jordaan (Eds.)
Ice-Structure Interaction
IUTAM-IAHR Symposium St. John's, Newfoundland, Canada

1. Introduction

This paper examines the flow of ice of most relevance to natural ice masses. The stress range considered is 0.02 to 0.2 MPa and the temperature range of interest is 0 to -50°C. Steady state tertiary ice flow (attained after deformation of initially isotropic ice to strains in excess of 20%) is also of most relevance to the physics of the natural ice masses. This is indicated by the large number of ice cores extracted from the ice masses which exhibit different anisotropic crystal structures with "easy glide" crystal orientations related to well understood stress configurations (e.g. Russell-Head and Budd, 1979; Thwaites and others, 1984; Gao and Jacka, 1987; Hooke, 1973; Herron, 1982; Dahl-Jensen, 1985; Gow and Williamson, 1976).

Figure 1 compares three crystal orientation fabrics (a small circle girdle, a single maximum and a multiple maxima fabric) from ice cores extracted from Law Dome, Antarctica with three fabrics developed in the laboratory. Although the stress configuration applied to the laboratory ice samples to develop these fabrics may not be unique (i.e. there may be other stress configurations which can also result in these crystal fabric patterns) it is instructive to outline the stress configurations actually applied.

The small circle girdle was developed in uniaxial compression to an octahedral strain of 32.5% (Jacka and Maccagnan, 1984). The single maximum pattern was developed in simple shear to an octahedral strain of 200% (Gao and others, in press). The multiple maxima fabric pattern (Gao, unpublished) of Figure 1 was developed as a result of slow stress release at -3.3 °C from an ice sample which had previously been strained to 27.6% octahedral in uniaxial compression, and which therefore had previously exhibited a small circle girdle fabric pattern. We know then, that at sufficiently high strains a small circle girdle crystal fabric pattern is generated by a compressive stress; that a single maximum fabric is generated by simple shear; and that a multiple maxima fabric is generated by stress release at warm temperatures, from a compressive stress.

These fabrics may also be generated by other stress configurations. Of particular concern is the fabric pattern and the associated steady state creep rates developed by a combination of compression and shear. Within natural ice masses, this combined stress configuration persists, compression dominating in the upper layers, and shear in lower layers. This is illustrated by the particle trajectories in the profile of Figure 1 which indi-

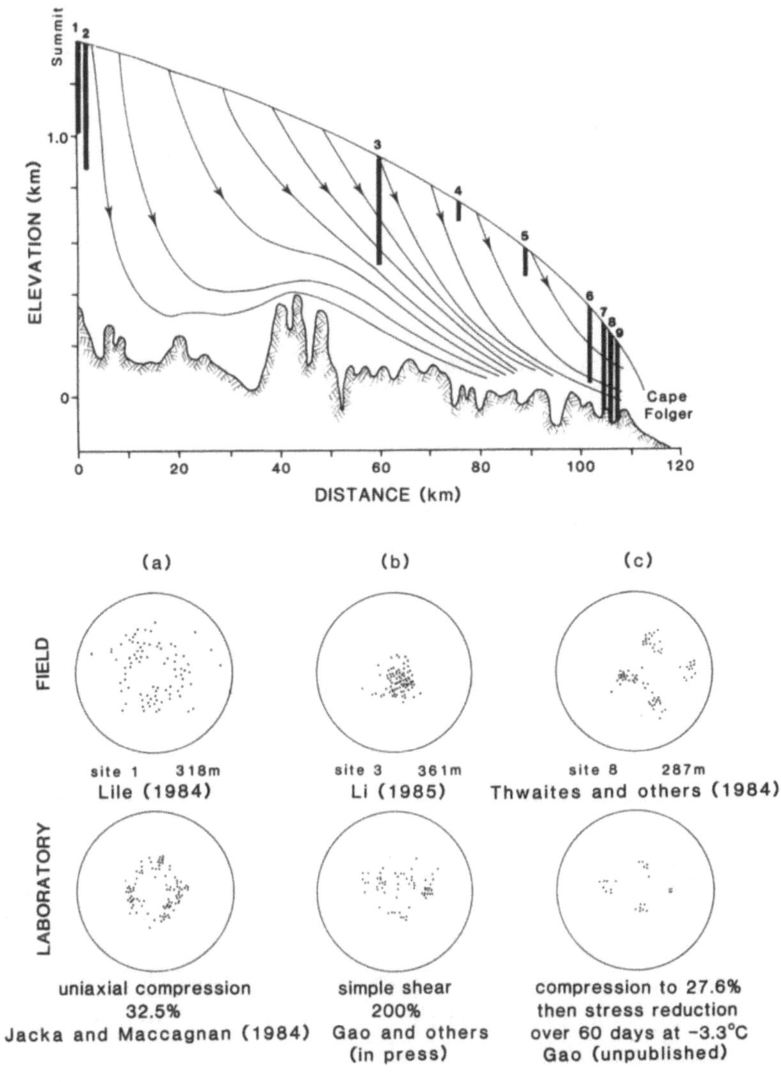

Figure 1. A cross-section of Law Dome, Antarctica from the summit to Cape Folger on the coast. Particle trajectories are shown along with locations of ice core drilling sites, numbered 1 to 9. Crystal orientation fabrics from three cores (from sites 1, 3 and 8) at marked depths illustrate (a) a small circle girdle fabric; (b) a single maximum fabric; and (c) a multiple maxima fabric. Also shown, are three similar orientation fabrics resulting from laboratory studies in different stress configurations.

cate predominantly vertical compression near the summit of Law Dome, shear in the lower layers, and more complex variable flow over the rough bedrock in the basal layers. In the basal ice, multiple maxima fabrics are found. In addition this ice is near or at the pressure melting point and often contains basal debris. This basal ice may therefore account for much of the movement of the whole ice mass and the deformation and sliding rates of this ice need close examination.

2. Steady State Ice Flow

Steady state ice flow is established at a strain which depends on the initial ice crystal structure, the stress configuration and the compatibility of the crystal structure with the stress configuration (i.e. whether the crystal orientation fabric is an easy glide or hard glide fabric for the given stress configuration). This concept is particularly important to understand and it is important to emphasize that a crystal orientation fabric can only be referred to as easy or hard glide with respect to a stress configuration and orientation.

If the initial crystal structure is not compatible with the stress configuration (hard glide), the strain required for the development of a compatible crystal structure will be greater than if the initial structure is compatible (easy glide). For an incompatible stress configuration/crystal structure relationship, the strain rate can be lower than even the isotropic minimum strain rate.

2.1 Steady State Ice Flow in Compression

The easy glide crystal orientation fabric for compression is a small circle girdle. This has been established by Jacka and Maccagnan (1984) who examined the development of this fabric pattern from initially isotropic ice and by Gao and Jacka (1987) who developed the same pattern through compression experiments to strains in excess of 60% on ice samples from an ice core from the summit (site 1 of Figure 1) of Law Dome, Antarctica and which initially exhibited various degrees of anisotropy. Alley (1988) and Azuma and Higashi (1985) propose from theoretical considerations of crystal rotations that in compression to very high strains, a single maximum pattern is developed. Their study assumed however, that crystal rotation alone generated the orientation fabric and that no recrystallisation took place. Gao (unpublished) carried out compression tests on ice from BHQ (site 3 of Figure 1), a high shear zone on Law Dome which

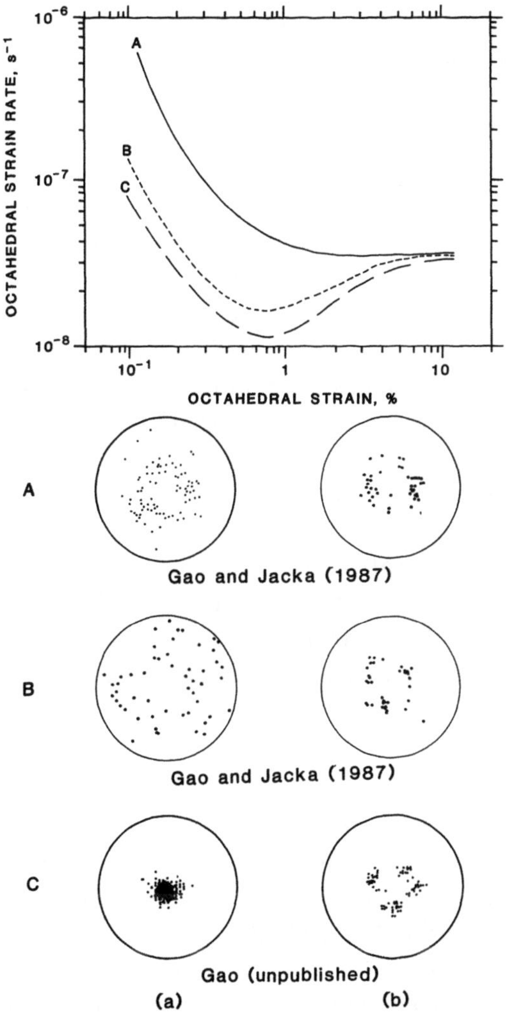

Figure 2. Creep curves (plots of octahedral strain rate as a function of octahedral strain) for three deformation tests in compression at -3.3°C and 0.2 MPa. Curve A: initially anisotropic ice exhibiting a small circle crystal orientation fabric; Curve B: initially isotropic ice; and Curve C: initially anisotropic ice exhibiting a single maximum fabric pattern. Curve A illustrates easy glide in compression; Curve C illustrates hard glide (at the minimum strain rate) in compression. Crystal orientation fabrics exhibited (a) prior to testing and (b) at test completion are also shown. Note that in all cases the final crystal fabric pattern was a small circle girdle.

initially exhibited single maximum fabrics. Gao's results indicated lower strain rates, even than the minimum isotropic strain rate. Figure 2 shows two creep curves (plots of strain rate as a function of total strain) from Gao and Jacka (1987) along with one from Gao (unpublished). Also shown are the crystal orientation fabrics prior to testing and at test conclusion. At octahedral strains of $>\sim$ 10%, the strain rate increased to the same tertiary value attained by initially isotropic ice and the crystal fabrics exhibited by Gao's samples at the test conclusions were small circle girdles (Figure 2). Huang and others (in press) also carried out repeated compression/annealing tests on ice from the BHQ core. They deformed samples cut parallel to, at 45° to and perpendicular to the long axis of the ice core. In all cases, a small circle girdle orientation fabric developed.

The small circle girdle fabric pattern is generated by and itself enhances ice flow in compression. This enhanced flow rate is a factor of ~ 3 higher than the minimum isotropic flow rate in compression. It appears that in steady state tertiary flow the rate of rotation and recrystallisation reach a balance with the mean angle of the girdle $\sim 25^o$ to the compression axis. In extension, the mean angle of the girdle pattern is $\sim 65^o$ (Jacka and Maccagnan, 1984).

2.2 Steady State Ice Flow in Shear

The steady state crystal orientation fabric pattern developed in parallel simple shear is a single maximum (e.g. Bouchez and Duval, 1982; Russell-Head and Budd, 1979; Gao and others, in press; from laboratory studies; and e.g. Russell-Head and Budd, 1979; Dahl-Jensen, 1985; Gow and Williamson, 1976; from field studies). This fabric is attained at an octahedral strain of $\sim$ 80% and has been found to persist to strains in excess of 200% (Bouchez and Duval, 1982; Gao and others, in press). The single maximum fabric develops in shear no matter what pattern is exhibited by the initial crystal fabric. The total strain required to develop the single maximum pattern however, may be greater if the initial pattern is a hard glide one for shear than if it is an easy glide pattern.

In shear, the enhanced flow rate for ice exhibiting a single maximum fabric pattern is a factor of ~ 5 (Gao and others, in press) to 8 (Russell-Head, 1985) faster than the minimum isotropic flow rate.

2.3 The Flow of Ice Exhibiting a Multiple Maxima Crystal Orientation Fabric Pattern

This paper is concerned with ice flow in compression and shear. It is worth noting however, that the basal ice exhibiting multiple maxima fabrics is usually relatively warm and thus pressure melting may be reached at the base where sliding may be significant. Multiple maxima fabrics have been developed in the laboratory by compressive deformation to large strains followed by stress reduction processes and thus the flow due to internal deformation in the field situation is likely to be small (due to the low stress) as found from the borehole results.

It would seem likely also that the multiple maxima fabric is an easy glide fabric for compression (more so at least than for shear) because the development through stress reduction processes of this fabric is more easily facilitated when the stress reduction follows a high compressive strain than when it follows a high shear strain.

3. Steady State Flow Laws For Compression and Shear

To compare flow properties of ice under different stress configurations, Nye (1953) and Budd (1968) have proposed the use of root mean square (octahedral) values of the principal stress and strain deviator tensors. The use of octahedral values however, implies flow dependence on the second invariant only of the stress and strain rate tensors. Thus, the application of octahedral values can be expected to apply to isotropic ice, but not to anisotropic ice.

Budd and Jacka (1989) conducted tests at 0.2 MPa shear and 0.2 MPa compression. They found that the minimum isotropic strain rates for shear were about an order of magnitude higher than those for compression in the temperature range 0 to -20°C. For 0.1 MPa octahedral over the same temperature range, the shear and compression results were closer, a factor of ~ 2 separating the minimum isotropic strain rates.

Gao (unpublished) compared octahedral minimum isotropic strain rates at -2.0°C over a range of octahedral stresses from 0.1 to 0.5 MPa. He also found that the octahedral strain rates for shear were a factor of ~ 2 higher than those for compression.

For the laboratory testing of ice flow, there remain problems concerning the scale and geometry of the ice samples in both compression (see e.g. Jacka, unpublished) and shear (see e.g. Russell-Head, 1985). These problems may account for the discrepancies in the octahedral results outlined above, and there is clearly a need to design test apparatus and techniques to overcome these problems.

For the current paper, our concern is with steady state anisotropic ice flow, for which the octahedral values do not directly apply. However, it has previously been established that in compression, the tertiary steady state flow rate is a factor of ~ 3 greater than the isotropic minimum

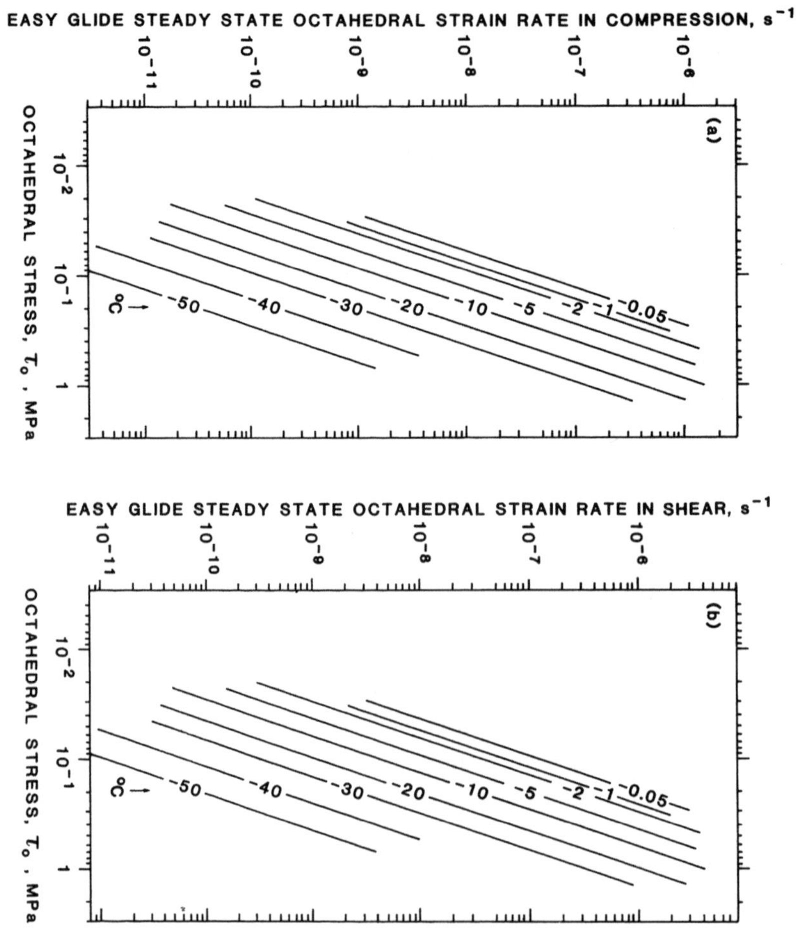

Figure 3. Flow law plots (octahedral strain rate as a function of octahedral stress) for easy glide steady state strain rate for (a) compression (of ice exhibiting a small circle girdle fabric pattern); and (b) shear (of ice exhibiting a single maximum fabric pattern).

octahedral flow rate. In shear, the tertiary steady state strain rate is ~ 5 to 8 times the isotropic minimum octahedral strain rate.

Budd and Jacka (1989) published a plot and table of isotropic minimum octahedral strain rates as a function of octahedral stress. Notwithstanding the above discrepancy of a factor of ~ 2, Budd and Jacka's table was compiled from data in both compression and shear.

It is now possible to construct two tables and two figures; one for compression (Table 1a, Figure 3a), and one for shear (Table 1b, Figure 3b). These tables and figures are constructed by multiplying the isotropic minimum octahedral values of Budd and Jacka (1989) by 3 for compression and by 8 for shear, and they provide flow laws for easy glide steady state ice flow.

The data of Tables 1a and 1b agree with the empirical relation provided by Glen (1958), with the addition of an extra quantity, E the enhancement of steady state easy glide strain rate, $\dot{\varepsilon}_{ss}$ over minimum isotropic octahedral strain rate, viz.,

$$\dot{\varepsilon}_{ss} = Ek_o\tau_o^n, \tag{1}$$

where τ_o is the octahedral stress, k_o is a function of temperature, and $n \simeq 3$. For compression, $E \simeq 3$. For shear, $E \simeq 8$. It is suggested that these flow laws provide a better estimate of steady state ice strain rates than the commonly used "Glen's Law" which is applicable only to minimum isotropic strain rate. The above flow laws apply to the bulk of the ice within polar ice masses. Furthermore, they apply to the deformation of ice tunnels, wharfs and other ice structures.

4. Steady State Crystal Size

In their experiments on the development of the small circle girdle crystal orientation fabric in compression, Jacka and Maccagnan (1984) noted that an equilibrium crystal size was established as tertiary steady state flow rates were attained. Jacka (1984) found that initial crystal size had little effect on ice flow rate (cf. Duval and LeGac, 1980; Jones and Chew, 1983) but that crystal size itself was affected by the steady state flow.

Table 1 (a) Steady state creep rates in compression for anisotropic ice in easy glide.

Temp °C	Octahedral stress, MPa 0.025	0.05	0.1	0.2	0.4
	Steady state octahedral creep rate, s^{-1}				
-0.05	7.2×10^{-10}	6.0×10^{-9}	4.2×10^{-8}	3.0×10^{-7}	2.7×10^{-6}
-1.0	3.3×10^{-10}	2.8×10^{-9}	2.0×10^{-8}	1.7×10^{-7}	1.3×10^{-6}
-2.0	2.0×10^{-10}	1.6×10^{-9}	1.2×10^{-8}	9.9×10^{-8}	7.2×10^{-7}
-5.0	7.5×10^{-11}	6.3×10^{-10}	4.5×10^{-9}	3.6×10^{-8}	2.6×10^{-7}
-10	2.7×10^{-11}	2.0×10^{-10}	1.6×10^{-9}	1.2×10^{-8}	9.6×10^{-8}
-20	6.0×10^{-12}	5.1×10^{-11}	3.9×10^{-10}	3.0×10^{-9}	2.4×10^{-8}
-30		1.4×10^{-11}	1.1×10^{-10}	9.0×10^{-10}	6.9×10^{-9}
-40		3.3×10^{-12}	2.4×10^{-11}	2.1×10^{-10}	1.7×10^{-9}
-50			4.5×10^{-12}	3.6×10^{-12}	2.6×10^{-10}

Table 1 (b) Steady state creep rates in shear for anisotropic ice in easy glide.

Temp °C	Octahedral stress, MPa 0.025	0.05	0.1	0.2	0.4
	Steady state octahedral creep rate, s^{-1}				
-0.05	1.9×10^{-9}	1.6×10^{-8}	1.1×10^{-7}	8.0×10^{-7}	7.2×10^{-6}
-1.0	8.8×10^{-10}	7.5×10^{-9}	5.4×10^{-8}	4.6×10^{-7}	3.4×10^{-6}
-2.0	5.4×10^{-10}	4.3×10^{-9}	3.3×10^{-8}	2.6×10^{-7}	1.9×10^{-6}
-5.0	2.0×10^{-10}	1.7×10^{-9}	1.2×10^{-8}	9.6×10^{-8}	6.8×10^{-7}
-10	7.2×10^{-11}	5.4×10^{-10}	4.2×10^{-9}	3.2×10^{-9}	2.6×10^{-7}
-20	1.6×10^{-11}	1.4×10^{-10}	1.0×10^{-9}	8.0×10^{-9}	6.4×10^{-8}
-30		3.8×10^{-11}	2.9×10^{-10}	2.4×10^{-9}	1.8×10^{-8}
-40		8.8×10^{-12}	6.4×10^{-11}	5.6×10^{-10}	4.4×10^{-9}
-50			1.2×10^{-11}	9.6×10^{-12}	6.8×10^{-10}

Ice samples initially exhibiting crystal sizes ranging from 0.7 to 4.5 mm diameter, after deformation in compression at the same temperature and stress to strains of 20% or more, exhibited the same final crystal size.

Gao and others (in press) carried out several experiments in shear to high strains (up to 200% octahedral), and again obtained an equilibrium crystal size along with approximately constant tertiary flow rates. In addition, compression tests on ice which initially exhibited single maximum orientation fabric patterns resulted in the same final crystal size as had resulted (at the same temperature and stress) in experiments commencing with isotropic ice (Gao, unpublished). Equilibrium crystal sizes were also attained by Gao and Jacka (1987) in compression tests on ice from Antarctic ice cores initially exhibiting various degrees of anisotropy.

The experiments outlined above suggest that an equilibrium crystal size is established along with the establishment of steady state ice flow. The actual crystal size seems independent of initial crystal size, independent of stress configuration, and independent of the initial crystal orientation fabric pattern. It seems to be dependent on stress, temperature and/or strain rate.

5. Recommendations for Further Study

Further study is clearly required of the comparisons of isotropic and anisotropic ice flow rate under different stress configurations. The usefulness of the octahedral relations for comparison of different stress configurations has still not been established satisfactorily. Of most need, because of the relevance to the flow of natural ice masses, are studies of ice flow in combined shear and compression.

Only after a clear understanding of the isotropic ice flow comparisons is obtained, can anisotropic ice flow in the different stress configurations be understood. There are fundamental questions to be answered. Does a sample (isotropic and anisotropic) undergoing shear deform more or less rapidly when subjected to a compressive stress? How does the crystal orientation fabric change from a single pole fabric (developed in shear) once a compressive stress is applied? How does it change from a small circle girdle pattern (developed in compression) if a shear stress is applied?

Crystal sizes in ice cores extracted from the natural ice masses are found to vary with depth. It has been recognised that the variations are associated with zones of high stress (Gow and Williamson, 1976). It has been suggested also that crystal size may be influenced by dust concentrations. Whatever factors affect crystal size, these effects must be added to the temperature dependent crystal growth with time. The laboratory experiments to high tertiary strains suggest that an equilibrium crystal size is established, the actual size, dependent on test temperature and stress. Further investigations are required of these equilibrium crystal sizes, in particular aimed towards improving our understanding of the crystal size profiles found in the ice sheets.

Comprehensive deformation and sliding experiments on ice with different concentrations of dirt are still required, particularly at temperatures near the melting point. They need to extend to high strains and they need to involve different stress configurations. Consideration must also be given to the physical and chemical properties of the dirt, to the grain size, to the percent of clay and to other soluble and insoluble impurities associated with the dirt. These experiments are particularly important because most of the ice mass movement may be due to the highly variable, warm basal layers.

To fully understand the development of crystal anisotropies, further data are required not only from laboratory experiments, but also from ice cores drilled at carefully selected locations in the ice sheets. The core drilling programs need to be planned to provide for "in the field" analysis of crystal fabrics and crystal size since these parameters may alter with time once the core is extracted. In addition to the core drilling itself, comprehensive glaciological data in the surrounding area is required. This data should include surface ice movement and strain rates, elevations, ice thicknesses and shear stresses. The borehole itself needs also to be studied for closure rates, inclination, vertical strain and temperature.

In recent years, comprehensive ice drilling programs as described above (most notably so far, the Greenland Dye 3 operation) have commenced. Along with the continuing laboratory studies, these drilling programs promise great hope for a better understanding of the relationships between crystal orientation fabrics and the creep of ice.

References

Alley, R.B. 1988. Fabrics in polar ice sheets: development and prediction. Science, Vol.240, No.4851, pp.493-495.

Azuma, N. and Higashi, A. 1985. Formation processes of ice fabric pattern in ice sheets. Annals of Glaciology, Vol.6, pp.130-135.

Bouchez, J.L. and Duval, P. 1982. The fabric of polycrystalline ice deformed in simple shear: experiments in torsion, natural deformation and geometrical interpretation. Textures and Microstructures, Vol.5, pp.171-190.

Budd, W.F. 1968. The dynamics of ice masses. ANARE Scientific Reports, Series A(IV), Pub. No.108, 216p.

Budd, W.F. and Jacka, T.H. 1989. A review of ice rheology for ice sheet modelling. Cold Regions Science and Technology, Vol.16, No.2, pp.107-144.

Dahl-Jensen, D. 1985. Determination of the flow properties at Dye 3, South Greenland, by bore-hole-tilting measurements and perturbation modelling. Journal of Glaciology, Vol.31, No.108, pp.92-98.

Duval, P. and LeGac, H. 1980. Does the permanent creep-rate of polycrystalline ice increase with crystal size? Journal of Glaciology, Vol.25, No.90, pp.151-157.

Gao, X.Q. unpublished. Laboratory studies of the development of anisotropic crystal structure and the flow properties of ice. Ph.D. thesis, 1989. Meteorology Department, The University of Melbourne, 167p.

Gao, X.Q. and Jacka, T.H. 1987. The approach to similar tertiary creep rates for Antarctic core ice and laboratory prepared ice. Journal de Physique, Colloque C1, Supplement to No.3, Vol.48, pp.C1-289-C1-296.

Gao, X.Q., Jacka, T.H. and Budd, W.F. in press. The development of ice crystal anisotropy in shear and comparisons of flow properties in shear and compression. Proceedings of International Symposium on Antarctic Research, Hangzhou, China, May, 1989.

Glen, J.W. 1958. The flow law of ice. IUGG/IAHS Symposium of Chamonix. IAHS Pub. No.47, pp.171-183.

Gow, A.J. and Williamson, T. 1976. Rheological implications of the internal structure and crystal fabrics of the West Antarctic ice sheet as revealed by deep core drilling at Byrd Station. CRREL Report 76-35, 25p.

Herron, S.L. 1982. Physical properties of the deep ice core from Camp Century, Greenland. Ph.D. thesis, State University of New York at Buffalo, 153p.

Hooke, R. LeB. 1973. Structure and flow in the margin of the Barnes Ice Cap, Baffin Island, N.W.T., Canada. Journal of Glaciology, Vol.12, No.66, pp.423-438.

Huang, M., Wang, W., Li, J. and Li, G. in press. Repeated compression-annealing experiments on anisotropic core ice. Proceedings of International Symposium on Antarctic Research, Hangzhou, China, May, 1989.

Jacka, T.H. 1984. Laboratory studies on relationships between ice crystal size and flow rate. Cold Regions Science and Technology, Vol.10, No.1, pp.31-42.

Jacka, T.H. unpublished. Experimental investigations of the flow of ice. Ph.D. thesis, Meteorology Department, The University of Melbourne, 1987, 236p.

Jacka, T.H. and Maccagnan, M. 1984. Ice crystallographic and strain rate changes with strain in compression and extension. Cold Regions Science and Technology, Vol.8, No.3, pp.269-286.

Jones, S.J. and Chew, H.A.M. 1983. Effect of sample and grain size on the compressive strength of ice. Annals of Glaciology, Vol.4, pp.129-132.

Li, J. 1985. Ice fabric pattern of the BHQ ice core from Law Dome Ice Cap, Antarctica. Science Bulletin, pp.1097-1100 (in Chinese).

Lile, R.C. 1984. The flow law for isotropic and anisotropic ice at low strain rates. ANARE Reports, No.132, 93p.

Nye, J.F. 1953. The flow of ice from measurements in glacier tunnels, laboratory experiments and the Jungfraufirn bore hole experiment. Proceedings of the Royal Society. Series A, Vol.219. No.1139, pp.477-489.

Russell-Head, D.S. 1985. Shear deformation of ice to large strains. (In: Australian Glaciological Research, 1982-83), ANARE Research Notes, No.28, pp.118-121.

Russell-Head, D.S. and Budd, W.F. 1979. Ice-sheet flow properties derived from bore-hole shear measurements combined with ice-core studies. Journal of Glaciology, Vol.24, No.90, pp.117-130.

Thwaites, R.J., Wilson, C.J.L. and McCray, A.P. 1984. Relationships between bore-hole closure and crystal fabrics in Antarctic ice core from Cape Folger. Journal of Glaciology, Vol.30, No.105, pp.171-179.

Discussion

Q: Duval, LGGE, CNRS

It seems that, for Antarctic ice, two physical mechanisms can induce anisotropic ice: 1) recrystallization processes, 2) rotation of crystals by dislocation glide. What do you think about the second mechanism, which seems to explain the fabrics of Vostok ice?

A: T.H. Jacka

Recrystallization processes certainly occur and have been exhibited in the laboratory by several researchers. Rotation of crystals may also occur, however I find it difficult to accept that rotation should occur without simultaneous recrystallization. It has been proposed that rotation alone may enhance anisotropic crystal structure at low temperatures (e.g. the Vostok ice core). This however cannot be conclusively proved or disproved in the laboratory because of the extremely long times involved. Since strain seems to be the controlling parameter for the creep properties of ice in the temperature range 0 to -32°C, I see no reason to assume any different processes at colder temperatures.

Q: S. Shyam Sunder, MIT

Do you observe significant grain size changes for strains up to 5% in comparison to the level of 20-30% presented in the lecture?

A: T. H. Jacka

The development of an equilibrium grain size is associated with the development of the preferred crystal orientation fabric and occurs over the same strain interval. That is, from ≈ 1% to 10 or 15% for initially isotropic ice. So by 5% strain, there is a small amount of grain size change. The rate of change of grain size is approximately linear (see Jacka, 1984; Jacka and Maccagnan, 1984) over the above strain interval.

Creep and Settlement of Ice Structures - A Case History

D.M. Masterson

Sandwell Swan Wooster Inc.
201-1167 Kensington Cres. N.W., Calgary, Alberta, Canada T2N 1X7

T. Smith

Thayer School of Engineering, Dartmouth College,
Hanover, NH 03755, U.S.A.

1. Introduction

During the late winter of 1986, BP Canada Inc. drilled three shallow, stratigraphic wells in the Cold Lake region of northern Alberta, Canada. The wells were related to the confirmation of heavy oil deposits beneath the bottom of a local lake called Marguerite Lake near BP's Wolf Lake development. The three wells were about 600 m deep and required approximately 3 days to drill. Light, truck mounted drilling rigs, weighing in total about 130 tonnes, are usually adequate to drill this type of hole.

BP decided to drill the wells in the winter using the ice cover which normally forms on top of the lake as a support for the drilling operation. This would simplify the operation and would be less costly and more environmentally acceptable than a summer floating or bottom founded operation. Access to the wells could be easily effected by means of a floating ice road.

Between early January and February 13, 1986 three ice pads were built by flooding and freezing water on top of the natural ice. The total ice thickness was, in this way, brought to about 1.5 m. An ice road system was built to connect the ice pads to shore. A map of the pads and roads on the lake is shown in Figure 1.

S. Jones · R. F. McKenna · J. Tillotson · I. Jordaan (Eds.)
Ice-Structure Interaction
IUTAM-IAHR Symposium St. John's, Newfoundland, Canada

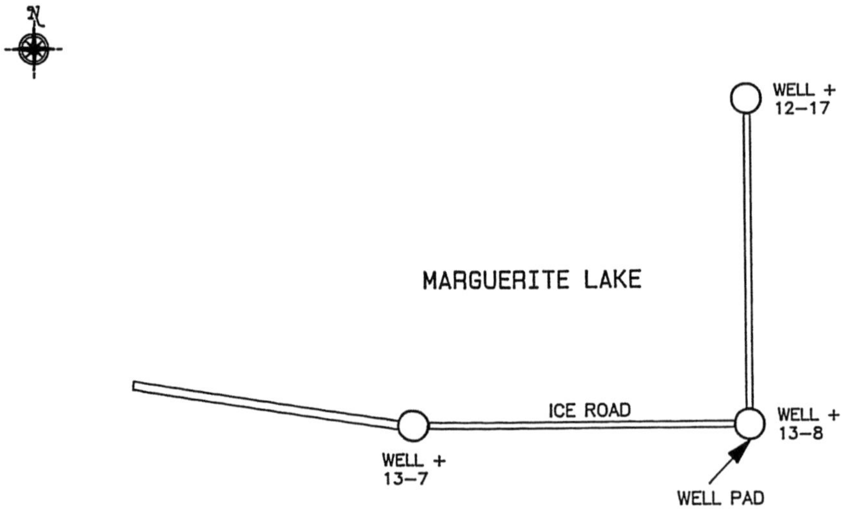

Figure 1. Ice platform layout Marguerite Lake.

While drilling the first well, all available freeboard of the floating ice was lost in the first few hours after rig up. By the end of drilling, the well centre was 120 mm under water and there was general flooding of the area around the rig. This was the first incidence of loss of freeboard during drilling of a well from a floating ice structure (Masterson et al., 1987).

It was subsequently determined that the rig weight reported to the BP and design engineers was considerably underestimated. A lighter rig was procured and the other two wells were drilled without incident. During drilling of the second well, there was an opportunity to monitor the effects of a smaller, short duration load increase and to gather useful information. This drilling project has yielded very useful information on ice structure design and performance and has helped substantiate that the assumptions used in current design practise are valid and not overly conservative (Masterson and Strandberg, 1980; Masterson and Gamble, 1986).

2. Ice Platform Design

The design of a floating ice platform is based on stress and creep or long term deflection criteria. The maximum, extreme fibre tensile stress is limited to a proven safe value of 520 kPa and the long term creep deflection must be less than the available freeboard of the ice. The available freeboard of the ice is normally about 10 percent of the total ice thickness, being dictated principally by buoyancy considerations. An engineering design report was prepared for BP by Sandwell Swan Wooster engineers. Since there is no code for ice platform design procedures, parameters were used which have proven to be safe in past projects.

The maximum, extreme fibre tensile stress is calculated using the methods developed by Westergaard (Timoshenko, 1940). An elastic homogeneous, isotropic plate resting on a perfectly elastic foundation is assumed. The maximum, extreme fibre tensile stress in the plate is

$$\sigma_{\max} = 0.275(1+\mu)\frac{P}{h^2}\log\left[\frac{Eh^3}{kb^4}\right], \tag{1}$$

where

$\sigma_{\max}$	= maximum, extreme fibre tensile stress,
P	= applied load,
h	= ice thickness under load,
E	= elastic modulus of ice,
k	= subgrade modulus = γ_{water},
b	= radius of loaded area, and
μ	= Poisson's ratio.

The effect of multiple loads is taken into account by using the principle of superposition and noting that the stress caused by a load applied a distance x from the point of interest is given by

$$\sigma_x = \sigma_{\max} \exp\left[\frac{-x}{0.691\ l}\right], \tag{2}$$

where

$$l = \left(\frac{Eh^3}{12(1-\mu^2)k}\right)^{1/4}. \tag{3}$$

The total stress at a point due to loads P_0, P_1, P_2, P_3 .... at distances x_0, x_1, x_2, x_3, ... is evaluated by the superposition equation

$$\sigma_{\text{Total}} = \sigma_0 + \sum_{n=1}^{i} \sigma_n \exp\left(-\frac{x_n}{0.691\ l}\right). \tag{4}$$

The maximum, instantaneous, extreme fibre tensile stress is limited to 520 kPa, a proven, safe design value (Masterson and Gamble, 1986). The calculation assumes the total load is applied instantaneously and that there are no creep effects, a conservative assumption.

The long term creep deflection is calculated using the reduced modulus method or exponential equations (Masterson and Strandberg, 1980; Masterson and Kivisild, 1980).

For these platforms the initial freeboard was small, being approximately 150 mm, and the margin for error in estimation of load or creep behaviour was correspondingly small. The available freeboard in larger platforms was generally of the order of 500 mm.

Long term creep can be calculated as follows:

$$\delta_{\text{Long Term}} = \frac{P}{8kl^{*2}}, \tag{5}$$

where

P = applied load,
k = subgrade modulus,
l^* = $\left(\frac{E^* h^3}{12(1-\mu^2)k}\right)^{1/4}$,
E^* = 0.1 E = E(t) = effective or reduced elastic modulus.

Effective or time dependent elastic modulus values for calculating the flexural creep of ice structures have been determined experimentally (Masterson et al., 1979). An expression for E as a function of time has been given in the above reference as

$$E(t) = 204.4t^{-0.64} \times 10^3, \tag{6}$$

where $E(t)$ is elastic modulus in units of MPa and t is time in days. The initial given loads and their distribution are shown in Figure 2. The stress and long term deflection equations were used to arrive at the design plan and cross section shown in Figures 3 and 4. An alternative load distribution for a second rig used for drilling the second and third wells is shown in Figure 5. This load distribution did not result in any changes in design geometry of the ice pads.

3. Construction

Construction of the ice pads began in early January and was completed February 13, 1986. Site visits were made by the engineers on four occasions. Ice thickness, freeboard and temperature were checked on these

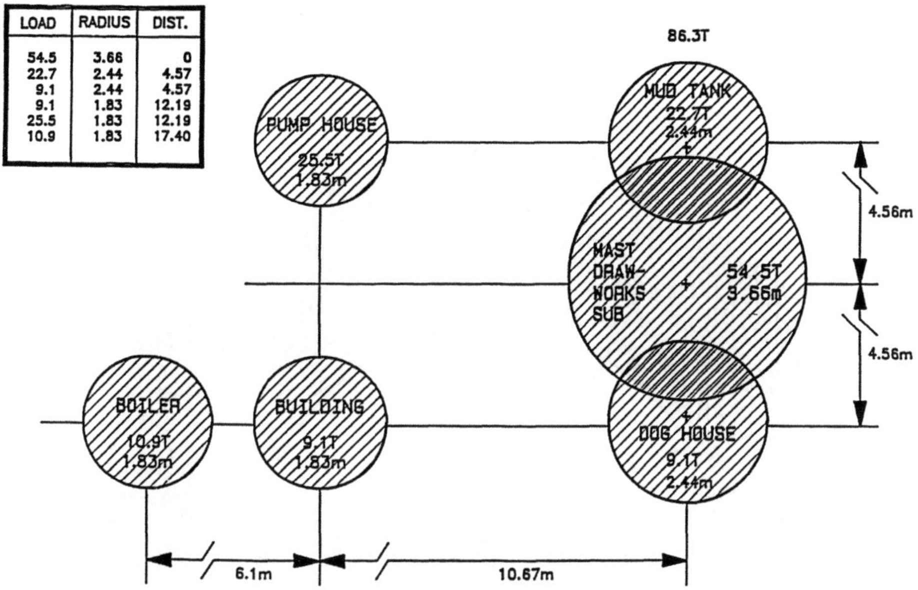

Figure 2. Design loads and configuration for Westburne Rig #37.

occasions. Small, 3 hp gasoline powered pumps were used to perform the flooding of the natural ice. The initial thickness of the ice at the start of construction was about 0.5 m. The temperatures during the construction period were unseasonably warm, slowing the construction progress and causing warm ice temperatures. This region of Alberta is noted for its cold winters and it was surprising to find temperatures generally in the range of -10 or -15°C or higher.

The following ice thicknesses were measured during construction

Date	Ice Thickness (m)
January 23	0.87 (all pads)
January 30	1.28 (all pads)
February 5	1.35 (all pads)
February 13	1.53 (Pad 12-17)

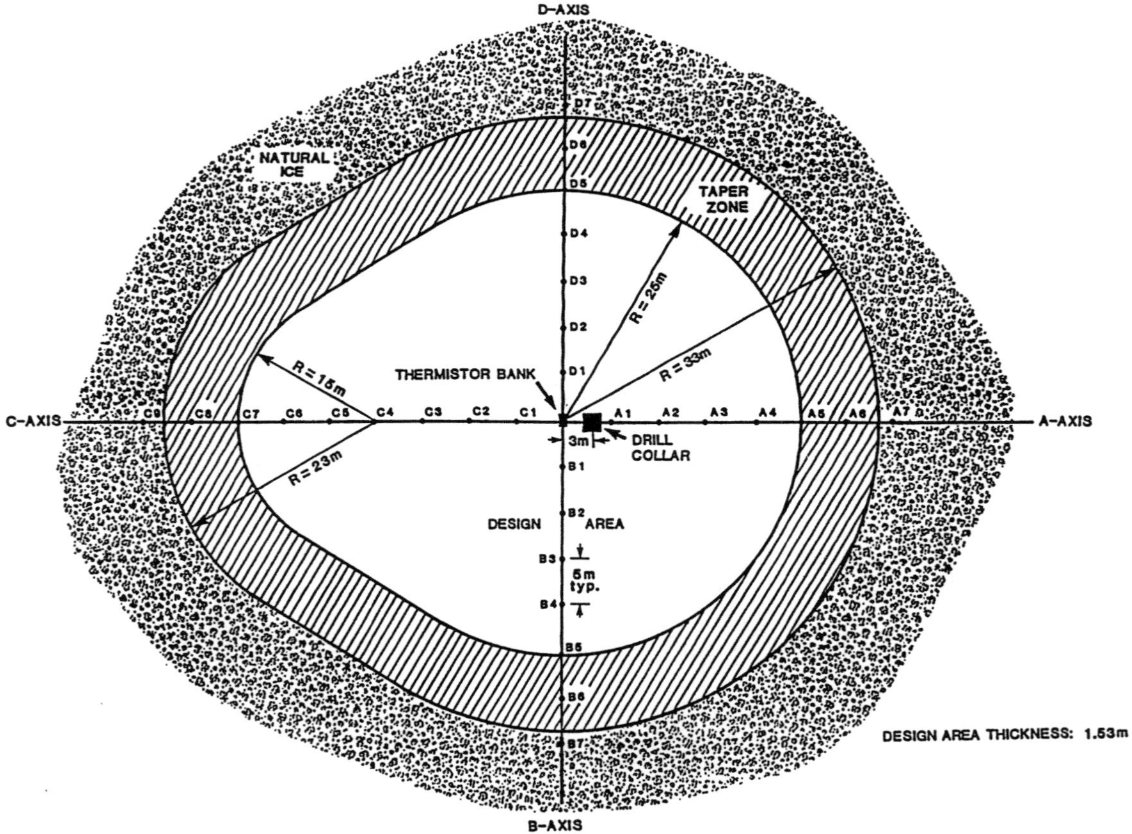

Figure 3. Ice pad layout (plan).

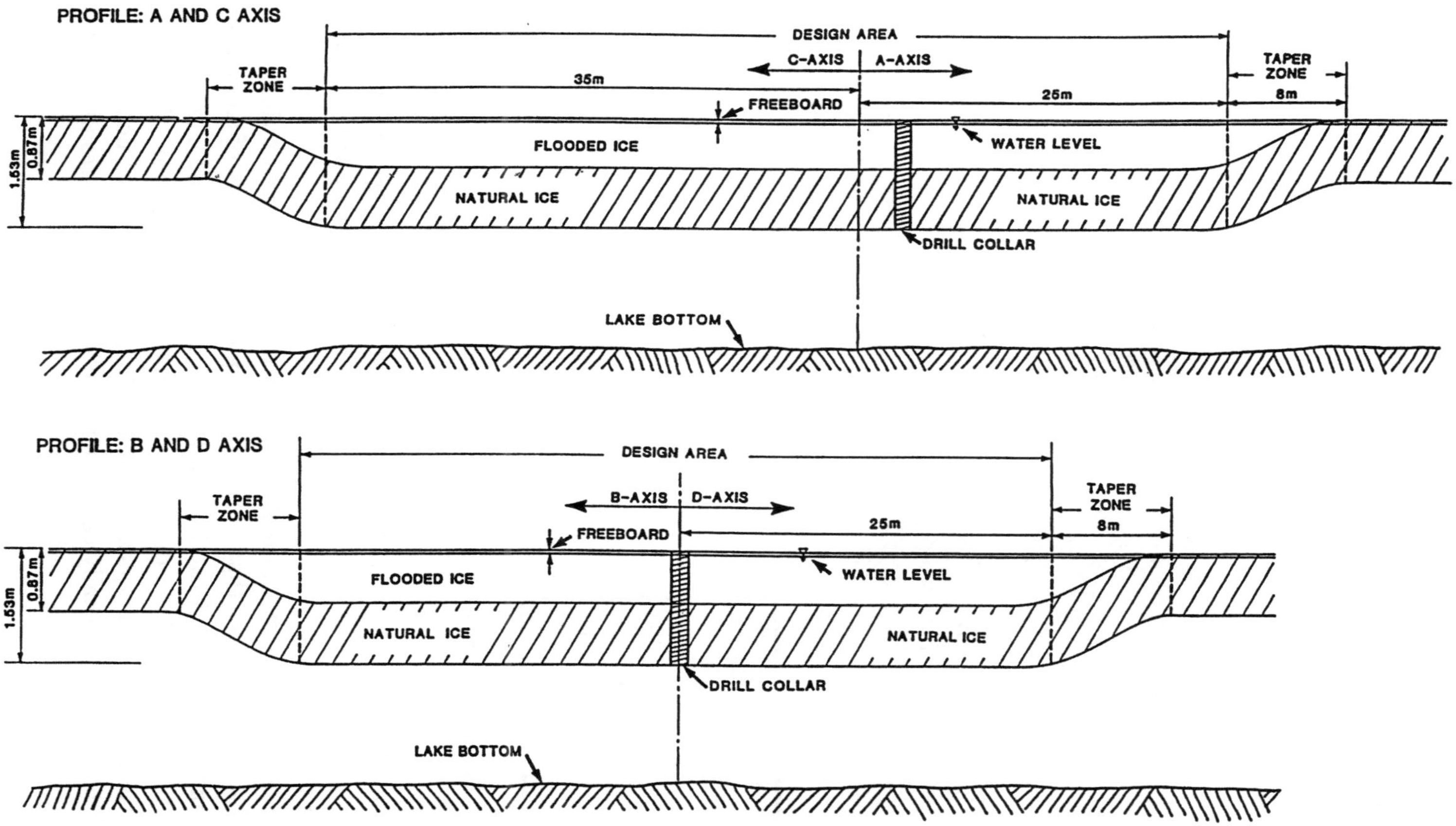

Figure 4. Ice pad profile.

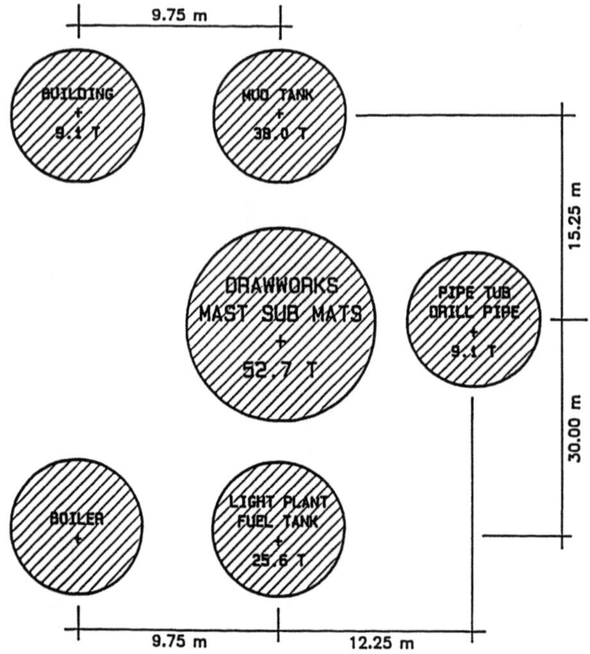

ITEM	LOAD (tonnes)	RADIUS (m)	DIST. (m)
DRAW WORKS SUB MAST MATS	52.7	2.0	0
PIPE TUB PIPE	9.1	3.0	12.25
LIGHT PLANT	25.6	3.0	15.25
BOILER	9.7	2.0	18.1
MUD TANK	33.0	2.0	30.0
BUILDING	9.1	2.0	31.5

Figure 5. Design loads and configuration for Westburne Rig #40.

The final thickness of pads 13-7 and 13-8 was 1.48 m, slightly less than the design thickness of 1.53 m, but construction was stopped to allow cooling of the pads.

The effects of warm ambient temperatures during construction were complicated by the fact that during flooding confining dikes of snow had been built and layers 75 to 100 mm thick had been flooded. These took much longer to freeze than do the normal flood layers of 25 mm thickness when free flooding techniques are used. Efforts to "conserve" flood water by dyking with snow usually cause more problems than any resulting savings in water warrant. At the end of construction the average freeboard of the ice was 75 to 90 mm instead of the planned 150 mm. There were water inclusions in the ice which resulted in the reduced freeboard. The ice temperature was uniformly 0°C. After cooling and complete freezing, the freeboard of the platforms reached the desired 150 mm.

4. Ice Strength and Stiffness

Ice strength and stiffness measurements of the ice pad were made after construction using a borehole jack (Masterson, 1983; Kivisild, 1975). Profiles of strength and elastic modulus were obtained at 150 mm intervals through the depth of the ice pads at various locations on the pads. The confined compressive strength of the ice varied between 20 and 30 MPa. A minimum of 10 MPa was required by the design. The measured elastic modulus was between 4 and 5 GPa and the design value was 4.8 GPa. Thus the mechanical properties of the ice were well within design specifications and met normal safe design criteria (Masterson and Gamble, 1986).

A load test was conducted on pad 13-7 using a 36.4 tonne truck to check the ice behaviour. This truck caused a maximum, extreme fibre tensile

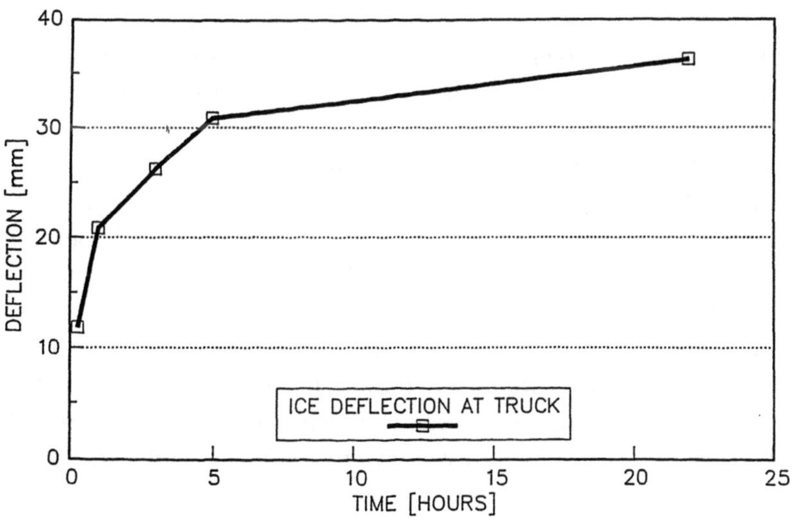

Figure 6. Truck deflection test, pad 13-8.

stress in the ice of 255 kPa, approximately one half the design value. Thus this load resulted in significant stress in the ice and was a realistic test of its competence to carry the desired loads. The instantaneous, elastic deflection and the long term creep deflection under the truck were measured and are plotted in Figure 6. The deflection rate was reduced to virtually zero after 24 hours and the instantaneous deflection yielded an elastic modulus of 6.26 GPa. The test thus confirmed the design assumptions and that the platforms were capable of carrying the specified loads.

5. Ice Pad Performance

5.1 Drilling from Pad 12-17

Before moving the rig onto the ice pads and beginning rig up, a truck mounted rathole drill was located at well centre to install a conductor pipe in the lake bed in preparation for drilling the hole. The weight of this rathole rig was about 75% of the design rig load. It was on location for several hours and caused 50 to 75 mm of deflection at the well centre, much of which was not recovered before the rig designated to drill the hole was brought in.

The rig up at 12-17 began at 1000 hrs. on February 23 and by 1600 hrs. all available freeboard at well centre had been lost. By 0400 hrs. on February 24 the ice had lost freeboard 10 m from well centre and by 1600 hrs. the bottom of the ice was in contact with the muddy bottom of the lake. A deflection vs. time graph for a survey point 10 m from the well centre is shown in Figure 7.

By February 28, 1986 the pad was more than 50 mm under water out to 10 m from well centre and the ice surface was 120 mm under water at well centre. Fortunately the lake was shallow at this well location and once the ice settled onto the bottom the deflection rate decreased dramatically. Unfortunately, the lake bottom was not stiff enough to stop all settlement. Had this well been in deeper water, as the other two sites were (Figure 1), it would have been necessary to abandon the well before finishing drilling. As it was the well was drilled as planned, although the crew had to work in a watery environment while drilling it. The rig was released on March 1, 1986 after 7 days on location.

There was no doubt that the ice had been overloaded. As pointed out in the previous section, the ice strengths and stiffnesses met or exceeded

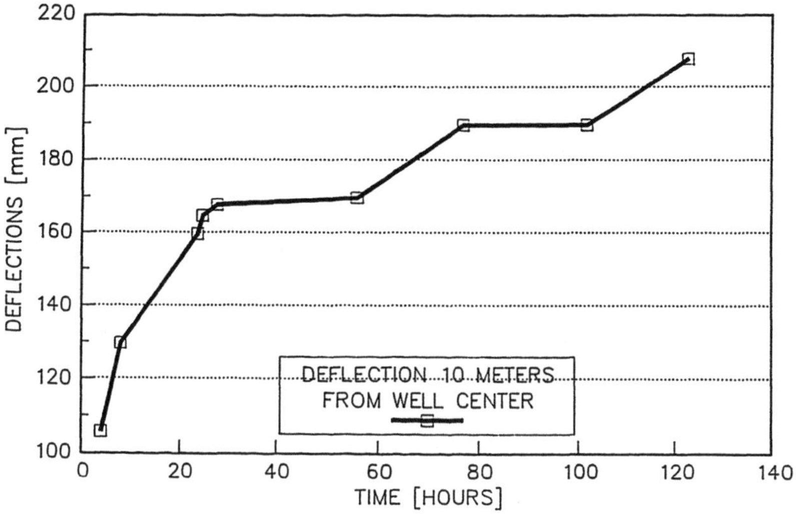

Figure 7. Deflection pad 12-17 at D10.

design specifications and pad 13-7 behaved as expected when tested with a long term load. On site inspection of the rig revealed it was not a truck mounted variety commonly used for drilling such holes but that it was a full, single stand rig with substructure, draw works and tower assembled on site during rig up. A very similar rig had been used to drill the first ice platform wells in the Arctic Islands by Panarctic Oils Ltd. (Baudais et al., 1974, 1976). A rig of this type would have to weigh at least 400 tonnes rigged up with full mud tank and basic supplies. The platform was designed for 132 tonnes (Figure 2).

No truck weigh bills were available to check weights and thus exact loadings on the ice could not be determined. It was decided that a different rig would be required to drill the remaining two holes. The loads and their distribution for this rig are shown in Figure 5. Each rig load was verified at a highway truck scale before it was allowed onto the floating ice of the lake. This was not done with the first rig because the loads given to the engineers by the drilling contractor were taken as fact. This proved to have been a mistake and loads were verified henceforth.

5.2 Drilling from Pads 13-8 and 13-9

The drilling went without serious incident on these pads. There were certain events of interest, one of which provides important information on the effects of load increase and on the adequacy of current design procedures and assumptions.

It was previously mentioned that, prior to rig up, a truck mounted drill called a rathole rig was located at well centre for several hours to set the conductor pipe for the well. This equipment weighed 75% of the design load of the pad and resulted in a loss of freeboard of 50 to 75 mm, only a portion of which might be recovered after removal of the load. In addition, cleanup around the well centre resulted in a loss of surface ice due to scraping, and thus a further loss of freeboard.

The load test, described in the previous section was conducted at well centre on pad 13-7. This resulted in some additional lost freeboard. It was thus decided to drill pad 13-8 first, allowing 13-7 to rebound.

Rig up at 13-8 began on March 2 and the rig was released March 7, 1986. There was virtually no loss of freeboard during the rig up, the freeboard being 60 mm before and after. Residual stress due to creep in the ice from previous loading by the rathole drill must have negated the initial effects of the stress due to the rig. That is, when the initial load was removed, the upward push by the water on the deflected shape would result in a compressive stress at the bottom of the ice and a tensile stress at the top fibre, opposite to the stress resulting from applied surface loading. Superposition of this stress plus that of opposite sense from the new surface load would result in a net lower stress than would be experienced in an initially unstressed ice. The result is low initial creep and deflection. With time this situation changes to a more normal stress regime.

A deflection vs. time graph for the main rig load and for the mud tank (see Figure 5) is shown in Figure 8. During the drilling of this well a loaded vacuum truck was located next to well centre for 5 hours from about hour 25 to hour 30. The vacuum truck weighed 25.5 tonnes or 50% of the rig load at well centre. This load, plus occasional visits from a 25 tonne water truck, caused the ice surface to sink to water level and the loss of all available freeboard at well centre for approximately a 36 hour period as shown in Figure 8. The situation was monitored very closely by engineers on site and no flooding of the ice surface resulted. The vacuum truck was removed and, as far as possible, no additional loads were permitted.

This additional load represented a 50% increase in the ice load and caused the loss of all freeboard. A brief analysis of this situation is

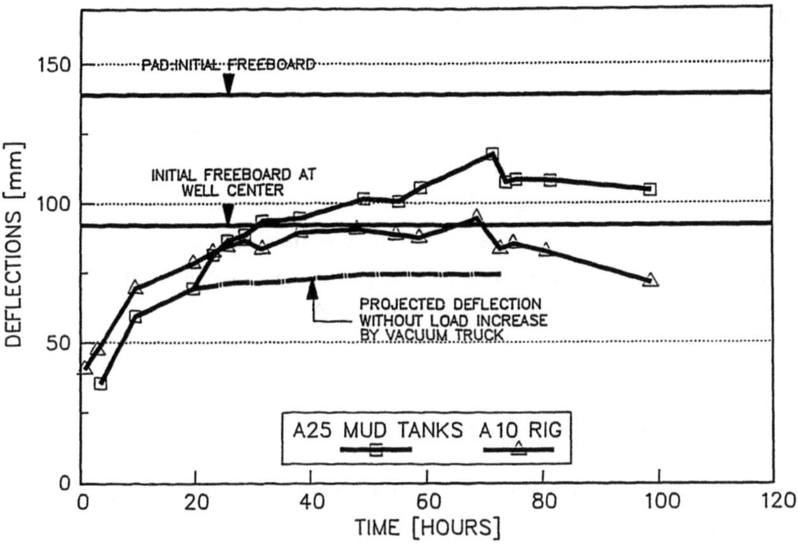

Figure 8. Deflection pad 13-8.

presented in the following section, showing the implications for future design of ice structures and the margins of safety provided by present guidelines.

The well at 13-8 was drilled without further incident. Cold weather did cause some surface cracking of the ice but these cracks remained dry and did not result in any weakening of the ice pad.

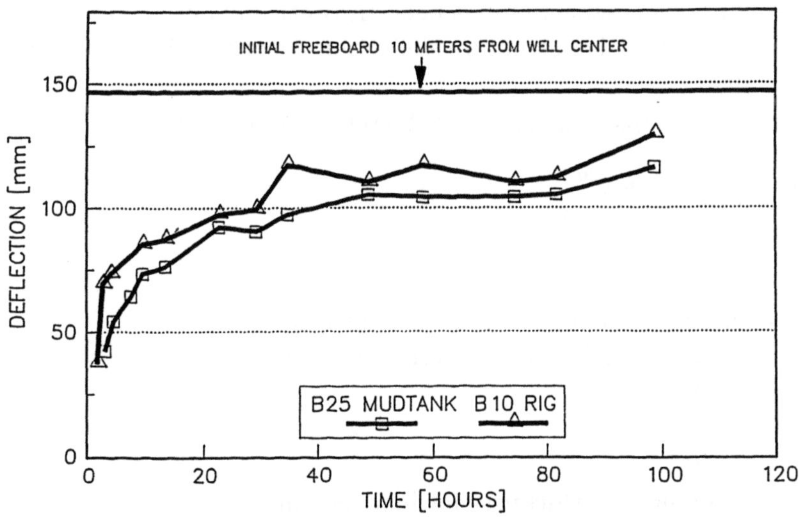

Figure 9. Deflection pad 13-7.

The well at pad 13-7 was drilled next. The final thickness of this pad was 1.48 m, slightly under the design thickness of 1.53 m. During the 6 day interval between the load test and rig up the pad rebounded to 75 mm freeboard. A deflection vs. time graph for this well is shown in Figure 9. Pad deflections and freeboard remained at acceptable levels throughout the period of loading. More than 15 mm of freeboard was available at the termination of drilling. The surface of the ice pad deteriorated significantly during the last two days. Warm temperatures caused considerable surface melting and ablation. No problems were encountered because of this.

6. Analysis of Creep Deflections

The loads which were placed on the ice pads during the drilling of these wells provided useful information on the adequacy of design assumptions used for load bearing ice structures. The overloading of pad 12-17 with the first drilling rig was so excessive and difficult to quantify, mainly because the loads were not known precisely enough, that little more can be said than has been. One can say that the state of the art is sufficiently developed that a structure designed for a particular load cannot be subjected to 3 times that load without severe consequences.

The load increase over design of pad 13-8 by the vacuum truck, and its effects, are, however, easily quantified. An exponential equation for the prediction of creep deflection has been theoretically derived and then calibrated using actual ice structure deflection records (Masterson and Strandberg, 1980):

The equation is written as follows (Masterson and Strandberg, 1980)

$$\frac{w(t)}{w_0(t)} = \left(\frac{P}{P_0}\right)^n \left(\frac{h_0}{h}\right)^{2n}, \tag{7}$$

where

$w(t)$ = unknown deflection of load P at time t,
$w_0(t)$ = reference or known deflection for load P_0,
P = load,
P_0 = reference load,
h = ice thickness,
h_0 = reference ice thickness for load P_0, and
n = flexural creep constant = 1.5.

If we investigate the effects of load variation only, then the ice thickness is constant, and $h = h_0$, and

$$w(t) = w_0(t) \quad (P/P_0)^{1.5}, \tag{8}$$

If the load is increased by 50%, as it was in the case of pad 13-8, then,

$$P = 1.5P_0, \tag{9}$$

and

$$w(t) = 1.84 \quad w_0(t). \tag{10}$$

Without the increase in load the deflection of pad 13-8 at well centre would have been less than the available freeboard, as is shown in Figure 8 by projecting the curve from the point where the additional load was applied to the pad. The increase in deflection was actually from 60 mm to 90 mm or an increase by a factor of 1.5. The equation, in this case, overpredicted the effect of a heavier load by 23%.

Within the limits to which loads and their distribution can be known and within the limits of measurement error, formulas based on basic elastic plate and material theory and calibrated by experience correctly predict that increasing the load on a floating ice structure above the design load by a factor of 1.5 will result in a corresponding increase in deflections which are likely to be excessive.

The design of the Marguerite Lake ice pads required an initial available freeboard of 150 mm. At design load, the deflection during drilling was calculated to be between 90 and 120 mm. According to current practise in design, increasing this deflection by 1/3 due to a 50% load increase would result in a situation where freeboard was lost and possibly where the pad was submerged at well centre, the point of maximum stress and load. Increasing the load on the pad by a factor of 3 over design would result in total submergence of the structure. Both of these predictions were verified.

From the above, it is seen that the current practise is reasonable and not overly conservative, especially when contingencies such as pre-deflection of the ice due to other loads and loss of ice due to preparatory activities are considered. It should be noted that where heavier rigs are used to drill wells or in any situation where the load of the basic equipment is in the hundreds or thousands of tonnes, then significant load increase by incidental equipment is less likely.

A 25 tonne vacuum truck would have gone unnoticed had it parked beside one of Panarctic's 1600 tonne rigs used for offshore drilling from floating ice platforms.

7. Summary and Conclusion

BP Canada successfully drilled three, shallow stratigraphic oil exploration wells into the bottom of Lake Marguerite in the late winter of 1986 using the natural ice cover as a support for drilling. The conclusions drawn from the experience and observations follow.

- Load increases experienced during the drilling of the wells verified current design procedures for predicting the long term creep deflection of floating ice structures.
- Increasing the applied load over design by a factor of 1.5 resulted in an increase in deflection by a factor of 1.5. Current power formulas in use predict an increase in deflection of 1.84, an overprediction of 23%. This is reasonable agreement considering errors involved in the estimation of load and its distribution plus variability in ice properties.
- Increasing the load over design by a factor of 3 resulted in loss of freeboard and flooding of the work area, something also predicted by current design formulas.

Acknowledgement

The authors wish to thank BP Canada Inc. for initiating this work and for granting permission to publish the results. The assistance and encouragement of Mr. Carl Merritt during the project were very much appreciated.

References

Baudais, D.J., Masterson, D.M. and Watts, J.S. 1974. A system for offshore drilling in the Arctic Islands. Journal of Canadian Petroleum Technology, July-September.

Baudais, D.J., Watts, J.S. and Masterson, D.M. 1976. A system for offshore drilling in the Arctic Islands. Offshore Technology Conference, Houston, Texas, Paper No.OTC 2622, Vol.3, pp.31-43.

Kivisild, H.R. 1975. Ice mechanics. Proc. Third International Conference on Port and Ocean Engineering under Arctic Conditions (POAC), Fairbanks, Alaska, Vol.2, pp.287-314.

Masterson, D.M. 1983. Interpretation of in situ borehole ice strength measurement tests. Proc. Symposium on the Mechanics of Ice, Rocks and Soils, ASME Applied Mechanics, Bioengineering and Fluids Engineering Conference, Houston, Texas.

Masterson, D.M. and Kivisild, H.R. 1980. Floating ice platforms: offshore oil exploration. ASCE Journal of the Structural Division, Vol.106, No.1, pp.133-143.

Masterson, D.M. and Strandberg A.G. 1980. Creep and relaxation in floating platforms - an analysis of case histories. Proc. Physics and Mechanics of Ice, IUTAM Symposium, Copenhagen, Denmark, 1979, pp.205-216.

Masterson, D.M. and Gamble, R.P. 1986. Current ice road and structure design and construction procedures. Proc. International Polar Transportation Conference, Vancouver, Canada, pp.602-648.

Masterson, D.M., Anderson, K.G. and Strandberg, A.G. 1979. Strain measurements in floating ice platforms and their application to platform design. Canadian Journal of Civil Engineering, Vol.6, pp.394-405.

Masterson, D.M., Baudais, D.J., Pare, A. and Bourns, M. 1987. Drilling of a well from a sprayed floating ice platform - Cape Allison C- 47. Proc. Sixth International Conference on Offshore Mechanics and Artic Engineering (OMAE), Houston, Texas, Vol.4, pp.9-16.

Timoshenko, S. 1940. Theory of plates and shells. McGraw Hill Book Co. Inc., New York, N.Y.

Discussion

Q: K.R. Croasdale, ESSO Resources Canada Ltd.

1. How are cracks accounted for in the empirical bearing capacity equations?

2. Were cracks present during the operation and did they have any noticeable effects?

3. A load of 3 to 4 times the design load would imply a bending stress of 3 to 4 times of 75 psi. Would you comment on why instantaneous failure of the ice sheet did not occur?

A: D. Masterson

1. The ice is assumed to have a "normal" amount of thermal cracking. Open leads require reassessment of stress and deflection.
2. The second well platform (13-8) had severe thermal cracking of the ice during drilling. No increase in deflection was noted because of this. The step increase was directly attributable to increased load by the vacuum truck.
3. Undoubtedly the stress was increased due to the load increase. The platform settled and contacted bottom before strains in the ice reached a level to cause failure. Had the water been deeper, the rig would have begun punching through. The ice was warm and ductile enough that brittle failure was precluded.

Q: L. Gold, NRC

Are the records of the performance of the Panarctic platforms now in the public domain? This data base would be excellent for the validation of theories for predicting the performance of ice platforms under static loads.

A: D.M. Masterson

The Panarctic ice platform data is in the public domain. Most of the key results have been published in journals and conference proceedings.

Q: R.P. Nordgren, Shell Development Company and Rice University

The test results could be compared with predictions of plate theory with a stress-strain relation for creep obtained from small scale experiments. This would provide a check on the empirical creep formula in the paper.

A: D. Masterson

The kind of comparison can be very useful. It has been attempted using finite element analysis in the past. Since stress relaxation as well as creep is involved, the process is not simple. We intend to publish more on this matter in the future.

Creep Constitutive Equations for Polycrystalline Ice and Effect of Microcracking

P. Duval

Laboratoire de Glaciologie et Géophysique de l'Environnement, CNRS
BP 96 38402 St. Martin d'Hères, Cedex, France

P. Kalifa

Institute for Marine Dynamics, N.R.C.,
P.O.Box 12093, Postal Station A, St. John's, Newfoundland, CANADA

J. Meyssonnier

Laboratoire de Glaciologie et Géophysique de l'Environnement, CNRS,
BP 96 38402 St. Martin d'Hères, Cedex, France

Abstract The deformation modes of polycrystalline ice are reviewed. Whereas single crystals of ice deform easily by shear on the basal plane, the minimum creep rate of polycrystalline ice is largely controlled by processes occuring on non-basal planes. Three main models describing the transient behaviour of polycrystalline ice are presented and discussed. Evidence that the occurrence of the first cracks, in the ductile-brittle transition, is related to a critical delayed elastic strain is presented. The mechanical behaviour of damaged ice is discussed. It is shown that damage enhances the creep rate notably.

1. Introduction

It is a matter of common experience that polycrystalline ice can be deformed up to large strains. In the flow of glaciers, plastic strains of more than 20% are easily achieved. For this to be possible, each ice crystal of the polycrystalline aggregate must have at least four independent systems of deformation available to it (Hutchinson, 1977). It is well known that the flow stress for slip on the basal plane of the ice crystal is very low. Dislocations on the basal planes can move for shear stresses lower than 0.01 MPa. This mode of deformation provides only two independent

S. Jones · R. F. McKenna · J. Tillotson · I. Jordaan (Eds.)
Ice-Structure Interaction
IUTAM-IAHR Symposium St. John's, Newfoundland, Canada

systems. Polycrystal plasticity must involve slip or climbing of dislocations on non-basal systems, and because of its difficulty, this process must play a major role in controlling the macroscopic behaviour (Duval et al., 1983). Grain boundary sliding, associated with intracrystalline dislocation glide, may make the deformation homogeneous. The nucleation of cracks at high stresses is probably caused by the highly plastic anisotropy of the ice crystals.

The aim of this paper is to examine the deformation modes of polycrystalline ice. Transient flow behaviour is especially analyzed with a particular emphasis given to the role of internal stresses in flow behaviour at low and high strain rates.

2. Deformation of Single Crystals of Ice

One of the main features of the plastic deformation of an ice crystal is its very strong anisotropy. The resistance to slip on the non-basal planes is much larger than that on the basal planes (Duval et al., 1983). A single crystal deforms easily by shear in the basal plane, i.e. along the systems (0001) - $< 11\bar{2}0 >$. The difficulty in producing non-basal glide appears to be surprising as edge dislocations in non-basal planes move faster than dislocations in the basal plane (Fukuda et al., 1987; Ahmad and Whitworth, 1988). The easier glide on the basal planes is attributed to a larger dislocation density in these planes than in the non-basal ones, and to the fact that screw dislocations are dissociated in the basal planes (Fukuda et al., 1987). In addition to the basal and prismatic slip systems, climbing of $< 0001 >$ or $1/3 < 11\bar{2}3 >$ dislocations can contribute to the deformation of the ice crystals along the c-axis. It appears that the contribution of this mechanism far exceeds that of the Nabarro diffusion process (Fukuda et al.).

3. Deformation of Polycrystalline ice

Extensive deformation of polycrystalline ice may occur if several systems of deformation are available. Results based on the self-consistent method show that extensive plasticity is possible with only four independent systems (Hutchinson, 1977). Shear may occur on the basal plane (giving two independent systems) and with a greater difficulty on the prismatic planes (giving two more independent systems). Since the ice resistance to creep depends greatly on the system involved, the plastic anisotropy is enormous. It follows that although the state of stresses within the polycrystalline ice aggregate may be seen as uniform on the very first stage of loading, the load is transferred to the harder systems as creep relaxes the resolved stress on the basal plane of each grain. Thus in an isotropic polycrystal, an increasingly non-uniform state of internal stresses develops, with the following consequences:

1. there is a geometric primary creep associated with a stress redistribution due to the relaxation of the softer systems, during which the creep rate decreases by a factor 100 or more;
2. the initial creep rate is largely determined by that of the softest system, but the steady-state creep rate is given by an average of the resistances of all systems;
3. there is a large anelastic or recoverable component of the creep strain which can be several times the initial elastic strain (Duval, 1978);
4. the internal stress distribution which develops during creep has a dominant wavelength in the order of the grain diameter and presents peaks whose amplitude can be several times the applied stress.

These internal stresses oppose further deformation giving directional or kinematic hardening. They can be relieved by reversing the strain, so they are responsible for the large recoverable strain which appears upon unloading. Superimposed on this kinematic hardening is the hardening caused by short-range interactions between dislocations on parallel or intersecting planes. This latter is considered as isotropic.

4. Constitutive Law

Le Gac and Duval (1980) proposed that the creep of polycrystalline ice is driven by an effective stress equal to the applied stress σ reduced by the back-stress S responsible for kinematic hardening and by the stress σ^* responsible for isotropic hardening. The creep rate for uniaxial compression or tension is thus given by:

$$\dot{\varepsilon} = A(|\sigma - S| - \sigma^*)^n, \tag{1}$$

where n is the power law exponent and A a temperature dependent factor. The internal stresses S and σ^* evolve with strain and their rate of change is a function of the strain-rate. Recovery processes depend on S and σ^* through a power law.

Recently Shyam Sunder and Wu (1989) have proposed a model based on the concept of reduced stress. The reduced stress $\sigma - S$ drives only the transient strain-rate. A non-dimensional drag stress B is introduced as well. The evolution equation for the transient creep is given by:

$$\dot{\varepsilon}_t = ((\sigma - S)/BV)^n, \tag{2}$$

where V is a temperature dependent factor. This model predicts that the transient strain is totally recoverable. With Le Gac and Duval's model a part of the transient strain is not recoverable. The very long time which would be necessary to achieve a totally recoverable strain during a creep recovery test makes it difficult to assess these predictions.

A non-linear visco-elastic model has been developed by Sinha (1979) to describe the transient and steady-state creep for transversely isotropic columnar ice. The recoverable delayed elastic strain represents the transient strain and is assumed to be associated with a grain boundary sliding mechanism. This time dependent strain component is assumed to be inversely proportional to the grain size. From Sinha (1979), the ratio of the maximum delayed elastic strain to the pure elastic strain lies between 3 and 9 for grain sizes varying from 3 to 1 mm. From Duval (1978), this ratio can be greater than 10 for granular ice. By adopting Sinha's relation, a good description of Jacka's (1984) creep data was obtained, but with a transient strain of 70 σ/E (Ashby and Duval, 1985).

5. Microcracking and Delayed Elasticity

Sinha (1984) assumed that the formation of a crack in columnar ice was related to a critical boundary-sliding displacement and hence to a critical delayed elastic strain. Experimental observations support this assumption of a critical delayed elastic strain. Kalifa et al. (1989) reached the same conclusion by determining the strain at first cracks for granular ice under triaxial compression. The effect of the confining pressure on the nucleation of cracks in an ice sample with a grain size of 5 mm is shown in Figure 1. It was demonstrated that the increase in the strain at the first cracks with the confining pressure was related to the increase of the delayed elastic strain. This reveals the predominant role of the transient strain in the crack nucleation process and this is consistent with other interpretations (Gold, 1972; Sinha, 1984; Cole, 1986). The stress concentration due to dislocation pile-ups or grain boundary sliding can induce crack nucleation.

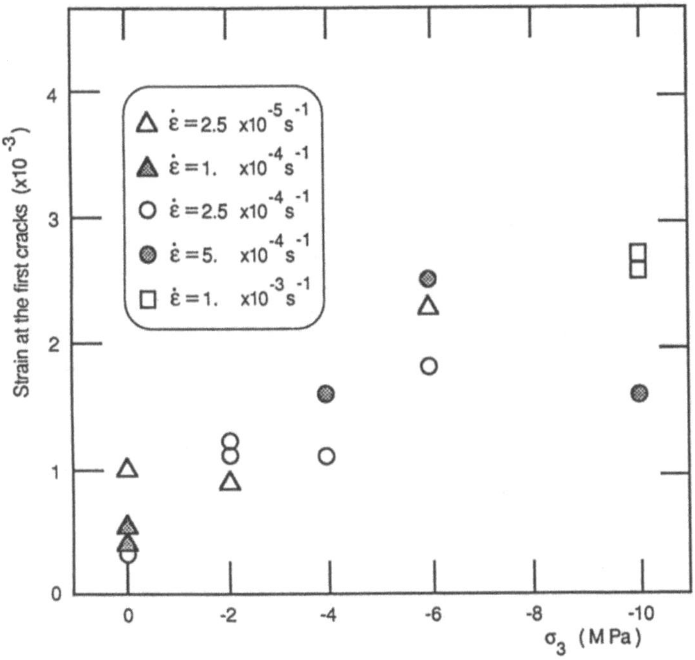

Figure 1. Effect of the confining pressure on the strain at first cracks ($T = -10^o$ C, grain size = 5 mm)

However, the time-dependent dislocation mechanisms can be questioned at high strain-rates since the time required for the formation of a dislocation pile-up may be too long with regards to the time observed for the initiation of the first cracks. From Cole (1988) and Shyam Sunder and Wu (1990) the elastic anisotropy of ice crystals can induce the nucleation of cracks at high rates of loading. But since this pure brittle mechanism does not require a critical delayed elastic strain, the dislocation mechanism cannot be ruled out for the nucleation of cracks in polycrystalline ice deformed up to $10^{-3}s^{-1}$ (Fig. 1).

The relative importance of the strain components in the deformation of damaged ice was analyzed by conducting mechanical tests under closed-loop controlled constant crosshead displacement velocity in uniaxial compression. Isotropic equiaxed polycrystalline ice with mean grain sizes of 1.5 mm and 4.5 mm were tested at -10°C ± 0.1°C. The height of the samples was 130 mm. A 500 kN hydraulic press provided the uniaxial force. The load was measured by means of a piezoelectric sensor and the strain was measured with a L.V.D.T. transducer mounted on the central part of the specimens. Figures 2 and 3 show the variation of both stress and strain with time for two ice samples with mean grain size of 1.5 and 4.5 mm. After unloading, the strain can be decomposed in an instantaneous elastic strain ε_e, a delayed elastic strain ε_d and a permanent viscous strain ε_v. Since ε_d increases with time after unloading, the values at t = infinity should be slightly higher than given in Figures 2 and 3. The elastic moduli of the damaged ice are respectively 7.1 GPa (grain diameter 4.5 mm) and 7.6 GPa (grain diameter 1.5 mm).

These values can be used to evaluate the state of damage of the samples (Chaboche, 1988). The evolution of damage is generally analyzed by

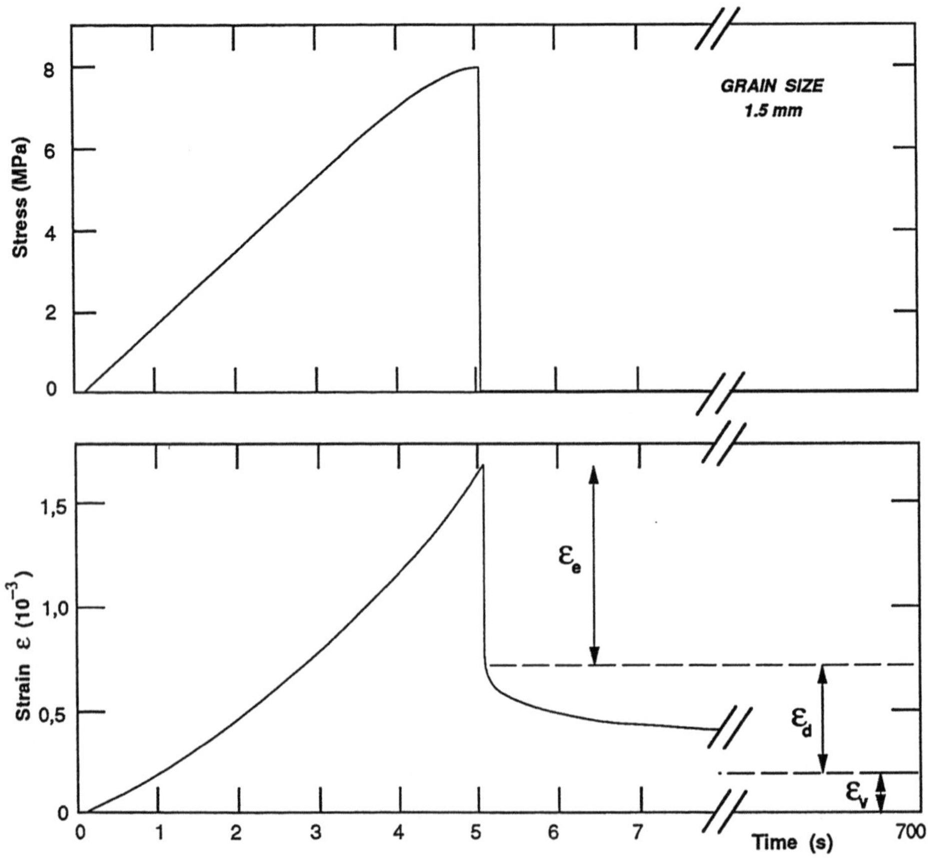

Figure 2. Stress and strain-time curves for isotropic polycrystalline ice under compression, upon loading and unloading ($T = -10°C$, grain size 1.5 mm, closed-loop controlled constant cross-head displacement velocity).

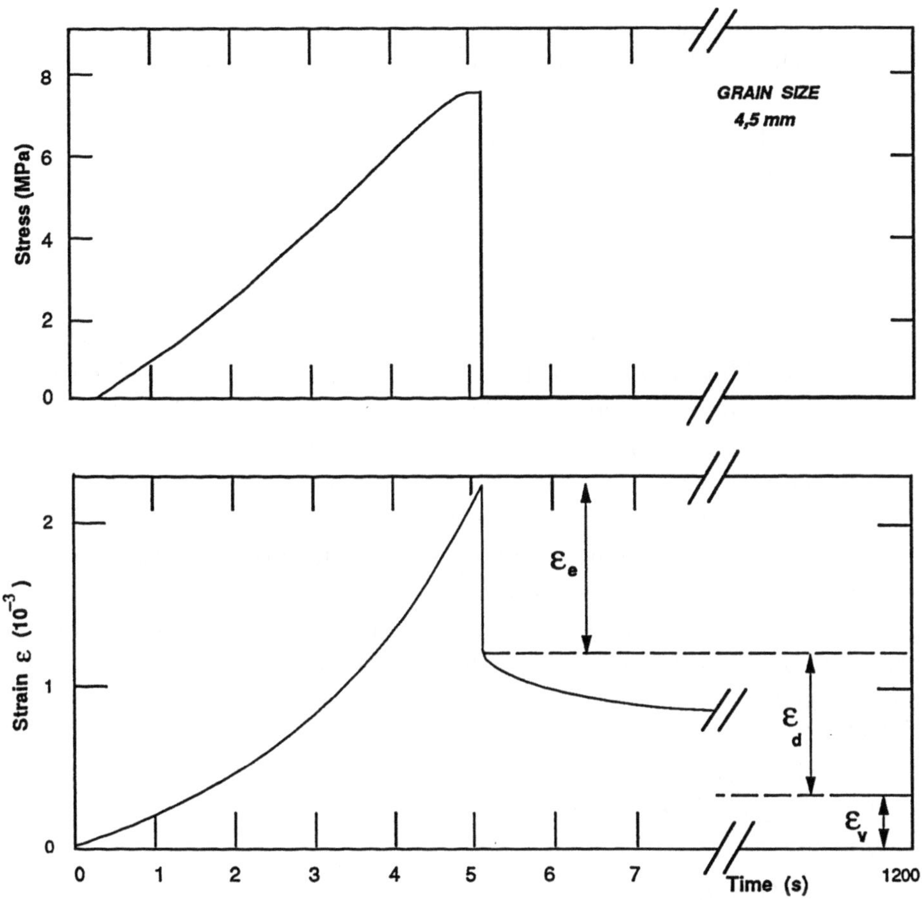

Figure 3. Stress and strain-time curves for isotropic polycrystalline ice under compression, upon loading and unloading ($T = -10°\mathrm{C}$, grain size 4.5 mm, closed-loop controlled constant cross-head displacement velocity).

adopting the concept of effective stress: the constitutive equations of the damaged material are deduced from the equations for the virgin material by replacing the stress with the effective stress (Chaboche, 1988; Sjölind, 1987).

The preliminary results shown in Figures 2 and 3 do not exhibit an obvious dependency of ε_d on the grain size. This raises the problem of the origin of the delayed elastic strain in polycrystalline ice. On the other hand, the visco-elastic model of Sinha (1979) specifies that the transient strain ε_t increases with decreasing grain size. On the other hand, a ratio of the maximum delayed elastic strain to elastic strain higher than 10 found by Duval (1978) cannot be induced only by grain boundary sliding (Raj and Ashby, 1971). It is therefore very important to initiate an experimental research program on the visco-elasticity of granular and columnar ice before and after the samples are damaged.

6. Creep Behaviour of Damaged Ice

A first analysis of the influence of damage on the creep rate of granular ice was done by Meyssonnier and Duval (1989). Ice samples were deformed in uniaxial compression at a constant strain rate of $5 \times 10^{-4}\mathrm{s}^{-1}$ until a certain amount of damage was reached. Then these samples underwent uniaxial compression creep tests, the level of stress being selected so that a further increase of microcracking was negligible. The main results of this study were that damage notably enhances the creep rate and that the exponent of the flow law keeps the same value (n=3) as at low stresses (see Figure 4). Up to the peak stress, the influence of damage seems rather well estimated by assuming that the ice exhibits the same response as undamaged ice submitted to an effective stress increasing with damage (Chaboche, 1988).

The effect of cracks on the creep rate was analyzed by Weertman (1969). For non-intersecting narrow cracks, this author gives the creep rate as:

$$\dot{\varepsilon} = \dot{\varepsilon}_u(1 + 2\pi N a^2 \sqrt{n}), \tag{3}$$

where $\dot{\varepsilon}_u$ is the creep rate of the undamaged ice, N is the number of cracks per unit area, a is the crack half-length and n the stress exponent of the flow law. This equation may be used as long as $a^2N \ll 1$. A crack density of 1 crack/grain was measured by Kalifa (1988) at the peak stress in the range 5×10^{-5} to $10^{-5}\mathrm{s}^{-1}$. By assuming that the crack length is about half the crystal size (Cole, 1986) the enhanced creep rate is about $1.6\,\dot{\varepsilon}_u$. A much higher factor was found by Meyssonnier and Duval (1989) at the peak stress.

The high plastic anisotropy described above must be taken into account to explain this large effect. From Weertman (1969) large effects on the creep-rate should be observed in anisotropic materials. For instance cracks increase the tensile creep-rate more than the shear creep-rate. On the other hand, cracks parallel to the basal plane can have a greater influ-

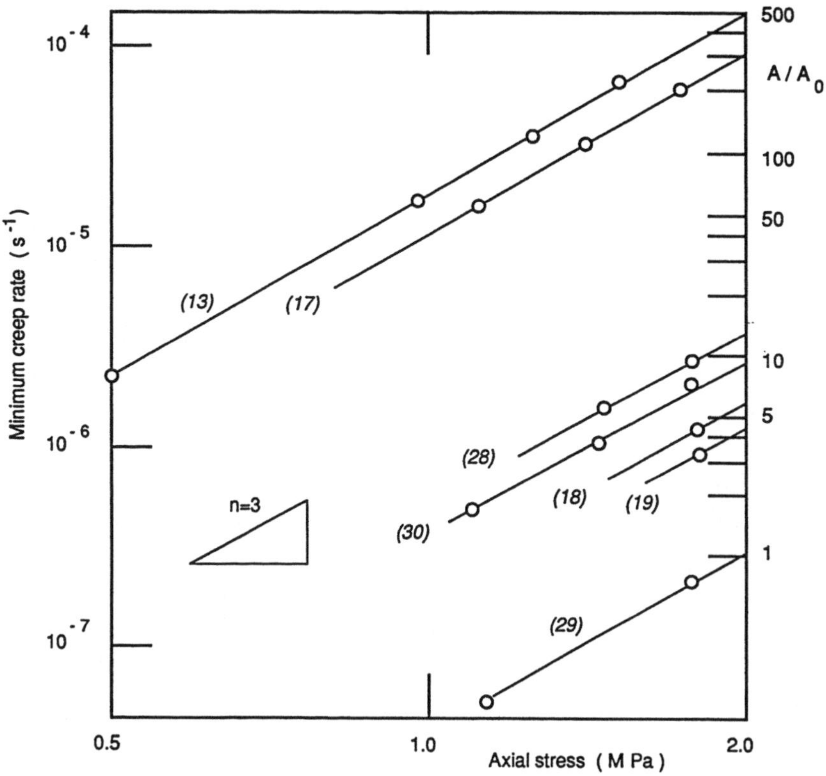

Figure 4. Minimum creep-rate for damaged samples, and ratios of the minimum creep rate to that of an undamaged sample, (No.29) ($T = -10°C$, grain size 4 mm, samples No. 18,19,28, and 30 were damaged up to the peak stress, samples No. 13 and 17 were damaged up to a strain of 0.9% and 0.5% respectively).

ence on the shear creep-rate than cracks perpendicular to it. In addition microcracking could affect primarily the grains which are misoriented for inelastic deformation, thus rendering the ice softer. A detailed analysis of the structure of the cracks is necessary to determine how microcracking really affects the creep-rate. For interacting cracks, friction across the compression cracks becomes the main process of the energy dissipation (Jordaan and McKenna, 1988; Jordaan and Timco, 1988).

7. Conclusion

The creep processes in polycrystalline ice have been reviewed. A general agreement exists on the processes involved in the steady-state creep deformation, and therefore on how they can be modelled. However, a great discrepancy is noted among the very few existing models for the transient behaviour of ice and their underlying physical mechanisms. Le Gac and Duval (1980) and Shyam Sunder and Wu's (1989) models are both based upon the concept of an internal stress involving intracrystalline processes: they differ mainly by the formulation adopted, the latter being perhaps more easy to deal with. To date, the conventional experiments do not allow us to come to a firm conclusion on this point. Sinha's (1979) widely used model for delayed elasticity is based upon the intercrystalline process of grain boundary sliding. Established for columnar ice, this model cannot predict the very large (compared to the pure elastic strain) recovered strain of granular ice. A great deal of experimental work should be performed in order to achieve a good description of transient creep. Such an effort will find its justification when modelling the behaviour of ice experiencing microcracking by permitting a better understanding of the nucleation process, a better accounting for the influence of the grain size on the peak stresses, and a correct estimate of the energy balance. Theoretical and experimental work should be done in parallel to understand the influence of a given population of microcracks on the steady-creep behaviour of ice and the possibly induced anisotropy. This work could help in understanding the nature of the peak observed in the stress-strain curves and in establishing the boundaries of this still hazy brittle-ductile transition zone.

Acknowledgements

This study was financially supported by the French "Centre National de la Recherche Scientifique" and "Mission de la Recherche et de la Technologie du Ministère de l'Education Nationale". The authors are especially grateful to Prof. E. Schulson, from Dartmouth College, Hanover-USA, for discussions and constructive comments. We wish to thank G. Marec for his technical suppport.

References

Ahmad, S. and Whitworth, R.W. 1988. Dislocation motion in ice: a study by synchrotron X-ray topography. Phil.Mag.A, Vol.57, No.5, pp.749-766.

Ashby, M.F. and Duval, P. 1985. The creep of polycrystalline ice. Cold Regions Science and Technology, Vol.11, pp.285-300.

Chaboche, J.L. 1988. Continuum damage mechanics: Part I, General concepts. Journal of Applied Mechanics, Vol.55, pp.59-64.

Cole, D.M. 1986. Effect of grain size on the internal fracturing of polycrystalline ice. CRREL Report 86-5, U.S. Army Cold Regions Research and Eng. Laboratory, Hanover, N.H., 79p.

Cole, D.M. 1988. Crack nucleation in polycrystalline ice. Cold Regions Science and Technology, Vol.15, pp.79-87.

Duval, P. 1978. Anelastic behaviour of polycrystalline ice. Journal of Glaciology, Vol.21, No.85, pp.621-628.

Duval, P., Ashby, M.F. and Andermann, I. 1983. Rate-controlling processes in the creep of polycrystalline ice. Journal of Physics and Chemistry, Vol.87, pp.4066-4074.

Fukuda, A., Hondoh, T. and Higashi, A. 1987. Dislocation mechanisms of plastic deformation of ice. Journal de Physique, C1, No.3, pp.163-173.

Gold, L.W. 1972. The failure process in columnar grained ice. National Research Council of Canada, Division of Building Res., N.R.C.C. 12637.

Hutchinson, J.W. 1977. Creep and plasticity of hexagonal polycrystals related to single crystal slip. Met. Transactions A, Vol.8A, pp.1465-1469.

Jacka, T.H. 1984. The time and strain required for the development of minimum strain-rates in ice. Cold Regions Science and Technology, Vol.8, pp.261-268.

Jordaan, I.J. and McKenna, R.F. 1988. Modelling of progressive damage in ice. Proc. 9th International Symposium on Ice, Sapporo, Japan, Vol.2, pp.585-624.

Jordaan, I.J. and Timco, G.W. 1988. Dynamics of the ice-crushing process. Journal of Glaciology, Vol.34, No.118, pp.318-326.

Kalifa, P. 1988. Contribution à l'étude de la fissuration dans la glace polycristalline en compression. Thése de Doctorat de l'Université Joseph Fourier, Grenoble, France, 149p.

Kalifa, P., Duval, P. and Ricard, M. 1989. Crack nucleation in polycrystalline ice under compressive stress states. Proc. 8th International Conference on Offshore Mechanics and Arctic Engineering, The Hague, The Netherlands, Vol.IV, pp.13-21.

Le Gac, H. and Duval, P. 1980. Constitutive relations for the non-elastic deformation of polycrystalline ice. Proc. IUTAM Symposium on Physics and Mechanics of Ice, Copenhagen, Denmark, pp.51-59.

Meyssonnier, J. and Duval, P. 1989. Creep behaviour of damaged ice under uniaxial compression: a preliminary study. Proc. 10th International Conference on Port and Ocean Engineering under Arctic Conditions, Lulea, Sweden, pp.225-234.

Raj, R. and Ashby, M.F. 1971. On grain boundary sliding and diffusional creep. Met. Trans., Vol.2, pp.1113-1127.

Shyam Sunder, S. and Wu, M.S. 1989. A differential flow model for polycrystalline ice. Cold Regions Science and Technology, Vol.16, pp.45-62.

Shyam Sunder, S. and Wu, M.S. 1990. Crack nucleation due to elastic anisotropy in polycrystalline ice, Cold Regions Science and Technology,Vol.18, pp.29-48.

Sinha, N.K. 1979. Grain boundary sliding in polycrystalline materials. Phil.Mag., Vol.40, No.6, pp.825-842.

Sinha, N.K. 1984. Intercrystalline cracking, grain boundary sliding and delayed elasticity at high temperatures. Journal of Materials Science, Vol.19, pp.359-376.

Sjölind, S.G. 1987. A constitutive model for ice as a damaging viscoelastic material. Cold Regions Science and Technology, Vol.14, pp.247-262.

Weertman, J. 1969. Effect of cracks on creep rate. Trans. of the A.S.M., Vol.62, pp.502-511.

Discussion

Q: L. Gold, NRC

This is only a comment and needs no response.

We have been considering a continuum of behaviour in the strain rate range of very low to very high e.g. 10^{-11} s^{-1} to 10^{-1} s^{-1}. Some of the features we have been considering are due to internal stress generated because of the inability of grains to conform with the strain of neighbouring grains. One deformation feature that has not been considered by the research community is the formation of voids, which occurs over a relatively narrow range of strain rate. Another is the collapse of a crack during deformation into a plane of voids, that occurs over an even smaller range of strain rate. A study of both mechanisms of volume increase could provide additional useful insight into the deformation behaviour of ice.

Q: O. Mahrenholtz

Equation (1) has two internal variables, mainly S and σ *. What are the corresponding equations of evolution for the analysis of the model?

A: P. Duval

The rates of change of S and σ * are given by:

$$\dot{S} = h\dot{\epsilon} - r \text{ and } \dot{\sigma} = h^* \; \dot{\epsilon} - r^*,$$

where h and h^* are hardening coefficients and r and r^* are recovery rates. For steady state we have $\dot{\epsilon}_{ss} = r/h = r^*/h^*$. If only static recovery occurs, we have: $r \propto \dot{S}^n$ and $r^* \propto (\sigma^*)^n$ (n is the exponent of the flow law) and h and h^* are independent of stresses.

Q: Ian Jordaan, Memorial University of Newfoundland

I was surprised by the difference between your results on the magnitude or delayed elastic strain and the prediction of Nirmal Sinha's model. Would you please comment further?

A: P. Duval

By assuming that the transient strain is fully recoverable, as predicted by the models of Sinha (1978) and Shyam Sunder and Wu (1989) the ratio of the maximum transient strain to elastic strain can be higher than 20 from creep data on granular ice. This result implies that only intragranular dislocation glide can explain this very high delayed elastic strain - more experimental data on granular ice are necessary to verify if the delayed elastic strain depends on grain size.

Kinetics of Microcracking and Dilatation in Polycrystalline Ice

Nirmal K. Sinha

Geotechnical Section
Institute for Research in Construction
National Research Council of Canada
Ottawa, Ontario, K1A 0R6

Abstract The paper presents a micromechanically based, structure sensitive constitutive equation for creep and dilatation for uniaxial stress application. The model can be applied for constant stress as well as constant strain-rate loading conditions. It is based on intragranular dislocation creep enhanced by grain-facet size cracks produced during deformation by the embrittlement process that is caused by an intergranular sliding mechanism. Incorporation of the kinetics of microcracking activity, enhancement of the matrix creep by these cracks and the associated generation of void volume is the foundation of the theory.

1. Introduction

When an ice body interacts with a structure, the three dimensional states of stress and strain that develop within the structure and in the ice are rather complicated. Elastic interactions take place momentarily in the beginning as the first contact is made. However, with time, which could last only a fraction of a second, inelastic interactions dominate the processes of deformation. This is primarily due to the dynamic response of the ice-structure system in conjunction with the nonlinear viscoelastic response of ice which is complicated further by the damage processes in the material that include micro- and macro-scale cracking activities. Inelastic response of the material of the structure could also play a role during these interactions. Analysis of such complicated processes requires a thorough knowledge of the mechanical properties of both ice and the material with which the structure is constructed.

S. Jones · R. F. McKenna · J. Tillotson · I. Jordaan (Eds.)
Ice-Structure Interaction
IUTAM-IAHR Symposium St. John's, Newfoundland, Canada

The lack of a reliable, microstructurally sensitive, three dimensional constitutive model of ice (and for that matter - of any material) is the stumbling block against the use of sophisticated computer modellings and numerical simulations available today. A successful constitutive equation should have a physical basis and be able to describe the features that can be readily seen in the real field conditions. Such a model for inelastic deformation must be able to predict the usual empirically derived relations and be adequate for extrapolation to account for the effects of microstructural alterations or loading conditions. To date most constitutive equations are highly empirical in nature and are useful for interpolation in the range investigated but inadequate for extrapolation.

Before trying to formulate the deformation and damage processes in ice under a complicated three-dimensional stress or strain field, it would be beneficial to understand first how cracks form and multiply in ice under simple loading conditions, such as uniaxial cases. The next problem is to understand how the kinetics of cracking influence the subsequent deformation processes in the matrix. Examinations should also be made on the effect of cracks in producing dilatations (or dilations) and hence changes in the volumetric strain. One advantage of examining the uniaxial response is the relative ease with which the theoretical predictions can be compared with those of experiments.

2. Elastic Response

Since the material responds elastically at the moment a load is applied and since elasticity plays, as will be shown here, a major role in shaping the viscoelastic response, it is appropriate to describe this response first. Following Hooke's linear elastic theory, the axial elastic strain, ε_e, under an uniaxial stress, σ, is given by,

$$\varepsilon_e = \sigma/E, \tag{1}$$

where E is the Young's modulus. The volumetric strain is given by,

$$\varepsilon_{vol}(\text{elastic}) = \varepsilon_e(1 - 2\mu), \tag{2}$$

where μ is the Poisson's ratio. Questions often raised are: What are the values of E and μ? How do they vary with ice types and loading directions? To obtain a satisfactory answer, one has to examine the lattice behaviour of ice. Ice belongs to the family of hexagonal crystals and the elastic response of a single crystal of ice can be presented by the following matrix:

$$\sigma_i = C_{i,j}\varepsilon_j \quad i,j = 1 \text{ to } 6, \tag{3}$$

or

$$\varepsilon_i = S_{i,j}\sigma_j \quad i,j = 1 \text{ to } 6. \tag{4}$$

There are 12 non-zero coefficients in the matrices C_{ij} and S_{ij} for ice. Out of these only five are completely independent. These are C_{11}, C_{12}, C_{13}, C_{33} and C_{44} for the stiffness constants and S_{11}, S_{12}, S_{13}, S_{33} and S_{44} for the compliances.

Some averaging principles are required in deriving the bulk elastic properties for polycrystalline mass from the single crystal elastic constants. Methods have been developed for calculating the Young's modulus, E, shear or rigidity modulus, G, and Poisson's ratio, μ, of polycrystalline ice - granular and columnar-grained - on the basis of single crystal data (Sinha, 1989c). The most useful aspect of the above exercise is the development of simple two term equations, readily available to the engineering needs, that describe the temperature dependence of the moduli. It is shown, for example, that Young's modulus and Poisson's ratio for isotropic, granular ice can be presented by,

$$E = 8.93 + 1.2 \times 10^{-2}(T_m - T), \tag{5a}$$

and

$$\mu = 0.308 + 7 \times 10^{-5}(T_m - T), \tag{5b}$$

where E is in $GN \cdot m^{-2}$ and the temperature, T, is in Kelvin and where T_m is the melting point. For transversely isotropic columnar-grained S-2 ice (see classification in Michel and Ramseier, 1971), the temperature dependence of Young's modulus in the horizontal plane (normal to the length of the columns) is given by,

$$E_{(h)} = 9.39 + 1.3 \times 10^{-2}(T_m - T). \tag{6}$$

Equations (5a) and (6) show that the elastic properties of granular ice and S-2 ice, in spite of their large differences in the fabric, are very similar. Note that the Poisson's ratio of granular ice is calculated to be 0.308 at the melting point and it increases only by about 1 percent in the temperature range 273 to 233 K, the range important to most engineering needs. In the same range of temperature, Young's modulus increases by about 5 percent.

It is shown further (Sinha, 1989c) that elastic response of S-2 ice is nearly isotropic; E in the vertical direction (growth direction) is only about two percent greater than that in the horizontal plane. S-3 type oriented sea ice shows behaviour similar to S-2 ice with slight anisotropy in the horizontal plane that depends on the degree of orientation. Columnar grained, S-1 type ice shows the maximum anisotropy; E in the vertical direction is 23 percent larger than that in the horizontal plane. The response of this ice in the horizontal plane, however, is the same as that of the S-2 ice in the vertical direction. Granular ice shows the least stiffness. Its Young's modulus is 5 percent lower than that of S-2 ice in the horizontal plane and 7 percent lower than that of S-1 ice in the same plane.

3. Kinetics of Microcracking

When a material is subjected to a tensile load, an immediate volume dilatation (Equation (2)) or a decrease in density occurs. A reverse phenomenon occurs under a compressive load and an immediate compaction or an increase in density occurs. The material is no longer in equilibrium and tends to re-establish its characteristic density by mobilizing additional mechanisms of deformation. Following elastic response, the time dependent or delayed elastic response, associated with the grain boundary sliding (gbs) mechanisms, takes a dominant role in shaping the deformation characteristics of polycrystalline materials, including ice, at homologous temperatures greater than about 0.4 T_m. For uniaxial constant stress, the delayed elastic strain, ε_d, is given by (Sinha, 1979),

$$\varepsilon_d = (c_1 d_1/d)(\sigma/E)^s[1 - exp\{-(a_T t)^b\}], \tag{7}$$

where d is the average grain diameter, b is a constant associated with the mobility of grain matrix dislocations and a_T is the inverse relaxation time at temperature, T. The stress exponent for gbs displacement is s

and c_1 is a constant corresponding to the unit grain size $d_1(= 1\text{mm})$ so that c_1/E^s gives the limiting delayed elastic strain (at infinite time) for unit grain size and under unit stress ($\sigma = 1MN.m^{-2}$).

The presence of Young's modulus, E, in Equation (7) indicates that delayed elasticity is related to the elastic strains associated with the grain matrix stress concentrations caused by grain boundary sliding activities. These stress concentrations often lead to the formation of wedge type microcracks, comparable to the length of the grain facets. If x_c is the critical gbs displacement corresponding to a critical gbs strain, ε^c_{gbs} and hence delayed elastic strain, ε^c_d (Sinha, 1979), and t_{fc} is the incubation time for the onset of formation of wedge type cracks, then (Sinha, 1984a),

$$\begin{aligned} x_c &= \varepsilon^c_{gbs}d/K = \varepsilon^c_d d/K \\ &= (c_1 d_1/K)(\sigma/E)^s[1 - \exp\{-(a_T t_{fc})^b\}], \end{aligned} \tag{8}$$

where K (assumed = 1) is a constant of proportionality between x_c and ε^c_{gbs} or ε^c_d.

Equation (8) proved useful in predicting the stress dependence of t_{fc} and the minimum stress, σ_m, that has to be exceeded before cracks form (Sinha, 1984a). Under a constant stress, the cracks form in ice when the stress exceeds about $5 \times 10^{-5}E$. For $\sigma > \sigma_m$ and $t > t_{fc}$, more and more sites of stress concentration reach criticality and the number of cracks increases (Gold, 1972, Sinha, 1982). This continued damage process could also be expressed in terms of gbs displacement (Sinha, 1984b),

$$N = N_c[\exp\{\Psi(x - x_c)\} - 1] = N_c[\exp\{\Psi(\varepsilon_d d/K - x_c)\} - 1], \tag{9}$$

where N_c is the crack density corresponding to the first cracks and Ψ is a constant that depends on temperature; negative values of N for $x < x_c$ are neglected. The second relation in Equation (9) shows that N is dependent on the delayed elastic strain, ε_d, given by Equation (7).

4. Kinetics of Dilatation

Assuming the average crack length, 2a, to be equal to the length of the average grain facet and assuming the cross-sectional geometry of the grains to be hexagonal, it can be shown that,

$$a = \left(\frac{\pi}{24(3)^{1/2}}\right)^{1/2} d \quad (or\ 2a \approx 0.55d), \tag{10}$$

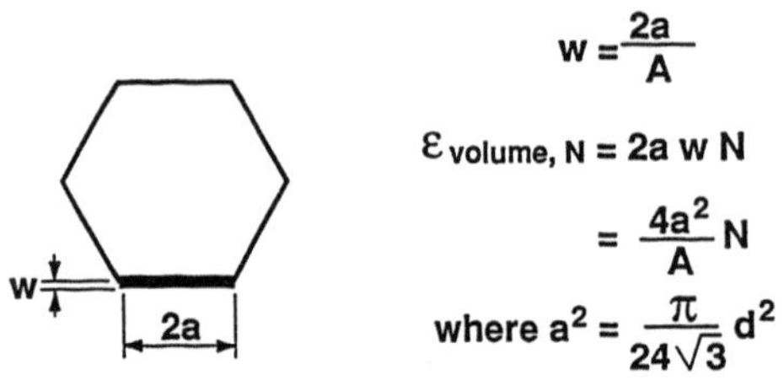

Figure 1. Geometry of the grains and the cracks.

where d is the average diameter of a circular grain having the same area ($\pi d^2/4$) as that of the hexagon (Figure 1).

Suppose w is the average depth (or opening) of the cracks. Then the average void volume per crack per unit volume of the material is $2aw$ if the cracks are assumed to be planar in geometry having their length infinitely larger than the other dimensions. The total porosity (void volume per unit volume) generated by a crack density of N is then given by,

$$\varepsilon_{vol}(\text{crack}) = 2awN = (\tfrac{\pi}{6(3)^{1/2}})^{1/2} dwN. \tag{11}$$

It should be pointed out that a simple crack geometry is assumed here to maintain simplicity in the formulations. For columnar-grained ice, which forms the bulk of lake ice as well as landfast sea ice, the assumed crack geometry is not far from reality (Sinha, 1988a). However, a "penny-shaped" crack geometry, to be discussed elsewhere, perhaps would be more suitable for granular ice.

The grain boundary sliding process produces stress concentrations at grain boundary jogs and triple boundaries, but the mechanism produces only negligible change in the volume, particularly in pure materials without any air or gas bubbles and other inclusions. Hence the material response, for simplicity, is considered to be incompressible due to this process (Sinha, 1987). The same assumptions can also be made for grain matrix viscous flow to be discussed later. Neglecting also the elastic creep or the effect of cracking activity on overall elastic strain (Sinha, 1988a), the total volumetric strain is given by the sum of the contributions made by the elastic response, Equation (2), and the cracks, Equation (11). Thus,

$$\begin{aligned} \varepsilon_{vol} &= \varepsilon_{vol}(\text{elastic}) + \varepsilon_{vol}(\text{cracks}) \\ &= \varepsilon_e(1-2\mu) + (\tfrac{\pi}{6(3)^{1/2}})^{1/2} dwN, \end{aligned} \tag{12}$$

Equations (9) and (12) give the kinetics of volumetric strain as,

$$\begin{aligned}\varepsilon_{vol} &= \varepsilon_e(1-2\mu) + \left(\tfrac{\pi}{6(3)^{1/2}}\right)^{1/2} dwN_c[\exp\{\Psi(x-x_c)\}-1] \\ &= \varepsilon_e(1-2\mu) + \left(\tfrac{\pi}{6(3)^{1/2}}\right)^{1/2} dwN_c[\exp\{\Psi(\varepsilon_d d/K-x_c)\}-1]. \quad (13)\end{aligned}$$

5. Crack Enhanced Matrix Creep

Formation of a crack intensifies the matrix creep rate. For dilute concentrations of non-interacting, long, grain-facet size cracks oriented with their major planes perpendicular to a uniaxially-applied tensile stress (or parallel to a uniaxially applied compressive load), and for materials obeying power law viscous flow, the enhanced creep rate at a constant temperature, $\dot{\varepsilon}_v$, is given by Sinha (1988a),

$$\dot{\varepsilon}_v = \dot{\varepsilon}_{Vo}(\sigma/\sigma_o)^n[1+\{\pi^2/(12(3)^{1/2})\}Nd^2n^{1/2}], \quad (14)$$

where $\dot{\varepsilon}_{Vo}$ is the viscous strain rate for unit stress, $\sigma_o = 1MN\cdot m^{-2}$ under conditions of no cracking, and n is the stress exponent for viscous flow.

6. Constitutive Equation

The total strain is given by the sum of the elastic, delayed elastic and viscous strains. From Equations (1), (7) and (14),

$$\begin{aligned}\varepsilon &= \varepsilon_e + \varepsilon_d + \varepsilon_v \\ &= (\sigma/E) + (c_1 d_1/d)(\sigma/E)^s[1-\exp\{-(a_T t)^b\}] \\ &+ \dot{\varepsilon}_{Vo}(\sigma/\sigma_o)^n \int_0^t [1+\{\pi^2/(12(3)^{1/2})\}Nd^2n^{1/2}]dt. \quad (15)\end{aligned}$$

7. Discussion

The applicability of the proposed Equations (9), (12) and (15) can be judged only if the material constants are known. Material constants should be determined by performing specific tests and after giving consideration to the micromechanics involved, as has been suggested in Sinha (1989a). In this respect considerable progress has been made in ice mechanics as may be seen from Table 1. Note that the values of E and μ given in the table, and determined experimentally, are compatible with theoretically developed Equations (5) and (6).

The simplest creep test is unconfined uniaxial in either tension or compression. The latter is preferred in testing geotechnical materials including ice. Usually a constant load is applied. For meaningful results to be obtained, the stress should be maintained constant or the strain level should be limited to avoid changes in stress level and nonuniform

Table 1. Material constants for creep and cracks for polycrystalline ice obtained from earlier experiments and analyses

μ	= (S-2 ice and S-3 sea ice)	Sinha (1988b)
$\dot{\varepsilon}_{Vo}$	$= 1.76 \times 10^{-7} s^{-1}$ ($T = 263$ K, $\sigma_o = 1$ MN m^{-2})	
E	= 9.5 GN m^{-2}(S-2 ice)	
K	= 1 (assumed), $s = 1$	
c_1	$= 9 \times 10^{-3}$, $d_1 = 1$ m (chosen unit)	Sinha (1979)
n	$= 3$, $b = 0.34 \approx 1/n$	
a_T	$= 2.5 \times 10^{-4}$ s^{-1} ($T = 263$ K)	
x_c	$= (M_1 - m_1T)d_1 / K$; where	Sinha (1984a)
M_1	$= 1.67 \times 10^{-6}$, $m_1 = 4.55 \times 10^{-9}$ K^{-1}	
N_c	= 550 m^{-2}, $\Psi = 1.33 \times 10^7$ m^{-1} ($T = 263$ K)	Sinha (1984b)

deformation of the specimen due to barreling effect. Often the current laboratory practice tends not to adhere to this ideal. Although the mathematical formulation of the constitutive equation is the issue here, it is good practice to be realistic. For this reason the following calculations are limited to only a few percent strain.

The kinetics of cracking, based on Equation (9) and Table 1, are illustrated in Figure 2; crack density per grain, corresponding to a grain diameter of 1 mm, is also shown in this figure (on the right hand side). These constant stress creep microfracturing calculations are consistent, as shown in Sinha (1984b), with the visual observations of Gold (1972), acoustic emission studies of Zaretsky et al. (1979) and dual acoustic emission/visual studies of Sinha (1982, 1989a) (all on columnar-grained ice) and acoustic emission studies of St. Lawrence and Cole (1982) (on granular ice).

Volumetric strains corresponding to the cracking activities can be calculated using Equation (13) provided the depth, w, of the cracks is known or assumed. To be realistic and to be simple, a width to depth ratio ($A = \frac{2a}{w}$) of 100 will be used here to illustrate the model. A set of calculations for volumetric strain during the early creep period is shown in

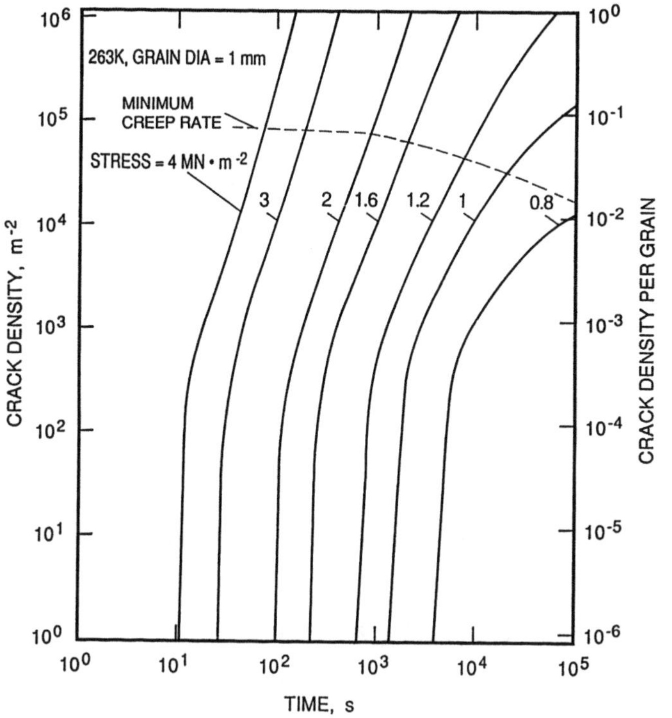

Figure 2. Stress dependence of the kinetics of microcracking under constant stress.

Figure 3. The small amount of initial elastic compaction might not be detected in most measurements. Figure 4 illustrates both axial strain and volumetric strain. In this figure the dilatation is shown as negative in order to be consistent with the chosen sign for the compressive axial creep curves. Examples of constant load creep and dilatation measurements carried out by the author are presented in Figure 5. Figures 5a and 5b are presented here to illustrate that the experimental results show the same characteristics as predicted theoretically. They show, however, that grain structure influences the behaviour. Since the assumed crack geometry is compatible to columnar-grained ice, Figures 4 and 5b are comparable provided the differences in the grain size are recognized. Although a detailed discussion is outside the scope of this paper, it should be pointed out that Figure 5a shows the relatively higher creep resistance of granular ice in comparison to columnar-grained ice in Figure 5b. The differences are predictable theoretically and will be presented elsewhere.

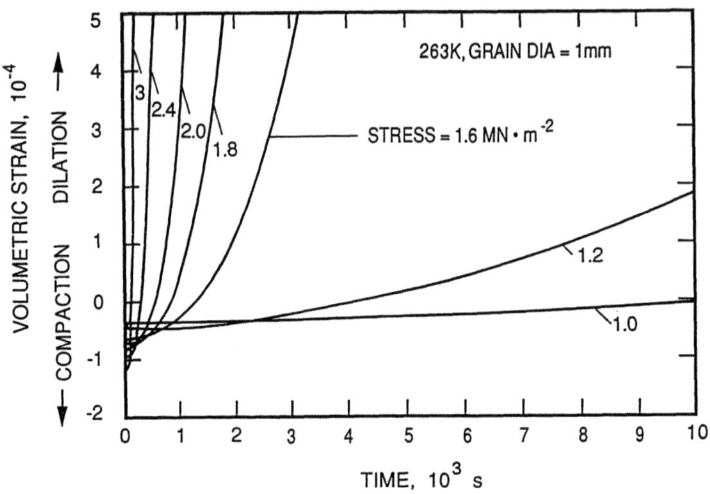

Figure 3. Stress and time dependence of volumetric strain under constant stress.

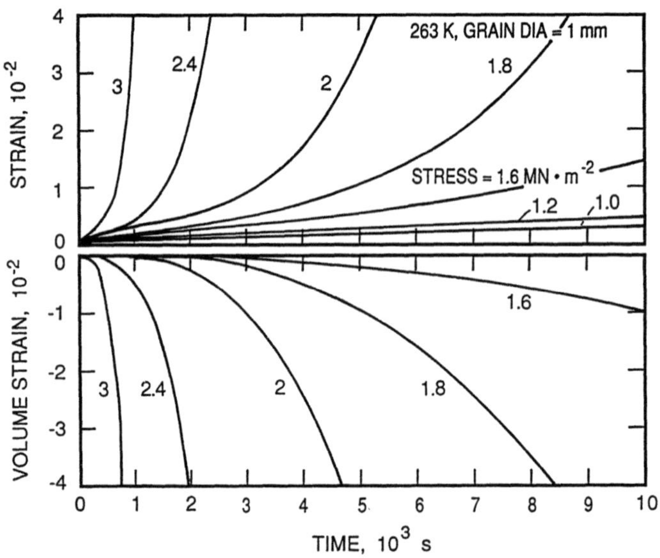

Figure 4. Theoretical stress and time dependence of axial and volumetric strain under constant stress.

Figure 4 shows the three pronounced stages of creep normally observed in polycrystalline materials in general including ice and frozen ground: the primary, the secondary and the tertiary (Glen, 1955; Goughnour and

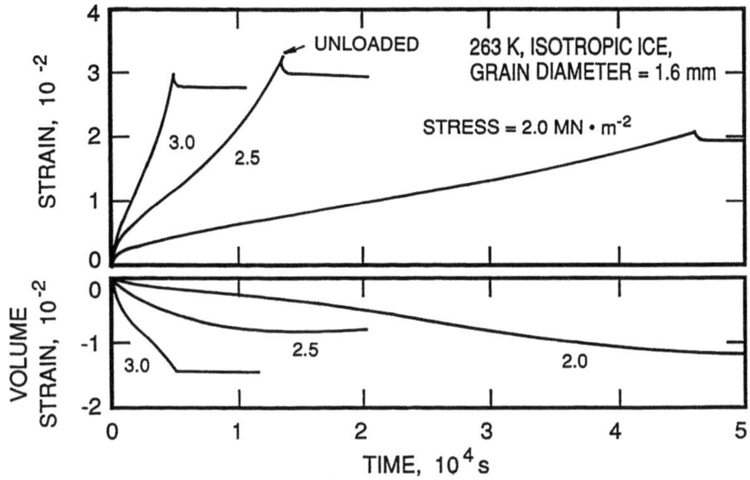

Figure 5a. Experimental observations on axial strain and volumetric strain under constant compressive load for isotropic ice.

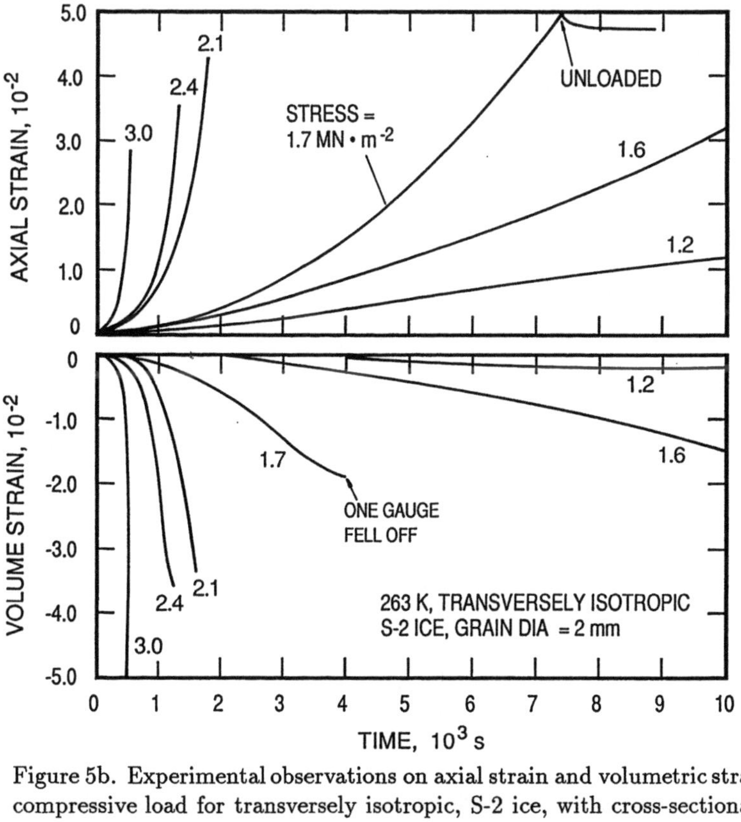

Figure 5b. Experimental observations on axial strain and volumetric strain under constant compressive load for transversely isotropic, S-2 ice, with cross-sectional grain diameters of 2mm.

Andersland, 1968; Barnes et al., 1971). The dependence of axial strain rate on time and stress is presented in Figure 6. These results bear striking agreement with experimental observations on granular ice (Mellor and Cole, 1982, 1983; Jacka, 1984) and frozen silt (Zhu and Carbee, 1987). Note that the time required to reach the minimum creep rate, t_m, is almost inversely proportional to the minimum strain rate, $\dot{\varepsilon}_m$, for stresses higher than about 1.6 $MN \cdot m^{-2}$ (broken line). This type of dependence is known as the Monkman-Grant type relation for brittle fracture in the field of metals and alloys, and may be presented in the form,

$$t_m \sigma^P = \text{constant}, \tag{16}$$

where P is a constant. Its value, from the calculations shown for stress range 2 to 4 $MN \cdot m^{-2}$ is 3.56 which agrees remarkably well with 3.59 for granular ice obtained experimentally by Mellor and Cole (1982) and the values obtained for metals and alloys (Sinha, 1989b). It should be mentioned here that the theory predicts a significant dependence of t_m on grain size and that this prediction has been confirmed by experiments on columnar-grained ice (Sinha, 1989a) and granular ice (Sinha, 1989b).

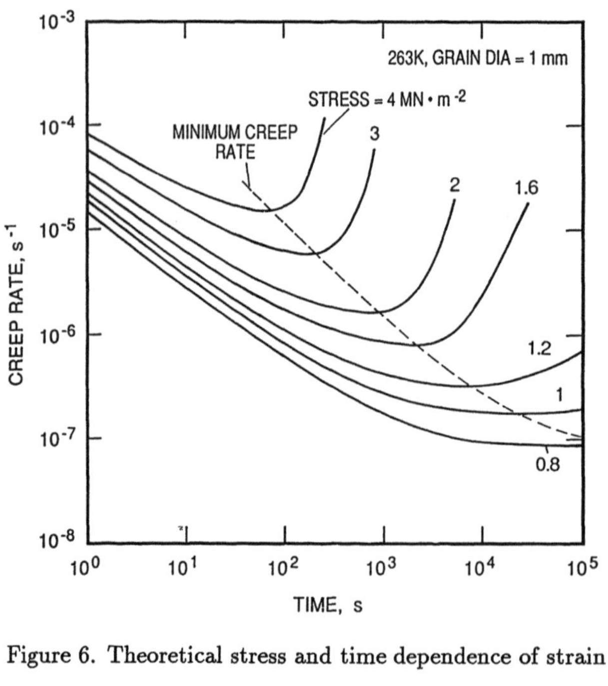

Figure 6. Theoretical stress and time dependence of strain rate under constant stress.

The theory predicts the crack density at minimum creep rate to increase rapidly with the increase in stress and then to approach a quasi-constant value (broken line in Figure 2) for the higher stress range. This is the range ($\sigma > 1.6 MNm^{-2}$) for which Equation (16) is also applicable. However, the crack densities are rather low at t_m as can be seen from the specific crack densities (less than one crack per grain). At these low crack densities, the dilatancy may not be noticeable, as may be seen in Figure 3 (and confirmed by experiments, Figure 5), unless very accurate measurements are made. Figure 4 indicates that the dilatation effect will not be pronounced until well into the tertiary stage.

The application of the proposed constitutive equations, (9) and (15), in formulating the material response under truly constant strain-rates has been presented in Sinha (1988a). Figure 7 shows a set of calculations for cracking activity at a few strain rates. Note that the crack density

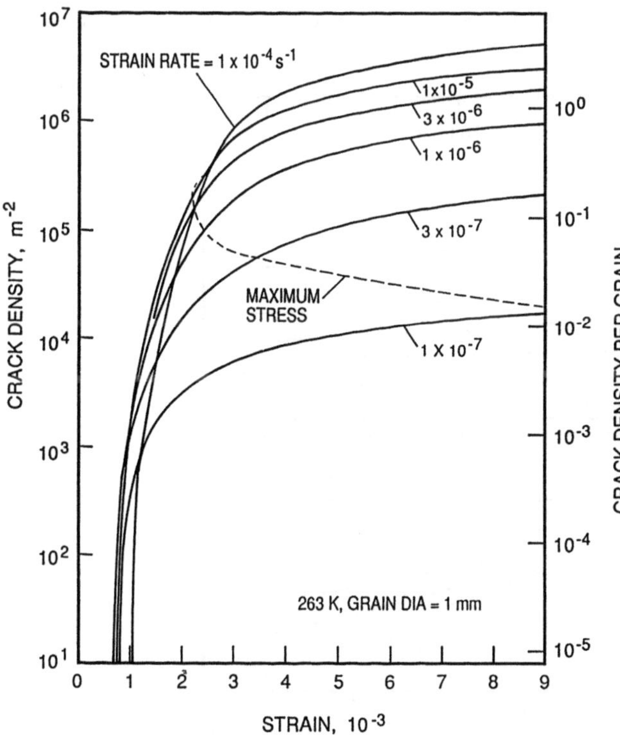

Figure 7. Strain and strain-rate dependence of cracking under constant strain-rate loading conditions.

at peak or upper-yield stress (broken line) increases rapidly with strain rate and then approaches a plateau (less than one crack per grain) at higher strain rates. The calculated stress and cracking histories, along with Equation (12), provide the stress-strain diagrams and the dilatancy under constant strain rates. Effect of strain-rates on the dependence of stress and volumetric strain on axial strain (or time, $t = \varepsilon/\dot{\varepsilon}$) are shown in Figure 8. Again the width to depth ratio of the cracks ($A = \frac{2a}{w}$) is assumed to be 100 and the compressive strain is given a positive sign. Calculations bear striking similarities with the experimental results for volumetric strain of frozen sand reported by Goughnour and Andersland (1968) and Baker et al. (1981).

Details of calculations for the pre-upper-yield period are shown in Figure 9, following usual convention for signs. As the compressive stress increases with strain, the elastic compaction increases until cracks start to form adding an effect of expansion. At some time, t_1(or $\varepsilon_1 = \dot{\varepsilon}t_1$), the rate of volumetric expansion due to the cracks balances the rate of volumetric elastic compaction. For $t > t_1$, the cracking activity dominates

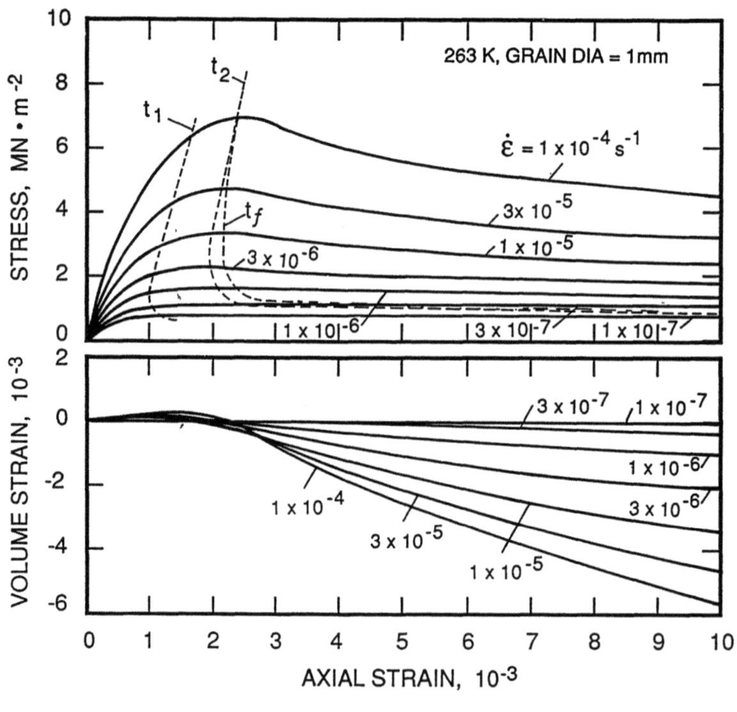

Figure 8. Stress-strain and volumetric stress-strain diagrams

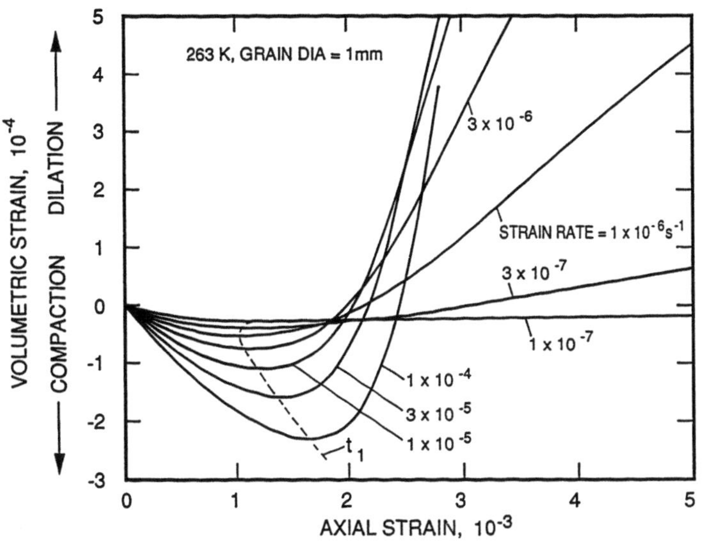

Figure 9. Theoretical strain dependence of volumetric strain as functions of constant strain rates.

the volumetric strain rate. Figure 8 shows that a net increase in volume occurs beyond t_2(or $\varepsilon_2 = \dot{\varepsilon} t_2$) when the volumetric elastic compaction balances the volume expansion due to the cracks. A net increase in volume occurs primarily during post peak stress period, or in other words, after failure. Note the correspondence between this behaviour and that for the constant stress response, that is, the predominance of dilatancy during tertiary creep period. Note also the similarities between the crack densities at upper-yield failure and those at t_m. Several other correspondences between the constant stress response and the constant strain rate behaviour, presented by Mellor and Cole (1983) can also be shown using the theory. These will be presented elsewhere.

The model can be used in predicting the mechanical behaviour under any prescribed strain or stress path. An example of its application in predicting the behaviour on repeated loadings is shown in Figure 10. In these cases instantaneous unloadings are made followed by complete recovery of delayed elasticity before loading again. The total recoverable strain is the total elastic and delayed elastic strain at the time of unloading; the permanent strain, at the beginning of a reloading cycle, is the accumulated viscous strain. Calculations used the crack density developed at the moment of unloading as the initial damage level during the reloading

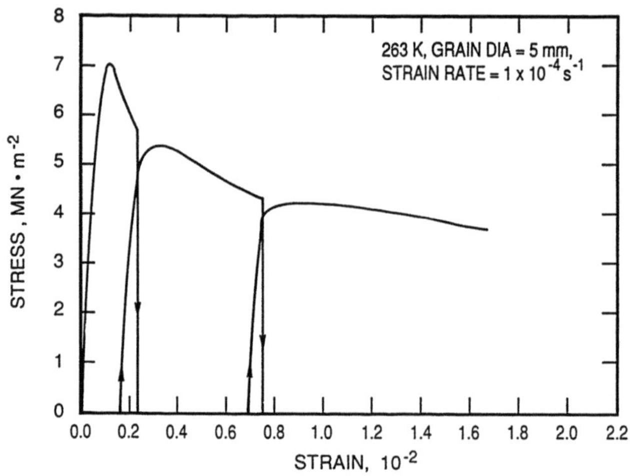

Figure 10. Theoretical predictions of stress-strain diagrams on repeat loadings.

period. No other assumptions in the computations were introduced. The computed results bear an extremely close resemblance to the experimental results in materials in general including those presented by Stone et al (1989) in ice. The theory, therefore, provides a quantitative explanation for the low ice forces during subsequent interactions in indentation tests reported by Sodhi and Nakazawa (1988). This exercise also shows that the effect of porosity on deformation and strength can also be formulated provided appropriate geometries of the inclusions are used. It would also be necessary to give consideration to the effect of the voids on the elastic and the delayed elastic effect.

This discussion would be incomplete without raising questions on what would happen to the kinetics of microfracturing and dilatation under triaxial loading conditions. It may be seen from the foregoing presentation that load application in other planes will either enhance or retard the crack initiation and multiplication processes, depending on the nature of the load, and hence the overall deformation mechanisms. The present model has been extended to three-dimensional loading conditions and will be presented elsewhere. Under a confining pressure, such as involved in most triaxial tests, longer times would be required before a critical gbs displacement is reached for crack initiation; thereby retarding the entire kinetics of cracking and resulting in higher strengths.

8. Conclusions

A rheological model applicable to any stress or strain path is presented. It describes the kinetics of microcracking, the microcrack enhanced creep and the dilatation effect produced by these cracks. The model is based on mechanics at the microscopic level, giving consideration to the geometry of the cracks in relation to the grain structure. Yet the methods of calculation are simple. It is shown that the model is capable of reproducing many commonly observed phenomena for both constant stress as well as constant strain-rate loading conditions. Grain facet size cracks develop when a critical delayed elastic strain corresponding to a critical grain-boundary sliding (gbs) displacement is reached. Further cracking is given in terms of the excess gbs displacement over its critical value. As cracks form, they enhance the matrix deformation and the volumetric strain. The crack enhancement leads, under constant stress, to minimum creep rate followed by accelerating or tertiary creep in association with the predominance of the dilatation effect. Under constant strain rates, the cracking activity leads to upper-yield failure followed by the post-failure period dominated by volumetric expansion. The model can also be applied to show the effect of reloading on strength.

References

Baker, T.H.W., Jones, S.J. and Parameswaran, V.R. 1981. Confined and unconfined compression tests on frozen sands. The Roger J.E. Brown Memorial Volume, Proc. 4th Canadian Permafrost Conference, Calgary, Alberta, March 2-6, Published by N.R.C. Canada, pp.387-393.

Barnes, P., Tabor, D. and Walker, J.C.F. 1971. The friction and creep of polycrystalline ice. Proc. Royal Soc. London, Ser. A, Vol.324, No.1557, pp.127-155.

Glen, J.W. 1955. The creep of polycrystalline ice. Proc. Royal Soc., London, Ser. A, Vol.228, No.1175, pp.519-538.

Gold, L.W. 1972. The process of failure of columnar-gained ice. Phil. Mag., Vol.26, No.2, pp. 311-328.

Goughnour, R.R. and Andersland, O.B. 1968. Mechanical properties of a sand- ice system. J. Soil Mechanics and Foundation Eng. Div., Proc. ASCE, Vol.94, No.SM4, pp.923-950.

Jacka, T.H. 1984. The time and strain required for the development of minimum strain-rates in ice. Cold Regions Sci. Tech., Vol.8, pp.261-268.

Michel, B. and Ramseier, R.O. 1971. Classification of river and lake ice. Canadian Geotechnical Journal, Vol.19, No.1, pp.36-45.

Mellor, M., and Cole, D.M. 1982. Deformation and failure of ice under constant stress or constant strain-rate. Cold Regions Science and Technology, Vol.5, pp.201-219.

Mellor, M., and Cole, D.M. 1983. Stress/strain/time relations for ice under uniaxial compression. Cold Regions Science and Technology, Vol.6, pp.207-230.

Sinha, N.K. 1979. Grain boundary sliding in polycrystalline materials. Phil. Mag.A., Vol.40, No.6, pp.825-842.

Sinha, N.K. 1982. Acoustic emission and microcracking in ice. Proc. 1982 Joint Conference on Experimental Mechanics, Society of Experimental Mechanics/Japan Society for Mechanical Engineers, Honolulu/Maui, Hawaii, May 1982, Society of Experimental Stress Analysis, Westport, CT, USA, Part 2, pp.767-772. (NRCC 20610)

Sinha, N.K. 1984a. Intercrystalline cracking, grain-boundary sliding, and delayed elasticity at high temperatures. Journal of Materials Science, Vol.19, No.2, February, pp.359-376.

Sinha, N.K. 1984b. Delayed elastic model for initiation and accumulation of creep cavitation at high temperatures. Advances in Fracture Research, Proc. 6th Int. Conf. on Fracture (ICF6), Pergamon Press, Oxford, Vol.3, pp.2295-2302.

Sinha, N.K. 1987. Effective Poisson's ratio of isotropic ice. Proceedings 6th International Symposium Offshore Mechanics and Arctic Engineering (ASME), Houston, Texas, USA, 1-5 March, Vol.4, pp.189-195.

Sinha, N.K. 1988a. Crack-enhanced creep in polycrystalline material: strain-rate sensitive strength and deformation of ice. J. Materials Science, Vol.23, No.12, pp.4415-4428.

Sinha, N.K. 1988b. Experiments on anisotropic and rate sensitive strain ratio and modulus of columnar-grained ice. Proc. 7th Int. Conf. on Offshore Mechanics and Arctic Engineering (OMAE/ASME), Houston, Texas, Feb. 7-12, 1988, Vol.4, pp.55-62. J. Offshore Mechanics and Arctic Eng., Trans. ASME, Vol.111, No.4, November, 1989, pp.354-360.

Sinha, N.K. 1989a. Microcrack-enhanced creep in polycrystalline material at elevated temperature. Acta Metallurgica, Vol.37, No.11, pp.3107-3118.

Sinha, N.K. 1989b. Ice and steel - a comparison of creep and failure. Mechanics of Creep Brittle Materials - 1, Proc. Euromech - 239, 15-17 August, 1988, Leicester, U.K., Edited by A.C.F. Cocks and A.R.S. Ponter, Elsevier Applied Science Publishers, London, pp.201-212.

Sinha, N.K. 1989c. Elasticity of natural types of polycrystalline ice. Cold Regions Science and Technology, Vol.17, No.2, pp.127-136.

Sodhi, D.S., and Nakazawa, N. 1988. Results from indentation tests on freshwater ice. Proc. 9th Int. Symp. on Ice, Sapporo, Int. Ass. on Hydraulic Research, Sapporo, Aug. 23-27, Vol.1, pp.341-350.

Stone, B.M., Jordaan, I.J., Jones, S.J. and McKenna, R.F. 1989. Damage of isotropic polycrystalline ice under moderate confining pressures. Proc. 10th Int. Conf. on Port and Ocean Eng. under Arctic Conditions (POAC), Luleå, Sweden, June 12-16, Vol.1, pp.408-419.

St.Lawrence, W.F. and Cole, D.M. 1982. Acoustic emissions from polycrystalline ice. US Army Cold Regions Research and Engineering Laboratory (CRREL), Report 82-21, 15p.

Zaretsky, Y.K., Chumichev, B.D. and Solomatin, V.I. 1979. Ice behaviour under load. Engineering Geology, Vol.13, No.1- 4, pp.299-309.

Zhu, Y. and Carbee, D.L. 1987. Creep and strength behavior of frozen silt in uniaxial compression. US Army Cold Regions Research and Engineering Laboratory (CRREL), Report 87-10, 67p.

CHAIRMAN'S REMARKS

L.W. Gold, National Research Council

This session had five papers that were an interesting and coherent set on the creep and viscoelasticity of ice. The information they contained covered the full range of stress associated with ductile behaviour and strain to more than 100%.

Ponter and Cocks reviewed information on the structural factors that cause the formation of internal cracks in ice and the resistance of plastic flow to be anisotropic. They extended the reference stress method for determining creep due to applied stresses to the state for which the resistance to flow of ice is significantly affected by the presence of internal cracks (i.e., cataclastic flow), and showed that the predicted behaviour is in agreement with field and laboratory measurements.

Jacka and Budd summarized observations made on the crystal fabrics that develop in polycrystalline ice during the tertiary creep stage at low stresses. This work has shown that the resistance to flow for a given strain in the tertiary region depends not only on the initial properties of the ice but also on the nature of the stress field under which it has been deformed.

Duval et al reviewed information on the factors controlling the deformation behaviour and development of internal stresses in polycrystalline ice. They presented and commented briefly on three models that have been proposed to describe the initial transient range in the creep behaviour of ice: that of Le Gac and Duval; Sunder and Wu; and Sinha. They reviewed progress that has been made in incorporating into the Le Gac and Duval model the effects of internal cracking activity on the creep behaviour.

After a review of the three term constitutive equation that he has proposed for ice, Sinha also presented a modification that incorporates the effect of internal cracking activity. He showed that the modified model is capable of reproducing response for both constant stress and constant strain rate conditions that is consistent with experimental observations, including the associated dilatation.

Masterson collaborated in the development of the analytical method used for the design of thick ice platforms (> 5 m) in Arctic sea ice. This method was essentially an empirical extension of elastic theory based on field measurements of the deflection rate of sea ice covers under relatively large load, and justified by observations on the performance of the platforms. In their paper, Masterson and Smith showed that the analytical model was satisfactory for the design of relatively thin platforms (about 1.5 m) incorporated into fresh water ice covers.

The papers presented in this session provide a good indication of the high level of analytical and experimental capabilities currently addressing time dependent ice deformation problems. It would now seem to be highly desirable to assemble good quality data bases that can be used to guide the development of and to validate analytical models of deformation behaviour, particularly their extension to the biaxial and triaxial stress fields that occur in nature. A good example is the information that has been compiled on the performance of the ice platforms used for offshore drilling in landfast ice. This is a small strain and strain rate problem. It should be an appropriate situation for identifying valid models of behaviour at the low end of the spectrum of deformation behaviour. Other examples are the information that now exists on the closure of tunnels and boreholes in ice, and on the interaction between ice covers and offshore structures. Such data sets, properly validated, would be very useful for the development of models covering the spectrum of deformation behaviour of ice and for the specification of limitations and simplifications in their application to specific problems.

SESSION II:

Fracture

Fracture Mechanics Models of Ice-Structure Interaction

Andrew C. Palmer

Andrew Palmer and Associates Limited
25 Victoria Street London SW1H OEX

Abstract Fracture mechanics models can explain some observed features of ice loading, among them anomalous "size effects" and the low forces seen in full-scale interactions (such as the Hans Island measurements, and some records of ice forces on gravel islands). A difficulty is that fracture mechanics can be made to explain too much, indeed almost anything, by appropriate hypothesis about initial defects and the development of additional defects by repeated fracture.

This paper attempts a critical assessment of the application of fracture mechanics, and identifies predictions which can be tested against observations in the field.

1. Introduction

A case for fracture mechanics as one of the components of an understanding of forces on offshore structures hardly needs to be re-stated. Ice is known to be an extremely brittle material, with a fracture toughness of the order of 0.1 $MN/m^{3/2}$, significantly smaller than glass (0.7 $MN/m^{3/2}$) and a thousand times less than steel (100 $MN/m^{3/2}$). Moreover, almost all ice deformation processes are characterised - at least on the level of superficial observation - by discontinuous breaking into fragments rather than by the smooth continuous deformation of creep and yield. Large deformations unaccompanied by fracture are the exception rather than the rule, and only occur in slow-moving ice sheets and glaciers (and in some carefully controlled small-scale tests).

Fracture mechanics as such can however be only part of the picture, and needs to be complemented by additional ideas from different disciplines. This paper explores the limitations of fracture mechanics and attempts to search out a way forward. It begins by contrasting an elemen-

S. Jones · R. F. McKenna · J. Tillotson · I. Jordaan (Eds.)
Ice-Structure Interaction
IUTAM-IAHR Symposium St. John's, Newfoundland, Canada

tary ("naive") fracture mechanics approach to ice forces with a plasticity approach.

2. Elementary Linear Elastic Fracture Mechanics Theory Approach to Ice Forces

A simple approach is straightforward, and its significance was recognized long ago by Goetze (1965). The analysis considers a cracked mass of ice, subjected to some external loading, and calculates the stress intensity factor induced by the loading. The crack propagates when the stress intensity factor reaches the fracture toughness (critical stress-intensity factor) for the ice. Figure 1 illustrates some of the geometries relevant to ice for which this has been done. Stress-intensity factors for a wider range of geometries can be found in standard references (see, for example, Hellan, 1984), and can be calculated numerically (Nemat-Nasser and Horii, 1982).

In the context of offshore structures, the forces required to propagate cracks are almost always small. In the simplest case, a vertical radial crack in a plane ice sheet is extended by a concentrated edge load (Figure 1a). The stress intensity factor is then,

$$K = 2.59F/t(\pi c)^{1/2}, \tag{1}$$

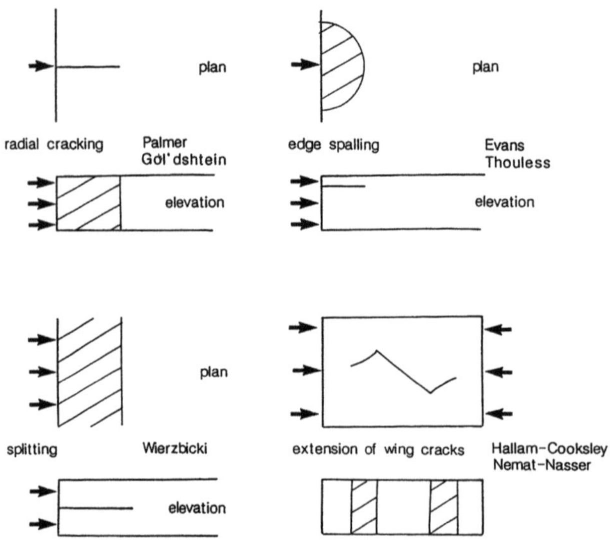

Figure 1. Ice crack propagation analyses.

where,

- F is the equivalent wedge-opening force applied at the end of the crack, taken as aP where P is the applied load and a is a multiplying factor determined by the local conditions under the applied load;
- t is the crack thickness; and
- c is the crack length.

The distance c to which a crack can be driven by a force P is therefore (Palmer et al., 1983),

$$c = 2.13(aP/tK_{1c})^2, \tag{2}$$

where K_{1c} is the fracture toughness. Taking t as 2 m, a as 0.5 and K_{1c} as 0.1 $MN/m^{3/2}$, the force to propagate a 10 m crack is about 1 MN and the force to propagate a 1000 m crack is about 10 MN.

Alternative geometries have less effective mechanisms for translating the externally-applied load into a crack-opening force. In the edge-spalling mode of Figure 1b, for instance, a crack radius of 10 m implies a force greater by a factor of about 10 (Evans et al., 1984; see also Thouless et al., 1987).

3. Elementary Plasticity Theory Approach to Ice Forces

The plasticity approach supposes that ice can be idealised as an elastic perfectly plastic material with a yield stress Y. The force under which an edge-loaded ice sheet can continue to deform ("collapse", in the sense of plasticity theory) is then,

$$P = IbtY, \tag{3}$$

where in addition,

- b is the contract breadth; and
- I is an indentation factor, related to the detailed geometry and the contact aspect ratio b/t.

I must be at least 1, but is generally not much more than 1. It can readily be determined through simple limit analysis.

Since ice creeps, the effective value of Y depends on the loading rate. Y is usually related to the uniaxial compressive stress that corresponds to a reference strain rate which is a multiple of U/b, where U is the ice velocity. The multiplier can be chosen empirically, or from reference stress theory (Ponter et al., 1983).

This gives much larger numbers. If b is 25 m, t is 2 m, I is 1, and Y is taken as 5 MN/m^2 (a reasonable value for sea ice deformed at a high velocity), P is 250 MN, more than two orders of magnitude greater than the force needed to propagate a crack 10 m long.

An attraction of plasticity theory is that it can readily be applied to complex geometries (Cammaert and Muggeridge, 1988), for which the limit theorems make it possible to determine close upper and lower bounds on ice forces.

4. Comparison Between Observation and Theories

Imagine that we had no way of observing what actually happens when an offshore structure interacts with ice, and had only the naive analyses outlined above; imagine in addition that other mechanisms such as creep buckling are not present.

If we only had the plasticity method, we would expect contact forces per unit area of the order of 5 MN/m^2, and forces corresponding to this value multiplied by the product of the structure breadth and the ice thickness. If we only had the fracture mechanics method, and supposed the ice initially to include crack defects with dimensions of 1 to 10 m, we would expect low forces, since applied forces of between 1 and 100 MN would induce widespread propagation of pre-existing cracks, and those cracks would link and lead to repeated fracture, which would progressively break the ice sheet into fragments.

Neither picture corresponds to what is observed, except in extreme cases. Something like the deformation that corresponds to plasticity theory is seen in some slow indentation tests on a laboratory scale, and in some indentations on a small field scale, on blocks of the order of 10 m across. In other field indentation tests, blocks obstinately split and break up.

Splitting deformations like those in simple fracture mechanics theory are indeed seen. In the Canadian Beaufort, for instance, satellite pictures show floating masses 10 km across split across by single cracks. This simple mode of deformation is not often seen close to fixed structures, because the transverse stresses tend to suppress propagation.

A second comparison can be made between predicted and observed total forces on offshore structures. Sanderson (1988a) has compiled all the data available into a plot of contact pressure (force/area) against contact area, and has carefully discussed each instance of full-scale measurement. One can argue about the detail of this compilation, and about its reliability as a predictor of ice forces, but several broad conclusions are plain. Whereas contact pressures that approach 10 MN/m^2 are often observed in laboratory tests and some intermediate-scale field tests, ice contact pressures observed on the scale of actual offshore structures are invariably smaller by a factor of between 5 and 100.

A fracture mechanics approach offers an appealing explanation. If the ice deforms by breaking up rather than by distorting continuously, the elementary fracture mechanics calculation shows that relatively small forces are required to extend the cracks. An explanation of the scale effects is that large volumes of ice include larger cracks which extend more easily.

This argument can be used to explain the low level of forces seen in the Hans Island measurements (Sanderson, 1988a), which were found to be unexpectedly low and first drew attention to the huge discrepancy between large-scale behaviour and the conclusions of plasticity-based theory. In the 80.08.15 event, for instance, the maximum force on a 5.6 m thickness ice floe reached 214 MN over a breadth of 279 m, corresponding to a 0.14 MN/m^2 contact pressure, which is inexplicable on a plasticity basis.

Total ice forces also include a clearing component, which is the force required to move fragments around the structure, by pushing them under the ice or heaping them on top. Sanderson's analysis (1988a) and similar analysis of ridge-building show that the clearing component is a small part of the total, and does not affect the conclusions reached here.

At a low level, fracture mechanics is a beginning of an explanation for observed behaviour, though it depends on a detailed analysis of the pre-existing fractures and the fracture process during the collision, which cannot be carried out from the available data and for which it would be extremely difficult to secure new data.

A useful and important independent conclusion from a fracture mechanics approach is that scale effects are to be expected, and that the results of model tests must be approached with extreme caution (Palmer and Rice, 1973; Atkins and Caddell, 1974; Kendall, 1978a,b; Carpinteri, 1983; Jones, 1986). If yield stress Y and fracture toughness K interact, there is a characteristic length K^2Y^2, and the ratio of physical dimensions to this length will not be correctly scaled in a model. This is likely to alter both the deformation mode and the forces.

Similarity requires the model and the prototype to have the same value of a brittleness number, defined by Carpinteri as $K_{1c}/YL^{1/2}$, where L is a length defining the system geometry. This number appears again and again in models of fracture/plasticity interaction. For example, in Evans et al.'s (1984) analysis of indentation spalling, the ratio of the indentation crack length to the plastic zone radius is $(K_{1c}/YL^{1/2})^{-1/2}$, where L is identified with the zone radius.

5. Limitations of Linear Elastic Fracture Mechanics

A simple fracture mechanics approach has many limitations. A particular weakness is that it explains too much, and indeed can be made to predict almost anything by a suitable postulate about pre-existing flaw numbers and dimensions. Until harder data can be found, a fracture mechanics explanation is somewhat vulnerable to Kendall's (1978a) reproaches - in another context - of size effects "rationalised by dubious statistical arguments involving invisible flaws".

A second limitation is that conventional fracture mechanics alone suggests too simple a picture of deformation modes. It suggests that ice should break into large fragments. This does happen sometimes, but by no means always. At Hans Island, for instance, the fragments seen in the videos of collision events are 5 to 10 m across, and similar fragment sizes are seen in rubble mounds and pressure ridges. In other instances, a crushing mode predominates, and the fragments are much smaller. This is often seen in the field and in the laboratory. In the 86.03.07 event at Amauligak I65, for instance, two heavily-ridged multi-year floes 6.5 m thick were driven against the Molikpaq caisson at up to 0.05 m/s, creating a peak horizontal force of 270 MN over a contact area 100 m wide, corresponding to a contact pressure of 0.42 MN/m^2 (Sanderson, 1988). The ice did not break into large fragments, but was "crushed and pulverised in contact with the structure and then extruded upwards and downwards as a fine powder". This and other large-scale behaviour dominated by crushing lead some engineers to conclude that offshore structures ought always to be designed for the forces corresponding to the crushing mode, and that it is not safe to rely on an empirical pressure/area curve (Dorris, 1989; Dorris and Winkler, 1989).

It is interesting to see what fracture mechanics can tell us about fragment size in crushing events. Kendall (1978a,b) examined a material which would either fracture or yield, and showed that large particles will break and small ones will distort plasticly. He found that a transition from fracture to yield will limit the minimum diameter of an isolated particle that can be crushed to

$$d_{min} = (32/3)ER/Y^2, \tag{4}$$

where

R is a fracture energy,
E is Young's modulus, and
Y is the yield stress.

Identifying R with the strain energy release rate G, equal to $K_{1C}^2/E(1-\nu^2)$, and taking K_{1c} as 0.1 $MN/m^{3/2}$ and Y as $5MN/m^2$, the minimum particle diameter is 4.7 mm, which is consistent with reports of the Amauligak event. Laboratory tests in the crushing mode appear to find rather smaller diameters, of the order of 1 mm: this may be because of limitations in the Kendall model, or because Y ought to be 10 MN/m^2, or because of additional triaxial constraint within a crushing zone.

6. Non-simultaneous Failure and Chaotic Behaviour

Conventional fracture mechanics asks questions about the initiation and growth of a single crack, or a small number of interacting cracks. In most engineering applications, propagation of a crack to a length of the same order as the dimensions of the structure almost invariably leads to a limit state, beyond which further fractures are of little concern.

Ice force applications are totally different. The very first crack is insignificant. Tens of thousands of fractures have taken place by the time the maximum force on the structure is reached, and the geometry of the fracture region is then a creation of myriad earlier fracture events. The ultimate challenge is to make sense of continued and repeated fracture.

A related complication is that movement of the ice against a fixed structure induces non-simultaneous failure, so that fractures are occurring at different locations and at different times, separated by time and space intervals in which no fractures are taking place, and the ice may not even be in contact with the structure. This can itself account for the observed reduction in force per unit area on large structures (Ashby et al., 1986), and Sanderson (1988a) has shown that a very simple non-simultaneous model can produce the same kind of irregular force/time history seen in field and laboratory investigations.

A difficulty is again that many models can be postulated, and that some of them mimic observed behaviour, but that it is difficult to know which models have worthwhile predictive power. Consider, for example, the model of repeated multiple non-simultaneous fracture illustrated in Figure 2. It imagines repeated failure to occur by linking of cracks, in such a way that repeated fracture creates a self-replicating geometry in which the average spacing of active cracks is a fixed multiple of the distance a crack has to be driven before it interacts with a neighbouring crack and a fragment breaks loose. An ice sheet is driven against a fixed structure by a force P over a total contact breadth b, and a series of cracks of length c are generated by equal forces applied close to the ends of the cracks, at a space λ c. The crack length is then related to the driving force for that crack by Equation (3), and so,

$$c = 2.13(a\lambda cP/btK_{1c})^2, \tag{5}$$

and therefore,

$$c = 0.47(K_{1c}bt/Pa\lambda)^2. \tag{6}$$

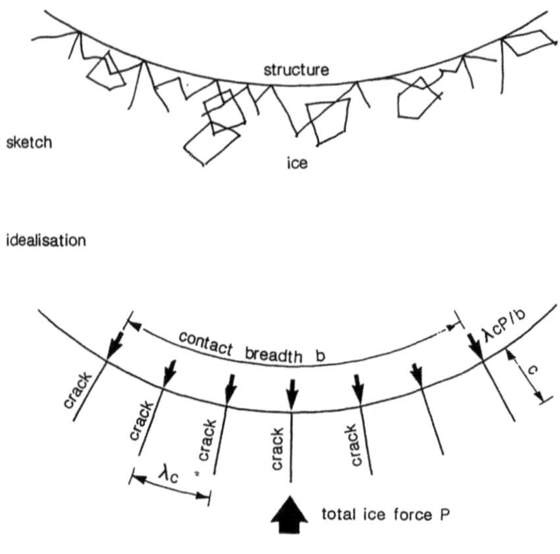

Figure 2. A model of multiple non-simultaneous loading.

Applying this to the Hans Island event described earlier, and taking λ and a each as 0.5 (so that a crack has to propagate twice the crack spacing before it links up with the next crack, which seems about right), c comes out to be 4 m, which agrees with the observed size of large fragments in Hans Island events. Equation (6) only relates the force to the fragment size, and does not predict either of them independently. It does predict that the fragment size ought to be inversely proportional to the square of the contact pressure P/bt. This is a testable prediction. If it turns out to be correct, it might be valuable in structural monitoring, since a progressive reduction in fragment size would be easy to see and would be a direct warning of increasing ice force.

Applied to the peak load in the Amauligak 86.03.07 event, with the same values of a and λ, c is 0.4 m, which is larger than observed.

The model just analysed is an incomplete model of non-simultaneous failure, since the instantaneous maximum force is divided equally between all the loading points which govern the fragment size. In reality, only some of those points will be active. This is shown clearly in Ashby et al.'s, (1986) analog model experiment, in which a floating sheet of brittle wax was pushed against a circular fixed structure. It can also be seen in experiments with brittle crispbread sheets. When a fragment breaks loose, it leaves a gap in the ice surface, and the ice has to move forward further before it contacts the corners left at the end of the first gap. A small number of long cracks advance ahead of the fracture zone. For a time they remain inactive, but they are later overtaken by the fracture zone and influence the breaking away of small fragments.

A further complication is that the ice is three-dimensional and non-homogeneous, above all in thick multi-year ridges which are most likely to induce the largest loads. Fragments will not only form by radial propagation of vertical cracks from contact points. There will also be horizontal-plane spalling and splitting cracks, which locally reduce the thickness of the ice and induce a series of spalling events. Some of the fragments will clear, but others will be jammed between the structure and the surface. The surface will evolve three-dimensionally. It appears to be worthwhile to study the evolution of the fracture surface predicted by simple models of fragment sizes, and this is in progress. Each fracture event modifies the surface, according to a transformation rule which reflects a simple fracture mechanics model of crack growth around a contact point. There are some loose parallels with fractal geometry, but the transformation rules are different, in that they do not correspond to infinitely-repeated division and refinement of a surface. Preliminary work suggests that many transformation rules are unstable, and lead to the interaction surface becoming progressively smoother.

The mode of non-simultaneous failure can be related to the loading conditions by a mode map (Palmer et al., 1983), though it is not certain what parameters determine the mode. The mode will almost certainly change during a loading event. The evolution of modes and forces can be followed in a phase diagram, which plots fragment size and force against time.

Figure 3a is one possible phase diagram for a Hans Island event, in which the initial mode at first contact is localised small scale crushing and fragment sizes increase until the floe comes to a stop. The Amauligak extreme events are quite different, and appear to show a progression from large fragments to small (Figure 3b). Any mode may be unstable: an initial large-fragment mode might be succeeded by a progressive blunting and smoothing of the ice face in contact with the structure, so that the load becomes shared over an increasing number of contact points and the fragment size continuously decreases. A crushing mode might also be unstable, if crushing forces rise to such a level that cracks outside the crushed region are driven ahead and linked up, so that the system reverts from crushing to large-scale repeated fracture.

Finally, there is no reason to suppose that continued movement necessarily leads to a stable mode of deformation, in which the deformation settles down to a replicating pattern with a roughly constant maximum force, even if the driving force and the ice properties are constant in space and time (as they rarely are in nature). The mode may oscillate, so that

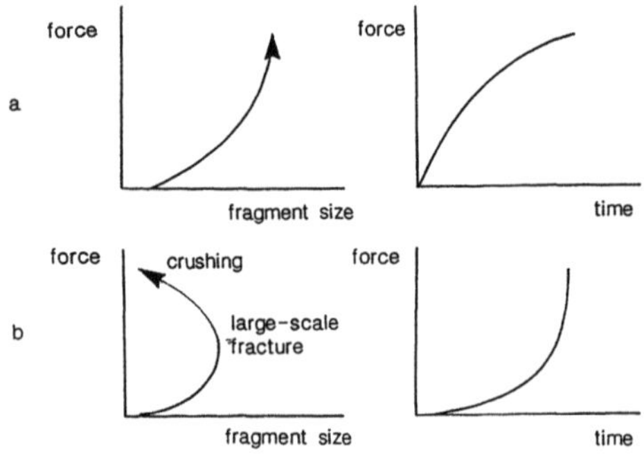

Figure 3. Phase diagrams for ice-structure interaction.

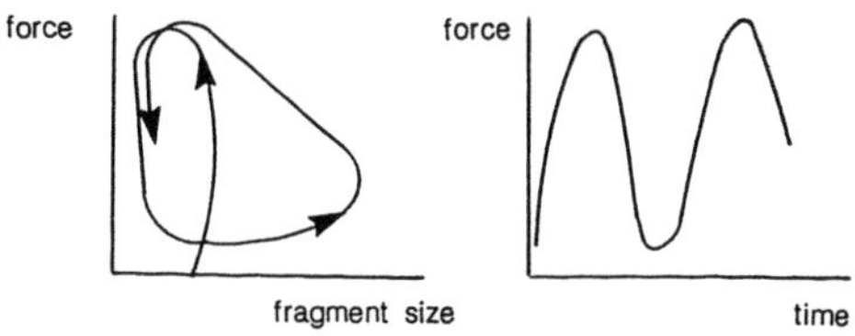

Figure 4. Phase diagram showing oscillatory response.

the system flips from crushing modes with large forces to large-fragment fracture modes with small forces (Figure 4).

These broad conclusions ought to be testable against observation. If it becomes possible to observe and record fragment sizes at the same time as ice forces, one would expect periods of low forces generally to be correlated with the formation of a small number of large fragments (but sometimes with continuous creep deformation in which large-scale fracture has only a secondary role). High forces would correspond to formation of a large number of small fragments, but this process might be unstable.

Recent developments in chaos theory (see for example, Gleick, 1988; Thompson and Stewart, 1986) have illustrated the extraordinary richness of the dynamic behaviour of very simple dynamic systems, characterised by chaotic attractors, period doubling, bifurcations, and strong sensitivity to initial conditions. An ice-structure system is far more complex, and the dynamics is further complicated by the flexibility of an offshore structure, which deflects significantly under ice forces and whose response can interact with non-simultaneous fracture. This interaction is likely to lead to chaotic responses.

One can conclude that simple responses are not to be expected, and that chaotic behaviour is likely.

7. A Way Forward

A theoretician hesitates to suggest observations that need to be made, fearing the exasperation of field workers who understand the extreme difficulties of making good measurements of ice behaviour. Tentatively, the present work suggests that:

1. Measurement of ice forces is essential for practical applications, and rightly has had high priority, but needs to be coupled with further observation of the geometry of deformation;

2. We need to know much more about non-simultaneous failure and detailed mechanisms; and
3. That numerical experimentation with rather simple models may throw light on the kind of complexity to be looked for in the field.

8. Conclusions

Fracture mechanics is but one of several theoretical tools that will need to be brought together before understanding is reached. Alone, it will not take us far, but in combination with other ideas such as chaos theory it ought to prove rewarding.

Acknowledgement

I would like to record my gratitude to British Petroleum, whose support of research in ice mechanics initiated some of the ideas developed here, and to several friends, particularly Tim Sanderson, Mike Jefferies and Dougal Goodman, for stimulating discussions.

References

Ashby, M.F., Palmer, A.C., Thouless, M., Howard, M., Hallam, S.D., Murrell, S.A.F., Jones, N., Sanderson, T.J.O. and Ponter, A.R.S. 1986. Non-simultaneous failure and ice loads on Arctic structures. Proceedings, Eighteenth Annual Offshore Technology Conference, Houston, paper OTC 5127, Vol.1, pp.399-404.

Atkins, A.G. and Caddell, R.M. 1974. The laws of similitude and crack propagation. International Journal of Mechanical Sciences, Vol.16, pp.541- 548.

Cammaert, A.B. and Muggeridge, D.B. 1988. Ice interaction with offshore structures. Van Nostrand Reinhold, New York.

Carpinteri, A. 1983. Plastic flow collapse vs. separation collapse (fracture) in elastic-plastic strain-hardening structures. Materiaux et Constructions, Vol. 16, pp.85-96.

Cooksley, S.D. 1984. Yield and fracture surfaces of brittle solids under multiaxial loading. Unpublished Ph.D. dissertation, University of Cambridge.

Dorris, J.F. 1989. The use of residual strength in the calculation of global crushing. Proceedings, Tenth International Conference on Port and Ocean Engineering under Arctic Conditions, Luleå, Vol.1, pp.467-480.

Dorris, J.F. and Winkler, M.M. 1989. On the relationship between global and local ice crushing loads. Proceedings, Tenth International Conference on Port and Ocean Engineering under Arctic Conditions, Luleå, Vol.1, pp.481-491.

Evans, A.G., Palmer, A.C., Goodman, D.J., Ashby, M.F., Hutchinson, J.W., Ponter, A.R.S., Williams, G.J. 1984. Indentation spalling of edge-loaded ice sheets. Proceedings, International Association for Hydraulic Research Ice Symposium, Hamburg, Vol.1, pp.113-121.

Gleick, J. 1988, Chaos. Heinemann, London.

Goetze, C.G. 1965. A study of brittle fracture as applied to ice. Technical Note, US Army Cold Regions Research and Engineering Laboratory.

Gol'dshtein, R.V. and Osipenko. N.M. 1983. Mekhanika razrusheniya i nektoriye voprosy razrusheniya l'da. Mekhanika i fizika l'da (ed. Gol'dshtein, R.V.), Izdatel'stvo Nauka, Moscow.

Hallam, S.D. 1986. The role of fracture in limiting ice forces. International Association for Hydraulic Research 86, Vol.2, pp.287-319.

Hellan, K. 1984. Introduction to fracture mechanics. McGraw-Hill, New York.

Jones, N. 1986. A note on ice scaling. Journal of Ship Research, Vol.30, pp.134-135.

Kendall, K. 1978a. Complexities of compression failure. Proceeding of the Royal Society, Series A, Vol.361, pp.245-263.

Kendall, K. 1978b. The impossibility of comminuting small particles by compression. Nature, Vol.272, pp.710, 711.

Nemat-Nasser, S. and Horii, H. 1982. Compression induced nonplanar crack extension with application to splitting, exfoliation and rockburst. Journal of Geophysical Research, Vol.87, pp.6805, 6821.

Palmer, A.C. and Rice, J.R. 1973. The growth of slip surfaces in the progressive failure of overconsolidated clay. Proceedings of the Royal Society, Series A, Vol.332, pp.527-548.

Palmer, A.C., Goodman, D.J., Ashby, M.F., Evans, A.G., Hutchinson, J.W. and Ponter, A.R.S. 1983. Fracture and its role in determining ice forces on offshore structures. Annals of Glaciology, Vol.4, pp.216-221.

Ponter, A.R.S., Palmer, A.C., Goodman, D.J., Ashby, M.F., Evans A.G., and Hutchinson, J.W. 1983. The force exerted by a moving ice sheet on an offshore structure. Part 1, The creep mode. Cold Regions Science and Technology, Vol.8, pp.109-118.

Sanderson, T.J.O. 1988a. Ice mechanics: risks to offshore structures. Graham and Trotman, London.

Sanderson, T.J.O. 1988b. The ice load question: some answers. IAHR Ice Symposium 1988, Sapporo.

Thompson, J.M.T. and Stewart, H.B. 1986. Nonlinear dynamics and chaos. John Wiley and Sons, Chichester.

Thouless M.D., Evans, A.G., Ashby, M.F., Hutchinson, J.W. 1987. The edge cracking and spalling of brittle plates. Acta Metallurgica, Vol.35, p.133.

Wierzbicki, T. 1985. Spalling and buckling of ice sheets. Proceedings, American Society of Civil Engineers Arctic '85 Conference, San Francisco, pp.953-961.

Discussion

Q: W. Nixon, Iowa Institute of Hydraulic Research

Three colleagues of mine at IIHR (R. Ettema, F. Stern, K. Georgakakos) have performed Chaotic Dynamic analysis on the behaviour of an ice-breaker in sheet ice. They found the behaviour to be deterministic with a fractal dimension between 4 and 5. But, having found this number, where do you go from here?

A: A.C. Palmer

I don't know that there is anything you can do with the fractal dimension. To me, the principal value of chaos theory up to this point is that it reminds us to look out for a high degree of complexity even in wholly deterministic systems.

I hope that there will be more quantitative results to be uncovered by future research.

Q: M. Metge, Canatec

In some of the ice crushing events on the Molikpaq, a major crack, several km long, would develop ahead of the structure and lead to a drop off in load. Would you comment on this from the fracture mechanics point of view.

A: A. Palmer

Fracture mechanics explains how a crack can be driven a long way. Depending on the far-field stress, a crack extension event can release some of the triaxial constraint in the immediate neighbourhood of the ice-structure contact, and allow a reduction in contact pressure. What is harder to understand in detail is the interaction between the crushing zone and the initiation of long cracks.

Q: T.J.O. Sanderson, BP International

I agree with Dr. Palmer's comments on chaos theory. One of the features of chaotic systems is that they produce apparently disordered behaviour within well-defined limits. Perhaps the lesson of chaos theory is that we should not seek to understand the details of every ice interaction, but we should be quite satisfied with the statistics of the behaviour within a probability distribution.

A: A. Palmer

I agree, except to say that we should seek to understand each detail of an observed event - and may indeed be successful - but that we cannot expect to predict each detail of future events. Weather forecasting is a useful analogy, which Gleick discusses in detail.

Q: E. Schulson, Dartmouth College

Cracks are characterized namely by size but also by thermo-mechanical history. What account is taken of this aspect when considering the application of fracture mechanics to the question of ice forces on structures?

A: A.C. Palmer

I completely agree that thermo-mechanical history is going to be an important component of any serious development of fracture-based analyses. The ice is often very close to its melting point. There is a lot of water about. There are often substantial time delays between successive fracture events. Blunting and healing will have a significant effect.

The Fracture Toughness of Ice

John P. Dempsey

Department of Civil and Environmental Engineering
Clarkson University
Potsdam, New York, 13676 USA

Abstract Topics to be addressed in this paper include

- the relevance of fracture mechanics to ice engineering
- the concept of fracture toughness
- the use of fracture toughness values for ice
- an examination and assessment of fracture toughness testing to date.

The last topic will comprise most of the paper. The author has intensively investigated the fracture toughness of freshwater ice. Insights and recommendations gained from this experience will be put forward. The discussion will focus on specimen size effects, the question of fracture toughness as a material property, fracture energy, and fracture mechanisms.

1. Relevance of Fracture Mechanics to Ice Engineering

The aim of fracture mechanics is to provide a quantitative description of the transformation of intact material into material broken and fragmented by crack growth. Griffith (1921, 1924) established a theoretical basis for fracture mechanics to explain results of fracture experiments on brittle materials. The development of this engineering discipline languished until the forties, when catastrophic accidents occurred to ships and storage tanks which could not be explained by the usual strength approach. Nowadays, fracture mechanics is an accepted tool in the safety assessment of aircraft, ships, pressure vessels, pipelines, and welded offshore steel structures. This approach has led to a rapid increase in the basic understanding of deformation processes active in a wide range of

S. Jones · R. F. McKenna · J. Tillotson · I. Jordaan (Eds.)
Ice-Structure Interaction
IUTAM-IAHR Symposium St. John's, Newfoundland, Canada

materials, particularly concerning the failure of brittle solids under compression. For this reason, materials such as concrete, rock and ceramics have been studied extensively since the sixties and seventies using fracture mechanics.

The relevance of fracture mechanics to ice engineering is intuitively acceptable once one has witnessed ice impact on a structure, an ice jam, an icebreaker at work, or a bearing capacity failure. For instance, the indentation and penetration of an ice sheet by a structure is typically fracture-dominated. The different failure modes observed and studied are crushing, crushing with spalling, crushing with radial cracking, crushing with radial and circumferential cracking, and radial/circumferential cracking with buckling (Sodhi, 1986; Timco, 1987; Blanchet et al., 1989). The mechanism by which the above failure modes are established is tensile fracture. To reliably predict ice forces on structures, and to better understand the role of fracture, the dependencies of the opening mode (Mode I) fracture toughness on such factors as loading rate, loading direction, specimen geometry and sample size, as well as the ice properties need to be investigated. The objectives of this paper are (1) to quantify the degree of applicability of LEFM (linear elastic fracture mechanics) to the determination of the resistance of ice to tensile fracture and (2) to forecast the necessary specimen size using the concept of an adequate specimen notch sensitivity.

2. Concept of Fracture Toughness

Griffith (1921, 1924) was the first to make a quantitative connection between strength and crack size. The fundamental importance of the origin, stable growth, and final propagation of crack-like flaws in material failure has since stimulated a generation of research into continuum mechanical descriptions of the stress and deformation fields in cracked bodies (Rice, 1968; Tada et al., 1985), including elastic-plastic and time-dependent fracture studies. The characteristic feature emerging from all linear elastic solutions to problems of cracks in homogeneous isotropic materials is the inverse square root of the crack tip stress singularity. To the extent that linear elasticity is appropriate, all crack tip stress fields are of identical functional form with the influence of the magnitude of loading, method of loading, and geometry of the cracked body being related in the crack tip region only through the stress intensity factor K_I (Irwin,

1957). Implicit in applications involving LEFM is the small scale yielding condition (Rice, 1968) which states that there is a region of inelastic deformation at the crack tip which must be small in comparison to the other geometrical dimensions.

The resistance of a material to crack extension, given by the value of K_1 required to advance the crack, depends on composition and microstructure, temperature, pressure, and chemical environment, and such factors as the loading rate, loading direction, specimen geometry, specimen thickness and sample size. The Mode I critical stress intensity factor is denoted here by K_{1c}, which is called the *fracture toughness*. If the plane strain fracture toughness has been established (ASTM E399), the fracture toughness is denoted by K_{Ic}. The thickness of the specimen (in other words, the length of the crack front) determines whether the specimen is in a state of plane stress, plane strain, or somewhere in between. Increasing thickness typically leads to first an increase in K_{1c}, then a decrease until the conservative (plane strain) plateau K_{Ic} is obtained.

The critical elastic energy release rate, G_{1c}, may also be used to characterize the resistance of a material to crack extension. In a linearly elastic homogeneous and isotropic material $G_1 = K_1^2/E'$, where $E' = E$ (plane stress) or $E/(1-\nu^2)$ (plane strain). If, in addition, $\Delta = CP$, where C denotes the compliance of the body and is a function of the geometry of the body and its elastic constants, and Δ is the load-point displacement separation, then $G_1 = (P^2/2h)C'(a)$. The value of K_1 or G_1 at which a crack extends in a given material should be independent of the size and shape of the test specimen and the method of loading.

In other words, if similarities exist in the local deformation fields for a class of configurations, experimental data on one configuration may then be rationally extended to the prediction of the behavior of the others.

3. Use of Fracture Toughness Values for Ice

Goetze (1965) was the first to speculate on the role of fracture in limiting ice forces. This topic was largely neglected until the eighties, being revived independently by Goldstein and Osipenko (1983) and Palmer et al. (1983). A representative survey of papers (from which a complete set of references may be found) that have further assessed the influence of fracture on ice forces includes those by Evans et al. (1984), Goldstein and Osipenko (1985a, b; 1989), Hallam (1986), Sanderson (1986), Timco and Frederking (1986), Schulson (1987), and Bhat et al. (1989).

The above papers reveal applications ranging from large scale tensile failure (spalling, radial/circumferential cracking), the dependence of ice forces on structures and an apparent dependence on the square root of the projected contact area, to the influence of microcracks on the fracture toughness. Certain topics draw directly upon the principles of fracture mechanics and a knowledge of fracture toughness *viz.*, (1) crack nucleation due to elastic anisotropy or dislocation pileup (Cole, 1986; Kalifa et al., 1989; Shyam Sunder and Wu, 1990); and (2) damage models for ice in compression (Pulkkinen, 1989; McKenna et al., 1989; Choi and Karr, 1989; Santaoja, 1989 and papers in this volume by Duval and Meyssonier, Sjölind, and Jordaan).

Clearly, the fracture toughness is not the only strength parameter needed; however, it may be more relevant than other strength measures when the integrity of an initially intact material is damaged by a few dominant cracks or by a system of microcracks. The importance of an understanding of fracture in ice is accentuated by the finding that the strength of ice under compression peaks at the transition from creep to brittle fracture. There is good reason to believe that the maximum force that can be exerted on a structure may be limited by a fracture-dominated failure mechanism.

Judging by the history of fracture mechanics applications to other heterogeneous solids, unanticipated benefits of a thorough knowledge of the fracture toughness of ice include:

- its value as a parameter for the classification of ice types;
- its appearance in dimensionless numbers important in small-scale modeling when the problems are governed by either inertia and fracture effects (Atkins, 1975) or gravitational and fracture forces (Jones, 1986);
- its value as an index of fragmentation processes.

4. Fracture Toughness Measurements to Date

The fracture toughness tests on ice reported to date have involved a number of ice types, including equiaxed granular freshwater ice, macrocrystalline unseeded S1 freshwater ice, columnar seeded S2 freshwater ice, sea ice, and model sea ices such as saline ice, and doped urea ice. Almost exclusively, these tests have concentrated on determining reproducible values for the Mode I (opening mode) critical-stress-intensity-factors association with the *initiation of a stationary macrocrack.* Considerable scatter in the fracture toughness results has been observed. Different factors cause this scatter, including ice type, crack length, loading rate, specimen size, test temperature, specimen geometry, crack orientation, specimen preparation, and crack-tip radius.

The measurements reported in the literature to date are summarized for freshwater ice in this paper. First, however, it is necessary to adopt a notational change (see also Dempsey et al., 1988). The ice fracture toughness testing community has placed a huge emphasis on determining, in particular, the Mode I critical-stress-intensity-factor (K_{Ic}). This notation implies a plane strain value, and all of the assumptions inherent in the ASTM E399-83. Certain of these assumptions are either unrealistic or patently inapplicable. For instance, the fracture toughness of ice has usually been computed by:

- using the maximum load (P_m) of the load versus time record;
- letting the effective crack length (a_e) equal the initial fabricated crack length (a_o);
- assuming that the sample thickness (h) is adequate;
- assuming that the crack is mathematically sharp;
- assuming that the material is homogeneous, isotropic, and linearly elastic.

Actually, the load at fracture (P_f) is not necessarily equal to the maximum load. The fabricated crack-tip radius (ρ) is not zero for a machined notch, in which case the effective crack length does not equal the machined length, and the crack is not then mathematically sharp. If a sharp crack is introduced at the tip of a band-saw cut by a scalpel, for instance, then the crack-tip radius is zero, but the length of this scribed crack is important, as is the variability and length of the associated crack front. Only for equiaxed granular freshwater ice is the fifth assumption above possibly applicable, and then only if the loading rate is high enough or the time-to-failure is short enough.

Typically, the material composition is anisotropic and nonhomogeneous. Furthermore, given the very high homologous test temperatures, the possibility of creep deformation at the crack-tip can be a significant factor. The time interval that the specimen is cracked prior to loading, any pre-loading interval, and the loading rate can influence the scale of deformation in the vicinity of the crack.

In view of the above remarks, a notational change in the symbol being used to represent the Mode I critical-stress-intensity-factor in ice is recommended, *viz.*,

$$K_Q = K_{apparent}^{initiation} = K_I(a_e, \rho, P_Q) \ . \tag{1}$$

That is, instead of K_{Ic}, the symbol K_Q is used here to symbolically represent both the inherent assumptions involved and the lack of a fracture toughness testing standard for each or any type of ice. Note that key features only are represented in Equation (1). The load P_Q is defined in ASTM E399; there are several different cases possible in the determination of P_Q from the opening load-CMOD (crack-mouth-opening-displacement) plot. A key requirement in terms of the nonlinearity of this curve is that $P_m/P_Q < 1.10$.

The notation in (1) has long been used in the fracture toughness testing of concrete and rock (Costin, 1981; Schmidt and Lutz, 1979). Wecharatana and Swartz (1988) recommend this notation in a report compiled on behalf of the Society of Experimental Mechanics Subdivision on Fracture of Concrete and Rock (112 references were surveyed).

To date, the *modus operandi* in the fracture toughness testing of ice has been to *assume* that $P_Q = P_m$. The effective crack length is a more difficult matter, because to date it has not been possible to fatigue crack notched ice specimens. It may be, however, that the precracking accomplished by scribing a sharp notch with a scalpel or razor blade is sufficient. This technique has been used extensively. Recently, Dempsey et al. (1989a) used the technique of freezing teflon coated razor blades in notched specimens with no precracking. The fracture toughness results were reported in terms of K_Q; however, with no precracking, it may have been advisable to introduce an even more tentative terminology

$$K_m = K_I(a_o, P_m) \ , \tag{2}$$

which can only be called an *approximate fracture toughness* (Ouchterlony, 1982).

In an attempt to quantitatively assess the influence of the microstructure on the observed fracture toughness values, macrofractography has recently been used by Dempsey and Wei (1989) to identify fracture mechanisms operative in the experiments reported by Dempsey et al. (1988) on macrocrystalline (S1) freshwater ice. A similar study of the fracture mechanisms operative during the tests reported by Dempsey et al. (1989b) using single-edge-notched-bend and wedge-loaded-compact-tension specimens on columnar (S2) freshwater ice is in progress. Earlier work of this type was reported by Kusumoto et al. (1985, 1986a, b).

5. Specimen Size Considerations

The criteria of validity for fracture toughness measurements of any type of ice are not yet established (Earle et al., 1984). The motivation behind the notation introduced in Equation (1) derives from the lack of a standard for determining the proper specimen size, loading rate, etc., for a valid K_{Ic} test of any ice material. The value of K_1 at which a stationary macrocrack extends in a given material should be independent of the size and shape of the test specimen and the method of loading. The critical-stress-intensity-factor (K_{Ic}) would then be an invariant quantity which could be regarded as a true *material property.*

The assumptions outlined in the above section are consistent with applications of linear-elastic-fracture mechanics. During a fracture test, LEFM requires that the zone of inelastic deformation in the vicinity of the crack tip (the process zone) be "small" with respect to the other dimensions of the specimen, usually the crack length (a) or the uncracked ligament (ℓ). These "small scale yielding" requirements (Rice, 1968) are well delineated in the LEFM standard for metals (ASTM E399) and are dictated by the ratio of toughness to yield strength, *viz.*,

$$a \ , \ \ell \ , \ h \geq 2.5(K_{Ic}/\sigma_y)^2 \ , \qquad (3)$$

where σ_y is the yield stress. All of the above dimensional criteria assume the presence of a plastically deformed zone (which has been observed in metals but not in ice) ahead of the crack.

These criteria ignore the extreme differences between ambient temperature metals testing and ice testing. The grain size of even the small-grained ice types differs by two to three orders of magnitude from that of most metals. Also, LEFM theory assumes an isotropic elastic medium.

In metals, while the individual grains are anisotropic they are oriented randomly; moreover, the metal grain size is very small compared with any flaw size. If the flaw size is comparable with grain size, the assumptions of continuum mechanics are no longer valid, and LEFM does not apply. In other words, true polycrystalline behavior would no longer be observed, but rather a pseudo-monocrystalline behavior. In addition to the much larger grain size, the fracture toughness testing of ice reported to date has been performed at very high homologous temperatures ($-10^oC \equiv 0.96T_m$).

To summarize briefly, the question of a suitable specimen size is complicated by the effects of grain size and high-temperature creep. The form of (3) is not convenient because the yield stress of ice (σ_y) is often unknown and not easily determined. The notation K_{Ic} is not ideal for the reasons that led to the introduction of K_Q in (1).

6. Grain Size Effects on Fracture Toughness Measurements

Ice strength has long been known to depend on the size of the ice samples tested. In the context of compression tests on seeded columnar (S2) freshwater ice prisms having an average grain size (d_{av}) equal to 4mm, Hirayama et al. (1974) found that the ice strength was independent of the size of the sample so long as the minimum cross-sectional dimension was greater than 25 d_{av}. In the context of tension tests, Schwarz et al. (1981) recommended that the same dimension be greater than 15 d_{av}. Earle et al. (1984) adopted the latter recommendation for plane strain fracture toughness testing, arguing that tensile strength is a closely related strength parameter. In the context of constant strain-rate compression tests on randomly oriented equiaxed freshwater ice, Jones and Chew (1981) determined that the minimum dimension must be greater than 12 d_{av} (otherwise the maximum yield stress would not be independent of the grain size).

Until recently there had been no systematic investigation of the grain size effect in terms of both crack length and specimen size. Dempsey et al. (1989a) tested seeded columnar freshwater ice (defined as S2 type ice by Michel and Ramseier, 1971) using the four-point-bend (4pt) and three-point-bend (3pt) loading configurations (Figure 1). S2 type ice is characterized by vertically-elongated columnar crystals exhibiting mainly horizontally oriented c-axes (compared to the plane of the ice sheet) with

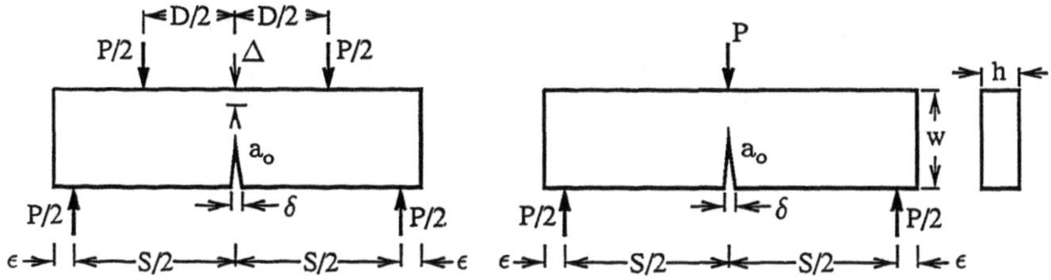

Figure 1. (a) Four-point-bend and (b) three-point-bend configuration.

random orientation (Figure 2a). The crack-microstructure orientation was kept fixed (Figure 2b), as was the grain size ($d_{av} \approx 3$mm). For this set of tests a tentative criterion was obtained:

$$a_o \geq 10 d_{av} \ , \quad \ell \geq 12 d_{av} \ . \tag{4}$$

In (4) a_o is the fabricated crack length, and ℓ is the length of the uncracked ligament ($\ell = w - a_o$) in the plane of the crack.

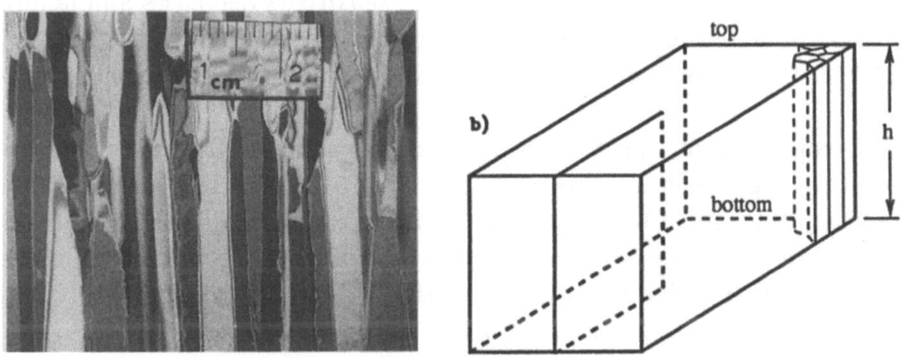

Figure 2. (a) Columnar S2 ice features and (b) radial crack orientation.

A constraint on the specimen thickness (h) is not included in Equation (4) for the following reasons. It turns out that for metals the plate thickness largely affects the state of stress at the crack tip. That is, in order to maintain plane strain along the greater part of the crack front, the plate thickness must be sufficiently large. A fairly thick specimen is required to determine the plane strain fracture toughness (the material property, in other words). For quasi-brittle materials such as rock, ceramics, and concrete, the specimen thickness requirement is not as severe. For instance, Mindess and Nadeau (1976) found for concrete that there was no dependence on the length of the crack front for thicknesses ranging between 4.7 d_a and 26.7 d_a, where d_a is the size of the aggregate (gravel).

Andrews (1983) stated a similar criterion to that in (4) for ice but also included a constraint that h $\geq 10d_{av}$; furthermore, he regarded his conditions as *additional* to those in (3). This author does not agree with the latter sentiment; furthermore, the constraint on h can only be established empirically and will surely differ for each ice type, and crack orientation combination.

7. Notch Sensitivity in Fracture Testing of S_2 Ice

For a specimen to be notch sensitive, the crack and uncracked ligament must both be long enough for the crack tip to be surrounded (except for the process zone) by a volume in which the elastostatic stress and displacement field is accurately portrayed by LEFM theory. For cracks that are too short or too long, the *notch sensitivity* decreases and failure in flexure or tension may preclude a true fracture failure in the sense of LEFM. "Too short" or "too long" must be found through experimentation.

Dempsey et al. (1989a) tested seeded columnar freshwater ice using three specimen sizes (Figure 1): w = 51mm $\approx 17d_{av}(4pt)$, 75mm $\approx 25d_{av}(4pt)$, and 102mm $\approx 34d_{av}(3pt)$. Correspondingly, three plateaus were observed in the apparent fracture toughness values obtained, for the following intermediate crack lengths (Figure 3): $5 \leq a_o/d_{av} \leq 10(0.3 \leq a_o/w \leq 0.6)$; $7 \leq a_o/d_{av} \leq 13(0.3 \leq a_o/w \leq 0.5)$, and $6 \leq a_o/d_{av} \leq 20(0.2 \leq a_o/d_{av} \leq 0.6)$, respectively.

The criterion in (4) was deduced from these results. There are, however, recurring features in these results that point to a variation in *notch sensitivity, viz.*,

- the increase or decrease of the fracture toughness K_Q with crack length;
- the overall increase of the plateau K_Q by increasing the specimen size;
- linear load-CMOD curves which negate the need for nonlinear fracture mechanics parameters.

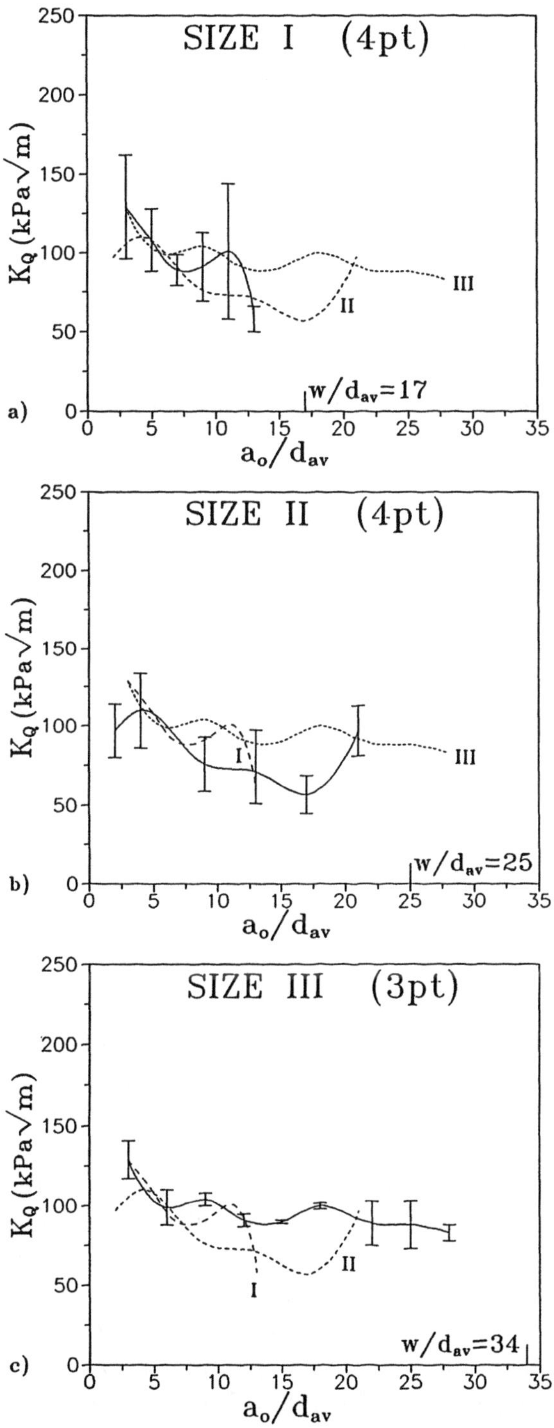

Figure 3. Apparent fracture toughness vs. ratio of crack length to grain size for sizes I-III.

The dependence of notch sensitivity on specimen size is discussed for concrete and aggregative materials by Carpinteri (1982a, b). Notch sensitivity, or the lack thereof, can be identified following the procedure used by Walsh (1972) and Carpinteri (1982b).

The stress-intensity-factor expressions for the single-edge-notched-bend (SENB) specimens tested by Dempsey et al. (1989a) are given by

$$\begin{aligned} K_Q &= (3P_f/hw)\sqrt{\pi a}F(a/w) \quad (4pt) \ , \\ K_Q &= \quad 3P_f G(a/w)/4h\sqrt{w} \quad (3pt) \ . \end{aligned} \tag{5}$$

The functions $F(a/w)$ and $G(a/w)$ are given by Dempsey et al. (1989a). The peak nominal tensile stress (σ_N) acting on the net sections at failure follows as

$$\begin{aligned} \sigma_N &= \ 3P_f w/h\ell^2 \quad (4pt) \ , \\ \sigma_N &= \ 9P_f w/2h\ell^2 \quad (3pt) \ , \end{aligned} \tag{6}$$

in which $\ell = w - a$. Using the value for P_f provided in (5), it now follows that

$$\sigma_N/\sigma_f = C_N \beta_f \ , \tag{7}$$

in which

$$\beta_f = K_Q/\sigma_f\sqrt{w} \ , \tag{8}$$

and

$$\begin{aligned} 1/C_N &= \ \sqrt{\pi u}F(u)(1-u)^2 \quad (4pt) \ , \\ 1/C_N &= \ \ G(u)(1-u)^2/6 \quad (3pt) \ . \end{aligned} \tag{9}$$

Following Carpinteri (1982b) the parameter β_f is called a "brittleness number", while σ_f is the flexural strength of an uncracked specimen. A plot of the σ_N/σ_f ratio versus the relative crack depth (varying the brittleness number β_f) for both the four-point- and three-point-bend configurations are shown in Figure 4. It is evident that fracture tests completely lose their meaning for β higher than $\beta_f^o \approx 0.528$ (4pt), 0.491 (3pt). Moreover, for values of $\beta < \beta_f^o$, fracture tests are significant only for cracks of intermediate length.

The parameter β_f introduced in Equation (8) is closely related to the characteristic length, introduced by Hillerborg (1985): $\ell_{ch} = EG_{1c}/\sigma_f^2$. It is useful to note that $\beta_f^2 \equiv \ell_{ch}/w$. A similar parameter has been introduced by Goldstein and Osipenko (1985a, 1989). Bazant (1987) has

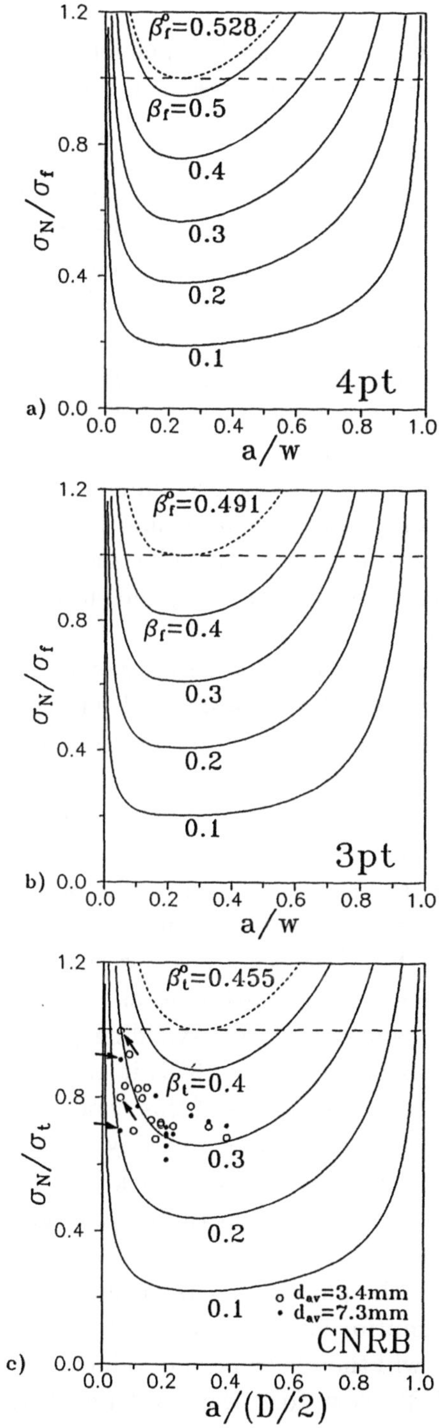

Figure 4. Brittleness plots for the 4pt, 3pt, and CNRB geometries.

criticized these parameters because they are not shape-independent. For this reason, the author has chosen to present his discussion in terms of the shape-independent ratio σ_N/σ_f.

Using the above information, the brittleness number β_f can be used to ensure that a particular series of fracture tests will be notch sensitive. For brittleness numbers that are too high ($\beta_f > \beta_f^o$) the fracture tests are meaningless, a strength failure occurring prior to the attainment of a critical-stress-intensity-factor. For lower β_f values the fracture tests are significant, but only if the fabricated crack lengths are within a certain interval. The plots in Figure 4 clearly reveal the optimum crack length to obtain the lowest degree of brittleness for a given specimen size. However, the essential importance of the brittleness number resides in the ability to forecast the necessary specimen size. A series of preliminary tests is first required to estimate the appropriate strength value and fracture toughness magnitude. It is then feasible to choose a particular specimen width w so that the tests will be significant (so that $\beta_f \leq \beta_f^o$) and suitably notch sensitive (which further restricts β_f).

Given the above information concerning the notch sensitivity of a particular fracture test, the fracture tests reported by Dempsey et al. (1989a) are re-examined. The ratios of the peak nominal outer fiber stress in the net section to the flexural strength (σ_N/σ_f) were plotted versus the ratio of the fabricated crack length to the specimen width (a_o/w) in Figure 5. The lines of constant brittleness numbers are also dashed in for comparison. Clearly, a feature common to each size portrayed in Figure 5 is the fact that a higher degree of brittleness (a lower brittleness number) is observed for intermediate crack lengths. Less scatter is observed for increasing specimen size. Guided by Dugdale type estimates as to when large-scale-yielding (l.s.y.) is observed versus small-scale-yielding (s.s.y) in terms of local to nominal stress ratios (Hutchinson, 1979) it is tentatively advisable to require that $\sigma_N/\sigma_f < 0.5$. In terms of this condition, it is evident that a sufficiently large specimen size was probably not examined by Dempsey et al. (1989a) since, for intermediate crack lengths, the specimens are not notch sensitive enough to keep the σ_N/σ_f ratio well below one half.

For a sufficiently large specimen size, the plateau K_Q value should no longer increase; in the sense of the assumptions involved in the statement of Equation (1), a "material property" will then have been obtained. Let this property be represented here by K_Q^∞. Now, from the development in

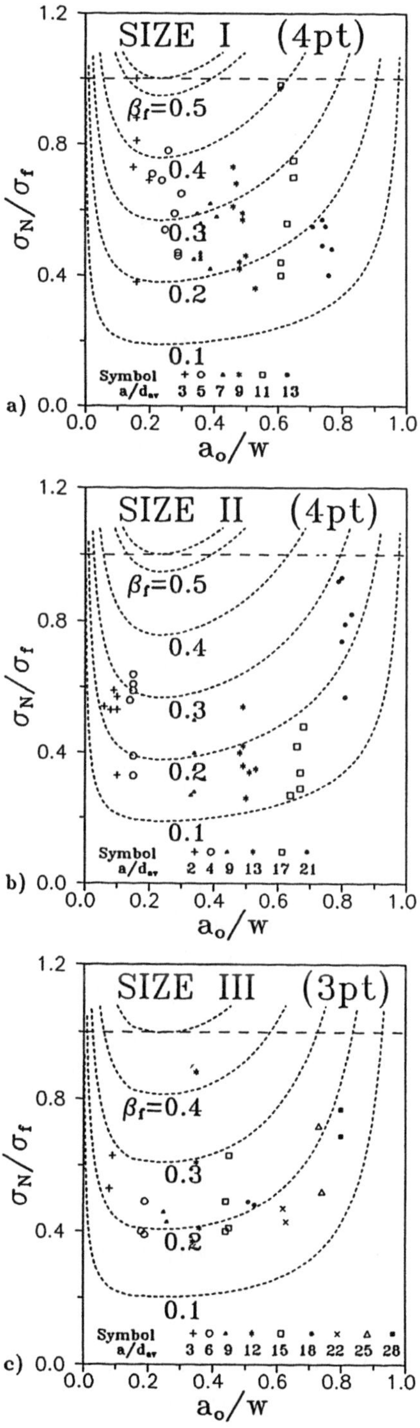

Figure 5. σ_N/σ_f vs. the ratio of crack length to specimen width for sizes I-III.

Equation (5) - (9), it is also instructive to note that $\sigma_N > \sigma_f$ provided that $w > w_{tr}$, where

$$\sigma_N = C_N K_Q / \sqrt{w} \text{ and } \sigma_f = C_N K_Q^{\infty} / \sqrt{w_{tr}} \,. \tag{10}$$

The specimen size w_{tr} is clearly the transitional size, for which if $w < w_{tr}$, a flexural failure would occur, and if $w > w_{tr}$, a fracture failure would occur. This argument is too simplistic (Walsh, 1972; Carpinteri, 1982a,b; Hillerborg, 1985; Bazant and Kim, 1985; Bazant and Pfeiffer, 1987; Shyam Sunder and Nanthikesan, 1987). Two factors are mainly involved: the size of the process zone and the length of the crack and the uncracked ligament as compared with the grain size. For these reasons, for widths approaching the transitional size, w_{tr}, the LEFM expressions become less and less applicable. It happens that the required width w^{∞} is many times w_{tr}. However, a quantitative estimate of the ratio w^{∞}/w_{tr} is still required.

Following from the flexural and fracture tests reported by Dempsey et al. (1989a) it is representative to assume that $\sigma_f = 1.5$ MPa and that $K_Q^{\infty} = 120$ kPa$\sqrt{\text{m}}$. Using the expressions in (10), the data reported by Dempsey et al. (1989a) are again re-examined, this time plotting for each size the apparent fracture toughness K_Q versus the ratio of the actual specimen width to the transitional width (w/w_{tr}). These plots are shown in Figure 6. The fracture specimens for Size I (Figure 6a) are clearly sub-size; the specimens for Size II and Size III show a significant improvement for the crack lengths which tend to be on the short side. From Figure 6b (Size II), the largest ratios of w/w_{tr} occur for $a/d_{av} \approx 3.9$ and 8.9; in other words, for $0.16 \leq a_o/w \leq 0.36$. From Figure 6c (Size III), the largest ratios of w/w_{tr} occur for $a/d_{av} \approx 6.4, 8.8$, and 12.0; that is, for $0.18 \leq a_o/w \leq 0.35$. For the latter range of crack lengths, the fracture toughness ranges between 87 and $141 kPa\sqrt{\text{m}}$. The optimum crack length intervals coincide with those which would be predicted with the brittleness plots in Figure 4. The overlap between the plateaus found from Figure 3 and Figure 6 suggest the optimum ranges for Sizes II and III to be, respectively, $0.3 \leq a_o/w \leq 0.36$ and $0.2 \leq a_o/w \leq 0.35$. The above trends are verified, especially for Size III, by replotting w/w_{tr} versus a_o/w for each size (Figure 7). The brittleness plot in Figure 5c also verifies the range stated for Size III.

The small scale yielding requirements stated in the LEFM standard for metals (ASTM E399) are quoted in (3). One objective here is to

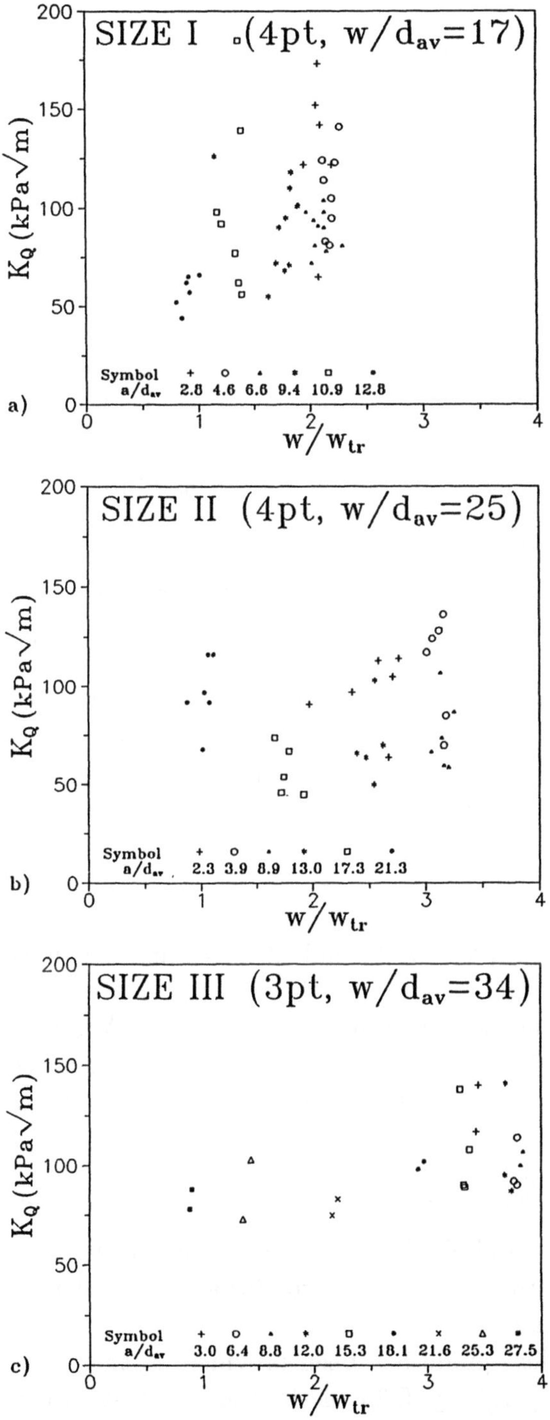

Figure 6. Apparent fracture toughness vs. the ratio of specimen width to transitional width for sizes I-III.

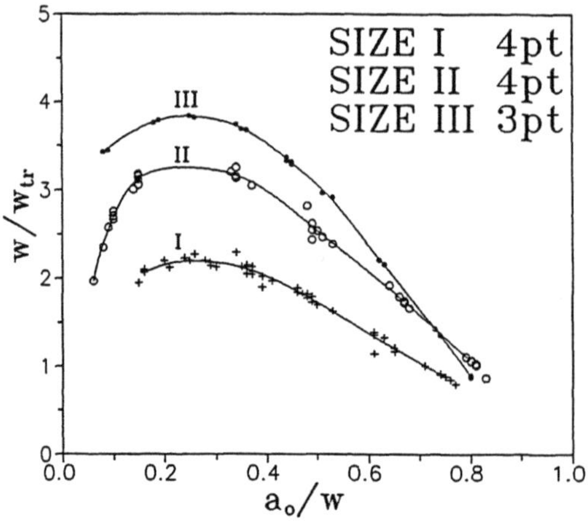

Figure 7. Ratio of specimen width to transitional width vs. a_o/w for sizes I-III.

derive a condition for the requisite specimen width w^∞ in terms of more easily attainable quantitites. For this purpose, it is proposed that

$$\sigma_N/\sigma_f < 0.4. \tag{11}$$

The results plotted in Figures 5 and 6 provide the impetus for this specimen shape-independent condition. In terms of the expressions in (10), the brittleness/size conditions in (11) is equivalent to requiring that $w^\infty > 25w_{tr}/4$ or, equivalently, that

$$w^\infty \geq (5C_N/2)^2(\frac{K_Q^\infty}{\sigma_f})^2 \ , \tag{12}$$

where C_N is the shape- and crack length-dependent factor and is defined in Equation (9). For the four-point-bend and three-point-bend configurations, $C_N \approx 2$ for the intermediate crack lengths of interest. Assuming $\sigma_f = 1.5$ MPa and $K_Q^\infty = 120$ kPa $\sqrt{\text{m}}$, then the required specimen size for the material property K_Q^∞ to be established using the 4pt or 3pt configuration is given by

$$w^\infty \geq 160\text{mm} \quad \text{or} \quad w^\infty/d_{av} \geq 53. \tag{13}$$

The condition stated in (12) parallels that in (3) but is now in terms of more readily attainable quantities. The same condition can be derived for tension tests, with σ_f replaced by the tensile strength, σ_t.

To summarize briefly, a clear guide has been provided as to the appropriate crack length ($0.2 \leq a_o/w \leq 0.35$) if the fracture tests are to be performed using either the four-point-bend or the three-point-bend configuration. The required width is given by (13) based on an *ad hoc* condition proposed in (11). The thickness h of each ice sheet was kept constant at 45mm. In other words, the length of the crack front (h) was not altered.

A survey of previous fracture tests on S2 freshwater ice is presented in Table 1. An attempt has been made to include all of the key factors that have undoubtedly contributed to the scatter in the reported apparent fracture toughness data. A certain amount of scatter in the results is expected, particularly if the crack length or the uncracked ligament is not significantly larger than the grain size, since the initial, fabricated crack front may terminate at a grain boundary or at some interior point of columnar crystal (or, partially, do both). The crack front "senses" at most only a few contiguous columnar crystals. In addition, note that different methods were used to fabricate and sharpen the cracks, and the time intervals between ice growth, crack formation and testing was certainly quite variable. If not radial (Figure 2a), crack growth was either vertically downwards or upwards. The fracture toughness results in Table 1 reveal a considerable degree of scatter, which is not surprising if one considers solely the range of specimen sizes, crack and ligament lengths, and grain sizes in the light of the results presented in Figures 3, 5, 6 and 7. Meaningful comparisons can be made with the work of Timco and Frederking (1986) and Bentley et al. (1988) since they used the same loading rate range, test temperature, loading conditions, and crack orientation (see Dempsey et al., 1989a).

The above crack length and specimen size considerations are primarily directed towards designing fracture toughness tests that both satisfy small scale yielding requirements and provide material properties (in the sense of Equation (11)) - toughness values independent of the size and geometry of the specimen. The results reported by Dempsey et al. (1989a) and re-examined in this paper suggest that the specimen sizes used in testing S2 ice to date have been *sub-size* (excluding the work by Bentley et al., 1988). Furthermore, the possibility of choosing a crack length to give a greater notch sensitivity has not been known or appreciated.

The above observations are in all likelihood applicable (in differing degrees) to other ice types, such as equiaxed-granular freshwater ice, macrocrystalline (S1) freshwater ice, doped models, or sea ice. This possibility is explored in the next section.

Table 1. Fracture toughness tests on S2 freshwater ice to date.

Authors (Year)	Test Type	a (*mm*) O'n[a]	d_{av} (*mm*)	a/d_{av}	[b] ℓ/d_{av}	Test Temp. (°C)	$\dot{K}$ (kPa$\sqrt{m}$/s)	K_Q (kPa$\sqrt{m}$)
Gold (1963)	Thermal shock	2.4-9	1.5-6	0.4-6				50-160[c]
Liu and Loop (1972)	CT	51 NI[d]	4.5	11	13	-3.5 -9.0	45 45	90-120 96-116
Goodman and Tabor (1978)	3pt Vickers Indenter Conical Indenter	7-10 1-10 1-10 NI	1*mm* to almost single crystals			-13 -38 -20 -20 -16	10-13	116 ±13 55-85 170-250 190-290 > 300(?)
Liu and Miller (1979)	CT	62-78 NI	5	12-16	34-38	-8 -9	1.67-3.32 38,46	113-327 110,130
Goodman (1979)	4pt	10 NI	10	1	1	-11	1000	85-153
Hamza and Muggeridge (1979)	3pt	25 25 C⊥	8 12	3 2	3 2	-3.9	33	42-82 116-164
Urabe and Yoshitake (1981)	3pt	50 C⊥,C∥	7.5	6	7	-20	80	90-170
Timco and Frederking (1982)	3pt	12 C⊥	2	6	24	-10	7-16	73-92
Andrews and Lockington (1983)	Pressurized Interface Crack	6.5	2-5	3.2-1.3				105 clear 50 bubbly
Andrews (1983,1985)	Radially cracked ring	6-13	0.91	7-14	8-15	-12		50-240
Nixon (1984)	Chevron notched round bar	112	1-5	22-112	8-38	-20	0.35 1.5 20.0	137-184 167-222 140-174
Danilenko (1985)	CT	8 to 40 C⊥,C∥	3-8	1-3 to 5-13	2-4 to 10-27	-10	0.1-2	110-320
Parsons and Snellen (1985)	CT	200-400 C⊥,C∥[g]	5-100 (top to bottom)	O'n dependent	O'n dependent	-15°C top surface (in situ)		100-400
Timco and Frederking (1986)	3pt 4pt	12 12 C⊥	2.5 2.5	5 5	15 15	-10	10-100	100-140
Kusumoto[e] et al. (1986a,b)	3pt 3pt 3pt	10 20[1] 20[2] NI[f]	10	1 2 2	1.5 3 3	-10	10^2-10^3	60-210 65-100 70-80
Bentley et al. (1988)	TDCB	165-561 165-561 C⊥	3 3	55-187 55-187	813-945 813-945	-2	200-1000	108-165
Parsons et al. (1989)	DT	250	1-5	250-50		-20°		124 ± 38 89 ± 14[c]

[a] Orientation of crack front perpendicular ($C\perp$) or parallel ($C\parallel$) to the c-axes.
[b] l is the length of the uncracked ligament in the plane of the crack
[c] Crack arrest toughness (K_Q^A)
[d] No information
[e] A sharp transition was observed in the range $10 \leq \dot{K} \leq 100$. Values of K_Q for $10^o \leq \dot{K} \leq 10^2$ were obtained but are not reported in the above table.
[f] The paper has not yet been translated; $^1h = 50$mm; $^2h = 200$mm
[g] Radial, spalling, and top downwards

8. Equiaxed and Macrocrystalline (S_1) Freshwater Ice and Sea Ice

8.1 Glacier Ice and Equiaxed Freshwater Ice

A survey of previous fracture toughness tests on glacier ice and randomly oriented equiaxed freshwater ice is presented in Table 2. This work is characterized by low crack length to grain size ratios and small specimen sizes given the average grain diameters. Not surprisingly, therefore, the apparent fracture toughness values reveal a substantial degree of scatter.

Table 2. Glacier ice and randomly oriented equiaxed freshwater ice.

Authors (Year)	Test Type	a (mm)	d_{av} (mm)	a/d_{av}	$^b\ell/d_{av}$	Test Temp. (°C)	$\dot{K}$ (kPa$\sqrt{m}$/s)	K_Q (kPa$\sqrt{m}$) $P_f = P_{max}$
Andrews[a] (1983,1985)	Radially cracked	4-11	9.6mm	0.4-1.1	1.9-2.6	–	3-9	31-93
Andrews[a] et al. (1984)	3pt	7-30	– (10)[c]	– (0.7-3)	– (2-4)	-1	0.1-7	70-180
Nixon and Schulson (1986)	CNRB[d]	9	2.2 8.2	4 1	17 4.5	-10 -10	$10,10^5$ $10,10^3,10^5$	~90 ~75
Nixon and Schulson (1988)	CNRB	8.5	1.6-9.3	1-5	23-4	-10	10.0	96-67
Nixon (1988a)	CNRB	5-18	3.4 7.3	1.5-5.2 0.7-2.4	11.8-8.1 5.5-3.8	-10 -10	10 10	68-100[e] 53-86[e]

[a]Glacier ice
[b]ℓ is the length of the uncracked ligament in the plane of the crack
[c] Assumed as representative
[d] Circumferentially notched round bar
[e] Re-computed using Tada's expression in (16)

Andrews (1983,1985) reported on the development of an internally or externally radially cracked ring fractured by internal pressure. For glacier ice their test specimen size (externally cracked) was very small compared to the grain size, and gave a mean apparent fracture toughness value of 58 kPa$\sqrt{\text{m}}$. This result caused Andrews and co-workers a lot of concern; as stated by Andrews (1983) ... "A value of around 115 kPa$\sqrt{\text{m}}$ would have been expected from these results. There are two possible causes of this discrepancy: the presence of air bubbles in the ice and the effect of grain size." A calculation of the σ_N/σ_t ratio using the information presented by Andrews (1983,1985) gives $\sigma_N/\sigma_t \approx 0.52$. This number is high compared to Equation (11); furthermore, the applicability of LEFM expressions when $0.4 \leq a_o/d_{av} \leq 1.1$ and $1.9 \leq \ell/d_{av} \leq 2.6$ is questionable. The specimen sizes were definitely not large enough to form an isotropic continuum in the vicinity of the cracks.

Nixon and Schulson (1986,1988) and Nixon (1988a) tested equiaxed granular freshwater ice using circumferentially-notched-round-bar (CNRB) specimens of constant outer diameter D and radius R. The equation analogous to that in (7) for the CNRB specimen is given by

$$\sigma_N/\sigma_t = C_N\beta_t \, , \tag{14}$$

in which

$$\beta_t = K_Q/\sigma_t\sqrt{D} \, , \tag{15}$$

and, given that $a_o/R = 1 - \xi$,

$$C_N = 4/\pi\xi^2 Y(\xi) \, , \tag{16}$$

where according to Brown and Srawley (1966)

$$Y(\xi) = 1.72/\xi - 1.27 \, , \tag{16a}$$

or according to Tada et al. (1985)

$$Y(\xi) = \frac{2}{\xi}G(\xi)\sqrt{\frac{2}{\pi}(\frac{1-\xi}{\xi})} \, , \tag{16b}$$

with

$$G(\xi) = [1 + \xi/2 + 3\xi^2/8 - 0.363\xi^3 + 0.731\xi^4]/2. \tag{16c}$$

The expression in (16a) provided by Brown and Srawley (1966) holds for $0.5 \leq \xi \leq 0.8$ or $0.2 \leq a_o/R \leq 0.5$. In terms of the diameter used in these CNRB tests (D = 91mm), (16a) is accurate within 1 percent for $9 \leq a_o \leq$22.5mm. The expression in (16b, c) from Tada et al. (1985) is accurate better than 1 percent over the whole range of crack lengths. In

(15), σ_t is the tensile strength of an unnotched sample; σ_N is the nominal tensile stress acting on the net section of the circumferentially notched specimen at failure.

A plot of the σ_N/σ_t ratio versus the relative crack depth, varying the brittleness number β_t, is shown in Figure 4c. Also shown in Figure 4c are the results from 29 fracture tests reported by Nixon (1988a). In this study, the notch length was varied between $0.06 \leq a_o/R \leq 0.39$ for two grain sizes, d_{av} = 3.4mm (16 tests) and 7.3mm (13 tests). For the shorter crack lengths ($a_o/R \leq 0.2$) the failure loads were re-computed using Tada's expression (16b) instead of Brown and Srawley's (16a) which is not valid in this range. The arrows (two for each grain size) in Figure 4c point out the samples *which did not fail at the notch.*

For d_{av} = 3.4mm, and assuming that σ_t= 0.94 MPa, observe that $0.68 \leq \sigma_N/\sigma_t \leq 1$. For d_{av} = 7.3mm and σ_t= 0.83 MPa, observe that $0.62 \leq \sigma_N/\sigma_t \leq 0.91$ (this tensile strength data was obtained from Currier and Schulson, 1982.) Clearly, these tests are very *notch insensitive.* The above observations, given the low values found by Andrews (1983, 1985), raise the possibility that the CNRB specimen size by Nixon and Schulson (1986, 1988) and Nixon (1988a) is relatively notch insensitive in which case the fracture toughness values reported by these authors are consistently too low (at least for test temperatures not too different from -10^oC).

The small scale yielding requirement stated in (11) is again used here (in terms of σ_N/σ_t) to estimate the requisite diameter D^∞ for the CNRB specimen. A similar expression to that in (12) is obtained for D^∞. The brittleness plot Figure 4c identifies the optimum crack lengths as being between $0.2 \leq a/R \leq 0.4$ for which C_N in (16) is approximately 2.26. Assuming that $\sigma_t \approx 1.0$ MPa, then the required specimen sizes for different estimates of the material property K_Q^∞ to be established using the CNRB configuration are given by

$$\begin{aligned} D^\infty &\geq \text{ 204mm for } K_Q^\infty = \text{ 80 kPa}\sqrt{\text{m}}\,, \\ D^\infty &\geq \text{ 319mm for } K_Q^\infty = \text{ 100 kPa}\sqrt{\text{m}}\,, \\ D^\infty &\geq \text{ 460mm for } K_Q^\infty = \text{ 120 kPa}\sqrt{\text{m}}\,. \end{aligned} \tag{17}$$

The 91mm diameter CNRB is severely sub-size if the size-independent apparent fracture toughness value is 80 kPa$\sqrt{\text{m}}$ or greater.

Nixon (1988a) chose to investigate the question of true polycrystalline behavior in terms of grain size versus both crack length and uncracked ligament length. The CNRB specimen size was kept fixed (D = 91mm). For an average grain size of 3.4mm, the notch depth to grain size varied between 1.5 and 5.2; for d_{av} = 7.3mm, $0.7 \leq a_o/d_{av} \leq 2.44$. Nixon notes that "no effect of notch depth was observed." This observation is consistent with the sample being notch insensitive. Furthermore, for very low a_o/d_{av} values (as above), anisotropy would be a factor. The K-calibration equations in (16) are for isotropic media as noted in the definition of K_Q in Equation (1).

The above discussion leads this author to question whether the grain size effect reported by Nixon and Schülson (1988) may be substantially influenced by the specimen size. Furthermore, the evidence strongly suggests that the observed notch insensitivity and inevitable anisotropy would have precluded true polycrystalline behavior. In all likelihood, K_Q^∞ is considerably higher than the values reported in Table 2.

Table 3. Fracture toughness tests on S1 ice to date.

Authors	Type	Dimensions w × h × S(mm)	O'n	°C	$\dot{K}$ kPa$\sqrt{m}$/s	K_Q (kPa$\sqrt{m}$)
Kollé	3pt	50 × 25 × 200	C∥	-17	1-10	240 ± 79
(1981)			C⊥			186 ± 82
Danilenko	CT	15 × 15 × 20	C∥	-10	0.1 - 1	335 ± 50
(1985)		to	C∥	-15		439 ± 72
		80 × 80 × 120	C⊥	-10		336 ± 94
			C⊥	-15		414 ± 85
Dempsey,	4pt	32 × 32 × 128	C∥	-10	100-250	465 ± 182
Nigam, and		48 × 48 × 192				280 ± 135
Cole (1988)		64 × 64 × 256				250 ± 72
		80 × 80 × 320				282 ± 136

8.2 Macrocrystalline (S1) Freshwater Ice

A survey of previous fracture toughness tests on macrocrystalline (S1) ice is presented in Table 3. S1 type ice is characterized by massive irregularly shaped crystals exhibiting, with respect to the plane of the ice sheet, vertical or near vertical c-axes (Michel and Ramseier, 1971). The fracture toughness results reported in Table 3 reveal a large degree of scatter. An investigative size effect study was reported by Dempsey et al. (1988) using the four-point-bend configuration and the same crack-orientation as in Figure 2b. Four specimen sizes were tested (see Table 3), with the band-sawed notch always equal to half the specimen depth ($a_o = w/2$). The grain sizes were not well controlled in the growth of this ice and varied between 10 and 40mm. The crack lengths to average grain size variations for sizes I-IV were as follows: $0.4 \leq a_o/d_{av} \leq 1.6; 0.6 \leq a_o/d_{av} \leq 2.4$; $0.8 \leq a_o/d_{av} \leq 3.2$; and $1 \leq a_o/d_{av} \leq 4$. Even for the largest specimen size, one would not expect to observe true polycrystalline behavior in these tests. Instead, the behavior measured could best be associated with multiple crystals in one extreme and probably either bicrystals or single crystals in the other.

In retrospect, the above specimen size study undertaken by Dempsey et al. (1988) was flawed in two respects. The variation in grain size was large, and the range of specimen sizes attempted was too small. Larger specimens or more finely grained S1 ice are needed. The results plotted by Dempsey et al. (1988) exhibited a large degree of scatter in both the flexural strength and the apparent fracture toughness. However, it is instructive to plot in the same figure (Figure 8) the actual flexural strengths (σ_f) and peak nominal stresses at fracture (σ_N) for each size. From Figure 8 it is evident, qualitatively at best, that these fracture tests were notch insensitive. In this instance, the notch insensitivity was almost totally caused by the multiple-cum-single crystal anisotropy of the test specimens.

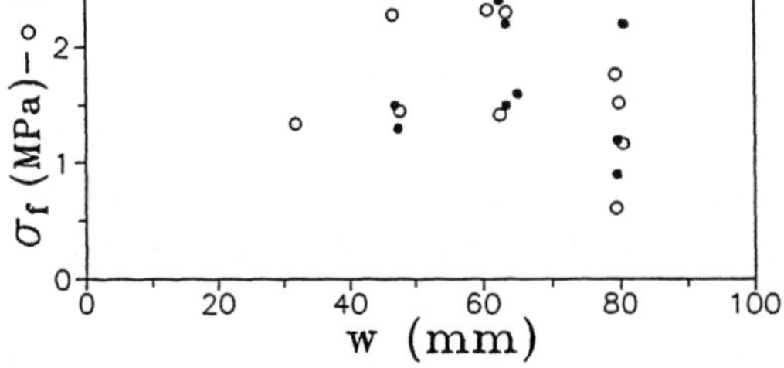

Figure 8. Comparison of nominal peak stress (σ_N) and flexural strength (σ_f) vs. specimen size.

8.3 Saline Ice

Nixon (1988b) reported four fracture toughness tests on saline ice using the CNRB geometry. Out of four tests, two broke away from the notch. The grain size of the saline ice was 10-20mm, while the outer diameter D was 100mm. Even for tension testing, this size would have less than 10 grain diameters across the diameter and would thus violate the $15d_{av}$ recommendation made by Schwarz et al. (1981). Anisotropy clearly was a major factor since $0.5 \leq a_o/d_{av} \leq 1$. To attempt fracture toughness tests with CNRB samples of 100mm diameter or less is inadvisable even for equiaxed granular freshwater ice as discussed above. This specimen size is not suitable for the fracture toughness testing of saline ice as the above notch insensitivity suggests.

8.4 Sea Ice

The effects of loading rate and specimen size on the fracture toughness of sea ice is a topic of utmost importance to those ice practitioners who wish to understand sea ice-structure interaction, the associated failure mechanisms and also the design of icebreakers. Space limitations in this paper do not permit a complete survey of previous fracture toughness tests on sea ice to be presented here. Instead, a brief analysis of results presented by Urabe et al. (1980) and Urabe and Yoshitake (1981) should reveal the crucial nature of the notch sensitivity question in the fracture toughness testing of sea ice.

Urabe et al. (1980) performed a series of three-point-bend fracture tests on sea ice, keeping the specimen size, temperature and crack length ($a_o/w = 0.19$) constant. The strain rate was varied through three decades (Table 4). Above a certain strain rate, the fracture toughness values reported were extremely strain rate-dependent. Unfortunately, all of the fracture tests reported can be shown to be notch insensitive. The loading rate effect discovered, therefore, is in all likelihood a feature of the small specimen size used. The correction factor introducted by Urabe and Yoshitake (1981) for even smaller specimen sizes has no theoretical basis and is perhaps too simplistic an approach to the difficulties posed by testing sub-size specimens.

The eleven fracture tests reported by Urabe et al. (1980) included flexural and fracture tests at each strain rate. Using their measure of the strain rate, their data is re-interpreted here in terms of the brittleness number β_f

Table 4. Fracture toughness tests on sea ice. (Urabe et al., 1980).

Test No.	$\dot{\varepsilon}$ (s^{-1})	σ_f (kgf/cm^2)	K_Q ($kgf\sqrt{cm}/cm^2$)	β_f	σ_N/σ_f
1	1.46×10^{-4}	3.79	8.00	0.35	0.72
2	5.40×10^{-4}	3.85	8.00	0.35	0.71
3	7.70×10^{-4}	3.72	8.00	0.36	0.73
4	8.97×10^{-4}	3.75	8.00	0.36	0.72
5	2.02×10^{-3}	3.50	7.20	0.34	0.70
6	2.91×10^{-3}	2.32	6.60	0.47	0.97
7	5.52×10^{-3}	1.75	5.58	0.53	1.08
8	8.34×10^{-3}	1.66	4.90	0.49	1.00
9	3.56×10^{-2}	1.11	2.60	0.39	0.80
10	3.64×10^{-2}	1.05	2.55	0.41	0.83
11	3.89×10^{-2}	1.11	2.40	0.36	0.74

and the ratio σ_N/σ_f (Table 4). The σ_N/σ_f condition in Equation (11) is violated for all tests. Interestingly, the transition from rate independence to rate dependence of the apparent fracture toughness values coincides directly with a sudden drop in brittleness and a further decrease in notch sensitivity.

9. Conclusions

The above discussion of the fracture toughness test results reported for freshwater ice in terms of notch sensitivity has been very instructive. In all likelihood, most specimen sizes used to date have been sub-size in terms of the small scale yielding required by linear elastic fracture mechanics (LEFM). An exact quantification of the degree of applicability of LEFM to tensile fracture in ice is made difficult by the coupled interaction of three major factors: grain size, specimen size, and loading rate. There appear to be two options available:

- upgrade the size of all specimen geometries until the notch sensitivity is such as to ensure LEFM;
- adopt the principles of nonlinear fracture mechanisms and concentrate on the associated toughness parameters.

It is highly likely that both approaches should be pursued. The first option will simply not be viable for certain types of ice and certain specimen geometries. Concerning the second option, size effects on fracture toughness and the use of nonlinear fracture mechanics has already received considerable attention for materials such as concrete and rock. It should be emphasized that an approach that is directly meaningful for anisotropic materials is one in which the resistance to crack growth is identified in terms of a critical value of the J-Integral, J_{1c}, (Miller, 1979) or in terms of the critical-energy-release-rate, G_{1c}, if the fracture event is linear elastic. The isotropy assumptions inherent in all of the work discussed in this paper are thereby removed. This approach has been followed by Dempsey et al. (1989b) in a recent paper using wedge-loaded-compact tension specimens and S2 freshwater ice.

Acknowledgement

This work has been supported in part by the U.S. National Science Foundation under grant no. MSM-86-18798 and in part by the U.S. Army under contract no. DACA 89-88-K-0013.

References

Andrews, R.M. 1983. Sheffield University Northeast Greenland Expedition, 1982. Final Report, pp. 27-59.

Andrews, R.M. 1985. Measurements of the fracture toughness of glacier ice. Journal of Glaciology, Vol. 31, pp. 171-176.

Andrews, E.H., and Lockington, N.A. 1983. The cohesive and adhesive strength of ice. Journal of Materials Science, Vol. 18, pp. 1455-1465.

Andrews, R.M., McGregor, A.R., and Miller, K.J. 1984. Fracture toughness of glacier ice. International Karakoram Project. (Edited by K.J. Miller), Cambridge University Press. Vol. 1, pp. 147-159.

ASTM E399-83. 1985. Standard test method for plane-strain fracture toughness of metallic materials. ASTM Standards, Vol. 03.01, pp. 547-582.

Atkins, A.G. 1975. Icebreaking modelling. Journal of Ship Research, Vol. 19, pp. 40-43.

Bazant, Z.P. 1987. Fracture energy of heterogeneous materials and similitude. Proceedings of the SEM/RILEM International Conference on Fracture of Concrete and Rock. (Edited by S.P. Shah and S.E. Swartz), Houston, Texas, pp. 390-402.

Bazant, Z.P., and Kim, J.-K. 1985. Fracture theory for nonhomogeneous brittle materials with application to ice. Proceedings. ARCTIC '85-Civil Engineering in the Arctic Offshore, ASCE, San Francisco, California, pp. 917-930.

Bazant, Z.P., and Pfeiffer, P.A. 1987. Determination of fracture energy from size effect and brittleness mumber. ACI Materials Journal, Vol. 84, pp. 463-480.

Bentley, D.L., Dempsey, J.P., Sodhi, D.S., and Wei, Y. 1988. Fracture of S2 columnar freshwater ice: floating double cantilever beam tests. Proceedings of the 9th International IAHR Symposium on Ice, Sapporo, Japan, Vol. 1, pp. 152-161.

Bhat, S.U., Choi, S.K., Wierzbicki, T., and Karr, D.G. 1989. Failure analysis of impacting ice floes. Proceedings of the 8th International OMAE Conference, The Hague, The Netherlands, Vol. IV, pp. 275-285.

Blanchet, D., Churcher, A., Fitzpatrick, J., and Badra-Blanchet, P. 1989. An analysis of observed failure mechanisms for laboratory, first-year and multi-year ice. Working Group on Ice Forces, Edited by G.W. Timco, Special Report 89-5, U.S. Army Cold Regions Research and Engineering Laboratory, Hanover, New Hampshire, pp. 77-124.

Brown, W.F., and Srawley, J.E. 1966. Plane strain crack toughness testing of high strength metallic materials. ASTM STP 410, pp. 15-16.

Carpinteri, A. 1982a. Application of fracture mechanics to concrete structures. ASCE Journal of the Structural Division, Vol. 108, pp. 833-843.

Carpinteri, A. 1982b. Notch sensitivity in fracture testing of aggregative materials. Engineering Fracture Mechanics, Vol. 16, pp. 467-481.

Choi, K., and Karr, D.G. 1989. A damage mechanics model for uniaxial creep and cyclic loading of polycrystalline ice. Proceedings of the 8th International OMAE Conference, The Hague, The Netherlands, Vol. IV, pp. 75-82.

Cole, D.M. 1986. Crack nucleation in polycrystalline ice. Cold Regions Science and Technology, Vol. 15, pp. 79-87.

Costin, L.S. 1981. Static and dynamic fracture behavior of oil shale. Fracture Mechanics Methods for Ceramics, Rocks, and Concrete, Edited by S.W. Freiman and E.R. Fuller, Jr., ASTM STP 745, 169-184.

Currier, J.H., and Schulson, E.M. 1982. The tensile strength of ice as a function of grain size. Acta Metallurgica, Vol. 30, pp. 1511-1514.

Danilenko, V.I. 1985. Determination of crack resistance (K_{1c},) of freshwater ice. Mechanics of Solids, Vol. 20, pp. 131-136.

Dempsey, J.P., and Wei, Y. 1989. Fracture toughness K_Q, and fractography of S1 type freshwater ice. Advances in Fracture Research, Edited by K. Salama, K. Ravi-Chandar, D.M.R. Taplin and P. Rama Rao, Pergamon Press, Vol. V, pp. 3421-3428.

Dempsey, J.P. Nigam, D., and Cole, D.M. 1988. The flexure and fracture of macrocrystalline S1 type freshwater ice. Proceedings of the 7th International OMAE Conference, Houston, Texas, Vol. IV, pp. 39-46.

Dempsey, J.P., Wei, Y., DeFranco, S., Ruben, R., and Frachetti, R. 1989a. Fracture toughness of S2 columnar freshwater ice: crack length and specimen size effects - part I. Proceedings of the 8th International OMAE Conference, The Hague, The Netherlands, Vol. IV, pp. 83-89.

Dempsey, J.P., Wei, Y., DeFranco, S., Ruben, R., and Frachetti, R. 1989b. Fracture toughness of S2 columnar freshwater ice: crack length and specimen size effects - part II. Proceedings of the 10th International POAC Conference, Lulea, Sweden, Vol. 1, pp. 199-207.

Earle, E.N., Frederking, R., Gavrilo, V.P., Goodman, D.J., Häusler, F.U., Mellor, M., Petrov, I.G., and Vaudrey, K. 1984. IAHR-recommendations on testing methods of ice. 4th Report of Working Group on Testing Methods in Ice, Proceedings of the 7th IAHR Symposium on Ice, Hamburg, Germany, Vol. IV, pp. 1-41.

Evans, A.G., Palmer, A.C., Goodman, D.J., Ashby, M.F., Hutchinson, J.W., Ponter, A.R.S., and Williams, G.J. (1984).Indentation spalling of edge loaded ice sheets. Proceedings of the 7th IAHR Symposium on Ice, Hamburg, Germany, Vol. I, pp. 113-121.

Goetze, C.G. 1965. A study of brittle fracture as applied to ice. Technical Note, unpublished, U.S. Army Cold Regions Research and Engineering Laboratory, Hanover, NH, 63p.

Gold, L.W. 1963. Crack formation in ice plates by thermal shocks. Canadian Journal of Physics, Vol. 41, pp. 1712-1728.

Goldstein, R.V. and Osipenko, N.M. 1983. Fracture mechanics and some questions of ice fracture. Mechanics and Physics of Ice, Nauka, Moscow, pp. 31-62 (in Russian).

Goldstein, R.V., and Osipenko, N.M. 1985a. Some mechanisms of localized fracture of ice cover under the action of compression. Proceedings of the 8th International POAC Conference, Narssarssuaq, Greenland, Vol. 3, pp. 1170-1188.

Goldstein, R.V., and Osipenko, N.M. 1985b. Fracture toughness and breaking of ice cover with ice breakers. Proceedings of the Arctic and Antarctic Scientific Research Institute (USSR), Vol. 391, pp. 137-158 (in Russian).

Goldstein, R.V., and Osipenko, N.M. 1989. Application of fracture mechanics in mathematical and physical modelling of ice breaking capability of icebreakers and ships. Proceedings of the 10th International POAC Conference, Lulea, Sweden, Vol. 3 (to appear).

Goodman, D.J., and Tabor, D. 1978. Fracture toughness of ice: a preliminary account of some new experiments. Journal of Glaciology, Vol. 21, pp. 651-660.

Goodman, D.J. 1979. Critical stress intensity factor (K_{1c},) measurements at high loading for polycrystalline ice. Proceedings of the IUTAM Symposium on the Physics and Mechanics of Ice, edited by P. Tryde, Copenhagen, Denmark, pp. 129-146.

Griffith, A.A. 1921. The phenomena of rupture and flow in solids. Proceedings of the Transactions of the Royal Society, Vol. A221, pp. 163-198.

Griffith, A.A. 1924. The theory of rupture. Proceedings of the 1st International Congress of Applied Mechanics, Delft, The Netherlands, pp. 55-63.

Hallam, S.D. 1986. The role of fracture in limiting ice forces. Proceedings of the 8th IAHR Symposium on Ice, Iowa City, Iowa, Vol. II, pp. 287-319.

Hamza, H., and Muggeridge, D.B. 1979. Plane strain fracture toughness (K_{1c}) of fresh water ice. Proceedings of the 5th International POAC Conference, Trondheim, Norway, Vol. 1, pp. 697-707.

Hillerborg, A. 1985. The theoretical basis of a method to determine the fracture energy G_F of concrete. Materials and Structures, Vol. 18, pp. 291-296.

Hirayama, K.-I., Schwarz, J., and Wu, H. C. 1974. Model technique for the investigation of ice forces on structures. Proceedings of the 2nd POAC Conference, Reykjavik, Iceland, pp. 332-344.

Hutchinson, J.W. 1979. Nonlinear Fracture Mechanics. Department of Solid Mechanics, Technical University of Denmark.

Irwin, G.R. 1957. Analysis of stress and strains near the end of a crack traversing a plate. Journal of Applied Mechanics, Vol. 24, pp. 361-364.

Jones, N. 1986. A note on ice-scaling. Journal of Ship Research, Vol. 30, pp. 134-135.

Jones, S.J., and Chew, H.A.M. 1981. On the grain-size dependence of secondary creep. Journal of Glaciology, Vol. 27, pp. 517-518.

Kalifa, P., Duval, P., and Ricard, M. 1989. Crack nucleation in polycrystalline ice under compressive stress states. Proceedings of the 8th International OMAE Conference, The Hague, The Netherlands, Vol. IV, pp. 13-21.

Kollé, J.J. (1981. Fracture toughness of ice; crystallographic anisotropy. Proceedings of the 6th International POAC Conference, Quebec, Canada, Vol. I, pp. 366-374.

Kusumoto, S., Kimura, N., Uchida, T., and Takase, T. 1986a. Study of fracture toughness of coarse grained ice and the effect of specimen size. Journal of the Society of Materials Science, Vol. 35, pp.659-663 (in Japanese).

Kusumoto, S., Takase, T., Uchida, T., Kidera, T., and Kaji, S. 1985. Fracture toughness of single crystal ice and columnar grained ice. Reports, Faculty of Engineering, Nagasaki University, Vol. 15, pp. 19-27.

Kusumoto, S., Kimura, N., Takase, T., and Kidera, T. 1986b. Study of fracture toughness of columnar grained ice using large specimens. Journal of the Society of Materials Science, Vol. 35, pp. 887-891 (in Japanese).

Liu, H.W., and Loop, L.W. 1972. Fracture toughness of freshwater ice. CRREL Technical Note, Hanover, N.H.

Liu, H.W., and Miller, K.J. 1979. Fracture toughness of fresh-water ice. Journal of Glaciology, Vol. 22, pp. 135-143.

McKenna, R.F., Meyssonier, J., and Jordaan, I.J., 1989. Peak pressures from a damage model for ice in compression. Proceedings of the 8th International OMAE Conference, The Hague, The Netherlands, Vol. IV, pp. 67-73.

Michel, B., and Ramseier, R.O., 1971. Classification of river and lake ice. Canadian Geotechnical Journal, Vol. 8, pp. 36-45.

Miller, K.J. 1979. The application of fracture mechanics to ice problems. Proceedings of the IUTAM Symposium on the Physics and Mechanics of Ice, edited by P. Tryde, Copenhagen, Denmark, pp. 265-277.

Mindess, S., and Nadeau, J.S. 1976. Effect of notch width on K_{Ic}, for mortar and concrete. Cement and Concrete Research, Vol. 6, pp. 529-534.

Nixon, W.A. 1984. Some Aspects of the Engineering Properties of Ice. Doctoral Thesis, Cambridge University, UK.

Nixon, W.A. 1988a. The effect of notch depth on the fracture toughness of freshwater ice. Cold Regions Science and Technology, Vol. 15, pp. 75-78.

Nixon, W.A. 1988b. Observations pertaining to the applicability of fracture mechanics to ice/structure interaction. ASCE Journal of Cold Regions Engineering, Vol. 2, pp. 1-12.

Nixon, W.A., and Schulson, E.M. 1986. Fracture toughness of freshwater ice as a function of loading rate. Ice Technology, Proceedings of the Ist International Conference, (Edited by T.K.S. Murphy, J.J. Connor, C.A. Brebbia), Cambridge, Massachusetts, pp. 287-296.

Nixon, W.A., and Schulson, E.M. 1988. The fracture toughness of ice over a range of grain sizes. ASME Journal of Offshore Mechanics and Arctic Engineering, Vol. 110, pp. 192-196.

Ouchterlony, F. 1982. Review of fracture toughness testing of rock. Solid Mechanics Archives, Vol. 7, pp. 131-211.

Palmer, A.C., Goodman, D.J., Ashby, M.F., Evans, A.G., Hutchinson, J.W., Ponter, A.R.S. 1983. Fracture and its role in determining ice forces on offshore structures. Annals of Glaciology, Vol. 4, pp. 216-221.

Parsons, B.L., and Snellen, J.B. 1985. Fracture toughness of freshwater prototype ice and carbamide model ice. Proceedings of the 8th International POAC Conference, Narssarssuaq, Greenland, Vol. 1, pp. 128-137.

Parsons, B.L., Snellen J.B., and Muggeridge, D.B. 1989. The double torsion test applied to fine grained freshwater columnar ice, and sea ice. The Mechanics of Creep Brittle Materials, (Edited by A.C. F. Cocks and A.R.S. Ponter), Elsevier Applied Science, London, pp.188-200.

Pulkkinen, E. 1989. Rate-sensitive damage and cracking model for ice. Proceedings of the 8th International OMAE Conference, The Hague, The Netherlands, Vol. IV, pp. 47-53.

Rice, J.R. 1968. Mathematical analysis in the mechanics of fracture. Fracture: An Advanced Treatise, (edited by H. Liebowitz), Academic Press, New York, Vol. 2, p. 191-311.

Sanderson, T.J.O. 1986. A pressure-area curve for ice. Proceedings of the 8th IAHR Symposium on Ice, Iowa City, Iowa, Vol. II, pp. 361-384.

Santaoja, K. 1989. Continuum damage mechanics approach to describe the multidirectional microcracking of ice. Proceedings of the 8th International OMAE Conference, The Hague, The Netherlands, Vol. IV, pp. 55-65.

Schmidt, R.A., and Lutz, T.J. (1979). K_{1c}, and J_{1c}, of Westerly Granite - effects of thickness and in-plane dimensions. Fracture Mechanics Applied to Brittle Materials, (Edited by S.W. Freiman). ASTM STP 678, pp. 166-182.

Schulson, E.M. 1987. The fracture of ice Ih. Journal de Physique, Vol. 48, pp. C1:207-220.

Schwarz, J., Frederking, R., Gavrilo, V., Petrov, I.G., Hirayama, K.-I., Mellor, M., Tryde, P., and Vaudrey, K.D. 1981. Standardized testing method for measuring mechanical properties of ice. Cold Regions Science and Technology, Vol. 4, pp. 245-253.

Shyam Sunder, S., and Nanthikesan, S. 1987. A tensile fracture model for ice. Proceedings of the 6th International OMAE Conference, Houston, Texas, Vol. IV, pp. 225-233.

Shyam Sunder, S., and Wu, M.S. 1990. Crack nucleation due to elastic anisotropy in polycrystalline ice. Cold Regions Science and Technology, Vol. 18, pp. 29-47.

Sodhi, D.S. 1986. Flexural and buckling failure of floating ice sheets against structures. Proceedings of the 8th IAHR Symposium on Ice, Iowa City, Iowa, Vol. II, pp. 339-359.

Tada, H., Paris, P.C., and Irwin, G.R. 1985. The Stress Analysis of Cracks. Handbook, Del Research Group, St. Louis.

Timco, G.W. 1987. Indentation and penetration of edge-loaded freshwater ice sheets in the brittle range. ASME Journal of Offshore Mechanics and Arctic Engineering, Vol. 109, pp. 287-294.

Timco, G.W., and Frederking, R.M.W., 1982. Comparative strengths of fresh water ice. Cold Regions Science and Technology, Vol. 6, pp. 21-27.

Timco, G.W., and Frederking, R.M.W. 1986. The effects of anisotropy and microcracks on the fracture toughness K_{1c}, of freshwater ice. Proceedings of the 5th International OMAE Conference, Tokyo, Japan, Vol. IV, pp. 341-348.

Urabe, N., and Yoshitake, A. 1981. Strain rate dependent fracture toughness (K_{1c},) of pure ice and sea ice. Proceedings of the 6th International IAHR Symposium on Ice, Quebec City, Canada, Vol.2, pp. 551-563.

Urabe, N., Iwasaki, T., and Yoshitake, A. 1980. Fracture toughness of sea ice. Cold Regions Science and Technology, Vol. 3, pp. 29-37.

Walsh, P.F. 1972. Fracture of plain concrete. Indian Concrete Journal, Vol. 46, pp. 469-470 and 476.

Wecharatana, M., and Swartz, S.E. 1988. Review of current notations, with recommendations, on fracture toughness testing of concrete and rock. Experimental Mechanics, Vol. 12, pp. 20-22.

Discussion

Q: E.M. Schulson, Dartmouth College

If the test specimens used by Nixon and Schulson (1986, 1988) and Nixon (1988 a,b) have generally been too small, then how would you explain the results (Nixon and Schulson, 1987) that the fracture toughness of ice of "infinitely large" grain size obtained by extrapolating to zero on a (grain size $^{-1/2}$) plot is very close to the value calculated from Griffith's theory?

A: J.P. Dempsey

The value calculated from Griffith theory is sensitive to the value used for the elastic modulus E and the surface energy which you chose to be the solid vapour value. The data (Nixon and Schulson, 1988) predicts this K_o term, to be between approximately 50-70 kPa $m^{1/2}$; your calculations, give for $\gamma_{sv} = 0.1$ J/m^2 and E between 7-11.8 GPa-a range of K_o between 30-51 kPa $m^{1/2}$. Your data would be better predicted with a higher value for the fracture energy, which would not be an unreasonable assumption.

To answer your question, the calculation you discuss, based as it is on assumptions and extrapolation, cannot be used to legitimize a particular specimen size.

Q: W.A. Nixon, IIHR

If you could make a randomly oriented polycrystalline freshwater ice sample "big enough", would you expect a grain size effect on K_Q^{∞}?

A: J.P. Dempsey

I do not disagree with the discussion on this topic presented by Nixon and Schulson (1988). Notably, there is more confusion in the ice fracture toughness literature concerning this question than for materials such as ceramics and steel.

Q: B.L. Parsons, IMD/NRCC

1. Could you comment on why you believe there is a grain size effect over the range of 1 mm to 100 mm when results from the field (the large grain size) are not significantly different from the small grain size (lab tests) results.

2. The only study in the literature on the J integral (Urabe and Yoshitake, 1981) toughness calculated from NLFM is the same as that calculated from LEFM. Why does this not convince you that LEFM is adequate for study of ice?

A: J.P. Dempsey

1. Details concerning a very interesting study you reported in 1985 are provided in Table 1 (Parsons and Snellen, 1985). Note that the grain size in this study varied between 5-100 mm. The fracture toughness results reported in this study vary hugely, from 100-500 kPa $m^{1/2}$. Because specific details concerning the microstructure and other factors in each test were not reported, this variation cannot be readily explained. However, your own previous work seems to refute the question you have just asked.

2. Urabe and Yoshitake (1981) did use J_{JC} to calculate a converted K_{IC} for $\dot{K} < 10$ KPA $m^{1/2}$ s^{-1}. Comparisons of K_{IC} in an LEFM sense in this regime are in all likelihood invalid because small scale yielding conditions would no longer be met. The lack of notch sensitivity and sub-size specimens have already been identified in the paper being discussed as the probable cause of a dramatic loading rate dependence exhibited by freshwater ice and sea ice fracture toughness data reported to date.

The application of NLFM must go hand-in-hand with a more sophisticated approach to fracture toughness testing. Ideally, future research will identify the particular deformation process active in the vicinity of the crack tip. There are as yet no quantitive estimates of either the shape or size of this process zone. The use of the J integral presupposes the existence of strain hardening in the vicinity of the crack tip characterized by a power law. In metals, the J_{IC} parameter has proved useful in describing stable elastic-plastic crack growth initiation. In ice, stable crack growth initiation is typically not observed.

Q: S.Shyam Sunder, MIT

Just a comment on your excellent presentation. The use of Rice and Reidel's criterion for estimating the process zone size in ice is incorrect. Under a tensile state of stress with a stress singularity at the crack-tip, the stress-strain law is elastic + transient creep prior to failure. The process zone size must be estimated using an accurate constitutive model of creep including transient creep. The elastic power-law model of Rice and Reidel is applicable to metals where considerable ductility is present in the process zone.

A: J.P. Dempsey

Thank you for your observation. However, let me briefly play the devil's advocate. Could it be possible that the Riedel-Rice expression might prove useful, at least in an order of magnitude sense, or better. A Discussion to this effect is presented in the paper in Bentley et al. (1989).

Q: G.W. Timco, NRC

Most of the fracture toughness values for ice measured to date rely on the recommendations laid down by Goodman.

1. How do your recommendations compare and differ from those of Goodman?

2. In your opinion, how many of the fracture toughness values in the literature are correct?

A: J.P. Dempsey

1. The extent of the Mode I creep zone was estimated by Riedel and Rice (1980) under conditions of elastic-nonlinear viscous material behaviour (uniaxially, $\dot{\varepsilon} = \dot{\sigma}$ /E + Bε^n. Goodman (1980) first used Riedel and Rice's expression, requiring that the creep zone be small compared to any of the dimensions of the specimen; 1/50 of the minimum dimension is an accepted measure. The Reidel and Rice (1980) expression has subsequently been used by Urabe and Yoshitake (1981), Nixon (1984), Timco and Frederking (1986), and Nixon and Schulson (1986, 1988).

   The adoption of this particular creep zone size estimate, which does not include the very active primary creep phase, has not met with unanimous support (see the comment by Shyam Sunder in this discussion). It is an overstatement to say that most of the fracture toughness values for ice measured to date rely on the creep zone size criterion adopted by Goodman (1980).

An important topic for future research is the degree of applicability of the Riedel and Rice (1980) creep zone size criterion to the fracture toughness testing of ice. To highlight this topic briefly, the paper by Nixon and Schulson (1986) determined using this criterion that their specimen/crack length size would not satisfy LEFM conditions if $\dot{K}$ (the loading rate) were less than 1.9 kPa $m^{1/2}$ s^{-1}. In fact, the transition from rate-dependent to rate- independent fracture toughness values occurred for $\dot{K} = 10$ KPa $m^{1/2}$ s^{-1}. At least in order to magnitude terms, the Riedel and Rice (1980) criterion apparently may be useful?.

2. In my opinion, there is no reason to doubt most of the fracture toughness values for ice reported in the literature. My concern, rather, is with values that may be regarded as material properties. For the fracture toughness to be a material property - given a particular ice type, growth history, crack orientation, method of macrocrack formation, temperature, and loading rate - the toughness value should be invariant with an increase in specimen size or change in specimen geometry. In terms of the fracture toughness investigations for ice reported to date, the results obtained by Dempsey et al. (1989 a,b) and discussed in this paper suggests that (1) almost all investigations have underestimated the necessary specimen size for small scale yielding conditions to apply, and (2) the importance of a sufficient notch sensitivity has been unknown until now in the ice fracture toughness field.

Riedel, H. and Rice, J.R. 1980. Tensile cracks in creeping solids. Fracture Mechanics: Twelfth Conference, ASTM STP 700, pp.112-130.

Goodman, D.J. 1980. The fracture toughness of Ice. Data report. Hokkaido University, Department of Applied Physics.

All other references are in the paper under discussion.

The Study of Fracture Toughness of Bohai Sea Ice

W.Shen

Dept. of Engineering Mechanics
Dalian University of Technology
Dalian, Liaoning, 116024, China

S.Z.Lin

Xiamen Branch Institute of China North-East
Construction Design Institute
Xiamen, China

P. Gu and X.A. Zhou

Dept. of Engineering Mechanics
Dalian University of Technology
Dalian, Liaoning, 116024, China

Abstract The evaluated method in engineering for ice load caused by the interaction of ice and structures is according to the compressive strength of ice. The test of interaction of ice and piles indicated that the majority of failures of ice sheets is mainly due to bending. Since the ice cannot avoid existing flaws and cracks, then if the loading rate is large the ice expresses brittle behavior, hence, it is reasonable to evaluate the strength of sea ice in accordance with the fracture toughness K_{IC} which is determined on the basis of linear elastic fracture mechanics. This paper states the principle and method of studies for mode I and mode II fracture toughness K_{IC} and K_{IIC}. The data obtained already by the author indicate that at 253 K, K_{IC} and K_{IIC} do not vary with the stress intensity factor rate, $\dot{K}_I$ and $\dot{K}_{II}$. Because the acting conditions of ice load are very complicated, in order to determine what fracture criterion would be suitable, the study of fracture problems of I-II mixed mode is necessary. The results obtained previously indicated that they coincide with the criterion established from the strain energy density factor theorem. The author had considered that during the interaction of ice and piles, the mode I fracture failure of ice is sometimes due to eccentric compression, so a concept of using the Compact Compression Test Specimen for the test of fracture toughness K_{IC} had been suggested, and some corresponding evaluation and analysis had been made. Some preliminary results had been obtained, which will be discussed.

S. Jones · R. F. McKenna · J. Tillotson · I. Jordaan (Eds.)
Ice-Structure Interaction
IUTAM-IAHR Symposium St. John's, Newfoundland, Canada

1. Introduction

For the last twenty years, the exploitation of petroleum resources under the Liaodong Gulf have been rapidly progressing. Through estimating, engineers unanimously agree that the ice load on a structure is the controlling load, and therefore should be considered seriously.

Last year a new oil and gas field was discovered in the Yellow River Mouth. It is of great value because it has the advantages of short distance from land, shallow water and convenient communication. In the demonstration of development plan, the ice load is one of the important problems. At present, the ice compressive strength is still taken as the main factor determining the ice load. The ice force acting on the platform had been evaluated by engineers, and it almost reached the earthquake force of intensity grade 8, and reached 2.3 times the total force of wave and running water. This shows the serious consequence of ignoring the ice force. In fact, sea ice unavoidably contains flaws, cracks, porosities and cavities, etc., and under the action of force, it may bend and twist rather than compress, hence, using only the compressive strength is not sufficient and will give a conservative estimate for the ice force. For more than twenty years, many ice mechanics specialists have attempted to study the fracture toughness of opening mode crack K_{IC} for ice, and have used this mechanical property to evaluate the ice load, in order to reduce the loads and hence the cost of the structure.

Those who have investigated fracture toughness K_{IC} for various kinds of ice (fresh water ice, sea ice, urea ice) are listed as follows: Gold (1963), Goodman and Tabor (1978), Liu and Miller (1979), Hamza and Muggeridge (1979), Goodman (1980), Urabe et al. (1980, 1981), Timco and Frederking (1983), Timco (1985), Shen and Lin (1986), Nixon and Schulson (1986), Dempsey et al. (1986), Bentley et al. (1987), Shen et al. (1987, 1988). They have used the different experimental arrangements and measuring methods such as 3-point bending, 4-point pure bending, tapered double cantilever beam, compact tension test specimens, indentation tests, etc., to study the fracture toughness K_{IC}. The theoretical foundation of these studies is linear elastic fracture mechanics which requires that the conditions involving uniform isotropic, linear elastic relation and plane strain state be satisfied. These conditions are approximately satisfied for metallic materials, but for ice the situation is uncertain. For polycrystalline ice under high loading rate and having sufficient thickness (larger than $2.5(K_{IC}/\sigma_y)^2$), the above three requirements may be satisfied. If one makes the crack plane parallel to the columnar crystal axis direction and has the bending plane perpendicular to it, then, sea ice will also satisfy the requirement of transverse isotropic. In the "IAHR- Recommendations on Testing Methods in Ice, 6th Report" (Häusler, 1988), the authors proposed a requirement that the diameter, or the least scale of the specimen, must be 10-20 times greater than the average diameter of the crystal grains. In order to satisfy this requirement, for the columnar grained ice of Bohai Sea with average diameter from 5mm at the upper layer to 12mm at the lower layer (measured in recent years), the thickness of specimen should be from a lower limit of 5cm to an upper limit of 24cm. According to this upper limit, for the 3-point bending test specimen, the required dimensions would be 24cm × 48cm × 200cm which is, obviously, difficult to realize. This paper adopted the lower limit. Some scholars have used the small scale yielding theorem of linear fracture mechanics applied to the ice specimen, i.e. adopted the size of plastic region $r_p = A(K_I/\sigma_y)^2$. If one applies this relation to a specimen of Bohai Sea ice, it will lead to following result: for the critical case, if one takes $K_{IC} = 80kPa \cdot m^{1/2}$ (from Shen and Lin, 1986), $\sigma_y = 1.5MPa$, coefficient $A = 0 \cdot 175$ (from Miller, 1980), a value of $r_p = 0.49mm$ is obtained. This result is smaller by one order of magnitude than the minimum average dimension of columnar grained ice, so it is of no significance. It also shows that using the LEFM in the study of fracture toughness should be done with care.

In calculating an ice load, the shear property of ice is also sometimes taken into consideration. So, the study of the problem of fracture toughness in the shearing mode (mode II) is necessary. During the interaction of ice sheets and piles, the shearing plane is parallel to the direction of the columnar grains. This paper also introduces this problem. Sometimes, the conditions of interaction between an ice sheet and piles due to the forces of wind, wave and tide are very complicated. In the majority of such cases, the failure of ice sheets probably belongs to the problem of mixed mode I and II. This paper indicates that the experimental results of sea ice for this mixed mode of fracture, coincide well with the curve obtained from the strain energy density factor theorem.

Generally, when an ice sheet acts against a pile, the ice sheet is subjected to eccentric compression, which leads to bending deformation. Owing to existing unavoidable flaws, it also leads to the possibility of opening mode fracture. Therefore, the authors suggest that using the compact compression test specimen (CCTS) to investigate the fracture toughness of mode I, may conform more to reality. This paper gives the primary results of an experimental investigation and evaluation for the CCTS. Further research still remains to be done, and we look forward to discussions with the specialists.

2. Collection, Transportation, Storage of Bohai Sea Ice and its Salinity, Density, Crystalline Fabric

The collection, packaging, transportation and storage of sea ice are very important matters. In the past few years, the authors have selected the coldest week (January 20-24) to collect the fast ice near the shore in following regions: Bayu Harbour, Liaohe River Mouth, Huludao and Jihua Island (situated to the east, north, northeast of Liaodong Gulf respectively). From January to February of 1988, the authors cooperated with Bohai Oil Company and used an icebreaker to collect the floating ice in Liaodong Gulf. The natural ice sheet was cut into ice blocks of dimensions 65-68cm $\times$ 25-38cm $\times$ the natural thickness (34-50cm) by labour power. Then, these blocks, which were without visible flaws, were packed in plastic bags in order to avoid sublimation, and put into wooden boxes which were filled with foamed plastic mats. The boxes were transported immediately by a refrigerated car, to the laboratory of Dalian University of Technology and were preserved in cold storage at a temperature lower than -15°C.

In order to determine the salinity and drainage characteristics of the sea ice, pieces of ice were picked up when the ice blocks were cut in the field, and sealed into bottles. After the blocks were put into cold storage, the collection of ice pieces were made once more every five days. The salinity measured for ice just taken from sea water was about 9%, but for ice transported to the laboratory, it became about 4-6%. The salinities vary with the thickness of the ice; at the upper layer it is about 2.3-3.0%, at the middle layer it is about 3.4-4.0%, at the lower layer it is about 3.6-5.5%. The salinity of ice collected from the Liaohe River Mouth belongs to the lower limit. The dimensions of the specimens were measured as soon as possible after recovery to determine volume, then the specimens were weighed on a platform scale with an accuracy of 2g. From these data, the density of sea ice is obtained as follows: at upper layer, it was 0.890-0.913 Mg/m^3, at middle layer it was 0.892 $\pm 0.011 Mg/m^3$ (for the sea ice of Huludao), at lower layer it was 0.871 $\pm 0.013 Mg/m^3$ (for the sea ice of Huludao). The density value of sea ice at Liaohe River Mouth belongs to the upper limit.

Thin sections of ice were cut from the ice blocks in both longitudinal and transverse orientation (in different layer). These sections were ground to 0.5mm in thickness, then attached on a glass slide. By observation between crossed polaroids, the crystalline fabric was seen to be columnar grained ice. The average diameter of the crystal grains was 4.7mm at upper layer, increasing gradually to 11.8mm at the lower layer.

There were two kinds of floating ice on the sea surface of Liaodong Gulf, i.e. single-layer and multi-layer ice. The average diameters of crystalline grain were 7.9-9.0mm and 2.0-2.5mm; the densities were 0.769-0.837 Mg/m^3 and 0.769- 0.840 Mg/m^3; the salinities were 4.9-8.0% and 2.2-9.3% for the single-layer and multi-layer sea ice respectively.

3. Critical Stress Intensity Factor of Mode I (K_{IC})

The critical stress intensity factor K_{IC} was measured by using the experimental arrangement in which the specimen was subjected to 3-point bending as shown in Fig. 1. (In Fig.1 as well as in Fig. 2, 3, 5, the specimen is prepared in such a way that the columnar axes are usually parallel to the z axis). In 1986, the dimensions of ice specimen were taken as B=8cm, W=16 cm, S=60 cm. The crack in the middle of the span was first cut by a steel saw, then the sharp crack with depth of 2mm was cut by a sharp saw-strip ground by emery wheel and then polished. The crack length "a" is generally $0.5W$. The orthogonality of four longitudinal lateral faces of specimen were closely checked by a right angle ruler

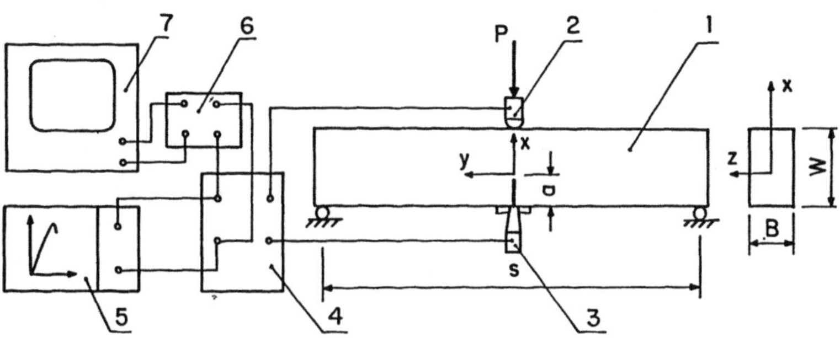

Figure 1. Experimental equipment setup for measuring fracture toughness. 1 – ice specimen for K_{IC}, 2 – load cell, 3 – crack mouth displacement transducer, 4 – dynamic strain indicator, 5 – X-Y recorder, 6 – A/D transformator, 7 – minicomputer Apple II.

after every specimen was prepared, they must be all within 1/300 in order to avoid the twisting as far as possible. The dimensions of specimen were measured carefully by a vernier calliper with an accuracy of 0.02mm. The crack length was measured once more after the specimen ruptured, and this length was taken for the calculation of K_{IC}. During the experiment a clip-on displacement gage was attached to the crack mouth, the crack opening displacement was input into a X-Y recorder.

Within the predetermined range of loading rate, the load P and opening displacement V appears as linear relation. So the fracture toughness K_{IC} may be obtained by substituting the maximum load P_{max} to the calibration of linear elastic stress intensity factor K_I. Shen and Lin (1986) used the calibration suggested by Chen Chi (1974).

$$K_{IC} = \frac{Ps}{4BW^{3/2}}(7.30 + 0.21\sqrt{\frac{s}{w}} - 2.9)\sec(\frac{\pi a}{2W})\sqrt{\tan(\frac{\pi a}{2W})}. \tag{1}$$

The results indicate that the fracture toughness K_{IC} is $7.9\times10^4 Pa\cdot m^{1/2}$ and seems independent of the stress intensity factor rate (SIFR) $\dot{K}_I$ over the range 0.1-4.0 $\times$ 10^4 $Pa \cdot m^{1/2}$ s^{-1} and at a temperature of T=253 K. In order to investigate the scale effect on fracture toughness of sea ice, Shen et al. (1987) prepared 3-point bending specimens of three different sizes: type A, 4cm $\times$ 8cm $\times$ 32cm; type B, 6cm $\times$ 12cm $\times$ 48cm; type C, 8cm $\times$ 16cm $\times$ 64cm.

As the types A, B, C of specimen are prepared from the same batch of ice blocks, so the grain sizes and forms in the vicinity of the crack are also the same, and all of these specimens have the same ratio a/W = 0.45-0.55. For the calculation of fracture toughness the authors used the calibration of China Test Standard YB 947-78.

$$K_{IC} = P_{max}s/BW^{3/2}[1.88 \times 0.75(\frac{a}{w} - 0.5)^2]\sec(\frac{\pi a}{2W})\sqrt{\tan(\frac{\pi a}{2W})}. \tag{2}$$

From the calculation it was found that: near the rate $\dot{K}_I$ = 12 $\times$ $10^4 Pa \cdot m^{1/2}s^{-1}$ and at the test temperature 253 K, the mean values of K_{IC} are obtained statistically: from sixteen specimens of type A was 5.2 $\times10^4 Pa \cdot m^{1/2}$; from fifteen specimens of type B was 4.7 $\times10^4 Pa \cdot m^{1/2}$ from seventeen specimens of type C was $4.4 \times 10^4 Pa \cdot m^{1/2}$. These results demonstrated a scale effect.

Shen et al. (March, 1988) performed a few fracture experiments for floating, single-layer sea ice collected from Liaodong Gulf again in the laboratory. The dimensions of 3-point bending specimen are 6cm $\times$ 12cm

$\times$ 48cm, crack length $a = 0.5W$, test temperature T = 263K. The authors used ASTM E399-81, and replaced P_Q by P_{max}, the expression for K_I and $f(a/W)$ may be written as follows:

$$K_{IC} = (P_{max}s/BW^{3/2})f(a/W)$$

$$f(a/W) = 3(a/W)^{1/2}\times \left[\frac{1.99 - (a/W)(1-(a/W))(2.15 - 3.93(a/W) + 2.7(a/W)^2)}{2[1+2(a/W)][1-(a/W)]^{3/2}}\right] \quad (3)$$

The values of K_{IC} under different stress intensity factor rate $\dot{K}_I$ are listed in Table 1.

Table 1 The fracture toughness of floating single-layer ice collected from Liaodong Gulf under different SIFR $\dot{K}_I$, 263K

$\dot{K}_I(kPa \times m^{1/2} \cdot s^{-1})$	5.548	7.118	5.038	5.636	6.328	5.178
$K_{IC}(\times 10^4 Pa \times m^{1/2}$	8.12	9.51	7.81	8.23	9.24	7.55

This table shows that the fracture toughness K_{IC} increases with the SIFR $\dot{K}_I$. However, these six specimens are too few to demonstrate this for certain. This problem will remain to be investigated further.

It should be pointed out that, although the authors used three different expressions for K_I above, the calculated results lead to the following conclusion: the non-dimensional terms $f(a/W)$ only differ by figures in the second decimal place. Hence, it is of no importance for using any one of them.

It remains a problem to study and discuss further whether or not the measured K_Q should be checked as is done for metallic materials. Till now, there is no international standard for measuring the fracture toughness of ice. This problem was a notable obstruction to the international comparison and exchanging of experience, it must be solved as far as possible.

4. Critical Stress Intensity Factor of Mode II (K_{IIC})

The measurement of the critical stress intensity factor of mode II (K_{IIC}) was done with an experimental arrangement shown in Fig.2. On the

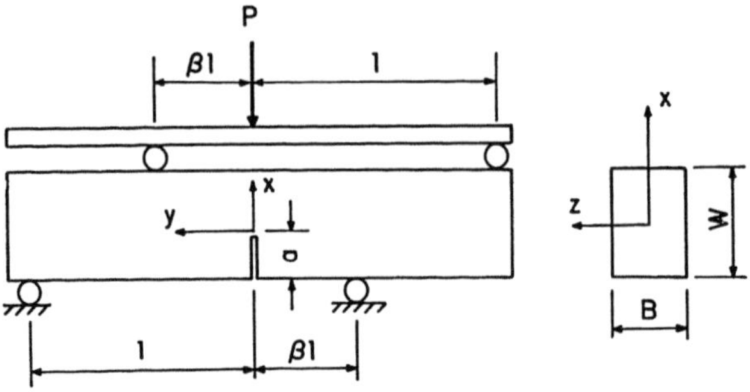

Figure 2. The test geometry for K_{IIC}.

section which coincides with the crack plane, the bending moment $M = 0$ and the shearing force $Q = \frac{1-\beta}{1+\beta}P$ (i.e. a pure shear). Therefore, $K_I = 0, K_{II} \neq 0$. Shen and Lin (1986) use the following data: $\beta = 1/2$ (so $Q = P/3$), $s = 2l = 40$cm, $W = 14$cm, $B = 8$cm, $a/W = 0.40\text{-}0.75$. The specimen was cut from the lower layer of Liaohe River Mouth sea ice, the X-Y plane of the specimen is parallel to the sea level. The K_{IIC} was determined by following expression for K_{II} derived by Chen Chi (1976), for $0.25 \leq a/W \leq 0.75$,

$$K_{IIc} = \frac{Q}{BW^{1/2}}[1.442 - 5.08((a/W) - 0.507)^2]\sec(\frac{\pi a}{2W})\sin(\frac{\pi a}{2W}), \quad (4)$$

where Q is the maximum shearing force. From the experimental results (1986), it had been found that over the stress intensity factor rate range, $\dot{K}_{II} = 0.01\text{--}2.00 \times 10^4 Pa{\cdot}m^{1/2}{\cdot}s^{-1}$ and at the temperature $T = 253K$, the fracture toughness, K_{IIC}, does not vary with the stress intensity factor rate $\dot{K}_{II}$.

5. Mixed Mode Fracture Criterion

5.1 Experimental Arrangement and Results

For the 3-point bending specimen, when the crack plane departs from the line along which the load acts, the bending moment M and shearing force Q exist simultaneously on this plane. Therefore $K_I \neq 0, K_{II} \neq 0$ and the $I - II$ mixed mode fracture is realized. For the dimensions of specimen, $s/W = 4, a/W = 0.40 - 0.60$ the investigated results obtained by Research Institute of Mechanics (Academia Sinica), (RIM, 1974) may be used and are listed in Table 2.

In this table, the calculated results of K_I and K_{II} for seven ratios of $2s_1/s$, and for five ratios of a/W, have been given.

Table 2 The numerical values of K_I and K_{II}

a/W	mode	2s/s						
		0	1/6	2/6	3/6	4/6	5/6	11/12
0.40	I	7.71	7.08	5.17	4.18	2.77	1.43	0.72
	II	0.00	0.516	0.70	0.675	0.649	0.688	0.822
0.45	I	8.86	8.07	6.49	4.65	3.16	1.59	
	II	0.00	0.517	0.781	0.744	0.733	0.732	
0.50	I	10.27	9.57	7.65	5.81	3.72	1.83	0.957
	II	0.00	0.705	0.932	0.920	0.832	0.830	0.880
0.55	I	12.11	10.96	9.08	6.51	4.31	2.25	
	II	0.00	0.794	0.990	1.103	0.988	1.05	
0.60	I	14.47	11.88	9.78	7.48	4.98	2.45	1.22
	II	0.00	1.174	1.124	1.138	1.160	1.147	1.045

mode I : $K_I BW^{3/2}/M$,. modeII : $K_{II}BW^{3/2}/M$

Shen and Lin (1986) performed fracture test in $I-II$ mixed mode. Their samples were cut from the lower layer of ice collected from the Liaoche River Mouth, its dimensions were $s = 60$cm, $B = 8$cm, $a/W = 0.40 - 0.60$, the ratio $2s_1/s$ is equal to $0, 1/6, 2/6, 3/6, 5/6, 11/12$ respectively. For every specimen, the failure load may be used to determine the K_I and K_{II} according to Table 2. It is evident that when $2s_1/s$ is equal to zero, the fracture toughness of mode I, K_{IC}, can be obtained. These results are plotted in Fig. 4 (every point corresponds to an experimental result except point $K_I = 0$ and $K_{II} = 0$).

5.2 The Strain Energy Density Factor Theorem

For the fracture problem of $I-II$ mixed mode the research was carried on from two respects: in what direction will the crack propagate during fracture; and what is the combination of K_I and K_{II} when the crack reaches the critical state? Shen and Lin (1986) used the strain energy density factor theorem to analyze the fracture problem of $I-II$ mixed mode. The combined stress field existing near the crack tip is (see Fig. 3)

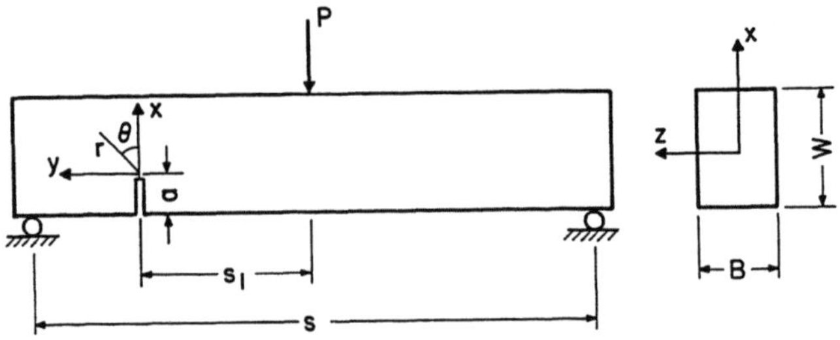

Figure 3. Experimental arrangement for mixed mode fracture.

$$\sigma_x = \frac{K_I}{(2\pi r)^{1/2}}\cos\frac{\theta}{2}(1-\sin\frac{\theta}{2}\sin\frac{3\theta}{2}) - \frac{K_{II}}{(2\pi r)^{1/2}}\sin\frac{\theta}{2}(2+\cos\frac{\theta}{2}\cos\frac{3\theta}{2}) + \dots$$

$$\sigma_y = \frac{K_I}{(2\pi r)^{1/2}}\cos\frac{\theta}{2}(1+\sin\frac{\theta}{2}\cos\frac{3\theta}{2}) + \frac{K_{II}}{(2\pi r)^{1/2}}\sin\frac{\theta}{2}\cos\frac{\theta}{2}\cos\frac{3\theta}{2} + \dots$$

$$\tau_{xy} = \frac{K_I}{(2\pi r)^{1/2}}\cos\frac{\theta}{2}\sin\frac{\theta}{2}\cos\frac{3\theta}{2} + \frac{K_{II}}{(2\pi r)^{1/2}}\cos\frac{\theta}{2}(1-\sin\frac{\theta}{2}\sin\frac{3\theta}{2}) + \dots$$

$$\sigma_z = 2\frac{K_I}{(2\pi r)^{1/2}}\cos\frac{\theta}{2} - 2\frac{K_{II}}{(2\pi r)^{1/2}}\sin\frac{\theta}{2} + \dots \quad (5)$$

The increment of strain energy in the elemental volume $dV = dxdydz$ is

$$dW = [\frac{1}{2E}(\sigma_x^2+\sigma_y^2+\sigma_z^2) - \frac{\nu}{E}(\sigma_x\sigma_y+\sigma_y\sigma_z+\sigma_z\sigma_x) + \frac{1}{2G}\tau_{xy}^2]dV. \quad (6)$$

Substituting equation (5) into equation (6), the expression of strain energy density function is

$$\frac{dW}{dV} = \frac{1}{r}(a_{11}K_I^2 + 2a_{12}K_IK_{II} + a_{22}K_{II}^2) + \dots, \quad (7)$$

The terms within the bracket are defined by Sih (1974) as the strain energy density factor

$$S = a_{11}K_I^2 + 2a_{12}K_IK_{II} + a_{22}K_{II}^2, \quad (8)$$

the individual terms are given by

$$
\begin{aligned}
a_{11} &= \frac{1}{16\pi G}(3 - 4\nu - \cos\theta)(1 + \cos\theta), \\
a_{12} &= \frac{1}{16\pi G} 2 \sin\theta[\cos\theta - (1 - 2\nu)], \\
a_{22} &= \frac{1}{16\pi G}[4(1-\nu)(1-\cos\theta) + (1+\cos\theta)(3\cos\theta - 1)].
\end{aligned} \tag{9}
$$

It is easy to prove that the strain energy density factor S depends mainly on the angle θ, rather than the stress intensity factor K_I, K_{II}. Sih deems that there exists a critical value, S_{CR}, which expresses the fracture toughness of materials and governs the crack propagation and supposes that the crack will propagate along such direction that the strain energy density possesses minimum value. That is, the crack propagating direction θ_0 should meet following conditions:

$$
\text{at } \theta - \theta_0, \qquad \frac{\partial S}{\partial \theta} = 0, \qquad \frac{\partial^2 S}{\partial \theta^2} > 0. \tag{10}
$$

Figure 4 shows the failure curve of $I - II$ mixed mode crack problem. It was plotted on the basis of strain energy density factor theorem, and this figure also shows the experimental results. From this diagram, it was found that the experimental results of sea ice for $I - II$ mixed mode fracture coincide well with the curve determined from the strain energy density factor theorem. It may be deemed that: the strain energy density factor theorem is able to ascertain more exactly than the maximum tension stress theorem that during the crack entering on the critical state the K_I, K_{II} present what combination.

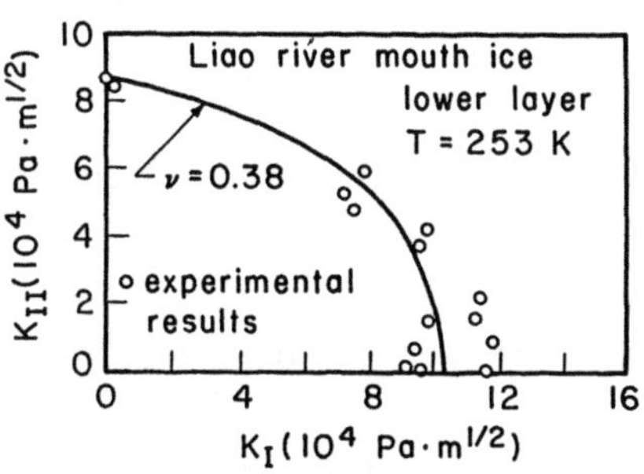

Figure 4. The criterion for mixed mode fracture.

For the mode I crack problem, $K_{II} = 0$, from equation (10), we obtain $\theta_0 = 0$, this is in accordance with the experimental results. From equation (8), if let $K_I = 0$ and then substitute it into equation (10), it will be the mode II problem. If the equation (8) is substituted into equation (10), it is mixed mode crack problem. For above two crack problems, the crack angle calculated by strain energy density factor does not coincide well with the experimental results. This shows that the mixed mode fracture problems of sea ice still should be researched further.

6. The Stress Intensity Factor K_I of Compact Compression Test Specimen

6.1. The Concept Formalized for the Compact Compression Test Specimen

In general, the samples that have been used for the measurement of fracture toughness, K_{IC}, are: the 3-point bending specimen; the compact tension specimen (Scrawley et al., 1967, 1972; Fan, 1965); the uniform axial tension specimen with an edge crack (Gross and Strawley, 1965; Brown and Strawley, 1967). Considering the loading condition and the situation of deformation and rupture (similar to an eccentric compression) of sea ice, Shen et al. (1988) had suggested a new concept concerning a compact compression test specimen (CCTS). The authors believe that the CCTS conforms better to the reality for ice in comparison than the other forms of specimen mentioned above. In 1988, Shen et al., also published some preliminary results about the CCTS obtained from experiment and finite element method, as well as the method of photoelasticity. However, these preliminary results are not ideal. In this paper, Shen et al., performed continuously some analysis and calculation. Considering the convenience for calculation the pattern of CCTS make some variation, see Fig.5.

6.2 Boundary Collocation Method

For the CCTS, only the symmetrical terms of Williams stress function (1957) are necessary, namely

$$\Phi(r,\theta) \cong \frac{PW}{B}\Sigma_{j=1}^{2m} X_j(\frac{r}{w})^{\frac{j}{2}+1}[-\cos(\frac{j}{2}-1)\theta\frac{\frac{j}{2}+(-1)^j}{\frac{j}{2}+1}\cos(\frac{j}{2}+1)\theta] \quad (11)$$

where, m is the number of collocation points. The rectangular and polar coordinates of these collocation points are (x_i, y_i), and (r_i, θ_i) respectively,

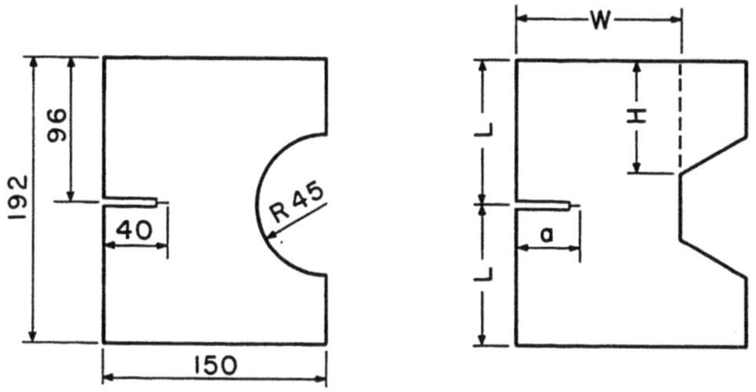

Figure 5. The geomety of CCTS a) initial size, thickness B=60mm b) present size, thickness B.

$i = 1, 2, 3, \ldots, m$. Let $\Phi_0(x_i, y_i)$ and $\frac{\partial \Phi_0(x_i, y_i)}{\partial_n}$ denote the values of $\Phi(r, \theta)$ and $\frac{\partial \Phi(r,\theta)}{\partial n}$ at the collocation points of the specimen without crack. In general, these values are taken as those at the corresponding points of the specimen which has the same dimensions but is possessed of crack. Therefore, we obtain the following coupled equations

$$\Phi(r_i, \theta_i) = \Phi_0(x_i, y_i)$$

$$i = 1, 2, 3, \ldots, m \tag{12}$$

$$\frac{\partial \Phi(r_i, \theta_i)}{\partial n} = \frac{\partial \Phi_0(x_i, y_i)}{\partial n}$$

or summarize into

$$\Sigma_{i=1}^{2m} A_{ij} X_j = B_J \qquad i = 1, 2, 3, \ldots, m. \tag{13}$$

There are 2m linear equations, from which the approximate value of K_I may be obtained

$$K_I = -\sqrt{\frac{2\pi}{W}} \frac{P}{W} X_1,$$

or write it as

$$\frac{K_I B W^{1/2}}{P} = -\sqrt{2\pi} X_1. \tag{14}$$

Considering the symmetrization, only the boundary ABCDE, that is, the half boundary of specimen is taken as the calculation boundary, see Fig. 6. On the stages AB, BC, DE of the boundary, the function $\Phi_0(x_i, y_i)$ is

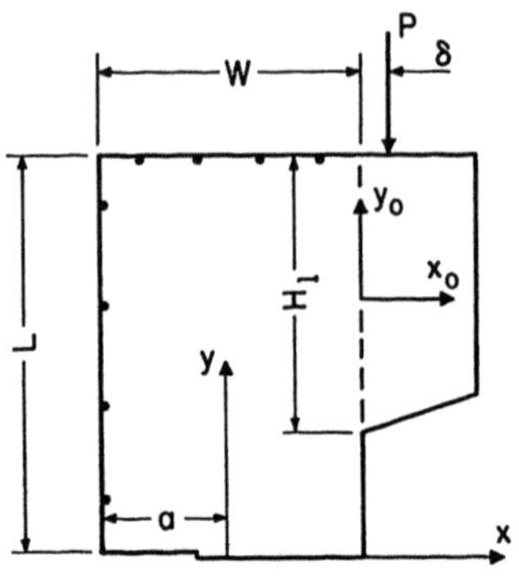

Figure 6. Boundary collocating point of CCTS.

accurate, but for the collocation points on the stage CD, the approximate values of function $\Phi_0(x_i, y_i)$ and $\frac{\partial \Phi_o(x_i,y_i)}{\partial n}$ are used, which are obtained by preliminary theorem of cantilever beam with one edge horizontal and the other edge tapered, that is,

$$\sigma(y_0) = \frac{My_0}{I},$$

$$\tau(Y_0) = \frac{VQ(y_0)}{BI} + \frac{MH_1}{4I}(1 - \frac{Q(y_0)H1}{I} + \frac{2y_0}{H_1})\frac{dH}{dx},$$

the expression for shear is derived by authors (referring to the theory as described in the "Strength of Materials" (Timoshenko and Gere, 1972)), where $\frac{dH}{dX}$ is the slope of the tapered edge. If the coordinate y_0 (with the middle point of edge CD as the origin) is translated into the coordinate y_ξ (with the crack tip as the origin), we obtain

$$\Phi_0(x_i, y_i) = \frac{\Delta M}{B} = \int_L^y \sigma(y_\xi)(y - y_\xi)dy_\xi \cdot B,$$

$$\frac{\partial \Phi_0(x_i, y_i)}{\partial n} = \frac{\Delta T}{B} = \int_L^y \tau(y_\xi)dy_\xi \cdot B.$$

6.3 Calculating Results

Twelve points are collocated on the boundary ABCDE as shown in Fig. 6. If let the distance $\delta \to 0$ (the δ is the deviation of the load P from line CD), then, $W_1 = W + \delta = 105$mm. The calculated results are listed in Table 3.

Table 3 The numerical results of $K_I B W^{1/2}/P$

L/W	a/W							
	0.1	0.2	0.3	0.4	0.5	0.6	0.7	0.8
3/5	1.473	2.007	2.634	3.412	4.385	5.540	6.770	6.862
2/3	2.124	2.849	3.503	4.307	5.248	6.389	7.754	9.391
4/5	2.905	3.859	4.578	5.388	6.320	7.447	8.769	10.000
1/1	3.948	5.243	6.222	7.271	8.482	9.956	11.490	11.450

The values which were obtained from the finite element method by Shen et al. (1988) are included in Table 1. So the data in this table is more widespread. By means of Table 3, and using the critical experimental values, P_{max} for three CCTS of Huludao sea ice (by Shen et al., 1988), the fracture toughness K_{IC} are calculated and listed in Table 4.

Table 4 The experimental values for the CCTS of Huludao sea ice

Specimen code	Maximum load $P_{max}(N)$	Specimen dimensions (mm)			$\frac{K_I B W^{1/2}}{P}$	K_{IC} $(\times 10^4 Pa \cdot m^{1/2})$
		W	B	a		
HuFP-2	443.2	105	60.2	40	4.760	13.08
HuFP-5	398.5	105	60.2	40	5.760	11.76
HuFP-10	461.9	105	60.2	41	5.839	13.82

From this table the average value of K_{IC} is 1.6 times larger than that obtained from the test of 3-point bending specimen by Shen and Lin (1986). Because there were only three specimens, and at that time the orientation of specimen is not in accordance with transverse isotropic, obviously conclusions should not be drawn. However, this makes clear that the study of the CCTS is still not enough. It is worthy of further research.

7. Discussion

From the study for the fracture toughness of Bohai Sea ice since 1984, the authors attempt to discuss the following problems with some specialists of this region:

1. Using the sea ice fracture toughness to determine the ice load will be probably more reasonable and economical than using the traditional calculation method, that is, using the compressive strength to determine the ice load.

2. There are already a lot of results of K_{IC} obtained from the experimental research for the 3-point bending specimen. It seems that within a certain range of stress intensity factor rate , $\dot{K}_I$, the K_{IC} is independent of the $\dot{K}_I$.

3. Using the theorem of $I - II$ mixed mode fracture to analyse and determine the ice load produced by interaction of sea ice and structures has practical meaning in engineering applications.

4. Measuring the fracture toughness, K_{IC}, of sea ice by means of CCTS is probably conforming to reality more than other forms of specimen (such as 3-point bending and compact compression, etc.).

Acknowledgement

The project was supported by National Natural Science Foundation of China.

References

Bentley, D.L., Dempsey, J.P. and Sodhi, D.S. 1987. The fracture toughness of urea model ice. ASCE Engineering Mechanics Division, 6th Specialty Conference, Buffalo, N.Y.

Brown, W.F. and Strawley, J.E. 1967. Current status of plane crack toughness testing. ASTM STP 410.

Chen, C. 1974. The stress intensity factor K_I of fracture toughness specimen subjected to 3-point bending. Journal of Mechanics (Chinese), No.3, p.121.

Chen, C. 1976. The research of K_{II} on 4-point bending beam with single-side crack. Proc. of 2nd Conference of Fracture Mechanics (Chinese).

Dempsey, J.P., Bentley, D.L., and Sodhi, D.S. 1986. Fracture toughness of model ice. Proc. 8th IAHR Symposium on Ice, Iowa City, IA, Vol.I, pp.365-376.

Fan, T.Y. 1976. The K_I of compact compression test specimen. Scientific Bulletin (Chinese), 21, p.494.

Gold, L.W. 1963. Crack formation of ice plate by thermal shock. Canadian J. Physics, 41, pp.1712-1238.

Goodman, D.J. 1980. Critical stress intensity factor (K_{IC}) measurements at high loading rates for polycrystalline ice. Physics and Mechanics of Ice, IUTAM Symposium, Copenhagen, Denmark, Springer-Verlag, pp.129-141.

Goodman, D.J. and Tabor, D. 1978. Fracture toughness of ice: a preliminary account of some new experiments. J. Glaciology, 21, No.85, pp.651-660.

Gross, B. and Strawley J.E. 1965. Stress Intensity Factors for Single-Edge-Notch Specimens in Bending or Combined Bending and Tension by Boundary Collocation of a Stress Function. NASA Technical Note D-2603, January.

Hamza, H. and Muggeridge, D.B. 1979. Plain strain fracture toughness (K_{IC}) of freshwater ice. Proc. POAC'79, Trondheim, Norway, Vol.I, pp.697-707.

Häusler, F.U. (ed.) 1988. IAHR Working Group on Testing Methods in Ice. IAHR recommendations on testing methods in ice, 6th report. Published by Hiroshi Saeki and Ken-ich Hirayama, Organizing Committee, IAHR Ice Symposium, Sapporo, Japan.

Liu, H.W. and Miller, K.J. 1979. Fracture toughness of fresh water ice. J. Glaciology, 22, No.86, pp.135-143.

Miller, K.J. 1980. The application of fracture mechanics to ice. Physics and Mechanics of Ice, IUTAM Symposium, Copenhagen, Denmark, Springer-Verlag, pp.265-277.

Nixon, W.A. and Schulson, E.M. 1986. Fracture toughness of freshwater ice as a function of loading rate. Proc. 1st International Ice Technology Conference, Cambridge, MA, pp.287-296.

RIM(Academia Sinica). 1976. Journal of Mechanics (Chinese), 3, p.168.

Scrawley, J.E. and Gross, B. 1967. Stress intensity factors for crackline-loaded edge-crack specimens. Mat. Res. and Standard, 77, p.155.

Scrawley, J.E. and Gross, B. 1972, Stress intensity factors for bend and compact specimens. Eng. Frac. Mech. 4, p.587.

Shen, W. and Lin, S.Z. 1986. Fracture toughness of Bohai Sea ice. Proc. OMAE'86, Tokyo, Japan, Vol.IV, pp.345-357.

Shen, W., Li, H.S., Zhang, X.P. and Bai, S.L. 1987. The experimental investigations on scale effect of Bohai Sea ice. Proc. OMAE'87, Houston, TX, Vol.IV, pp.173-179.

Shen. W., Li, H.S., Zhang, X.P. and Ji, X.H. 1988. A new concept of compact compression test specimen to study the K_I and K_{IC} for Bohai Sea ice. Proc. OMAE'88, Houston, TX, Vol.IV, pp.25-29.

Sih, G.C. 1974. Strain energy density factor applied to mixed mode crack problems. International Journal of Fracture, 10, p.305.

Timco, G.W. 1985. Flexural strength and fracture toughness of urea model ice. Proc. OMAE'85, Dallas, TX, Vol. II, pp.199-208.

Timco, G.W. and Frederking, R.M.W. 1983. Flexural strength and fracture toughness of sea ice. Cold Regions Science and Technology, 8, pp.35-41.

Timoshenko, S. and Gere, J. 1972. Strength of Materials. Van Nostrand Reinhold Company, p.135.

Urabe, N., Iwasake, T. and Yoshitake, A. 1980. Fracture toughness of sea ice. Cold Regions Science and Technology, 3, pp.29-37.

Urabe, N. and Yoshitake, A. 1981. Strain-rate dependent fracture toughness (K_{IC})of pure ice and sea ice. Proc. IAHR Symposium on Ice, Quebec City, PQ, Vol.II, pp.551-563.

Williams, M.L. 1957. On the stress distribution at the base of a stationary crack. J. Appl. Mech, 24, p.109.

Discussion

Q: B.L. Parsons (IMD/NRCC)

To my knowledge you have reported the only results for K_{IIc} of ice. Have you attempted a K_{IIIc} tests? Our preliminary work on K_{IIIc} using a tearing force, indicates that it is approximately 10% of K_{Ic}, K_{IIc}. Could you speculate on why this should be so low?

A: W. Shen

We haven't, because this is an out-of-plane shear. We cannot speculate on it, since we have not done it.

The Tensile and Compressive Fracture of Ice

E. M. Schulson

Thayer School of Engineering, Dartmouth College
Hanover, NH 03755

Abstract The brittle fracture of ice under uniaxial tension and compression is discussed in terms of the nucleation and the propagation of internal cracks. The discussion reveals that fracture under both applied stress states can be understood in terms of theories previously developed for the fracture of brittle solids. Also, it shows that the applied strain rate which marks the transition from ductile to brittle behavior under compression can be expressed quantitatively by invoking a fracture model which incorporates crack-tip creep and frictional crack sliding.

1. Introduction

This paper considers the brittle fracture of ice. The discussion is limited to fracture under uniaxial loading and focuses on the underlying physical processes. Experimental details are omitted, in the interests of space, but may be obtained from the literature cited where fuller discussions are given.

The failure of ice through brittle fracture, unlike that through ductile yielding, is a relatively sudden process. It involves the propagation of cracks, either pre-existing or nucleated under the applied load. Thus, whether the applied loading is largely tensile, such as the bending of an ice sheet, or mainly compressive, such as the brittle crushing of an ice feature, cracks are the dominant microstructural element governing the failure. The discussion, therefore, centers on the fundamental question: Under what conditions will cracks propagate?

S. Jones · R. F. McKenna · J. Tillotson · I. Jordaan (Eds.)
Ice-Structure Interaction
IUTAM-IAHR Symposium St. John's, Newfoundland, Canada

2. Tensile Fracture

Brittle fracture under uniaxial tension occurs over essentially all strain rates of practical interest, i.e., at rates above $\approx 10^{-8}\ \mathrm{s}^{-1}$. Three cases are worth exploring: virgin ice, cracked ice, and saline ice.

2.1 Virgin Ice

Two criteria must be met before fracture can occur (Currier and Schulson, 1982): the inelastic strain must be sufficient to nucleate cracks; and the tensile stress normal to the plane of the cracks must be sufficient to propagate them. In terms of the axial tensile stress, the criterion for crack nucleation is given by (Smith and Barnby, 1967):

$$\sigma_T^N = \sigma_0 + kd^{-\frac{1}{2}}, \tag{1}$$

where σ_0 and k are material constants and where d is the grain size. For crack propagation the criterion is given by:

$$\sigma_T^P = \frac{YK_{IC}}{(\pi C)^{\frac{1}{2}}}, \tag{2}$$

where K_{IC} in the critical stress intensity factor for mode I (i.e. crack opening) loading and C is the crack half-length; Y is a geometric parameter and equals $\pi/2$ for non-interactive, penny-shaped, internal cracks of radius C oriented normal to the applied stress. The crack length is set by the grain size (i.e. $2C = \alpha d$, where α is another constant, Lee and Schulson, 1988) and so the second criterion, like the first, can be written in terms of this parameter; i.e.:

$$\sigma_T^P = \frac{YK_{IC}d^{-\frac{1}{2}}}{(\pi\alpha/2)^{\frac{1}{2}}}. \tag{3}$$

Both criteria are presented graphically in Figure 1. They lead to the expectations that the fracture strength of coarsely-grained ice is controlled by the nucleation of cracks, and that the strength of finely-grained ice is controlled by the propagation of cracks. "Coarse" and "fine" mean coarser than and finer than a critical size, d_c, which corresponds to the cross-over point in Figure 1. It is given by:

$$d_c = \left(\frac{\left(\frac{2}{\pi\alpha}\right)^{\frac{1}{2}} YK_{IC} - k}{\sigma_o}\right)^2. \tag{4}$$

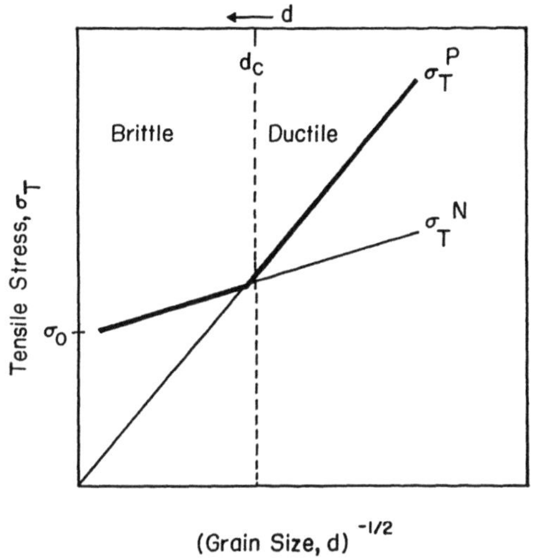

Figure 1. Schematic sketch of the effect of grain size, d, on the applied tensile stress, σ_T^N, to nucleate a crack and on the stress, σ_T^P, to propagate a crack. The crack size is proportional to the grain size. The critical grain size, d_c, marks a transition from brittle ($d > d_c$) to ductile ($d < d_c$) behavior. (After Schulson, 1987).

Owing to the dependence on temperature and strain rate of the material constants K_{IC}, k and σ_o, the critical grain size is expected to vary with loading conditions.

Figures 2 and 3 show two sets of experimental results (Lee and Schulson, 1988) obtained from relatively bubble-free, fresh-water granular ice fractured at $-10°$C, at a relatively high strain rate of 10^{-3} s^{-1} and at a relatively low rate of 10^{-7} s^{-1}. The grain size (taken as the average spacing between grain boundaries as seen on a thin section) spanned the range which most commonly occurs in nature (1mm to $\approx$ 10mm). Both sets show that the strength is proportional to $d^{-1/2}$. The first set is described by Equation (1) and the second, by Equation (2), suggesting that crack nucleation controlled the strength of the more rapidly strained material and that crack propagation controlled the strength of the more slowly strained ice. In other words, the first cracks to form in the more rapidly strained ice appear to have propagated immediately, because the applied tensile stress upon nucleation was more than enough for propagation (i.e. the grain sizes of all aggregates was above d_c). On the other hand, the first cracks to form in the more slowly strained ice appear to have been stable, because the applied tensile stress for nucleation in this case was

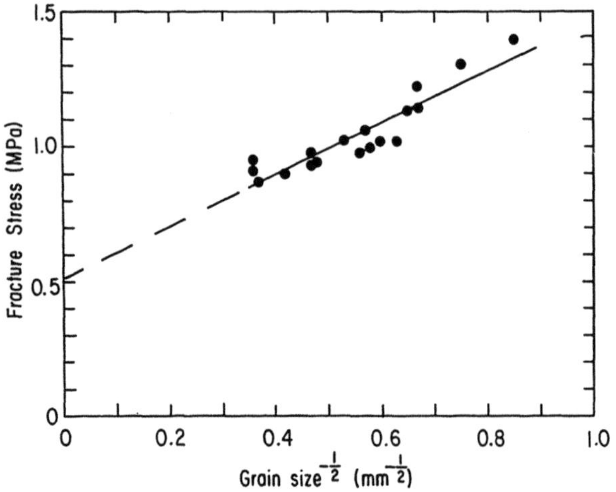

Figure 2. The tensile strength versus (grain size)$^{-\frac{1}{2}}$ of granular, fresh-water ice strained at a constant rate of 10^{-3} s^{-1} at $-10°$C. The data are described by Equation 1. (After Lee and Schulson, 1988).

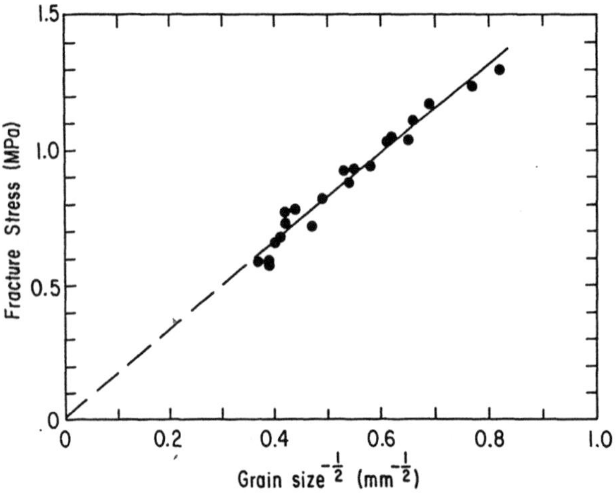

Figure 3. The tensile strength versus (grain size)$^{-\frac{1}{2}}$ for granular, fresh-water ice strained at a constant rate of 10^{-7} s^{-1} at $-10°$C. The data are described by Equation 2. (After Lee and Schulson, 1988).

less than that needed for propagation (i.e. for all aggregates $d < d_c$). Additional stress was needed to fracture the aggregates, resulting in inelastic strain additional to that needed to nucleate cracks. These results suggest that d_c increases with decreasing strain rate.

Consistent with the above views are three additional observations (Schulson, 1987; Lee and Schulson, 1988): the two parts from specimens fractured at the higher rate were crack free, while those from the specimens fractured at the lower rate contained internal cracks of grain-size dimensions; granular ice strained at $-10°C$ at the intermediate strain rate of 10^{-6} s^{-1} exhibits an intermediate critical grain size of ≈ 1.5mm (Schulson et al., 1984); and the slopes of the curves in Figures 2 and 3 are in fair accord with calculations, i.e.:

$$k_{\text{meas.}} = 0.3\text{MPa} \cdot \text{m}^{1/2} \quad \text{vs.} \quad k_{\text{calc.}} = 0.06\text{MPa} \cdot \text{m}^{1/2}, \text{ and}$$

$$k_{\text{meas.}} = 0.52\text{MPa} \cdot \text{m}^{1/2} \quad \text{vs.} \quad k_{\text{calc.}} = 0.05\text{MPa} \cdot \text{m}^{1/2}.$$

The fracture mode, incidentally, is transgranular cleavage. Cleavage steps in coarsely grained ice can be seen by the naked eye. In finely grained ice they can be seen using the environmental scanning electron microscope, (Schulson et al., 1989a).

It appears, therefore, that the tensile fracture of virgin ice can be understood in terms of the nucleation and the transgranular propagation of internal cracks.

2.2 Cracked Ice

When long, sharp cracks are present, they are expected to propagate and to reduce the strength of the ice. On the other hand, when either short cracks or long, blunt ones are present, they are expected to be stable and to have no effect at all on the tensile strength. "Long" and "short" refer to the crack length relative to a critical size noted below.

The basis for these views is shown in Figure 4 (Schulson et al., 1989c). The curve describes the critical crack size vs. the grain size of the matrix in which the cracks are imbedded. It was obtained by equating the stress to nucleate a crack (Equation 1) with the stress to propagate a crack of length $2C$ (Equation 3). The neighboring curves denote an uncertainty in K_{IC} of $\pm 6\%$. The numbers are specific to fresh-water granular ice at $-10°C$, but the message is general: sharp cracks which fall above the curve are expected to be unstable, while those which fall below are expected to be stable. In reasonable keeping with these expectations, with two notable exceptions, are data obtained from pre-cracked material which was deformed at $-10°C$ at 10^{-3} s^{-1} (where, from Section 2.1,

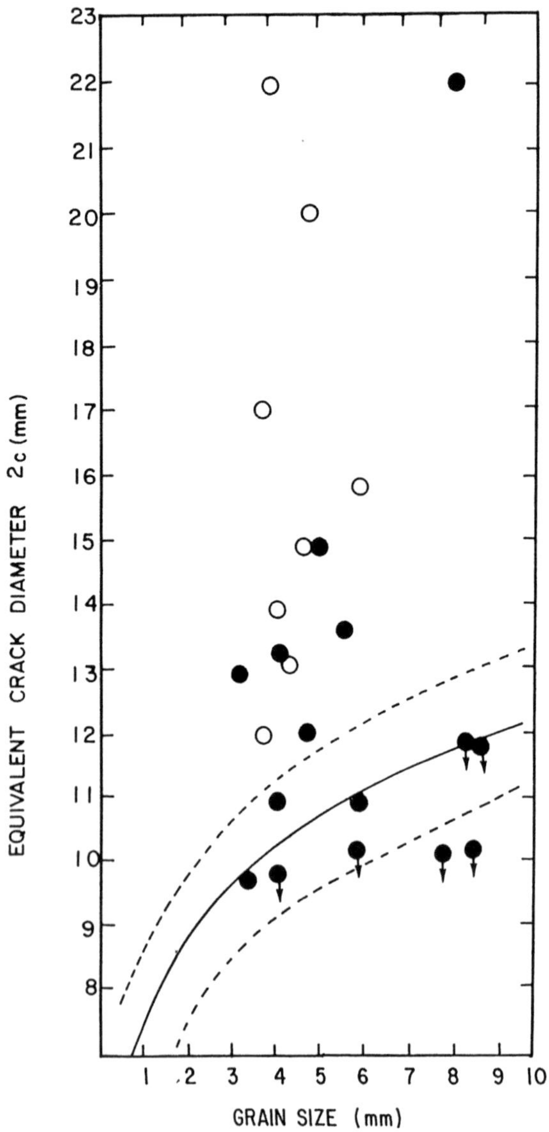

Figure 4. The equivalent crack size versus grain size for granular, fresh-water ice strained in tension at −10°C at 10^{-3} s^{-1}. The open and closed points, respectively, denote experimentally observed unstable and stable cracks. The arrows on some of the closed points indicate that the crack size was smaller than indicated. The large (22mm dia.) crack contained within the coarsely-grained (d = 8.3mm) matrix was stabilized through crack-tip creep. (After Schulson et al., 1989c).

virgin ice fractured through the nucleation of cracks). They are shown in Figure 4. The open points denote pre-existing cracks which propagated; correspondingly, the tensile strength was lower than it would have been had new cracks nucleated, by an amount in agreement with calculations based upon fracture mechanics. On the other hand, the closed points denote pre-existing cracks which did not propagate; correspondingly, the tensile strength was the same as that for uncracked ice.

The notable exceptions were two large (22mm and 15mm) cracks. In these cases only, the cracked aggregate was crept (at 0.6 to 0.7 MPa for about 3 min. at −10°C) before it was rapidly loaded to fracture. During this deformation the creep appeared to be sufficient (see calculation, Schulson et al., 1989c) to stabilize the cracks. These results are notable, because they suggest that the thermal-mechanical history of a crack is as important as its size in determining its propensity to propagate.

Cracked ice is thus more complicated than virgin ice. Yet, an understanding of its tensile fracture is possible within the context of crack mechanics, provided that care is taken to incorporate crack-tip creep when appropriate.

2.3 Saline Ice

Crack nucleation may not be an issue in saline ice. Pores, most of which are brine-filled at temperatures between initial freezing (−2°)C and final freezing (≈ −22°)C, are part of the microstructure of this material and are organized in plate-like arrays which may act as cracks (Weeks and Assur, 1967). Instead, crack propagation may control the fracture process. Supporting this possibility are the results from ring tensile tests which show that the strength is directly proportional to (brine volume)$^{-1/2}$, Figure 5. Goetze (1965) predicted this relationship on the basis of the kind of theory described in Section 2.1. In this case, the "crack size" was assumed to be equivalent to the pore size.

Mechanistic studies are required before the physical processes can be better understood. Such studies are still in the early stages. It is apparent, however, that saline ice is a complex material and that unambiguous conclusions will be more difficult to draw.

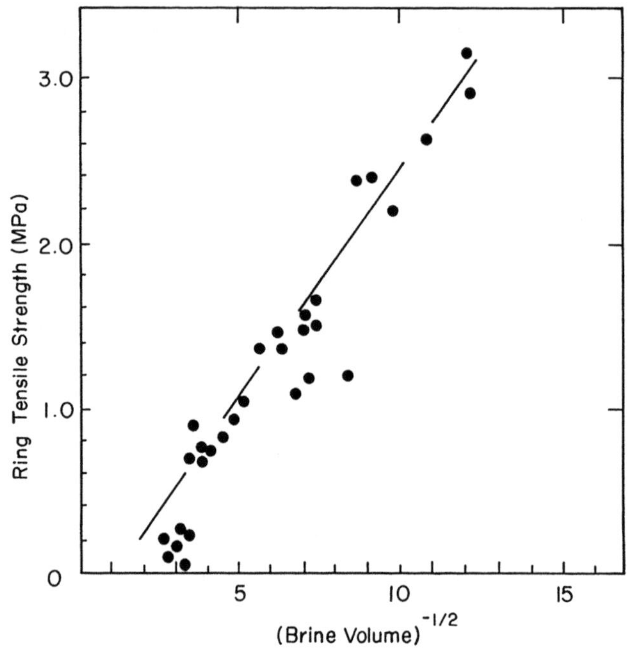

Figure 5. The ring tensile strength of saline ice versus (brine volume)$^{-1/2}$. (After Goetze 1965).

3. Compressive Fracture

Brittle failure under compression occurs when ice is strained rapidly; i.e., at rates in excess of around 10^{-3} s^{-1} at temperatures around -10°C (Hawkes and Mellor, 1972). When strained more slowly, the ice is ductile. The strain rate that marks the transition increases with confinement (Jones, 1982), with grain refinement (Schulson and Cannon, 1984; Cole, 1987) and with pre-strain (Kuehn et al., 1988) and, for columnar ice, with a change in the direction of loading from parallel to perpendicular to the columns. The transition strain rate is important, because it sets the conditions under which the compressive strength is a maximum.

Compressive fracture is more complicated than tensile fracture. The nucleation of cracks is still important, but additional processes such as frictional crack sliding, stable crack propagation, and crack interaction may occur.

The discussion which follows should be viewed as an attempt to unveil the underlying physical process, rather than as a summary of an accepted position.

3.1 Fracture Modes

Three fracture modes have been identified: shear faulting, axial splitting and explosive failure. Figure 6 illustrates these modes schematically. The first two are commonly observed in other brittle solids, such as rock and concrete (Jaeger and Cook, 1979).

Shear faulting occurs under uniaxial loading when the ends of test pieces are confined (Schulson et al., 1989b) either through ice-platen interaction or through specimen shape (i.e. dumb-bell). The fracture surfaces are inclined $\approx 30°$ to the loading direction and owing to the sliding of one piece of ice over the other are covered with finely granulated powder. Also, the fracture surfaces are characterized by a roughness of the scale of the grain size. Axial splitting occurs either when the ice-platen interface generates lateral tensile stresses or when the platens have little interaction with the ice. Ice fractured in this manner is characterized usually by several long axial cracks and, unlike the shear faulted material, by relatively smooth fracture surfaces free from powder. Explosive fracture has been reported for relatively warm (-10 to $-5°$) ice, and appears to be characterized in part by grain-sized fragments (Cole, 1987).

Cracks again are the key elements. Pre-existing cracks, however, are probably less of a factor in compressive failure than in tensile fracture. Instead, internal microcracks generated during loading are the ones of primary interest. They are about equal in size to the grain size (at least for fresh-water ice) (Gold, 1966, 1970, 1972; Kalifa et al., 1989; Cole, 1986), are contained within individual grains and/or along grain boundaries and terminate at grain boundaries. They are aligned more or less along the direction of maximum compressive load (albeit spread about the loading direction) and beyond a critical strain (Sinha, 1982) increase in number density with increasing strain. Once a sufficient number links up within a localized shear band, shear faulting appears to occur. When the cracks extend instead, axial splitting occurs. Explosive fracture appears to reflect the highest internal crack density, since it corresponds to the highest strengths observed under a given set of conditions (Cole, 1987). This mode implies that the internal cracks neither linked up nor prop-

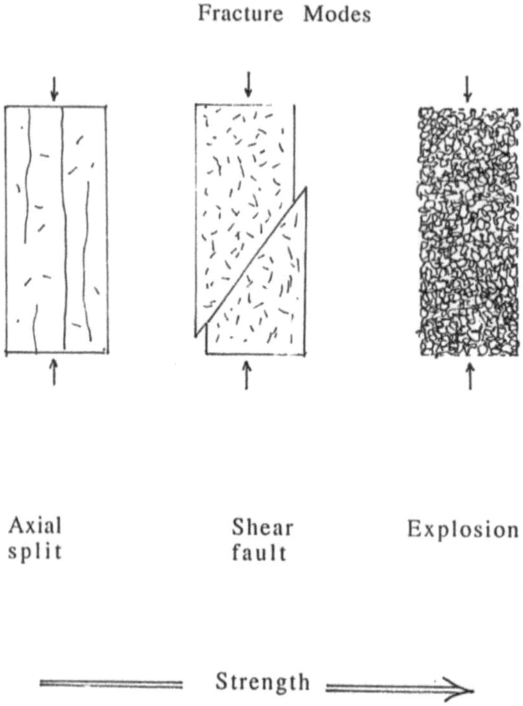

Figure 6. Schematic sketch of brittle compressive fracture modes under compression.

agated, but owing to unusually high stability against propagation along the direction of loading, reached a number density sufficiently great to allow the material to disintegrate.

3.2 A Physical Model

Barring explosive fracture, a physical model (McClintock and Walsh, 1963; Nemat-Nasser and Horii, 1982; Ashby and Hallam, 1986; Horii and Nemat-Nasser, 1986) developed for brittle solids and based upon frictional crack sliding can explain a number of the features of the brittle compressive fracture of ice. Another model (Costin, 1983, 1985), based upon the view that internal heterogeneities generate localized tensile stresses and that the magnitude of these stresses is proportional to the applied deviatoric stress, might also apply. It will not be considered further, however, because this model is more phenomenological than physical.

The idea of the frictional sliding model is that out-of-plane extensions or wings form within the tensile zone of microcracks which are inclined to the far-field load, Figure 7. The wings initiate and then grow along the direction of loading through frictional sliding across the opposing faces of the parent crack. With increasing applied load, the sliding increases and the wings lengthen. If microcracks within an echelon array sprout wings, then the wings interact and a shear fault develops (see Figure 15, Schulson, 1987). Otherwise, the wings lengthen on just a few cracks and an axial split develops. The wings grow when the tensile stress localized at their tips is such that the attendant mode I stress intensity factor reaches the critical level; i.e., $K_I = K_{IC}$. Under far-field compressive loading, unlike tensile loading, the axial crack growth is stable; i.e., the length increases as the load increases.

What evidence supports this model? The most direct is wing-cracks themselves. Figure 8 (Cannon et al., 1989) shows such cracks entrapped

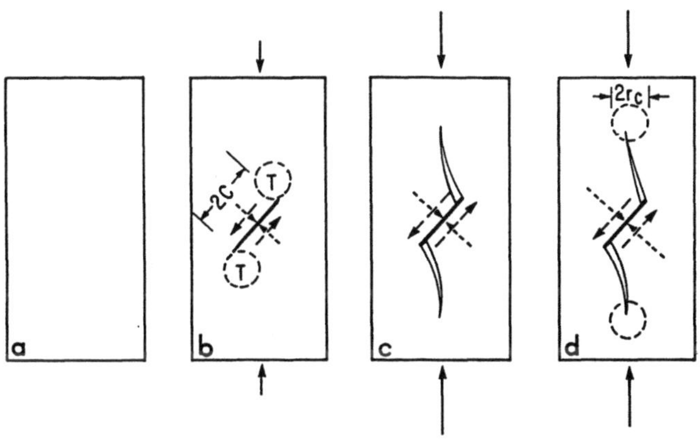

Figure 7. Schematic sketch of the frictional sliding, wing crack model at different stages of loading under uniaxial compression:

(a) no load;

(b) small load, such that crack is nucleated; the T regions at the crack tips above and below the crack plane denote regions of tensile stress; the broken arrows denote induced shear stress which drives the sliding;

(c) higher load, such that out-of-plane extensions, or wings, have formed, aligned along the loading direction;

(d) same state as (c), but including crack-tip creep within zones at ends of wing cracks.

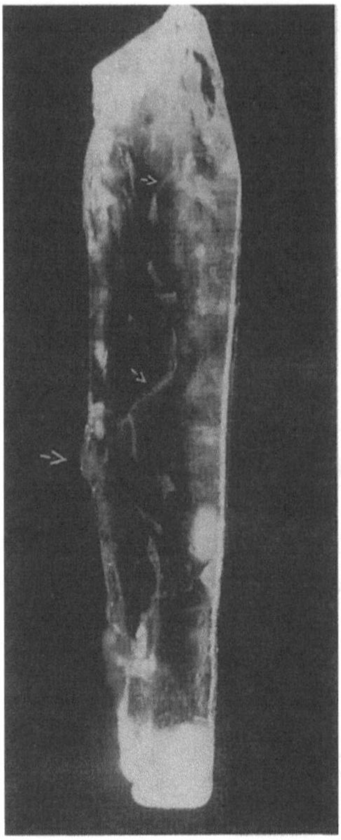

Figure 8. Wing-cracks (arrowed) in $S2$ columnar, fresh-water ice strained at 10^{-2} s^{-1} at $-20°C$ in a direction normal to the columns. The load was applied along the direction of the shard shown (vertical in figure) and the columns were perpendicular to the plane of the image. Note the large wing crack in the center and the smaller wing crack near the top of the image. The step (arrowed) on the axially-split fracture surface appears to have formed through the propagation of a wing crack. The width of the shard is about 12mm. (After Cannon et al., 1989).

within a shard which broke free from a test specimen. The examples shown were formed within coarsely-grained (d = 5mm), $S2$ fresh-water columnar ice loaded under uniaxial compression in a direction normal to the columns, at $-20°C$ at 10^{-2} s^{-1}. The parent crack nucleated along a grain boundary, which was inclined about 45° to the loading direction and was about as long as the boundary, as evident from post-test, thin-section analysis under crossed polarized filters; the two wings propagated transgranularly a short distance into two adjacent grains. As the model predicts, the wings are aligned more or less along the direction of loading.

A fuller account of wing cracking will be given elsewhere, showing that the cracks lengthen with increasing stress in reasonable quantitative accord with theory.

Other evidence is in the form of a quantitative comparison between theory and measurement and was introduced earlier (Schulson, 1987). A detailed discussion will be given elsewhere. Briefly, the theory dictates that the brittle compressive fracture strength is directly proportional to K_{IC} and to (crack size ≈ grain size)$^{-1/2}$ and is inversely proportional to either $(1-\mu)$ or $[(1+\mu^2)^{1/2}-\mu]$, where μ is the coefficient of sliding friction for ice on ice. Measurements on granular, fresh-water ice, Figure 9, show that the strength, although scattered, is consistent with the $d^{-1/2}$

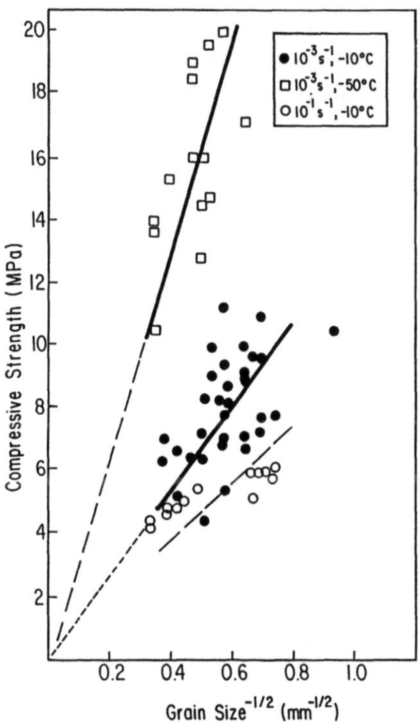

Figure 9. The uniaxial compressive fracture strength of granular fresh-water ice versus (grain size)$^{-1/2}$. The data were obtained from material strained at −10°C and at −50°C at strain rates of 10^{-3} s^{-1} and 10^{-1} s^{-1}. Note the marked increase in strength with decrease in temperature, and the decrease in strength with increase in strain rate. The latter point is apparent only in the more finely-grained ice where the difference in strength is greater than the scatter in the data. (After Schulson, 1987).

dictate, that the marked ($\approx$ 250%) increase in strength upon reducing the temperature from $-10°$C to $-50°$C is consistent with the increase in friction from $\mu \approx 0.5$ at $-10°$C to $\mu = 0.8$ at $-40°$C (Jones, 1989), and that the decrease ($\approx$ 30%) in strength at $-10°$C upon raising the strain rate from 10^{-3} s^{-1} to 10^{-1} s^{-1} is consistent with the decrease in friction from $\mu \approx 0.5$ to $\mu \approx 0.3$ (Jones, 1989) upon raising the sliding velocity from around $10^{-6} - 10^{-5}$ m $\cdot$ s^{-1} to $10^{-4} - 10^{-3}$ m $\cdot$ s^{-1} (μ is a function of sliding velocity, v, which was computed from the relationship $v \approx 2\dot{\varepsilon}d$ where $\dot{\varepsilon}$ is the applied strain rate).

The other relevant point is the brittle compressive strength of warm $S2$ columnar saline ice. Over the range of strain rates from 10^{-3} to 10^{-1} s^{-1} where this ice, too, is brittle, the strength at $-10°$C of aggregates loaded perpendicular to the columns appears to remain constant, although scattered, Figure 10 (Kuehn and Schulson, 1989). In this case, the ice contains brine-filled pores which fracture and thus provide a water lubricant for sliding. In the case of the fresh-water ice, frictional heating

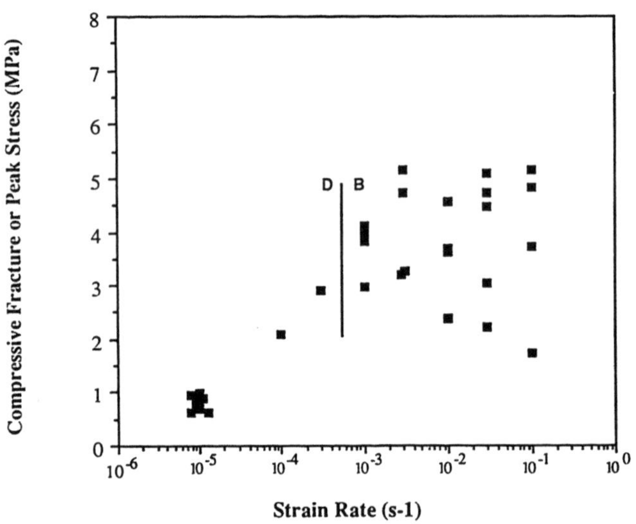

Figure 10. The compressive strength versus strain rate of $S2$ columnar, saline ice ($\approx$ 4 ppt salinity) loaded normal to the columns ($\approx$ 4mm dia.) at $-10°$C. A ductile-to-brittle transition ($D \Rightarrow B$) occurs at a strain rate of 0.5×10^{-3} s^{-1}, where the stress-strain curve changes from one exhibiting a peak to one exhibiting only a rising branch. The brittle fracture strength, although scattered, appears not to decrease with increasing strain rate. (After Kuehn and Schulson, 1989).

was needed to produce the lubricant, which explains the reduction in the compressive fracture strength of that material with increasing strain rate.

Appealing though the crack sliding model appears to be, caution is warranted before adopting it without question. Wings, while seen in both fresh-water columnar ice (reported here) and in fresh-water granular ice (to be discussed elsewhere), have been reported only under one set of conditions (−20°C, 10^{-2} s^{-1}), and the $d^{-1/2}$ functionality characterizes any model based upon fracture mechanics. Nevertheless, the model appears to merit further consideration.

3.3 Ductile-to-Brittle Transition: A Criterion

In the light of the above discussion, consider the ductile-to-brittle transition. The inclined crack from which the wings form is assumed to have nucleated during loading, but at an applied stress lower than that for wing initiation and growth: for this reason, nucleation, although a necessary prelude, does not appear explicitly in the model. Two crack-zone processes are thought to be important. One is the intensification of stress at the crack tip, and the other is the relaxation of stress. When the rate of loading (which is set by the applied strain rate) is high, K_I eventually reaches K_{IC}. On the other hand, when the loading rate is low, K_I does not reach K_{IC}. The question is: how high is high?

Assume that stress relaxation occurs through power-law creep, and let r_c, Figure 7d, be the radius of the creep zone within the vicinity of the loaded crack. When r_c exceeds a critical size, K_I no longer characterizes the stress field at the crack tip. The critical size is difficult to specify: for the purposes of introducing the criterion it is assumed here to be a small fraction, f, of the smallest dimension of the cracked aggregate. This dimension for the rather densely cracked condition which characterizes failed ice at the transition point is assumed to be equal to the microstructural scale, λ. Thus, a criterion for the ductile-to-brittle transition may be written as:

$$r_c = f\lambda. \tag{5}$$

In the subsequent development it is assumed, albeit rather arbitrarily, that $f \approx 1/50$. A similar fraction relates the size of the plastic zone to the minimum dimension of a test specimen needed to obtain a valid measure of K_{IC} (ASTM - E399, 1981).

Riedel and Rice (1980) have modelled r_c for pre-existing cracks under a constant tensile stress. In the present context the stress, and hence K_I, is not constant but is rising. Yet, K_I does not rise gradually from zero: it presumably exists as soon as the crack is nucleated. An upper limit on r_c may then be estimated from Riedel and Rice's analysis by writing $K_I \approx K_{IC}$. Thus:

$$r_c \cong \frac{K_{IC}^2}{2\pi E^2}\left[\frac{(n+1)^2 E^n B t}{2n\alpha^{n+1}}\right]^{2/n-1} F, \tag{6}$$

where E is Young's modulus, α is a dimensionless parameter of order unity, and F is a parameter which specifies the shape of the creep zone and is of order unity; B and n are constants in the power law (Glen, 1953):

$$\dot{\varepsilon}_C = B\sigma^n, \tag{7}$$

where $\dot{\varepsilon}_C$ is the steady-state creep rate and σ is the applied stress. The time, t, may be approximated by the ratio:

$$t = \frac{K_{IC}}{\dot{K}_I}, \tag{8}$$

where $\dot{K}_I$ is the rate of increase in the mode I stress intensity factor and may be given by the product:

$$\frac{\delta K_I}{\delta t} = \frac{\delta K_I}{\delta \sigma}\frac{\delta \sigma}{\delta \varepsilon}\frac{\delta \varepsilon}{\delta t}, \tag{9}$$

where $\delta\sigma/\delta\varepsilon$ is a modified Young's modulus (i.e., reduced by damage), and where $\delta\varepsilon/\delta t = \dot{\varepsilon}_a$, the applied strain rate. The other derivative, $\delta K_I/\delta\sigma$, requires a model of the crack-tip stress field and for consistency we use the one for a short, loaded wing crack of length 2λ; i.e. (Ashby and Hallam, 1986):

$$\frac{\delta K_I}{\delta \sigma} = \left(\frac{\pi\lambda}{6}\right)^{1/2}\left[(1+\mu^2)^{1/2} - \mu\right]. \tag{10}$$

Upon collecting terms and substituting the values $\alpha = 1$, $F = 1$, and $n = 3$ (Barnes et al., 1971), upon equating Equations 5 and 6, and upon neglecting the reduction in Young's modulus owing to damage, the applied strain rate corresponding to the ductile-to-brittle transition rate may be expressed as:

$$\dot{\varepsilon}_{D/B} = \frac{28K_{IC}^{3}B}{\lambda^{3/2}[(1+\mu^{2})^{1/2}-\mu]}. \tag{11}$$

Letting $B = 1.6 \times 10^{-7}\text{MPa}^{-3}\ \text{s}^{-1}$ (Barnes et al., 1971), $K_{IC} = 0.1\text{MPa}\ \text{m}^{1/2}$ (Nixon and Schulson, 1987) and $\mu = 0.5$ (Jones, 1989), Equation (11) gives $\dot{\varepsilon}_{D/B} = 0.1 \times 10^{-3}\ \text{s}^{-1}$ at $-10°\text{C}$ for granular, fresh-water ice of 3mm ($= 2\lambda$) grain size. This result is within an order of magnitude of the rate observed. The equation also incorporates the observations that both grain refinement (through reducing λ) and pre-strain (presumably through increasing B) increase $\dot{\varepsilon}_{D/B}$.

While encouraging, the agreement between theory and experiment is better than it deserves to be, given all the assumptions and the approximations that were made in the development. The analysis, nevertheless, offers a guide to experiment. Further work, it is hoped, will lead to improvements.

4. Conclusion

The brittle fracture of ice can be understood in terms of the nucleation and the propagation of internal cracks. Fracture under tensile loading is relatively well explored, and appears to conform to theory. The outstanding issues here relate to saline ice. Fracture under compressive loading is more complicated and less well explored, but shows some evidence of conforming to the dictates of a theory based upon the frictional crack-sliding model developed for brittle solids.

Acknowledgements

I would like to thank Mr. D. M. Cole, Dr. P. Duval, Prof. H.J. Frost and Dr. J. Meysonnier for valuable discussions. Also, I would like to thank Mr. G.A. Kuehn for technical assistance, Messrs. N.P. Cannon, and T.R. Smith for their search for wing cracks, and Ms. D. Withrow for typing the manuscript.

The work was funded by the Office of Naval Research through a University Research Initiative Program, grant number N00014-86K-0695.

References

ASTM - E 399. 1981. Standard test method for plane-strain fracture toughness of metallic materials. Amer. Soc. for Test. and Mater.

Ashby, M.F. and Hallam, S.D. 1986. The failure of brittle solids containing small cracks under compressive stress states. Acta Met. Vol. 34, pp. 497-510.

Barnes, P., Tabor, D., and Walker, J.F.C. 1971. The friction and creep of polycrystalline ice. Proc. Royal Society of London, A. Vol. 324, pp. 127-155.

Cannon, N.P., Schulson, E.M., and Smith, T.R. 1989. Wing cracks in ice. (to be submitted for publication).

Cole, D.M. 1986. Effect of grain size on the internal fracturing of polycrystalline ice. USA-CRREL Rpt. 86-5.

Cole, D.M. 1987. Strain-rate and grain-size effects in ice. J. Glaciology. Vol. 33, pp. 274-280.

Costin, L.S. 1983. A microcrack model for the deformation and failure of brittle rock. J. Geophys. Res. Vol. 88, pp. 9485-9492.

Costin, L.S. 1985. Damage mechanics in the post-failure region. Mech. of Mater. Vol. 4, pp. 149-160.

Currier, J.H. and Schulson, E.M. 1982. The tensile strength of ice as a function of grain size. Acta Met. Vol. 30, pp. 1511-1514.

Glen, J.W. 1953. The creep of polycrystalline ice. Proc. Roy. Soc. A. Vol. 228, pp. 519-538.

Goetze, C.G. 1965. A study of brittle fracture as applied to ice. USA-CRREL Tech. Note (unpublished).

Gold, L.W. 1966. Dependence of crack formation on crystallographic orientation for ice. Can. J. Phys. Vol. 44, pp. 2757-2764.

Gold, L.W. 1970. Process of failure in ice. Can. Geotech. J. Vol. 7, pp. 405-413.

Gold, L.W. 1972. The process of failure of columnar-grained ice. Phil. Mag. Vol. 26, pp. 311-328.

Hawkes, I. and Mellor, M. 1972. Deformation and fracture of ice under uniaxial stress. J. Glaciology. Vol. 11, pp. 103-129.

Horii, H. and Nemat-Nasser, S. 1986. Brittle fracture in compression: splitting, faulting and brittle-ductile transition. Phil. Trans. Roy. Soc. London. A. Vol. 319, pp. 337-374.

Jaeger, C. and Cook, N.G.W. 1979. Fundamentals of Rock Mechanics (3rd ed). London: Chapman Hall.

Jones, D.E. 1989. An experimental investigation of low-speed, ice — ice friction. M.E. Thesis, Thayer School of Engineering, Dartmouth College.

Jones, S.J. 1982. The confined compressive strength of polycrystalline ice. J. Glaciology. Vol. 28, pp. 171-177.

Kalifa, P., Duval, P., and Ricard, M. 1989. Nucleation of cracks in polycrystalline ice under compression. Proc. OMAE, The Hague, The Netherlands, Vol. 4, pp. 13-21.

Kuehn, G.A., Schulson, E.M. and Nixon, W.A. 1988. The effects of prestrain on the compressive ductile-to-brittle transition in ice. Proc. IAHR Ice Symp., Sapporo, Japan. pp. 109-117.

Kuehn, G.A. and Schulson, E.M. 1989. (unpublished results).

Lee, R.W. and Schulson, E.M. 1988. The strength and ductility of ice under tension. J. Offshore Mechanics Arctic Engg. Vol. 110, pp. 187-191.

McClintock, F.A. and Walsh, J.B. 1963. Friction on Griffith cracks in rock under pressure. Proc. 4th U.S. Nat. Cong. Appl. Mech. ASME. Vol. 2. pp. 1015-1021.

Nemat-Nasser, S. and Horii, H. 1982. Compression-induced nonplanar crack extension with application to splitting, exfoliation and rockburst. J. Geophys. Res. Vol. 87, pp. 6805-6821.

Nixon, W.A. and Schulson, E.M. 1987. A micromechanical model of the fracture toughness of ice. J. de Physique 48:C1,pp. 313-319.

Riedel, H. and Rice, J.R. 1980. Tensile cracks in creeping solids. Fracture mechanics, 12th Symp., ASTM STP 700, American Society for Testing and Materials. pp. 112-130.

Schulson, E.M. 1987. The fracture of ice *Ih*. J. de Physique. 48:C1,pp. 207-220.

Schulson, E.M. and Cannon, N.P. 1984. The effect of grain size on the compressive strength of ice. IAHR Ice Symp., Hamburg, Germany. pp. 24-38.

Schulson, E.M., Lim, P.N. and Lee, R.W. 1984. A brittle to ductile transition in ice under tension. Phil. Mag. Vol. 49, pp. 353-363.

Schulson, E.M., Baker, I., Robertson, C.D., Bolon, R.B., and Harnimon, R.J. 1989a. Fractography of ice. J. Mat. Sci. Letts. Vol. 8, pp. 1193-1194.

Schulson, E.M., Gies, M.C. and Lasonde, G.J. 1989b. The effect of the specimen-platen interface on internal cracking and brittle fracture of ice under compression: high-speed photography. J. Glaciology, Vol. 35, pp. 378-382.

Schulson, E.M., Hoxie, S.G., and Nixon, W.A. 1989c. The tensile strength of cracked ice. Phil. Mag. A. Vol. 59, pp. 303-311.

Sinha, N.K. 1982. Delayed elastic strain criterion for first cracks in ice. Proc. IUTAM Symp. of Deformation and Failure of Granular Materials. pp. 323-330.

Smith, E. and Barnby, J.T. 1967. Crack nucleation in crystalline solids. Metal Sci. J. Vol. 1, pp. 56-64.

Weeks, W. and Assur, A. 1967. The mechanical properties of sea ice. USA-CRREL Monograph. II-C3.

Discussion

Q: W.A. Nixon, IIHR

You identified three types of brittle compressive failure in ice. Is your model of the transitional strain rate applicable to all three types of failure?

A: E.M. Schulson

I don't understand explosive failure very well, and so I don't know whether the model works in that case. It should, however, be applicable to brittle fracture through either of the other two modes.

Q: Ian Jordaan, Memorial University of Newfoundland

I have also been investigating the role of friction between crack surfaces and feel that it is important to consider this, and is an important part of your contribution. With regard to wing cracks, if the initiation is by shear in grain boundaries, it might not be possible to observe this movement. Could you please comment?

A: E.M. Schulson

I don't know exactly how the "parent crack" (i.e. the inclined one at the grain boundary) forms. I suspect elastic anisotropy is important.

Q: T.J.O. Sanderson, BP International

Thank you for an excellent paper – it is a delight to see real evidence of wing – crack formation. What, however, do you think is the mechanism of formation of the nucleation cracks? They are at $\approx 45^o$ to the principal axes - are they due to mode II (shear) opening forces, or is there a local mode I tensile stress due to, for instance, grain anisotropy?

A: E.M. Schulson

I don't know the mechanism of crack nucleation. I suspect, however, that it has more to do with elastic anisotropy than with dislocations at the relatively high strain rate and low temperature where the cracks were observed.

Q: B.L. Parsons, IMD/NRC

In the spirit in which your paper was given, I would like to suggest two refinements to the calculation of the brittle to ductile transition strain rate, both in Equation (6). As you point out the cracks are nucleated under load, but I would suggest the use of $K_{Ia} = 0.09 \mathrm{MPam}^{1/2}$ (Parsons, Snellen and Muggeridge, 1989), as in the original work of Rice and Riedel the applied K_I is assumed to be a step function. If K_c was used, the crack would grow, while the arrest toughness is what one obtains at the crack tip after growth.

Also with respect to Equation (6), the modulus in the original theory is a constant, whereas in ice it is a more linear function of time. This is true also of the approximation $\frac{\delta T}{\delta \varepsilon} = E$.

A: E.M. Schulson

The model is really a first cut at the problem. Its merit is that it brings together material properties in a manner that appears at least qualitatively consistent with experience. Quantitatively, it also seems to work, at least roughly. But, as I note, it should not be expected to do too much, given the assumptions and approximations. Refinements of the kind you mentioned could improve it.

Q: L. Gold, NRC

You stated that crack nucleation was due primarily to elastic anisotropy. Am I correct in assuming you meant for high rates of strain?

A: E.M. Schulson

Yes.

Q: Robert Gagnon, Institute for Marine Dynamics, NRC

Are you convinced that bubbles are not playing a role?

A: E.M. Schulson

No.

Q: R. Frederking, NRC

1. First an observation: G. Timco and I have done very short term loading and observed grain boundary sliding in S2 ice after 1ms of loading. Could you comment on this?

2. What was the grain orientation with respect to the wing crack orientation you showed (observed) in S2 ice?

A: E.M. Schulson

1. GBS cannot be ruled out. It occurs in other materials at high temperatures and might well occur in ice. Quite possibly, your conditions were such that GBS did occur.

2. The grains were columnar in shape and were aligned in a direction normal to the viewing screen. In one case, the c-axis in one grain was parallel to the wing and in the other grain was normal to it.

Q: D.M. Cole, USA CRREL

1. What were the grain size, $\dot{\varepsilon}$, and L/D ratio of your specimen in which you observed the wing cracks?

2. These conditions are somewhat exotic. Would you please comment on the fact that you have observed wing cracking only under fairly unusual experimental conditions.

A: E.M. Schulson

1. $\approx$ 7 to 8mm, $\approx 10^{-2}$ s^{-1}, $\approx$ 1.5 in cylindrical samples.

2. I don't think the conditions were exotic. The growing wing cracks I showed during my presentations were generated within specimens compressed between stainless steel platens at -20°C at $\approx 10^{-2}$ s^{-1}.

Q: S. Sunder, MIT

1. Just a comment. We have done an analysis of an isotropic polycrystal with elastically anisotropic crystals. The results show that in

compression, first cracks nucleate at an angle about 30°. These can serve as the cracks for wing crack initiation.

2. A similar analysis with bubbles shows that bubbles influence the results for smaller grain size and more in tension than in compression. Papers concerned with these analyses are to be published in Cold Regions Science and Technology.

A: E. Schulson

1. Yes, I would agree.
2. If an isolated bubble is relatively large w.r.t. the grain size, and it is girdled by a sharp crack, then it could reduce the tensile strength σ_T. On the other hand, if it is not girdled by a sharp crack, then I question whether it would affect σ_T. I base this on our observation (reported here) that small, sharp cracks do not affect σ_T and on the fact that cracks intensify stress more than bubbles. In compression, I would agree with your comment.

SESSION III:

Damage and Compressive Fracture

Models for the Deformation Behavior of Viscoelastic Media with Distributed Damage and Their Applicability to Ice

R. A. Schapery

Department of Civil Engineering, Texas A & M University
College Station, Texas 77843

Abstract A constitutive theory that was originally developed for nonlinear viscoelastic composite materials with damage is described. It is shown that the predicted behavior is consistent with that of polycrystalline ice. This theory is quite different from those developed for ice by other investigators. Therefore, besides reviewing the results, we give the underlying mathematical and physical arguments. First, some aspects of linear and nonlinear viscoelastic constitutive theory without damage are reviewed. Then, previously developed results for uniaxial loading are described and shown to be consistent with the behavior of ice under conditions of constant compressive stress and strain rate. Encouraged by these results we review the basis for the theory for uniaxial and other types of proportional loading, starting with a multiaxial constitutive equation without damage and an equation for microcrack growth rate. Finally, some aspects of the nonproportional loading problem are discussed.

1. Introduction

Polycrystalline ice behaves as a brittle, nonlinear viscoelastic material under a wide range of conditions of engineering interest (Sanderson, 1988). The brittleness and polycrystalline structure combine to produce a significant level of distributed microcracking under many compressive stress states and histories. Typically, as loading or time increases under moderate to high stresses, an increasing number of grain facet-size microcracks, both transgranular and along grain boundaries, appear (e.g. Gold, 1972 and Sinha, 1988). In controlled load tests the increasing density of microcracks adds considerably to the overall strain and strain rate. For controlled deformation tests the resulting load may be much less than would be the case without microcracking. In contrast, under tensile loading the

S. Jones · R. F. McKenna · J. Tillotson · I. Jordaan (Eds.)
Ice-Structure Interaction
IUTAM-IAHR Symposium St. John's, Newfoundland, Canada

cracking process is dominated by one rapidly growing macrocrack which leads to test specimen failure at relatively small strains.

Here we shall consider only the case of globally stable, distributed microcracking. Effects of nonlinear viscoelasticity and microcracking will be combined to give a mathematical description of the overall or average stress-strain behavior of ice. The effect of individual microcracks is smeared out in that the microcracked material will be modelled as a homogeneous continuum on a scale that is much larger than the grain size. The study of media which can be idealized in this way falls under the heading of "continuum damage mechanics" (CDM); for a general overview see Krajcinovic and Lemaitre (1987), and for studies of nonlinear viscoelastic media with emphasis on the tensorial description of damage see Onat and Leckie (1988) and Weitsman (1988). Jordaan and McKenna (1988) discuss some fundamental problems of damage and CDM for ice. We shall not attempt here to review the many different aspects of CDM. Rather, only those that are considered to be especially relevant to viscoelastic ice behavior will be discussed.

The assumption of a homogeneous continuum is introduced for the purpose of describing the deformation behavior of a damaged material element through a constitutive equation. This equation, as used here, relates the overall or average stresses and strains for a material element; they are defined in the usual way for polycrystalline and composite materials, whether elastic or viscoelastic, without or with microcracks (e.g. Hashin, 1965; Schapery, 1967; Schapery, 1981; and Horii and Nemat-Nasser, 1983). In applications, such as prediction of the forces between ice and engineering structures, the damage will vary spatially due to stress gradients. Of course, the validity of a constitutive equation (based on homogeneous damage) becomes questionable wherever the stress variation is not very small over distances that are comparable to grain size and the local average crack spacing and size. As long as the error arising from a stress gradient is not important, one may use the same constitutive equation from point-to-point while allowing for the nonuniform damage through numerical values of the parameters in the equation that accounts for the damage. As discussed by Jordaan (1986) direct numerical methods of analysis, such as the finite element method, usually will be needed to predict ice-structure interactions partially because of the nonhomogeneity of the damage and the mutual dependencies of the damage and stress distributions.

In this paper we shall discuss primarily a constitutive theory that was originally developed for viscoelastic composite materials with microcracking (Schapery, 1981). However, the theory appears to be valid for ice (Harper, 1986, 1989), and therefore much of the paper is an exposition of the earlier work along with a description of its use in the characterization of ice. Uniaxial compression is considered first in order to bring out the essential effects of stress or strain history with a minimum of complexity. Next, multiaxial stress effects are introduced for the case of proportional loading, wherein all overall stress or strain components have the same time-dependence. The form of the constitutive equation for this case is the same as for uniaxial loading for materials with intrinsic power law nonlinearities (i.e. nonlinearities apart from effects of damage). Finally, certain aspects of the much more complex problem of nonproportional loading are discussed. Some new results are given which follow from the recent work of Schapery (1990).

Our approach to developing constitutive equation for ice with damage is considerably different from that followed by Sjölind (1987) and Karr and Choi (1989). In these two studies some information on the crystalline microstructure and microcrack state are combined with continuum mechanics to develop a multiaxial constitutive theory. Sjölind assumes the ice undergoes only secondary creep at constant damage and expresses damage growth rates in terms of thermodynamically-based dissipation potentials. The effect of damage on creep is introduced by replacing the actual stress by the effective stress; the effective stress tensor is defined as the actual stress tensor multiplied by a damage-dependent tensor. The elastic moduli also vary with damage. Karr and Choi add the delayed elastic strain (primary creep) to secondary creep and incorporate the effects of damage through the effective stress. Their damage evolution law distinguishes between the effects of grain boundary cracking and transgranular cracking; this damage is incorporated in the constitutive equation by means of a second-order damage tensor that accounts for the crack orientation distribution. The damage growth rate is postulated to depend on the damage and the resolved normal stresses and strain rates; delayed elastic strain rate is used for damage due to grain boundary cracks while total strain rate is used for transgranular cracks. In both of the cited studies satisfactory agreement with limited experimental data is achieved. Another physically-motivated theory has been proposed by Sinha (1984, 1988) for uniaxial compression, as guided by his extensive experimental studies.

While physically motivated, these studies use neither micro-crack growth equations based on viscoelastic fracture mechanics nor a rigorous viscoelastic analysis which accounts directly for growing and stationary cracks in an otherwise continuous material. The shortcomings are understandable in view of the complexity of the problem. However, as we have shown (Schapery, 1981), by using a so-called modified superposition principle to characterize the intrinsic nonlinear viscoelastic behavior of the continuum, the effect of microcracking may be rigorously and easily included starting with an elastic solution. This version of a viscoelastic constitutive equation will be defined and used with the damage analysis in Sections 2 and 3. First, however, as background for the nonlinear theory we want to comment on certain mathematical representations of linear viscoelastic behavior for undamaged ice.

2. Uniaxial Constitutive Equations

2.1 Linear Theory

Consider a uniaxial stress state of tension or compression. The most general form of a linear, nonaging viscoelastic constitutive equation is

$$\varepsilon = \int_{0^-}^{t} D(t-\tau)\frac{d\sigma}{d\tau}d\tau, \tag{1}$$

where ε and σ are the axial strain and stress, respectively; it is assumed that $\sigma = \varepsilon = 0$ when $t < 0$. The lower limit is 0^-, rather than 0, in order to allow for a discontinuity in stress at $\tau = 0$, such as in a creep test. The quantity $D(t)$ is the uniaxial creep compliance, which is defined as the ratio of strain-to-stress when σ is a constant σ_o applied at $t = 0$ (a creep test),

$$D(t) \equiv \varepsilon(t)/\sigma_0 \qquad (t > 0). \tag{2}$$

Also, $D(t) \equiv 0$ when $t < 0$. Given $D(t)$ and any stress history $\sigma(t)$ which vanishes when $t < 0$, Eq. (1) enables the prediction of strain history. Note that one is to use $\sigma = \sigma(\tau)$ in Eq. (1), where τ is the dummy time variable of integration.

Equation (2) may be recovered from Eq. (1) by first observing that for a creep test $\sigma(\tau) = \sigma_0 H(\tau)$, where $H(\tau)$ is the unit step function ($H(\tau) = 0$ for $\tau < 0$ and $H(\tau) = 1$ for $\tau > 0$), and then using $d\sigma/d\tau = \sigma_0\delta(\tau)$ where $\delta(\tau)$ is the Dirac delta function.

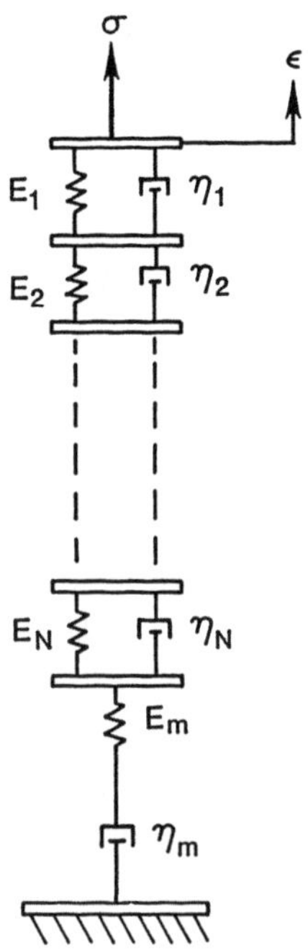

Figure 1. Generalized Voigt model.

Viscoelastic behavior is often described in terms of the behavior of a collection of Hookean springs (with moduli E_i) and Newtonian dashpots (with viscosities η_i) such as that in Figure 1. Solution of the system of differential equations for this model when stress is constant yields

$$D \equiv \frac{\varepsilon}{\sigma_0} = \frac{1}{E_m} + \sum_{i=1}^{N} \frac{1}{E_i}[1 - \exp(-t/\rho_i)] + \frac{t}{\eta_m}, \tag{3}$$

where $\rho_i \equiv \eta_i/E_i$ is the ith "retardation time". Biot (1954) has used irreversible thermodynamics to show that Eq. (3) is the most general form of time dependence physically possible. The three different terms in Eq. (3), counting $\sum$ as one term, are the elastic, transient and steady-flow compliances, respectively. When multiplied by stress, they are usually called, respectively, the elastic strain, delayed elastic strain (or primary creep), and viscous creep strain (or secondary creep).

In contrast to Eq. (3), for many viscoelastic materials,

$$D = \frac{1}{E_m} + D_1 t^n + \frac{t}{\eta_m}, \tag{4}$$

where D_1 and n are nonnegative constants; also $0 < n < 1$. The special case $\eta_m = \infty$ gives a creep strain rate which vanishes at long times. Eq. (4) is not inconsistent with Eq. (3). Rather, the term $D_1 t^n$ may be derived from the summation in Eq. (3) by a limiting process in which the summation is approximated by an integral and the coefficients $1/E_i$ are replaced by a power law distribution function of ρ (e.g. Ferry, 1970).

The time-dependence in Sinha's (1978) equation for creep strain of ice corresponds to the creep compliance

$$D = \frac{1}{E_m} + D_c[1 - \exp(-t/t_c)^{n'}] + t/\eta_m, \tag{5}$$

where $n' \simeq 1/3$ and t_c is constant; D_c and η_m may obey power laws in stress, and thus nonlinear behavior is not excluded in Sinha's equation. When $t/t_c \ll 1$ the second term reduces to the power law in Eq. (4) if we use $n' = n$. Moreover, just as for the power law, the second term in Eq. (5) may be represented by the model in Fig. 1 in which there is an infinite number of springs and dashpots and a distribution function is used in place of $1/E_i$.

The inverse of Eq. (1) is

$$\sigma = \int_{0-}^{t} E(t-\tau)\frac{d\varepsilon}{d\tau}d\tau, \tag{6}$$

where $E(t)$ is the relaxation modulus, which is the ratio of stress-to-strain when ε is a constant ε_0 applied at $t = 0$ (relaxation test),

$$E(t) = \sigma(t)/\varepsilon_0. \tag{7}$$

An implicit relationship between $D(t)$ and $E(t)$ may be derived by Laplace transforming Eqs. (1) and (6) and requiring both equations to describe the same stress-strain behavior; thus

$$s\bar{E} = (s\bar{D})^{-1}, \tag{8}$$

where the overbar denotes the Laplace transform of a function $f(t)$,

$$\bar{f} = \int_{0-}^{\infty} f(t)e^{-st}dt, \tag{9}$$

in which s is the transform parameter; the lower limit in Eq. (9) is 0^-, just as in Eqs. (1) and (6), in order to allow for contributions to the integral at $t = 0$ due to a possible Dirac delta singularity in f such as when it represents the stress rate in a creep test.

For the special case of a power law creep compliance

$$D = D_1 t^n, \tag{10}$$

Eq. (8) yields

$$E = \frac{\sin n\pi}{D_1 n\pi} t^{-n} \tag{11}$$

or

$$ED = \sin\ n\pi / n\pi, \tag{12}$$

showing that $ED \neq 1$ unless $n = 0$. Finally, we observe for a constant strain rate $\dot{\varepsilon}$ that Eq. (6) reduces to (after a change of variable, $u = t - \tau$),

$$\sigma = \dot{\varepsilon} \int_0^t E(u) du, \tag{13}$$

where the lower limit of 0, instead of 0^-, can be used in Eq. (6) because the given strain history $\varepsilon(\tau)$ is continuous at $\tau = 0$. For the power law Eq. (11),

$$\sigma = \frac{\sin n\pi\dot{\varepsilon}}{D_1(1-n)n\pi} t^{1-n}. \tag{14}$$

Let us introduce the secant modulus for the constant strain rate test, $E_s \equiv \sigma/\varepsilon$, and then obtain from Eqs. (10) and (14) a simple relationship between creep and constant strain rate response,

$$E_s D = \frac{\sin n\pi}{(1-n)n\pi}. \tag{15}$$

Typically for ice $n = 1/3$, and therefore $E_s D \simeq 1.24$.

2.2 Nonlinear Theory Without Damage

A simple generalization of Eq. (1) to nonlinear behavior consists of replacing σ by a function of stress $F(\sigma)$,

$$\varepsilon = \int_{0^-}^t D(t-\tau) \frac{dF}{d\tau} d\tau, \tag{16}$$

where $F(0) = 0$ and $dF/d\tau = (dF/d\sigma)(d\sigma/d\tau)$. This equation, sometimes called a "modified superposition principle", is introduced here because of its simplicity and, when damage effects are introduced, appears to be general enough to correctly predict viscoelastic behavior of ice. Also, after generalization to three-dimensional loading on specimens or structures, it leads to simple methods for predicting viscoelastic responses from elastic solutions. For other, more general single-integral representations which have been used for ice see Williams (1984).

Two special stress inputs are of interest. For a creep test at stress σ_0,

$$\varepsilon = D(t)F(\sigma_0). \tag{17}$$

For a creep and recovery test, $\sigma(t) = \sigma_0[H(t) - H(t - t_1)]$, where $t_1 > 0$ is the time at which the stress is removed, we may write $F(\sigma(\tau)) = F(\sigma_0)[H(\tau) - H(\tau - t_1)]$, and thus obtain from Eq. (16)

$$\varepsilon = F(\sigma_0)[D(t) - D(t - t_1)] \tag{18}$$

or, equivalently,

$$\varepsilon = \varepsilon_c(t) - \varepsilon_c(t - t_1), \tag{19}$$

where $\varepsilon_c(t)$ is the creep strain, Eq. (17). Figure 2 illustrates Eq. (19) in which $A = \varepsilon_c(\Delta t)$ and $\Delta t = t - t_1$. The simple relationship between the recovery strain ($t > t_1$) and creep strain shown in this figure may be established directly from the principle of superposition for a linear viscoelastic material. It is interesting therefore that the relationship also holds for the nonlinear material defined by Eq. (16).

The model in Fig. 1 is helpful in contrasting the present formulation and terminology with that used for ice by some other investigators. The spring (E_m) and dashpot (η_m) at the bottom of the model provide the elastic and viscous creep strain, and comprise the so-called Maxwell unit or model. The delayed elastic strain in creep (primary creep) is often

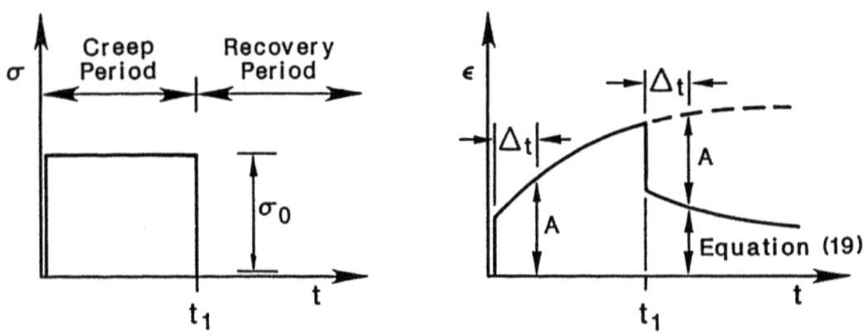

Figure 2. Creep and recovery test.

accounted for through one of the Voigt units consisting of a spring and dashpot in parallel (say E_1 and η_1). For nonlinear materials one or more of the spring moduli and dashpot viscosities vary with internal or external stress or strain or possibly other variables such as strain rate; the terms modulus and viscosity obviously are used rather loosely here. By allowing for nonlinearity through the moduli and viscosities and suitable generalization to a three-dimensional state of stress, we may consider Sjölind's (1987) representation of viscoelasticity to be a two-element Maxwell unit, and we obtain Karr and Choi's (1989) representation by adding one Voigt unit to provide the delayed elastic strain. Sunder and Wu (1989a) have given a description of viscoelastic behavior without damage that may also be viewed as a four-element model comprised of the nonlinear Maxwell and Voigt units, where the spring in the Voigt unit produces the so-called back-stress.

In our approach many springs and dashpots may be needed, but all nonlinearity, apart from damage, is in the one function $F(\sigma)$ (which replaces σ in Fig. 1). Alternatively, by using two springs and two dashpots of arbitrary complexity one can represent very general nonlinear behavior. However, we have chosen to introduce all nonlinearity in the one function $F(\sigma)$ and account for arbitrary time-dependence in creep through the function selected for $D(t)$; one may use this latter function, such as that in Eq. (4), directly in the analysis, or one may decide to use the equivalent spring and dashpot formulation to express the behavior in terms of differential equations. Our choice (in which the nonlinearity is simple and the time-dependence is general) turns out to allow for a simple but rigorous microcrack growth analysis, and seems preferable if supported by experimental results. Experience with nonlinear viscoelastic materials has shown that significantly different forms of constitutive equations are often in agreement with the same limited experimental data; consequently it would not be surprising to find that this is also true for ice.

2.3 Nonlinear Theory With Damage

Consider a continuum whose constitutive equation without damage is a three-dimensional generalization of Eq. (16) (cf. Eqs. (71)). Now, suppose that randomly or regularly distributed microcracks pop-in suddenly and/or grow quasi-statically. For a specimen under an applied axial stress σ, and with this type of distributed damage, the axial strain may be written in the form

$$\varepsilon = E_R \int_{0^-}^{t} D(t-\tau)\frac{df}{d\tau}d\tau \tag{20}$$

where

$$f = f(\sigma, S_k), \tag{21}$$

and the quantities $S_k (k = 1, 2, \ldots)$ are time-varying parameters that account for the effect of microcracking on the overall axial strain ε. (This result and much of the following uniaxial theory are taken from Schapery (1981)). The coefficient E_R is a free constant called the "reference modulus". The material with or without damage is not necessarily isotropic. For an elastic material D is a constant; if we take $E_R = D^{-1}$ then Eq. (20) reduces to $\varepsilon = f(\sigma, S_k)$. Thus, f is the strain that exists in an elastic material with the same instantaneous set of damage parameters S_k as the viscoelastic material.

For uniaxial or more general proportional stressing of a material characterized by a nonlinear function $F(\sigma)$ and crack growth rate behavior which obey power laws in stress, the quantities f and S_k are

$$f = \left|\frac{\sigma}{\sigma_2}\right|^r g(S)\mathrm{sgn}(\sigma), \tag{22}$$

$$S = \int_0^t \left|\frac{\sigma}{\sigma_1}\right|^q f_1 d\tau, \tag{23}$$

where σ_1, σ_2, r, and q are positive constants, $\mathrm{sgn}(\sigma)$ is the sign of σ ($\mathrm{sgn}(\sigma) = -1$ for compression), $|\cdot|$ denotes absolute value, and f_1 accounts for certain complications that may arise in characterizing crack-tip material such as that due to temperature changes; it is often permissible to use $f_1 = 1$, and we shall do so here in applications. For simplicity we have shown only one damage parameter in the presently arbitrary but positive function g. Different S_k would be distinguished by using different values of q or functions f_1; whether or not it is necessary to use two separate damage parameters to account for grain boundary and transgranular cracks in ice remains to be seen. Loading may act on the microcrack faces, such as normal and frictional stresses that arise from an axial compression load; if the friction is Coulombic (i.e., the local shear stress is proportional to the local normal stress acting across the crack faces) then the form of Eqs. (22) and (23) is the same with or without friction.

For most materials, including ice, there is considerable difference in the effect of tension compared to compression loading on the type and amount of damage that develops; as noted previously, tensile behavior of ice is dominated by a single crack. The intrinsic nonlinearity, as reflected by the constants r and σ_2, may also be different. Although it is not exhibited in the notation in Eqs. (22) and (23), if different amounts and types of distributed microcracking were to exist in tension and compression, we could use different functions g and f_1 and different constants in tension and compression. It should also be recognized that when $r \neq 1$ there is no _linear_ viscoelastic range of behavior, even without damage ($g = 1$). Thus, one should view this power law behavior as an approximation which does not necessarily apply at very low stresses. Similarly, $D(t)$ does _not_ have to be the linear viscoelastic creep compliance; rather, it is assumed only that $D(t)$ reflects the creep behavior over some useful stress range.

If the strain history is specified, as in a constant strain rate test, then we would like to be able to invert Eq. (20) to predict stress. To do so first solve Eq. (20) for f by using Eq. (6) and the similarity between Eqs. (1) and (20), and then introduce the power law, Eq. (22). There results

$$\sigma = \frac{\sigma_2 |\varepsilon^0|^{1/r}}{g^{1/r}} \operatorname{sgn}(\varepsilon^0) \tag{24}$$

where

$$\varepsilon^0 = \frac{1}{E_R} \int_{0^-}^{t} E(t-\tau) \frac{d\varepsilon}{d\tau} d\tau, \tag{25}$$

which we call the _pseudo strain_; observe that for an elastic material we may use $E = E_R$ and thus $\varepsilon^0 = \varepsilon$. The right-hand side of Eq. (24) depends on stress through $g = g(S)$, and thus we still need to express S in terms of strain. This may be done by substituting Eq. (24) into the time-derivative of Eq. (23) and then rearranging the result to obtain

$$\int_0^S g^{q/r} dS = S_\varepsilon \tag{26}$$

where

$$S_\varepsilon \equiv \left(\frac{\sigma_2}{\sigma_1}\right)^q \int_0^t |\varepsilon^0|^{q/r} f_1 d\tau. \tag{27}$$

The damage parameter S_ε replaces S, where Eq. (26) is an implicit relationship for finding S in terms of S_ε; $S = S(S_\varepsilon)$ may then be substituted into g in Eq. (24) to obtain

$$\sigma = \sigma_2 g_\varepsilon |\varepsilon^0|^{1/r} \operatorname{sgn}(\varepsilon^0) \tag{28}$$

where

$$g_\varepsilon \equiv g^{-1/r}. \tag{29}$$

Had we started with Eq. (28), in which g_ε had been found from tests under controlled strain history, then the steps could be reversed to predict strain history in terms of stress history from Eq. (20). Instead of Eq. (26) we find that

$$\int_0^{S_\varepsilon} g_\varepsilon^q dS_\varepsilon = S, \tag{30}$$

which may be used to solve for $S_\varepsilon = S_\varepsilon(S)$, and then $g = g_\varepsilon^{-r}(S_\varepsilon)$ would be used in Eq. (22).

In a subsequent discussion on applications of this theory to ice, we shall consider two special functions $g(S)$ which reflect softening due to damage,

$$g = \exp(\lambda S) \tag{31}$$

and

$$g = [1 + (S/S_1)^\beta]^\alpha, \tag{32}$$

where λ, α, β, and S_1 are positive constants. Substitution of Eq. (31) into Eq. (26) and then using Eq. (29) yields

$$g_\varepsilon = (1 + \lambda q S_\varepsilon / r)^{-1/q}. \tag{33}$$

For the case of $\beta = 1$ in Eq. (32) an explicit result for g_ε may also be found,

$$g_\varepsilon = \left(1 + \frac{\gamma}{r}\frac{S_\varepsilon}{S_1}\right)^{-\alpha/\gamma} \tag{34}$$

where $\gamma \equiv \alpha q + r$.

Equations (31) and (33) were used previously in an example problem by Schapery (1981). Harper (1986) then showed that the theory based on these damage functions predicts realistic behavior for ice under constant stress and constant strain rate, as shown in Figs. 3 and 4.

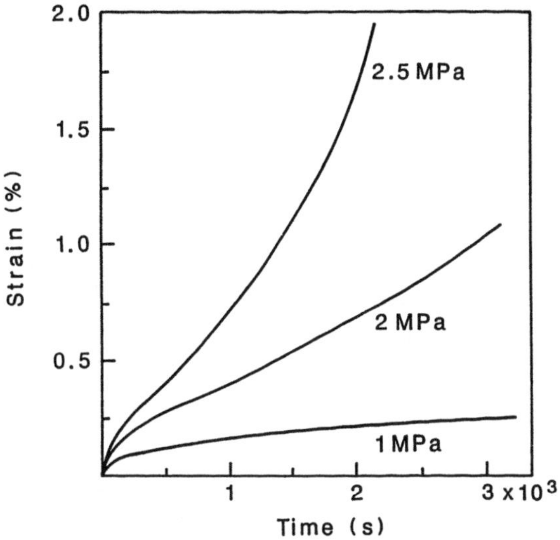

Figure 3. Typical creep curves predicted by Eq.(20). After Harper (1986).

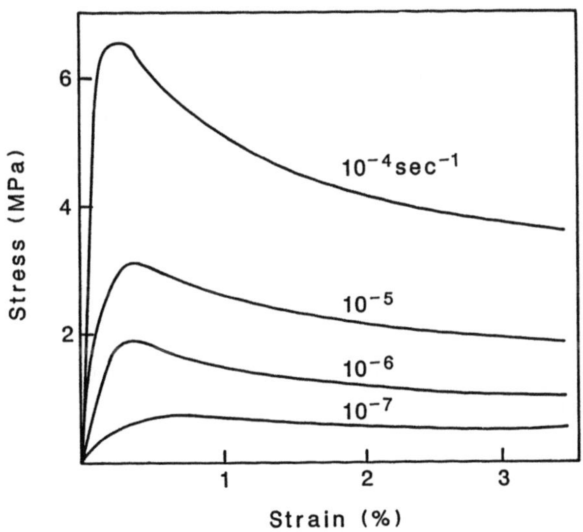

Figure 4. Typical stress-strain curves predicted by Eq.(28) for constant strain-rate loading. After Harper (1986).

Equation (20), with f given in Eq. (22), and its inverse Eq. (28), enable either strain or stress to be predicted in terms of the other variable. In the next section we shall specialize these equations to creep and constant strain rate and then show how the functions g and g_ε may be found from experimental data without using any a prior assumption on the form of their dependence on the damage parameter S or S_ε.

2.4 Compression Creep Test: Determination of $g(S)$

In considering *compression only*, it is helpful to simplify the notation by a sign change in which σ and ε^0 are specified as positive rather than negative quantities; the absolute value and sgn notation may then be omitted. In order to find g from creep test data, it is useful to substitute the data in a rearranged form of Eq. (24),

$$g = \left(\frac{\sigma_2}{\sigma}\right)^r \varepsilon^0. \tag{35}$$

For creep and $f_1 = 1$, Eq. (23) becomes

$$S = \left(\frac{\sigma}{\sigma_1}\right)^q t. \tag{36}$$

Let us suppose that σ_2, r and $D(t)$ have been obtained from creep tests at low enough stresses and/or times to not introduce significant damage, so that $g = 1$. (Observe that Eq. (17) applies in this case, with $F = f$, and so it may be used with creep data to characterize the undamaged material.) Then, in the testing range for which $g \neq 1$, we may use experimental measurements of $\varepsilon(t)$ to evaluate ε^0 from Eq. (25); recall that the modulus E_R is an arbitrarily specified constant, which is introduced to make ε^0 dimensionless, and that $E(t)$ may be found from $D(t)$. The right side of Eq. (35) is thus known, which is also g, and we may plot it as in Fig. 5. If the theory is correct, g is a function of only S, Eq. (36), which may be written as

$$\log S = q \log(\sigma/\sigma_1) + \log t. \tag{37}$$

The constant σ_1 may be selected as any one of the stresses used in the creep tests. The curve of g or $\log g$ against $\log t$ for each of the other stresses differs from the g for $\sigma = \sigma_1$ by a horizontal translation of an amount $-q \log(\sigma/\sigma_1)$. Hence, by measuring these distances one may determine the exponent q; how well the individual curves superpose by horizontal shifting to form a single, so-called *master curve* of g vs. S (which is the same as the curve for $\sigma = \sigma_1$) provides an indication of the accuracy of this characterization.

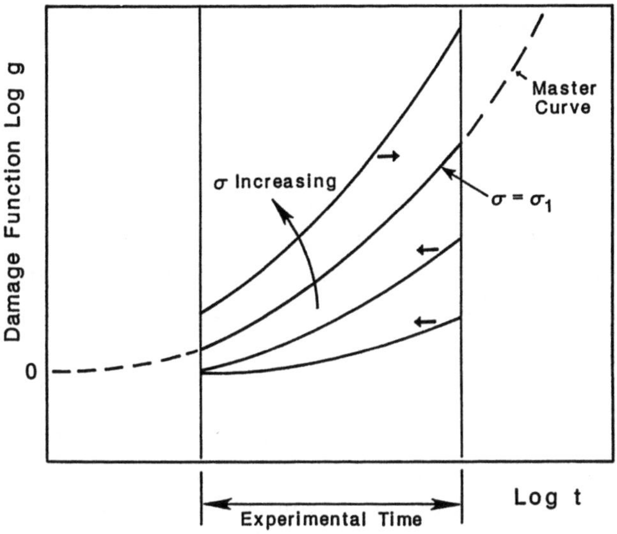

Figure 5. Diagram showing the use of creep data to obtain the damage function.

2.5 Constant Strain Rate Test in Compression: Determination of $g_\varepsilon(S_\varepsilon)$

Constant strain rate test data may be used to find g_ε. From Eq. (28),

$$g_\varepsilon = \left(\frac{\sigma}{\sigma_2}\right)(\varepsilon^0)^{-1/r} \tag{38}$$

and from Eqs. (25) and (27) with $\varepsilon = \dot{\varepsilon}t$ and $f_1 = 1$,

$$\varepsilon^0 = \dot{\varepsilon}h_1 \tag{39}$$

$$S_\varepsilon = \left(\frac{\sigma_2}{\sigma_1}\right)^q \dot{\varepsilon}^{q/r}h_2 \tag{40}$$

where

$$h_1 \equiv \int_0^t Edt/E_R \tag{41}$$

and

$$h_2 \equiv \int_0^t h_1^{q/r}dt. \tag{42}$$

Assuming the material has been characterized in the undamaged state, so that σ_2, r and h_1 are known, we may obtain g_ε as a function of time from Eq. (38), where σ comes from test data. Observe that

$$\log S_\varepsilon = \log\left[\left(\frac{\sigma_2}{\sigma_1}\right)^q \dot{\varepsilon}^{q/r}\right] + \log h_2. \tag{43}$$

Thus, a curve of g_ε or $\log g_\varepsilon$ versus $\log h_2$, for each strain rate, will differ from another by only a horizontal translation, similar to the creep results in Fig. 5 (although g_ε has a negative rather than positive slope). Let $\dot{\varepsilon}_R$ be the strain rate used for the curve to which all of the others are referred, and select σ_1 such that the first term on the right side of Eq. (43) vanishes; i.e.,

$$\sigma_1 = \sigma_2\, \dot{\varepsilon}_R{}^{1/r}. \tag{44}$$

Then this curve (or better, the master curve coming from an average of all the shifted data) is $g_\varepsilon(S_\varepsilon)$. The one difficulty here, compared to creep, is that h_2 depends on the unknown q, and thus iteration or a numerical method method is needed. The best q is the one that leads to a *master curve* with the smallest data spread.

2.6 Simultaneous Determination of Nonlinear Properties and the Damage Function

Recall that if either $g(S)$ or $g_\varepsilon(S_\varepsilon)$ is known, Eq. (29) and either Eq. (26) or (30) may be used to obtain one function from the other one. Also, in our discussion of both creep and constant strain rate tests it was assumed that r, σ_2, and $E(t)$ or $D(t)$ could be found with sufficient accuracy from data on undamaged specimens. This may not be possible if damage develops at very low stresses. In such a case, these quantities would have to be found simultaneously with the determination of g or g_ε.

In the next section special but realistic functional forms for $E(t)$ and g will be used to predict behavior. These forms may be useful as at least first approximations when data from the damaged state has to be used to find all material constants and functions. However, results from creep tests may be used to determine g and the nonlinear properties r and σ_2 simultaneously, without iterating. First, $E(t)$ may be obtained from $D(t)$ by using Eq. (11) in the case of a power law or by exact Eq. (8), or by approximate methods. Then, after selecting a value for E_R, ε^0 may be found from its definition, Eq. (25), and measured values of strain.

In order to see how ε^0 should be used in the data analysis, we observe from Eq. (35) that

$$\log \varepsilon^0 = r \log(\sigma/\sigma_2) + \log g. \tag{45}$$

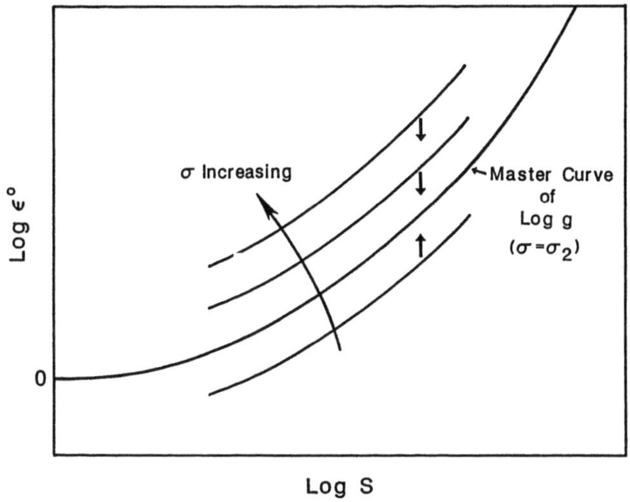

Figure 6. Diagram showing the use of creep data to find the effect of stress level at constant damage.

If S were known at each stress and time, we could plot experimental values of $\log \varepsilon^0$ against S and obtain curves as illustrated in Fig. 6. Namely, according to Eq. (45), the curves for constant stress are shifted *vertically* with respect to the master curve by an amount $r \log(\sigma/\sigma_2)$. By measuring the amount of vertical shift needed for each stress to form a master curve passing through the point ($g = 1$, $S = 0$) we deduce r and σ_2; when $\sigma = \sigma_2$ the $\log \varepsilon^0$ curve is *log g*. Although S is not really known initially (since q and σ_1 are not known) we may combine the horizontal and vertical shifting of Figs. 5 and 6 to find simultaneously all constants and $g(S)$. Specifically, plot $\log \varepsilon^0$ versus $\log t$ for each stress level. Then, the individual curves will differ from the (unknown) master curve by a combination of a vertical and horizontal translation. If the theory is valid, one may trace over each curve on a separate piece of paper and construct the master curve; any one stress may be selected as σ_1, and the corresponding (reference) curve is the one to which all curves are shifted. We may then plot the vertical and horizontal translations against $\log \sigma$ to obtain r and q as the slopes of these lines; if one or both of these lines is not straight we may still characterize the material using appropriate curve fits as long as a master curve can be formed. The value of σ_2 is found from the amount of vertical shift (cf. Eq. (45)) needed to made this reference curve asymptotic to $\log \varepsilon^0 = 0$ as $\log S \to -\infty$. This resulting curve is the function $g(S)$; the last vertical shift makes $g = 1$ at $S = 0$.

It has been assumed that $E(t)$ is known in order to construct $\log \varepsilon^0$ curves. If one cannot measure creep at low enough stress levels for the material to be in its linear viscoelastic range (or if $E(t)$ is different from the linear viscoelastic relaxation modulus) then some numerical work or iteration appears to be needed. If a power law $E(t) = E_1 t^{-n}$ can be used, then we may select $E_R = E_1$ (it is not necessary that E_R have dimensions of modulus), so that the only unknown in ε^0 is the value of n; one would then select the one n that is found to produce the master curve with the least data spread.

2.7 Prediction of Compression Creep

In order to demonstrate the use and the validity of the theory, Eqs. (20), (22) and (23) will be used to predict creep strain and show its agreement with experimental data on ice. As the creep compliance we shall use that in Eq. (4), but without the steady flow term; however, if desired one may derive the steady flow strain from the results for the transient component by letting $D_1 \rightarrow \eta_m^{-1}$ and $n \rightarrow 1$ in the latter prediction.

The strain may be written in the form

$$\varepsilon = \varepsilon_e + \varepsilon_t \tag{46}$$

with elastic strain,

$$\varepsilon_e = E_R E_m^{-1} \left(\frac{\sigma}{\sigma_2}\right)^r g \tag{47}$$

and transient strain,

$$\varepsilon_t = E_R D_1 \int_{0^-}^{t} (t-\tau)^n \frac{df}{d\tau} d\tau. \tag{48}$$

The damage function $g = g(S)$ is considered to be known. Equations (46) - (48) are not limited to constant stress (applied at $t = 0$), but here we shall use this special input. In evaluating the integral in Eq. (48) one must account for the fact that f is discontinuous at the time the constant stress is applied. This may be done by first writing Eq. (22) with the unit step function $H(t)$,

$$f = H(t) \left(\frac{\sigma}{\sigma_2}\right)^r g, \tag{49}$$

where σ is now totally independent of time. Then,

$$df/dt = \delta(t)\left(\frac{\sigma}{\sigma_2}\right)^r g + H(t)\left(\frac{\sigma}{\sigma_2}\right)^r \frac{dg}{dt}, \tag{50}$$

in which $\delta(t)$ is the Dirac delta function. For creep the damage parameter is given by Eq. (36). Substitute Eq. (50) into Eq. (48) after replacing t by τ,

$$\varepsilon_t = E_R D_1 \left(\frac{\sigma}{\sigma_2}\right)^r \left[t^n + \int_0^t (t-\tau)^n \frac{dg}{d\tau} d\tau\right], \tag{51}$$

where we have used the fact that $S = 0$ initially and therefore $g = 1$ at $\tau = 0$; as the discontinuity in f at $\tau = 0$ has already been accounted for we may use 0, instead of 0^-, as the lower limit.

It will be helpful to rewrite Eq. (51) by making a change of variables in which t and τ are replaced by the damage parameter using Eq. (36); i.e.

$$t = \left(\frac{\sigma_1}{\sigma}\right)^q S, \qquad \tau = \left(\frac{\sigma_1}{\sigma}\right)^q S'. \tag{52}$$

Equation (51) becomes

$$\varepsilon_t = E_R D_1 \left(\frac{\sigma}{\sigma_2}\right)^r \left(\frac{\sigma_1}{\sigma}\right)^{qn} G, \tag{53}$$

where $G = G(S)$ is defined by

$$G \equiv \int_0^S (S - S')^n \frac{dg}{dS'} dS' + S^n. \tag{54}$$

Also of interest are the elastic and transient strain rates. We find from Eqs.(36), (47), and (53),

$$\dot{\varepsilon}_e = E_R E_m^{-1} \left(\frac{\sigma}{\sigma_2}\right)^r \left(\frac{\sigma}{\sigma_1}\right)^q \frac{dg}{dS} \tag{55}$$

and

$$\dot{\varepsilon}_t = E_R D_1 \left(\frac{\sigma}{\sigma_2}\right)^r \left(\frac{\sigma}{\sigma_1}\right)^{q(1-n)} \frac{dG}{dS}, \tag{56}$$

where the overdot denotes the time derivative. If the steady flow term t/η_m in Eq. (4) had been included, it would produce an additional strain rate

$$\dot{\varepsilon}_S = E_R \eta_m^{-1} \left(\frac{\sigma}{\sigma_2}\right)^r g, \tag{57}$$

which comes from Eq. (56) by using the substitutions $n \to 1$ and $D_1 \to \eta_m^{-1}$. As expected, with damage growth both the elastic and steady-flow strain rates vary with time. Moreover, the variation with time and stress is different for each of the three components.

Ashby and Duval (1985) in a study of ice showed that the effect of stress and time on total strain rate, until the minimum strain rate $\dot{\varepsilon}_{min}$ is reached, may be expressed in terms of a single function (or master curve) using normalized variables; the experimental data and different theories for a range of stresses are shown in Fig. 7. The present theory is consistent with these data if the elastic and steady-flow strain rates are negligible compared to $\dot{\varepsilon}_t$. From Eq. (56) for this case,

$$\frac{\dot{\varepsilon}}{\dot{\varepsilon}_{\min}} = \frac{dG/dS}{(dG/dS)_{\min}}. \tag{58}$$

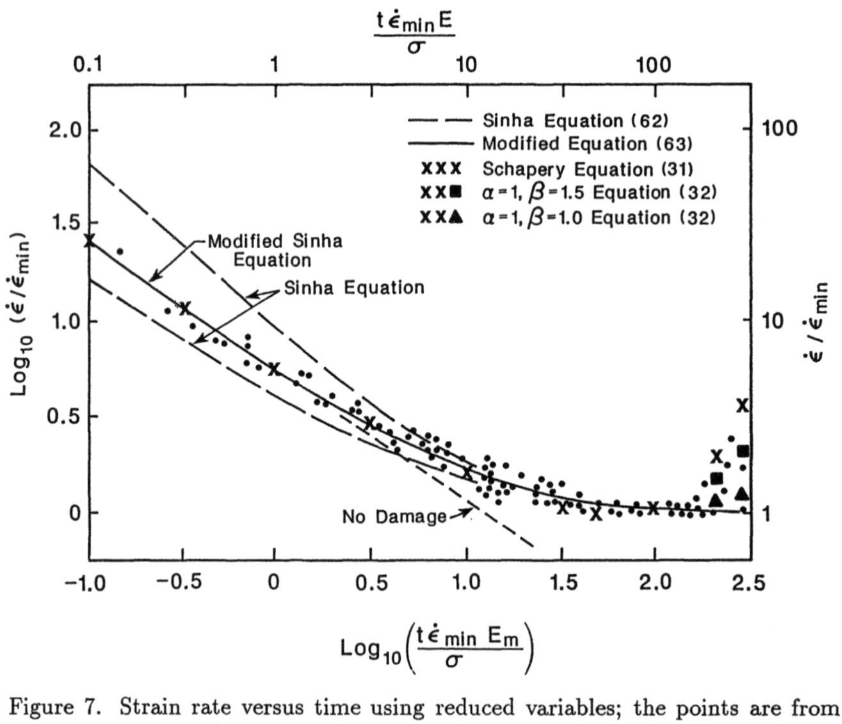

Figure 7. Strain rate versus time using reduced variables; the points are from the ice creep data of Jacka (1984). After Ashby and Duval (1985).

This ratio is a function of only S, and therefore we need to be able to express S in terms of only the abscissa in Fig. 7. Denote the abscissa variable by $\hat{t}$,

$$\hat{t} \equiv t\dot{\varepsilon}_{\min} E_m/\sigma. \tag{59}$$

Then from Eqs. (52 and 56),

$$\hat{t} \sim \sigma^{(r-qn-1)} S. \tag{60}$$

Thus, if

$$q = (r-1)/n, \tag{61}$$

the abscissa, $\log \hat{t}$, is independent of stress and is proportional to $\log S$, apart from an additive constant. The additive constant may be adjusted to match the minimum in Eq. (58) with that of the experimental data. However, further study of experimental data is needed to determine if $\hat{t}$ is really independent of stress.

The prediction (X) in Fig. 7 uses the exponential damage function Eq. (31) and $n = 1/3$ in Eq. (54); numerical integration was employed. Analytical integration may be used along with Gamma functions to evaluate G based on the damage function in Eq. (32) if $\alpha = 1$; results for the two cases $\beta = 1$ and $\beta = 1.5$ are approximately the same as for Eq. (31) (if the location of the minimum strain rate is suitably adjusted) except at times beyond the minimum strain rate. Thus, experimental data from the tertiary creep region apparently is most important in establishing the appropriate form of damage function. However, the comparison in Fig. 7 of the theories with and without damage shows that damage growth *does* affect creep behavior prior to reaching the minimum rate; indeed, without it a minimum rate would not exist.

Also drawn in Fig. 7 are two curves based on Sinha's equation for creep of ice without damage, which Ashby and Duval wrote in the dimensionless form

$$\hat{\varepsilon} \equiv \frac{\varepsilon}{\sigma/E_m} = 1 + A\left\{1 - \exp\left[-\frac{E_m}{\sigma}\left(\frac{c'\sigma}{E_m}\hat{t}\right)^n\right]\right\} + \hat{t}, \tag{62}$$

where A and c' are dimensionless constants; the upper and lower curves reflect the stress range of the experimental data. They show that by modifying this equation so that the stress-dependence disappears,

$$\hat{\varepsilon} = 1 + A\{1 - \exp(-c''\hat{t}^n)\} + \hat{t}, \tag{63}$$

in which c'' is a constant, then there is a good agreement with the strain-rate data in Fig. 7. If $\dot{\varepsilon}_{\min}$ is proportional to stress, $\hat{t}$ will be independent of stress and Eq. (63) characterizes <u>*linear*</u> viscoelastic behavior. When $\dot{\varepsilon}_{\min}$ is nonlinear in stress, Eq. (63) is like that for a linear material except the time-scale is a function of stress; this type of behavior has been reported for some solid polymers (Schapery, 1971). In contrast to Eq. (63), the damage model based on Eq. (31) gives comparable agreement with experiment and accounts for behavior in the tertiary creep region.

2.8 Minimum Creep Rate and Strength Relationships

Harper (1989) used the exponential damage function, Eq. (31), and the power law creep compliance $D(t) = D_1 t^n$ in deriving the following relationships for creep,

$$\dot{\varepsilon}_{\min} \sim \sigma^a, \tag{64}$$

$$t_{\min} \sim \sigma^{-q}, \tag{65}$$

$$\varepsilon_{\min} \sim \sigma^{r-nq}, \tag{66}$$

where

$$a \equiv r + q(1-n). \tag{67}$$

Also, $\varepsilon_{\min}$ and $t_{\min}$ are the strain and time when the strain rate is $\dot{\varepsilon}_{\min}$. Relationships for the maximum stress $\sigma_{\max}$ in a constant strain rate test are similar,

$$\sigma_{\max} \sim \dot{\varepsilon}^{1/a}, \tag{68}$$

$$t_{\max} \sim \sigma_{\max}^{-q}, \tag{69}$$

$$\varepsilon_{\max} \sim \sigma_{\max}^{r-nq}, \tag{70}$$

where $\varepsilon_{\max}$ and $t_{\max}$ are the strain and time when $\sigma_{\max}$ is reached. The power law behavior in Eqs. (64) - (70) and relationships among some of the coefficients in the power laws for the two types of tests were observed by Harper (1989) to be consistent with Mellor and Cole's (1982, 1983) results for isotropic ice in uniaxial compression.

One may show that Eqs. (64) - (70) do not depend on the use of the exponential form of the damage function. Rather, any function $g(S)$ for which a minimum creep strain rate exists leads to these power laws with the same exponents.

For ice, $a \simeq 3$ in Eq. (64) for minimum creep rate, which together with Eq. (61) yields $q = 2$ and $r = 2n + 1$. However, it should be recalled that Eq. (61) is based on the assumption that $\hat{t}$ is independent of stress. Rather than plotting data as shown in Fig. 7, the present damage theory implies one should determine if a master curve can be found as shown in Fig. 5; if so, the resulting value of q and the commonly reported value of $a = 3$ may be used with Eq. (67) to determine if the stress exponent in Eq. (60) vanishes.

3. Multiaxial Constitutive Equations

The uniaxial constitutive theory with damage growth, Eqs. (20) - (23), was stated without proof, although evidence of its validity for ice was given. Here we shall fill in some of the mathematical and physical background to these and more general equations for multiaxial loading. See Schapery (1981, 1984, 1987, 1990) for additional details.

3.1 Nonlinear Theory Without Damage

A three-dimensional generalization of the nonlinear viscoelastic constitutive Eq. (16) may be written in the form

$$\varepsilon_{ij} = E_R \int_{0^-}^{t} D(t-\tau) \frac{\partial \varepsilon_{ij}^e}{\partial \tau} d\tau, \tag{71}$$

where E_R and $D(t)$ have the same significance as previously introduced, ε_{ij} are the components of the strain tensor in a rectangular Cartesian coordinate system $x_i (i = 1, 2, 3)$, and ε_{ij}^e are dimensionless (if E_R has the dimensions of modulus) tensor functions of the stresses, $\varepsilon_{ij}^e = \varepsilon_{ij}^e(\sigma_{k\ell})$. The functions $E_R \varepsilon_{ij}^e$, which replace $F(\sigma)$ in Eq. (16), must be such that Eq. (71a) reduces to a thermodynamically valid form for the elastic response at short times. This condition implies ε_{ij}^e must be expressible as the derivative of a potential function, say $\widehat{W}_c = \widehat{W}_c(\sigma_{ij})$; i.e.

$$\varepsilon_{ij}^e = \partial \widehat{W}_c / \partial \sigma_{ij}. \tag{71b}$$

If we select E_R as the initial modulus $1/D(0)$, then $\widehat{W}_c$ is the so-called complementary strain energy density (energy/initial volume) for the elastic response. The behavior characterized by Eqs. (71) may be isotropic or anisotropic, depending on the form of the function $\widehat{W}_c$. However, all time-dependence is expressed through one function, $D(t)$; this function has no effect on $\widehat{W}_c$, given the stresses. It is assumed Eqs. (71) characterize the behavior of polycrystalline ice _without_ damage.

The power-law stress dependence in the uniaxial Eq. (22) may be generalized to the multiaxial stress formulation by using a function $\widehat{W}_c$ which has the mathematical form of a so-called homogeneous function of degree $1 + r$. Specifically, we assume

$$\widehat{W}_c(c\sigma_{ij}) = |c|^{r+1}\widehat{W}_c(\sigma_{ij}) \tag{72}$$

where c is any stress factor.

3.2 Nonlinear Theory With Damage

The form of Eqs. (71) is such that it leads to a relatively simple method for predicting _viscoelastic_ structural response from an _elastic_ analysis based on the function $\widehat{W}_c$. The structure may be simply a statistically homogeneous macroscopic material element with growing microcracks, which is what will be considered here; the overall or average stresses and strains will be denoted by σ_{ij} and ε_{ij}, respectively. The micro-stresses within the material element are of course nonuniform because of the microcracks (and polycrystalline structure of ice). One important consequence of Eqs. (71) is that the stress distribution throughout the viscoelastic material element is the _same_ as for an elastic element under the same applied σ_{ij} with or without crack growth. The response analysis is made even simpler when Eq. (72) is valid and all of the applied stresses vary with time in the same way (which is called proportional stressing). After introducing a crack growth equation in the next section, it will be shown that the theory leads to Eq. (22) and (23) for proportional stressing.

3.3 The Crack Growth Equation

Consider any one of many cracks or crack-like defects in a material element for which the area growth in the time dt along any infinitesimal segment of its edge is given by dA and the movement normal to the edge is da. Then, if the constitutive equation for the surrounding intact material is that in Eqs. (71), and the crack-tip failure process zone is small compared to crack length, the equation for quasi-static growth is often expressible as a power law in the complementary strain-energy release rate, $\partial W_c/\partial A$,

$$\dot{a} = f_c(\partial W_c/\partial A)^k, \tag{73}$$

where k is a positive constant and f_c reflects properties of the crack tip material; f_c may vary with temperature and with the ratio of shearing to normal stresses at the tip, i.e. the so-called mode ratio. It should be added that W_c denotes the _total_ complementary strain energy for an elastic material element with cracks; without cracks, W_c/V is the complementary strain energy density $\widehat{W}_c$ appearing in Eq. (71b), where V is the initial volume of the macroscopic material element. It may be shown that $\partial W_c/\partial A = -\partial W/\partial A$, where W is the strain energy of the cracked material expressed in terms of pseudo strains ε_{ij}^0 and crack geometry, and is related to W_c through $W = V\sum \sigma_{ij}\varepsilon_{ij}^0 - W_c$. Also,

$$\varepsilon_{ij}^0 \equiv \frac{1}{E_R}\int_{0^-}^{t} E(t-\tau)\frac{\partial \varepsilon_{ij}}{\partial \tau}d\tau, \tag{74}$$

which is the three-dimensional version of the pseudo strain in Eq. (25). Strictly speaking, the material is not elastic, but when the ε_{ij}^0 are used instead of ε_{ij} the constitutive equations are identical to those for an elastic body; on this basis we call W the strain energy of an elastic material.

Equation (73) predicts crack growth for even very small values of $\partial W_c/\partial A$. In reality, for most materials $\dot{a} = 0$ when $\partial W_c/\partial A$ is less than some material-dependent value. If $k \gg 1$ this modification to Eq. (73) does not have a significant effect on the constitutive equation for the damaged material, and it will not be used here. At the expense of additional complexity in the theory, a low-speed cut-off could be introduced.

3.4 Proportional Stressing of Power Law Materials with Damage Growth

Proportional stressing is defined as

$$\sigma_{ij} = \sigma\sigma'_{ij}, \tag{75}$$

where $\sigma = \sigma(t)$ and the σ'_{ij} are the components of a constant symmetric tensor. The complementary strain energy density W'_c for a microcracked material whose uncracked state is characterized by a power-law $\widehat{W}_c$, Eq. (72), can be shown to obey a similar equation,

$$W'_c(\sigma\sigma'_{ij}) = |\sigma|^{r+1}W'_c(\sigma'_{ij}). \tag{76}$$

Although not explicitly shown, W_c' depends on the geometry of the time-varying microcracks; note that $W_c' \equiv W_c/V$. Eq. (76) is valid even if tractions act on crack faces if these tractions are proportional to σ; such will be the case for closed cracks with crack-face sliding if the frictional traction is zero or proportional to the contact pressure. Let us multiply Eq. (76) by V to obtain W_c, and then substitute it into the crack growth law Eq. (73) to find

$$\dot{a} = cf_1 \left|\frac{\sigma}{\sigma_1}\right|^q (\partial W_{c1}/\partial A)^k, \tag{77}$$

where $q = k(r+1)$, $W_{c1} \equiv W_c$ for $\sigma = 1$, $f_1 \equiv f_c\sigma_1^q/c$, and σ_1 and c are arbitrary constants with dimensions which make σ/σ_1 and f_1 dimensionless.

For an isolated penny-shaped crack of radius a, $\partial W_{c1}/\partial A = c_p a$ where c_p is a constant. For this case Eq. (77) may be integrated; thus,

$$a = \left[a_0^{1-k} - (k-1)cc_p^k \int_0^t \left|\frac{\sigma}{\sigma_1}\right|^q f_1 dt\right]^{1/(1-k)}, \tag{78}$$

where a_0 is the initial crack radius. If $k > 1$, then $a \to \infty$ when the quantity in the brackets vanishes. We may think of the time at which this happens as the "failure time", denoted by t_f, and therefore

$$a_0^{1-k}/(k-1)cc_p^k = \int_0^{t_f} \left|\frac{\sigma}{\sigma_1}\right|^q f_1 dt, \tag{79}$$

which is an implicit equation for finding t_f. Figure 8 illustrates the microcrack growth, in which the "sudden growth" occurs essentially at the time t_f.

For ice in compression, the microcracks are arrested at a size comparable to the grain facet length. Equation (78) does not reflect the arrest, which is possibly produced by a rapid drop-off in the right-hand side of Eq. (77) when the crack tip reaches a grain that lies across the crack plane. There may be a period in which the crack speed is so high it is limited by inertial effects. We assume the amount of time spent in the high-speed propagation phase is negligible compared to the time t_f of Eq. (79) even though most of the growth may occur at high speed. It should also be added that during the high-speed growth phase and in the arrested state no assumptions are needed concerning the crack's geometry and interactions with other cracks and other microstructural details, as these interactions will not appreciably affect t_f.

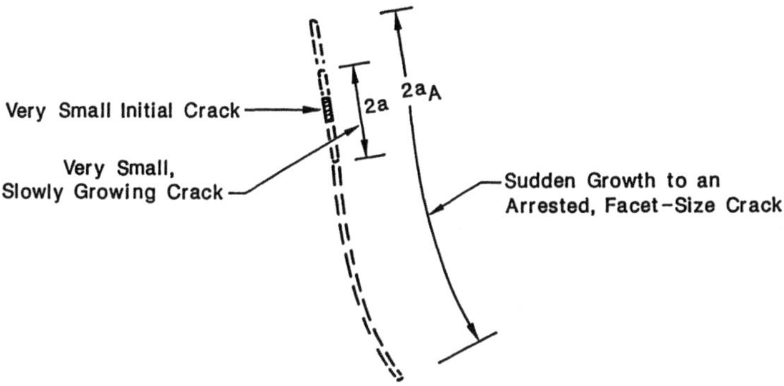

Figure 8. Growth and arrest of an idealized microcrack.

The left side of Eq. (79) may vary from crack-to-crack, producing different values of t_f. Let us denote the value of the left side for the mth crack by S_m and the resulting failure time by t_m. Then, assuming q and f_1 are the same for all cracks,

$$S_m = \int_0^{t_m} \left|\frac{\sigma}{\sigma_1}\right|^q f_1 dt. \tag{80}$$

In order to account for the effect of the microcracks on overall response of a macroscopic material element, let us first assume each crack has no effect until it has reached its arrested size. Namely, during the time $0 \leq t < t_m$, the mth crack is too small to soften the material; but softening does occur at t_m when the crack "pops in". The final size, shape and orientation of each crack is assumed to be independent of the loading history (as defined by $\sigma(t)$), and thus may be identified in terms the index m or, equivalently, in terms of S_m. Thus, to account for the effect of microcracking on the macroscopic behavior, we may write $W_c' = W_c'(\sigma_{ij}, S_m)$. As S_m increases so does W_c', which is illustrated in Fig. 9. It is reasonable to assume there are a great many S_m which are spaced closely, and therefore we may smooth out the curve in Fig. 9 and consider W_c' to be a continuous function of the variable S. After dropping the subscript m in Eq. (80), we arrive at Eq. (23) for S.

There may be some cracks for which slow crack growth affects W_c', whether or not the bracketed quantity in Eq. (78) eventually vanishes. Even in this case we may use the one parameter S to reflect all damage because the instantaneous crack size, Eq. (78), is expressed in terms of S, Eq. (23).

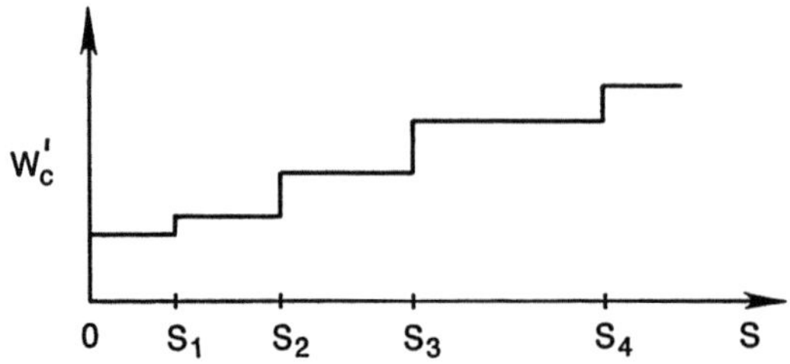

Figure 9. Effect of the sudden growth and arrest of several microcracks on the complementary strain energy density.

Having established that the function $W_c' = W_c'(\sigma_{ij}, S)$ accounts for microcracking, we now want to predict the overall strains ε_{ij}. The correspondence principle established by Schapery (1981) yields the simple result,

$$\varepsilon_{ij} = E_R \int_{0-}^{t} D(t-\tau) \frac{\partial}{\partial \tau} \left(\frac{\partial W_c'}{\partial \sigma_{ij}} \right) d\tau, \tag{81}$$

which is identical to Eqs. (71), except W_c' appears in place of $\widehat{W}_c$. From Eqs. (75) and (76),

$$\frac{\partial W_c'}{\partial \sigma_{ij}} = |\sigma|^{r+1} \frac{\partial W_c'}{\partial \sigma_{ij}'} \frac{1}{\sigma} = |\sigma|^r \frac{\partial W_c'}{\partial \sigma_{ij}'} \mathrm{sgn}(\sigma), \tag{82}$$

where the tensor $\partial W_c'/\partial \sigma_{ij}' \equiv \partial W_c'(\sigma_{k\ell}', S)/\partial \sigma_{ij}'$ is independent of σ. Hence, the equation for each strain component is identical in form to Eq. (20) if Eq. (22) is used for f.

Equation (81) does not in general predict proportional straining even though proportional stressing, Eq. (75), has been used. If, however, $\partial W_c'/\partial \sigma_{ij}' = \varepsilon_{ij}' \, g(S)$, where the ε_{ij}' are constants, then there is proportional straining since Eqs. (81) and (82) yield

$$\varepsilon_{ij} = \hat{\varepsilon} \varepsilon_{ij}', \tag{83}$$

where

$$\hat{\varepsilon} = E_R \int_{0-}^{t} D(t-\tau) \frac{\partial}{\partial \tau} [|\sigma|^r g \mathrm{sgn}(\sigma)] d\tau. \tag{84}$$

It should be added that Eq. (81) is <u>*not*</u> limited to the power law form of $\widehat{W}_c$ and W_c', Eqs. (72) and (76). However, this form is needed to arrive at a damage parameter that is a power law in the applied stress, Eq. (23), where σ is the factor in Eq. (75).

As another generalization, recall that the form for S came from a crack growth equation which is a power law in energy release rate. Suppose we had <u>*not*</u> introduced crack growth as the mechanism for at least the early stage of damage, but instead had simply used a phenomenological law for the mth source of damage,

$$\dot{a}_m = f_1 \left| \frac{\sigma}{\sigma_1} \right|^q F_m, \tag{85}$$

where a_m represents any parameter that accounts for material failure, such as the number of broken bonds, size of a cavity, etc., and $F_m = F_m(a_m)$, then,

$$\int_{a_{mo}}^{a_m} \frac{da_m}{F_m} = \int_0^t \left| \frac{\sigma}{\sigma_1} \right|^q f_1 dt, \tag{86}$$

showing that, as before, $a_m = a_m(S)$ and thus S, Eq. (23), still serves as the measure of damage. The way in which F_m varies with a_m determines whether or not pop-in or smooth growth behavior occurs. If, instead of the power law in stress in Eq. (85), a more general function of stress $F_\sigma(\sigma)$ were used, then a single damage parameter could still be used in which $|\sigma/\sigma_1|^q$ is replaced by F_σ (assuming, of course, F_σ is the same for all cracks or sources of damage). However, the power law behavior in Eqs. (64) - (70) does depend on the use of a power law in stress in S; since these equations are consistent with observations for ice, it is believed that the stress power law in Eq. (23) is sufficiently general for ice.

3.5 Nonproportional Loading

In the remainder of Section 3 we shall discuss an approach to developing constitutive equations for a damaged material under nonproportional loading (stressing or straining), while allowing for frictional contact between crack faces. The essential ideas will be brought out without accounting for intrinsic viscoelasticity. Then, as long as Eqs. (71) are applicable to all of the material comprising a microcracked body, except possibly for that at the crack tips, the extension of the equations to intrinsic viscoelasticity is easily made. On the basis of the elastic-viscoelastic correspondence principle (Schapery, 1981), all one has to do is replace ε_{ij}

by ε_{ij}^0, where ε_{ij}^0 is defined in Eq. (74). A similar substitution applies to the generalized displacements q_i discussed below.

Because of the desire to account for crack-face friction, we shall begin with a notation using generalized forces Q_i and generalized displacements q_i, where $i = 1, 2, \ldots, N$. Some of the q_i may represent the average strains or boundary displacements of a material element with distributed cracks, while others may be needed to describe crack-face displacements when there are work-related tractions on crack faces such as friction. The generalized forces are defined in the usual manner wherein $Q_i \delta q_i$, for each i, is the virtual work done on the body when it undergoes a virtual displacement δq_i. Some of these generalized forces may be average stresses or forces acting on the material's outer boundary, while others may be tractions or forces acting on crack faces. For an elastic body with strain energy W,

$$Q_i = \partial W / \partial q_i, \tag{87}$$

with the inverse expressed in terms of complementary energy

$$q_i = \partial W_c / \partial Q_i, \tag{88}$$

where $W_c = Q_i q_i - W$; *from now on in this section repeated indices imply summation over their range, unless stated otherwise.* In Eqs. (87) and (88) $W = W(q_i, S_p)$ and $W_c = W_c(Q_i, S_p)$ respectively; these energies depend on a set of parameters $S_p (p = 1, 2, \ldots, P)$ that serve to define the state of damage; there may be as many as needed to give the detailed geometry of each crack or there may be a small number of parameters which characterize the statistics of the crack distribution or other quantities needed to account for changes in the microstructure. If Eqs. (87) and (88) relate stresses and strains (rather than forces and displacements) the W and W_c are energies for a unit initial volume, denoted previously by W' and W_c'.

3.6 Effect of Friction

The development of an explicit constitutive theory is complicated by crack-face friction. In order to illustrate this and motivate an approach that simplifies the problem, consider the geometry in Fig. 10. A slab of material in plane strain with one crack of length "a" is loaded in compression and shear between non-rotating, rigid plates. Assuming the

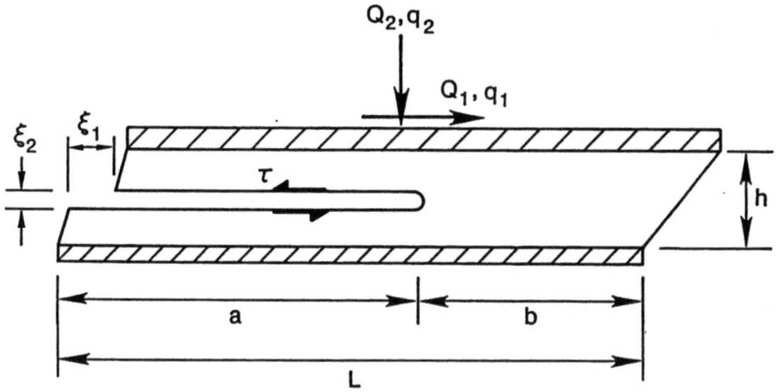

Figure 10. Cracked layer subjected to compression and shear.

crack faces are in contact ($\xi_2 = 0$), a normal compressive traction σ_n (not shown) and a shear traction τ act on the crack faces. Assuming $h \ll a$ and $h \ll b$, the total strain energy in a linear elastic, isotropic material which has unit thickness (normal to the page) is found to be

$$W = \frac{hG}{2}\left[(L-a)\left(\frac{q_1}{h}\right)^2 + a\left(\frac{q_1+q_3}{h}\right)^2\right] + \frac{LhE_1}{2}\left(\frac{q_2}{h}\right)^2, \tag{89}$$

in which $q_3 \equiv -\xi_1$ and

$$E_1 = \frac{(1-\nu)E}{(1+\nu)(1-2\nu)}, \tag{90}$$

where E is Young's modulus, G is the shear modulus, and ν is Poisson's ratio. From Eq. (87),

$$\begin{aligned} Q_1 &= \frac{G}{h}[Lq_1 + aq_3], \\ Q_2 &= \frac{LE_1}{h}q_2\,, \quad Q_3 = \frac{Ga}{h}(q_1+q_3). \end{aligned} \tag{91}$$

It should be noted that $Q_3 = \tau a$ because $Q_3\delta q_3 = -Q_3\delta\xi_1$ is the virtual work due to τ.

Let us assume $\dot{\xi}_1 > 0$ and introduce the Coulomb law for friction $\tau = \mu\sigma_n$ where μ is the friction coefficient for sliding; then $Q_3 = \mu a\sigma_n = \mu a E_1 q_2/h$. Use this result in the last of Eq. (91) to solve for q_3 and then substitute q_3 into the first of Eq. (91). The result may be written in terms of nominal stresses σ_{12} and σ_{22} and nominal strains γ_{12} and ε_{22},

$$\begin{aligned}\sigma_{12} &= C_{11}\gamma_{12} + C_{12}\varepsilon_{22}, \\ \sigma_{22} &= C_{21}\gamma_{12} + C_{22}\varepsilon_{22},\end{aligned} \tag{92}$$

where

$$\begin{aligned}\sigma_{12} &\equiv Q_1/L, & \sigma_{22} &\equiv Q_{22}/L, \\ \gamma_{12} &\equiv q_1/h, & \varepsilon_{22} &\equiv q_2/h,\end{aligned} \tag{93}$$

and

$$\begin{aligned}C_{11} &= Gb/L, & C_{12} &= \mu a E_1/L, \\ C_{21} &= 0, & C_{22} &= E_1.\end{aligned} \tag{94}$$

It is important to recognize that the matrix C_{ij} is not symmetric unless $\mu = 0$ or q_3 is constant; if $\mu \neq 0$ and σ_n is sufficently large, then q_3 will be constant and we must return to the first of Eq. (91) to obtain the shear stress-strain equation.

The loss of symmetry implies the overall force-displacement or stress-strain equations cannot be derived directly from a strain energy function. However, if the slip displacement q_3 is used as an independent variable, as in Eq. (89), a strain energy function does exist in the sense that Eq. (87) is applicable.

Analogous results, but expressed in terms of increments in overall stress and strain, are obtained for a general nonlinear elastic body with one or more cracks. If these are widely spaced, flat cracks in a homogeneous material, only one slip displacement vector (which gives a measure of the interface sliding displacement)is needed in the strain energy for each crack if the crack surface shear traction is uniformly distributed (or its distribution is specified), just as in the example in Fig. 10. A complementary energy formulation would lead to similar results. It should be added that Dougill (1975) used a sliding block problem as a model of a frictional material to show how friction leads to unsymmetric coefficients in incremental force-displacement equations.

Suppose that instead of the friction law $\tau = \mu\sigma_n$, the tractions τ and σ_n develop from a thin layer of damaged material which, for example, is in the gap of width "a" and height ξ_2 in Fig. 10. This situation may be realistic for ice as viewed by Jordaan and McKenna (1989). Let us assume the crack layer material for the example in Fig. 10 can be characterized by a strain energy function $W_\ell = W_\ell(\xi_1, \xi_2, S_q)$, in that

$$\sigma a = -\partial W_\ell / \partial \xi_2, \quad \tau a = -\partial W_\ell / \partial \xi_1, \tag{95}$$

where the set of S_q represents all parameters needed to account for damage in the thin layer. Thus, the cracked body is, in effect, replaced by a nonhomogeneous elastic *continuum* with strain energy $W + W_\ell$; this means there is no need to explicitly include ξ_1 and ξ_2 as independent variables in Eqs. (87) and (88). This conclusion obviously carries over to bodies with any number of cracks as long as the cracks are replaced by continuous material characterized by a strain energy function. The crack layer material may have a large effect on overall behavior. For example, nonlinearity of the layers could produce significant nonlinear behavior of a body which is otherwise linearly elastic. Changes in S_p, like changes in "a" in Fig. 10 (but now representing changes in the geometry of each layer which are analogous to crack growth or healing) produce additional overall nonlinearity.

Whether one idealizes crack face interactions through a layer of (damaged) elastic material or uses contact mechanics and a friction equation such as Coulomb's law, there is a need to introduce additional equations for predicting changes in S_p. If the layer idealization is used, then the set S_p includes those parameters that are needed to account for changing damage in the layers.

3.7 A Model Based on Elastic-Like Damage Growth

In this section we shall discuss some equations and results given by Schapery (1990). The evolution equation for each S_p which changes at any particular time is specified as

$$-\frac{\partial W}{\partial S_p} = \frac{\partial W_s}{\partial S_p}, \tag{96}$$

where $W_s = W_s(S_p)$ is a potential with the physical significance of being the total work of crack or layer growth. If, for example, the *required* work per unit of new area, G_c, is constant and the same for all cracks, then $W_s = G_c \times$ total crack surface area; in the case of layers rather than true cracks, this surface area is the area of the layers, such as $a \times 1$ in Fig. 10.

The left side of Eq. (96) is the rate of change of *available* work (like the so-called "energy release rate" used in fracture mechanics) while the right side is the rate of change of *required* work. When this familiar fracture-type of equation is applicable, the total work input in any process,

$$W_T = \int Q_i dq_i, \tag{97}$$

is a function of state as defined by the values of q_i and S_p; it is given by

$$W_T = W + W_s. \tag{98}$$

Moreover, when we use Eq. (96) to solve for all of those S_p (in terms ofq_i) which change in a process, then

$$Q_i = \partial W_T / \partial q_i, \tag{99}$$

in which $W_T = W_T(q_i, S_r)$; the set S_r represents those parameters which are constant during the process under consideration. According to Eq. (99), the material behaves as a hyperelastic body, in which the total work replaces the strain energy function. Thus, when Eq. (96) is applicable for a particular set of varying S_p, one may characterize the material as one would characterize a nonlinear elastic material without damage. Moreover, for incremental changes ΔQ_i and Δq_i, the incremental stiffness matrix C_{ij} defined in $\Delta Q_i = C_{ij} \Delta q_j$ is symmetric. Conversely, if it is known (from experiment, say) that $C_{ij} = C_{ji}$, then Eq. (96) is obeyed.

Suppose that instead of Eq. (96), equations for $\dot{S}_p$ like that for $\dot{a}$ in Eq. (73) are applicable. Then, if $k \gg 1$ and $\ddot{S}_p \geq 0$, it can be shown Eq. (96) may be used in place of the equations for $\dot{S}_p$ by simply employing a time-dependent work of damage, $W_s = W_s(S_p, t)$; namely, as an approximation, one may use a work of damage which is like that for an ageing elastic material.

3.8 A Model Based on Viscoelastic Damage Growth

Next, we want to give an idealization that leads to a constitutive equation for nonproportional loading which exhibits damage parameters analogous to S and S_ε in Eqs. (23) and (27), and which allows one to draw upon results from proportional loading to predict response to more general loading. Suppose there are M penny-shaped cracks which are widely spaced and parallel. Let $a_m (m = 1, \ldots, M)$ be the radius of the mth crack, and assume there are no relevant parameters involving a length scale other than a_m. By dimensional analysis one may show that the strain energy has the form

$$W = W_0 - w_1 \sum a_m^3, \tag{100}$$

where W_0 and w_1 are positive functions of only q_i. The viscoelastic crack growth law Eq. (73) may be written in the form

$$\dot{a}_m = f_c(-\partial W/\partial A_m)^k, \tag{101}$$

where $A_m = \pi a_m^2$ and we have used the identity $-\partial W/\partial A_m = \partial W_c/\partial A_m$. Eqs. (100) and (101) yield

$$\dot{a}_m = f_c \left(\frac{3w_1}{2\pi}\right)^k a_m^k. \tag{102}$$

The solution to this differential equation is analogous to Eq. (78), in which a_m replaces a and ${w_1}^k$ replaces σ^q. By following earlier arguments it may be concluded that the effect of all parallel cracks is accounted for through one damage parameter of the form

$$S_1 \equiv \int_0^t {w_1}^k f_c dt. \tag{103}$$

From Eq. (100) the strain energy has the form

$$W = W_0 - w_1 h_1(S_1), \tag{104}$$

where the function $h_1(S_1)$ is related to the (statistical) distributon of the current crack sizes.

For widely-spaced cracks which have $b = 1, 2, \ldots, B$ different orientations, we have only to use b as the subscript in place of 1 in Eqs. (103) and (104), and use in Eq. (104) a sum over all orientations,

$$W = W_0 - \sum_b w_b h_b(S_b). \tag{105}$$

If, for a given set of deformation histories, the w_b and h_b functions are at least approximately independent of orientation, apart from constant factors, then we may omit the subscript and write,

$$W \simeq W_0 - wh(\hat{S}), \tag{106}$$

where

$$\hat{S} = \int_0^t w f_c dt. \tag{107}$$

With only one damage parameter $\hat{S}$, Eq. (87) yields

$$Q_i = \frac{\partial W_0}{\partial q_i} - \frac{\partial w}{\partial q_i} h(\hat{S}). \tag{108}$$

If the Q_i and q_i are replaced by stresses and strains, then

$$\sigma_{ij} = \frac{\partial W_0'}{\partial \varepsilon_{ij}} - \frac{\partial w'}{\partial \varepsilon_{ij}} h(\hat{S}), \tag{109}$$

where W_0' and w' are energies for an initial unit volume of material. For a power law material with proportional straining through the factor $\varepsilon(t)$, Eq. (109) must reduce to Eq. (28); thus, apart from constant factors, $\hat{S} = S_\varepsilon$ and $h = 1 - g_\varepsilon$.

Use of the complementary energy, instead of W, leads to analogous results. Of course, in this case a comparison to proportional stressing, rather than proportional straining, would be made.

4. Conclusions

The problem of developing a realistic viscoelastic constitutive equation for polycrystalline ice with distributed damage is clearly a difficult one. For this reason we have concentrated on proportional loading and, as a special case, uniaxial compression. If the theory can be experimentally verified for this type of loading using a variety of histories such as constant load, constant displacement rate, etc., then there is hope for predicting behavior under more general types of loading.

One of the difficulties is that there is still not a generally accepted viscoelastic constitutive theory for ice without damage. Here, we selected a constitutive equation for the undamaged state which appears to account for at least the major features of ice behavior and which allows for a relatively simple analysis of the microcracking. For proportional loading the effect of damage growth enters through a simple function of the applied stress history, Eq. (23), or strain history, Eq. (27). A procedure for using experimental data to fully characterize the behavior was described for uniaxial compression and then shown to be applicable to other types of proportional loading. For nonproportional loading, only highly idealized types of damage were addressed. The effect of crack-face friction was briefly discussed and an idealized model was proposed.

At this time it is not clear which details of the microstructure of ice have to be accounted for to develop a realistic constitutive equation for ice under stress states and histories of engineering interest. Presumably, the more one relies on experimental results using stress states and histories which simulate those in the applications of interest the less demand there is for modeling of physical details. In contrast to the work of others (e.g. Sinha (1988) and Sunder and Wu (1989b)) we did not explicitly use information on the microstructure, such as the size of grains and their anisotropic properties. However, such information could be introduced through the equation for microcrack growth and the strain energy and thus enable prediction of the effect of changes in microstructural parameters on overall stress-strain behavior.

Temperature could also be introduced to account for its effect on the viscoelastic behavior and for thermal expansion. Sinha (1978) found in his studies of creep of ice that the viscoelastic behaviour depends on temperature in the same way as many polymeric materials do, in that ice is thermorheologically simple. This behavior is characterized by using a temperature-modified time scale. Using standard notation, for undamaged material we simply replace the time argument in the relaxation modulus and creep compliance by t/a_T, where a_T is a function of temperature. Thermorheologically simple behavior will be exhibited by the theory with damage growth if, in addition, we let f_1 in Eq. (23) and f_c in Eq. (73) depend on temperature through the factor $1/a_T$.

References

Ashby, M.F. and Duval, P. 1985. The creep of polycrystalline ice. Cold Regions Science and Technology, Vol. 11, pp. 285-300.

Biot, M.A. 1954. Theory of stress-strain relations in anisotropic viscoelasticity and relaxation phenomena. Journal of Applied Physics, Vol. 25, pp. 1385-91.

Dougill, J.W. 1975. Some remarks on path dependence in the small in plasticity. Quarterly of Applied Mathematics, Vol.32, pp.233-243.

Ferry, J.D. 1970. Viscoelastic Properties of Polymers. John Wiley & Sons, Inc., New York.

Gold, L.W. 1972. The failure process in columnar-grained ice. National Research Council, Canada, Technical Paper No. 369.

Harper, B.D. 1986. A uniaxial nonlinear viscoelastic constitutive relation for ice. Journal of Energy Resources Technology, Vol.108, pp.156-160.

Harper, B.D. 1989. Some implications of a nonlinear viscoelastic constitutive theory regarding interrelationships between creep and strength behavior of ice at failure. Journal of Offshore Mechanics and Arctic Engineering, Vol.111, pp.144-148.

Hashin, Z. 1965. Viscoelastic behavior of heterogeneous media. Journal of Applied Mechanics, Vol.32, pp.630-636.

Horii, H. and Nemat-Nasser, S. 1983. Overall moduli of solids with microcracks: load-induced anisotropy. Journal of the Mechanics and Physics of Solids, Vol.31, pp.155-171.

Jacka, T.H. 1984. The time and strain required for the development of minimum strain-rates in ice. Cold Regions Science and Technology, Vol.8, pp. 261-268.

Jordaan, I.J. 1986. Numerical and finite element techniques in calculation of ice-structure interaction. Proceedings International Association for Hydraulic Research (IAHR) Symposium on Ice, Iowa City, Iowa, Unversité Laval, Québec, Vol.II, pp.405-441.

Jordaan, I.J. and McKenna, R.F. 1988. Modelling of progressive damage in ice. IAHR Symposium on Ice, Sapparo, Japan, Vol.II, pp.585-624.

Jordaan, I.J. and McKenna, R.F. 1989. Processes of deformation and fracture of ice in compression. Proceedings IUTAM/IAHR Symposium on Ice-Structure Interaction, St. John's, Newfoundland.

Karr, D.G. and Choi, K. 1989. A three-dimensional constitutive damage model for polycrystalline ice. Mechanics of Materials, Vol.8, pp.55-66.

Krajcinovic, D. and Lemaitre, J., Editors 1987. Continuum Damage Mechanics Theory and Applications. Springer-Verlag, New York – Wien.

Mellor, M. and Cole, D.M. 1982. Deformation and failure of ice under constant stress or constant strain-rate. Cold Regions Science and Technology, Vol.5, pp.201-219.

Mellor, M. and Cole, D.M. 1983. Stress/strain/time relations for ice under uniaxial compression. Cold Regions Science and Technology, Vol.6, pp.207-230.

Onat, E.T. and Leckie, F.A. 1988. Representation of mechanical behavior in the presence of changing internal structure. Journal of Applied Mechanics, Vol.55, pp.1-10.

Sanderson, T.J.O. 1988. Ice Mechanics: Risks to Offshore Structures. Graham and Trotman, London, UK.

Schapery, R.A. 1967. Stress analysis of viscoelastic composite materials. Journal of Composite Materials, Vol.1, pp.228-267.

Schapery, R.A. 1971. On the characterization of nonlinear viscoelastic materials. Journal of Polymer Engineering and Science, Vol.11, pp.295-310.

Schapery, R.A. 1981. On viscoelastic deformation and failure behavior of composite materials with distributed flaws. 1981 Advances in Aerospace Structures and Materials-AD-01, ASME, Edited by S.S. Wang and W.J. Renton, pp. 5-20.

Schapery, R.A. 1984. Correspondence principles and a generalized J integral for large deformation and fracture analysis of viscoelastic media. International Journal of Fracture, Vol.25, pp.195-223.

Schapery, R.A. 1987. Deformation and fracture characterization of inelastic composite materials using potentials. Polymer Engineering and Science, Vol.27, pp.63-76.

Schapery, R.A. 1990. A theory of mechanical behavior of elastic media with growing damage and other changes in structure. Journal of the Mechanics and Physics of Solids, Vol.38, pp.215-253.

Sinha, N.K. 1978. Rheology of columnar-grained ice. Experimental Mechanics, Vol.18, pp.464-470.

Sinha, N.K. 1984. Intercrystalline cracking, grain-boundary sliding, and delayed elasticity at high temperatures. Journal of Materials Science, Vol.19, pp.359-376.

Sinha, N.K. 1988. Crack-enhanced creep in polycrystalline material: strain-rate sensitive strength and deformation of ice. Philosophical Magazine, Vol.A43, pp.4415-4428.

Sjölind, S.G. 1987. A constitutive model for ice as a damaging viscoelastic material. Cold Regions Science and Technology, Vol.41, pp.247-262.

Weitsman, Y. 1988. A continuum damage model for viscoelastic materials. Journal of Applied Mechanics, Vol.55, pp.773-780.

Sunder, S. and Wu, M.S. 1989a. A differential flow model for polycrystalline ice. Cold Regions Science and Technology, Vol.16, pp.45-62.

Sunder, S. and Wu, M.S. 1989b. Crack nucleation due to elastic anisotropy in polycrystalline ice. Cold Regions Science and Technology, In Press.

Williams, H.T. 1984. Single integral representations in ice mechanics. Cold Regions Science and Technology, Vol. 9, pp. 89-95.

Discussion

Q: J. Dempsey (Clarkson University)

In terms of tensile fracture initiation from a fabricated macrocrack, we have found a great deal of difficulty in trying to initiate stable cracking. The best we have achieved so far is crack growth by discrete crack jumps. Given your knowledge and viscoelastic fracture, could you comment on this difficulty.

A: R.A. Schapery

Instability of crack growth may arise from elastic energy in the specimen and loading device or from the fracture toughness, G_c, versus crack speed relationship, or both. If, for example, the fracture toughness is constant, but the elastic energy release rate, G, increases with crack length at any given applied grip displacement, then the crack growth will be unstable. Here the instability arises from the specimen's elastic energy. On the other hand, if there is significant energy stored in the assembly that connects the point of controlled displacement application and the specimen grip or end plate, an otherwise stable growth may be unstable. In other words, for stable crack growth one needs $\frac{\partial G}{\partial a} < 0$ where $G = G_{\text{spec}} + G_{\text{loading assembly}}$ and a = crack length.

The discrete crack jump behaviour may exist even when $\frac{\partial G}{\partial a} < 0$ if $\frac{\partial G_c}{\partial \dot{a}} < 0$, i.e. if the fracture toughness decreases with crack speed $\dot{a}$. Suppose for example that the initiation fracture toughness is larger than that for continuing growth and that $\frac{\partial G}{\partial a} < 0$. Then, when the specimen is loaded and the initiation G_c is exceeded, the crack will propagate dynamically; but it stops when the available energy G falls below the propagation value of G_c. Obviously, the larger $|\frac{\partial G}{\partial a}|$ is, the shorter the jumps in crack growth will be.

Discrete crack jumps often are observed in fracture testing of plastics over some ranges of temperature and machine crosshead speed. Bulk or global viscoelastic behaviour tends to stabilize crack growth.

Microfracture and The Compressive Failure of Polycrystalline Ice

David M. Cole

U.S. Army Cold Regions Research and Engineering Laboratory
Hanover, N.H.

Abstract This paper discusses strain rate and grain size effects on the process of microcrack nucleation in freshwater polycrystalline ice. It also examines crack nucleation mechanisms that are likely to operate under conditions of practical concern. Special attention is paid to the ductile-to-brittle transition and the role of grain size in the transition region.

Experimental results on granular ice having bonded end caps revealed that there was a distinct beginning and end to the crack nucleation stage prior to brittle compressive failure. Straining at a rate of 10^{-3} s^{-1} generated a population of stable microcracks over a certain range of stress; after this nucleation stage, the specimen continued to sustain increased loading up to the point of failure without extension of the existing cracks or nucleation of additional cracks. Several attempts to characterize the additional cracking associated with the ultimate collapse of the specimen proved fruitless. Additional topics include the influence of grain boundary ledges on the internal stress field in polycrystalline ice, grain and crack size measurements and some important aspects of experimental technique.

1. Introduction

This paper examines the experimental evidence for and theoretical explanations of microcracking in ice and frames the nucleation problem in light of experimental findings of the past few years. Microcracks nucleate in ice under conditions of interest in engineering problems, and their occurrence is intimately related to the important transition from ductile to brittle behavior. A detailed understanding of the underlying micromechanics of the ductile-to-brittle transition will clarify our view of processes observed in the laboratory and provide a sound basis for interpreting larger-scale observations.

S. Jones · R. F. McKenna · J. Tillotson · I. Jordaan (Eds.)
Ice-Structure Interaction
IUTAM-IAHR Symposium St. John's, Newfoundland, Canada

The following discussions concentrate on observations of microcracks in laboratory-prepared freshwater ice, theoretical aspects of crack nucleation, and grain size effects on the ductile-to-brittle transition. The theoretical discussions incorporate the notion that grain boundary ledges are significant stress concentrators and sources of dislocations. Finally, some observations on the microcracking processes leading up to brittle compressive failure are presented and discussed. Important aspects of experimental technique are stressed throughout.

2. Observations of Microcracks in Ice

Gold (1972) made extensive observations on microcracking in columnar-grained freshwater ice, paying special attention to the time and strain at which cracks began to appear and tracking the crack population during creep straining. Zaretsky et al. (1979) also studied the evolution of internal cracking in freshwater ice, incorporating the use of acoustic emission monitoring techniques.

More recently, Cole (1986) investigated the crack size distribution as a function of grain size as well as the crack population as a function of strain and load level. Cole (1986) also contains a detailed review of laboratory studies on the topic of internal cracking in ice. Hallam et al. (1987) observed microfractures in several laboratory-prepared freshwater ice specimens with findings that were in general agreement with those of Cole (1986).

Recent work by Kalifa et al. (1989) demonstrates hydrostatic pressure effects on crack nucleation in freshwater polycrystalline ice. The experimental results quantified the increase in stress and strain levels for crack nucleation with increasing confining pressure. The size and orientation of the microcracks did not change with pressure, the plane of the fractures had a strong tendency to align with the compressive axis, and their size scaled with the grain size. Kalifa et al. note the general agreement of their observations with those reported by Cole (1986): the average crack size was found to be approximately 0.6 times the average grain size, and the maximum crack size was typically 2 to 3 times the average grain size. Theoretical considerations presented by Cole (1986) led to a crack size-grain size ratio of 0.65, which is in reasonable agreement with experimental observations. Despite the wide range in experimental conditions, differences in ice structure and specimen preparation methods in the crack nucleation studies mentioned above, there are some common trends in the findings: microfracturing begins at strains on the order of 10^{-4} and peak fracturing rates occur at strains in the range of 1×10^{-3} to 4×10^{-3}.

Figure 1 shows examples of typical microcracks nucleated during the creep of laboratory-prepared freshwater ice. The photographs are from sections of the test material which have been trimmed on a microtome to 50 - 100 mm thickness and are referred to as thick sections. Such sections taken parallel and perpendicular to the compressive axis allow measurements to determine crack size distributions and give insight regarding the aspect ratio of the nucleated cracks, as seen in Figure 2. The ratio of the average crack length projected in the horizontal plane to the average length projected in the vertical plane is 0.93 for the results shown in Figure 2. Figure 3, from Cole (1986), shows the distribution of over 2.2×10^3 microcrack observations from 25 specimens. The crack lengths were measured along the direction of compressive stress and have been normalized to the average grain size. The form of this distribution reflects well the distribution found for each specimen individually.

Since the study of microcracking in ice relies very heavily on detailed experimental observations, a brief discussion of several relevant aspects

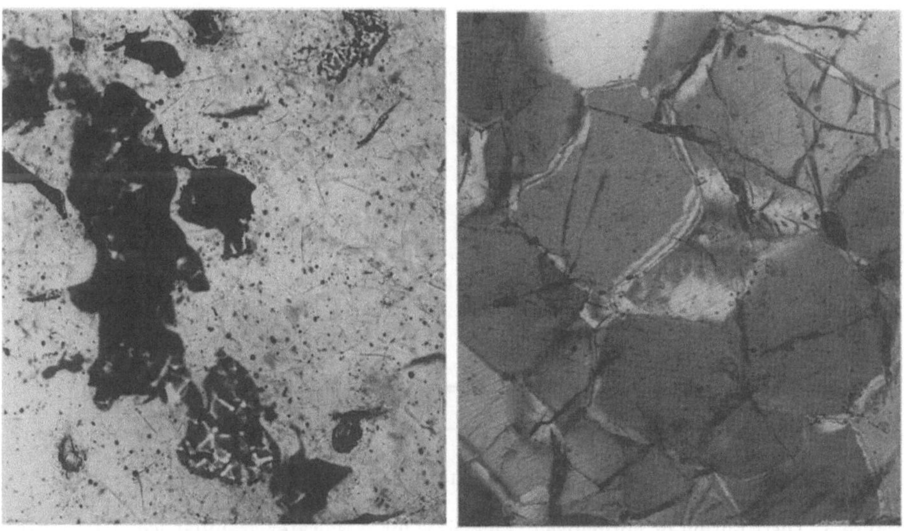

Figure 1. Typical microcracks nucleated during the creep of granular ice as viewed a) perpendicular and b) parallel to the compressive stress axis.

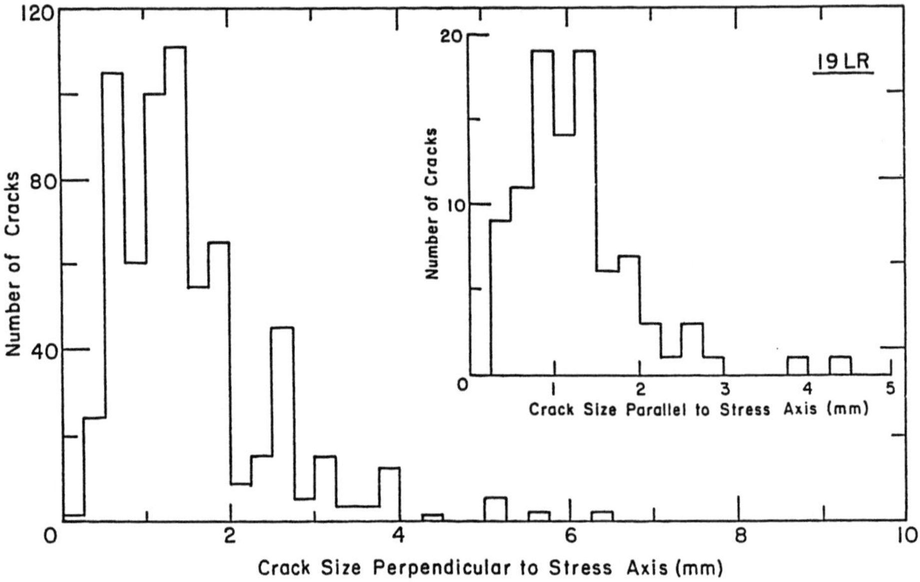

Figure 2. Crack size distributions as measured in the direction of and perpendicular to the compressive stress axis, $d = 1.5$mm.

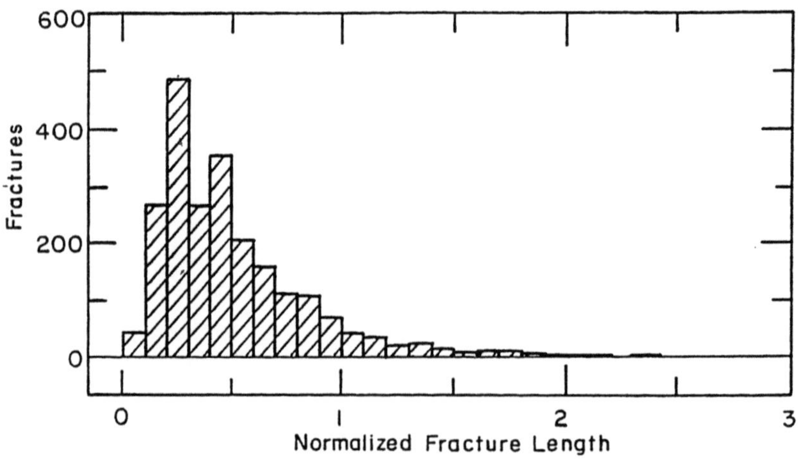

Figure 3. Distribution of normalized crack sizes for freshwater granular ice. Crack sizes have been normalized to the average grain size, from Cole (1986).

of experimental methodology is in order. Studies relating microcrack size to grain size demand careful estimates of both quantities. Measurements on cracks isolated from the atmosphere should be made within hours of the termination of the experiment in order to avoid the effects of crack healing. On the other hand, cracks open to the atmosphere require a significantly greater time to heal owing to a shift of the rate controlling process of healing brought about by the presence of gas in the void (Colbeck, 1986).

The topic of grain size measurement deserves attention as there are several methods currently in use. The frequently used mean intercept length method is known to underestimate the average size of equiaxial grains (Dieter, 1976), but is expected to give good results for horizontal sections of columnar ice. In the present case of granular ice formed from equiaxial seed grains, the author has found that grain sizes calculated using $d = (6/\pi N_A)^{1/2}$, where N_A is the number of grains per unit area, are in very good agreement with the size of the seed grains used to form the specimen. Cole (1986), examining material up to a grain size of 5 mm, indicates that the intercept method of estimating grain size produces averages that are approximately 30 percent lower than the above method. Currier (1981) presented data which show that for grain sizes near 10 mm, the intercept method yields values which are as low as 61 percent of the average size of the seed grains used to form the specimen. Such variations are significant, especially when the measurements are to be used in a micromechanical context, as they carry over into calculations of slip line length, grain facet size, grain volume and the ratio of microcrack size to grain size. Jacka (1984) and Cole (1986) present comparisons of several methods used to estimate grain size.

With regard to presenting information on the extent of microcracking, the number of cracks per unit area or volume, quantities that are easily measured, can give a misleading indication of the extent of microstructural damage when the specimen grain size varies. Normalizing the number of cracks to the number of grains in the volume under consideration gives a clearer indication of the extent of damage to the structure and provides a dimensionless quantity that approaches the order of unity for high crack densities. Note that the method used to determine grain size will have a significant effect on the calculated number of grains per unit volume and hence on the estimate of the number of cracks per grain.

3. Nucleation Mechanisms and Modeling

A stress concentration within the material must reach a certain level (nucleation stress) for a microcrack to form. Dislocation pileups provide a mechanism to achieve the nucleation stress when plastic flow occurs. Elastically generated stress concentrations can achieve the nucleation stress at deformation rates which are too high to allow dislocation pileups to form.

The dislocation pileup mechanism has been successfully applied to ice in the work of Schulson and Cannon (1984) and Cole (1986). The form of the model generally used is from Smith and Barnby (1967) who modified the original Stroh (1957) model to account for the influence of shear stress. Such models require the formation of a pileup of approximately 10^3 dislocations under the conditions of interest. Cole (1988) pointed out that the Frank-Read source, often assumed to operate when the pileup model is invoked, cannot generate dislocations at a sufficient rate to explain the occurrence of microcracking by this mechanism in times less than approximately 1.5×10^3 s. An attractive alternative to explain the more rapid formation of pileups is the grain boundary ledge source mechanism. Work by Hondoh and Higashi (1983) clearly indicates that grain boundaries in ice contain ledges capable of producing significant stress concentrations when undergoing shear deformation. The ledges were observed to act as efficient sources of dislocations under very low applied stress levels. Under this mechanism, dislocations are emitted directly from the stress field at the grain boundary ledge and travel at significantly higher velocities than those generated by a Frank-Read source operating under the background stress. It is the feeling of the present writer that the dislocation sources continue to operate, but produce fewer and fewer dislocations, as the deformation or loading rate increases to the onset of brittle behavior.

Although the role of dislocation pileups diminishes as the transition to brittle behavior is approached, some recent experimental results indicate that dislocation motion may play a role in the stress-strain behavior up to relatively high loading rates. Work by Cole and Gould (1989) on laboratory-prepared granular ice under fully reversed cyclic loading between ±1 MPa (which is below the crack nucleation threshold) indicated that damping attributable to dislocations increased with decreasing grain size, suggesting that an increase in grain boundary area is responsible for generating an increased level of dislocation motion. In addition, the damping remained noticeable throughout the nominally ductile range and was observed for loading times as low as 0.25 s. These observations support the notion that grain boundaries in general, and thus in the present argument, grain boundary ledges, remain effective sources of stress concentrations at the higher loading rates and would then presumably act as preferred sites for microcrack nucleation.

An argument can also be put forth for microcracking at grain boundaries based on the elastic anisotropy of the ice lattice. A simple model to demonstrate the anisotropy effect was presented by Cole (1988). That model considers only an isolated pair of grains and neglects the character of the far field stress.

Very recent work by Shyam Sunder and Wu (1989) presents a much more detailed model of crack nucleation due to elastic anisotropy. That model accounts for the nature of the loading state in a polycrystal and predicts a significant difference in the nucleation stress levels for tensile and compressive loading states. The latter trend is borne out in results discussed below, where under nominally brittle conditions, the nucleation stress in compression was approximately three times the tensile strength found for similar material under nucleation-controlled conditions. This line of study will be of great use in clarifying compressive failure processes.

4. Microcracking and the Ductile-To-Brittle Transition

It is important to ask whether and to what extent microcracking participates in the overall compressive failure processes in ice. An increase in the rate of cracking has been noted for certain structural types of ice entering tertiary creep (Gold, 1972), but it is generally not required that microcracks develop for accelerating creep to begin, as evidenced by the acoustic emission results of St. Lawrence and Cole (1982). However, microcracking clearly affects the high-strain creep behavior. A significant amount of microcracking promotes the eventual catastrophic failure of a specimen. In the absence of a significant level of microcracking, dynamic recrystallization dominates the high-strain creep behavior as observed in the experiments of Jacka (1984).

In the macroscopically ductile regime, the extent of microcracking varies greatly with structure, strain, strain rate, temperature and grain size. For the purposes of this discussion, material which exhibits a significant amount of inelastic behavior and continues to resist deformation after the peak stress has been reached will be considered nominally ductile. Nominally brittle behavior, on the other hand, is characterized by very little inelastic strain and a catastrophic failure at maximum stress which disintegrates the specimen.

Experimental results indicate that when polycrystalline ice is tested under conditions which lead to microcracking, the peak stress decreases with increasing grain size. This remains true throughout the ductile-to-brittle transition as well as in the brittle regime. In addition, the microcrack density is closely linked to power law breakdown: as the microcrack density approaches one crack per grain, the shift toward rate-insensitive behavior is evident and the failure strain decreases from approximately 0.01 to the range of 0.001 to 0.004 (Cole, 1987).

For well-controlled laboratory experiments, fine-grained ice ($d \approx 1$ mm) exhibits a roughly 30 percent drop in peak stress with the onset of brittle behavior near a strain rate of $10^{-3}s^{-1}$. The peak stress vs. strain rate curve for coarse-grained ice ($d > 5$ mm) follows the power law curve until reaching a plateau at a strain rate somewhat greater than $10^{-4}s^{-1}$ (Cole, 1987). Thus, the increased propensity for microcracking in the larger-grained material causes a shift to brittle behavior at a significantly lower strain rate than found for the fine-grained material.

Figure 4 shows a plot of peak stress vs. $d^{-1/2}$ using the data of Schulson and Cannon (1984), Cole (1987), and Schulson (1987). Interestingly, this function linearized the results for both the ductile and brittle regimes reasonably well despite the significant differences in the macroscopic behav-

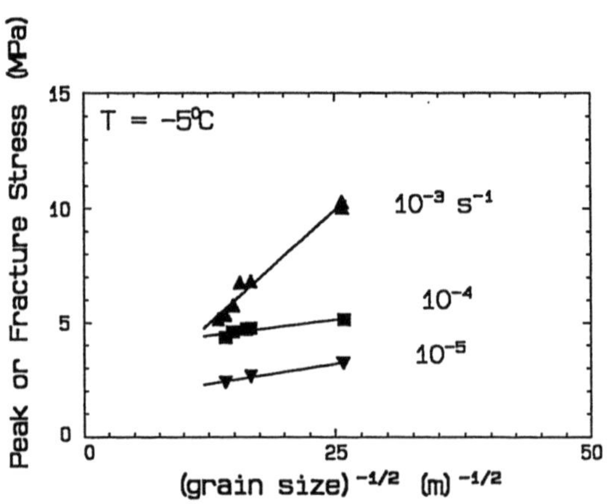

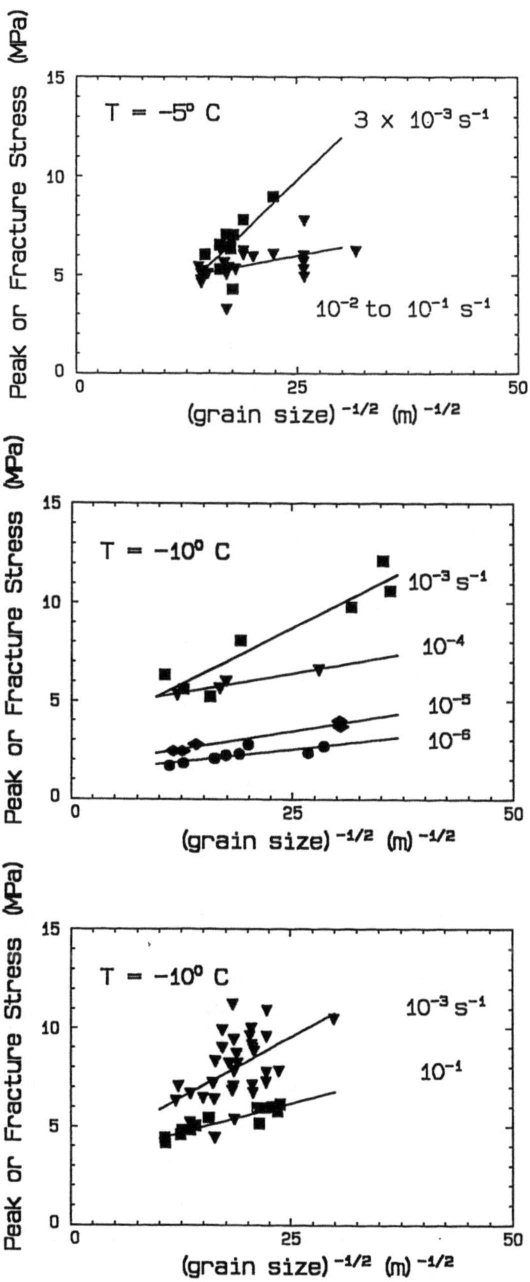

Figure 4. Peak or fracture stress vs $d^{-1/2}$ for laboratory-prepared granular ice. Data in a) and b) from Cole (1987), c) from Schulson and Cannon (1984), and d) from Schulson (1987).

ior. Excluding the results at 10^{-3} and $3 \times 10^{-3}s^{-1}$, each curve represents experimental results that are macroscopically either brittle or nominally ductile, each exhibits a positive stress axis intercept, and the slopes are uniform. The two exceptions to this (seen in Fig. 4a and 4b) have relatively steep slopes and zero or near-zero intercepts. Examination of the stress-strain curves associated with these data, particularly for a strain rate of $10^{-3}s^{-1}$, indicate that the fine-grained specimens were behaving in a nominally ductile manner whereas the coarse-grained material was in the brittle regime.

The -5^oC and -10^oC results in Figure 4 are not strictly comparable owing to differences in the methods used to calculate the average grain sizes. For the -5^oC results, grain size was determined using the alternative formula discussed above, and the linear intercept method was used for the -10^oC results.

An important aspect of the above observations is that strain rate cannot be used as the sole indicator of whether ice is brittle or ductile. Grain size must also be considered. The abnormally strong dependence of peak stress on grain size at a strain rate of $10^{-3}s^{-1}$ is a consequence of the transition to brittle behavior rather than a result of a grain size effect on a particular mechanism of deformation. With regard to the modeling of either the ductile or brittle regime, it seems appropriate to first define the range of the regime in terms of actual behavior, considering both strain rate and grain size.

5. Some Observations of Microcracking During Brittle Failure

The following experiments were performed in order to gain insight in to the compressive failure process in the brittle regime. Specimens having an average grain size of approximately 5 mm were deformed to various levels of axial strain at a strain rate of $10^{-3}s^{-1}$ at -10^oC. A triangular waveform served as the command signal for the closed-loop testing machine, so the specimen was loaded and unloaded at the same strain rate. The testing system was operated in strain control, and an extensometer mounted on the ice provided deformation measurement and served as the feedback source for the machine. The rigid clamping system used in these experiments (Cole and Gould, 1989) functions in both tension and compression and allowed the controlled reversal of the strain rate without overshoot. The nominal maximum strain applied in a given cycle of loading was increased from an initial value of 5×10^{-4} to 7.5, 9.0 and 10.0×10^{-4} with several minutes between successive loadings. Figure 5 presents stress-strain plots and modulus calculations from these experiments.

There is a noticeable curvature in the stress-strain data prior to the onset of internal cracking, indicating that microcracking is not necessary for ice to deviate from linear elastic behavior. Additionally, although the method of control in these experiments was not particularly well suited to monitoring strain recovery after each cycle, it appeared that most of the offset evident between the start and finish of each load cycle (Fig. 5) was recovered within a short time.

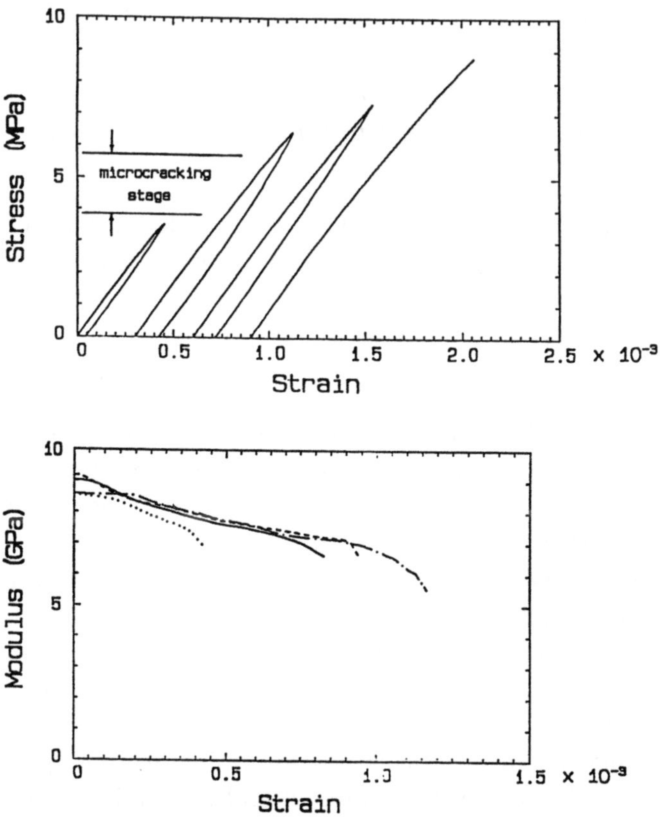

Figure 5. Stress-strain curves for a 5 mm grain size specimen subjected to increasing compressive strains. a) Representative curves from the reloading experiments. Stress levels for the microcracking stage are indicated. b) Effective modulus vs axial strain for the loading portions of the curves in a).

The intent of this series of experiments was to halt the deformation at appropriate intervals to examine the various stages in the failure process. Four stages became evident:

1. The first stage was nominally elastic, but the stress-strain behavior was not perfectly linear and no visible microcracking occurred. The stress and strain limits of this stage were 3.29 MPa and 4.4×10^{-4}.

2. The onset of microcrack nucleation marked the beginning of the second stage. A network of stable microcracks began to nucleate between 3.29 MPa and 3.95 MPa. The crack population increased noticeably with increasing stress up to 5.65 MPa.

3. The third stage commenced with the end of the crack nucleation stage, and its distinguishing feature was the absence of further visible microstructural damage. The increasing peak stress caused no additional microcracking, and the existing microcracks appeared stable.

4. The fourth and final stage was the sudden and complete failure of the specimen. The debris consisted primarily of grain-sized fragments, with evidence of large vertical cracks which extended into the relatively intact but highly damaged material adhering to the end caps.

The transformation from the dense array of stable microcracks produced in stage 2 to complete failure occurred over a very brief strain interval. Several attempts to reverse straining during this stage, and thus capture additional details of the fracturing, proved fruitless as a consequence of the instability of the process. In one experiment, the failure process was arrested prior to the disintegration of the specimen. However, the highly damaged specimen completely fell apart when an attempt was made to remove it from the loading apparatus, precluding any detailed observations.

It is worth mentioning that in the stages of deformation leading up to failure, the microcracks did not appear to interact and no stable tension cracking was observed. The process of stable tension crack extension from a shear crack in a compressive stress field has been considered by Nemat-Nasser and Horii (1982), who use the term “crack kinking”, and by Ashby and Hallam (1986), who use the term “wing crack” in referring to the configuration. The wing cracking process has been incorporated in a model of compressive failure of freshwater ice by Schulson (1987), who employed it as a means to address the strong temperature effect he noted on the compressive strength. The absence of wing crack development in freshwater granular ice was also noted by Kalifa et al. (1989).

Figure 6 shows micrographs taken under polarized light from these experiments. The microcrack population in the material that could be recovered consisted of a combination of grain boundary and intracrystalline cracks. Figure 6a shows part of a large vertical crack which ran into the less-damaged material near the end caps. The path of such cracks consisted of segments of grain boundary and transcrystalline cracks. Figure 6b shows a typical microcracking pattern as viewed along the compressive stress axis.

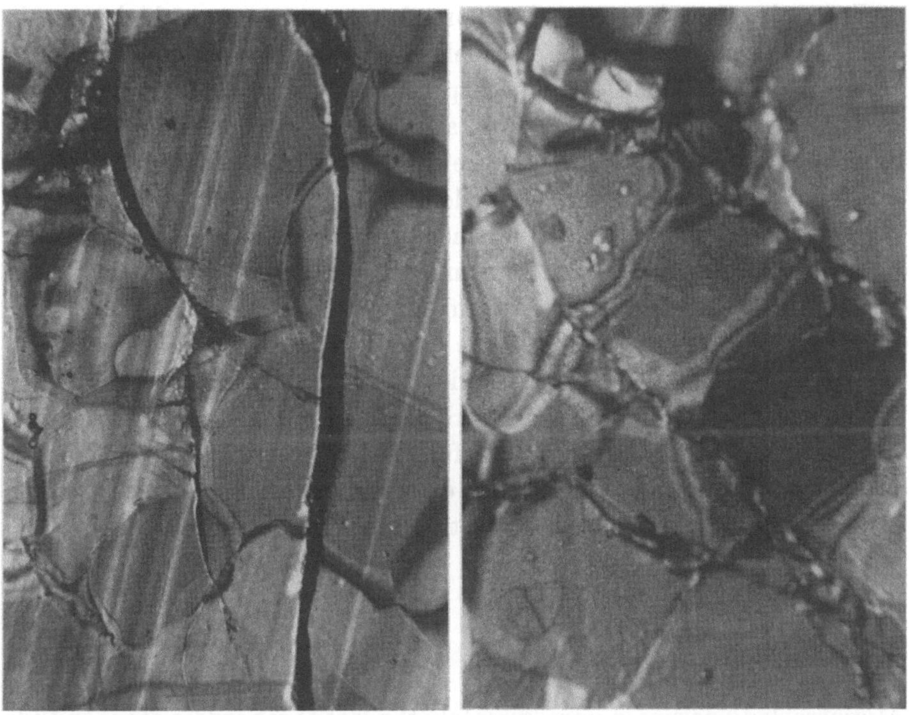

Figure 6. Micrographs from the brittle compressive failure of a 5 mm grain size specimen. Microcracking pattern as viewed a) perpendicular to the axis of loading and b) along the axis of loading. Both are of the relatively intact material immediately adjacent to the failure zone. The large crack on the right in a) is a segment of a long vertical crack which extended into the material adhering to the end cap.

There is a steady decrease in the slope of the stress-strain curve up to failure, with no apparent changes when microcracking occurs (Figure 5a). Additionally, at strains where comparison is possible, the results indicate that the unloading and reloading moduli were unaffected by the presence of the microcrack array. The observation that crack nucleation does not occur continuously through the final stages of failure, but rather has a definite beginning and end, gives some interesting insight to the failure process, as discussed in the following section.

Although it is possible that the present method of reloading the specimen altered the failure process, the findings are in general agreement with the behavior observed by the author in numerous experiments in this regime. Additional experiments of this type will clarify the trends noted above and enhance our understanding of the brittle compressive failure process.

Before leaving the topic of compression experiments at higher strain rates, the importance of the parallelism of the specimen end planes deserves mention. Ductile regime experiments are very forgiving in this regard, but in the brittle regime, slight stress concentrations brought about by non-uniform loading of the end planes can result in abnormally low failure stresses or alter the mode of failure. Such non-uniform loading can cause apparently brittle behavior under conditions which, under uniform loading, would lead to nominally ductile behavior. From the author's experience, there are two manifestations of this problem in the microcracking regime for specimens having bonded end caps. Gross errors produce macroscopic cracks prior to the nucleation of a significant number of microcracks. Minor errors result in a non-symmetric array of microcracks. It is important to carefully assess every experiment for such problems. The remedy lies in both extremely careful specimen preparation and in diligent attention to testing machine alignment.

6. Discussion

Microcracking occurs in approximately the last three orders of magnitude of strain rate in the ductile regime. When microcracking levels become significant, a drop in the strain level for peak stress or the strain rate minimum is observed. This drop in the critical strain level is the first macroscopic evidence of the beginning of the transition to brittle behavior and precedes the onset of rate-insensitive behavior by at least an order of magnitude of strain rate.

Although grain size may appear to be a second-order variable with regard to its effect on strength, studies of grain size effects contribute significantly to our understanding of the micro-mechanics of ice and thus help guide the modeling process. Additionally, the transition to brittle behavior has been shown to be very sensitive to grain size. The observation that plots of σ_{max}vs $d^{-1/2}$ generally exhibit a positive stress-axis intercept for conditions that produce microcracking, provided that the data do not straddle the ductile-to-brittle transition, is an important feature of the data and should be accounted for in any micromechanical model of the process. The stress-axis intercept is rate sensitive in the nominally ductile regime and not noticeably so in the brittle regime. For the ductile regime, the relationship between the intercept values obtained from Fig.4 and the applicable strain rates follows approximately an $n=3$ power law. A possible exception to the non-zero intercept observation occurs in results presented by Schulson (1987) for compressive strengths at -50^oC. Additional experimental results are certainly required, however, to further clarify the trends just noted.

The experimental evidence discussed above makes a reasonable case for considering grain boundary ledges to be a significant source of stress concentrations within polycrystalline ice. This premise has appealing implications with regard to deformational processes in ice in general and to the process of microcracking in particular. When deformation is sufficiently slow, the variations in strain energy density associated with such grain boundary features, along with the variations caused by dislocation pileups, can drive the grain boundary migration process. Additionally, slip can be propagated through favorably oriented grains under the action of such stress concentrations. This slip process can in turn generate dislocation pileups which nucleate cracks by the mechanism described earlier. At rates which disallow alternative mechanisms to relax stress concentrations, or at sites which are not favorably oriented for slip, microcracks begin to form in direct association with the stress fields at the grain boundaries. From the foregoing discussions, it appears that local grain boundary geometry and orientation to the stress field will dictate the preferred mechanism of crack nucleation under nominally brittle conditions. Ledges will be important to the nucleation process when they are present and favorably oriented, and elastic anisotropy would presumably drive the process in the absence of ledges.

Basically, the process of microcracking in ice serves the function of relieving internal stress concentrations. With regard to brittle failure, the observation that the microcrack nucleation process ends prior to failure indicates that the microcracking effectively exhausts all the important stress concentrations in the material before the failure stress is reached. Thus, the microcracking process has the effect of producing a more homogeneous internal stress state prior to failure and establishes an array of defects which link up at failure.

Interestingly, no significant change in the stress-strain curve appears to coincide with the formation of the microcrack array. This suggests that microcracking *perse* is not of primary importance in the observed deviations from linear elastic behavior. Since the effects of crack healing were to be avoided in the reloading experiments, there was insufficient time for full recovery of the anelastic strain between loading cycles. As noted above, however, a significant portion of the inelastic strain appeared to be recoverable. It seems likely that dislocation processes are responsible for the bulk of the anelastic strains observed in the reloading experiments and hence for the deviations from linear elastic behavior. Allowing for full recovery would provide a means to assess directly any permanent strain associated with the microcrack population. Future work will address this matter.

7. Concluding Remarks

The foregoing discussions and observations clearly demonstrate that microcracking is intimately related to the transition from ductile to brittle behavior and is the major agent of structural damage prior to brittle compressive failure. Consideration of the possible mechanisms of deformation led to a line of thought which holds that grain boundary ledges are significant in the deformational regimes under consideration.

A brief series of experiments indicated the existence of a discrete microcracking phase prior to brittle compressive failure. A population of stable microcracks nucleated over a certain range of stress. Continued increases in stress apparently did not further damage the specimen until the point of catastrophic failure, which occurred in an unstable manner.

The inescapable link between theory and experiment in the field of ice mechanics is an underlying theme of this work. Insight regarding the complex process of compressive failure can only come about through exhaustive, direct examination of specimens at all stages of the failure process. Furthermore, such micro-level observations should be carried out for as large an experimental scale as practical. With continued study, microcracking will prove to be a predictable function of ice structure, temperature and loading conditions. Thus, for processes which engage ice in compression, microcracking can serve as a unique link between fundamental laboratory work and studies at larger scales.

Acknowledgements

The author expresses his appreciation to Glenn Durell, operator of CRREL's materials testing facility, to Alan Richard who prepared ice specimens and to Nancy Perron for her thin section and photography work. The author gratefully acknowledges the support of CRREL and the Office of Naval Research.

References

Ashby, M.F. and Hallam, S D. 1986. The failure of brittle solids containing small cracks under compressive stress states. Acta Metall., Vol.34, No.3, pp.497-510.

Colbeck, S.C. 1986. Theory of microfracture healing in ice. Acta Metall., Vol.34, No.1, pp.89-95.

Cole, D.M. 1986. Effect of grain size on the internal fracturing of polycrystalline ice. CRREL Report 86-5.

Cole, D.M. 1987. Strain rate and grain size effects in ice. J. Glaciol., Vol.33, No.115, pp.274-280.

Cole, D.M. 1988. Crack nucleation in polycrystalline ice. Cold Regions Sci. Technol., Vol.15, pp.79-87.

Cole, D.M. and Gould, L.D. 1989. Uniaxial tension/compression tests on ice - Preliminary results. Proceedings of the 8th Intl. Conf. on Offshore Mechanics and Arctic Engineering, The Hague, The Netherlands, Vol.4, pp.37-42.

Currier, J.H. 1981. The brittle to ductile transition in polycrystalline ice under tension. M. Sc. Thesis, Dartmouth College, Hanover, NH., June.

Dieter, G.E. 1976. Mechanical Metallurgy. McGraw-Hill, New York, Second ed.

Gold, L.W. 1972. The process of failure in columnar-grained ice. Phil. Mag., Vol.26, No.2, pp.311-328.

Hallam, S.D., Duval, P. and Ashby M.F. 1987. A study of cracks in polycrystalline ice under uniaxial compression. J. de Physique, Vol.48, No.3, pp.C1-303 - C1-311.

Hondoh, T. and Higashi, A. 1983. Generation and absorption of dislocations at large-angle grain boundaries in deformed ice crystals. J. Phys. Chem., Vol.87, No.21, pp.4044-4050.

Jacka, T.H. 1984. Laboratory studies on relationships between ice crystal size and flow rate. Cold Regions Science and Technology, Vol.10, pp.31-42.

Kalifa, P., Duval, P. and Ricard, M. 1989. Crack nucleation in polycrystalline ice under compressive stress states. Proceedings of the 8th Intl. Conf. on Offshore Mechanics and Arctic Engineering, The Hague, The Netherlands, Vol.4, pp. 13-21.

Nemat-Nasser, S. and Horii, H. 1982. Compression-induced nonplanar crack extension with application to splitting, exfoliation, and rockburst. J. Geophys. Res., Vol.87, No.B8, pp.6805-6821.

St. Lawrence, W.F. and Cole, D.M. 1982. Acoustic emissions from polycrystalline ice. Cold Regions Sci. and Technol., Vol.5, pp.183-199.

Schulson, E.M. 1987. The fracture of ice Ih. J. de Physique, supp. 3, Tome 48, March, pp. 207-220.

Schulson, E.M. and Cannon, N.P. 1984. The effect of grain size on the compressive strength of ice. Proceedings of the IAHR Ice Symposium, Hamburg, Vol. 1, pp.29-38.

Shyam Sunder, S. and Wu, M.S. 1989. Crack nucleation due to elastic anisotropy in polycrystalline ice. Submitted to Cold Regions Science and Technology.

Smith, E. and Barnby, J.T. 1967. Crack nucleation in crystalline solids. Metal Science Journal, Vol.1, pp.56-64.

Stroh, A.N. 1957. A theory of the fracture of metals. Advances in Phys., Vol.6, No.24, pp.418-465.

Zaretsky, Y.K., Chumichev, B.D. and Solomatin, V.I. 1979. Ice behavior under load. Engineering Geology, Vol.13, pp.299-302.

Discussion

Q: L.Gold, NRC

I have found that a specimen can be kept intact during brittle failure by wrapping it in plastic film.

Was strain in your tests measured by strain gauges mounted on specimen?

A: D.M. Cole

The deformation was measured using a strain gauge extensometer that monitored the relative displacement of two circumferential clamps mounted on the ice.

Q: E.M. Schulson, Dartmouth College

You noted that the density of microcracks (within ice strained rapidly at -10°C) saturates before microscopic fracture occurs. Would you please comment on how certain you are of this point? I'm wondering whether the saturation is only an apparent effect, owing to the possible shielding of new cracks by old ones once the density reaches a sufficiently high level.

A: D.M. Cole

As a consequence of the relatively large grain size (d = 5.0 mm, specimen diameter was 50 mm), the microcracks were clearly discernable and the development of the crack population could be visually monitored with reasonable confidence. This would not be possible in fine-grained ice since the high number of small cracks renders that material nearly opaque under similar conditions. In addition, I have noticed this behaviour in numerous monotonically-loaded specimens as well as in ice-structure interaction tests on freshwater columnar ice. It has only now been possible to examine these stages of compressive failure in detail because of the loading fixture which we have recently developed.

Some Questions on Ice and Ice Cover Fracture in Compression

R.V. Goldstein and N.M. Osipenko

Institute of Mechanical Problems,
Academy of Science of USSR,
Prospect Vernadskogo 101, 117526 Moscow, USSR

Abstract In the paper there are considered some mechanical models of fracture processes in ice and ice cover on various scales under compression, including the dependence of ice compression strength on its porosity and strain rate, growth of ridges as cracklike faults with overlapping surfaces in ice cover under compression, the possibilities for improving capacity of ships or icebreakers to fracture ice cover in conditions of ice compression.

1. Ice Compression Strength and Its Porosity

Firstly let us consider ice compression strength on the scale of the laboratory specimen. The value mainly depends upon ice structure, salinity and so on (Mellor, 1980; Bogorodsky and Gavrilo, 1980). For characteristic columnar-grained sea ice specimens, orientation is also essential. Under compression in the direction perpendicular to the columns the microfracture mechanism is connected with the activity of structural elements transforming external compressive stress into local tensile or shear stresses. Shear stresses play the main role at small strain rates. At large strain rates brittle fracture prevails connected with growth of mode I microcracks.

Presence of voids leads to a dependence of compressive ice strength on its porosity (Weeks and Assur, 1977). The dependence is characteristic for columnar-grained sea ice, when compression acts perpendicular to the direction of the columns. Let us consider the porosity dependence of compressive columnar-grained ice strength using the model proposed in Goldstein et al. (1974). We shall model a block of ice as a medium

S. Jones · R. F. McKenna · J. Tillotson · I. Jordaan (Eds.)
Ice-Structure Interaction
IUTAM-IAHR Symposium St. John's, Newfoundland, Canada

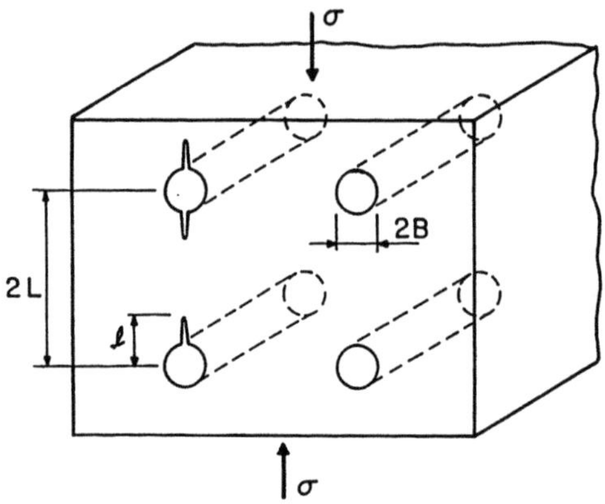

Figure 1. Ice block model with parallel cylindrical voids.

containing parallel cylindrical voids of characteristic radius with characteristic distance 2L from one axis to another (Fig. 1).

The stress state in a block under compressive loads acting perpendicular to the axes of the voids can be represented by the stress state in an elastic plane with circular voids. It is well known that near the void contour under axial compression there arise tensile stresses normal to the compression direction. The tension leads to mode I crack formation growing from the void contour in the compression direction. For a single void in a plane the crack growth is stable. In a real case of a great number of voids, initial stable crack growth near each void is followed by unstable blending and generation of main ruptures also oriented along the compression direction.

Further we shall use for convenience the following expression approximating numerical results for the stress intensity factor K_I at the ends of the cracks growing symmetrically from the circular void contour in the compression direction (Goldstein et al., 1974).

$$\frac{K_I}{\sigma\sqrt{R}} \approx \frac{8-(\ell/R)}{80[(\ell/R)-1]}, 1.2 < (\ell/R) < 4. \tag{1}$$

The condition of cracks blending (and formation of a main rupture) under the axial compression takes the form

$$\ell \sim L. \tag{2}$$

In the model, which we consider, medium porosity n is connected with (R/L)

$$(R/L) \sim n^{1/2}. \tag{3}$$

Finally from (1)-(3) we obtain

$$\sigma_p \approx \frac{K^*_{IC}}{\pi\sqrt{L}} \; \frac{80(1-n^{1/2})}{n^{1/4}(8n^{1/2}-1)}. \tag{4}$$

Here K^*_{IC} - ice fracture toughness on the microcracks scale and it is taken into account that for ice the characteristic distance between voids L is a more stable parameter than the voids radius R. Formula (4) determines ice compression strength dependence on its porosity for a faintly porous ice.

Comparison of (4) with experimental results (Weeks and Assur, 1977; Nadreau and Michel, 1984) shows good agreement. For convenience we take $R \sim 2L\sqrt{\nu/\pi}, \nu-$ brine volume in ice. Then from (4)

$$\sigma_p \approx \frac{24K^*_{IC}}{L^{1/2}\nu^{1/4}} \frac{(1-1.12\nu^{1/2})}{(9\nu^{1/2}-1)}. \tag{5}$$

According to (Weeks and Assur, 1977) $2L \approx$ const, $2L \approx 0.25mm$. Then the best agreement with experimental results (Fig. 2) we obtain taking $K^*_{IC} \approx 0.2kg/cm^{3/2}$. The K^*_{IC} value is about an order less than the value of macroscopic ice fracture toughness (Goodman, 1979; Goldstein and Osipenko, 1983).

Notice that (5) is more adequate for relatively high strain rates and ice strength.

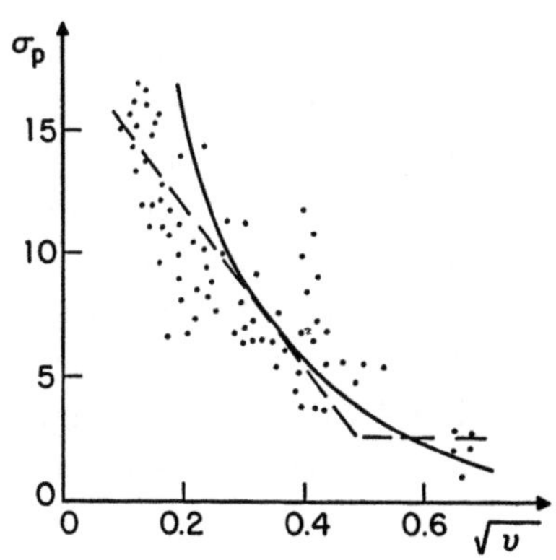

Figure 2. σ_p vs. brine volume ν.

2. Influence of Strain Rate in Compressive Strength

In this connection it is interesting to estimate an influence of a strain rate in the limits of the presented model. Ice compressive strength σ_p depends upon strain rate nonmonotonically (Mellor, 1980; Bogorodsky and Gavrilo, 1980; Sunder, 1986). The maximum value corresponds to $\dot{\varepsilon} \sim 10^{-2}S^{-1}$, for subsequent growth of $\dot{\varepsilon}$ the value of σ_p drops and then stabilization of its level occurs.

Fracture at low strain rates $\dot{\varepsilon} < 10^{-4}s^{-1}$ is well described by a phenomenological dependences of the types: $\dot{\varepsilon} \sim \sigma^m, \dot{\varepsilon} \sim sh(\sigma/\sigma_0)$ and so on (Nadreau and Michel, 1984). As shear types of deformation have the main role, stress averaging occurs because of creep and therefore the sensitivity of the medium to defects is small. Localization of strain and brittle fracture elements growth occurs in the final deformation stages only.

Another picture takes place in the region of visco-brittle transition and with further $\dot{\varepsilon}$ growth. Here is observed a large acoustic activity (Zarezkij and Chymichev, 1972) numerous microfractures and so on. We suppose that the presented model of ice fracture in compression is valid for just that case.

Let us consider at the limits of the model the influence of $\dot{\varepsilon}$ on the ice microfracture toughness K^*_{IC}.

Firstly we note that, contrary to such structural materials as metals, ice fracture occurs without appreciable traces of plastic strains on micro- and macroscales. This circumstance and the obvious influence of nonlinear ice properties on fracture (Mellor, 1980) suggest an idea about the possibility of describing the phenomenon using two scale models of fracture (Goldstein and Osipenko, 1985b). In accordance with the model, two zones are distinguished near the crack tip - outer and inner. In the outer zone strain distribution is quasihomogeneous without localization of deformation. In the inner zone local strains are very high. It is possible to connect the viscobrittle transition with deformation peculiarities in the inner zone, more sensitive to strain rate change. Let us identify the inner zone dimension in ice with the characteristic thickness of the quasiliquid layer (Savel'ev, 1980), covering every ice surface at least at temperatures $\geq -15^oC$. Water as a polar liquid has near a wettable surface an ordered structure with properties of a viscoelastic liquid and with great strength in thin layers. The model allows us to explain some observable effects.

As usually in fracture mechanics we can write approximately, that

$$K_{IC} \sim \sigma\sqrt{r}, \tag{6}$$

where σ and r are specific stress and structure size of the material characterizing fracture processes in the end region of the crack.

In the model we identify r with quasiliquid layer thickness and σ - with its strength. Let us consider the layer as a Maxwell medium with the relation for $\dot{\varepsilon}$

$$\dot{\varepsilon} = \frac{1}{E}\dot{\sigma} + \frac{1}{\eta}\sigma. \tag{7}$$

Here η - viscosity coefficient, E - Young's modulus. As a characteristic time moment in a porous medium fracture process, we take the moment of blending of the cracks growing from adjacent voids. For the moment $\dot{\varepsilon}$ at the crack tip is proportional to $\dot{\varepsilon}$ of the whole body. Then from (7) for fixed $\dot{\varepsilon}$

$$\sigma = \dot{\varepsilon}\eta(1 - \exp(-(E/\eta)t)). \tag{8}$$

For small $\dot{\varepsilon}$ the quasiliquid layer has approximately constant thickness and lines all crack surfaces completely. It is convenient to offer loading time t in (8) as the characteristic time of crack length extension on the value of the layer thickness. At constant layer thickness the time is inversely proportional to $\dot{\varepsilon} : t \sim \varepsilon^{-1}$. We obtain from (6) and (8)

$$K^*_{IC} \approx \dot{\varepsilon}\eta(1 - \exp(-E/\eta\dot{\varepsilon}))\sqrt{r}. \tag{9}$$

After expansion of the exponent the series in (9) takes the form

$$K^*_{IC} \approx E\sqrt{r}(1 - \frac{1}{2}\frac{E}{\eta\dot{\varepsilon}}).$$

It is seen from (9) that K^*_{IC} tends to constant level with $\dot{\varepsilon}$ growth, if the quasiliquid layer has a constant thickness. However full restoration of the layer thickness to equilibrium level during crack propagation is possible only at very low local strain rates.

Increasing the strain rate can lead to a local decrease of the layer thickness because the balance between the processes of restoration and fracture of the quasiliquid layer is upset. Micromechanics of such phenomena up to date has not worked (Deriagin et al., 1985).

Suppose that the extension of viscous layer is proportional to strain rate. Then the thickness of the layer is also $\sim \dot{\varepsilon}^{-1}$. Thus from (9) (with accuracy of a coefficient C)

$$K_{IC}^{*} \approx C\dot{\varepsilon}^{-1/2}\eta(1 - \exp(-E/\eta\dot{\varepsilon})). \tag{10}$$

Consider for simplicity the case $(E/\eta) << 1$. Then the function $K_{IC}^{*}(\dot{\varepsilon})$ has a maximum at $\dot{\varepsilon} \approx 3/2(E/\eta)$. Further increase of $\dot{\varepsilon}$ leads to a decrease of effective K_{IC} value.

Let us make some estimates. Maximum ice strength value corresponds (according to experimental results) to $\dot{\varepsilon} \sim 10^{-2} - 10^{-3}s^{-1}$. If we take the viscosity of surface layer approximately equal to ice viscosity at -0°C, i.e. $\eta \sim 10^{10}Pa \cdot s^2$, then we obtain for $E \sim 10^2 - 10^3 kg/cm^2$. From (9) taking $r \sim 10^{-2}Nm$ (Savel'ev, 1980) it follows that

$$(K_{IC}^{*})_{max} \approx 0.03 \div 0.3 kg/cm^{3/2}.$$

The value is close to the K_{IC}^{*} estimate given above in section 1. The described model allows us to explain the mechanism of rate dependence of ice strength and to understand qualitatively the effect of an increase in $\dot{\varepsilon}$ corresponding to viscobrittle transition at the ice prestrain (Kuehn et al., 1988).

3. Ice Cover Resistance to Ridge Growth

The process of fracture in compression takes place at a larger scale in ice cover during ridge formation.

Let us consider the ice cover as a thin elastic plate on a hydraulic foundation. Suppose that the ice plate has a cracklike defect or breach and the compression acts perpendicularly to the breach direction. In the end regions of the breach, one specific form of localized fracture can take place. Since the ice plate thickness is small compared with the other plate dimensions, then the fracture products can be displaced out of the plane of the plate and surfaces of growing cracklike defect can overlap. Such a form of fracture has been discussed by the authors (Goldstein and Osipenko, 1985a). In the end regions of the described "crack of compression" various mechanisms of fracture at the compression and bearing-out of fracture products can be seen.

Because of the obvious analogy of the problem of the crack of compression growth and the problem of tensile crack growth it is possible to introduce a special material parameter characterizing the resistibility of plate material to the propagation of the crack of compression. The parameter is the critical value of the compressive stress intensity factor at the end regions of a crack with overlapping surfaces in a plate. For ice we call this parameter - hummock formation resistance K_h.

Value of K_h, like K_{IC}, can be determined experimentally. For thin ice plates in particular $K_h \approx 200 kg/cm^{3/2}$ (i.e. K_h is an order higher than K_{IC}).

It is possible to determine some analytical dependencies of K_h considering various mechanisms of fracture in the end regions of the crack of compression (cleavage, combined shear, system of microrupture and so on) (Goldstein and Osipenko; 1985). All fracture criteria in the inner problem (in the end regions) contain strength characteristics of the material depending on loading parameters, in particular on strain rate. This leads to the existence of various regimes of the crack of compression propagation.

As an example we now consider the case when in the end regions of the crack of compression local loss of stability of a thin ice plate takes place (some estimates show that the mechanism can work up to 30-50 cm ice plate thickness h). Local loss of stability leads to formation of a segment of individual fold in the end region. On the fold crest (and then on its sides) there are derived cracks because of bending stress action or local fracture takes place at the compression (Fig. 3). For quantitative analysis of the mechanism we shall use an end region model like the Leonov-Panasuk-Dagdel (LPD) model.

Suppose that the unfractured part of the loss of stability fold is an analog of narrow plastic zone-thin strip, where average stresses σ_y normal to the fold axis are constant in the plate plane. The estimate of the stress (from the stability condition of the part of the end region considered as

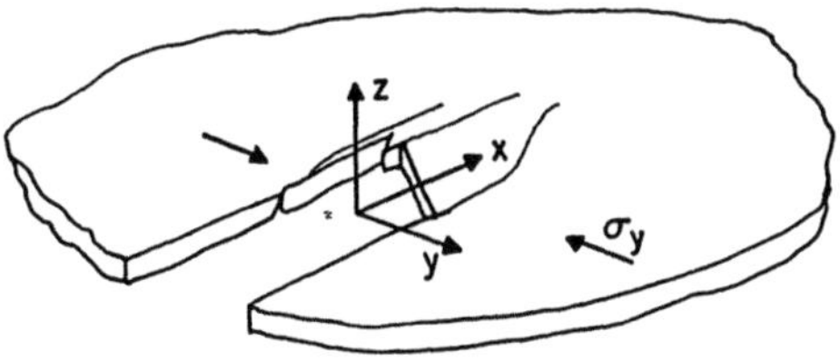

Figure 3. Formation of cracks at a fold crest.

a beam of unit width on elastic foundation) leads to

$$\sigma_y \approx 1/3\sqrt{Ehk}. \tag{11}$$

Where k is the coefficient of elasticity of foundation. Let the fracture condition have the form $|\sigma_{max}| = \sigma^*$, where σ_{max} is the maximum stress in the bending beam section, σ^* the effective strength of the plate material. It is possible to find the values of σ_{max} considering elastic equilibrium of the fold element as an effective beam of the length H (half of the wave of loss of stability of the beam on an elastic foundation), loaded on the end section with the force $F = \sigma_y h$

$$\sigma_{max} = \frac{6M}{h^2} \pm \sigma_y. \tag{12}$$

Here signs $\pm$ correspond to compression or tension respectively, M - bending moment on the fold crest ($M \sim F d_z, d_z$ - vertical displacement of the fold crest). Consider now a limiting rapprochement of fold edges in the ice cover plane at folds bending in the moment, when fracture condition takes place. This limiting reproachment can be treated as analogue of the critical crack opening displacement in model LPD. It is possible to present for small flexure the relation between longitudinal d_y (along σ_y direction) and transverse d_z displacements on the fold crest in the form

$$d_y \sim \frac{\pi^2}{2} \cdot \frac{d_z^2}{H}, \qquad H = \pi \left(\frac{Eh^3}{12k}\right)^{1/4}, \tag{13}$$

where H is a half wave length of a loss of stability fold. Hence, we obtain using intercommunication between the crack opening and fracture toughness in LPD model

$$K_h \approx \sqrt{d_y \sigma_y E} \approx A E^{5/8} (kh)^{3/8} (\sigma^* \sqrt{\frac{3k}{E}} \pm \sqrt{h}) \tag{14}$$

$$A \approx 1.$$

For small ice thicknesses [1]

$$K_h \approx A E^{1/8} k^{-1/8} h^{3/8} \sigma^*. \tag{15}$$

[1] Note the analogous result we can derive from (6), supposing, that fracture takes place, when stresses on the contour of a notch of radius $\sim H$ achieve limiting value σ^*.

We can now estimate a connection between the velocity of local strain in the fracture region and the velocity of compression crack motion

$$\frac{\partial \varepsilon}{\partial t} = \frac{\partial \varepsilon}{\partial x} \cdot \frac{\partial x}{\partial t}. \tag{16}$$

Further, $(\partial x/\partial t) = \dot{\ell}, (\partial \varepsilon/\partial x) \sim \varepsilon_{max}/\Delta, \varepsilon_{max} \sim (d_y/H)$, where Δ - longitudinal fold dimension. According to LPD model

$$\frac{d_y}{\Delta} = \frac{8\sigma^*}{\pi E}, \tag{17}$$

hence

$$\dot{\varepsilon} \approx B\dot{\ell}\sigma^* \left(\frac{k}{E^5 h^3}\right)^{1/4}, B = \frac{8}{\pi^2}(12)^{1/4}.$$

Suppose that ice rheological properties become apparent mainly in the compressive crack end region and affect ice strength only (here we have in mind compressive strength σ_p).

Estimate for example the influence rheology, using Glen presentation (Assur, 1980)

$$\dot{\varepsilon} = D\sigma^n. \tag{18}$$

From (15) and (17) we find

$$K_h \approx D^*(\dot{\ell}\sigma_o^n)^{1/(n-1)} \left[\frac{k^{(3-n)}}{E^{(11-n)} h^{3(3-n)}}\right]^{1/8(n-1)}. \tag{19}$$

Comparing the value of K_h with the stress intensity factor values in various loading schemes it is possible to derive laws of compressive crack growth. Particularly, for quasistatic growth of a crack in ice cover, loaded at the infinity with uniform compression acting along the normal to the crack axis, we can write the condition of limiting equilibrium in the form

$$K_I \sim \sigma_\infty \sqrt{\ell}. \tag{20}$$

Hence

$$\dot{\ell} \sim \ell^{1/2(n-1)} (k^{(n-3)} E^{(11-n)} h^{(9-3n)})^{1/8} \sigma_\infty^{(n-1)} \sigma_o^{-n}. \tag{21}$$

It is known from experiments (Mellor, 1980; Nadreau and Michel, 1986) that there exist three characteristic ranges of n values. At very low strain

rates ($\dot{\varepsilon} < 10^{-8}s^{-1}$) $n \sim 2$. The value of n increases as $\dot{\varepsilon}$ grows. In the range of $\dot{\varepsilon} \sim 10^{-6} - 10^{-3}s^{-1}$ usually $n \sim 5$ or 6. At the strain constraint and small pressure $n \sim 3$, if $\dot{\varepsilon} \sim 10^{-6}s^{-1}$.
According to these possibilities we obtain from (21)

$$\begin{aligned} n &= 2 \quad \dot{\ell} \sim \ell^{1/2}h^{3/8} \quad K_h \sim \dot{\ell}h^{-3/8} \\ n &= 3 \quad \dot{\ell} \sim \ell \quad K_h \sim \dot{\ell} \\ n &= 5 \quad \dot{\ell} \sim \ell^2 h^{-3/4} \quad K_h \sim \dot{\ell}^{1/4}h^{3/16}. \end{aligned} \tag{22}$$

It is seen, that hummock growth resistance K_h increases with increase of compression crack propagation in all cases considered. This leads to a stable regime of hummock development according to K_h. Indeed an increase of stress intensity factor at the extension of a compression crack is compensated for by the growth of effective K_h at the appropriate increase in velocity of compression crack propagation. Unstable, catastrophic regimes come at strain rates, which correspond to maximum strength ($n \to \infty$). It is easy to draw a conclusion about instability of compression cracks in the range of maximum strength using (10), connecting σ_p and $\dot{\varepsilon}$ in the fracture model considered in the previous section.

Note that the influence of various parameters in (22) changes depending on the n -value. In particular, the ice thickness increases crack velocity at small $\dot{\varepsilon}$, and reduces it at large $\dot{\varepsilon}$. On the whole the velocity of hummock (compression crack) propagation characterizes stress level and mechanical properties of ice. At the initial state of slow propagation $\dot{\ell} \sim \sqrt{\ell}$, then compression crack accelerates and a transition to instability occurs. The estimates show that the velocity of the compression crack tip $\sim (1-5).10^{-3}m/s$ take place i.e. for $\sigma_\infty \sim 0.2kg/cm^2, \ell \sim 1km$, $n = 5, h \sim 10cm$, $\sigma_o \sim 100 \pm 50kg/cm^2, E = 10^5kg/cm^2$. In the case $\dot{\varepsilon}$ max in the tip equals $\dot{\varepsilon} \sim 10^{-6} - 10^{-5}s^{-1}$.

This corresponds to the range, where $n = 5$. The system behaviour can be different at the other variants of loading. For example, under the action of concentrated loads at the centre of the compression crack $K_I = P\sqrt{\pi\ell}$ and for $n = 5$ we have

$$\dot{\ell} \approx \frac{P^4}{\ell^2}\frac{1}{h^{3/4}}. \tag{23}$$

We have a stable regime of the compression crack propagation with reduction of the velocity.

4. Artificial Hummock Formation Near Structures

More complicated forms of fracture in compression take place in processes of ice cover interaction with the piles of an ice-resistant structure. The compression cracks considered above can serve for unloading of piles from the ice cover pressure. In the case when the embryo of hummock is initiated, i.e. with an artificial cut, it can play a role of an initial defect. The state of limiting equilibrium of the compression crack is connected with the condition $K_I = K_h$.

Suppose, for example, that a compression crack having a length larger than the diameter of a single pile is disposed near the pile and is oriented transverse to the direction of ice motion. The ice cover can be compressed with the stresses σ_∞. If the ice cover moves relative to the pile, then a system of loads arises, which can be presented in the outer problem as a pile reaction P (Fig. 4).

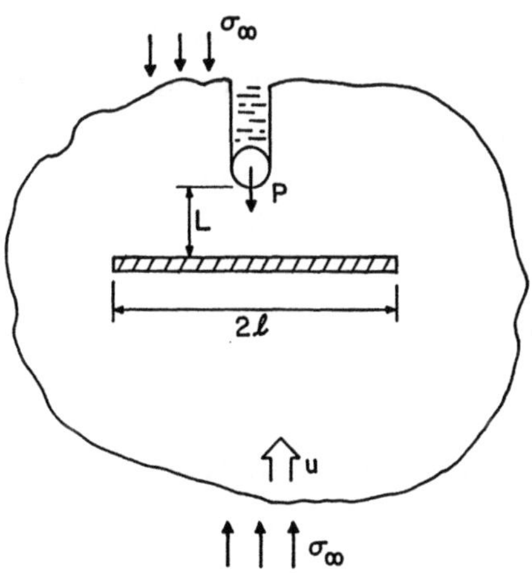

Figure 4. Pile reaction due to relative motion between ice sheet and pile.

Let us consider for definiteness, that ice compression and ice motion take place in the direction perpendicular to the direction of a compression crack, having length (2ℓ). Suppose that the pile is disposed in the middle range of the crack at a distance L from the crack.

In this case the following expressions of the stress intensity factors K_I, K_{II} can be obtained (Murakami et al., 1987)

$$K_I = \frac{P\sqrt{\ell(1+\nu)}}{8\sqrt{\pi(\ell^2+L^2)}}\left(\frac{4}{1+\nu}+\frac{2L^2}{\ell^2+L^2}\right)+\sigma_\infty\sqrt{\pi\ell}, \tag{24}$$

$$K_{II} = \frac{P(1+\nu)}{8\sqrt{\pi\ell}}\left[\frac{L}{\sqrt{\ell^2+L^2}}\left(\frac{4}{1+\nu}+\frac{2\ell^2}{\ell^2+L^2}\right)+2\frac{1-\nu}{1+\nu}\right], \tag{25}$$

here ν - the Poisson ratio.

The following variants of the system reaction are possible. At small crack lengths $\ell < L$ value of $|K_I|$ increases as the length of compression crack increases. For further crack length growth, $|K_I|$ decreases. In the case $K_h = const$, then we will have a stable regime of compression crack growth, when for crack length growth it is necessary to increase P value.

This regime changes into $|K_I|$ growth under the action of ice compression σ_∞. An active artificial hummock formation can serve as an element of pile protection at various regimes of the system.

Observe now the connection between the values of pile reaction P and compression crack velocity $\dot{\ell}$ in the case when $K_h = K_h(\dot{\ell})$, using e.g. (22) at $n = 5$. Suppose that the ice cover displacement velocity $\dot{u} = const$ and take into account that the average displacements of the crack contour of a uniformly loaded crack are proportional to the crack length. Then we derive for the load on the pile from (22), (24) at $\ell < L$

$$P \sim \frac{m\dot{\ell}^{1/4}L}{\ell^{1/2}}, m = c^*\sigma_o^{5/4}\left(\frac{h^3}{k^4E^3}\right)^{1/16}. \tag{26}$$

Here it is supposed that $K_h \sim \dot{\ell}^{1/4}$. Let us assume $\dot{\ell} \sim \dot{u}(E\ell/P)$, then

$$P \sim \left(\frac{m^4\dot{u}EL^4}{\ell}\right)^{1/5}. \tag{27}$$

If the pressure P on the pile is maintained constant (because of ice fracture in the neighbourhood of the pile), then

$$\dot{\ell} \sim \frac{P^4\ell^2}{m^4L^4}. \tag{28}$$

When the compression crack length increases $\ell >> L$

$$m(\dot{\ell})^{1/4} \sim \frac{P}{\sqrt{\ell}} + \sigma_\infty\sqrt{\ell}. \tag{29}$$

At small σ_∞ and $P = const$ the crack braking is possible. If $\dot{\ell} \approx const$, then P has the maximum at $\ell \approx (m/2\sigma_\infty)^2\sqrt{\ell}$.

The above considered example demonstrates that there exists a possibility of control of loads on an engineering structure by means of artificial hummock formation.

Note that in the case, when $K_{II} \neq 0$, the compression crack trajectory will be curvilinear. Now according to experiments (Goldstein and Osipenko, 1985) we use the hypothesis of generalized normal compression for compression cracks (by analogy with local symmetry criterion for cracks in combined tension and shear). By means of the hypothesis we estimate range of deflection angles φ of the compression crack from the rectilinear trajectory. For small φ from (Erdogan and Sih, 1963) it follows, that

$$\varphi \sim -2K_{II}/K_I. \tag{30}$$

The bounds show that in the problem being considered a hummock grows enveloping the pile.

5. Reduction of Ice Cover Compression Action on Ships

Under the action of ice compression the strain constraint at the fracture increases reducing icebreaking capacity of ships (Kashteljan et al., 1968). It is possible to suggest some schemes of motion in conditions of ice compression for improving the ice breaking capacity of ships or some schemes of reducing the action of ice compression (see in this connection Goldstein and Osipenko, 1983). For those purposes it is necessary from the point of view of crack theory to decrease the intensity of compressive stress near the channel tip by means of definite changes of the channel shape (considered as a cut in ice cover).

We can pick out some classes of problems concerning various aspects of ice navigation, for example, increase of icebreaking capability; optimization of the energy expenditures of ships during the action of ice compression (also if standing still for long periods of time) etc. Fracture mechanics application allows us to find nontrivial solutions of the problems (Goldstein and Osipenko, 1982).

Let us consider for example the problem of the arrangement of ships in compressed ice cover. It is convenient in particular, to use the solution (Antipov and Popov, 1988) of the plane elasticity problem for interconnecting cracks (Fig. 5a). If the cracks are lying symmetrically relative to the direction of compression, then the stress intensity factor value K_I diminishes as the value of angle θ decreases from $(\pi/2)$ to 0 and K_I changes the sign at $\theta \sim 11^o$. A positive value of K_I in the crack tips achieves its maximum for $\theta_o \sim 5 - 7^o$.

In application to the problem of ship arrangement in ice compression conditions this means that by making two intersecting channels, as is shown in Fig. 5, with $\theta = \theta_o$ and arranging the ships near the channel tips, we shall provide locally near the ships the action of tensile stress instead of compression of ice cover.

Consider now the case when the axis of symmetry of two intersecting channels (cuts) is inclined at an angle $\sim \pi/4$ to the compression direction (Fig. 5b). When θ increases for the cut approaching the direction perpendicular to the compression direction, the K_I value increases, and for the other cut decreases at the same time.

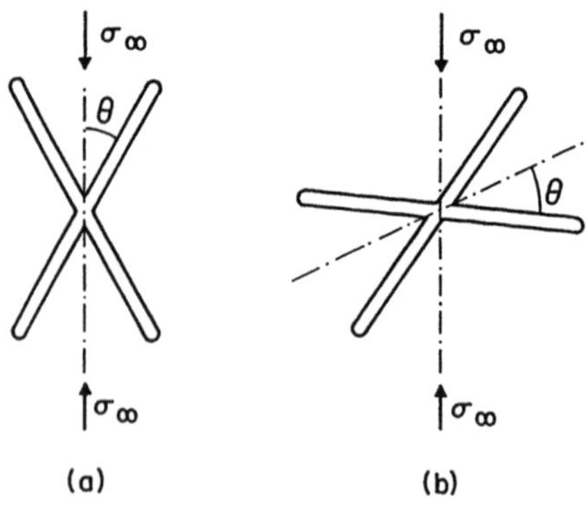

Figure 5. Intersecting channnels under ice sheet compression conditions: a) axis of symetry of channels corresponding to σ_∞, b) axis of symetry of channels inclined by $\sim \pi/4$ to σ_∞.

The tension takes place near the tip of this cut at $\theta > 5^o$. The effect of tension is stronger for cruciform or T-shaped location of the intersecting cuts (Murakami et al., 1987). This result leads us to some suggestion about possible trajectory of icebreaker motion for reducing the action of ice compression (Fig. 6). At the first stage the icebreaker moves across the compression direction at some distance S from the ship, so that $\ell \geq S$. Then the icebreaker can get out along the channel in the region, from which it can make its way towards the ship along the direction of the

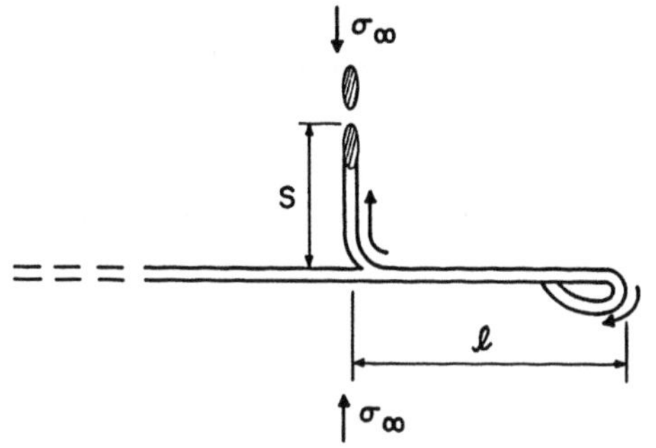

Figure 6. Proposed icebreaker trajectory.

compression. In this case both icebreaker and ship will be in conditions of local tension. It has been considered in (Goldstein and Osipenko, 1986) the other possible type of curvelinear trajectory of the icebreaker motion for reducing compression near the ship.

References

Antipov, Y.A., Popov, G.J. 1988. Plane stress of elastic plane with two intersecting cuts. Mechanica Tverdogo Tela, Vol.52, No.4, pp.617-627 (in Russian).

Assur, A. 1980. Some promising trends in ice mechanics. Physics and Mechanics of Ice. Springer-Verlag, Berlin, Heidelberg, New York, pp.1-15.

Bogorodsky, V.V., Gavrilo, V.P. 1980. Ice. Gydrometeoisdat, Leningrad, 384 p. (in Russian).

Deriagin, B.V., Churaev, N.V., Muller, V.M. 1985. Surface Forces. Nauka, Moskwa, 399 p.

Erdogan, F., Sih, G.S. 1963. On the crack extension in planes under plane loading and transverse shear. Trans. ASME, Ser D, Vol.85, No.4, pp.519-527.

Goldstein, R.V., Osipenko, N.M. 1982. Fracture mechanics of ice cover. Preprint, Instituta Problem Mechaneke Akademee Nauk SSSR, No. 200, 66 p. (in Russian).

Goldstein, R.V., Osipenko, N.M. 1983. Fracture mechanics and some questions of ice fracture. In: Mechanika i Phisika L'da.Nauka, Moskwa, pp.31-62.

Goldstein, R.V., Osipenko, N.M. 1983. Some aspects of fracture mechanics of ice cover. Proc. 7th. International Conference on Port and Ocean Engineering Under Arctic Conditions (POAC 1983), Helsinki, Finland, Vol. 3, pp.132-143.

Goldstein, R.V., Osipenko, N.M. 1985a. Some mechanisms of localized fracture of ice cover under the action of compression. Proc. 8th International Conference on Port and Ocean Engineering Under Arctic Conditions (POAC 85), Narssarssuaq, Greenland, Vol. 3, pp.1170-1188.

Goldstein, R.V., Osipenko, N.M. 1985b. Two scale models of low alloy steel fracture toughness. II Vsesouznyi Simposium po Mechanike Razrushenija. Gitomir. Abstracts., Vol.II, p.72.

Goldstein, R.V., Osipenko, N.M. 1986. Ice fracture mechanics and some of its applications. Proc. POLARTECH 86, VTT Symp. 70, Helsinki, Espoo, Finland, Vol.1 pp.197- 210.

Goldstein, R.V., Ladugin, V.M., Osipenko, N.M. 1974. Model of fracture of low porosity material under compression and tension. Phisico-technicheskie Problemy Rasrabotki Poleznych Eskopaemych, No.1, pp.3-13 (in Russian).

Goodman, D.J. 1979. Fracture toughness measurements for polycrystalline ice of high strain rates. Proc. Ice Phys. Workshop, University of Hokkaido, Japan.

Kashtalijan, V.I., Posnyk, T.N., Ryvlin, A.Y. 1968. Ice resistance to ship motion. Sudostroenie, Leningrad, 238 p.

Kuehn, G.A., Schulson, E.M., Nixon, W.A. 1988. The effects of prestrain on the compressive ductile-to-brittle transition in ice. Proc. IAHR Ice Symp., Sapporo, Japan, Vol.1, pp.109-117.

Määttänen, M. 1987. Ten years of ice-induced vibration isolation in lighthouses. Proc. 6th. Int. Offshore Mechanics and Arctic Engineering Symp.,Houston, TX, Vol.IV, pp.261-266.

Mellor, M. 1980. Mechanical properties of polycrystalline ice. Physics and Mechanics of Ice. Springer-Verlag, Berlin, Heidelberg, New York, pp.217-245.

Nadreau, J.-P., Michel, B. 1984. Ice properties in relation to ice forces. Proc. IAHR Ice Symp., Hamburg, Germany, Vol.4, pp.63-115.

Nadreau, J.-P., Michel, B. 1986. Secondary creep in confined ice samples. In Proc. IAHR Ice Symp., Iowa City, Iowa, Vol.I, pp.307-318.

Savel'ev, B.A. 1980. Structure and composition of natural ice. Izdatel'stvo Moskowskogo Universiteta, 280 p. (in Russian).

Stress Intensity Factors Handbook. 1987. Pergamon Press, Oxford, Vol.I, p.640.

Sunder, S.S. 1986. An integrated constitutive theory for the mechanical behaviour of sea ice: Experimental verification. Proc. IAHR Ice Symp., Iowa City, Iowa, Vol.I, pp.253-264.

Weeks, W.F., Assur, A. 1977. Fracture of lake and sea ice. Fracture, Vol.7, Pt. I, Mir, Moskwa, pp.513-623 (in Russian).

Zarezkij, Y.K., Chymichev, B.D.1972. Creep of ice of short duration. Nauka, Novosibirsk. 120 p. (in Russian).

Seasonal Compressive Strength of Beaufort Sea Ice Sheets

G. W. Timco and R.M.W. Frederking

National Research Council of Canada
Ottawa, Ontario, Canada

Abstract An analysis has been performed to calculate the uniaxial compressive strength of first-year sea ice sheets in the Canadian Beaufort Sea. The analysis makes use of a recently-developed model by the authors. It was found that the ice sheet reaches a maximum strength in March/April of each year. At that time typical strength values are 4.5 MPa, 1.5 MPa and 0.5 MPa at strain rates of 10^{-3} s^{-1}, 10^{-5} s^{-1} and 10^{-7} s^{-1}. In general, the strength values for the sea ice are reasonably consistent throughout the Beaufort Sea region, regardless of the severity of the winter. The information supplied in this paper should provide the design engineer with useful information on the strength properties necessary for designing offshore structures in this region.

1. Introduction

In recent years there has been considerable offshore exploration in the Canadian Beaufort Sea (see Figure 1). A number of different types of offshore structures have been designed and built including artificial berm islands, caisson-retained islands and caisson-type structures. For these structures, the ice load was the predominant design criterion. During the autumn, winter and spring months when first-year sea ice surrounds the structures, there can be considerable ice movement. This leads to ice forces on the structures. While first-year sea ice may not represent the most extreme ice condition, it does represent the case for frequently generated ice loads. During an ice loading event, the strength of the ice sheet plays a key role in determining the ultimate load on the structure. When the ice sheet fails by a crushing process, the ice load is assumed to be related to the uniaxial compressive strength of the ice sheet. The nature of this relation is still subject to some questions; however, a cal-

S. Jones · R. F. McKenna · J. Tillotson · I. Jordaan (Eds.)
Ice-Structure Interaction
IUTAM-IAHR Symposium St. John's, Newfoundland, Canada

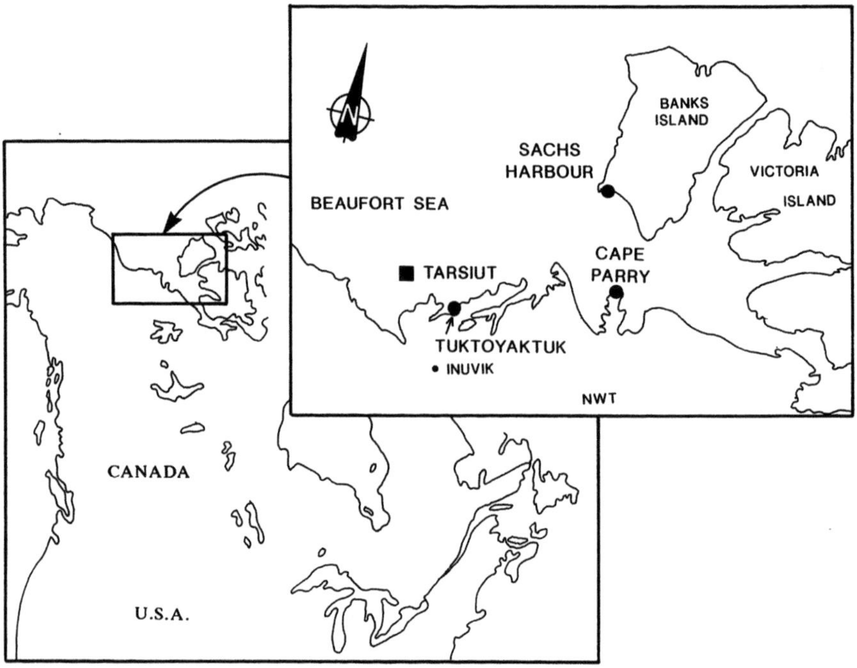

Figure 1. Map showing the location of the Canadian Beaufort Sea, Tuktoyaktuk, Cape Parry and Sachs Harbour.

culated uniaxial compressive strength for the ice sheet provides a basis for comparing ice loads measured on offshore structures at various times during a winter season.

This seasonal variation of sea ice sheet compressive strength for the Canadian Beaufort Sea region is developed in this paper. Using available meteorological data and a model recently developed by the authors (Timco and Frederking, 1990), the spatial and temporal variation of the thickness and compressive strength of first-year sea ice is detailed. This is done for both extreme and average conditions. This information should be useful for interpreting ice loads already measured and for future designs of Beaufort Sea structures.

2. A Model for the Compressive Strength

A model has been developed for predicting the compressive strength of sea ice sheets (Timco and Frederking, 1990). This model makes use of the results of numerous small-scale compression tests on sea ice, and the relationship between the intrinsic and extrinsic properties of the ice sheet. A summary of the steps used in developing the model follows.

The model is based on two assumptions. One, there is no size-effect in compressive strength for sea ice. This assumption appears to be a reasonable one, based on the large-scale compressive strength tests performed by Exxon in two multi-million dollar projects (Chen and Lee, 1986; Lee et al., 1986; Petrie and Poplin, 1986; Wang and Poplin, 1986). Exxon showed that the compressive strength of the full ice sheet was very well represented by the average of the results of small-scale strength tests. For this to be true, the environmental conditions for the small-scale tests have to be the same as if the samples were in situ in the ice sheet. Two, the strength of the ice can be related to the environmental conditions of the ice. This assumption also is reasonable for sea ice (Weeks and Ackley, 1982).

The first step in developing this procedure was to compile the existing data on small-scale strength tests. It was found that the results of 195 small-scale strength tests of the ice were well represented by

$$\sigma_c = 37(\dot{\varepsilon})^{0.22}[1 - (\nu_T/270)^{0.5}], \tag{1}$$

for horizontally-loaded columnar ice, and

$$\sigma_c = 49(\dot{\varepsilon})^{0.22}[1 - (\nu_T/280)^{0.5}], \tag{2}$$

for granular ice where $\dot{\varepsilon}$ is the strain rate and ν_T is the total porosity (brine + air) of the ice in parts per thousand (ppt). The strain rate range for these equations is from 10^{-7} s^{-1} to 10^{-4} s^{-1}. Above this strain rate, premature (brittle) failure of the ice can occur in some, but not all instances. Thus the strain rate range can be extended to 10^{-3} s^{-1} whilst keeping in mind that the ice may fail at lower stresses than that predicted for this higher strain rate range. These equations relate the horizontally-loaded uniaxial compressive strength of sea ice explicitly to grain type, loading strain rate and total porosity, and implicitly to ice salinity, temperature and density.

These ice test conditions were then reformulated in terms of the extrinsic parameters appropriate for the whole ice sheet. For this, the total porosity of the ice (ν_T) was assumed to be the sum of the brine volume (ν_b) and the air porosity (ν_a) in the ice. It was found that the total porosity could be expressed in terms of the ice temperature (T_i), ice thickness

(h_i) and bulk ice density (ρ) as

$$\begin{aligned} \nu_T &= 1 - [\rho/0.917 - 1.403 \times 10^{-4} T_i] \\ &\quad + \rho\, [F_3(h_i)/F_1(T_i)]/[1 + F_2(T_i)], \end{aligned} \tag{3}$$

where the salinity (S_i) of the ice is directly related to the thickness of the ice sheet by

$$\begin{aligned} S_i &= 13.4 - 17.4\, h_i \qquad \text{for} \quad h_i \leq 0.34 \text{ m} \\ &= 8.0 - 1.62\, h_i \qquad \text{for} \quad h_i \geq 0.34 \text{ m} \\ &= F_3(h_i). \end{aligned} \tag{4}$$

The functions $F_1(T_i)$ and $F_2(T_i)$ are given as (from Cox and Weeks, 1982)

$$\begin{aligned} F_1(T_i) &= -4.732 - 22.45\, T_i - 0.6397\, T_i^2 - 0.01074\, T_i^3 \\ &\qquad \text{for } -2 \geq T_i \geq -22.9 \\ &= 9899 + 1309\, T_i + 55.27\, T_i^2 + 0.716\, T_i^3 \\ &\qquad \text{for } -22.9 \geq T_i \geq -30, \end{aligned} \tag{5}$$

and

$$\begin{aligned} F_2(T_i) &= 8.903 \times 10^{-2} - 1.763 \times 10^{-2}\, T_i - 5.33 \times 10^{-4}\, T_i^2 \\ &\quad -8.801 \times 10^{-6}\, T_i^3 \qquad \text{for } -2 \geq T_i \geq -22.9 \\ &= 8.547 + 1.089\, T_i + 4.518 \times 10^{-2}\, T_i^2 + 5.819 \times 10^{-4}\, T_i^3 \\ &\qquad \text{for } -22.9 \geq T_i \geq -30, \end{aligned} \tag{6}$$

where T_i is in ^oC.

For sea ice there is usually a temperature gradient in the ice and the temperature and porosity of the ice sheet vary significantly with depth. In order to ensure that the model calculated the ice strength at the correct environmental conditions, the ice sheet was separated into nine individual layers. The average ice temperature (T_i) of each layer was related to the temperature at the ice surface (T_s) and ice bottom (T_B) by

$$T_{\text{layer } n} = T_s - (2n - 1)[(T_s - T_B)/18], \tag{7}$$

where T_B is approximately -2^oC. The ice surface temperature (T_s) is generally warmer than the air temperature (T_a) due to the presence of snow. The following empirical relation between the ice surface temperature (T_s) and the air temperature (T_a) has been found

$$\begin{aligned} T_s &= T_a \qquad \text{for } -2 \geq T_a \geq -10^oC, \\ T_s &= 0.6T_a - 4 \quad \text{for } T_a < -10^\circ C. \end{aligned} \tag{8}$$

The porosity for each layer was calculated based on the local temperature, salinity and density. Then, using the defining equations from the small-scale tests, the ice strength was determined for each layer. The average of these calculated compressive strengths was used to give the large-scale compressive strength for the full ice sheet. The reader is referred to Timco and Frederking (1990) for full details of the procedure and the development of the defining equations.

In Timco and Frederking (1990) a comparison was made to the measured large-scale strength values reported by Exxon. As shown in Figure 2, there is excellent agreement between the measured values and those calculated using our model. With this approach it is possible to predict the compressive strength of a sea ice sheet. The only two environmental parameters necessary to define the strength are the ice thickness and the air temperature. The average and extreme values of these parameters are easily obtained from climatic data. Hence, average and extreme values of the ice sheet compressive strength can also be obtained.

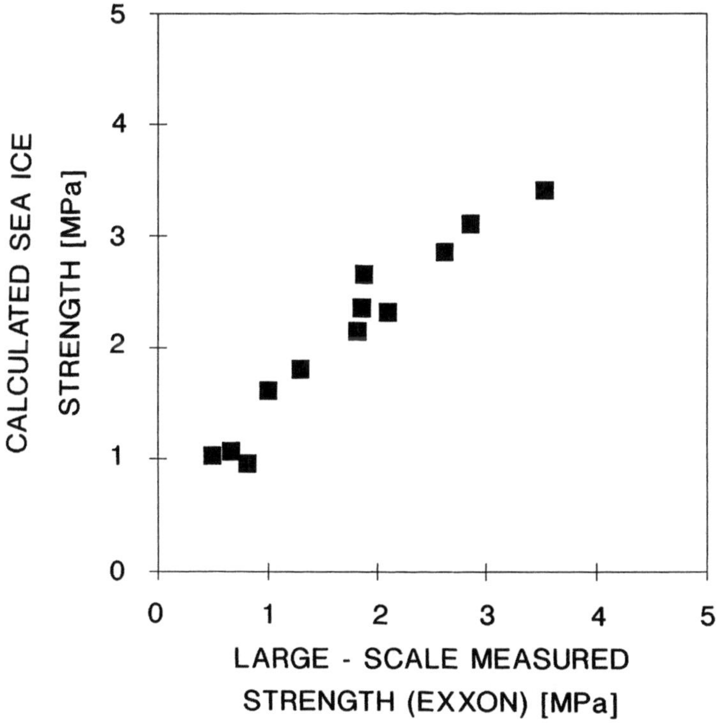

Figure 2. Comparison of the calculated compressive strengths to those measured by Exxon in large-scale tests.

3. Meteorological Information for the Beaufort Sea

In order to use the model it is necessary to have meteorological data from the Beaufort region. These can be obtained from the Atmospheric Environment Service (AES) of Canada. Details of ice thickness and air temperatures are available for Tuktoyaktuk, Cape Parry and Sachs Harbour (see Figure 1). These three sites surround the main drilling region in the Canadian Beaufort Sea. For all of these sites, the average and the upper and lower extremal values can be obtained.

Figures 3 and 4 give the annual air temperature and ice thickness for Tuktoyaktuk. From the figures it is seen that the air temperature reaches a minimum in February and the ice thickness reaches a maximum in April/May of each year. Similar information is given for Cape Parry in Figures 5 and 6, and for Sachs Harbour in Figures 7 and 8. Comparing the figures indicates that there are some regional differences in both the air temperature and ice thickness for these regions. In general the differences are not great. Their influence on the compressive strength will be investigated in the next section.

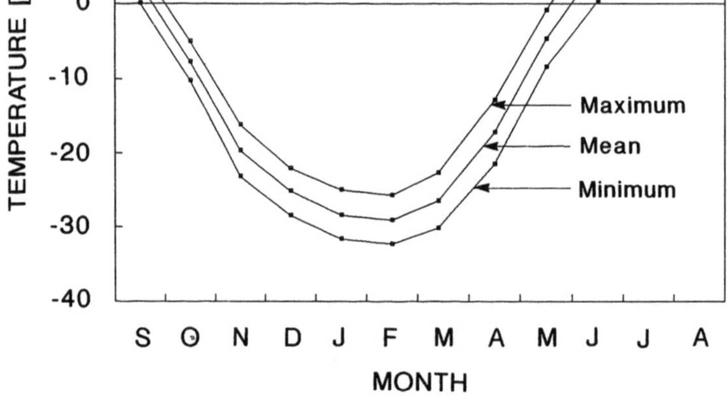

Figure 3. Annual air temperature for Tuktoyaktuk showing both average and extreme temperature conditions.

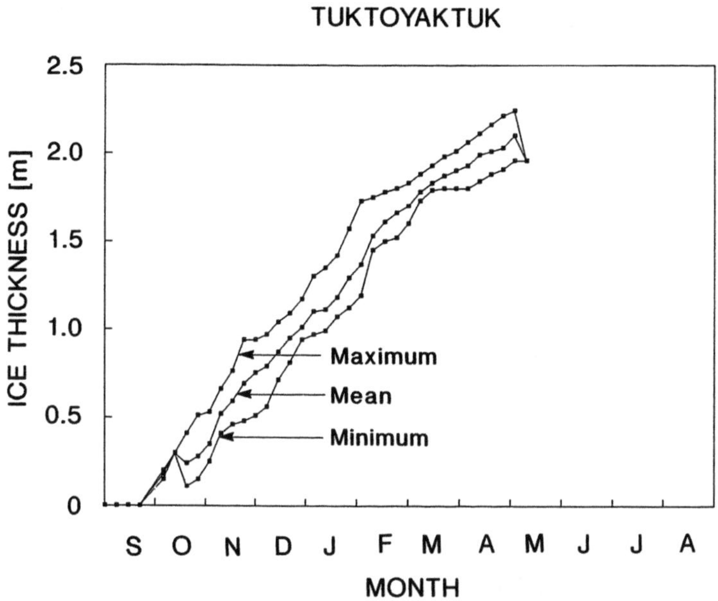

Figure 4. Annual ice thickness at Tuktoyaktuk showing both average and extreme conditions.

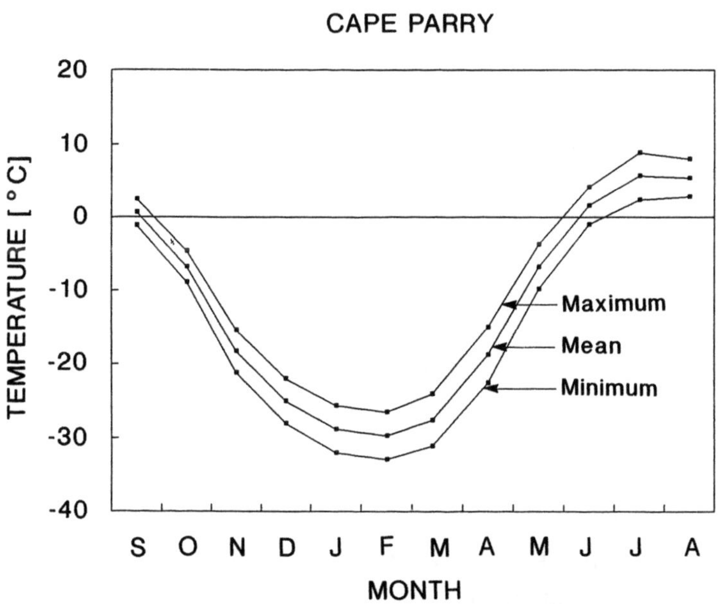

Figure 5. Annual air temperature for Cape Parry showing both average and extreme temperature conditions.

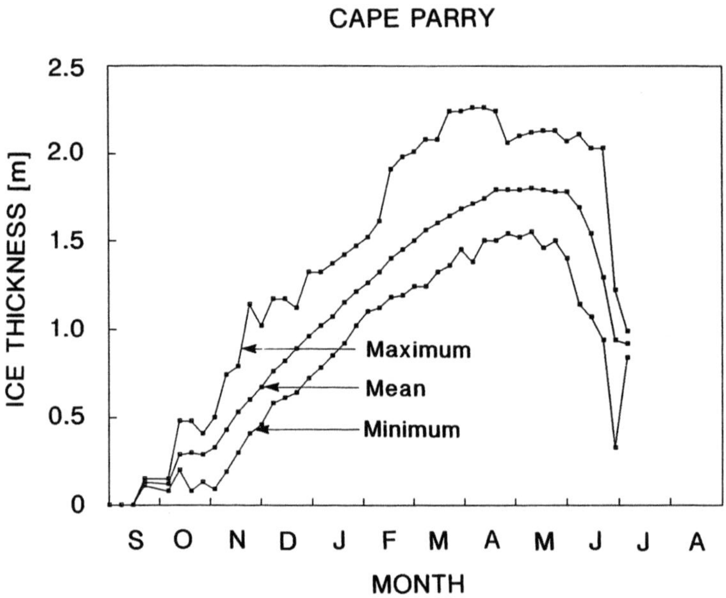

Figure 6. Annual ice thickness at Cape Parry showing both average and extreme conditions.

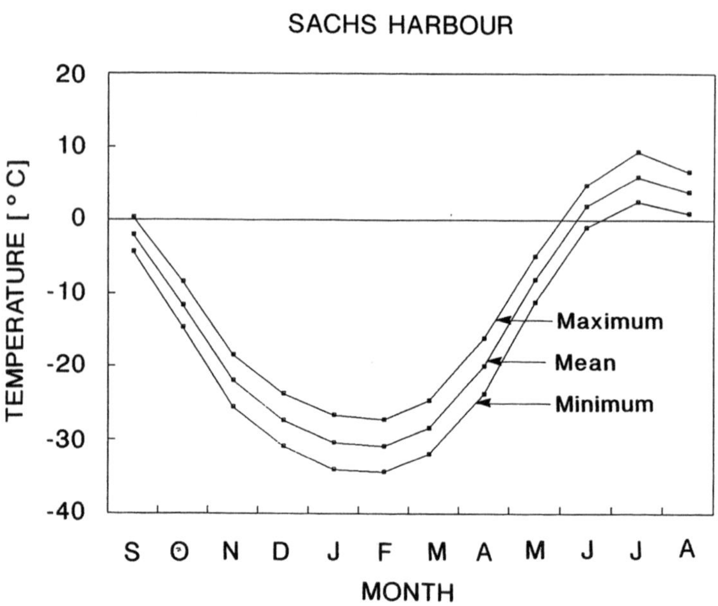

Figure 7. Annual air temperature for Sachs Harbour showing both average and extreme temperature conditions.

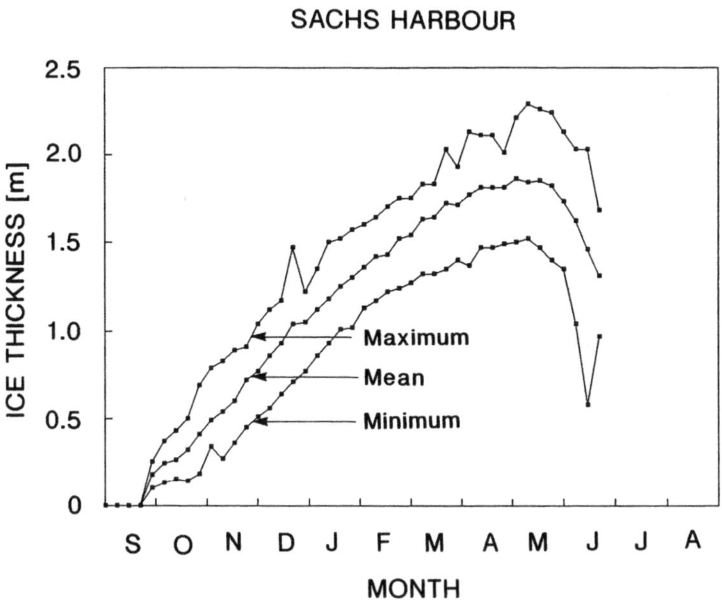

Figure 8. Annual ice thickness at Sachs Harbour showing both average and extreme conditions.

4. Seasonal Compressive Strength

Figure 9 gives the seasonal variation of the compressive strength of first-year sea ice sheets at Tuktoyaktuk. The strength values for an "average" winter and extremal winters are shown. The strength values were determined for a strain rate of 10^{-4} s^{-1}. The average values were determined by using both average ice thickness and average air temperature values in the model. The upper extremal values were calculated using the maximum ice thickness (which has the lowest ice salinity) and minimum air temperatures for each month. The lower extremal values were calculated using the minimum ice thickness and maximum air temperatures for each month. From the figure it is seen that the compressive strength reaches a maximum value in March/April each year. At a strain rate of 10^{-4} s^{-1}, the maximum strength value is approximately 3 MPa. There are some differences between the average values and the extremal values of the compressive strength, but these are only significant at the beginning and end of the season (October and May). For the remainder of the winter, the range of strength is, perhaps surprisingly, not very great.

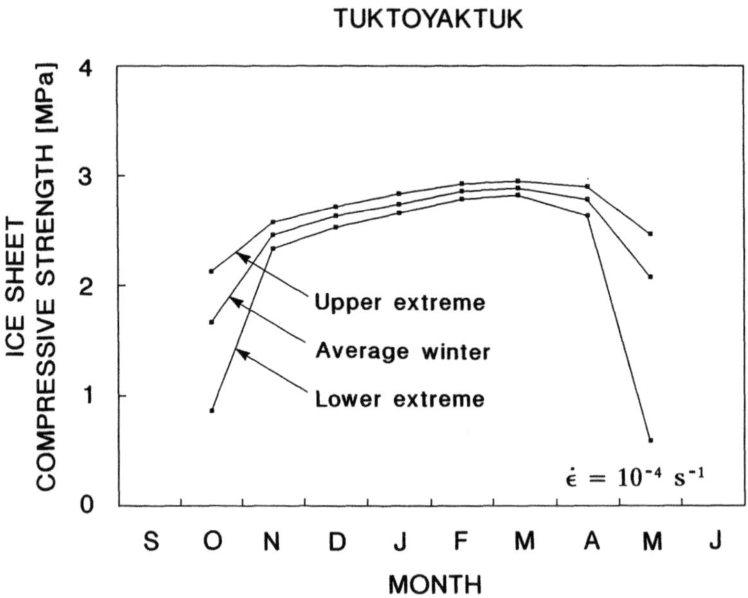

Figure 9. Seasonal variation of the compressive strength of the first-year ice sheets at Tuktoyaktuk.

Figures 10 and 11 show the comparable graph to Figure 9 for both Cape Parry and Sachs Harbour respectively. In comparing these figures for the three sites it is clear that there is reasonable uniformity in the ice sheet strength values in the Beaufort Sea.

Because of this consistency it is possible to generate curves which represent the ice sheet strength of Beaufort Sea ice for both average and severe winter conditions. This is shown in Figures 12 and 13 for strain rates from $10^{-3}\ s^{-1}$ to $10^{-7}\ s^{-1}$. The curves show a general uniformity of the compressive strength at any strain rate throughout the winter. As would be expected, the strain rate has a significant influence on the strength of the ice. At the highest strain rate of $10^{-3}\ s^{-1}$, the ice reaches a strength of approximately 5 MPa in the early spring (March/April). After that, both the strength and the ice thickness decrease. Figures 12 and 13 can be used as good representations of the compressive strength of first-year Beaufort sea ice.

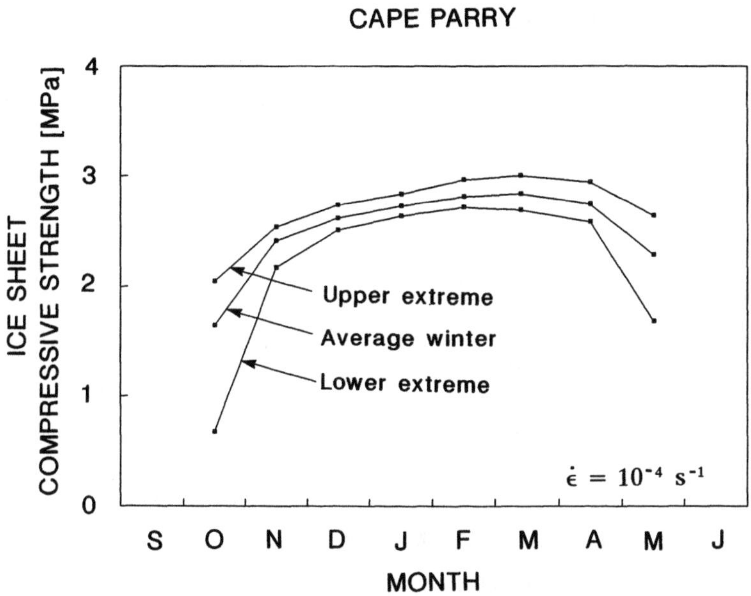

Figure 10. Seasonal variation of the compressive strength of the first-year ice sheets at Cape Parry.

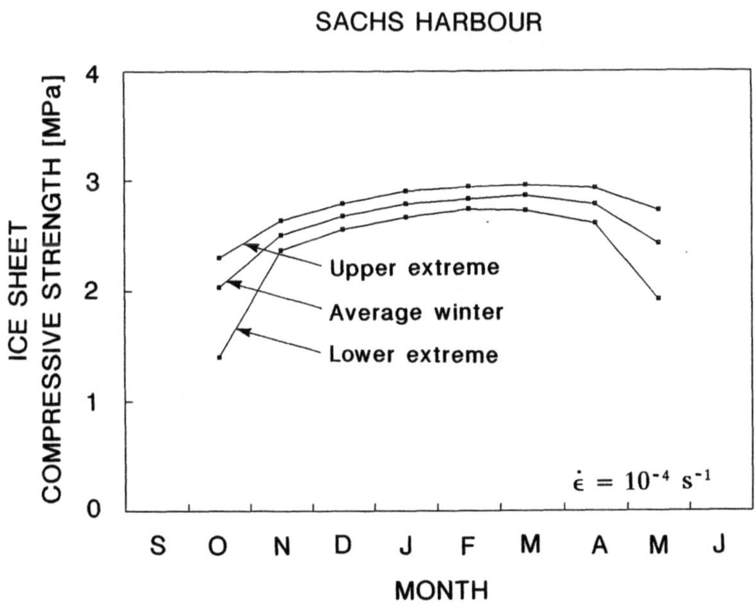

Figure 11. Seasonal variation of the compressive strength of the first-year ice sheets at Sachs Harbour.

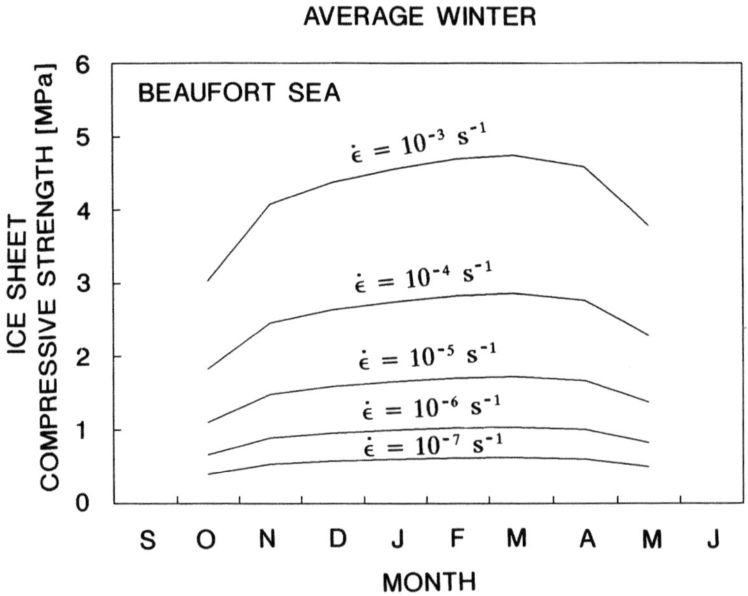

Figure 12. Seasonal variation of the uniaxial compressive strength of first-year sea ice sheets in the Canadian Beaufort Sea for average winter conditions.

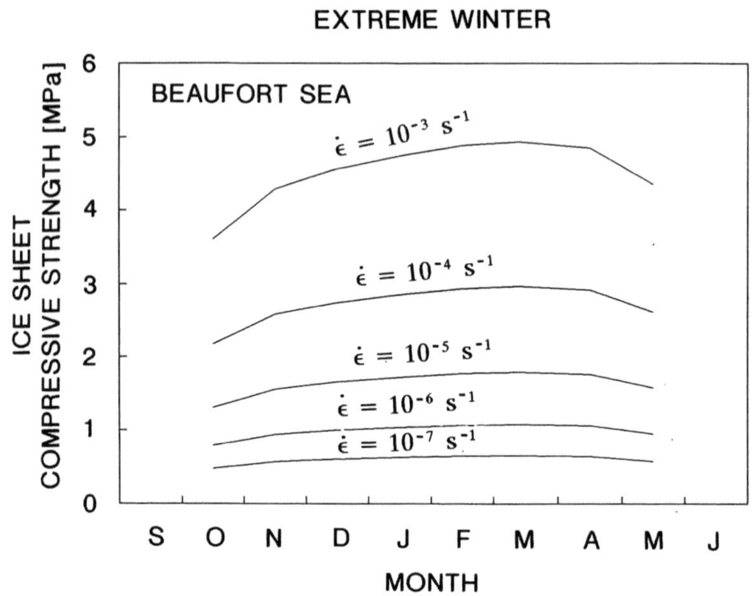

Figure 13. Seasonal variation of the uniaxial compressive strength of first-year sea ice sheets in the Canadian Beaufort Sea for extreme winter conditions.

5. Ice Loads on Molikpaq at the Tarsiut Location

As was mentioned in the introduction, the compressive strength of sea ice sheets can be related to the ice loads measured on offshore structures for evaluating performance experience. The Molikpaq is one structure for which first-year sea ice loading information is available. Hnatiuk and Felzien (1985) have reported measurement results for seven loading events for the 1984/85 winter at the Tarsiut P-45 location (see Figure 1). Four events involving first year sea ice are summarized in Table 1. No ice thickness data were given by Hnatiuk and Felzien (1985). Meteorological data recorded at Tuktoyaktuk, Sachs Harbour and Cape Parry for December 1984 and January 1985 were examined to obtain an average ice thickness, h_i, and air temperature, T_a, for the dates of the 4 events. The values obtained are presented in Table 1. Note that the air temperature, T_a, is the mean of the 5 preceding days. Average ice pressure was calculated from the measured load using the ice thickness h_i and assuming that the ice load was uniformly applied across the full 100 m width of the structure. The strain rate of the interaction was estimated from the empirical formula, $\dot{\varepsilon} = V/4d$, where V is velocity and d is the structure width. Using the known air temperature, the ice thickness and the estimated strain rate for the interaction, ice sheet strengths were calculated based on the model of section 2. The results are given in Table 1. It can be seen that the average ice pressure is about one-half of the ice sheet strength.

Tablè 1. First-Year Sea Ice Loading Events on Molikpaq at the Tarsiut P-45 Location

Date	Ice Load′	Ice Thickness	Ice Velocity	Strain Rate	T_{air}	Pressure	Ice Strength
	(MN)	(m)	(m/s)	(s^{-1})	(°C)	(MPa)	(MPa)
84.12.08	120	0.82	0.10	2.50E-04	-27.7	1.46	3.24
84.12.14	120	0.91	0.05	1.25E-04	-31.8	1.32	2.86
85.01.02	140	1.09	0.05	1.25E-04	-25.7	1.28	2.82
85.01.26	120	1.27	0.05	1.25E-04	-25.4	0.94	2.86

6. Summary and Conclusions

An analysis has been performed to present the meteorological information for the Canadian Beaufort Sea. Using this as input into a model developed to predict the compressive strength of sea ice sheets, representative compressive strength values have been determined for this area. These curves should provide design engineers with information important for the safe and efficient design of offshore structures for the Beaufort Sea.

References

AES (Atmospheric Environment Service) 1982. Updated ice thickness climatology for Canadian stations, Ottawa, Ont., Canada (unpublished).

Chen, A.C.T. and Lee, J. 1986. Large-scale ice strength tests at slow strain rates. Proc. OMAE 86, Tokyo, Japan, Vol.4, pp.374-378.

Cox, G.F.N. and Weeks, W.F. 1982. Equations for determining the gas and brine volumes in sea ice samples. CRREL Report 82-30, Hanover, N.H., U.S.A.

Hnatiuk, J. and Felzien, E.E. 1985. Molikpaq - an integrated mobile Arctic drilling caisson, 17th Annual Offshore Technology Conference, May 6-9, 1985, Houston, Texas, OTC 4940.

Lee, J., Ralston, T.D. and Petrie, D.H. 1986. Full-thickness sea ice strength tests. Proc. IAHR Ice Symposium, Iowa City, Iowa, U.S.A., Vol.1, pp.293-306.

Petrie, D.H. and Poplin, J.P. 1986. Comparison of small-scale and large-scale sea ice strengths. Proc. IAHR Ice Symposium, Iowa City, Iowa, U.S.A., Vol.1, pp. 265-277.

Timco, G.W. and Frederking, R.M.W. 1986. Confined compression tests: outlining the failure envelope of columnar sea ice. Cold Regions Sci. and Tech. Vol.12, No.1, pp.13-28.

Timco, G.W. and Frederking, R.M.W. 1990. Compressive strength of sea ice sheets. Cold Regions Science and Technology. Vol.17, No.3, pp.227-240.

Wang, Y.S. and Poplin, J.P. 1986. Laboratory compressive tests of sea ice at slow strain rates from a field test program. Proc. OMAE 86, Tokyo, Japan, Vol. 4, pp.379-384.

Weeks, W.F. and Ackley, S.F. 1982. The growth, structure and properties of sea ice. U.S.A. CRREL Monograph 82-1 Hanover, N.H., U.S.A.

Discussion

Q: T.J.O. Sanderson, BP London

I disagree with some of the conclusions drawn from the Exxon large-scale tests. These tests were carried out in the ductile deformation regime (less than about 10^{-4} s^{-1}) and might have yielded quite different conclusions if carried out in the brittle strain rate regime (10^{-3} s^{-1} or 10^{-2} s^{-1}). Under these conditions of brittle behaviour I do not accept that "strength" is a true material property.

A: G.Timco and R. Frederking

You are mistaken in your quotation of the range of strain rates for the uni-axial tests performed by Exxon in their large-scale tests. For these tests the strain rate range extended from 8×10^{-7} s^{-1} to 8×10^{-4} s^{-1}. This strain rate range is virtually the same as that described in the current paper and in the model developed by the authors for predicting the compressive strength of sea ice sheets. Therefore both the model and the predicted strengths for the Beaufort Sea are supported by the results of the Exxon tests. However, your statements regarding the concept of "strength" at higher rates is in agreement with our own views. In the paper there is a discussion of the applicable range for which this model should be used.

Q: T.H. Jacka, Australian Antarctic Division

Your annual air temperature data appear to consist of curves of long term mean, long term mean maximum and long term mean minimum data. Your quoted compressive strengths however apply to an average winter and a severe winter. The temperature data you should use then is the mean temperature over e.g. the warmest 5 years and the coldest 5 years.

A: G. Timco and R. Frederking

The temperature data presented in Figures 3, 5 and 7 represent the means for each month of daily maximum temperature, daily minimum temperature and the average of the daily maximum and minimum temperature. We have defined "average" and "extreme" winters in a different manner than that suggested by you; i.e. we have taken a collection of average months and extreme months which would not necessarily correspond collectively to a whole average or extreme winter. Your definition of temperature for an average winter and an extreme winter is perhaps more realistic and could be incorporated into the model. I expect, however that we would not see much different between the two approaches.

Q: L. Gold, NRC Ottawa

One of the problems in comparing the results of field measurements is to obtain reasonable measure of the strength of the ice. Perhaps one should look at the present work not as a means of determining the strength of the ice cover but as a simple means that can readily be standardized for determining a strength index that can be used for correlating results of field measurements and associated failure mechanisms at the structure.

A: G. Timco and R. Frederking

We agree that the results of the present model and the predicted strength curves presented in this paper can be used as a good strength index for correlating past measurements of ice loading events on Beaufort Sea structures. The excellent agreement with the Exxon large-scale data, however suggests more than this. The correlation instills more confidence in the data as a measure of the true strength of the ice cover.

Processes of Deformation and Fracture of Ice in Compression

Ian J. Jordaan and Richard F. McKenna

Faculty of Engineering and Applied Science,
Memorial University of Newfoundland,
St. John's, Newfoundland, Canada A1B 3X5

Abstract The analysis of crushing of ice in compression is addressed. As background, a test result from the Pond Inlet spherical indenter experiments is presented. This shows the dramatic fluctuations in load experienced when crushing is the dominant mechanism of ice fracture. In order to analyse the degradation of strength of ice in compression, damage mechanics is used. Starting with combined Maxwell and Kelvin elements to model undamaged ice, a rate-theory approach is used to generate microcracks. Separate sets are developed, one for the Maxwell and the other for the Kelvin element. The effect of microcracks is to degrade the elastic potential of both elements and to cause an enhancement of viscous movements, taken as being stress-dependent, also for both elements. The analysis is developed in the context of thermodynamic theory and the results are compared to a set of experiments. Good agreement is obtained. Phenomena which lie outside the framework of the viscoelastic elements are discussed. These relate primarily to frictional movements across newly developed crack faces. Stick-slip movement is a possible mechanism in this regard. Sintering, which can be seen as the reverse of the damage process, is also potentially of importance. Immediately following rapid release of strain energy, kinetic effects may need to be considered.

1. Introduction

Ice in compression will initially deform elastically, followed by creep if the load is sustained: at higher loading rates, microcracks appear and the material eventually disintegrates. In the case of indentation of ice sheets or masses, this can be manifested by crushed ice in the form of a powder along a narrow layer adjacent to the loading device.

Other brittle materials, for example concrete and rocks, also show a similar tendency to disintegrate by microcracking under compressive

S. Jones · R. F. McKenna · J. Tillotson · I. Jordaan (Eds.)
Ice-Structure Interaction
IUTAM-IAHR Symposium St. John's, Newfoundland, Canada

states of stress. In designing an engineered system, it is usual practice to analyse the response of the system including only a limited degree of damage and very often only elastic response is considered. Those states corresponding to a high degree of damage have a low probability of occurrence, in practice (if well engineered!) and are not given much attention in terms of analysis.

The same luxury is not permitted the ice engineer or scientist given the task of estimating the loads, as well as any possible variation of the loads with time, acting on a man-made structure. In the crushing mode of failure, repeated cycles of variation in load and consequent ice-induced vibrations are observed. At the same time, the crushed ice is ejected in the form of distinct particles. A full analysis of this process has to address the initial degradation of elasticity associated with the formation of microcracks, as well as the transition to, and the establishment of flow in the crushed ice. An outline of the process as it relates to the dynamics of ice crushing has been given in (Timco and Jordaan, 1987; Jordaan and Timco, 1988). The dynamic behaviour was associated with finite jumps in the thickness of the crushed layer. The peak stresses in the ice were associated with pulverisation events. As an example, to motivate the work, some results of field tests using a spherical indenter will be presented.

2. Medium Scale Indentation: the Process of Damage and Pulverisation

A single result from a series of tests carried out in a grounded iceberg at Pond Inlet (Geotech, 1985) will be reported. The spherical indenter segment of area 1.0 m^2 with a radius 560 mm was used to indent a plane surface in a tunnel excavated in the iceberg. The indenter speed was servo-controlled and varied from a maximum of 100 mm/s at the start of the indentation to zero at the end, according to a cosine curve. The end of the test was at an indention of about 130 mm, or complete penetration of the spherical segment.

The result is shown in Figure 1. The force-time curve shows a sharp variation, typically at 28 Hz during the middle period of the test. The crushed layer is shown in Figure 1(b); generally the thickness of the layer was of the order of 40 mm, although at certain positions it is quite narrow. No doubt this is associated with areas of local high pressure at the

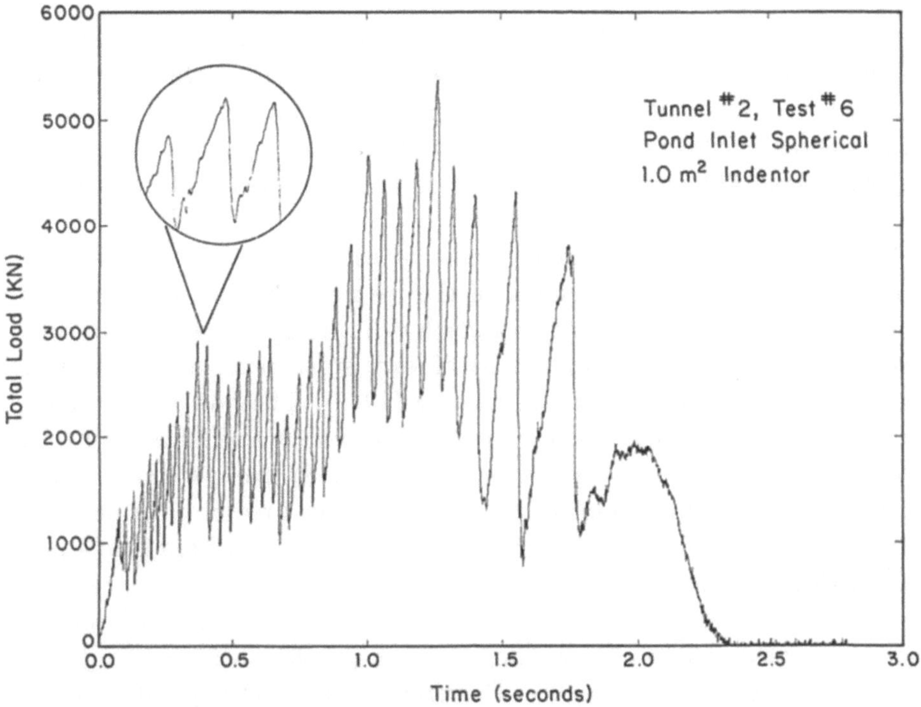

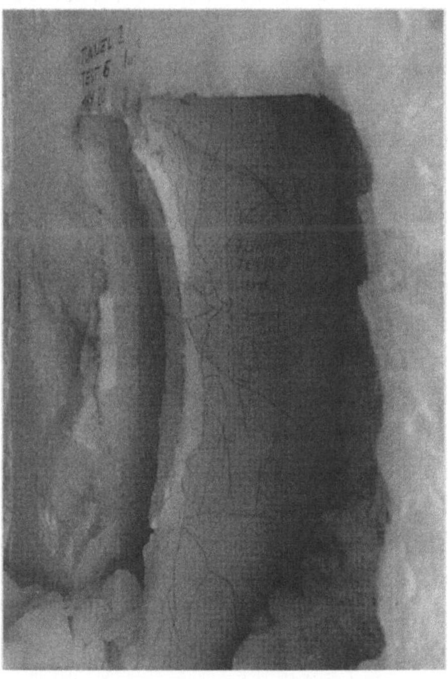

Figure 1. Spherical indenter:
(a) time series of total
load exerted on the indenter;
(b) photograph of cross-section
of the crushed layer.

interface. The period of fluctuation in load corresponds to a distance of about 2 mm, so this is a small fraction of the average layer thickness. The crushed layer was quite distinct and tended to be in a sintered condition after the tests. This raises the possibility that sintering might occur during the tests. The cycles of pulverisation and clearing could be followed by sintering and by repulverisation.

Repeated pulverisation could also occur within a layer if sintering occurs, given the high confinement of the ice near the axis of symmetry in the present tests. It is also noteworthy that the pulverisation in such layers has been observed to consist of very small particles (parts of a grain); see also Kheisin and Cherepanov (1970). The flow of pulverised ice was treated as a slow viscous flow by Kurdyumov and Kheisin (1976); this was also adopted in Jordaan and Timco (1988). No systematic study of the rheological behaviour of crushed ice under pressure has been reported and this subject needs further study. The object here is to study the process leading to disintegration.

Figure 2 shows schematically the development of a crushed layer. In spite of considerable microcracking, ice can preserve a skeletal rigidity, and this is termed damaging ice. Pulverisation marks the point at which the particles of ice move relative to each other and the skeletal structure is lost. Account must be taken of the fact that at the moment when the load drops in the cycles of Figure 1(a), a substantial amount of energy flows from the surrounding ice and in appropriate cases, from the structure and indenter as well, into the pulverising ice. Thus the local strain rates in the transition $A \rightarrow B \rightarrow C$ (Figure 2) will likely be increasing dramatically. The main purpose of our research is to study this transition, using appropriate constitutive theory for an element of ice coupled with numerical analysis based on (for example) the finite element method (e.g. Cormeau et al., 1986) to facilitate numerical computations. The process of damage, pulverisation and extrusion is clearly linked to shear, since the extruding ice has little shear strength compared to the virgin ice, yet it can transmit pressure.

In order to study the behavior of an element of ice and in particular the process of damage, a series of uni- and triaxial tests have been conducted in the laboratory on granulated ice (Stone et al., 1989) together with accompanying analysis (McKenna et al., 1989). In the present paper, the initial matching of theory and experiment will be presented and discussed in relation to the damage and pulverisation process.

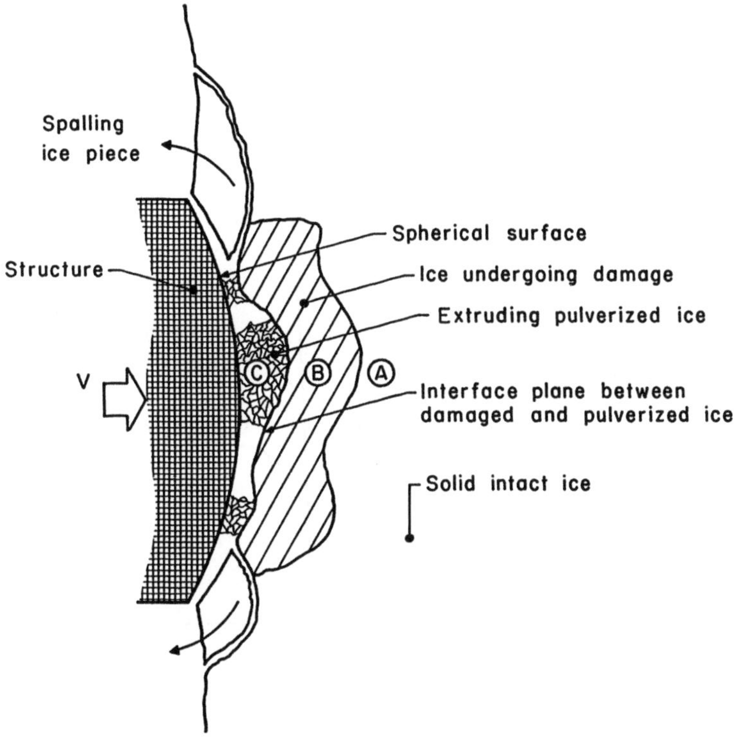

Figure 2. Schematic representation of the development of the crushed ice layer adjacent to the spherical indenter.

A typical set of results from the laboratory test program is shown in Figure 3. An important feature is the fact that the plateau after the initial peak is almost horizontal. This is indicative of continuing energy dissipation, and subsequent reloading confirms that the stress-strain curve becomes stabilised, i.e. repeats itself if the strain rate is maintained in the subsequent cycle (Plateau X in Figure 3). If the strain rate is increased on a subsequent reloading cycle, the stress attains a new peak (Curve Z) with accompanying damage and subsequent re-stabilisation; on the other hand, a decrease in strain rate on the second loading results in the material absorbing energy in dissipative mechanisms at a lower rate, Plateau Y, than in Plateau X. From many experiments such as Y in which the samples were reloaded at the same or lower rates, the axial stress stabilizes to a plateau with little or no decrease even for large strains. The stress values were found to be consistent with crack enhanced creep at a constant value of damage (Stone et al., 1989).

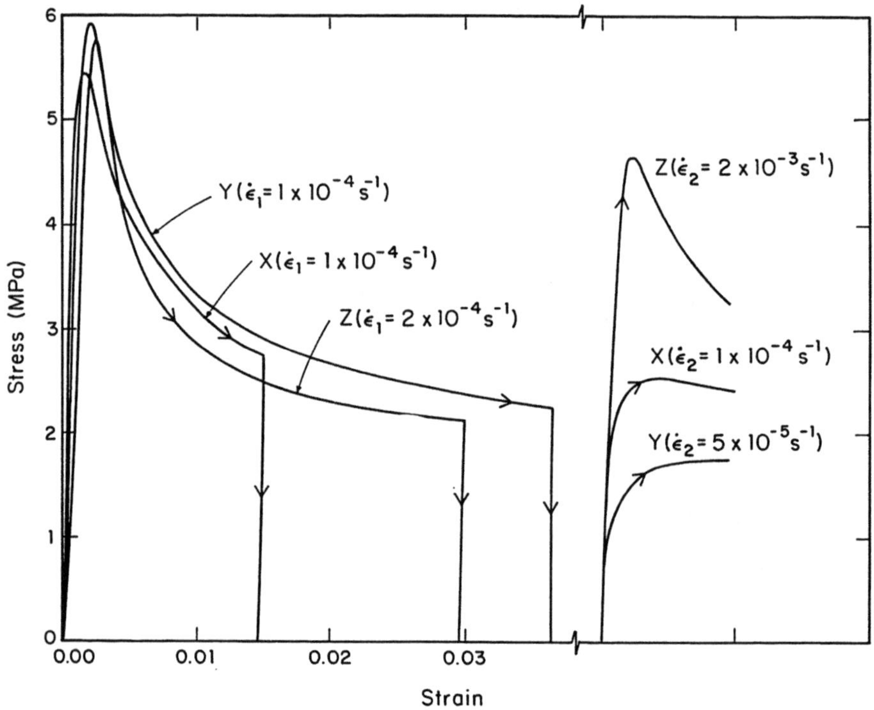

Figure 3. Constant strain rate uniaxial compression test results: reloading at same rate X, slower rate Y and faster rate Z. Reloading strains are offset in the illustration.

3. Overview of Problem

In the absence of cracks ice can be adequately characterised as a viscoelastic material with elastic, delayed elastic and flow components (Sinha, 1978). Thus the total strain, ϵ, is written as the sum of these components, i.e.

$$\epsilon = \epsilon_e + \epsilon_f + \epsilon_d, \tag{1}$$

where ϵ_e = elastic strain, ϵ_f = flow and ϵ_d = delayed elastic strain.

These three terms of equation (1) can be represented by means of the components of a mechanical model (Figure 4), the first two terms corresponding to the Maxwell unit and the last one to the Kelvin unit. The viscous movement in the Maxwell unit has been attributed to slip in the basal plane of grains of ice, whereas delayed elastic movement has been associated with grain boundary sliding.

The stress concentration in the spring of the Kelvin unit is associated with stress concentrations at grain boundary junctions. This has been discussed extensively by Sinha (1984, for example) who did not use formal rheological models as in Figure 4.

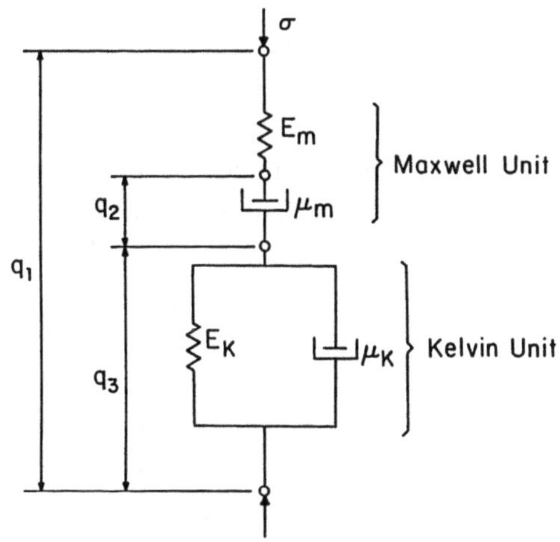

Figure 4. Rheological model for ice.

We are dealing with a heterogeneous material, with the anisotropic grains oriented randomly in various directions, and with grain boundaries of random length, orientation and number per grain. Provided the distance over which strain is measured is large compared to the grain size, equation (1) and Figure 4 will provide a reasonable idealisation: in this case, the distance must be large compared to the grain size (i.e. several times).

For strain rates greater than about $10^{-7}s^{-1}$, cracks will form; these will be mainly but not exclusively in the direction of the stress under uniaxial stress (Hallam, 1984), and in the direction of the largest principal stress under triaxial conditions (Kalifa et al., 1989). This process is not uncommon in brittle materials and can lead to failure of a specimen in compression by axial splitting (Ashby and Hallam, 1986; Brace and Bombolakis, 1963; Nemat-Nasser, 1989). We will ignore this phenomenon in the following: it does not seem relevant to the crushing process in ice since it can be suppressed by the imposition of a small confining pressure.

As loading progresses, and particularly at high rates, cracks in ice join together and there will be a tendency for more inclined cracking. The ultimate state of crushed ice has been observed to be a powder and the size of the particles can be smaller than the parent grains.

The mechanism of formation of the initial cracks (oriented in the direction of the main principal stress) could possibly be associated with energy of lattice deformation. Even under uniaxial conditions there is axial shortening and a lateral lengthening of lattice dimensions with energy associated with the latter, which would be released by an axial crack. There are also elastic misfits between grains, which should be analysed taking into account the anisotropic nature of the ice crystal (e.g. Cole, 1988). Soon after loading, grain boundary sliding commences: as Cottrell (1984) points out, the shear strength of the boundary is small and it acts as a shear crack. Figure 5 shows two possible mechanisms of crack formation due to grain boundary sliding; see also the work of Nemat-Nasser (1989).

The mechanism of crack formation in ice in compression needs further study. It should be emphasized that cracks in ice in compression initiate, grow a short distance (of the order of a grain facet), then stop. The propagation phase is not dominant; rather the rate of occurrence would be the dominant aspect. In a given volume of ice, it is convenient to think of "arrivals" in time with a certain rate, much like a Poisson process in probability theory.

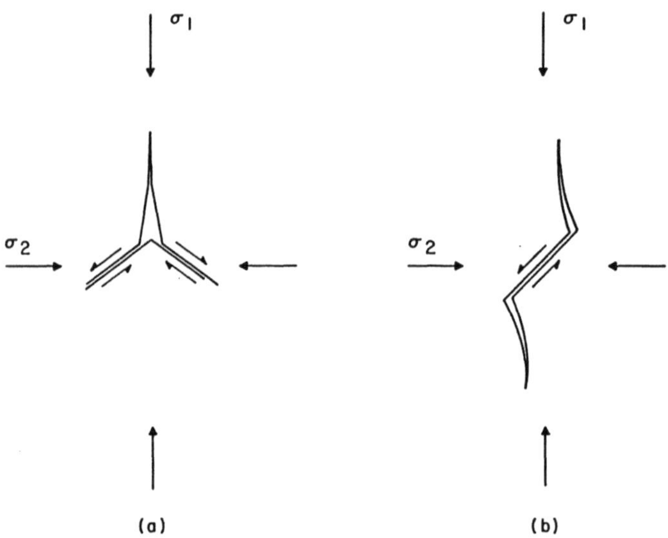

Figure 5. Crack formation resulting from grain boundary sliding.

4. General Considerations

To formulate the problem we use the concept of generalised forces Q_i with associated displacements q_i (see the works of Biot and Schapery). The product of each pair, $Q_i \delta q_i$ (no summation), where δq_i denotes an increment in displacement, represents an increment of work. The displacements q_i are taken in the following as the independent state variables, or coordinates. Generally, the variables will consist of extensive (q_i) and intensive (Q_i) pairs in dual vector spaces, with a scalar product representing work. This general notation is retained for three reasons: first, the analysis of viscoelasticity requires hidden coordinates, defined by the condition that the associated force variables are zero. These hidden coordinates pertain essentially to the specification of components of strain in viscoelasticity; for example delayed elastic as against flow. They are therefore related to internal mechanisms, such as grain boundary sliding, internal stress concentrations, and slip on basal planes. The second reason is the need to specify tractions and displacements across crack faces. These can be handled quite well by a subset of pairs Q_j, q_j within the total set of generalized forces and displacements. The forces Q_j are not zero in this case. The third reason is the need to specify the developing damage within the specimen; for example a crack of the Griffith type in a tensile field could be modelled using crack area as the extensive variable. The appropriate conjugate variable in this case is the energy release rate, G_c (per unit area), but it is not a generalized force in the sense that the product $Q_j \delta q_j$ does not do work on the body.

The measure of damage, for example, crack area, or crack density (number per unit volume) represents an internal state variable since we cannot control it in the same way as external variables such as displacement, or load. As we shall see, for microcracking in compression, the volume of material should be large enough so that we can consider the "average" effect of the individual cracks, i.e. ultimately we will not wish to model each crack, only the average effect by means of some measure of damage.

It was noted above that some of the force-displacement pairs will represent tractions and displacements across crack faces. As the crack initially propagates, and in subsequent deformation such as might result from viscoelastic distortion of the crack, frictional dissipation, accompanied by high stresses at asperities, by adhesion, by bond breaking and possibly by local crushing of the ice, can occur in zones near the crack faces. These local events and processes occur in a thin layer adjacent to the crack faces, and are not characteristic of the parent material, although as noted, the tractions across them must be included in the analysis (see also Figure 6).

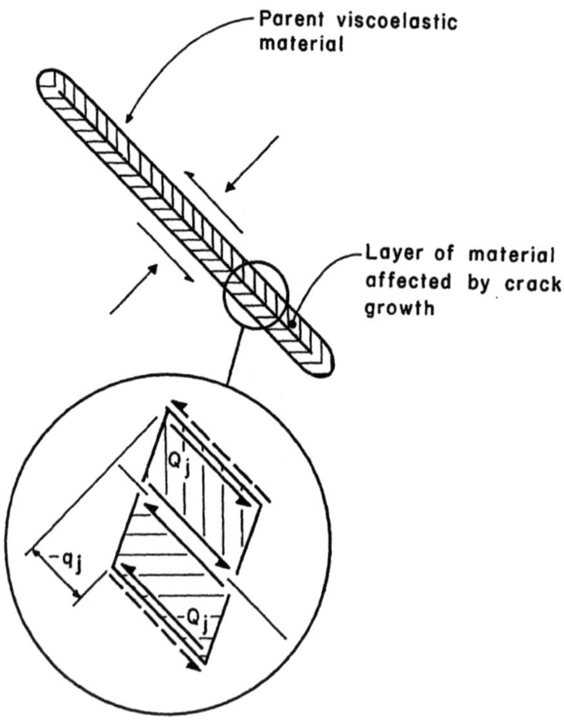

Figure 6. Schematic of a crack in compression.

It is important at the outset to define the system of interest. This is specified as the solid body excluding the cracks and the thin layer of adjacent material, i.e. the parent viscoelastic material, and is denoted "system I". The thin layer must also be analyzed; note that this is very thin in case of tension cracks of an extremely brittle material, and vanishingly thin for a Griffith crack (Rice, 1978). We treat those thin layers which transmit tractions as an additional system denoted "system II", and surround both systems I and II by a heat reservoir, which maintains isothermal conditions. Heat generated by viscoelastic movements and by friction in the layers at crack faces is immediately absorbed at constant temperature into the reservoir in a similar manner to that proposed by Bridgman (1950) for plastic flow.

To summarise the situation with regard to coordinates and forces, we specify the following sets for system I:

External Set:	$\{q_i, Q_i\}, i = 1, 2, \ \ldots, k$	External coordinates and forces, excluding tractions on cracks.
Internal Hidden Set:	$\{q_i, Q_i\}, i = k+1, \ldots, l$	Internal (hidden) coordinates in visco-elasticity; Q's are zero.
Internal Damage Set:	$\{q_i, Q_i\}, i = l+1, \ \ldots, m$	Internal coordinates and associated forces in damage characterisation; the actual work at the crack tip can be expressed as an energy release rate of damage variables such as crack area (see for example Schapery, 1981)
Internal Traction Set:	$\{q_i, Q_i\}, i = m+1, \ \ldots, n$	Internal coordinates and forces tractions on cracks.

Together, these constitute the set $\{q_i, Q_i\}, i = 1, 2, \ \ldots, n$.

It is reasonable to assume that irreversible processes take place in both systems I and II: following Prirogine (1967) we postulate a non-negative entropy increase with time in both systems. In order to allow for elastic movements, we consider systems I and II separately, each in conjunction with the heat reservoir, since a compressive stress in an elastic body under isothermal conditions will result in a decrease in entropy of the material with a corresponding increase of entropy in the reservoir as heat flows out of the body.

Considering first system I,

$$\dot{S} = \dot{S}_I + \dot{S}_R \geq 0, \tag{2}$$

where S = entropy of system I (S_I) plus that of the reservoir (S_R), and superimposed dots indicate rates of change with respect to time. Conservation of energy requires that

$$\dot{U}_I = Q_i\dot{q}_i - \dot{h}, \tag{3}$$

where U_I = internal energy material in system I and h = heat leaving system I. Summation is implied in the term $Q_i\dot{q}_i$, and will be in the following unless otherwise stated. It should be noted the tractions Q_j in Figure 6 should be interpreted so that they act on the body of system I. They are in the opposite direction to the displacements q_i, hence the value of q_i in Figure 6 is shown as negative, which will always be the case for positive Q_j. Put another way, the product $Q_j\dot{q}_j$ for the shearing tractions is always negative, say $-\dot{W}_d$ where $\dot{W}_d \geq 0$. The work associated with normal forces in addition to the shears shown in Figure 6 should also be considered but is omitted for simplicity. Since $\dot{S}_R = \dot{h}/T$,

$$T\dot{S} = Q_i\dot{q}_i - \dot{H} \geq 0, \tag{4}$$

where $H = U_I - TS_I$, the Helmholtz free energy.

Further analysis requires the consideration of the individual terms in equation (4). Following the subdivision of work terms for system I above, this will be done under the headings of viscoelastic characterisation, progress of damage, effect of cracks on creep and internal tractions. Before commencing these topics, it is worth noting that for a relatively "soft" layer, the free energy in system II is negligible and the system may be characterized by

$$T\dot{S} = \sum_{j=m+1}^{n} Q_j\dot{q}_j = \dot{W}_d. \tag{5}$$

5. Viscoelastic Characterisation of Ice

We shall consider in this section only those coordinates corresponding to the "external" and "hidden" sets of the preceding section, i.e. undamaged ice with the coordinates for internal damage and tractions as zero. Following a previous discussion (Jordaan and McKenna, 1988b), the creep will be considered in terms of the model of Figure 4; the external coordinate (considering uniaxial loading, for example) corresponds to the strain q_i across the assembly of units, with two internal components q_2 and q_3 as shown. A convenient way to deal with non-linearities is to make the viscosity of the dashpots dependent on the current stress in the dashpot. Such a relationship can be derived from equation (4), using theory based on the work of Biot (1954) and Schapery (1968); see also Jordaan and McKenna (1988b). In brief, we write equation (4) as

$$T\dot{S} = (Q_i - \frac{\partial H}{\partial q_i})\dot{q}_i = g(\sigma)b_{ij}\dot{q}_i\dot{q}_j, \tag{6}$$

in which b_{ij} is a constant matrix, given that we are considering conditions close to equilibrium, and $g(\sigma)$ is a nonlinear function of stress (indeed, functions of other variables could be considered as well: we do not do so here, for simplicity). One can think of the product $g(\sigma)b_{ij}$ as stress-dependent viscosities. For the isothermal conditions of the present approach, we take the Helmholtz free energy as being the strain energy W (in both springs of Figure 4) and thus

$$\frac{\partial H}{\partial q_i} = \frac{\partial W}{\partial q_i} = a_{ij}q_j, \tag{7}$$

where a_{ij} is the matrix of elastic coefficients. Note that in the case where $\dot{S} = 0$ (elastic system), from equation (5), $Q_i = \partial W/\partial q_i$ as dealt with by Jordaan and McKenna (1988a) in discussing elastic damaging materials. In solving for stresses, attention is focussed on the following identity, based on equations (6) and (7):

$$Q_i - a_{ij}q_j = g(\sigma)b_{ij}\dot{q}_j, \quad i = 1, 2, \ldots k, k+1, \ldots, l. \tag{8}$$

The hidden coordinates can be eliminated from this set of equations by solving for q_i in those for which $i = k+1, \ldots l$ (noting that $Q_i = 0$ in this set) and substituting back into the first k equations: see Biot (1954) and Schapery (1968) for details. The result is a set of constitutive equations which may be represented physically by springs and dashpots in arrays such as Figure 4. The dashpots may be taken as having nonlinear stress-dependent viscosities.

The two dashpots of Figure 4 will have different non-linearities, and these are associated with entropy generated in different regions - within grains, in the case of the Maxwell unit and in grain boundaries in the case of the Kelvin unit. For uniaxial stress σ, the strain in the Maxwell unit is as follows, taking into account equation (1):

$$\epsilon_e + \epsilon_f = \sigma \left[\frac{1}{E_0} + \frac{t}{\mu(\sigma)} \right], \tag{9}$$

where E_0 is the elastic modulus ($\simeq$ 9.5 GPa) and $\mu(\sigma) = 1/(\propto \sigma^{n-1})$, with $\propto$ and n constant, $\propto \simeq 1.76 \times 10^{-7} s^{-1}$ at T=263K, $n \simeq 3$.

The delayed elastic term presents some difficulty, because the usual forms presented in the literature are empirical fits to date. Sinha (1984) used the form:

$$\epsilon_d = \frac{\sigma}{E_K}[1 - \exp - (a_T t)^b], \tag{10}$$

where $E_K = (d/c)E$, d = grain size and c is a constant (9mm for d in mm), $a_T = (2.5 \times 10^{-4} s^{-1}$ at 263K) and $b \simeq (1/n)$ are constants.

The expression for a Kelvin unit with stress-dependent dashpot is identical to equation (10) except that the argument of the exponent is $-\int_0^t [E_K/\mu_K(\sigma_d)] d\tau$ where E_k is the spring constant for the Kelvin unit and σ_d and μ_k are the dashpot stress and viscosity and τ is an intermediate time. For a step application of stress σ maintained from t=0, this can be made identical to equation (10) provided that

$$\frac{\mu_k(\sigma, \sigma_d)}{E_k} = \frac{n}{a_T}[l_n(\frac{\sigma}{\sigma_d})]^{n-1}. \tag{11}$$

The stress-dependent viscosity of equation (11) will give identical results to Sinha's equation for the case considered, i.e. compliance due to a constant stress. For loading at a constant strain rate, expression (11) will yield similar but not identical delayed elastic strains to those calculated using (10). It is noteworthy that the use of a stress-dependent viscosity permits one to avoid the necessity to make current stress and strain under variable stress conditions a function of the entire history of past stress and strain. Thus in incremental analysis, the strains in the next step depend only on the current σ and σ_d, as in the viscoelastic model of Figure 4. On the other hand, the use of equation (10) requires the storage of stress and strain at all previous points in time.

It should not be expected that the simple rheological model of Figure 4 remains unchanged, even in structure, during the damage process. The parameters, both for elasticity and creep will change, and the possibility of different phenomena entering the scene must be considered.

6. Progress of Damage

The effect of tractions across crack faces will be ignored in this and in the following sections. In the earlier stages of damage in ice in compression, a series of distinct cracks propagate and then stop, the length being the order of a grain facet. The set of coordinates $q_i, i = l+1, \ldots m$ can, as noted, be considered to model planar microcracks with deterministic or random areas and directions, the latter defining the orientation of the plane (direction). It is the essence of a fracture or damage process that the energy released for cracking activity comes from strain energy W within the body (equivalent to the Helmholtz free energy H in equation (4)). We therefore investigate the strain energy release terms $\dot{H} = (\partial W / \partial q_i)\dot{q}_i$ for $i = l+1, \ldots m$ in equation (4) in the present section. The second effect of microcracking is the increase in creep rate for external coordinates $\dot{q}_i, i = 1, \ldots k$, in effect a modification of the function $g(\sigma)b_{ij}$ in equation (6) –this is deferred to the next section.

The strain energy release terms mentioned in the preceding paragraph are considered as expected values, in the probabilistic sense: it is a basic postulate of damage mechanics that the volume of material under consideration is large compared to the individual cracks. Thus, if $N(t)$ is the crack density at any time t, we may write

$$\dot{W} = \sum_{i=\ell+1}^{m} \frac{\partial W}{\partial q_i}\dot{q}_i = -A_m G \dot{N}, \tag{12}$$

recognising that the summation extends over many cracks with mean area A_m, and G=energy release per unit crack area. Only the "damage" q's have been varied in equation (12).

There will be terms corresponding to equation (12) which will contribute to the overall production of entropy in equation (4) (Schapery, 1988). Applying again the extensive property of entropy, we can regard this as additional to the viscoelastic production of entropy, equation (6). For reversible work involved in the creation of surfaces (surface tension) a term may be added to the free energy if this surface is part of the system (Rice, 1978): the additional value of entropy production per unit area is $(G - 2\gamma)/T$, where γ is the reversible work per unit area.

Another measure of damage is the average value of an elastic constant: we will use the uniaxial Young's modulus E, for example. Thus, at any time t the damage is given by $\lambda(t)$ and $E = E_0(1-\lambda)$ where E_0 is the initial elastic modulus. In obtaining equation (12), we held the coordinates (strains) other than those corresponding to damage progression constant, and the equivalent case for λ (see McKenna et al., 1989) is

$$\dot{W} = \frac{\partial W}{\partial \lambda}\dot{\lambda} = -\frac{1}{2}\frac{\bar{\sigma}^2}{E_0}\dot{\lambda}, \tag{13}$$

where, $\bar{\sigma} = \sigma/(1-\lambda)$ and σ = current stress. The value $\bar{\sigma}$ is sometimes termed the "effective stress". Combining equations (12) and (13):

$$\frac{\partial \lambda}{\partial N} = \frac{2E_0 A_m G}{\bar{\sigma}^2}. \tag{14}$$

It is clearly desirable to have a unique relationship between λ and N, i.e. a given crack density should result in a unique value of elastic modulus. Since $\bar{\sigma} = \sigma/(1-\lambda)$ in equation (14), and since the variation of σ with time or strain is not unique (consider for example constant strain-rate tests), the value of $(A_m G) \equiv F$ in equation (14) must be a function of σ in order to achieve the desired result. The simplest way to achieve this is take F as a function of σ and λ in the following form: $F = C_1[\sigma/(1-\lambda)]^2 = C_1\bar{\sigma}$. Then $\partial\lambda/\partial N = 2E_0C_1$ from equation (14). This relationship was used in the sequel.

The rheological model developed for ice actually contains two springs (Figure 4) and it is necessary to develop expressions for two rates of occurrence, denoted $\dot{N}_M$ and $\dot{N}_K$ for the Maxwell and Kelvin units, respectively. A reasonable approach to this problem is based on rate theory, first applied to fracture problems by Tobolsky and Eyring (1943). This has subsequently applied by Raj and Ashby (1975) and to ice by McKenna et al. (1989).

The rate expression proposed is as follows

$$\dot{N} = \dot{N}_0[exp(\frac{\bar{\sigma}-\sigma_c}{\sigma_0}) - 1], \tag{15}$$

where $\dot{N}_0$ = reference rate, σ_c = threshold stress, and σ_0 = constant (units of stress). Separate expressions were developed for the two springs: for the delayed elastic strain (Kelvin unit), we note that a critical delayed elastic strain was developed by Sinha (1984) for initiation of first cracks. This is completely equivalent to a critical stress in the spring in the Kelvin unit; taking the uniaxial case and the constants of Sinha (1984), we find $\sigma_c = 0.50$ MPa. This value is adopted here, recognising that there is a need to investigate further crack formation in granular ice. In creep tests conducted by Gold (1972), a threshold stress of 0.59 MPa was required before cracks formed, which is consistent with the above result.

7. Changes in Geometry: Effect of Cracks on Creep

The presence of cracks, even without considering crack growth or new crack nucleation, will cause an increase in the overall creep rate of a specimen.

At the simplest level, if one considers the parameter λ as being the area "lost" due to damage, then the effective stress is $\bar{\sigma} = \sigma/(1-\lambda)$, for uniaxial stress σ. The strain rate at any time t, from equation (9) is $\propto \sigma^n$. Substituting $\bar{\sigma}$ for σ in this equation one obtains a strain rate of $[\propto /(1-\lambda)^n]\sigma^n$. Tests to date (e.g. Stone et al., 1989) suggest that the form of this is correct but that the coefficient in square brackets underestimates the creep rate.

Weertman (1969) studied the effect of cracks on the creep rate using dislocation theory. For a low density of cracks he found a result of the kind $\dot{\epsilon} = \dot{\epsilon}_0(1 + 2\pi a^2 N n^{1/2})$ for power-law creep, in damaged material where $\dot{\epsilon}_0 = \propto \sigma^n$ in the undamaged material. Here a = half crack length and N = crack density per unit area. The relationship just quoted was used by Sinha (1988) in analyzing crack-enhanced creep. For higher crack densities, Weertman noted that the creep rates were proportional to σ^n but the proportionality constant involved higher powers of N, e.g.,$\propto (N)^{n+1}$ or $(N^2)^{n+1}$. This suggests that an exponential form might be worth exploring; thus

$$\dot{\epsilon} = \dot{\epsilon}_0 exp(\beta N), \tag{16}$$

where β is a constant.

The effect that we are discussing in the present section arises from changes in geometry of the specimen. The introduction of a discontinuity in the specimen causes stress concentrations and consequent magnification of the creep rate near the crack tips and a reduction of (tensile or shear) stress near the middle of the crack, with corresponding increases elsewhere. The results quoted above suggest that the effect of these cracks can be modelled by adjusting the viscosity coefficient $g(\sigma)$ in equation (6). For a complete three-dimensional analysis, dilatations due to microcracking must also be included, and this will require additional terms.

8. Internal Tractions

Considerable internal dissipation of energy will occur in mechanisms such as those illustrated in Figure 5. As damage progresses there will be increasing numbers of cracks with faces in contact and with compression and shear transmitted from one face to the other (Figure 6). Eventually the pulverised ice extrudes and these frictional processes dominate. The mechanics of kinetic friction must then be considered (Molgaard, 1988).

In a previous review (Jordaan and McKenna, 1988a) attention was drawn to the fact that many continuum formulations do not account for such cracks as a considerable dissipator of energy. This aspect was addressed previously in the present paper by the set of internal coordinates $\{Q_j, q_j\} j = m+1, \ldots, n$. The movement q_j is always in the opposite direction of Q_j resulting in a net removal of mechanical energy from the specimen (System I) at a rate $\dot{W}_d$.

Data on friction of ice on ice at slow speeds with relatively high normal stresses is not readily available. Figure 7 shows a composite (log-log) plot of shear strain versus shear stress for secondary creep, for crack-enhanced creep (note the slopes are similar, indicating $n \simeq 3$), and for sliding friction of ice on granite. Barnes et al. (1971) analysed the latter movement in terms of creep of a thin layer ($10^{-4}m$) of material and deduced that the creep in this layer was much enhanced as a result of recrystallisation. No doubt recrystallisation takes place; in addition, at the higher stresses there could also be local crushing at asperities. There was considerable evidence of cracking in the works reported by Barnes et al. In any event, the stress-dependence of local frictional movement at lower strain rates, zone S in Figure 7, would be similar to that of the specimen as a whole, i.e. $n \simeq 3$, but at an enhanced rate.

It is suggested that the lower part of the sliding friction curve (i.e. at lower rates) could well apply to slow movements of ice on ice where adhesion is important. It is tempting also to suggest that the shear stresses across crack faces or zones such as those in Figure 5 are proportional to the appropriate deviatoric component of stress s_{ab} i.e. $Q_j \propto s_{ab}$ (possibly averaged). In the uniaxial case, where the applied stress σ is proportional to the stress deviators for slow movement, since $\dot{W}_d \propto \sigma\dot{\epsilon}$ and $\dot{\epsilon} \propto \sigma^n$ (Figure 7)

$$\dot{W}_d = - \sum_{j=m+1}^{n} Q_j \dot{q}_j \propto \sigma^{n+1}, \tag{17}$$

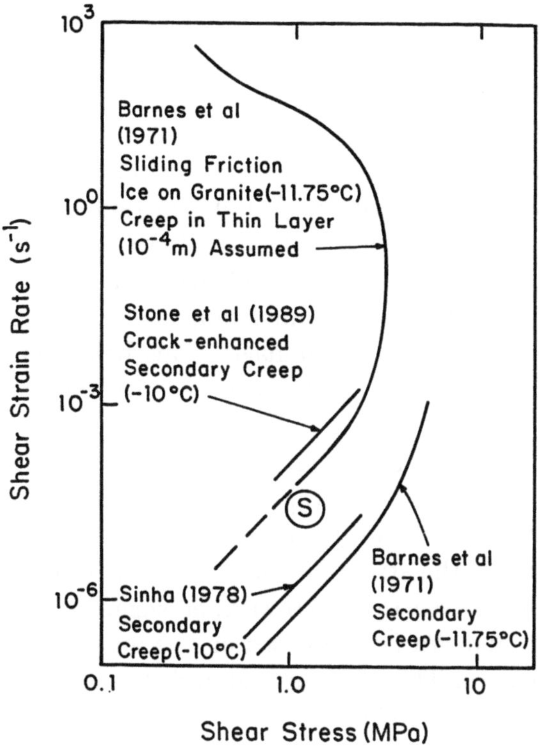

Figure 7. Experimental results for shear strain rate versus shear stress for secondary creep.

which will be averaged over the appropriate volume, large enough to smooth local "details". Thus slow creep movement across cracks can be modelled as a contributor to creep of the specimen as a whole.

The relatively slow creep movements across crack faces would not be expected to affect significantly the strain energy, W; in effect, external work $Q_i\dot{q}(i = 1\ to\ k)$ will flow through the body and leave as $\dot{W}_d$. This situation will not prevail during the dramatic fluctuations in load shown in Figure 1, at least not in the zones where the pulverisation of the ice is taking place. Figure 7 shows that as the shear strain rate reaches a certain value ($\simeq 10^{-3}s^{-1}$) the shear stress ceases to increase and eventually declines. This could be associated with stick-slip movements within the ice, as previously proposed (Jordaan and McKenna, 1988a). Measurements of ice friction under triaxial stress conditions have demonstrated stick-slip (Beeman et al, 1988), albeit that these measurements were taken at very low temperatures.

9. Incremental Analysis

The experiments conducted at -10^0C by Stone et al. (1989) on bubble-free, granular ice with a grain size of 3mm provide a means of testing the theory described above. Particular emphasis was placed on the loading and reloading of ice samples in uniaxial compression at constant strain rates. As noted earlier, the experiments show that distributed damage increases the ability of the ice to dissipate energy. Once damage has stabilised (the plateau in the stress-strain curve), unloading and reloading at a lower rate does not result in any further damage. This behaviour is described by secondary creep, enhanced by the presence of cracks as described by equation (16).

The progress of damage follows the rate relation (14) for both the Maxwell and Kelvin springs in Figure 4. Note that damage of the main spring results from the total stress while damage in the Kelvin unit is only a function of the elastic portion of the stress in that unit. The critical stress σ_c for the main spring was taken to be 3.0 MPa as suggested by Cole (1988) and Shyam Sunder and Wu (1990) based on the anisotropic nature of the elastic grain properties. The rate constant and stresses used for fitting the model to the data are as follows:

Parameter	Kelvin Unit	Maxwell Unit
$\dot{N}_0[m^{-3}s^{-1}]$	1.67×10^5	5×10^5
$\sigma_c[MP_a]$	0.25	3
$\sigma_0[MP_a]$	5	5
$\beta[m^3]$	$5d^3$	$5d^3$
C_1	7.5×10^{-8}	2.5×10^{-8}

An incremental analysis was used to solve equations (1), (9), (10), (11), (14), (15) and (16) for the damaging viscoelastic material. This formulation takes advantage of the distinction between the instantaneous elastic and the time-dependent dissipative portions of the model. For suitably small time increments, the ice can be seen to creep (primary and secondary creep) under a constant stress σ and then damage to decrease the stress. The total stress change over the time increment is the sum of the decrease due to damage and an elastic adjustment based on the current elastic properties of the main spring and the difference between the applied and creep strain increments.

The experimental and model results for loading at a strain rate of $10^{-4}s^{-1}$ are shown in Figure 8. The combination of damage to the main spring and to the delayed elastic spring allows for an accurate representation of the test results, both for the initial loading and reloading at lower strain rates. Model results assume a complete relaxation of the delayed elastic strain.

Unfortunately, no crack densities were available to validate the calculated values. The average energy required for the formation of a crack (A_mG in (14)) was estimated to result in crack densities in the order of the grain density.

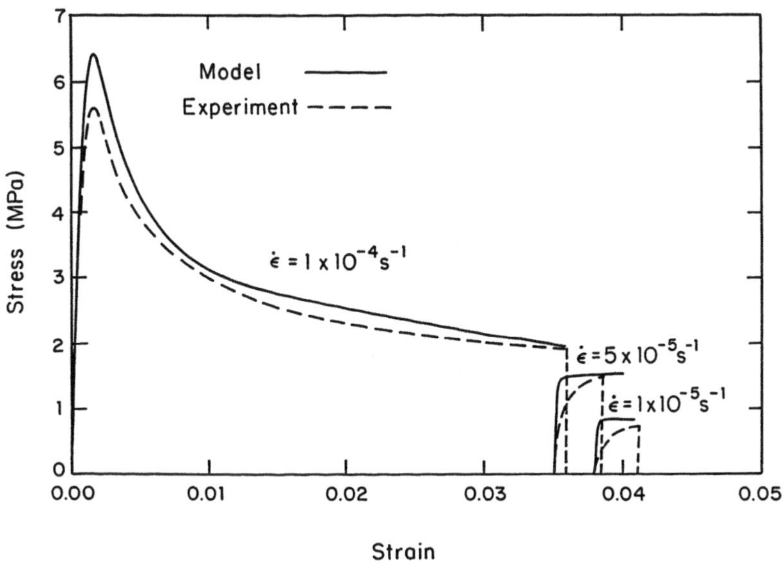

Figure 8. Uniaxial compression of granular ice (d = 3 mm): model results for reloading at lower strain rates.

Reloading at higher rates in Figure 9 showed that the model has potential for balancing crack enhanced creep and further damage. The model gives reasonable values for the magnitude of stress of the peak on reloading, but creep dominates further damage resulting in a broader peak than observed for the second reloading. One possible cause is that crack formation in (15) is sensitive to stress only. Further work will address whether equation (15) is appropriate.

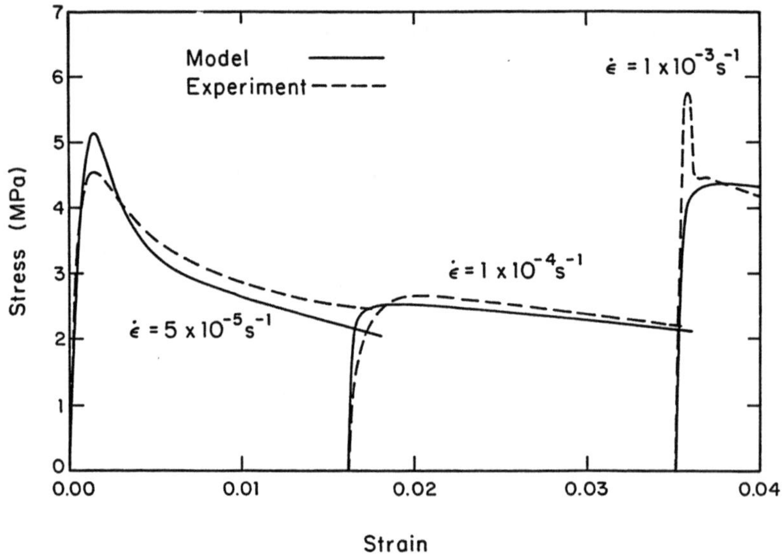

Figure 9. Uniaxial compression of granular ice (d = 3 mm): reloading at higher strain rate.

Initial results also show that the present model is a reasonable representation for ice under uniaxial compression at strain rates between $10^{-6}s^{-1}$ and $10^{-3}s^{-1}$. Nevertheless, it should be emphasised that the results presented are preliminary, because they deal exclusively with the uniaxial situation. The analysis of recent triaxial test results, including measurements of crack densities, will assist with the modelling.

10. Concluding Remarks

The overall objective of the analysis above has been to put the process of damage into the perspective of ice-structure interaction, where, in the crushing mode, significant fluctuations in load occur. The transition from zones A to B to C of Figure 2 will hopefully be the focus of considerable further research: progress in the early stages of the damage process, leading to peak stresses, has been reported in the paper. The declining part of the load-time curve of Figure 1, in particular, needs further work. No doubt, the process of softening leads to a flux of energy into the damaging zone with consequent further damage; this is aided by stress gradients. Models of stick-slip processes (Figure 10 for example) abound in the literature but a more fundamental understanding of the process is

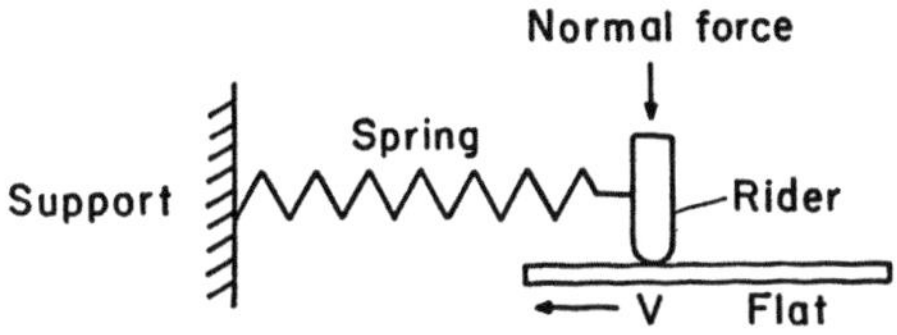

Figure 10. Simple model for stick-slip process.

needed. The dramatic drops in load would release considerable energy and kinetic effects may need to be considered during this phase.

Two processes for crack formation have been suggested in this paper: as a result of the elastic properties of the grains and due to sliding along grain boundaries. Experiments are underway to isolate these processes and with a view of improving the rate relation proposed here for describing crack formation.

Under the high confining pressures that would be attained particularly in the central zone behind the spherical indenter, evidence of sintering was observed. This type of process could contribute to the stiffening of the ascending branch of the force-time curves of Figure 1. Sintering can be considered as a reverse of the damage process described: see, for example, Schapery (1980).

Acknowledgements

Financial support for this research from the Natural Sciences and Engineering Research Council of Canada, as well as Mobil Oil Canada Properties, is gratefully acknowledged. Permission granted by the Hibernia Joint Venture Participants, that is, Mobil Oil Canada Properties (Operator), Chevron Canada Resources Ltd, Gulf Canada Resources, Inc., Petro-Canada Resources, to publish the results embodied in Figure 1 from the Pond Inlet spherical indentor tests is especially appreciated. Discussions with Dr. J. Molgaard on friction were most helpful.

References

Ashby, M.F. and Hallam, S.D. 1986. The fracture of brittle solids containing small cracks under compressive stress states. Acta Metallurgica, Vol.34, No.3, pp.497-510.

Barnes, P., Tabor, D. and Walker, J.C.F. 1971. The friction and creep of polycrystalline ice. Proceedings, Royal Society London, A, Vol.324, pp.127- 155.

Beeman, M., Durham, W.B., and Kirby, S.H. 1988. Friction of ice. Journal of Geophysical Research, Vol.93, No.B7, pp.7625-7633.

Biot, M.A. 1954. Theory of stress-strain relations in anisotropic viscoelasticity and relaxation phenomena. Journal of Applied Physics, Vol.25, No.11, pp.1385-1391.

Brace, W.Q. and Bombolakis, E.G. 1963. A note on brittle crack growth in compression. Journal of Geophysical Research, Vol.68, No.12, pp.3709-3713.

Bridgman, P.W. 1950. The thermodynamics of plastic deformation and generalised entropy. Reviews of Modern Physics, Vol.22, No.1, pp.56-63.

Cole, D.M. 1988. Crack nucleation in polycrystalline ice. Cold Regions Science and Technology, Vol.15, pp.79-87.

Cormeau, A., Jordaan, I.J. and Maes, M. 1986. Development of a model for progressive damage of ice. Report submitted to the Department of Public Works, Ottawa, by Det norske Veritas (Canada) Ltd.

Cottrell, A.H. 1964. The mechanical properties of matter. Wiley.

Geotech, 1985. Medium scale iceberg impact simulation test program. Report prepared for Mobil Oil Canada, Ltd.

Gold, L.W. 1972. The failure process in columnar grained ice. Technical Paper No. 369 of the Division of Building Research, National Research Council of Canada, 103 p.

Hallam, S.D. 1984. Yield and fracture surfaces of brittle solids under multi-axial loading. Ph.D. Thesis, Cambridge University Engineering Dept., Cambridge, England.

Jordaan, I.J. and McKenna, R.F. 1988a. Modelling of progressive damage in ice. Proceedings International Association for Hydraulic Research (IAHR) Symposium on Ice, Sapporo, Japan, August, Vol.II, pp.585-624.

Jordaan, I.J. and McKenna, R.F. 1988b. Constitutive relations for the creep of ice. Proceedings, International Association for Hydraulic Research (IAHR) Symposium on Ice, Sapporo, Japan, August, Vol.III, pp.47-58.

Jordaan, I.J. and Timco, G.W. 1988. Dynamics of the ice-crushing process. Journal of Glaciology, Vol.34, No.118, pp.318-326.

Kalifa, P., Duval, P. and Ricard, M. 1989. Crack nucleation in polycrystalline ice under compressive states. Proceedings Offshore Mechanics and Arctic Engineering (OMAE) Symposium, A.S.M.E., Vol.4, pp.13-21.

Kheisin, D.E. and Cherepanov, N.V. 1970. Change of ice structure in the zone of impact of a solid body against the ice cover surface. Problems of the Arctic and Antaratic, Issues 33-35, pp. 239-245.

Kurdyumov, V.A. and Kheisin, D.E. 1976. Hydrodynamic model of the impact of a solid on ice. Prikladnaya Mekhanika, Vol.12, No.10. pp.103-109.

McKenna, R.F., Meyssonnier, J. and Jordaan, I.J. 1989. Peak pressures from a damage model for ice in compression. Proceedings of OMAE, The Hague, The Netherlands, Vol.IV, pp.67-73.

Molgaard, J. 1988. On the nature of the process of kinetic friction. Unpublished manuscript.

Nemat-Nasser, S. 1989. Compression-induced ductile flow of brittle materials and brittle fracturing of ductile materials. Invited Lecture, Proceedings of the 12th Canadian Congress of Applied Mechanics, Carleton University, Ottawa, Vol.1, pp.10-21.

Prirogine, I. 1967. Introduction to Thermodynamics of Irreversible Processes. Third Edition, Interscience, New York.

Raj, R. and Ashby, M.F. 1975. Intergranular fracture at elevated temperature. Acta Metallurgica. Vol.23, pp.653-666.

Rice, J.R. 1978. Thermodynamics of quasi-static growth of Griffith cracks. Journal of the Mechanics and Physics of Solids, Vol.26, pp.61-78.

Schapery, R.A. 1968. On a thermodynamic constitutive theory and its application to various nonlinear materials. Proc. IUTAM Symposium, East Kilbride, Springer, New York, pp.249-285.

Schapery, R.A. 1980. On constitutive equations for viscoelastic composite materials with damage. Workshop on a Continuum Mechanics Approach to Damage and Life Prediction, National Science Foundation, May 4-7, pp.119-133.

Schapery, R.A. 1981. On viscoelastic deformation and failure behaviour of composite materials with distributed flaws. Advances in Aerospace Structures and Materials, ASME.

Schapery, R.A. 1988. A theory of mechanical behaviour of elastic media with growing damage and other changes in structure. Mechanics and Materials Center, Texas A and M University, College Station, Texas, Report No. MM.5762-88-1.

Shyam Sunder, S. and Wu, M.S. 1990. Crack nucleation due to elastic anisotropy in polycrystalline ice. Cold Regions Science and Technology, Vol.18, pp.29-48.

Sinha, N.K. 1978. Short-term rheology of polycrystalline ice. Journal of Glaciology, Vol.21, No.85, pp.457-473.

Sinha, N.K. 1984. Intercrystalline cracking, grain-boundary sliding, and delayed elasticity at high temperatures. Journal of Materials Science, Vol.19, pp.359-376.

Sinha, N.K. 1988. Crack-enhanced creep in polycrystalline material: strain- rate sensitive strength and deformation of ice. Journal of Materials Science, Vol.23, No.12, p.4415.

Stone, B.M., Jordaan, I.J., Jones, S.J. and McKenna, R.F. 1989. Damage of isotropic polycrystalline ice under moderate confining pressures. Proceedings 10th International Conference on Port and Ocean Engineering under Arctic Conditions, Luleå, Sweden, Vol.1, pp.408-419.

Timco, G.W. and Jordaan, I.J. 1987. Time series variations in ice crushing. Proceedings 9th International Conference on Port and Ocean Engineering Under Arctic Conditions, Fairbanks, Alaska, Vol.1, pp.13-20.

Tobolsky, A. and Eyring, H. 1943. Mechanical properties of polymeric materials. Journal of Chemical Physics, Vol.11, pp.125-134.

Weertman, J. 1969. Effect of cracks on creep rate. Transactions of the ASM, Vol.62, pp.502-511.

Discussion

Q: T.J.O. Sanderson, BP International

Your comments about microscopic processes at sliding cracks are interesting. If local stress concentrations cause pressure melting, then the possibility exists that such cracks might refreeze on unloading. This may explain why diagonal cracks involved in wing crack formation are very hard to observe. Does this appear likely?

A: Ian J. Jordaan

We consider that pressure melting may well be an important part of the process. Refreezing on unloading may then have the effect noted and also "stiffen" the structure of the ice on the next load increase. Diagonal cracks need not be present; grain boundaries may serve this purpose.

Q: S. Shyam Sunder, MIT

I am happy to see that you have included (somewhat indirectly) the Helmholtz Free Energy contribution from both instantaneous elasticity (E_M) and delayed elasticity (E_K). This question is with regard to the number of material parameters required specifically to model damage processes (not viscoelasticity), their physical basis, and how the individual material parameters are determined from experimental data.

A: Ian J. Jordaan

Two damage mechanisms have been described in the model: due to strictly elastic processes and due to sliding along grain boundaries. The damage progression for each mechanism was described using a rate relation that includes 3 parameters, a threshold stress for initiation σ_c, a normalising stress σ_0, and a rate constant $\dot{N}_0$. Enhancement for the delayed elastic and permanent creep strains was described using the enhancement parameters, β. If all of these parameters are independent, then there is a total of 8 constants. In practice, the rate of loading will establish which of these are relevant in a given situation. The enhancement parameters, β, have been calculated from a sequence of constant loads applied to damaged ice. Damage initiation stresses are now being addressed by several research groups while the remaining parameters have been derived from constant strain rate experiments. Schapery (this volume) shows a physical basis for using a power law rate relation for damage progression which behaves in a similar manner to the exponential form presented here.

Ice crushing by microcracking is a complex process. The present model identifies some key processes that require further analysis and experimentation.

A Plasticity Model for the Crushing of Ice

J. F. Dorris

Offshore Engineering Research Department
Shell Development Company
Houston, Texas, USA

Abstract A plasticity formulation is used to develop a material model for ductile ice crushing. The model includes rate and pressure sensitivity, strain softening, and dependence on volumetric deformations. A nonassociated flow rule is adopted for the plastic potential. A relationship is then proposed which relates the plastic potential to the yield function. This relationship permits volumetric deformations to be calculated directly from measured stress strain curves. These calculations show good qualitative agreement with experimental measurements of volume change. A brittle failure criterion based on dilatant deformation follows naturally from the plastic potential. The model is intended for application to ice indentation problems where crushing is the dominant failure mode.

1. Introduction

Ice crushing loads are of considerable importance to the design of Arctic offshore structures. On a local scale, crushing pressure governs the design of shell thickness, stiffeners, and bulkhead spacing while, on a global scale, crushing loads influence the sliding resistance of the structure. Measured contact stresses, when plotted as a function of contact area, show a significant decrease with increasing area (Sanderson, 1986). Laboratory and field indentation tests provide data for small areas, while global loads on structures are used to infer effective contact pressure for large areas. The lack of data for intermediate areas and the large drop in pressure with increasing area can give the false impression that local and global contact pressures are unrelated phenomena. The large disparity in these pressures is often explained by size effects and nonsimultaneous failure.

S. Jones · R. F. McKenna · J. Tillotson · I. Jordaan (Eds.)
Ice-Structure Interaction
IUTAM-IAHR Symposium St. John's, Newfoundland, Canada

Dorris and Winkler (1989) propose a conceptual pressure vs. area model for crushing which provides a connection between local and global pressures. Their model is based on heuristic arguments regarding the loading conditions which can be expected for small and large areas. Local pressures on small areas are normally associated with the initial contact of undamaged ice, small strain, and high confining pressure. Global pressures are generally associated with damaged ice, large strain, and low confinement. Thus, it seems natural to identify local pressures with confined compressive strengths and global pressures with the unconfined post peak (residual) compressive strength. In fact, the ratio of these strength parameters is similar to the ratio of local to global indentation pressures.

Based on this discussion, Dorris and Winkler argue that the connection between local and global pressures can be explained by small-scale mechanical tests. Thus, the development of a general material model for crushing which covers the range of engineering interest should describe all of the mechanical properties observed in conventional triaxial tests. These properties are rate and pressure sensitivity, strain softening (hardening), and volumetric deformations. Rate and pressure sensitivity have long been recognized as necessary requirements for a realistic mechanical description of ice, but relatively little attention has been given to the effects of strain softening and volumetric deformations. Strain softening is required to describe the variation in stress between confined peak strengths and unconfined residual strength. Volumetric deformations are a direct consequence of pressure sensitivity and must be considered when developing multiaxial stress strain relationships.

This paper presents a material model for ice crushing which contains all of the features listed above. Triaxial test data describing the stress strain response and volumetric deformation of multiyear ice are first presented to provide an experimental basis for the development of the model. The crushing model is then formulated from principles of plasticity. The formulation begins by assuming a nonassociated flow rule for the plastic potential. A relationship is then proposed which relates the plastic potential to the yield function. This relationship recognizes the residual strength as a fundamental material property and provides a connection between the stress strain response of ice and its accompanying volumetric deformation. Thus, volumetric deformations can be calculated directly from measured stress strain curves. These calculations show good qualitative agreement with experimental measurements of volumetric deformation.

The crushing model assumes that the ice remains in the ductile regime. Thus, the solution of practical problems such as ice indentation would require a failure criterion which distinguishes between ductile and brittle responses. It will be shown that a pressure and rate dependent failure criterion follows naturally from the plastic potential and is associated with dilatant volumetric deformations.

Finally, a first-order differential equation is derived which describes the evolution of the effective plastic strain rate. This equation replaces the consistency condition found in plasticity models. In principle, the differential equation for effective plastic strain rate, the constitutive equations, and the equations of equilibrium will provide solutions to boundary value problems of general interest. This system of equations can be solved by a finite element formulation.

2. The Compression of Ice and Indentation Tests

To motivate the development of the crushing model, we begin with a brief review of the mechanical response of ice under compressive stress states and ice indentation tests.

The mechanical response of ice as observed in triaxial compression tests covers the whole range of material response, depending on its rate of loading. At low strain rates, the response is ductile, and the stress strain curve is characterized by a rise in stress to a peak value followed by a decrease in stress to a fairly constant postpeak value (i.e., residual strength). The peak strength and residual strength depend on pressure and strain rate. The nonmonotonic nature of the stress strain curve is due to softening of the ice from progressive damage of its microstructure. A dominant softening mechanism is microcracking. At high strain rates, the response is brittle, and the stress strain curve is characterized by a sharp rise to a peak value where catastrophic failure of the sample occurs. A critical strain rate marks the transition from ductile to brittle behavior, where peak strength obtains its maximum value. By increasing the confining pressure, the critical strain rate moves to a higher value. The mechanical properties of ice are also sensitive to temperature. However, the data and applications considered here are restricted to a constant temperature, but temperature effects can easily be included by making model parameters depend on temperature.

The behavior of ice during indentation tests is analogous to its behavior seen in triaxial compression tests. That is, the force required for a rigid indentor to penetrate an ice feature depends on the velocity of indentation. At low velocities, the ice response is ductile, and the force displacement curve is smooth. Although a damaged zone of highly fissured ice occurs in the vicinity of the indentor, the ice essentially remains a continuum under ductile conditions. At high velocities, the ice response is brittle, and the force displacement curve is jagged. The peaks and valleys correspond to the initiation and propagation of large cracks which cause spalling of the ice surface. The maximum indentation pressures occur at a critical velocity which marks the transition from a ductile to brittle response.

Ice indentation tests are of fundamental interest, since they provide an empirical basis for estimating ice crushing loads. Ralston (1979) predicts global ice loads on gravel islands by applying a crushing model developed from ice sheet indentation tests. Johnson and Benoit (1987) develop a pressure vs. area curve from spherical indentation tests which can be used to estimate local contact stresses. However, the extension of these empirical approaches to applications with different geometries, boundary conditions, loading conditions, etc. is not obvious. Of particular interest is the extrapolation of current pressure vs. area curves to larger areas where field indentation tests are not feasible.

The most notable approach to modeling the indentation process has been the application of plastic limit analysis. Ralston (1978) provides analytic support for the indentation factors derived from sheet indentation tests, and Nordgren (1988) provides insight into the different failure modes which can be expected from local contact stresses on conical structures. However, indentation pressures from limit analysis are normalized with respect to unconfined compressive strength so that the determination of actual pressure requires an a priori assumption regarding an effective strain rate. Also, there is no strain softening effect in limit analysis models.

In principle, the finite element method can be applied to any ice indentation problem where the force displacement curve is smooth and the ice remains a continuum. Thus, its range of application would cover the ductile regime and could even predict maximum indentation pressures if a suitable failure criterion existed. Unfortunately, the use of the finite element method has been hindered by the lack of a realistic constitutive model for ice. The crushing model presented in the following is intended to fill this need. Its development requires triaxial tests conducted under ductile conditions to calibrate the sensitivity of constitutive parameters to pressure and strain rate as well as triaxial tests conducted at high strain rates to define the brittle failure criterion.

3. Triaxial Tests

3.1 Test Conditions and Ice Properties

A program of conventional triaxial tests on multiyear ice samples was conducted to produce the data required for a general crushing model. All tests were conducted at a temperature of −5°C. Test conditions included axial strain rates up to 10^{-1} sec^{-1} and confining pressures up to 10 MPa. Test samples were right circular cylinders with a length of 250 mm and a diameter of 100 mm. Each sample was initially loaded with a hydrostatic pressure (σ_{33}). An axial load ($\sigma_{11} - \sigma_{33}$) was then applied. During the axial loading portion of the test, axial strain rate ($\dot{\epsilon}_{11}$) and radial confining pressure (σ_{33}) were held constant by closed loop feedback control. All tests were loaded until the sample failed or 10 percent axial strain was reached.

The ice samples were taken from a multiyear floe located near Melville Island in the Canadian Arctic. After testing, crystallographic thin sections were made for each test sample and classified according to the scheme developed by Richter and Cox (1984). All test samples were classified as either Type I (granular), Type III (mixed granular and columnar), or Type IIIB (largely Type I with columnar inclusions). The Type IIIB samples had a small percentage of columnar ice and, in some cases, could be classified as Type I. The average salinity of all test samples was 0.26 ± 0.19 o/oo and the average density was 0.902 ± 0.014 g/cm^3. As a whole, the physical properties were more uniform than the properties reported in other test programs on multiyear ice (e.g., Cox et al., 1984; Cox et al., 1985). Consequently, the measured mechanical properties in the present program showed less scatter than is normally seen with multiyear ice samples.

In the following, four tests from the above program will be presented. The test conditions were selected for two levels of constant strain rate and constant confining pressure to illustrate the effect of those parameters on the stress strain response and volumetric deformation. The test conditions and sample designation for the four tests are listed in Table 1. The particular test samples at each test condition were selected to match as closely as possible the failure mode and physical properties. The physical

Table 1. Test conditions for each ice sample.

σ_{33} \ $\dot{\epsilon}_{11}$	.001/sec	.01/sec
2.62 MPa	T494	T481
10 MPa	T485	T466

Table 2. Physical Properties of ice samples listed in Table 1.

Test Sample	Salinity (0/00)	Density (g/cm³)	Crystal* Structure
T494	0.1	.898	I
T485	0.1	.898	I
T481	0.3	.912	IIIB
T466	0.3	.886	I

*from Richter and Cox (1984)

properties for the four samples are listed in Table 2. The failure mode of each sample was primarily uniform barreling. However, tests T481 and T485 showed evidence of the initiation of a shear plane. A fairly uniform distribution of microcracks throughout these two samples indicates that the shear planes probably occurred late in the test and did not contribute significantly to the overall deformation.

3.2 Stress Strain Response

Figure 1 illustrates the deviatoric stress strain curves for the four tests listed in Table 1. All curves show a stable postpeak response typical of ductile test conditions. The shapes of the stress strain curves consist of a transient response characterized by the peak strength and a steady state response characterized by the residual strength. During deformation from a natural state, ice undergoes structural changes by means of micromechanical mechanisms such as dislocation motion and microcrack initiation. For a constant temperature, the relative importance of each

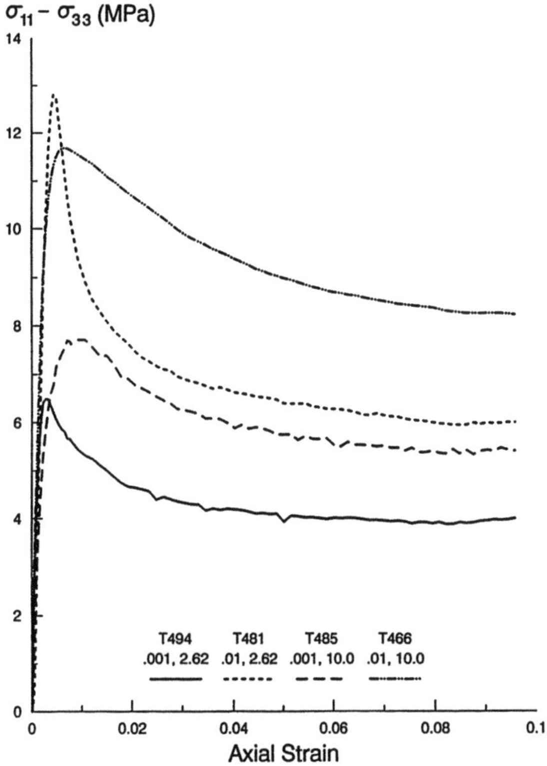

Figure 1. Stress strain curves for the four tests listed in Table 1.

mechanism and, hence, the energy requirements to deform the material depend on confining pressure and strain rate.

The effects of confining pressure and strain rate on the peak and residual strengths are shown in Figure 2. The mean unconfined peak and residual strengths reported by Dorris (1989) are included in this figure to illustrate the development of a failure surface. Uniaxial tests loaded at strain rates above 10^{-3} sec^{-1} do not show a stable postpeak response. However, when a confining pressure of $\sigma_{33} = 2.62$ MPa is applied, residual strength values can be observed at $\dot{\epsilon}_{11} = 10^{-2}$ sec^{-1}. The dashed lines in Figure 2 are drawn to show an approximate failure envelope which separates ductile and brittle responses. A more accurate shape of the failure surface would require further testing.

The solid lines in Figure 2 represent pressure and rate dependent yield surfaces. A general crushing model for ice should describe the evolution of the surfaces from Figure 2a to Figure 2b as deformation proceeds.

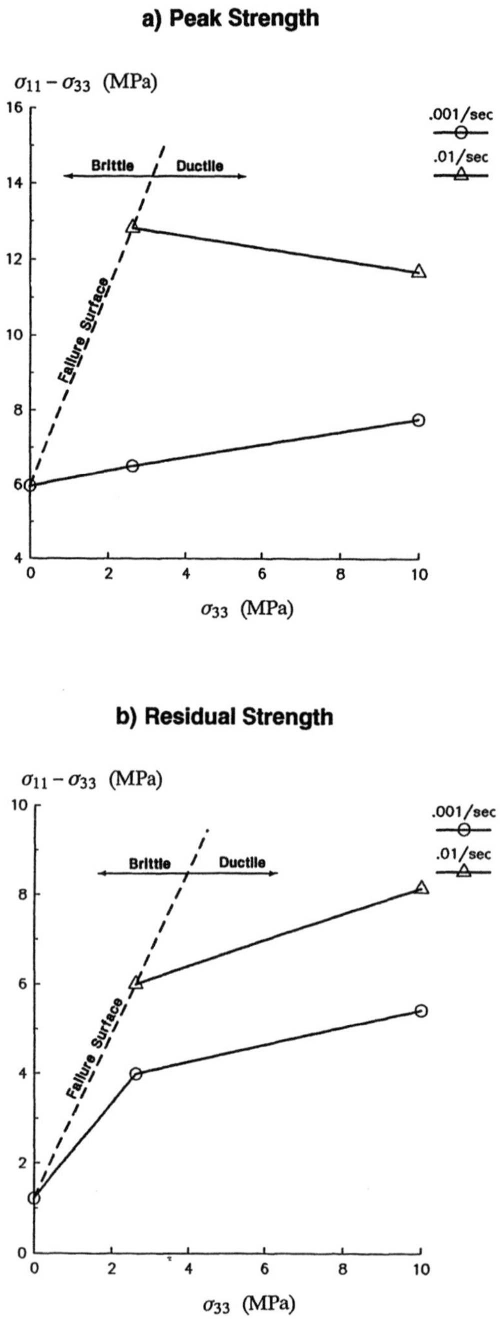

Figure 2. The dependence of strength on confining pressure and strain rate: (a) peak strength, (b) residual strength.

In plasticity models, the evolution of yield surfaces is accounted for by adopting a hardening (or softening) rule. However, the change in the shapes of the surfaces in Figure 2 indicates that the isotropic or kinematic hardening rules commonly found in plasticity models are not applicable for ice.

3.3 Volumetric Deformation

Two techniques were used to measure volumetric deformations during each triaxial test. The first technique measured the diametral expansion of the sample in two orthogonal directions at its midpoint. The second technique related the fluid displacement within the confining cell to the sample deformation. Complete details of both techniques are given by Spencer et al. (1989).

The sign convention used throughout this paper assumes compaction (compression) is positive and dilation (tension) is negative. Thus, the volumetric deformation inferred from the two pairs of diametral gages is given by

$$\left(\frac{V_0 - V}{V_0}\right)_d = 1 - \left(\frac{D}{D_0}\right)^2 \left(\frac{L}{L_0}\right) = 1 - (1 - \bar{\epsilon}_{33})^2(1 - \bar{\epsilon}_{11}). \tag{1}$$

This formulation assumes that the test sample remains a right circular cylinder throughout deformation, i.e., there is no barrelling. Here, V_0, D_0, and L_0 denote initial values of volume, diameter, and length, while V, D, and L denote current values. The quantities $\bar{\epsilon}_{22} = \bar{\epsilon}_{33} = (D_0 - D)/D_0$ and $\bar{\epsilon}_{11} = (L_0 - L)/L_0$ denote the principal components of engineering strain. The two pairs of diametral gages generally yield different values due to local anisotropy of the ice so that the average of the two gages are used in the calculation of $\bar{\epsilon}_{33}$. The volumetric deformation inferred from the confining cell intensifier is the difference in the volumes displaced by the main actuator and the confining cell intensifier, i.e.,

$$\left(\frac{V_0 - V}{V_0}\right)_i = \frac{A_a d_a - A_i d_i}{V_0}. \tag{2}$$

Here, A_a and d_a are the cross-sectional area and the displacement of the main actuator, while A_i and d_i refer to the intensifier.

Figure 3 illustrates the volumetric deformation determined from each technique for the four tests listed in Table 1. Both techniques generally show good agreement at small strain but begin to diverge beyond about 1 percent axial strain. The divergence is due to barrelling of the test sample.

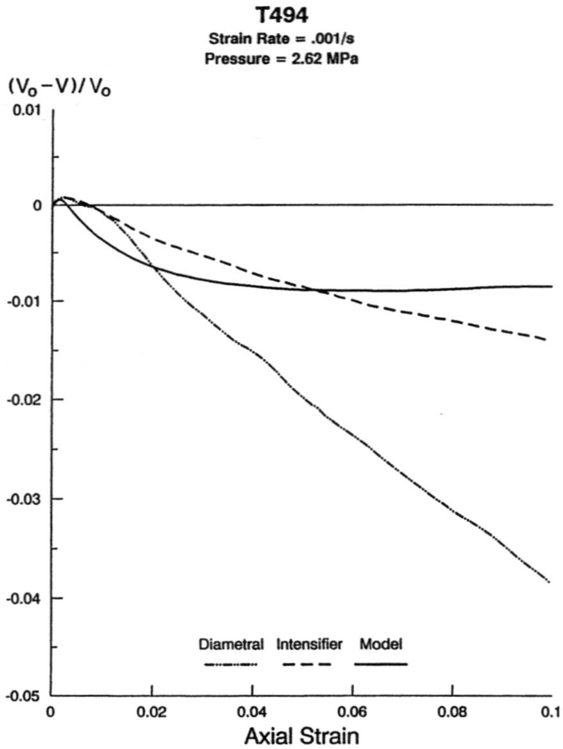
T494
Strain Rate = .001/s
Pressure = 2.62 MPa
(Vo−V)/Vo
0.01
0
-0.01
-0.02
-0.03
-0.04
-0.05
0
0.02
0.04
0.06
0.08
0.1
Axial Strain
Diametral
Intensifier
Model

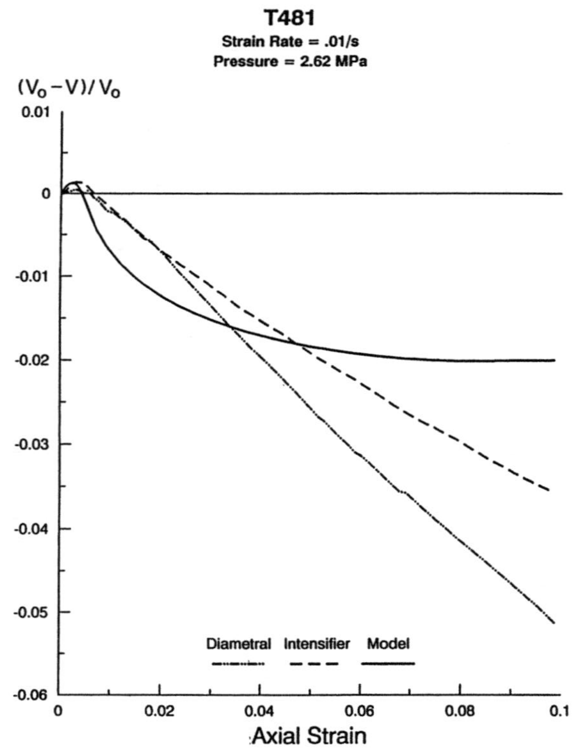
T481
Strain Rate = .01/s
Pressure = 2.62 MPa
(Vo−V)/Vo
0.01
0
-0.01
-0.02
-0.03
-0.04
-0.05
-0.06
0
0.02
0.04
0.06
0.08
0.1
Axial Strain
Diametral
Intensifier
Model

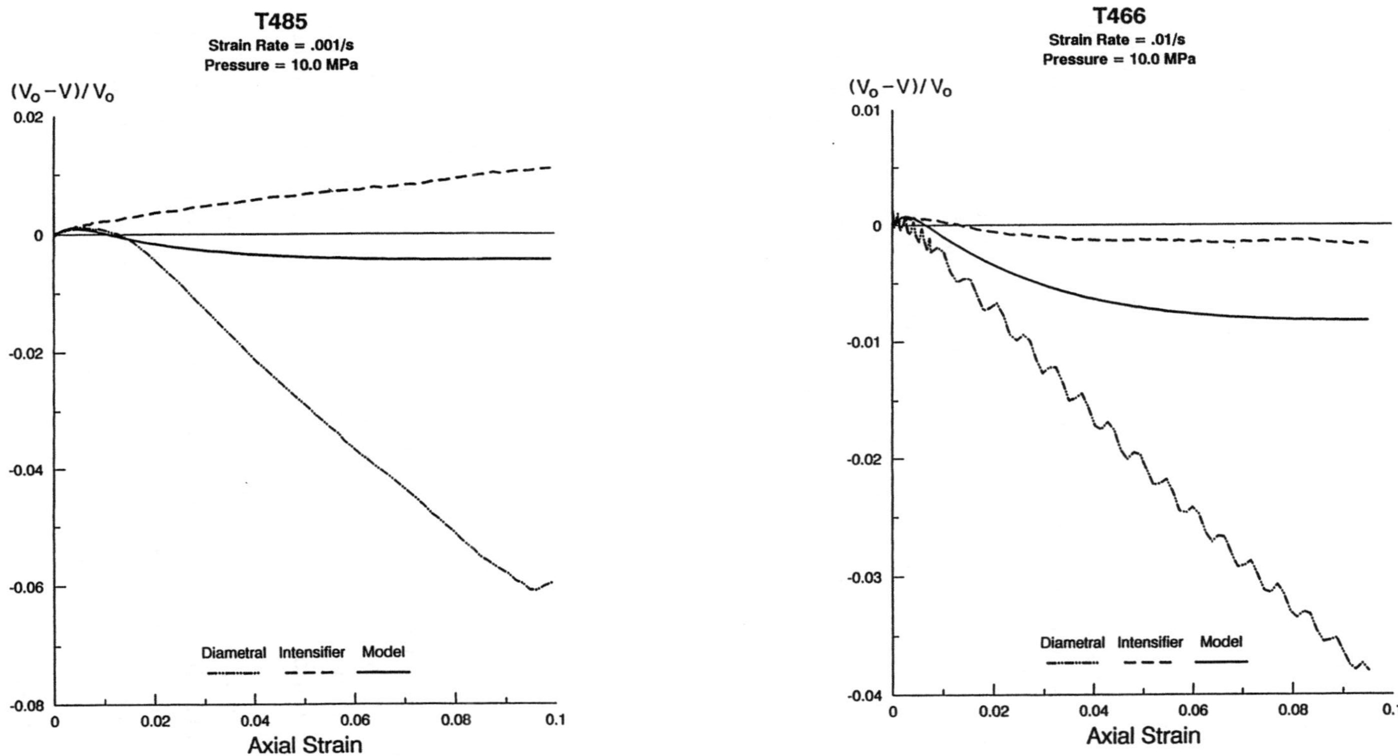

Figure 3. Volumetric deformation of the four tests listed in Table 1.

Friction at the ice platen interface constrains the sample at both ends and produces a nonhomogeneous state of strain along its length. With the onset of barrelling, the volumetric deformations from the diametral gages become questionable. Although the diametral gages provide a good estimate of the radial strain at the midpoint, the axial strain there is unknown. The actual axial strain at the midpoint is probably greater than the measured axial strain, since the measured strain is really a mean value averaged along the sample length. Consequently, the diametral gages probably overestimate volumetric deformation at large strain. The volumetric deformation inferred from the intensifier is not affected by nonhomogeneous deformations. However, the intensifier measurements are sensitive to the confining cell stiffness and any air entrained in the test system. Corrections for these factors are made during data acquisition.

Finally, dilatant volume change in ice is attributed to the creation of void space by microcrack initiaion. Intuitively one would expect that dilatancy would increase with increasing strain rate and decreasing pressure. In Figure 3, the final volume strain from the intensifier follow these trends for each level of constant strain rate and constant pressure. The diametral gages generally follow these trends at small strain.

4. Ice Crushing Model

4.1 Formation

Formulation of the crushing model begins by assuming a fixed rectangular Cartesian coordinate system with coordinate axes, x_i, i =1,2,3. Time is denoted by the quantity t, and a superposed dot denotes a material time derivative. The applications considered here are restricted to small strain triaxial tests so that material time derivatives are acceptable. A general application of the model for finite strain would require properly invariant rates of change.

The velocity field is v_i, and $D_{ij} = (v_{i,j} + v_{j,i})/2$ denotes the rate of deformation tensor, where a comma followed by an index represents partial differentiation with respect to the corresponding coordinate. Following the common practice in plasticity, we assume that the rate of deformation obeys an exact additive decomposition. That is,

$$D_{ij} = D^e_{ij} + D^p_{ij}, \tag{3}$$

where the superscript e denotes the contribution from the elastic deformation and superscript p denotes the plastic contribution. Creep strain rates can also be accounted for by adding a power law creep term to the right hand side of Equation (3), but for simplicity that term is omitted here.

The elastic strain rate is related to stress rate by Hooke's Law,

$$D_{ij}^{e'} = \frac{\dot{\sigma}'_{ij}}{2\mu}, \qquad D_{kk}^{e} = \frac{\dot{\sigma}_{kk}}{3K}, \tag{4}$$

where repeated indices are summed, a prime denotes the deviatoric part, σ_{ij} are the components of Cauchy stress, μ is the shear modulus, and K is the bulk modulus. The plastic strain rate is expressed in terms of a plastic potential, g, as

$$D_{ij}^{p} = \lambda \frac{\partial g}{\partial \sigma_{ij}}, \tag{5}$$

where λ is a scalar function. Plastic stress states are distinguished from elastic states by the yield function, f. A stress state on the surface $f = 0$ defines a plastic state, while the condition $f < 0$ defines an elastic state. A complete plasticity model requires conditions which define loading and unloading from a plastic state. Explicit conditions exist for a rate independent material with strain hardening, but none exist for a rate dependent material with strain softening. At this time, experimental data do not exist which can provide guidance in developing these conditions. Consequently, this issue is deferred by restricting loading to paths where the plastic strain increases monotonically throughout deformation.

The yield function, f, and plastic potential, g, are assumed to depend on the state of stress, strain history, and strain rate. The dependence on shear stress and pressure are separated in the following manner:

$$\begin{aligned} f &= \tau - F(I, \gamma_p, \dot{\gamma}_p) \qquad \text{(Yield Function)}, \\ g &= \tau + G(I, \gamma_p, \dot{\gamma}_p) \qquad \text{(Plastic Potential)}, \end{aligned} \tag{6}$$

where $\tau = \left(\frac{3}{2}\sigma'_{ij}, \sigma'_{ij},\right)^{1/2}$ is the effective shear stress, $I = \sigma_{kk}$ is a measure of hydrostatic pressure, and $\gamma_p = \int \left(\frac{2}{3} D_{ij}^{p'} D_{ij}^{p'}\right)^{1/2} dt$ is effective plastic shear strain. The ice is assumed to be initially isotropic, and loading is restricted to monotonic paths from an undeformed state. The yield function and plastic potential depend on only two stress invariants so that the model is two-dimensional in stress space. The functional forms for f and g show an explicit dependence on the effective plastic shear strain and its rate but no dependence on the plastic volume strain and its rate. An implicit dependence is assumed by introducing the plastic volume strain, $\Delta_p = \Delta_p(\tau, I, \gamma_p, \dot{\gamma}_p) = \int D^p_{kk} dt$. The validity of this assumption is demonstrated later when a relationship between F and G is proposed.

It is easily shown that $\lambda = \dot{\gamma}_p$. The substitution of this identity and the plastic potential from Equation (6) into Equation (5) yields

$$D^p_{ij} = \dot{\gamma}_p \left(\frac{3}{2} \frac{\sigma'_{ij}}{\tau} + \frac{\partial G}{\partial I} \delta_{ij} \right), \tag{7}$$

where δ_{ij} is the identity tensor. By defining a total effective shear strain and total volume strain, Equations (4) and (7) can be used to obtain two scalar constitutive equations:

$$\dot{\gamma} = \frac{\dot{\tau}}{3\mu} + \dot{\gamma}_p, \qquad \dot{\Delta} = \frac{\dot{I}}{3K} + 3\frac{\partial G}{\partial I} \dot{\gamma}_p, \tag{8}$$

where $\dot{\gamma} = \left(\frac{2}{3} D'_{ij} D'_{ij} \right)^{1/2}$ and $\dot{\Delta} = D_{kk}$. These equations are valid as long as the yield condition, $f = \tau - F = 0$, is satisfied.

It is apparent that, as far as the plastic potential is concerned, only $\partial G / \partial I$ appears in the constitutive Equations (8). The physical significance of this parameter is seen by taking the trace of Equation (7) and rearranging,

$$3\frac{\partial G}{\partial I} = \frac{\dot{\Delta}_p}{\dot{\gamma}_p} = \frac{d\Delta_p}{d\gamma_p}. \tag{9}$$

Hence, $\partial G / \partial I$ is the increment of volumetric plastic strain per unit increment of effective plastic strain.

The model formulated above accounts for pressure dependence, strain softening (hardening), and rate dependence through the dependence of F and G on I, γ_p, and $\dot{\gamma}_p$. Dependence on volumetric deformations is obtained through the parameter $\partial G / \partial I$. Although the model is developed from principles of plasticity, it should not be considered a true plasticity model, since unloading is not defined.

4.2 A Relationship between Stress and Volumetric Deformation

Triaxial tests conducted with constant axial strain rate and constant radial pressure boundary conditions provide two independent measurements: (1) the axial or deviatoric stress, and (2) the radial or volumetric deformation. By assuming the yield function and plastic potential to be distinct (i.e., $F \neq G$), the proposed model provides two data functions to describe these measurements. The function F represents the strength and can be constructed from the data in Figure 1. The function G is related to plastic volume change and can be constructed from the data in Figure 3. Although F and G can be constructed from independent experimental measurements, it is intuitively reasonable that strength and volumetric deformation should be related. This is certainly true for soils and is clearly illustrated by critical state soil mechanics (Schofield and Wroth, 1968) which states that the continuous shearing of a soil will lead to a critical state defined by a constant flow stress and current values of pressure and volume.

The idea of a critical state is adopted here by assuming

$$\frac{\partial G}{\partial I} = \frac{F_\infty - F}{I} = \frac{1}{3}\frac{d\Delta_p}{d\gamma_p}. \tag{10}$$

Here, $F_\infty = \lim_{\gamma_p \to \infty} F = F_\infty(I, \dot{\gamma}_p)$ is the residual strength and is identified as the critical state for ice which depends on pressure and strain rate. Implicit in this assumption is the notion that under continuous deformation, ice will re-organize its internal structure through a variety of micromechanisms so that it will flow under a constant shear stress. This point of view sees the peak strength as an artifact of the internal structural changes and the residual strength as a fundamental material property.

In addition to the two constitutive relations in Equation (8), Equation (10) presents a third relation which defines plastic volume strain and permits the formulation of F and G as functions of $(I, \gamma_p, \dot{\gamma}_p)$ only. A comparison of the current state of stress, (F, I), with residual strength determines the plastic volumetric deformation. The condition $F < F_\infty$ yields compaction and is satisfied when the stress strain curve approaches F_∞ monotonically from below. The condition $F > F_\infty$ yields a dilatant response and is satisfied for nonmonotonic stress strain curves which approach F_∞ from above. The inverse relationship between $\partial G/\partial I$ and I implies little plastic volumetric deformation is expected for high confining pressures. At large strain, $F \to F_\infty$ and $\partial G/\partial I \to 0$. Thus the assumed relationship between F and G associates incompressible plastic flow with the critical or steady state response.

The integration of Equation (10) would yield an explicit expression for G within some constant of integration. This constant would depend on F and involve either the initial yield condition or the large strain response of F. As an example, the constant could be determined by requiring $\lim_{\gamma_p \to \infty} G = F_\infty$ which would yield an associated flow for the model at large strain. Details such as this can only be resolved with the actual construction of F and G.

4.3 Application to Triaxial Tests

Consider the triaxial compression of a right circular cylinder. Again let σ_{11} denote axial stress and $\sigma_{22} = \sigma_{33}$ denote radial confining pressure with ϵ_{11} and $\epsilon_{22} = \epsilon_{33}$ referring to the corresponding components of logarithmic strain. With these definitions, the constitutive Equations (8) become

$$\begin{aligned} \dot{\gamma} &= \frac{2}{3}(\dot{\epsilon}_{11} - \dot{\epsilon}_{33}) = \frac{\dot{\tau}}{3\mu} + \dot{\gamma}_p, \\ \dot{\Delta} &= \dot{\epsilon}_{11} + 2\dot{\epsilon}_{33} = \frac{\dot{I}}{3K} + 3\frac{\partial G}{\partial I}\dot{\gamma}_p, \end{aligned} \tag{11}$$

where $\tau = (\sigma_{11} - \sigma_{33})$ and $I = (\sigma_{11} + 2\sigma_{33})$. The boundary conditions for the triaxial test are assumed to be a constant axial strain rate and constant radial confining pressure. Constant radial confinement implies $\dot{I} = \dot{\tau}$.

The Equations (11) can be combined to eliminate $\dot{\gamma}_p$. This operation yields a differential equation for ϵ_{33}, i.e.,

$$\frac{d\epsilon_{33}}{d\epsilon_{11}} = \frac{\frac{1}{3K}\frac{d\tau}{d\epsilon_{11}} + \left(2 - \frac{1}{\mu}\frac{d\tau}{d\epsilon_{11}}\right)\frac{\partial G}{\partial I} - 1}{2(1 + \frac{\partial G}{\partial I})}, \tag{12}$$

where $d\tau/d\epsilon_{11}$ is the tangent modulus of the stress strain curve. Equation (12) is a generalized Poisson's ratio for ice which accounts for plastic deformation. In the limit of large strain, $d\tau/d\epsilon_{11} \to 0$ and $\partial G/\partial I \to 0$. Thus, $\lim_{\epsilon_{11} \to \infty} d\epsilon_{33}/d\epsilon_{11} = -1/2$, which is the condition for incompressible flow.

Equation (12) can be integrated to yield ϵ_{33}. For simplicity, the Euler method with a strain step of $\Delta\epsilon_{11} \approx 0.0001$ is used for the integration. The tangent modulus in Equation (12) is calculated directly from the stress strain curve by the first forward difference formula. Values of $\mu = 4000$ MPa and K = 6667 MPa are used for the elastic constants.

The initial condition is $\epsilon_{33}(0) = 0$. For our sign convention, the principal components of engineering strain are related to the logarithmic strain components by the relationship, $\epsilon_{ij} = -\ln(1 - \bar{\epsilon}_{ij})$ for $i = j = 1, 2, 3$. Thus, the volumetric deformation predicted by the model is

$$\frac{V_0 - V}{V_0} = 1 - \exp(-2\epsilon_{33})\exp(-\epsilon_{11}). \tag{13}$$

The volumetric deformation predicted by the model for each test is shown in Figure 4. In all cases, initial compaction is followed by a net dilatancy. This response is to be expected from the assumed relationship between F and G. At small strain, the condition for compaction, $F < F_\infty$, is satisfied, while near the peak stress and beyond, the condition for dilatancy, $F > F_\infty$, is satisfied. Thus, the model associates strain softening with dilatant deformations. Finally, the responses in Figure 4 show an

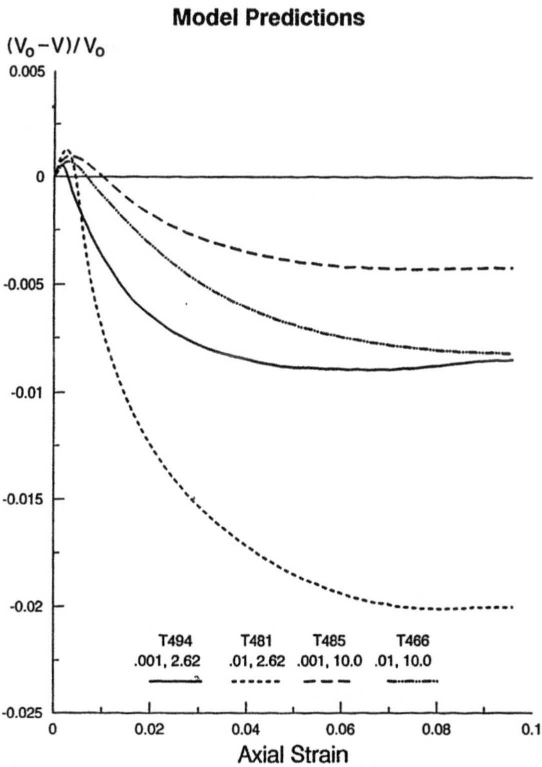

Figure 4. Model predictions for the volumetric deformation of each test listed in Table 1.

increase in dilatancy with increasing strain rate and decreasing pressure in accordance with physical intuition. These trends can be anticipated by the shape of the stress strain curves in Figure 1. Sharp stress strain curves which are characterized by a rapid loss in strength are associated with high rates of dilatancy, whereas blunt curves with a slower loss in strength are associated with lower dilatancy rates.

Figure 3 provides a quantative comparison between the model and experimental measurement. In all cases, the model underpredicts the net dilatancy measured from the diametral gages, which is expected from previous discussion. Similarly, differences between the model and the intensifier should be expected, since the model assumes a homogeneous state of strain.

Greater confidence in the model predictions can be obtained by studying several repetitions of the same test condition. Three tests were conducted for $\sigma_{33} = 2.62$ MPa and $\epsilon_{11} = 10^{-2}$ sec^{-1} including test T481. The physical properties of each test sample were very similar and yielded maximum and residual strengths within 10 percent. Figure 5 illustrates polygonal envelopes constructed to bound the diametral and intensifier measurements along with the model predictions obtained from the three stress strain curves. The diametral gages show a large amount of scatter, since they provide point measurements and, hence, are very sensitive to local deformations. The intensifier which provides volume averaged measurements shows less scatter. The model shows the least amount of scatter which is on the order of magnitude of that seen in the stress strain curves. Futhermore, the model favors the diametral measurements in the early stages of deformation and the intensifier in the later stages. This trend holds for the other test conditions.

The volumetric deformations predicted by the model show a good qualitative correlation with experimental measurements. These predictions are obtained directly from the stress strain curves without the benefit of a model constant to adjust the results. The introduction of such a constant to the relationship between F and G (Equation (10)) would improve the quantitative comparison between the model and the data. However, the data scatter, the experimental problem with nonhomogeneous deformations, and the fundamental differences between the two types of measurements do not support making this adjustment at this time.

The calculation of $d\epsilon_{33}/d\epsilon_{11}$ is not necessary for the solution of boundary value problems but is presented here to demonstrate the validity of

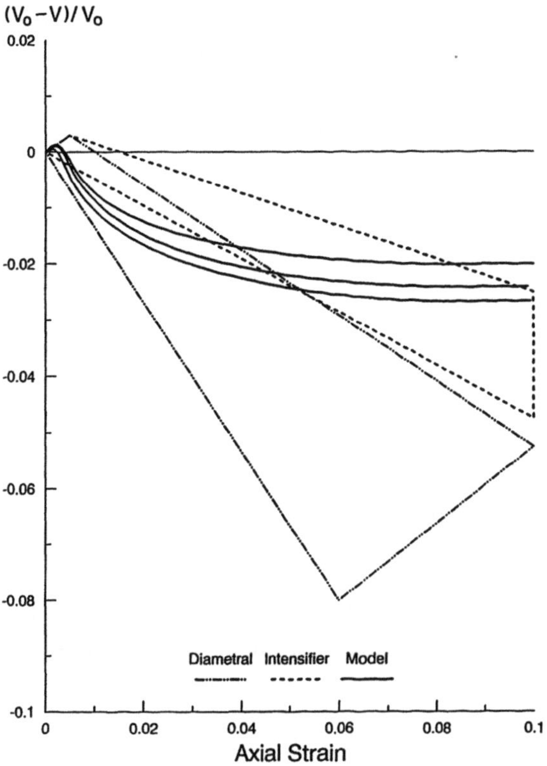

Figure 5. A comparison between model predictions and experimental measurements for three tests conducted at $\dot{\epsilon}_{11} = 0.01/\text{sec}$ and $\sigma_{33} = 2.62\text{MPa}$.

the relationship between F and G. The important quantity for problem solving is the effective plastic strain rate, $\dot{\gamma}_p$, which governs the magnitude of plastic deformation. A differential equation for $\dot{\gamma}_p$ is obtained by combining Equations (11) and taking the material time derivative. The resulting equation is

$$\left(1+\frac{\partial G}{\partial I}\right)\ddot{\gamma}_p+\left(\frac{\partial \dot{G}}{\partial I}\right)\dot{\gamma}_p=\ddot{\epsilon}_{11}-\frac{\ddot{\tau}}{3\mu}-\frac{\ddot{I}}{9K}, \tag{14}$$

where

$$\left(\frac{\partial \dot{G}}{\partial I}\right)=-\frac{1}{I}\left(\dot{F}+\frac{\partial G}{\partial I}\dot{I}\right)$$

Equation (14) is a first-order differential equation which describes the evolution of $\dot{\gamma}_p$ with time. This equation replaces the consistency condition found in conventional plasticity models and incorporates rate dependence in the model. Except for $\ddot{\epsilon}_{11}$, all terms in Equation (14) are independent of the coordinate system. The interpretation of this frame dependent term in the context of a general boundary value problem will be discussed in a later section.

For the triaxial tests considered here, Equation (14) reduces to

$$\left(1+\frac{\partial G}{\partial I}\right)\ddot{\gamma}_p+\left(\frac{\partial \dot{G}}{\partial I}\right)\dot{\gamma}_p=-\frac{\ddot{\tau}}{E}, \tag{15}$$

where E denotes Young's modulus. All terms in Equation (15) depend on F and, hence, can be obtained directly from the stress strain curves. However, the noise from the numerical calculation of $\ddot{\tau}$ is unacceptable. As an alternative, F is constructed analytically by assuming

$$\begin{aligned} F(\sigma_{33},\epsilon_{11},\dot{\epsilon}_{11}) &= F_1\exp(-\eta/\eta_1)+F_2\exp(-\eta/\eta_2) \\ &\qquad +F_3\exp(-\eta/\eta_3)+F_\infty, \end{aligned} \tag{16}$$

where $\eta = (\epsilon_{11}-\epsilon_y)/(\epsilon_m-\epsilon_y)$. The exponential terms represent the transient response, and F_∞ is the steady state response. The coefficients F_1, F_2, and F_3 are found by requiring

$$\begin{aligned} F(\sigma_{33},\epsilon_y,\dot{\epsilon}_{11}) &= \tau_y \\ F(\sigma_{33},\epsilon_m,\dot{\epsilon}_{11}) &= \tau_m(\sigma_{33},\dot{\epsilon}_{11}) \\ \frac{dF}{d\eta}(\sigma_{33},\epsilon_m,\dot{\epsilon}_{11}) &= 0 \end{aligned} \tag{17}$$

where (ϵ_m,τ_m) is the peak value of the stress strain curve with test conditions $(\sigma_{33},\dot{\epsilon}_{11})$, and $(\epsilon_y,\tau_y) = (0.000025, 0.25)$ is arbitrarily chosen as a rate and pressure independent yield point. By trial and error, $\eta_1 = 0.5, \eta_2 = 1.0$, and $\eta_3 = 5.0$ with suitable F_1, F_2, F_3, and F_∞ are found to yield reasonable approximations of the stress strain curves in Figure 1. Note that temperature effects can easily be incorporated into this construction of F by making (ϵ_m,τ_m) and F_∞ depend on temperature.

The function F is constructed for each test, and Equation (15) is integrated by the Euler method with the initial condition $\dot{\gamma}_p(\epsilon_y) = \frac{2}{3}(1+\nu)\dot{\epsilon}_{11} - \frac{1}{3\mu}\dot{\tau}$ where ν is the elastic Poisson's ratio. Figure 6 illustrates $\dot{\gamma}_p$ normalized with respect to $\dot{\epsilon}_{11}$ for each test. The evolution of $\dot{\gamma}_p$ consists of a transient response which depends on $\dot{\epsilon}_{11}$ and σ_{33} and a steady state response. In all cases, the transient response reaches a peak value greater

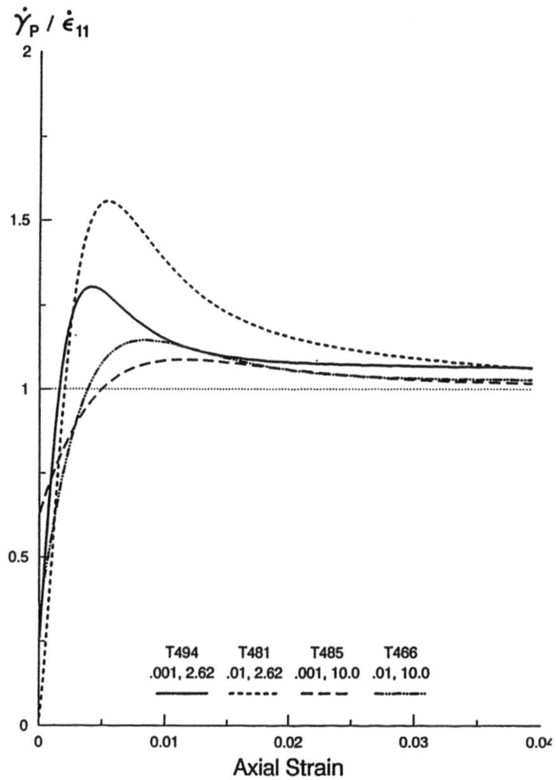

Figure 6. Normalized effective plastic strain rate for the four tests listed in Table 1.

than one. The magnitude of the peak value depends on the degree of dilatancy. At large strain, $\dot{\gamma}_p/\dot{\epsilon}_{11}$ approaches one from above. This result together with the condition of incompressibility at large strain indicates that the steady state response of the model is perfectly plastic.

4.4 A Rate and Pressure Dependent Failure Criterion

In previous sections, it was implied that microcrack initiation and dilatancy are related phenomena and are associated with the strain softening (i.e., reduction in strength) observed in tests conducted under ductile conditions. This connection between crack initiation and dilatancy can be extended to predict the total loss of strength (i.e., brittle failure).

Under ductile conditions, microcracks initiate but do not propagate. The existence of the microcracks creates void space and the dilatant deformations seen in Figure 3. The degree of cracking and, hence, the amount of dilatancy depend on the confining pressure and strain rate. For critical combinations of pressure and strain rate, the cracks will propagate in an unstable manner producing a critical dilatancy rate where total loss of strength is imminent. Equations (12) and (14) suggest that the critical dilatancy rate is $(\partial G/\partial I)_{cr}$ = -1, since, at that value, $d\epsilon_{33}/d\epsilon_{11}$ and $\ddot{\gamma}_p$ become unbounded.

This criterion with Equation (10) yields an approximate expression for points near the failure surface. If F^* denotes this approximate surface, then the failure criterion is

$$F^* = I + F_\infty(I, \dot{\gamma}_p). \tag{18}$$

If F_∞ is rewritten as a function of $(\sigma_{33}, \dot{\epsilon}_{11})$, then the dashed lines in Figure 2 offer a rough representation of $F*$ in the (σ_{33}, τ) plane. However, the dependence of F_∞ on pressure indicates that the failure surface in this plane is likely to be nonlinear.

4.5 Application to Indentation Problems

The application of the model has been restricted to triaxial tests where the stress state is assumed to be homogeneous. Indentation problems generally involve nonhomogeneous states of stress and, hence, require the simultaneous solution of the equations of equilibrium and the constitutive equations. This can be accomplished by incorporating the model into a finite element formulation.

For indentation problems, there will, in general, be 10 unknowns: three components of displacement ($u_i = \int v_i dt$), six components of stress (σ_{ij}), and the effective plastic strain rate ($\dot{\gamma}_p$). The three equations of equilibrium, the six equations for D_{ij} (i.e., Equation (3)), and the differential equation for $\dot{\gamma}_p$ (i.e., Equation (14)) are sufficient to solve for these unknowns. Since deformations and local material rotations can be large for indentation problems, all material time derivatives must be replaced by properly invariant rates of change.

Finally, the application of Equation (14) to indentation problems requires an interpretation of the coordinate system associated with $\ddot{\epsilon}_{11}$. The appearance of $\ddot{\epsilon}_{11}$ introduces a local coordinate system into the calculation of $\dot{\gamma}_p$ and is attributed to microcracking. The presence of microcracks creates a fabric within an originally isotropic material. For triaxial tests, the cracks tend to initiate randomly in planes normal to the direction of least principal stress. Since the stress state is axially symmetric, the fabric has a plane of symmetry whose normal is parallel to the direction of maximum principal stress. To maintain consistency with triaxial test data, the local x_1 direction should be interpreted as the direction of maximum principal stress.

A specific application of the model is shown schematically in Figure 7 which illustrates the penetration of ice by a rigid sphere with velocity, v. At any point P, a local coordinate system (x, z) is established which aligns with the direction of maximum principal stress. The local system is located with respect to the global system (X, Z) by the angle θ and distance R. For $\theta \approx \pm\pi/2$ and R slightly greater than the contact radius, the point P is located near a free surface where the confinement is small. For a small indentation velocity, the pressure and strain rate at P are likely to yield a ductile response. If the indentation velocity is sufficiently high, then the pressure and strain rate at P are likely to yield $\partial G/\partial I = -1$. At this point, the transition from a ductile to a brittle response occurs, where the contact pressure is expected to be maximum.

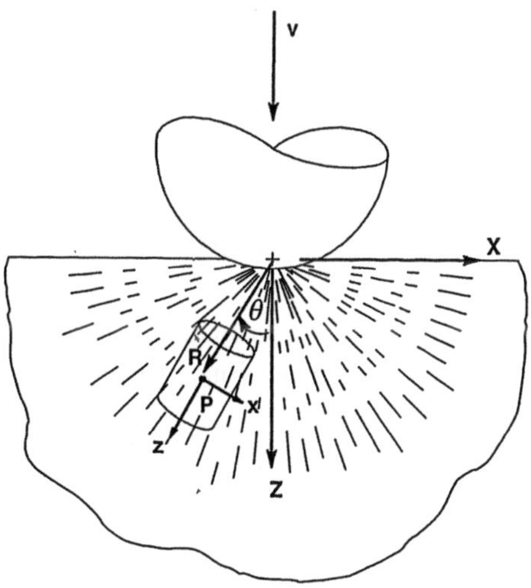

Figure 7. Application of the model to the indentation of ice by a spherical indentor.

5. Conclusion

A material model for the crushing of ice has been formulated to include pressure and rate sensitivity, strain softening, and volumetric deformations. The model is restricted to the ductile regime and is complemented with a failure criterion which predicts the onset of brittle failure. The model is in the early stage of development and requires further testing. A critical test is the model's sensitivity to load path. This work along with the construction of F and G is in progress.

Application of the model to the solution of meaningful problems would require a finite element formulation. The completion of this step would provide an analytical tool for the estimation of ice crushing pressures over the range of engineering interest. Model predictions for small and intermediate areas are expected to be accurate, since crushing is the predominant failure mode at those areas. At large areas, where other failure modes (e.g., buckling, fracture) are likely to occur, model predictions are expected to provide an upper bound.

References

Cox, G. F. N., Richter-Menge, J. A., Weeks, W. F., Mellor, M., and Bosworth, H. W. 1984. The mechanical properties of multi-year sea ice, phase I: test results. Report 84-9. Cold Regions Res. Eng. Lab., Hanover, NH, U.S.A.

Cox, G. F. N., Richter-Menge, J. A., Weeks, W. F., Bosworth, H., Perron, N., Mellor, M., and Durrel, G. 1985. The mechanical properties of multi-year ice, phase II: test results. Report 85-16. Cold Regions Res. Eng. Lab., Hanover, NH, U.S. A.

Dorris, J.F. 1989. The use of residual strength in the calculation of crushing loads on wide structures. Proc. 10th Int. Conf. on Port and Ocean Eng. Under Arctic Conditions, Lulea, Sweden, pp.467-480.

Dorris, J.F. and Winkler, M.M. 1989. On the relationship between global and local ice crushing loads. Proc. 10th Int. Conf. on Port and Ocean Eng. Under Arctic Conditions, Lulea, Sweden, pp.481-491.

Johnson, R.C. and Benoit, J.R. 1987. Iceberg impact strength. Proc. Offshore Tech. Conf., OTC Paper 5599, Houston, TX, Vol. 4, pp.417-423.

Nordgren, R.P. 1988. Plastic analysis of ice contact problems. ASME J. App. Mech., Vol 55, pp.73-80.

Ralston, T.D. 1978. An analysis of ice sheet indentation. Proc. Fourth IAHR Sym. on Ice Problems, Lulea, Sweden, Pt. 1, pp.13-31.

Ralston, T.D. 1979. Sea ice loads. Tech. Seminar on Alaskan Beaufort Sea Gravel Island Design, Presented by Exxon Company, U.S.A.

Richter, J.A. and Cox, G.F.N. 1984. A preliminary examination of the effect of structure on the strength of ice samples from multi-year pressure ridges. Proc. ASME 3rd

Int. Symposium on Offshore Mech. and Arctic Eng., New Orleans, LA, U.S.A., Vol. 3, pp.140-144.

Sanderson, T.J.O. 1986. A pressure-area curve for ice. Proc. 8th International Symposium on Ice, IAHR, Iowa City, Iowa, U.S.A, Vol. 2, pp.361-384.

Schofield, A. and Wroth, P. 1968. Critical State Soil Mechanics. McGraw-Hill, London.

Spencer, P.A., Masterson, D.M. and Dorris, J.F. 1989. Volumetric deformation of multiyear ice samples. To be submitted to Cold Regions Science and Technology.

Discussion

Q: T.J.O. Sanderson, BP International

You commented that for larger-scale interactions it is only necessary to consider residual stresses. Would you please explain how the necessary damage occurs to enable this? Ice may approach a structure in an undamaged state, and I think you need to explain why it does not pass through a state of maximum stress.

A: J. Dorris

There are two cases to consider: 1) good contact along the entire structure width in the frozen-in condition, and 2) impact conditions. In the first case, the far field stress, which drives the ice, builds up over a period of hours or even days. During this time prior to breakout the ice will experience a period of strain acceleration or strain history. Because of this it is expected that the ice will not go through the peaks seen in a constant strain rate compression test. This reasoning is supported by the strengths measured by a soft test machine. There the strains are accelerating and the measured strengths are typically much less than the strengths measured in a stiff machine. In the second case of impact, the initial contact area is expected to be small and the ice undamaged. This is a local contact problem which should be associated with peak strength values. Although the contact stress is high, the total force or integrated pressure is low. By the time the structure has fully penetrated the ice feature, large strains and significant damage of the ice has occurred. At this time the residual strength is the appropriate strength value. Although these stresses are much lower than peak, their integration across the contact width are expected to yield the global load on the structure. Both of the above arguments rely on strain history. The model depends on strain and should be able to model both cases.

Q: R.P. Nordgren, Shell Development Co.

The cyclic unloading-reloading tests reported at this symposium by Cole and by Jordaan show permanent strain and reyielding at a critical stress level. It should be possible to model such behaviour by the plasticity model proposed by Dorris upon the addition of suitable loading and unloading criteria. Further experiments with cyclic non-proportional loading would be of interest to confirm the existence of a rate-dependent yield function.

A: J.F. Dorris

I agree with you completely. As stated in my paper, the model is restricted to monotonic levels and strain paths due to a lack of data on unloading-reloading cycles. The type of data presented by Cole and Jordaan should lead to a generalization of the model which would include non-monotonic paths.

Q: A. Porter, University of Leicester

Rate dependent plasticity theory seems not to include a significant features of ice behaviour. For $\dot{\sigma} < 0$, if the rate is sufficiently large, plasticity theory gives a zero inelastic strain rate, whereas ice shows significant inelastic strain recovery. Would you like to comment?

A: J.F. Dorris

I am not sure I understand the question. I believe that the data shown by Cole and Jordaan show that ice undergoes significant permanent deformation upon load removal. I have also seen indentation tests where the ice contains a permanent impression of the indentor after it has been removed. This indicated to me that a plasticity approach is appropriate for ice.

Q:

You are treating ice as a classic plastic material in terms of a yield surface. Would you comment on the types of problems (e.g. monotonic loadings and proportional loading you intend to solve with your model).

A: J.F. Dorris

The paper mentions that the model is currently restricted to monotonic load or strain paths. Indentors penetrating ice under a constant velocity would be an example of this type of problem. In order to solve problems of unloading and reloading the model would have to be generalized. This would require test data on unloading. These data do not currently exist.

Q: Oskar H. Mahrenholtz, TUHH, FRG

I have doubts about the existence of a plastic potential in the mentioned way. The Von Mises type describes an inviscid material $D^p_{ij} = \lambda \frac{\partial_g}{\partial \sigma_{ij}}$ is not explicit. Objectivity, rotation, Clausius-Duheim where inequality should be regarded.

A: J.F. Dorris

Permanent deformation seen in indentation tests indicates that a plasticity approach is suitable. It is mentioned in the paper that for application to large deformation that all material time derivatives should be replaced by objective time derivatives. The material time derivative was used in the paper to facilitate calculations and was restricted to small strain triaxial tests.

Strength and Failure Modes of Pure Ice and Multi-Year Sea Ice Under Triaxial Loading

S.A.F. Murrell, P.R. Sammonds and M.A. Rist

Rock and Ice Physics Laboratory
Department of Geological Sciences, University College London
Gower Street, London WC1E 6BT, England UK

Abstract Following a brief description of a new servo-controlled triaxial testing system for large ice samples (ca. 40mm diameter) with facilities for acoustic emission (AE) observations, and of a technique for manufacturing pure ice samples, laboratory results are presented for the strength of ice under uniaxial tension, and under uniaxial and triaxial compression. Data has been gathered using strain rates up to 10^{-2}/s, confining pressures up to 30MPa and temperatures down to -40°C. The pure ice had a uniform grain size of about 1mm and a high density. The sea ice, collected from the Canadian High Arctic, had a large natural variability of structure, salinity and density. The test conditions straddle the boundary between brittle cracking and fracture behaviour, and high temperature plasticity. The behaviour shows some similarities with that of silicate rocks and covalent oxides. At low temperatures and high strain rates ice fracture exhibits a pressure dependence similar to that of brittle rocks, with strength increasing as confining pressure increases. AE is also similar to that of brittle rocks. However, at higher temperatures and pressures, where ductile behaviour occurs, increasing pressure leads to decreasing strength. The peak strengths observed during ductile behaviour are not yet fully understood. The results are discussed in terms of Griffith crack theory.

1. Introduction

In this paper we shall compare data from pure laboratory-grown ice of uniform 1mm grain size with data from multi-year sea ice which has considerable characteristic heterogeneity.

The experiments on sea-ice were performed because of a requirement for better estimates, based on appropriate experimental data, of the maximum likely ice loads on engineering structures (oil exploration and production platforms) in offshore cold regions (e.g. the Beaufort Sea). Apart

S. Jones · R. F. McKenna · J. Tillotson · I. Jordaan (Eds.)
Ice-Structure Interaction
IUTAM-IAHR Symposium St. John's, Newfoundland, Canada

from such engineering uses, ice mechanics is of fundamental significance also in the geological sciences in connection with the geophysical behaviour of polar and mountain-glacier ice masses.

Croasdale (1980) showed that the forces available from high (15-30 m/s) wind shear over moderately large (400-2000 km^2) expanses of ice are sufficient to cause very large loads (200-7000 MN). However ice-structure interactions are usually at loading rates where fracture is the dominant deformation mode and ice loads are limited by the strength of ice locally failing against the structure (Sanderson, 1984). In this so-called stress-limit analysis the ice-structure interaction is considered to be an indentation problem. Early analyses (Michel, 1970) used a crushing strength for ice and an empirical factor to describe the interaction. Later, plastic limit analysis was applied to the problem (Ralston, 1978), where ice strength is limited by a plastic yield. The reference stress method was also applied (Ponter et al., 1983), where ice failure by creep deformation is assumed.

However the indentation of ice involves a complex triaxial stress field, and a number of deformation modes occur, ranging from tensile fracture at the edge of the indenter to confined compressive fracture behind the indenter and creep deformation where the confinement is sufficient to prevent crack nucleation and growth. A number of experimental studies have been carried out to investigate the problem. These range from full-scale measurements of impact forces between multi-year ice and rock islands (Danielewicz and Cornett, 1984) to laboratory indentation tests (Timco, 1986) where crushing, crushing and spalling, crushing with radial cracks, crushing with radial and circumferential cracks have been noted as deformation modes. Ice fracture has been introduced into the ice-structure indentation analysis by applying a scale-dependent fracture cut-off to the reference stress method (Walden et al., 1987). Hallam (1986) has examined the problem more rigorously by assuming an elastic-brittle material response. However a complete treatment of the problem is still required.

This paper is concerned with the strength of ice under triaxial loading and under conditions of temperature and strain rate giving rise to behaviour ranging from brittle (involving cracking and fracture) to fully ductile (involving crystal plasticity alone). A new servo-controlled triaxial testing system for large ice samples (40mm diameter) is described, and results are presented from tests on samples of pure ice (laboratory grown) and of multi-year sea ice. In a number of tests acoustic emission was measured, illustrating a clear distinction between brittle and ductile behaviour. Uniaxial tension and compression tests were also performed.

It is well known that triaxial loading strongly affects brittle behaviour, whereas it has little effect on crystal plasticity (Murrell, 1964; Murrell and Digby, 1970; Paterson, 1978; Jaeger and Cook, 1979; Horii and Nemat-Nasser, 1986; Ashby and Hallam, 1986). Using the notation $\sigma_1, \sigma_2, \sigma_3$ for the principal stresses, and taking tensile stresses as positive, then in a conventional triaxial compressional test in which cylindrical samples are subjected to a confining pressure ($P = \sigma_1 = \sigma_2$) and to a separate axial load ($R = \sigma_3$) the distinction between brittle and crystal plastic behaviour is that the strength $(R - P)$ where $(R - P) = (\sigma_3 - \sigma_1)$ is strongly dependent on confining pressure (P) in the brittle case, but is almost independent of it in the crystal plastic case. This is well demonstrated in a Mohr diagram (Figure 1).

Plastic yield in crystal plasticity generally follows the von Mises criterion,

$$\sigma_y = [\Sigma(\sigma_i - \sigma_j)^2/6]^{\frac{1}{2}}. \tag{1}$$

For brittle materials containing many randomly-oriented identical cracks, as envisaged in Griffith's theory (1920), the initiation of crack growth (the prelude to crack interaction and fracture) occurs when:

$$(\sigma_1 - \sigma_3)^2 = (2 - \nu)^2 \sigma_T [\sigma_T(4 - \nu) - 2(\sigma_1 + \sigma_3)], \tag{2}$$

(Murrell and Digby, 1970; taking the simplest case of penny-shaped cracks, where σ_T = uniaxial tensile fracture strength, and ν = Poisson's ratio). Murrell and Digby (1970) found that the crack-initiation criterion is independent of the intermediate principal stress ($\sigma_2 = Q$). It will be noted that the uniaxial brittle compressive strength is an order of magnitude larger than the tensile strength.

The two criteria contained in Equations 1 and 2 are illustrated in Figure 1. An important fundamental difference between crystal plasticity and brittle fracture is that the plastic yield or flow stress σ_y is strongly dependent on homologous temperature (T/T_m) (where T and T_m are the test temperature and the melting temperature respectively), and on strain-rate ($\dot{\varepsilon}$), whereas the brittle tensile strength (σ_T) and Poisson's ratio (ν) are almost independent of T and $\dot{\varepsilon}$ for a Griffith material with pre-existing cracks (see Murrell, 1990, for further discussion of this).

The initiation of crack growth does not immediately lead to brittle

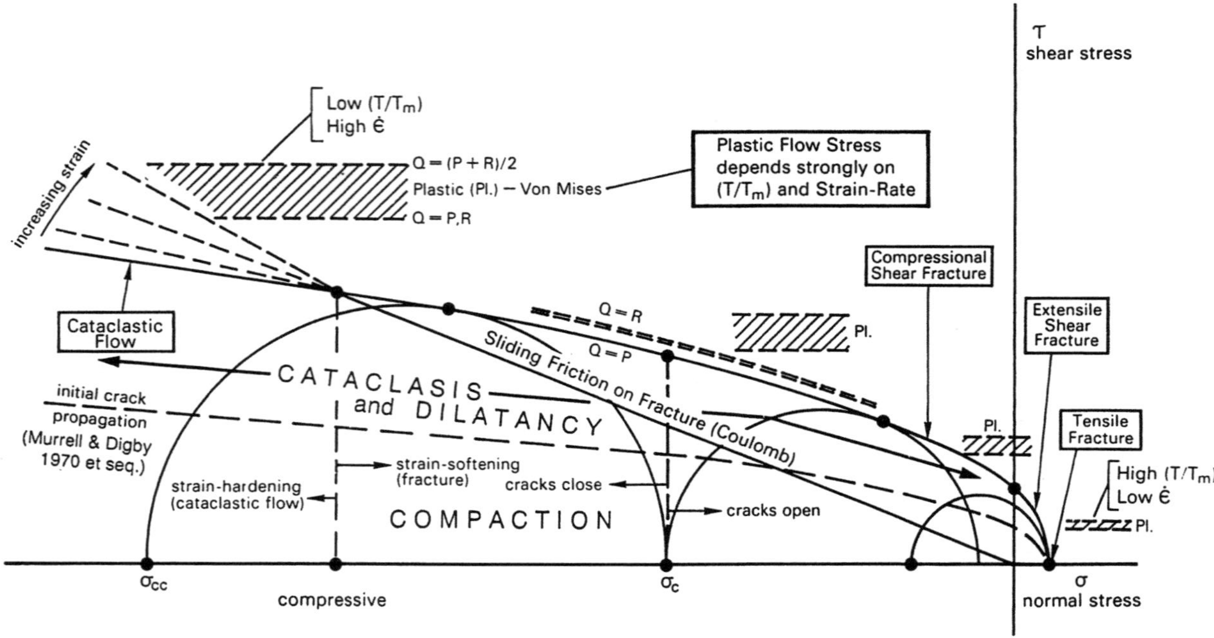

Figure 1. Mohr diagram with tensile stresses taken as positive showing Mohr envelopes for crack propagation, fracture and sliding friction, and for plastic and cataclastic flow. Q = intermediate principal stress. σ_c is the normal pressure required to close a "penny-shaped" crack, and σ_{cc} is the pressure above which all such cracks are closed. Compaction involving crack closure occurs under stress conditions below the crack propagation envelope, and cataclasis with dilatancy associated with new crack propagation occurs under stress conditions lying between the fracture and crack propagation envelopes and in the cataclastic flow region of the diagram.

fracture except in the case of the most tensile of triaxial stress states (e.g. uniaxial tension). As confining pressure increases there is a transition from tensile to shear (extensional or compressive) fracture, and in the case of shear fracture crack initiation is followed by stable crack growth requiring an increasing differential load until crack interaction results in the formation of a fracture which becomes unstable when it reaches the sample boundary. (At very low confining pressures axial splitting caused by buckling stresses is an alternative failure mode, see Horii and Nemat-Nasser, 1986; Ashby and Hallam, 1986). During the period of stable crack growth more and more cracks are nucleated and grow, so that crack damage increases, and this can be observed through its effect on acoustic emission (see below) and on elastic wave propagation (Jones and Murrell, 1989; Sammonds et al., 1989a). We describe this process as cataclasis (Figure 1). Cataclasis is a process of general fragmentation, (Murrell, 1990). It results in either the formation of a layer of powdered material in shear fracture zones (called fault gouge in rock mechanics), or the occurrence of cataclastic flow under conditions in which formation of a shear fracture is inhibited (e.g. at very high confining pressures, see Figure 1, or in fluid-saturated rocks under undrained conditions when dilatancy-hardening occurs, see Ismail and Murrell, 1976). Shear fracture occurs at differential stresses of approximately twice those required to cause the initiation of crack growth. Scanning electron microscope (SEM) images of brittle crack growth leading to fragmentation in rocks have been published by Fonseka et al. (1985).

Figure 1 also shows the conditions under which a transition occurs from brittle to ductile behaviour. It can be seen that these conditions depend on both the triaxial stress system (in particular the confining pressure and the tensile or compressive nature of the stresses) and the temperature and strain-rate (grain-size is also important, see Murrell, 1990). Few triaxial mechanical tests have been reported on ice, and the great majority have been limited to relatively high temperatures and relatively low strain-rates, under which conditions crystal plastic creep has been the dominant deformation mechanism (Jones, 1982; Nadreau and Michel, 1986; Richter-Menge et al., 1986; Cox and Richter-Menge, 1986). However, Durham et al., (1983) have studied the confined brittle fracture of pure ice at temperatures below 195K. The temperatures in large bodies of natural ice on Earth are high relative to the melting temperature of ice, so that typical ice temperatures of 223 - 273K correspond to homologous temperatures of 0.82 -1.0, but we shall show that the brittle-ductile transition in natural ice does occur under tensile stress or low confining pressure conditions when the strain rate takes values that are of interest under natural or engineering conditions (e.g. 10^{-7}/s to 10^{-2}/s).

2. Apparatus

A schematic diagram of the triaxial testing apparatus used is shown in Figure 2. A 100kN servo-controlled actuator loads a ram assembly contained in a large pressure vessel (1 tonne mass). The apparatus is located inside a -20°C armoured cold room, but is independently cooled by refrigerant circulating in a coil around the vessel. A test specimen 100mm long by 40mm diameter is used, which is enclosed in an indium metal jacket to exclude the confining pressure medium (nitrogen gas) from cracks and pores. The apparatus has a capability for performing tests at pressures up to 300MPa, with temperatures down to -90°C and strain-rates in the range 10^{-9}/s to 10^{-1}/s.

The vessel design incorporates a pressure-balanced ram whereby the high pressure confining fluid acts on the ram shoulder (below seal A) whilst the space below the shoulder (between seals B and C) is vented to atmosphere. The actuator is therefore only required to apply the differential load to the specimen. Differential load is measured with an external load cell, and specimen deformation is measured externally using a displacement transducer mounted on the ram. The sum of specimen and ram deformation is measured, but the latter, being elastic and proportional to the applied load is continuously subtracted from the displacement signal within the servo-control loop so that control is based on the specimen deformation alone (for detail see Sammonds et al., 1989b). The system has a very high effective stiffness. Although the axial elastic deformation of the apparatus measured at the specimen position is ca. 0.02mm/tonne (compared with about 0.075mm/tonne for ice specimens), this is subtracted from the control signal so that the effective system elastic deformation is close to zero. Rist (1989 and Rist et al 1990) has carried out detailed comparisons of measurements made by this method with measurements made using an LVDT mounted directly on calibration specimens, and has confirmed the efficacy of the system. The standard triaxial test procedure was employed, setting the confining pressure at a fixed value and advancing the actuator at a constant rate. Jacketed specimens were checked for leaks before and after testing.

For tensile tests the same apparatus was used, with dumb-bell shaped

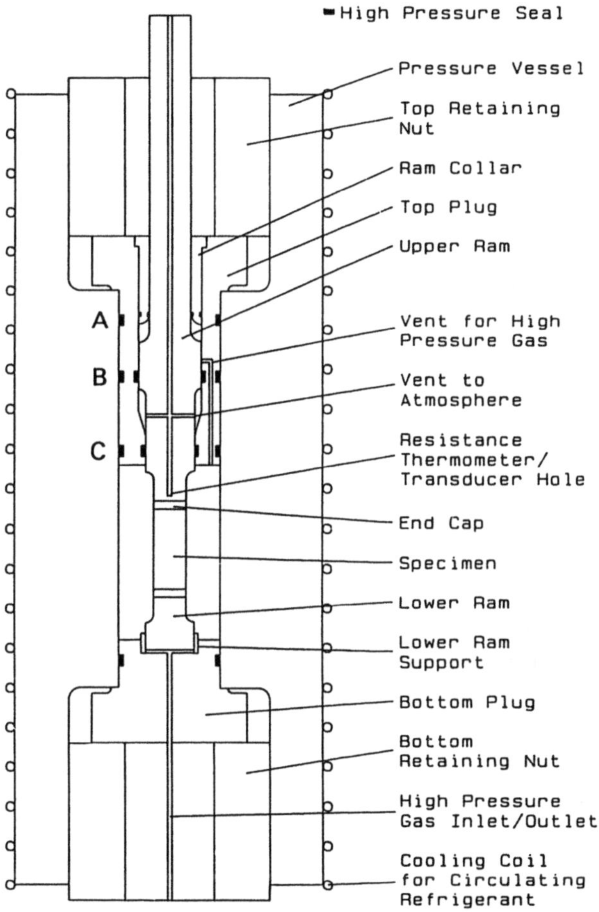

Figure 2. The triaxial testing cell.

specimens (Figure 3) cemented in aluminium end caps attached by pins to motorcycle chains which were in turn attached to the load cell and to the bottom of the triaxial vessel respectively. Expanded polystyrene foam was packed into the top and bottom of the vessel to help stabilise the temperature around the specimen. For tensile tests servo-control was based on specimen deformation measured directly using an LVDT clamped on a 70mm gauge length at the centre of the specimen. Calibration showed that good strain rate control was achieved in both tensile and triaxial compression tests. In both compression and tension tests, temperature was measured by platinum resistance thermometers placed close to the ice specimen.

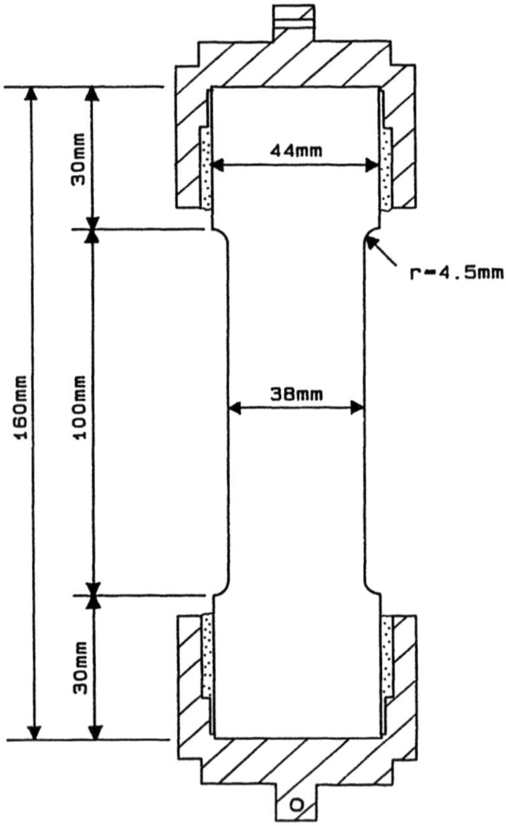

Figure 3. Tensile specimen shape.

3. Ice Tested

Floating ice is a hazard to offshore operations, and this hazard increases with increasing ice thickness and mass. In the Beaufort Sea multi-year sea ice floes are the most common thick ice feature, and therefore present the greatest ice hazard. Nevertheless few tests have been made on multi-year sea ice (Sinha, 1984, 1985; Cox et al., 1984; Cox and Richter-Menge, 1986). Because of the gross structural heterogeneity of this ice type the interpretation of strength data obtained on it in the laboratory is difficult. A comparative study with laboratory grown pure ice is useful.

As a basis for a fundamental study of ice strength under triaxial stress conditions a series of tests has been carried out on laboratory grown pure ice of uniform 1mm grain size. It is intended to extend this study to include the effect of grain-size, salinity, and air and mineral content. The preparation technique used is similar to that described by Jacka and Lile (1984). Ice seeds were first produced from frozen distilled water which was crushed and sieved to the required size. These grains were then added to distilled, de-aerated water at just below 0°C to form an ice-water mixture which was subsequently poured into a tall perspex cylinder with an internal diameter equal to the required diameter of the ice specimen. Obvious bubbles were removed by agitation before the mixture was compressed using a screw press in order to squeeze out excess pore water in which further bubbles could form. Finally the samples were allowed to solidify. Some very fine bubbles remained, but the amount of included air was small, the mean bulk density being 915 ± 1 kg/m^3, and specimen quality was highly reproducible. Meltwater conductivity of the specimens was lower than $10 \times 10^{-6}\Omega^{-1}\text{cm}^{-1}$.

The multi-year sea ice tested in this programme came from a single floe off Buckingham Island in the Canadian Arctic archipelago, collected during a British Petroleum expedition. Cores 1m long by 325mm diameter were flown back to London and kept in a coldstore at -28°C. The thickness of the floe was found (by coring) to be 7m. The top 200mm consisted of bubbly ice (average density 850 kg/m^3), with predominantly fine-grained polycrystalline ice (grain size 1-2 mm) in the uppermost portion. In the next 50mm a columnar structure was rapidly established with a grain size (column width) of 3 to 7mm. The average density of test specimens taken from the floe was 899 ± 2kg/m^3. Further details will be published at a later date (see Sammonds, 1988). For the present tests, specimens were taken mostly from the same location at an unridged portion of the floe, and mostly at the same depth (0.8 -1.6 m) below the surface. Using a pillar drill, horizontal experimental specimens were cored in the plane of the ice floe and in the direction of magnetic north.

All ice specimens were machined in a lathe to the correct length and with their ends flat and parallel to within ± 0.03mm.

4. Results

The laboratory grown pure ice is isotropic. However, the sea ice is anisotropic. Most of our tests on sea ice were on cores drilled in a fixed horizontal direction in the ice mass, but vertically oriented cores were also tested, and the latter are stronger. A few experiments were performed on first-year ice. The dependence of strength on the inverse of the square root of brine content (Assur, 1958) was confirmed, and a tendency was also found for strength (measured under fixed conditions) to increase roughly linearly with depth in the ice mass measured relative to a fixed level surface (sea level).

4.1. Deformation Modes and Strength

Brittle failure was characterized by a predominantly linear increase of stress with strain climaxed by an audible report from the specimen at peak stress, followed by a stress drop.

Brittle and ductile tensile fractures in sea ice specimens are shown in Figures 4a,b. Fractures in pure ice are similar. For tests on both ice types at -10°C the transition from ductile to brittle fracture occurred when strain-rate was increased from 1.0×10^{-7}/s to 1.0×10^{-4}/s.

Typical tensile stress-strain curves for pure ice at this temperature are given in Figure 5. The average strength of the two ice types in the two different deformation modes is given in Table 1.

Specimens which failed in tension in a brittle manner produced fractures with a smooth surface normal to the tensile stress. Those which failed in a ductile manner showed evidence of plastic deformation and necking, with ragged fracture surfaces inclined (at about 70°) to the tension axis.

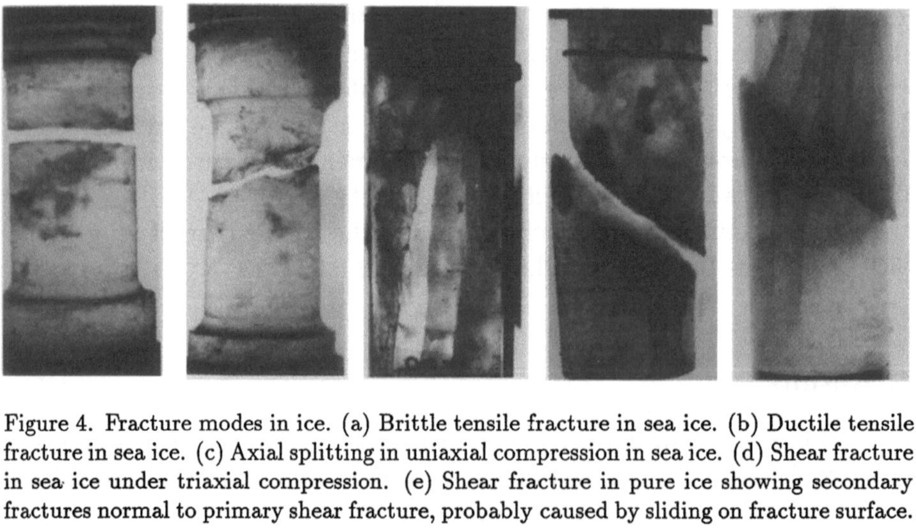

Figure 4. Fracture modes in ice. (a) Brittle tensile fracture in sea ice. (b) Ductile tensile fracture in sea ice. (c) Axial splitting in uniaxial compression in sea ice. (d) Shear fracture in sea ice under triaxial compression. (e) Shear fracture in pure ice showing secondary fractures normal to primary shear fracture, probably caused by sliding on fracture surface.

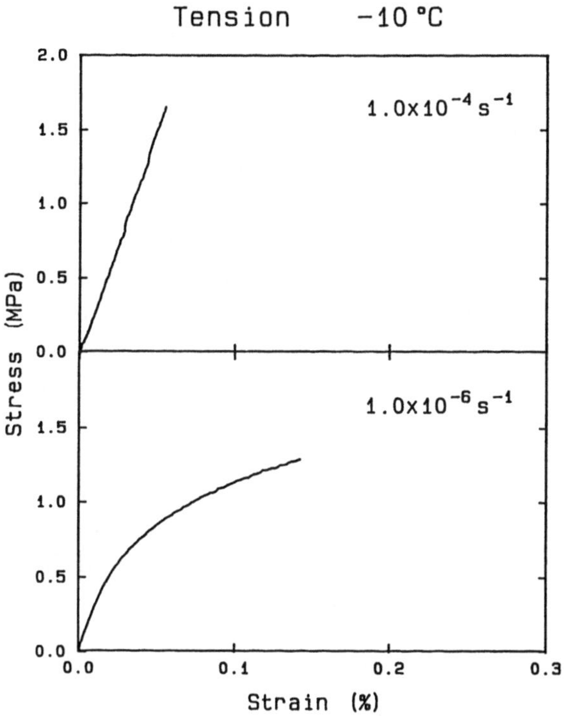

Figure 5. Tensile stress-strain curves for pure ice specimens at -10° C and various strain-rates.

Table 1. Uniaxial Tensile Strength (mean values in MPa) at −10°C

		Sea Ice	Pure Ice
Brittle	hor.	0.95	1.43
	vert.	1.39	
Ductile	hor.	0.47	1.05
	vert.	1.44	

Under uniaxial compression completely brittle failure in ice is characterized by multiple axial splitting (Figure 4c), though in pure ice this is accompanied by the formation of a relatively wide shear band, consisting of smaller fractures inclined at about 30° to σ_3. Axial splitting is inhibited by confining pressure however, and brittle failure takes place then by the formation of a narrow shear fracture inclined at an angle of about 45° (Figure 4d). Occasionally large secondary fractures were seen to extend perpendicular to the principal, shear fracture surface (Figure 4e), apparently formed as the two halves of the specimen slid over each other after failure. Such failures have also been noted by Durham et al., (1983). Uniaxial compression tests at -10°C showed a transition from ductile behaviour (though with some cracking) at a strain rate of 10^{-6}/s to brittle fracture by axial splitting at a strain rate of 10^{-2}/s. Results of such uniaxial tests at -10°C, -20°C, and -40°C are given in Table 2. They show that the compressive strength is substantially higher than the tensile strength under otherwise similar conditions. The uniaxial compressive strength of pure ice is strongly temperature dependent.

Under triaxial compression at low temperatures and high strain rates shear fracture can occur, but is eliminated as the confining pressure increases. This is caused by two distinct effects, firstly the increased diffi-

Table 2. Uniaxial Compressive Strength (mean values in MPa)

	$\dot{\varepsilon}$ (s^{-1})		Sea Ice		Pure Ice
				-10°C	
Brittle	10^{-2}	hor.	5.78		9.08
		vert.	18.4		
		**	9.11		
Ductile	10^{-5}	hor.	0.66		2.49
		vert.	5.12		
		**	0.49		
				-20°C	
Brittle	10^{-2}	hor.	6.70		5.80
Semi-Ductile	10^{-5}	hor.	1.77		
	10^{-4}	hor.			6.50
				-40°C	
Brittle	10^{-2}	hor.	10.99		27.5
Semi-Ductile	10^{-5}	hor.	1.98		
	10^{-4}	hor.			16.6

(Note ** refers to data for first year sea ice specimens)

Table 3. Triaxial Compressive Strength (mean values in MPa)

	$\dot{\varepsilon}$		Sea Ice	Pure Ice
			-10°C $P = 7$ MPa	
Brittle	10^{-2}/s	hor.	16.85	
		vert.	22.36	
		**	9.66	
Semi-Ductile	10^{-2}/s	hor.		24.5
	10^{-3}/s		4.53	13.0
Ductile	10^{-3}/s	hor.	4.62	
	10^{-4}/s			8.0
			-20°C $P = 14$ MPa	
Brittle	10^{-2}/s	hor.	19.43	
Semi-Ductile	10^{-3}/s	hor.	9.33	18.5
	10^{-2}/s			34.0
Ductile	10^{-5}/s	hor.	1.23	8.85
			-20°C $P = 30$ MPa	
Semi-Ductile	10^{-2}/s	hor.	25.32	32.4
Ductile	10^{-4}/s	hor.	6.43	12.0
			-40°C $P = 14$ MPa	
Brittle	10^{-2}/s	hor.	36.25	55.9
Ductile	10^{-4}/s	hor.	15.55	22.5

culty of nucleating or propagating cracks in a triaxial compressive stress field, and secondly the greater ease of crystal plastic flow as the melting point is lowered under pressure. Results from a range of triaxial compression tests at temperatures of -10°C, -20°C, and -40°C and at confining pressures of 7 MPa, 14 MPa, and 30 MPa are given in Table 3.

4.2. Role of Cracking in Ductile Failure

One of the most interesting features of these experiments, which can be seen most clearly in the pure ice specimens, concerns the complex role of cracking in ductile failure. Except at temperatures higher than -10°C a typical stress-strain curve for ductile deformation of ice shows initial linear-elastic deformation, then a period of anelastic strain-hardening up to a peak failure stress, followed by strain-softening at a rate that de-

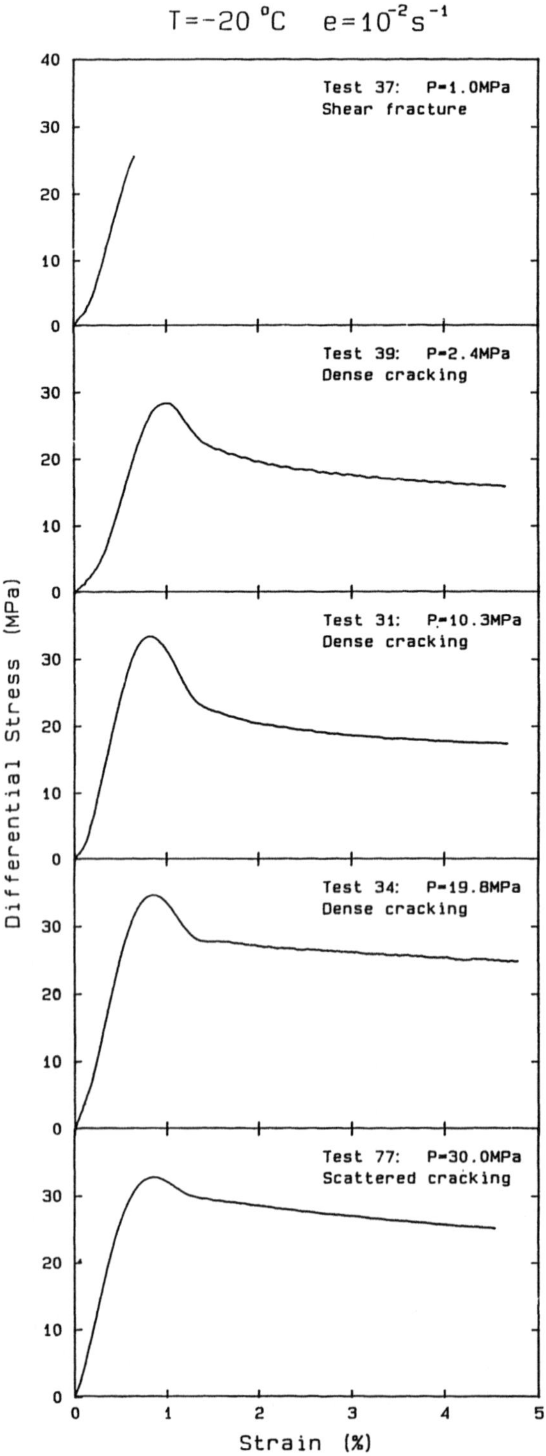

Figure 6. Stress-strain curves for pure ice specimens under triaxial compression at -20°C and a strain-rate of 10^{-2}/s.

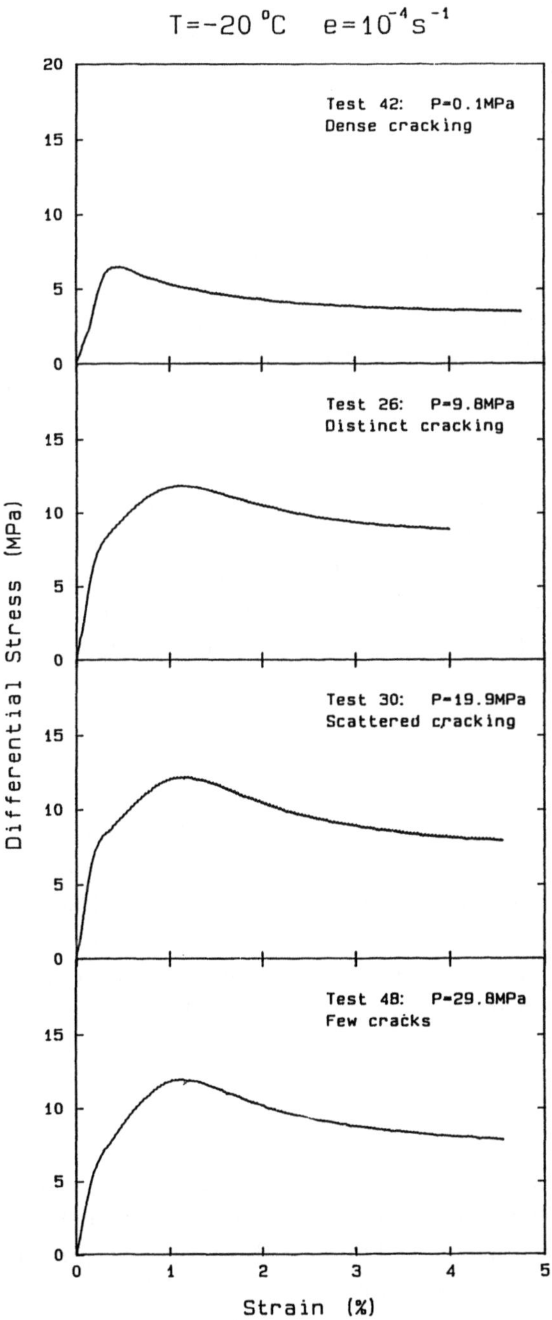

Figure 7. Stress-strain curves for pure ice in triaxial compression at -20°C and a strain rate of 10^{-4}/s, P =confining pressure.

creases with strain. The extent of all these features depends strongly on temperature, strain rate, and confining pressure. Examples are shown in Figure 6 for tests at -20°C and a strain rate of 10^{-2}/s. It was found possible to distinguish five levels of cracking associated with such ductile deformation as follows: (i) dense cracking; (ii) distinct cracking; (iii) scattered cracking; (iv) few cracks; (v) no cracks. Where pressure melting has apparently occurred the crack distribution is uneven.

Test conditions giving rise to some of these levels of cracking are shown in Figure 6, and the range is extended in Figure 7, which shows data for tests at -20°C and a lower strain rate (10^{-4}/s). The latter figure also shows examples of double-yield behaviour (noted earlier by Mellor and Cole, 1982 in pure ice, and by Cox and Richter-Menge, 1986 in sea ice) occurring at confining pressures of 10 MPa and above, with a lower yield point at a stress difference of about 7 MPa (and a strain of about 0.002), which is presumably associated with the initiation of crystal plasticity (and perhaps of crack nucleation), followed by a peak stress of about 12 MPa (at a strain of about 0.01) when strain softening is initiated. No cracks were observed in tests at -5°C, a strain-rate of 10^{-3}/s, and a confining pressure of 30 MPa.

Where cracks could be individually distinguished in the pure ice specimens they were approximately 1-2mm long, close to the grain-size of the undeformed ice. These cracks were aligned predominantly in the direction of maximum compression ($R = \sigma_3$). Both grain-boundary and trans-granular cracks are found. In sea ice, which had grain sizes of 3 to 7mm and columnar structures in many cases, axial cracks ranging in size from a few mm to tens of mm formed in horizontal multi-year specimens, apparently along grain boundaries. For vertical specimens the first cracks to form were a few mm long in the horizontal plane, but these were soon dominated by axial cracking.

4.3. Acoustic Emission (AE) in Ductile and Brittle Failure

By incorporating a piezoelectric (lithium niobate) transducer in the upper ram close to the specimen it was possible to observe AE (at frequencies ≈1 MHz) associated with cracking, using Dunegan-Endevco equipment. Figure 8 shows results from four tests, ranging from "no crack" conditions (Figure 8a) to "dense cracking" conditions (Figure 8d). It is noteworthy that when cracking is intense then AE remains at a high level at strains well beyond the peak stress point (which occurs at strains ≈0.01). This

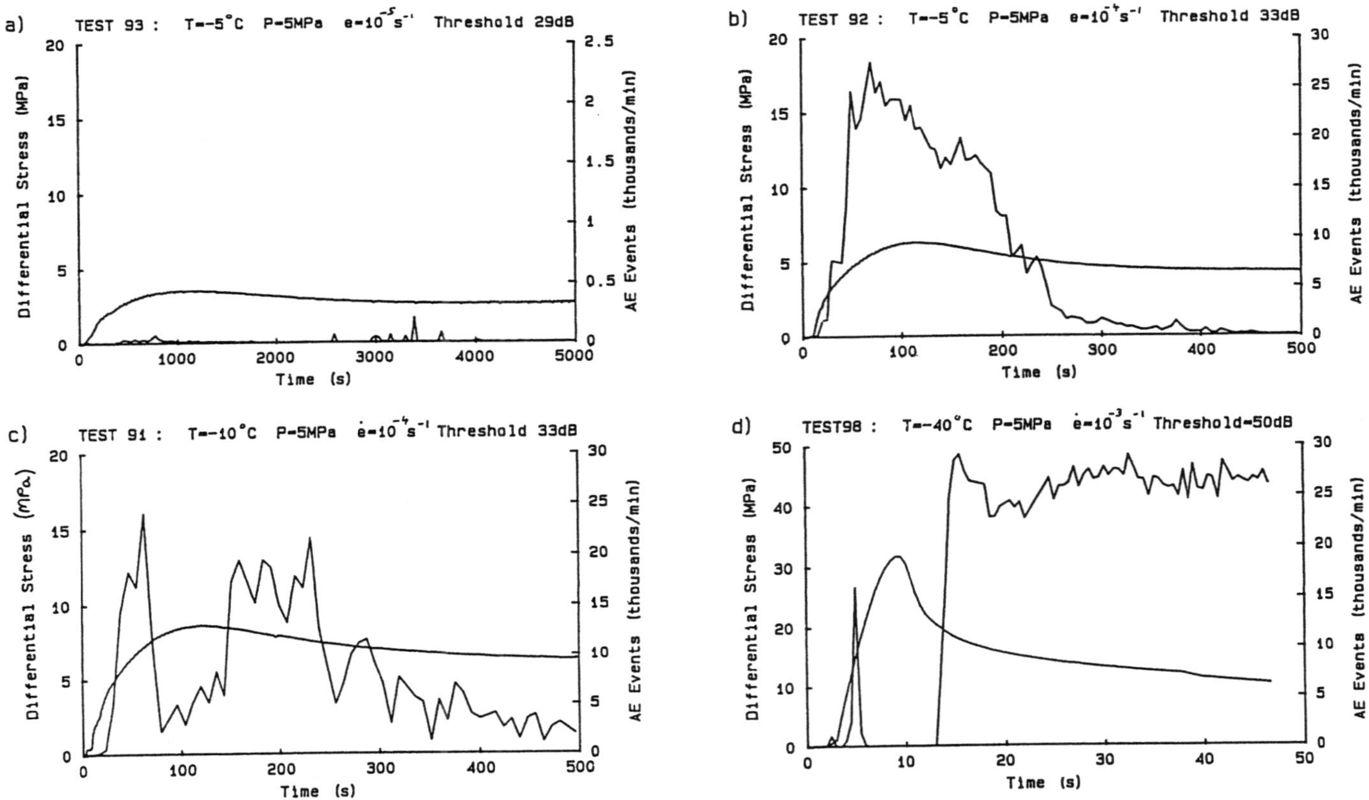

Figure 8. Acoustic emission event-rate history under various test conditions, together with stress-time curves (the smooth ones). (a) no cracks. (b) scattered cracks. (c) distinct cracking. (d) dense cracking.

is a very good example of cataclastic deformation as discussed in the introduction.

4.4. Strength as a Function of Confining Pressure

Figure 9 shows strength at -40°C as a function of confining pressure and of strain rate for pure ice. There is a strong effect of pressure on strength at the highest strain rate of 10^{-2}/s, when brittle behaviour is found. At lower strain rates, where behaviour is more ductile, there is little effect of pressure. At temperatures below -20°C pressure weakening occurs at the higher confining pressures, and is particularly marked at -5°C. Similar effects are found in sea ice.

In Figure 10 strength is shown as a function of confining pressure, and a mapping of cracking activity is also shown.

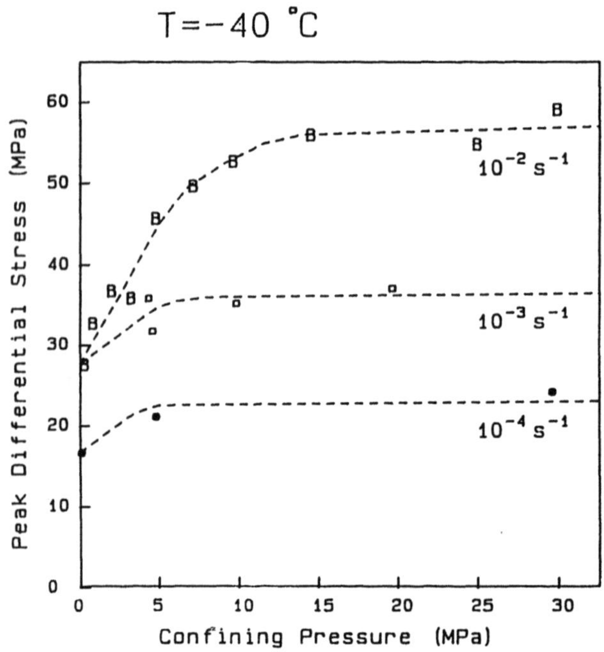

Figure 9. Peak differential stress ($\sigma_3 - \sigma_1$) versus confining pressure for pure ice under triaxial compression at -40°C. B=brittle fracture.

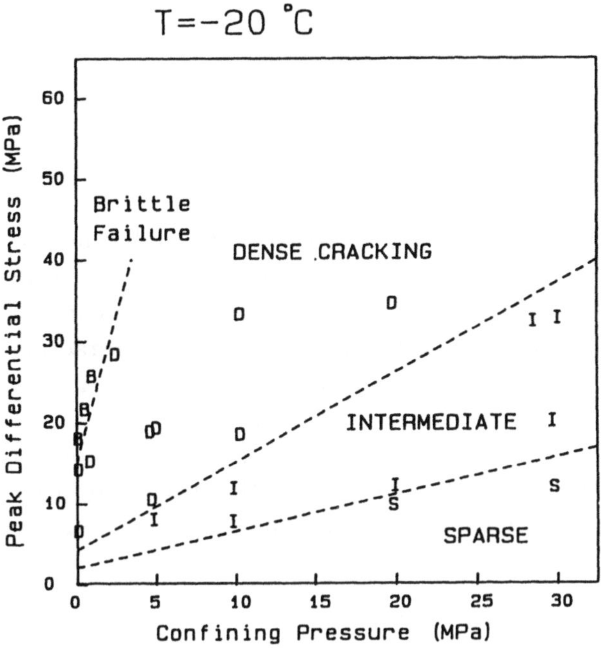

Figure 10. Relationship between pure ice strength, failure mode, and cracking activity.

5. Discussion and Conclusions

The data presented in this paper show the strength of ice ranging from less than 1 MPa at low strain rates in tension to as much as 60 MPa under triaxial compression at 30 MPa confining pressure, high strain rates and a temperature of -40°C. The coarser grained and more heterogeneous sea ice tends to have a lower strength than pure ice of finer grain size when measured in the horizontal direction, but vertically oriented sea ice has a strength similar to pure ice measured under the same conditions. Higher strength is associated with brittle or semiductile behaviour as determined not only by macroscopic behaviour but also by observations of microcracking activity, through direct visual observations of cracks and also through measurement of AE.

The uniaxial compressive strength data in Table 2 show the strong effect of temperature on strength, which is particularly striking in pure ice at high strain-rates. This indicates an effect of temperature on crack nucleation by microplasticity, showing that pre-existing cracks as envisaged in Griffith's (1920) theory are not the cause of failure, since the fracture strength under this theory is not strongly temperature dependent (Murrell, 1990). The temperature dependence is less marked in sea ice, showing that pre-existing cracks are probably important in this case. Concomitantly the increase of strength under triaxial compression is more marked in sea ice, where pre-existing cracks are present, than in pure ice.

The behaviour of ice in the transitional conditions between ductile and brittle behaviour is clearly extremely complex. The role of cracking in semi-ductile behaviour is not yet well understood, and the consequences of a requirement for crack nucleation preceding crack growth and interaction under compression have not yet been theoretically and quantitatively studied. Without further studies on laboratory grown saline ice and on ice with columnar structure it is not possible to model the behaviour of sea ice more precisely. Nevertheless, it is clear that laboratory grown ice is an excellent subject for deformation studies. Space precludes further discussion.

Acknowledgments

This research was supported by British Petroleum, by the NERC (studentship awards to PRS and MAR), and by University College London and especially the Geological Sciences Department. We are grateful for the strong support received from Professor M.G. Audley-Charles.

References

Ashby, M.F. and Hallam, S.D. 1986. The failure of brittle solids containing small cracks under compressive stress states. Acta Metall. Vol.34, pp.497-510.

Assur, A. 1958. Composition of sea ice and its tensile strength. NAS, NRC Publication 598. Arctic Sea Ice. pp.106-138.

Cox, G.F.N. et al. 1984. Mechanical properties of multi-year sea ice. Phase I and Phase II: Test Results. CRREL Report 84-8, pp.84-89.

Cox, G.F.N. and Richter-Menge, J.A. 1986. Confined compressive strength of multi-year pressure ridge sea ice samples. Proc. 5th Int. OMAE Symp. Tokyo. Vol.4, pp.365-371.

Croasdale, K.R. 1980. Ice forces on fixed rigid structures. In: IAHR Working Group on ice forces on structures, a state-of-the-art report, ed. T.Carstens, CRREL Special Report 80-26, pp.34-106.

Danielewicz, B.W. and Cornett, S. 1984. Ice forces on Hans Island 1983. APOA Project No.202.

Durham, W.B., Heard, H.C. and Kirby, S.H. 1983. Experimental deformation of polycrystalline H_2O ice at high pressure and low temperature: preliminary results. J.Geophys.Res. Vol.88, pp.B377-B392.

Fonseka, G.M., Murrell, S.A.F. and Barnes, P. 1985. Scanning electron microscope and acoustic emission studies of crack development in rocks. Int.J.Rock Mechs.Min.Sci. and Geomech.Abstr. Vol.22, pp.273-285.

Griffith, A.A. 1920. The phenomena of rupture and flow in solids. Phil. Trans. Roy. Soc. Lond. Vol.A221, pp.163-198.

Hallam, S.D. 1986. The role of fracture in limiting ice forces. Proc. IAHR Ice Symposium, Iowa City, Iowa. Vol.2, pp.287-320.

Horii, H. and Nemat-Nasser, S. 1986. Brittle failure in compression: splitting, faulting and brittle-ductile transition. Phil. Trans. Roy. Soc. Lond. Vol.A319, pp.337-374.

Ismail, I.A.H. and Murrell, S.A.F. 1976. Dilatancy and the strength of rock under undrained conditions. Geophys.J.R.Astr.Soc. Vol.44, pp.107-134.

Jacka, T.H. and Lile, R.C. 1984. Sample preparation techniques and compression apparatus for ice flow studies. Cold Regions Sci.Technol. Vol.8, pp.235-240.

Jaeger, J.C. and Cook, N.G.W. 1979. Fundamentals of Rock Mechanics, 3rd Edition. Chapman and Hall: New York.

Jones, C. and Murrell, S.A.F. 1989. Acoustic compressional wave velocity and dilatancy in triaxially stressed rock. In V. Maury and D. Fourmaintraux eds. Rock mechanics and rock physics at great depth, Proc. ISRM-SPE Rock at Great Depth - International Symposium, Pau, France, Aug. 1989. Vol.1, pp.241-247.

Jones, S.J. 1982. The confined compressive strength of polycrystalline ice. J.Glaciol. Vol.28, No.98, pp.171-177.

Mellor, M. and Cole, D.M. 1982. Deformation and failure of ice under constant stress or constant strain-rate. Cold Regions Sci.Technol. Vol.5, pp.201-219.

Michel, B. 1970. Ice pressure on engineering structures. CRREL Monograph III- B1b

Murrell, S.A.F. 1964. The theory of the propagation of elliptical Griffith cracks under various conditions of plane strain or plane stress: part I and parts II and III. Brit.J.Appl.Phys. Vol.15, pp.1195-1210 and 1211-1223.

Murrell, S.A.F. 1990. Brittle to ductile transitions in polycrystalline non-metallic materials. In D.J.Barber and P.G.Meredith eds. Deformation Processes in Minerals, Ceramics and Rocks. Mineralogical Society Special Volume. Unwin-Hyman: London. pp. 109-137.

Murrell, S.A.F. and Digby, P.J. 1970. The theory of brittle fracture initiation under triaxial stress conditions: part I and part II. Geophys.J.R.Astr.Soc. Vol. 19, pp.309-334 and 499-512.

Nadreau, J.P. and Michel, B. 1986. Courbe intrinsèque de la glace en compression. J.de Phys., 7th Symp. on the Physics and Chemistry of Ice, Grenoble.

Paterson, M.S. 1978. Experimental Rock Deformation - the Brittle Field. Springer-Verlag: Berlin.

Ponter, A.R.S. et al. 1983. The force exerted by a moving ice sheet on an offshore structure. Part I. The creep mode. Cold Regions Sci.Technol. Vol.8, pp.109-118.

Ralston, T.D. 1978. An analysis of ice sheet indentation. Proc. 5th IAHR Symp. on Ice Problems, Luleå, August 1978, Part I, pp.13-31.

Richter-Menge, J.A. et al. 1986. Triaxial testing of first-year sea ice. CRREL Report 86-16.

Rist, M.A. 1989. The effect of temperature and strain-rate on the triaxial deformation and strength of pure polycrystalline ice. Ph.D. Thesis, University of London.

Rist, M.A., Murrell, S.A.F. and Sammonds, P.R. 1990. Strain-rate control during deformation of ice: an assessment of the performance of a new servo-controlled triaxial testing system. Cold Regions Science and Technology (In press).

Sammonds, P.R. 1988. Triaxial deformation experiments on natural sea ice as a function of temperature and strain-rate. Ph.D. Thesis, University of London.

Sammonds, P.R., Ayling, M.R., Jones, C., Meredith, P.G. and Murrell, S.A.F. 1989a. A laboratory investigation of acoustic emission and elastic wave velocity changes during rock failure under triaxial stress. In V. Maury and D. Fourmaintraux eds. Rock mechanics and rock physics at great depth, Proceedings ISRM-SPE Rock at Great Depth - International Symposium, Pau, France, August 1989. Vol. 1, pp. 233-240.

Sammonds, P.R., Murrell, S.A.F. and Rist, M.A. 1989b. Fracture of multi-year sea ice under triaxial stresses: apparatus description and preliminary results. Trans. ASME, JOMAE, Vol.111, pp.258-263.

Sanderson, T.J.O. 1984. Theoretical and measured forces on wide structures. Proc. IAHR Ice Forces Symp., Hamburg, Aug. 1984, Vol.4, pp.151-207.

Sinha, N.K. 1984. Uniaxial compressive strength of first-year and multi-year sea ice. Can.J.Civil Eng. Vol.11, pp.82-91.

Sinha, N.K. 1985. Confined strength and deformation of second-year columnar-grained sea ice in Mould Bay. Proc. 4th Int.OMAE Symp., Dallas. Vol.2, pp.209-219.

Timco, G.W. 1986. Indentation penetration of edge-loaded freshwater ice sheets in the brittle range. Proc. 5th Int. OMAE Symp., Tokyo, Vol.4, pp.444-452.

Walden, J.T., Hallam, S.D. and Baldwin, J.T. 1987. An explicit technique for calculating first-year ice loads on structures. Proc. 6th Int. OMAE Symp., Houston. Vol.4, pp.267-272.

Discussion

Q: Oskar H. Mahrenholtz, TUHH

Under your extreme experimental conditions with high power density (60 MPA triaxial pressure) there may occur shear bands with shear localization, even local melting.

A: S.A.F. Murrell

We observe shear fracture, as described in our paper, and I agree that because of the low thermal conductivity of ice it is quite possible that localized melting on such shear fracture may well occur once the fracture has formed and movement against friction occurs. This is well-known to occur in the shearing of silicate rocks.

Q: Ian Jordaan, Memorial University

Regarding the cessation of acoustic emission during your tests, is it possibly related to a cessation also of cracking? Could you please comment?

A: S.A.F. Murrell

I think it relates to the falling rate of stress-increase as the peak load is closely approached and to the deformation-mode transition that occurs at the peak load. I believe these factors do lead to a cessation of growth of many cracks just before peak load, but this is presumably succeeded by a new form of crack growth beyond peak load. Where shear fracture occurs this process can be understood. However, where ductile fracture occurs at about peak load, or semi-brittle deformation is occurring beyond peak load, the cause of the renewed acoustic emission (and new form of cracking) is not clear at present.

SEA ICE UNDER COMPLEX STRESS STATES: CONSTITUTIVE MODELLING AND TESTING

F.U. Häusler

Hamburgische Schiffbau-
Vesuchsanstalt GmbH,
Hamburg, F.R. Germany

H.G. Matthies, C.S. Moore

Germanischer Lloyd
Hamburg, F.R. Germany

Abstract Based on typical ice–structure interaction scenarios the importance of complex stress states for ice load evaluation is emphasized. The thermodynamic basis for a constitutive model for the description of sea ice is formulated, taking into account complex stress states and damage. It is suggested that the effect of total porosity on sea ice strength be considered in a constitutive model as some kind of existing damage. Procedures for determining the strength characteristics of sea ice under complex stress states are described.

1. Introduction

Design ice loads are governed by failure processes in the ice. This paper presents a basis for the formulation of constitutive equations for ice under multiaxial stress states including the effects of damage. Section 2 describes some typical ice–structure interaction scenarios emphasizing the importance of complex stress states for ice load evaluation. In section 3 the thermodynamic basis for a constitutive model for the description of sea ice is formulated. Porosity of sea ice is considered in section 4 as a form of existing damage. The reference strength concept is discussed in the context of continuum damage mechanics. Finally, procedures for

S. Jones · R. F. McKenna · J. Tillotson · I. Jordaan (Eds.)
Ice-Structure Interaction
IUTAM-IAHR Symposium St. John's, Newfoundland, Canada

determining the strength characteristics of sea ice under complex stress states are described in section 5.

2. Stress States in Ice Sheets as Produced by Ice–Structure Interaction

Due to constraints given by the ice sheet itself and by the structure with which it interacts, the stress states which build up within the ice body usually are rather complex. Since ice strength is known to depend on hydrostatic stress, ice load determination based on uniaxial strengths of ice may lead to erroneous results. In Figure 1 three basic ice–structure interaction scenarios are shown together with the stresses which qualitatively act at characteristic material points. Although other ice–structure interaction scenarios can be thought of, most of them can in principle be reduced to the three given here:

- narrow indenter, e.g. cylindrical column (Figure 1a)
- wide indenter, e.g. straight vertical wall (Figure 1b),
- sloping indenter, e.g. upward breaking cone (Figure 1c).

In the "narrow indenter" case the lateral compressive stresses near the indenter are balanced by tensile stresses at a certain distance. These tensile stresses usually lead to failure in the form of radial cracking. But the formation of radial cracks does not mean that the ice sheet has lost its load bearing capacity. Even if affected by the already existing radial cracks, final failure is usually governed by mechanisms like buckling or local crushing. The latter occurs at higher stress rates in close vicinity to the ice–structure interface, i.e. in the regime where the ice is under biaxial compression (see Figure 1a). The shear stresses which may act due to friction between ice and structure shall not be considered here. Since a rise in hydrostatic pressure (= decrease of the mean stress) results in increasing suppression of crack formation (Frost and Ashby, 1982; Jones, 1982), and as such in increasing ice strength (very high hydrostatic pressures excluded), an analysis of the ice-failure processes near the ice–structure interface should consider the multiaxial strength properties of the ice. This applies in particular to columnar grained ice loaded horizontally, i.e. in the plane of the ice cover as given here, since its structural anisotropy acts like a vertical confinement.

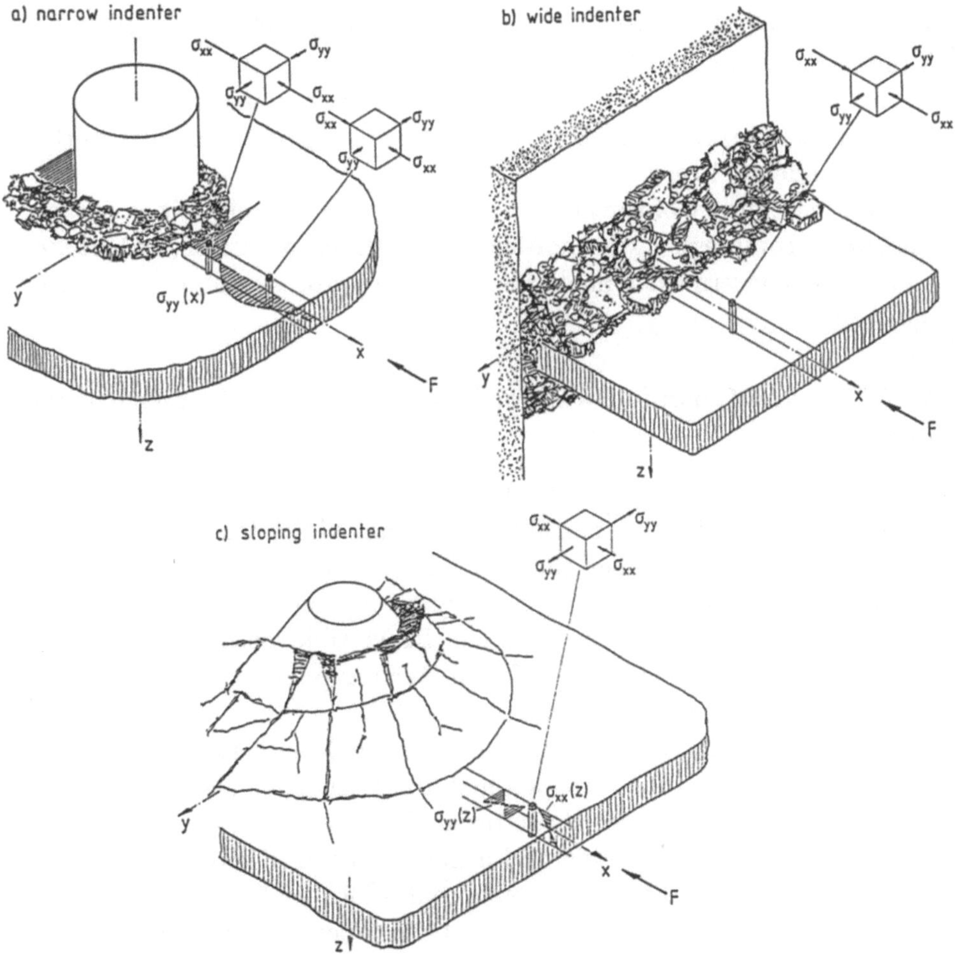

Figure 1. Typical ice-structure interaction scenarios.

The above considerations on biaxial compression apply to the case of the "wide indenter" (Figure 1b) as well. In the case of the "sloping indenter" (Figure 1c) the principal stresses are also found to be in the plane of the ice cover, balancing both the normal and the in plane components of the ice–structure interaction force. In the given example of a wider conical indenter the variation of ice strength due to the multiaxiality of the stress state is not so much a topic of concern because the tensile strength of ice appears not to be very much affected by transverse tensile or compressive stresses - see failure envelopes established by Timco and Frederking (1984, 1986) or by Häusler (1983, 1986). Nevertheless, if the sloping structure is rather narrow, as for example the stem of an icebreaker, local crushing may occur due to high compressive stresses in the contact zone, and these stresses usually are multiaxial, similar to the "narrow indenter" case.

As a consequence of the above considerations it is emphasized that at least under compression, the multiaxial strength properties of (sea) ice are a matter of concern when analysing the failure processes in ice–structure interactions.

3. A Thermodynamic Basis for Phenomenological Constitutive Models

In our opinion a phenomenological constitutive law should be subject to some general restrictions, namely it must not violate the laws of thermodynamics, it should obey the principle of material frame indifference and be invariant under the symmetry group actions appropriate for the particular material under consideration. Irreversible thermodynamics (de Groot and Masur, 1984) is therefore a starting point well suited to the development of constitutive laws.

We shall assume the material to be a simple material (Noll, 1972; Krawietz, 1986), and the state of a material point to be completely specified by the Green-Lagrange strain tensor $\mathbf{E}$, the absolute temperature T, the temperature gradient ∇T and a finite number of internal variables which are hidden and not observable, but which represent the memory of the material and through which the history dependence is manifested. These internal variables will be collectively denoted Q for the time being. They will be assumed to be independent and not include $\mathbf{E}$, T or ∇T.

3.1 Entropy Production

By combining the first and second law of thermodynamics, we arrive at the dissipation inequality

$$\mathbf{S} : \dot{\mathbf{E}} - (\dot{\varphi} + s\dot{T}) - \frac{1}{T}\mathbf{q}^T \cdot \nabla T \geq 0. \tag{1}$$

Here $\mathbf{S}$ is the second Piola-Kirchhoff stress tensor, φ and s the specific free energy and entropy per volume and $\mathbf{q}$ the heat flux vector (Halphen and Nguyen, 1975; Marsden and Hughes, 1983; Krawietz, 1986). φ is assumed differentiable and is a function of $\mathbf{E}$, T and the internal parameters

Q. From Equation 1 we deduce with standard arguments the following relations:

$$\mathbf{S} = \frac{\partial}{\partial \mathbf{E}}\varphi(\mathbf{E}, T, Q), \qquad s = -\frac{\partial}{\partial T}\varphi(\mathbf{E}, T, Q),$$

$$< P, \dot{Q} > \geq 0, \quad \text{and} \qquad -\frac{1}{T}\mathbf{q}^T \cdot \nabla T \geq 0, \tag{2}$$

where $P = -\frac{\partial}{\partial Q}\varphi(\mathbf{E}, T, Q)$ is the thermodynamic force associated with the fluxes $\dot{Q}$ and $< \cdot, \cdot >$ is a suitable duality pairing or inner product.

The second law is reduced to the last two relations. While the second of these restricts the direction of heat flow, the first requires the intrinsic dissipation to be non-negative. We shall now concentrate on the mechanical aspect. In view of the Onsager reciprocity relations (de Groot and Masur, 1984), this suggests the existence of a convex "dissipation potential" (Halphen and Nguyen, 1975) Ω $(P, \mathbf{E}, T, Q)$ with its minimum at $P = 0$. The fluxes will be the derivative of Ω in some generalised sense. This generalisation is necessary for rate-independent processes, where the dissipation potential is not continuously differentiable, and is provided by the notion of subdifferential (Oden, 1986) $\partial_p\Omega$ with respect to the variable P. The evolution law or "flow rule" for the internal parameters is given by

$$\dot{Q} \ \varepsilon \ \partial_p\Omega(P) <=> \forall R :< R - P, \dot{Q} > +\Omega(P) \leq \Omega(R). \tag{3}$$

The first relation is termed a differential inclusion and the second a variational inequality. Actually the existence of the dissipation potential implies that the dissipation is maximal, i.e. any other choice than (3) for the evolution of Q would give lower dissipation. This condition of maximum dissipation is connected to the Onsager relations and has been termed an extremal principle of irreversible thermodynamics (Ziegler, 1961). In other words, entropy not only increases under non-stationary conditions, but it increases as quickly as possible.

Since $\Omega(P)$ is a scalar, which has to be invariant under the action of the material symmetry group (Noll, 1972; Krawietz, 1986), the Curie principle (de Groot and Masur, 1984) implies that P is composed only of tensors of even order and hence by duality so is Q.

These principles have been used by several workers to construct phenomenological laws (cf. Halphen and Nguyen, 1975; Lemaître and Marquis, 1988), and in particular also for ice (Sjölind, 1987; Matthies, 1989).

3.2 Specific Form of Phenomenological Constitutive Relations

To fix ideas, we first consider a simple rheological model given in Figure 2, where we have elastic, delayed elastic or primary creep, visco-plastic or secondary creep and inviscid plastic mechanisms, and all elements are subject to damage. We shall generalise this model to anisotropic situations and multiaxial stress states. Similar approaches have been employed by several workers (e.g. Tomin et al., 1986; Sjölind, 1987; Pulkinnen, 1989). We shall follow Krajcinovic (1985) in the treatment of damage. Similar, but isotropic models are described by Choï and Karr (1989).

According to the theory in the preceding sections, we have to specify internal parameters Q, the Helmholtz free energy φ and the dissipation potential W, all other relations then being given by the thermodynamic laws.

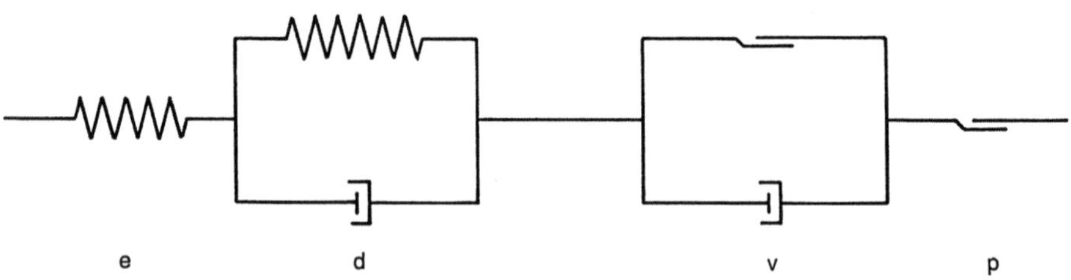

Figure 2. Rheological model.

3.3 Internal Variables

With reference to the rheological model, we decompose the total strain $\mathbf{E}$

$$\mathbf{E} = \mathbf{E}_e + \mathbf{E}_d + \mathbf{E}_v + \mathbf{E}_p, \tag{4}$$

into elastic strain $\mathbf{E}_e$, delayed or visco-elastic strain $\mathbf{E}_d$, visco-plastic or creep strain $\mathbf{E}_v$ and inviscid plastic strain $\mathbf{E}_p$. The inviscid plastic mechanism serves mainly as an ultimate stress delimiter. The decomposition Equation 4 is appropriate if the total strain is not too large (Krawietz, 1986).

Similarly to Choi and Karr (1989), we will assume two damage mechanisms, one dependent on delayed elastic strain (Sinha, 1982), and one dependent on total strain (Gold, 1972). The damage will be described by two symmetric tensors $\mathbf{Y}_1$ and $\mathbf{Y}_2$ (Leckie and Onat, 1980), such that the amount of damage $\omega_i(\mathbf{n})$ in the direction with normal $\mathbf{n}$ is given by $\omega_i(\mathbf{n}) = \mathbf{n}^T \cdot \mathbf{Y}_i \cdot \mathbf{n},\ i = 1, 2.$

Our set of internal variables then is

$$Q = (\mathbf{E}_d, \mathbf{E}_v, \mathbf{E}_p, \mathbf{Y}_1, \mathbf{Y}_2). \tag{5}$$

Observe that $\mathbf{E}_e$ was not included in the set of internal variables, since the total strain $\mathbf{E}$ would then be fixed through Equation 4.

3.4 Helmholtz Free Energy

The Helmholtz free energy $\varphi(\mathbf{E}, T, Q)$ will be assumed to be given in the form

$$\varphi(\mathbf{E}, T, Q) = \varphi_e(\mathbf{E}_e, T, \mathbf{Y}_i) + \varphi_d(\mathbf{E}_d, T, \mathbf{Y}_i) + \varphi_T(\mathbf{E}_e, T, \mathbf{E}_d). \tag{6}$$

We will not further specify φ_T, which is responsible for thermal stresses (Lemaître and Marquis, 1988). The simplest anisotropic forms are

$$\varphi_e = 1/2\mathbf{S}(\mathbf{E}_e, \mathbf{Y}_1, \mathbf{Y}_2) : \mathbf{E}_e \quad \text{and} \quad \varphi_d = 1/2\mathbf{S}_d(\mathbf{E}_d, \mathbf{Y}_1, \mathbf{Y}_2) : \mathbf{E}_d, \tag{7}$$

where $\mathbf{S}$ is the stress and $\mathbf{S}_d$ the delayed elastic stress. Both are a linear function of the net or effective stresses $\mathbf{S}_n = A_e : \mathbf{E}_e$ and $\mathbf{S}_{nd} = A_d : \mathbf{E}_d$; A_e and A_d are the fourth order tensors for the instantaneous and delayed elasticity, and therefore we set

$$\mathbf{S} = L(\mathbf{Y}_1, \mathbf{Y}_2)(A_e : \mathbf{E}_e), \quad \text{and} \quad \mathbf{S}_d = L(\mathbf{Y}_1, \mathbf{Y}_2)(A_d : \mathbf{E}_d). \tag{8}$$

3.5 Thermodynamic Forces

The associated thermodynamic forces are given by

$$\begin{aligned} -\frac{\partial \varphi}{\partial \mathbf{E}_d} &= \mathbf{S}_d, & -\frac{\partial \varphi}{\partial \mathbf{E}_v} &= \mathbf{S}, \\ -\frac{\partial \varphi}{\partial \mathbf{E}_p} &= \mathbf{S}, & -\frac{\partial \varphi}{\partial \mathbf{Y}_i} &= \mathbf{Z}_i, \quad i - 1, 2. \end{aligned} \tag{9}$$

The forces $\mathbf{Z}_i$ need further specification. If we know the linear mapping $L(\mathbf{Y}_1, \mathbf{Y}_2)$, $\mathbf{Z}_1$ and $\mathbf{Z}_2$ can be computed from Equations 7-9 with the chain rule.

3.6 Continuum Damage Theory

Following Krajcinovic (1985) we distinguish active and passive micro-cracks, which influences the effective damage. L depends on $\mathbf{Y}_1, \mathbf{Y}_2$ through the effective damage tensor $\mathbf{D}$, where

$$\mathbf{D} = \frac{1}{4\pi} \left(\int_{\{\mathbf{n}\varepsilon S^2 \mid \mathbf{n}^T \cdot \mathbf{E}_d \cdot \mathbf{n} \geq \varepsilon_{crd}\}} \omega_1(\mathbf{n})\, \mathbf{n} \cdot \mathbf{n}^T \, dA + \int_{\{\mathbf{n}\varepsilon S^2 \mid \mathbf{n}^T \cdot \mathbf{E} \cdot \mathbf{n} \geq \varepsilon_{cr}\}} \omega_2(\mathbf{n}) \mathbf{n} \cdot \mathbf{n}^T \, dA \right) \tag{10}$$

and the specific form of dependence of L or rather L^{-1} on $\mathbf{D}$ will be discussed in the next section. In Equation 10 we see that damage is only taken into account if the resolved strain or delayed elastic strain in the direction $\mathbf{n}$ is larger than some threshold value.

3.7 Dissipation Potential

For the dissipation potential we will also assume a decomposition into different mechanisms, where the dependence on stress is through the net stress $\mathbf{S}_n$ or $\mathbf{S}_{nd}$:

$$\Omega(P) = \Omega_d(\mathbf{S}_d) + \Omega_v(\mathbf{S}) + \Omega_p(\mathbf{S}) + \Omega_Y(\mathbf{Z}_1, \mathbf{Z}_2). \tag{11}$$

The first two terms are differentiable and give the rate-dependent behaviour, while the last two are not differentiable - inviscid plasticity is not rate-dependent - and a similar assumption is used for the damage evolution (Lemaître, 1985). We set

$$\Omega_d(\mathbf{S}_d) = k_d/2\ \mathbf{S}_{nd} : C_d : \mathbf{S}_{nd}, \quad \text{and}$$
$$\Omega_v(\mathbf{S}) \ = k_v/(m+1)(S_{eff-v}(\mathbf{S}_n)/S_v)^{m+1}, \tag{12}$$

where k_d is a constant and C_d a symmetric positive definite fourth order tensor, k_v is a constant, m the power law exponent, S_v a scaling parameter and S_{eff-v} a non-isotropic analogue of effective or equivalent stress (e.g. Sanderson, 1988). All the 'constants' may be temperature dependent.

The non-differentiable part is given by indicator functions, modelling the rate independent behaviour (Halphen and Nguyen, 1975; Matthies,

1989):

$$\Omega_p(\mathbf{S}) = \begin{cases} 0 & \text{if } S_{eff-p}(\mathbf{S}_n) \leq S_p \\ +\infty & \text{otherwise,} \end{cases}$$

$$\Omega_y(\mathbf{Z}_1, \mathbf{Z}_2) = \begin{cases} 0 & \text{if } Z_{eff-i}(\mathbf{Z}_i) \leq Z_{e-i}, i = 1, 2. \\ +\infty & \text{otherwise,} \end{cases} \tag{13}$$

Again S_{eff-p} and Z_{eff-i}, i=1,2 are non-isotropic analogues of effective stress or effective strain energy density and S_p and Z_{e-i} are threshold values.

3.8 Evolution Laws

In general, if Ω is composed of a differentiable component Ω_1 and a non-differentiable indicator function Ω_2, relation (3) may be simplified (e.g. Oden, 1986):

$$\forall\ R \text{ such that } \Omega_2(R) = 0: \quad < R - P, \dot{Q} - \frac{\partial \Omega_1}{\partial P} > \leq 0, \tag{14}$$

which is reminiscent of flow rules in plasticity. Our evolution laws therefore are:

$$\dot{\mathbf{E}}_d = k_d C_d : \mathbf{S}_{nd} \quad \dot{\mathbf{E}}_v = \frac{k_v}{S_v} \left(\frac{S_{eff-v}(\mathbf{S}_n)}{S_v} \right)^m \frac{\partial S_{eff-v}(\mathbf{S}_n)}{\partial \mathbf{S}},$$

$$\forall\ \mathbf{S}',\ \mathbf{Z}_1{}', \mathbf{Z}_2{}' \text{ such that } \Omega_p(\mathbf{S}) = 0 \text{ and } \Omega_y(\mathbf{Z}_1{}', \mathbf{Z}_2{}') = 0:$$

$$(\mathbf{S}' - \mathbf{S}) : \dot{\mathbf{E}}_p + (\mathbf{Z}_1{}' - \mathbf{Z}_1) : \dot{\mathbf{Y}}_1 + (\mathbf{Z}_2{}' - \mathbf{Z}_2) : \dot{\mathbf{Y}}_2 \leq 0. \tag{15}$$

The variational inequality may also be formulated in a more conventional form with Lagrange multipliers at some loss of generality (Halphen and Nguyen, 1975).

3.9 Numerical Implementation

The mathematical structure of these evolution equations is the same as for combined visco- and inviscid plasticity, and similar algorithms may be used, where projection methods are preferred for reasons of efficiency and numerical stability (Matthies, 1989).

4. Porosity Considered as Existing Damage in Sea Ice

Damage of ice can be related to the nucleation, growth and coalescence of defects. On a microscopic scale, various mechanisms such as grain boundary dislocation pile-up, trans-granular cracking, and also porosity are responsible. To take into account the resulting density and orientation of microcracks or microvoids, a number of continuous variables representative of microscopic discontinuities in the ice are introduced. The damage variables can be defined at a point in direction **n** by a volume average of the area of intersections of microdefects with a plane normal to the unit vector **n** (Kachanov, 1958; Hult, 1979; Lemaître and Marquis, 1988). The reduction in material effectiveness can be related to the reduction of the area A of a surface of the elementary volume by the effective area of the defects in the plane, δA. This reduction should be corrected to take into account stress concentrations and microcrack interaction. Damage D for this surface can be defined then by the ratio of these two areas:

$$D = \delta A / A \tag{16}$$

where $D = 0$ corresponds to undamaged material and $D = 1$ corresponds to completely ineffective material. The theory of isotropic damage assumes that D does not depend upon **n** and thus is represented by a scalar. The anisotropy of damage has been taken into account by the introduction of second order tensors as described above.

4.1 Effective Stress Tensors

Numerical application of continuum damage utilizes the concept of net or effective stress. Let σ be the Cauchy stress tensor acting on the tetrahedron $OPQR$, then the force vector acting on the surface element can be expressed as (Murakami and Ohno, 1981)

$$A\tau = A\sigma \cdot \mathbf{n} \tag{17}$$

where τ is the corresponding stress vector, as shown in Figure 3. The surface element $P^*Q^*R^*$ is subjected to the same force which can be written in terms of the '*' quantities through

$$\begin{aligned} A^*\tau^* &= A\tau \\ &= \sigma \cdot (\mathbf{I} - \mathbf{D})^{-1} \cdot (A^*\mathbf{n}^*) \\ &= \sigma^* \cdot (A^*\mathbf{n}^*), \end{aligned} \tag{18}$$

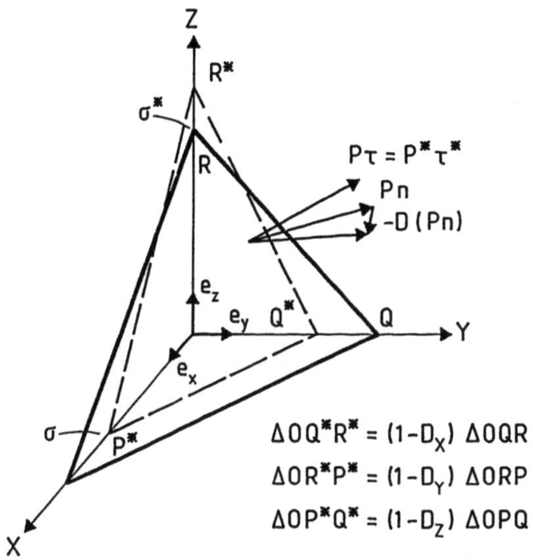

Figure 3. Net area and force vector in damaged and undamaged body.

where $\mathbf{n}^*$ is the unit normal of surface, τ^* is the stress vector on $P^*Q^*R^*$, $\mathbf{I}$ is the unit tensor, and $\mathbf{D}$ the damage tensor. The net or effective stress is then defined in terms of the Cauchy stress and the damage tensor by

$$\sigma^* = \sigma \cdot \mathbf{\Theta}, \quad \mathbf{\Theta} = (\mathbf{I} - \mathbf{D})^{-1}. \tag{19}$$

The effective stress tensor σ^* will not be symmetric if the damage is anisotropic. For use in analysis this effective stress tensor is often replaced by a symmetric tensor of the form

$$\sigma^{**} = 1/2(\sigma \cdot \mathbf{\Theta} + \mathbf{\Theta} \cdot \sigma). \tag{20}$$

This approach does not result in a symmetric transformation tensor between the stress forms. A pseudo-net stress tensor introduced by Betten (1981) enforcing symmetry of the transformation tensor between Cauchy and net stress, is equivalent to using

$$\sigma^{**} = \mathbf{\Theta} \cdot \sigma \cdot \mathbf{\Theta}^T, \tag{21}$$

which for the case of isotropic damage where the transformation tensor can be replaced by a scalar is

$$\sigma^{**} = (1 - D)^{-2} \sigma. \tag{22}$$

4.2 Relation to Total Porosity/Reference Strength Method

In the total porosity/reference strength method, ice strength is expressed as a function of the reference strength σ_o , total porosity ν, temperature T and strain rate (Häusler, 1988) in the following form

$$\sigma = \sigma_o(1-\sqrt{\nu})^2 f(T,\dot{\varepsilon}), \tag{23}$$

where the particular form of the function f depends upon whether brittle or ductile failure is being considered. This relation can be rewritten in the net stress form of continuum damage methods by identifying the reference strength with the net or effective stress tensor. Then,

$$\sigma_o = \sigma(1-\sqrt{\nu})^{-2}, \tag{24}$$

where the effects of temperature and strain rate have been dropped from further consideration. One can immediately identify this form with that in Equation 22. However, for isotropic damage one would expect the dependence upon porosity to be $\nu^{2/3}$ rather than $\nu^{1/2}$. This is apparently inconsistent.

To resolve this it is worth considering the development of the reference strength form. Weeks and Assur (1968) proposed this form for brine volume based upon geometric considerations and upon curve fitting to experimental data. The assumption is that the columnar growth of the sea ice leads to parallel rows of cylindrical gas/brine volumes which effectively extend vertically the full depth of the grains.

From a continuum damage perspective this leads to two descriptions of damage, one parallel to the columnar grains and one perpendicular (i.e. parallel to the plane of the ice cover). Figure 4 shows an idealization of the geometry. The porosity is equal to the relative volume of the brine and gas pocket with radius a_o and is equal to

$$\nu = \pi a_o^2 g/(a\ b\ g)\ (= \pi a_o^2/a^2 \text{ for } a = b). \tag{25}$$

Consider now the reduction in effective area in a vertical plane (e.g. the xz plane in Figure 4). Here

$$\delta A_{xz} = 2a_o g/(a\ g) = (2a_o/a)\ \propto \sqrt{\nu}. \tag{26}$$

One then obtains the dependence upon $\nu^{1/2}$ given in Equation 23. However, if one considers the effective area reduction in the horizontal plane

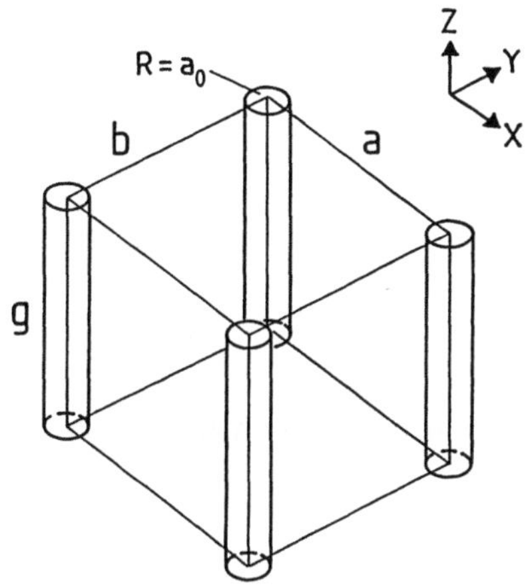

Figure 4. Cylindrical geometry of sea ice porosity.

one obtains

$$\delta A_{xy} = \pi a_o^2/(a\ b) = (\pi a_o^2/a^2)\ \propto\ \nu. \tag{27}$$

Thus the total porosity approach to reference strength for ice is consistent with continuum damage approaches when a specific geometry of the columnar orientation of sea ice with its brine channels and gas volumes is applied. The specific form stated in Equation 23 is applicable for ice strength in the plane of the ice cover for columnar sea ice. Vertical strengths should depend upon porosity through the form

$$\sigma_o = \sigma(1-\nu)^2 \quad f(T, \dot{\varepsilon}). \tag{28}$$

This is supported by experimental results (Häusler, 1989) where the accuracy of the failure envelopes established based upon the $(1\text{-}\sqrt{\nu})^2$ form is good for strengths perpendicular to the columnar grains but not so good for strengths parallel to the grains.

A critical part of continuum damage mechanics is the evolution of damage. The thermodynamic aspects of this have been discussed above, here a few points regarding porosity of sea ice are mentioned. In porous, columnar grained sea ice, the evolution of damage can be associated with three physical phenomena. One is grain boundary dislocation pile-up such as described by Sinha (1982) leading to microcrack nucleation. These piled up dislocations should cause cracks with a similar geometry to the existing porosity and can be treated as an increase in porosity. Secondly, when loaded in compression in the ice cover plane, microcracks will propagate from the pores in a direction more-or-less parallel to the direction of maximum compressive stress as described by Sammis and Ashby (1986). Thirdly, trans-granular cracking due to expansion in the vertical direction under horizontal compressive loads may occur. This has been discussed by Ashby and Hallam (1986) and Karr and Choi (1989).

5. Methods of Evaluating Ice Strength under Complex Stress States

In order to fill the above material model considerations for ice with life, uniaxial and multiaxial strength testing is required. The technology for such tests is available, because multiaxial strength testing is well established for other materials, e.g. concrete (e.g. Gerstle et al., 1976), and adaptation to ice has been performed mainly within the past decade. At present two methods represent the state-of-the-art of multiaxial testing of ice: the "standard triaxial test" (Figure 5a) and the true triaxial loading of cubic samples by means of brush platens (Figure 5b). When applied to the same type of multi-year pressure ridge ice both methods yield compatible results (Häusler et al., 1988). Uniaxial testing methods are not considered here.

The design of the "standard triaxial test" originates from von Kármán (1911). A cylindrical sample is placed inside a pressure vessel. While axial load is applied to the circular end surfaces, as in an ordinary uniaxial test, confining hydrostatic pressure acts on the cylindrical surface of the specimen. Thus the stress state in the "standard triaxial test" is limited to stresses of the type $\sigma_1 \neq 0$; $\sigma_2 = \sigma_3 \leq 0$. When investigating anisotropic columnar grained (sea) ice, this limitation makes it necessary to prepare cylindrical specimens, the axis of which is oriented in various directions with respect to the parental ice cover. The sample preparation itself is simple, and requires high precision for the end surfaces only. A disadvantage of the "standard triaxial test" is that strain measurements on the sample's surface require pressure tight strain transducers. Another is that the axial force reading must be corrected, because it is corrupted to a certain degree by friction between the piston and the side wall of the test cell, and at the linear bushing. An advantage is that - in principle - tensile axial loading is possible.

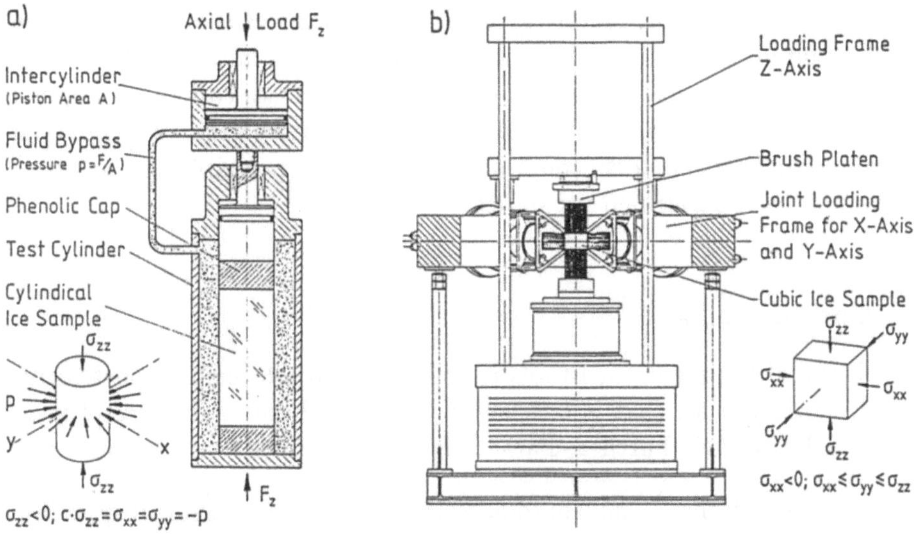

Figure 5. Multiaxial testing devices for ice: a) Modified triaxial cell (after Mellor et al., 1984); b) True triaxial loading frame with brush-platens (after Häusler, 1989)

The true triaxial loading of cubic ice samples with brush platens allows for stress states of the type $\sigma_1 < 0$; $\sigma_1 \leq \sigma_2 \leq \sigma_3$. The brush platens, originally designed by Hilsdorf (1965) for concrete testing, are - in principle - rigid metal platens, which are subdivided into an array of slender rods separated from each other and clamped at the foot. In their longitudinal direction the rods are stiff, and so can transmit normal stresses to the cube's faces. In their transverse direction the rods are flexible. Being separated, they can follow lateral deformations of the sample, while applying only a negligible amount of lateral constraint to it. Also, because the brush platens follow the sample's deformation they do not interact with each other, even if acting on all six faces of the cube simultaneously. Doubts on the brush platen technique when applied to ice discussed by Sanderson (1988) with respect to possible crack initiation at the edges of the individual rods, have been cleared up by Lasonde et al. (1988). The possibility with the true triaxial loading to individually control each of the three load directions makes this technique convenient to study anisotropic ice. In addition, there is always access to the 12 edges of the cube for the purpose of strain measurements directly on the sample. Furthermore, properly oriented cubic samples can easily be cut from a vertical ice core. Of course, the accuracy required in machining the ice cubes is unfortunately rather high, because of the rigidity of the brush platens in the load direction. Another disadvantage of the brush platens in connection with ice is that tension stresses cannot be applied (yet) to a sample, due to the lack of an appropriate glue. Thus biaxial tension-compression tests (pure shear), which in principle are possible, have not been performed yet on ice.

6. Conclusion

With the above mentioned tools it is possible to establish a constitutive model for (sea) ice that is consistent with the thermodynamics of solid body deformation. Further characteristics of the model are as follows:

- it contains the formalism to describe various mechanisms responsible for damage;
- the observable variables such as stress and strain are obtainable from free energy and dissipation potentials;
- multiaxial behaviour and anisotropy are included for the solid body deformation as well as for damage.

The consideration of brine and air porosity as existing damage is shown to be consistent with the reference strength method when a specific pore geometry is assumed. The technical means for determining the model coefficients are available. The numerical implementation and experimental verification of the model are tasks which remain to be performed.

Acknowledgement

This work was partly supported by the Federal Ministry for Research and Technology (BMFT).

References

Ashby, M.F. and Hallam, S.D. 1986. The failure of brittle solids containing small cracks under compressive stress states. Acta Metall., Vol.34, pp.497-510.

Betten, J. 1983. Damage tensors in continuum mechanics. J. de Mécanique, Vol.22, pp.13-32.

Choi, K. and Karr, D. 1989. A damage mechanics model for uniaxial creep and cyclic loading of polycrystalline ice. Proc. 8th OMAE - The Hague 1989, ed. N.K. Sinha, D.S. Sodhi and J.S. Chung, ASME, New York, Vol.4, pp.75-82.

de Groot, S.R. and Masur, P. 1984. Non-equilibrium thermodynamics. Dover, New York.

Frost, H.J. and Ashby, M.F. 1982. Deformation mechanism maps - The plasticity and creep of metals and ceramics. Pergamon Press, Oxford.

Gerstle, K.H., Linse, D.H., Bertacci, P., Kotsovos, M.D., Ko, H.-Y., Newman, J.B., Rossi, P., Schickert, G., Taylor, M.A., Traina, L.A. and Zimmerman, R.M. 1976. Strength of concrete under multiaxial stress states. Proc. McHenry Symposium, Mexico City, October 1976, pp.103-131.

Gold, L.W., 1972. The failure process in columnar-grained ice. National Research Council, Canada, Technical paper No.369.

Halphen, B. and Nguyen, Q.S. 1975. Sur les matériaux standards généralisés. J. de Mécanique, 14, pp.39-63.

Häusler, F.U. 1989. Beitrag zur Ermittlung der Kräfte beim Eisbrechen unter besonderer Berücksichtigung der Anisotropie des Eises und seiner Versagenseigenschaften unter mehrachsiger Beanspruchung. Diss. TU Hamburg-Harburg. Institut für Schiffbau der Universität Hamburg, Bericht Nr. 494.

Häusler, F.U. 1983. Comparison between different yield functions for saline ice. Annals of Glaciology, Vol.4, pp.105-109.

Häusler, F.U. 1986. Multiaxial mechanical properties of urea doped ice. Proc. IAHR - Symposium on Ice, Iowa City, August 18-22, 1986, Vol.I, pp.349-363.

Häusler, F.U. 1988. Reference strengths based on total porosity - a tool for the description of the ductile and brittle strength of ice frozen from sea water or other aqueous solutions. Proc. Ninth IAHR Ice Symposium, ed. H. Saeki and K. Hirayama, Sapporo, Vol.1, pp.77-91.

Häusler, F.U., Earle, E.N. and Gerchow, P. 1988. Uniaxial and biaxial compressive strength of ice sampled from multi-year pressure ridges. In: Port and Ocean Engineering under Arctic Conditions, ed. W.M. Sackinger, and M.O. Jeffries, Vol.I, pp.1-12.

Hilsdorf, H. 1965. Bestimmung der zweiachsigen Festigkeit des Betons. Deutscher Ausschuss für Stahlbeton, Heft 173, Wilhelm Ernst und Sohn, Berlin.

Hult, J. 1979. CDM capabilities, limitation and promises. In: Mechanisms of Deformation and Fracture, ed. K.E. Easterling, Pergamon Press, Oxford,

Jones, S.J. 1982. The confined compressive strength of polycrystalline ice. Journal of Glaciology 28(98), pp.171-177.

Kachanov, L.M. 1958. Time of the rupture process under creep conditions. Izv. Akad. Nauk SSR, Otd. Tech. Nauk, 8, pp.26-31.

von Kármán, Th. 1911. Festigkeitsversuche unter allseitigem Druck. Zeitschrift des VDI, Heft 42, pp.37-68.

Karr, D. and Choi, K. 1989. A three dimensional constitutive damage model for polycrystalline ice. Mechanics of Materials, Vol.8, No.1, pp.55-66.

Krajcinovic, D. 1985. Continuous damage mechanics revisited: Basic concepts and definitions. J. of Applied Mechanics, Vol.52, pp.829-834.

Krawietz, A. 1986. Materialtheorie. Springer Verlag, Berlin.

Lasonde, G.J., Gies, M.C. and Schulson, E.M. 1988. The effects of end conditions on the strength and fracture of ice under compression. Proc. Ninth IAHR Ice Symposium, ed. H. Saeki and K. Hirayama, Sapporo, Vol.1, pp.99-108.

Leckie, F.A. and Onat, E.T. 1981. Tensorial nature of damage measuring internal variables. Physical Non-Linearities in Structural Analysis, ed. H. Hult and J. Lemaître, Springer-Verlag, Berlin, pp.140-155.

Lemaître, J. 1985. Coupled elasto-plasticity and damage constitutive equations. Comp. Meth. Appl. Mech. Eng., Vol.51, pp.31-50.

Lemaître, J. and Marquis, D. 1988. Modelling elasto-plasticity, damage and ageing as coupled behaviours in engineering materials. Applied Solid Mechanics-2, ed. A.S. Tooth and J. Spence, Elsevier Applied Science, London, p.277-302.

Marsden, J.E. and Hughes, T.J.R. 1983. Mathematical foundations of elasticity. Prentice Hall, Englewood Cliffs, N.J.

Matthies, H.G. 1989. The rate problem for complex material behaviour with internal variables. Proc. Computational Plasticity - COMPLAS II - Barcelona 1989, ed. D.R.J. Owen et al., Pineridge Press, Swansea, pp.27-48.

Mellor, M., Cox, G.F.N. and Bosworth, H. 1984. Mechanical properties of multi-year sea ice - testing techniques. CRREL Report 84-8.

Murakami, S. and Ohno, N. 1981. A continuum theory of creep and creep damage. in: Creep in Structures, ed. A.R.S. Ponter and D.R. Hayhurst, Springer Verlag, Berlin, pp.422-444.

Noll, W. 1972. A new mathematical theory of simple materials. Arch. Rat. Mech. Anal., Vol.48, p.1-50.

Oden, J.T. 1986. Qualitative methods in nonlinear mechanics. Prentice Hall, Englewood Cliffs, N.J.

Pulkkinen, E. 1989. Rate-sensitive damage and cracking model for ice. Proc. Eighth OMAE - The Hague, ed. N.K. Sinha, D.S. Sodhi and J.A. Chung, ASME, New York, Vol.4, pp.47-53.

Sammis, C.G. and Ashby, M.F. 1986. The failure of brittle porous solids under compressive stress states. Acta Metall., Vol.34, No.3, pp.511-526.

Sanderson, T.J.O. 1988. Ice mechanics - risks to offshore structures. Graham and Trotman, London.

Santaoja, K. 1989. Continuum damage mechanics approach to describe the multidirectional microcracking of ice. Proc. Eighth OMAE - The Hague, ed. N.K. Sinha, D.S. Sodhi, J.S. Chung, ASME, New York, Vol.4, pp.55-65.

Sinha, N.K. 1982. Delayed elastic strain criterion for first cracks in ice. Proc. IUTAM Symposium on Deformation and Failure of Granular Materials, ed. P.A. Vermeer, and H.J.A. Luger, A.A. Balkema, Rotterdam, pp.323-330.

Sjölind, S.G. 1987. A constitutive model for ice as a damaging visco-elastic material. Cold Regions Science and Technology, Vol.14, pp.247-262.

Timco, G.W. and Frederking, R.M.W. 1984. An investigation of the failure envelope of granular/discontinuous columnar sea ice. Cold Regions Science and Technology, Vol.9, pp.17-27.

Timco, G.W. and Frederking, R.M.W. 1986. Confined compression tests: Outlining the failure envelope of columnar sea ice. Cold Regions Science and Technology, Vol.12, pp.13-28.

Tomin, M.J., Cheung, M., Cormeau, A. and Jordaan, I.J. 1986. Analysis of failure modes and damage processes of freshwater ice in indentation tests. Proc. Fifth OMAE, ed. V.J. Lunardini, et al., ASME, New York, Vol.IV, pp.453-460.

Weeks, W.F. and Assur, A. 1968. The mechanical properties of sea ice. Proc. Ice Pressures against Structures, Québec, Canada, 1966. National Research Council of Canada, Associate Committe on Geotechnical Research, Technical Memorandum No. 92, pp.25-78.

Ziegler, H. 1961. Zwei Extremalprinzipien der irreversiblen Thermodynamik. Ing. Archiv, Vol.30, pp.410-416.

SESSION IV:

Friction and Broken Ice Masses

Mechanisms of Ice Friction

J. Mølgaard

Faculty of Engineering and Applied Science
Memorial University of Newfoundland
St. John's, Newfoundland, Canada A1B 3X5

Abstract Friction involving ice is reviewed in both steady and transient conditions, considering contact mechanics, adhesion, fracture, crushing, elastic and plastic deformation, recrystallization, and melting as co-operating or competing steps and mechanisms in the overall process, with reference to recent experimental work.

1. Friction as a Process

Young's modulus and yield stress are examples of material properties definable for a single given material and specified test conditions. They are parameters derived from a particular deformation process, carried out in each case on a single material under defined conditions. They are material properties in the sense that, while they are determined on samples of particular dimensions, they are expected to apply in the deformation of any object, of any dimensions, made from the same material in the same state.

A coefficient of friction is not a material property in the usual sense as it involves at least two objects, i.e. in general at least two materials, and depends on the whole system. The coefficient characterizes the process involved, not just the materials.

In kinetic friction the concern is not only with a force resisting motion but also with the dissipation of mechanical work in this process through a number of separate steps and mechanisms, such as elastic deformation, plastic deformation, creep, repeated adhesion and fracture, transfer of solid material between the surfaces, as well as modification of the surface

S. Jones · R. F. McKenna · J. Tillotson · I. Jordaan (Eds.)
Ice-Structure Interaction
IUTAM-IAHR Symposium St. John's, Newfoundland, Canada

material by deformation, chemical reaction, and modification by phase changes (which may include melting).

In Figure 1, following Czichos and Mølgaard (1977), a ship and the ice it encounters are represented as elements 1 and 2, respectively, of a system, and the transactions which can occur between these elements indicated in a tabular fashion. A third element is the material, including water, air, and ice debris, in the interface between 1 and 2. These items in the interface can play a significant role in the friction process both in the contact mechanics and in the dissipation of mechanical work. A coefficient of kinetic friction is thus an indicator of the intensity of the friction process (Czichos, 1978).

The dissipation of mechanical work in kinetic friction need not be confined to, or occur primarily at, a geometrical plane of interfacial contact between objects sliding or rolling on each other. At least on an atomic scale the process can involve a considerable volume of material. Portions of the regions involved can have properties quite different from the rest of the objects, the differences in part due to the friction process itself. For instance, the dissipation of mechanical work produces heat, thus raising the temperatures of parts of the system and affecting physical and chemical mechanisms including the mechanical deformation, phase changes and any chemical reactions. These temperature changes depend on the thermal resistances, or thermal impedance, of the system.

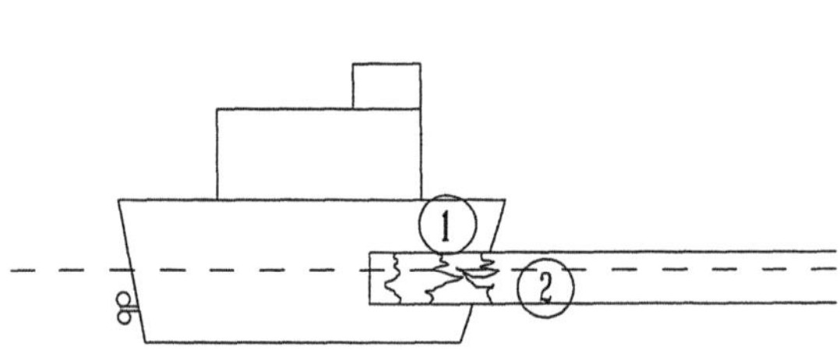

Possible transactions between elements 1 and 2

	On 1	On 2	Between 1 and 2
Elastic deformation	x	x	
Plastic deformation		x	
Creep		x	
Adhesion			x
Fracture		x	
Transfer			x
Material modification		x	

Figure 1. Ship-ice system; elements and transactions.

Kinetic friction can also be affected profoundly by mechanical impedance which, like thermal impedance, is a property of the whole system, not the surfaces. There can be considerable differences in friction under field conditions and laboratory tests because the systems involved are different. Figure 2 illustrates the wide range of coeffcients of friction, both in their levels and their dependence on temperature obtained in different systems, with examples taken from Schulz and Knappwost (1968) for tests at 1 mm/s, from Tusima (1977) at 0.074 mm/s and from Roberts and Richardson (1981) at 0.1 mm/s, the first two for ice on steel and the third for rubber on ice, and also for rubber on wet glass, as discussed below. Roberts and Richardson, and also Tusima, employed spherical specimens of the test material, while Schulz and Knappwost used a thrust-ring configuration.

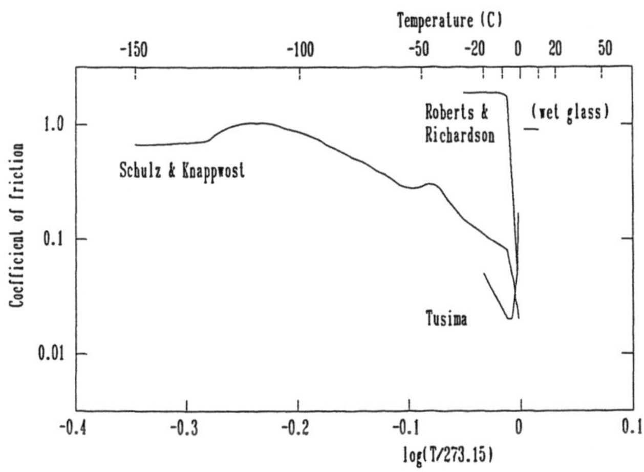

Figure 2. Effect of temperature on friction of ice in different systems.

In Figure 3 the differing effect of velocity is illustrated with results from Roberts and Richardson (1981) at −15°C, Tusima (1977) at −10°C, Akkok et al. (1987) at −20°C, Barnes et al. (1971) at −11.75°C, and Oksanen (1983) at −15°C. While the results from Roberts and Richardson are for rubber on ice, the other data is for steel on ice. Each source of data differs from the others in the mechanical system and specimen surface finish employed.

Some of the stages in kinetic friction involving ice are discussed in the following. The main theoretical issue in many instances is the possible

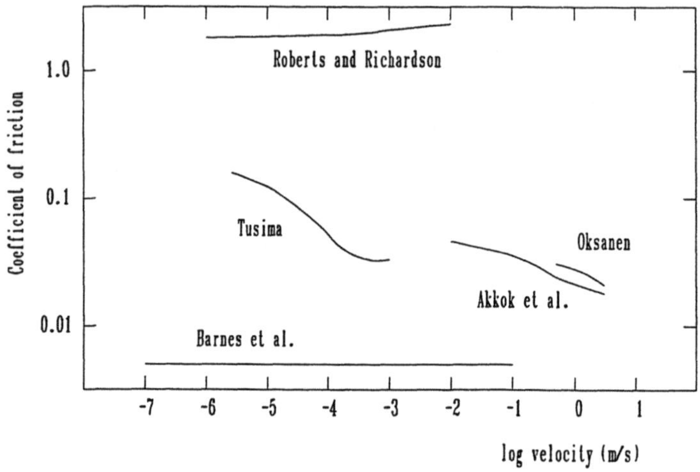

Figure 3. Effect of velocity changes on friction of ice in different systems.

presence and role of water at the sliding interface. The main practical issue is usually the prediction of the coefficient of friction as a function of the parameters of the system in question. The discussion indicates the complexity of the process which must be understood in order to resolve either issue.

2. Contact and Deformation

Theories of friction involve estimates of the total area at which there is intimate or "real" contact, i.e. where interatomic bonds can form across the contact, and also involve estimates of stresses at these contacts.

For many laboratory tests it is usual to prepare ice surfaces as flat and smooth as possible. This obviously is not always the case for ice in the field or for surfaces on laboratory specimens affected by wear or fracture, so for friction we must, in general, consider contact involving rough surfaces. A standard approach in theories of friction is to assume that the deformation at the points of contact on a rough surface will be plastic, if the loads are sufficiently large and radii of curvature sufficiently small at the contacts.

The stress patterns at such contact points will be similar to those at a hardness test indentor during a standard hardness test, with a mean normal stress indicated by the hardness number arrived at in Vickers or

Brinell tests. See Bowden and Tabor (1950 and 1964). The temperature and the duration of the contact can have a significant effect on the contact pressure as noted by Barnes et al. (1971) for hardness tests on ice in which a steel ball was loaded against ice at a variety of loads and durations of contact with the area of the indentations measured after removing the load. It was assumed by Barnes et al. that these indentations were produced by plastic deformation in the short duration tests or by creep in the case of loads applied for some time. While polycrystalline ice is not known to exhibit much, if any, plastic (i.e. permanent) deformation in uniaxial loading, creep has been given considerable study under those conditions. Deformation in uniaxial loading is usually accompanied by the formation of cracks which obviously can facilitate deformation on a macroscale as noted, for instance, by Gold (1972) and Sinha (1981).

Substantial permanent deformation is possible at sufficiently high deviatoric stresses in confined multiaxial loading of ice. This may, at least in part, be due to the formation and healing of microcracks as, for instance, studied by Jones (1982). It seems to be common to refer to a large permanent strain achieved through the microcracking as plastic deformation, like the permanent deformation achieved solely by dislocation movement in more ductile materials. The implications for contact mechanics are similar although the effect upon the microstructure and mechanical properties differ.

There is some evidence in the photographs published by Barnes et al. of fractures at their hardness indentations to confirm the brittle nature of ice even under the conditions under the spherical hardness test indentor. Goodman and Tabor (1978) also noted cracks at indentations produced in Vickers microhardness tests on ice. However, whatever the detailed interpretation of the deformation in these tests they do provide a measure of the ability of an ice surface to support a load applied to a confined area in a manner similar to that occurring in friction.

3. Material Modification

In general, the modification of surface material by processes such as work hardening, cracking, chemical reactions, and phase changes is likely to occur in a friction process involving materials capable of modification by these means.

Appreciable plastic deformation, without microcracking, seems to have only been observed in single crystals of ice. The high homologous temperature in question in most cases with ice would tend to modify or anneal out deformation structures, or lead to the recrystallisation of a deformed structure, as noted by Barnes et al.(1971) and by Tusima (1977). The formation of microcracks must involve the expenditure of some mechanical work. Some of this work is considered to be stored as surface energy at the crack surfaces and at the surfaces of particles formed as debris. This work is thereby removed from work transmitted through the system. The transformation of energy from one form to another depends on the appropriate process occurring, and the surface energy at crack or particle surfaces is not necessarily recoverable as mechanical work when cracks are healed. Ice is not chemically reactive, unlike metals which usually form films of reaction products on their surfaces in processes which are enhanced by frictional heat. However, there is a possible role for contaminants in the water from which ice may have been formed. These may concentrate as a brine on the surface and at grain boundaries. The lower the temperature the smaller the quantity of brine produced and the higher the salt concentration in the brine, down to a temperature at which the salt precipitates.

A liquid phase can also be produced by melting. An increase of hydrostatic pressure (up to about 200 MPa) lowers the melting point (Hobbs, 1974), and this may be sufficient for melting to occur below the ambient temperature. However, melting requires heat transfer to the site at which melting occurs, and so is not necessarily an instantaneous event.

4. Work Dissipation

In Figure 4 some of the available explanations for kinetic friction are illustrated schematically. Static friction is relatively simple and is solely concerned with a force resisting a tendency to produce motion but with no continuous motion. It can be analyzed solely in terms of contact mechanics.

In kinetic friction we must account for work dissipation and must consider the appropriate mechanisms at the microscopic and atomic scale at and close to the contacting surfaces.

Sliding at the surface could be equivalent to grain boundary sliding, where the lattice effects are limited to a region a few atomic distances thick on both sides of the boundary (Poirier, 1985). The dissipation could also be due to plastic deformation of a thicker layer of material on

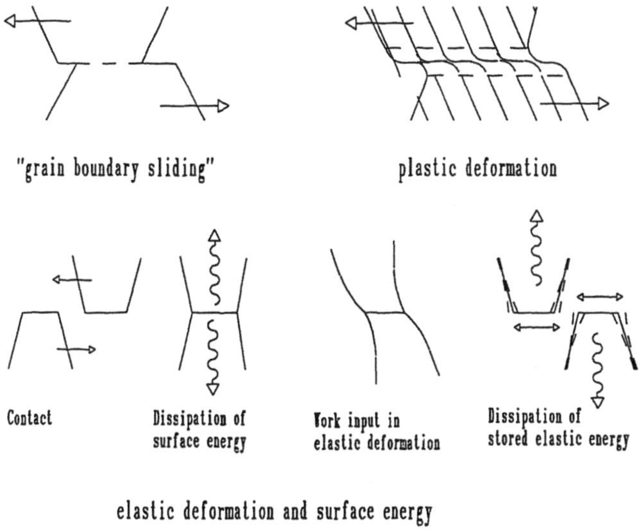

Figure 4. Some mechanisms in kinetic friction.

either side of an adhesion preceding fracture of the adhesion, or in the deformation of impinging asperities.

We must also consider the effect of cycles of repeated adhesion at a contact and subsequent fracture. Dissipation may be partly due to plastic deformation associated with fracture propagation, particularly at the tips of cracks, or as dissipation of the work involved in elastic deformation of the bulk of the objects prior to separation or fracture which may exceed that accountable for as additional surface energy. The excess must be dissipated in some manner, presumably involving internal friction. There must also be dissipation in irreversible migration of atoms accompanying solid surface formation and elimination as part of the process of maintaining thermodynamic equilibrium during these processes (Gurney, 1949).

Mechanical relaxation within ice has been advocated by Schulz and Knappwost (1968) as a mechanism for the dissipation of mechanical work in friction involving ice at speeds and loads low enough to avoid frictional melting. Using the activation energy and relaxation times associated with proton reorientation they were able to correlate maxima in an empirical relationship of the coefficient of friction with temperature down to -170°C with mechanical relaxation processes. As illustrated in Figure 2, a modest peak at around -50°C is attributed to internal mechanical resonance parallel to the c-axis, while a major peak at around -120°C is attributed to a response to periodic deformation driven by features on the counter-surface in their experiments, i.e. a feature dependent on the system.

5. The Ice Surface

Water may form at the sliding interface, due to frictional heat, or due to pressure melting, or by a combination of both mechanisms. The question of whether this occurs or not is a central issue in a consideration of ice friction. However, there are two other possible reasons for a liquid at the surface. One is the formation of a brine from impurities, as mentioned above, the other follows from the thermodynamics of the interface. Neither mechanism requires friction for the presence of water on the surface, though both may affect the frictional process.

Faraday considered the nature of the surface of ice in connection with the observation that pieces of ice will freeze together upon contact at temperatures above the freezing point, an effect subsequently referred to as regelation (Faraday, 1860). Weyl (1951), Jellinek (1967), and Fletcher (1968), have provided a modern structural and thermodynamic analysis predicting a liquid-like surface layer of the order of 10 to 40Å at temperatures down to a few degrees below 0°C. This liquid-like layer is sometimes referred to as the Faraday layer.

If a thin layer of liquid is present on ice an adhesion force is likely to be caused by the surface tension of the liquid and a friction force will result from shear of the liquid layer. The smaller the thickness of the liquid layer, the greater the adhesion and friction forces. In the case of the Faraday layer, adhesion and friction might also involve a refreezing of the liquid in the interface, as postulated by Faraday.

Experiments by Roberts and Richardson (1981) of the friction of rubber on ice at a range of temperatures, which they compared with the friction of the same rubber object sliding on glass covered by a thin film of water at just above 0°C, throw some doubt on melt water as a cause of the low friction in their experiments with ice just below 0°C. The friction coefficient was much higher for rubber on wet glass at sliding speeds up to 0.1 mm/s than for rubber sliding on ice at the same speeds at comparable temperatures in the same equipment, as shown in Figure 2. However, the soft rubber hemisphere used in these tests left a noticeable mark on the ice surfaces at temperatures above -10°C, indicating that the ice surface was softer than the bulk of the ice in that temperature range. Jellinek (1967) concluded from adhesion experiments that the viscosity of the liquid-like 'Faraday layer' formed without frictional melting or contamination on ice surfaces is probably somewhere in the region of 1 to 100 Ns/m^2. The viscosity of normal water is about 0.002 Ns/m^2 so that the Faraday layer offers a transition from the solid to the liquid state different from the usual liquid form of water. In the experiments by Roberts and Richardson the low friction of ice may be due to the low shear strength of a Faraday layer rather than due to a layer of melt water maintained hydrodynamically (or elastohydrodynamically) between the surfaces, which only seemed to be possible at the highest speed used, 10 mm/s, with rubber on wet glass.

6. Frictional Melting

While it is indicated in the preceding section how a liquid, or liquid-like, layer may be present on the surface of ice below the bulk melting point, there is no doubt that at a sufficiently high rate of generation of frictional heat there will be water at the interface due to melting. A task in friction theory is to establish the conditions under which this occurs and predict the resulting coefficient of friction.

Several analyses of friction involving ice have been based on this possible mechanism. Two of the most recent are by Oksanen (1983) and Akkok et al. (1987).

The proposition is simple. A layer of melt water, once formed, will become the source of friction since shear will then be limited to the water layer. The mechanism will be self balancing at a particular film thickness in a steady state, since no more heat will be generated than that required to maintain the melting surfaces at the melting point.

There are some points which do not seem to have been given enough attention in treatments of this process. For instance, it is likely that the melt water film thickness increases from the leading edge to the trailing edge of a contact region as the water produced at the forward end is added to that produced at the trailing end. A diverging liquid film will produce a negative hydrostatic pressure pulling contacting surfaces together. Separation will depend on meltwater being produced at a rate greater than that at which it can escape from the interface. It is therefore likely that there will always be some solid to solid contact at the leading edge of a melting contact area, with the load increased somewhat by negative pressure in the melt water. A frictional force will be generated both at the solid to solid contact at the leading edge as well as in the liquid film behind it, a form of mixed lubrication.

Estimates of the temperature reached in frictional contacts involve estimates of the resistance presented to the frictional heat flowing away from the contacts. This aspect still has to be included in a theory for circumstances in which the contact temperature is fixed at the melting point.

In the following the problem is reviewed through some example calculations, indicating the effect of varying the velocity, load, coefficient of friction and number of simultaneous contact points. It should also be noted that the contact temperature depends on whether the points of contact move on the ice or the other materials, and therefore depends on the topography of the surfaces.

Carslaw and Jaeger (1959) treated the flow of heat into a semi-infinite solid through a small stationary region on the surface and through a region moving on the surface, and Archard (1959) simplified the relationships which apply to contact temperatures developed by frictional heat.

The procedure is to estimate the thermal resistance presented in each object to the flow of that heat from the contact, combine all the resistances, and multiply by the heat flow rate.

Denoting the thermal resistances in object I and II by R_I and R_{II} the combined total resistance, R, is given by

$$R = R_I \cdot R_{II}/(R_I + R_{II}). \tag{1}$$

If we have, say, a hemispherical steel slider moving on an ice surface, the contact is stationary on the slider and moving on the ice. The resistance is greater for a source of heat stationary on the surface than for a moving source. A contact is considered moving if the dimensionless variable, ξ, is greater than 0.1, and is moving fast if ξ is greater than 5.0, where

$$\xi = V \cdot a/2\kappa, \tag{2}$$

V is the velocity, a the radius of the contact area, and κ is the thermal diffusivity.

For a stationary circular contact the contact resistance is

$$0.25/K \cdot a, \tag{3}$$

where K is the thermal conductivity.

Table 1 : Estimated Contact temperature increase (°C), for ice against the types of material listed

Velocity (m/s)	Load N	Coefficient of friction	Number of asperities	Contact temperature rise Moving contact on: ice	other	both (°C)
Metals:						
0.1	10	0.10	10	1	1	1
1.0	1	0.10	3	4	2	2
1.0	10	0.10	10	6	3	3
1.0	100	0.10	10	16	5	5
10.0	1	0.10	3	26	7	6
10.0	1	0.10	100	6	3	3
10.0	10	0.10	10	39	9	8
Ceramics:						
0.1	10	0.10	10	3	2	2
0.1	100	0.10	10	7	4	3
1.0	1	0.10	3	10	6	4
1.0	10	0.10	10	15	7	6
1.0	100	0.10	10	33	13	10
Glass:						
0.1	10	0.10	3	10	13	7
0.1	10	0.10	10	7	9	5
0.1	100	0.10	10	13	18	9
1.0	1	0.10	3	18	26	12
1.0	10	0.10	10	24	36	15
Plastics:						
0.10	100	0.10	3	11	24	8
0.10	100	0.10	10	8	16	6
0.10	1000	0.10	10	14	35	11
1.00	10	0.10	3	19	52	15
Ice:						
0.10	100	0.10	3	9	9	5
0.10	100	0.10	10	6	6	4
0.10	1000	0.10	10	12	12	7
1.00	10	0.10	3	17	17	10
1.00	100	0.10	10	23	23	13
10.00	10	0.10	100	23	23	13

For a fast moving contact the thermal resistance is

$$0.219\xi^{-1/2}/K \cdot a. \tag{4}$$

Resistances for moving contacts for which ξ is between 0.1 and 5 are given by Archard, as given by the dimensionless function in Table 2.

Table 2 : Resistance, R, as a function of ξ, for intermediate ξ

ξ:	0.1	0.5	1.0	2.0	3.0	4.0	5.0
R[c2aK]ξ:	0.3	1.1	1.9	2.9	3.7	4.3	4.9

The total resistance also depends on the number of simultaneous contacts and any surface layers with low thermal conductivity (Grosberg and Mølgaard, 1966), which in the present context might be rust or paint films on metal. These expressions applied to ice in contact with other types of materials at typical loads and velocities give the temperature rise at the contact points indicated in Table 1.

The representative values in Table 3 for the relevant materials properties were used in the calculations.

The contact area was assumed to be determined by yield, with the mean contact stress equal to three times the uniaxial yield stress, as is the case under the typical hardness test indentor on most materials (Bowden and Tabor, 1950). A nominal value of 0.1 is assumed here for the coefficient of friction. The calculated temperature rise can be scaled up or down, in direct proportion to the actual coefficient of friction in any particular instance. The number of contact points assumed is either 3, 10, or 100.

The three right hand columns of Table 1 are for the three likely alternative configurations: moving transient contacts on the ice; moving transient contacts on the other material (which in the lowest section of the table is also ice); and the case where both surfaces experience transient moving contacts.

Table 3 : Property values assumed in estimates of contact temperature increase

	Thermal Conductivity W/(m.K)	Density kg/m^3	Specific Heat J/(kg.K)	Yield Stress MPa
Ice	2.3	920	2090	10
Metals	100.0	8000	400	200
Ceramics	20.0	3000	800	4000
Glass	1.0	2500	700	3600
Plastics	0.20	1100	1400	50

Some of these estimates are illustrated in Figure 5, for the effect of velocity on the contact temperature rise with different materials sliding on ice. Ten simultaneous contact points which are transient on both objects in contact are assumed. A coeffcient of friction of 0.1 and a load of 100N was assumed in all instances, except for rubber for which the load is 10N.

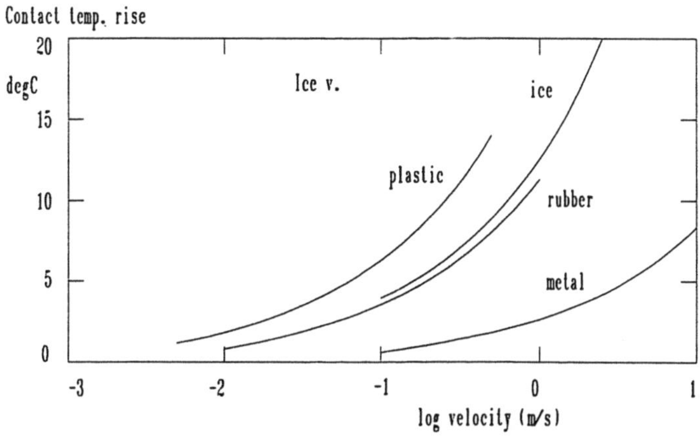

Figure 5. Contact temperature rise v. velocity for ice aliding on various materials effect of velocity changes on friction of ice in different systems.

In Figure 6 the effect of load on contact temperature rise is shown for plastic and metal, along with the differing role of the configuration involved, assuming ten simultaneous contact points, at 1 m/s for the metal and 0.1 m/s for the plastic and a coefficient of friction of 0.1 in both cases. For materials with relatively high thermal conductivity sliding on ice the contact temperatures are highest when the moving contacts are on the ice, as for a metal slider moving on a smooth ice sheet, a common configuration in laboratory experiments. However the opposite configuration is likely for rough ice moving against a smoother ship's hull. On the other hand, for ice contacting plastics the higher temperatures occur with the moving contact on the plastic, compared with the opposite configuration, an aspect relevant to studies of ice sliding on hull coatings.

Unless a single contact point follows from the shape of one of the objects, one can assume at least three points of contact, and somewhere between 10 and 100 points of contact is often quite plausible. Increasing the number of contact points decreases the contact temperature, so this

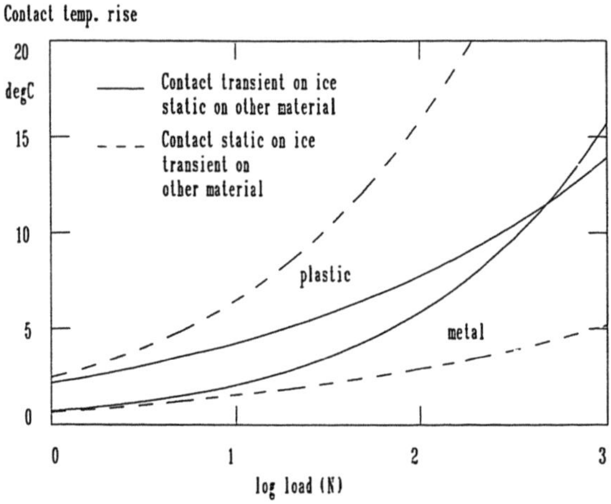

Figure 6. Contact temperature rise v. load for metal and plastic sliding on ice.

is an aspect to be considered for nominally flat ice sliding against rough surfaces.

Table 1, and Figures 5 and 6 serve mainly to indicate whether, under any given experimental or field conditions, the local temperature rise due to friction is likely to bring the contacts up from the ambient temperature to the melting point. If the temperature rise indicated is much above the difference between zero and the ambient temperature, melt water is likely to be present at the contacts and will at least contribute to the friction process, and perhaps dominate it.

For instance, Oksanen (1983), presents friction coefficients for ice on ice at speeds between 1 and 3 m/s. The values are between 0.01 and 0.04. At -15°C the coefficient increases with increasing speed but from Table 1 it is clear that melting is unlikely to have occurred. A similar dependence of the coefficient of friction on speed was observed by Tusima (1977) for steel on ice under conditions where melting clearly could not have occurred. At -1°C, however, Oksanen concluded that melting occurred, in accord with Table 1. At that temperature an increasing velocity increases the coefficient of friction, as one might expect with a liquid lubricant film.

Evans et al.(1976) claimed that melting took place in their experiments at a load of about 45 N and speeds between 0.2 and 10 m/s at temperatures around -11.5°C with four points of contact in most tests in which the coefficient of friction was about 0.02 at the highest speeds and 0.05 at the lowest speeds. The contacts were stationary on the metal and plastic test pieces used. From the estimates in Table 1 melting seems unlikely.

7. Transient Friction at High Loads

Most of the work done hitherto on friction involving ice has been at loads less than the yield or fracture load of the ice. Also, most experiments have been taken to a steady state, upon which the studies have concentrated.

In icebreaking and other field conditions loads reach the crushing strength of the ice at the interface and conditions are likely to be transient. Both aspects are being reproduced in friction studies in the laboratories of Memorial University on which a separate paper is presented here (Gagnon and Mølgaard, 1989).

Surfaces in general, including ice surfaces in the field, are rough and initial stress at points of contact are likely to exceed the yield stress. In the short time involved neither creep nor pure plastic flow is likely. Since the points of contact are confined the contact stresses may exceed those normally measured in uniaxial compression tests and it is likely that cracks are initiated from the contact location. The other surface may abrade the ice, or wear it by adhesion plus shear. Either this, or plastic flow, will increase the contact area. A liquid film on the ice at this initial stage could be the Faraday type of film or could be liquid already present at grain boundaries in ice formed from impure water, as is obviously the case for ice formed from seawater.

If sufficient frictional heat is developed, the contact surface of the ice will melt. It is also possible that the surface temperature is merely increased to the point where a Faraday film is present below the melting point. In the case of melt water forming there may be a form of mixed lubrication combining solid-solid contact with shear of a liquid film. Melting will increase the real area of contact and given sufficient time could produce a self-regulating situation at the contact with a load below the failure stress of ice.

A continuously increasing load on the ice may cause much of the surface layer of the ice to shatter into small pieces, as has been observed in indentation tests, in friction studies here, and in some circumstances in ice breaking. Jordaan and Timco (1987), and Jordaan and McKenna (1988) have recently reviewed and studied a number of such observations, with some suggestions on the process involved. They note the work needed to create the extra surface area as crack surface is a small portion of the total work expended in producing and expelling the crushed ice. Their estimate of the work to produce crack surface may be an underestimate if it is based on mechanical work in crack formation in tensile conditions. Crack formation under shear and compression could be accompanied by considerable shear, and also friction involving repeated fracture and adhesion between the crack surfaces. Nevertheless, at present it seems that the major part of the work in the process of crushing and expulsion of the crushed ice is dissipated in friction between the small pieces of ice.

The freshly formed surfaces on crushed particles will on the whole be clean. If any melt water has been produced at the original surface at the start of the friction process, or continues to be produced, that water has a large surface area to cover once crushing has occurred, and will likely be much thinner than on the initial surface. In any event the small radii of curvature on the particles formed by crushing is likely to allow the particles to penetrate any liquid film, permitting solid-solid contact.

An important aspect of the process involving sliding, and possibly rolling, between particles of freshly crushed ice is the short time involved, typically milliseconds. The mechanisms and effects involved are therefore transient. The sliding distances in each event are also short, typically no more than one millimetre. Steady state analysis hardly can apply.

The most likely processes of friction at the surfaces of crushed particles will be shearing of inter-particle bonds and mechanical relaxation processes, and thus of the dry contact type rather than the wet contact type. The dissipation associated with repeated adhesion and fracture, where much of the mechanical work associated with fracture is not recovered in subsequent adhesion, may also play a part. At temperatures close to the melting point, as may be the case temporarily in the crushed ice immediately after crushing, there may be appreciable dissipation involving the rolling and melting of fine debris between the larger particles. A friction force peak was observed by Schulz and Knappwost (1968) at around -0.2°C when fine debris of ice was present in their work.

8. Empirical Support For Theory

Relatively little in the way of supplementary work is usually done in ice friction studies to confirm the postulated mechanisms directly by independent observation, nor is it easy to do so though some examples of supplementary observations can be mentioned.

Tusima (1977) observed different friction coefficients for steel on the surfaces of single crystals of ice of different crystallographic orientation at low speeds not involving melting, confirming the role of the solid crystal structure in that case since the coefficient of friction differed for different orientations. He also took into account evidence of ploughing at temperatures close to the melting point. Schulz and Knappwost (1968) varied the periodicity of surface features on their specimens and thereby found support for their hypothesis of mechanical relaxation as the mechanism of friction without melting.

Roberts and Richardson (1981) observed the interface between transparent rubber and ice for evidence of adhesion in the form of 'Schallamach waves' (Schallamach, 1971) in the rubber and some evidence of a water film from interference colours. These authors differ from many others studying friction of ice in being concerned as much, or more, with the processes in the rubber and its properties contributing to the total friction process as with the properties of ice and the processes in it.

9. Conclusion

This paper is a review of some the significant mechanisms known to play a role in friction involving ice. The friction of ice is a complex process and the correlation of laboratory and field work, with a good understanding of the friction process, is not necessarily straightforward and has yet to be achieved.

References

Akkok, M., Ettles, C.M.McC., and Calabrese, S.J. 1987. Parameters affecting the kinetic friction of ice. Journal of Tribology, Transactions of the ASME. Vol.109, pp.552-561.

Archard, J.F. 1959. Temperature of rubbing surfaces. Wear. Vol.2, pp.438-455.

Barnes, P., Tabor, D., and Walker, J.C.F. 1971. Friction and creep of polycrystalline ice.. Proc. Royal Soc. London. Vol.324, pp.127-155.

Bowden, F.P. and Tabor, D. 1950. The friction and lubrication of solids. Part I. Oxford: Clarendon Press.

Bowden, F.P. and Tabor, D. 1964. The friction and lubrication of solids. Part II. Oxford: Clarendon Press.

Carslaw, H.S. and Jaeger, J.C. 1959. Conduction of heat in solids. (Second Edition). Oxford: Clarendon Press.

Czichos, H. 1978. Tribology: a systems approach to the science and technology of friction, lubrication and wear. Amsterdam: Elsevier. Chs.3 and 4.

Czichos, H., and Mølgaard, J. 1977. Towards a general theory of tribological systems. Wear. Vol.44, pp.247-264.

Evans, D.C.B., Nye, J.F. and Cheeseman, K.J. 1976. The kinetic friction of ice. Proc. Royal Soc. London, Vol. A 437, pp.493-512.

Faraday, M. 1860. Note on regelation. Proc. Royal Society of London, Vol.10, p.440.

Fletcher, N.H. 1968. Surface structure of water and ice. II. A revised model. Philosophical Mag. (London) Vol.18 (8th series). pp.1287-1300.

Gagnon, R.E. and Mølgaard, J. 1989. Crushing friction experiments on freshwater ice. IUTAM/IAHR Symposium on Ice/Structure Interaction. St. John's, Newfoundland.

Gold, L.W. 1972. The process of failure of columnar-grained ice. Phil Mag. Vol.26, pp.311-328.

Goodman, D.J. and Tabor, D. 1978. Fracture toughness of ice: a preliminary account of some new experiments. Journal of Glaciology. Vol.21, pp.651-660.

Grosberg, P. and Mølgaard, J. 1966. Aspects of the wear of spinning travellers: the division of heat at rubbing surfaces. Proc. Institution of Mechanical Engineers (U.K.) Vol.181 Pt.3L, pp.16-23.

Gurney, C. 1949. Surface forces in liquids and solids. Proc. Physical Society. London. Vol.62. pp.639-648.

Hobbs, P.V. 1974. Ice Physics. Oxford: Clarendon Press. p.61.

Jellinek, H.H.G. 1967. Liquid-like (transition) Layer on Ice. Journal of Colloid and Interface Science. Vol.25, pp.192-205.

Jones, S.J. 1982. The confined compressive strength of polycrystalline ice. Journal of Glaciology. Vol.28, pp.171- 177.

Jordaan, I.J. and McKenna, R.F. 1988. Ice crushing by impact and indentation: a literature review. Report prepared for the Institute for Marine Dynamics, National Research Council Canada, CR-AVR-02.

Jordaan, I.J. and Timco, G.W. 1988. Dynamics of the ice crushing process. Journal of Glaciology. Vol.34, pp.318-326.

Oksanen, P. 1983. Friction and adhesion of ice. Technical Research Centre of Finland, Espoo, 36pp.

Poirier, J.-P. 1985. Creep of crystals. Cambridge: Cambridge University Press. Section 2.4.

Raraty, L.E. and Tabor, D. 1958. The adhesion and strength properties of ice. Proc. Royal Soc. London. Vol.A 245, pp.184-201.

Roberts, A.D. and Richardson, J.C. 1981. Interface study of rubber-ice friction. Wear, Vol.67, pp.55-69.

Schallamach, A. 1971. How does rubber slide? Wear, Vol.17, pp.301-312.

Schulz, H.H. and Knappwost, A. 1968. Die Festkörperreibung des Eises als Relaxionseffekt". Wear. Vol.11, pp.3-20.

Sinha, N.K. 1981. Rate sensitivity of compressive strength of columnar-grained ice. Experimental Mechanics. Vol.21, pp.209-218.

Tusima, K. 1977. Friction of a steel ball on a single crystal of ice. Journal of Glaciology, Vol.19, pp.225-235.

Weyl, W.A. 1951. Surface structures of water and some of its physical and chemical manifestations. Journal Colloid Science, Vol.6, pp.389.

Crushing Friction Experiments on Freshwater Ice

R. E. Gagnon

Institute for Marine Dynamics, National Research Council of Canada
St. John's, Newfoundland, Canada, A1B 3T5

J. Mølgaard

Faculty of Engineering and Applied Science, Memorial University of Newfoundland
St. John's, Newfoundland, Canada, A1B 3X5

Abstract A unique hydraulically driven friction apparatus has been used to measure friction forces on freshwater ice specimens against a steel surface under crushing normal loads at various temperatures. The friction coefficient was found to be a decreasing function of increasing sliding speed and temperature and did not appear to be influenced by fabric, texture or the presence of bubbles. There was no correlation between normal load and the coefficient of friction. There was evidence that crushing friction has a higher coefficient than non-crushing friction, and possible mechanisms to explain this behavior are discussed. Periodic crushing was evident.

1. Introduction

A major component of the power required for the movement of ships in ice is due to frictional forces generated between the ice and the ship's hull, both above and below the water line. The friction of ice is also significant in the complex of forces occurring between stationary structures and moving ice fields. While a considerable amount of research has been done on ice friction (for example Evans et al., 1976; Akkok et al., 1986; Forland and Tatinclaux, 1985), no laboratory experiments have been conducted using crushing normal loads, though ship-based measurements of full-scale friction forces, which probably involved some crushing, have been made (Hoffman, 1985; Enkvist and Mustamaeki, 1986). The

S. Jones · R. F. McKenna · J. Tillotson · I. Jordaan (Eds.)
Ice-Structure Interaction
IUTAM-IAHR Symposium St. John's, Newfoundland, Canada

present work reports the preliminary results from the first series of tests performed using an apparatus designed specifically for laboratory crushing friction experiments on ice.

2. Apparatus

The apparatus used for the crushing friction experiments essentially consists of a steel wheel (Figure 1, A) with a radius of 28 cm driven by a hydraulic winch at speeds in the range 0-40 rpm with torques up to 5 kN-m. Ice specimens (B) are loaded in a holder (C) suspended by two stainless steel arms (D) tangentially oriented to the outer wheel surface. The ends of the arms are connected to the rigid frame of the apparatus by a shaft (F) and bearings. The holder is connected to a platform by another shaft with bearings (H) which passes through both. The platform is attached by four steel rods to a hydraulic ram (G) which can pull the holder and ice specimen against the curved peripheral outer surface of the rotating wheel with loads up to 50 kN. The shafts and bearings (F, H) facilitate the movement of the holder with minimum resistance. The pulling force is monitored by a load cell (I) located between the ram and the rods. The internal LVDT of the hydraulic ram measures the posi-

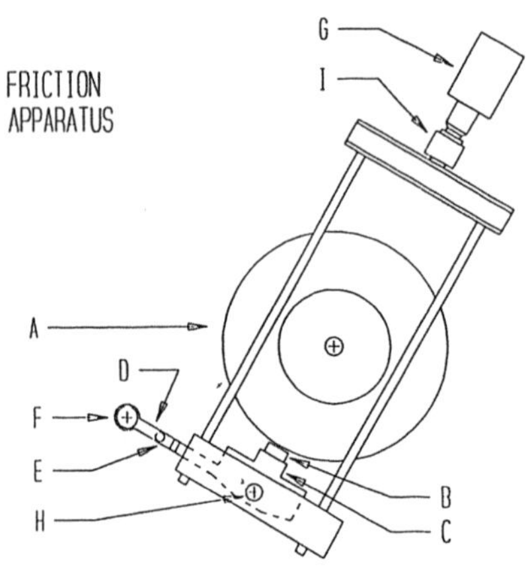

Figure 1. Crushing friction apparatus: A, mild steel wheel; B, ice specimen; C, specimen holder; D, support and sensor arms of the sample holder; E, strain gages for measuring tension and bending; F, support shaft and bearings for sample holder; G, hydraulic ram; H, shaft and bearings for transmitting the pulling force of the hydraulic ram to the specimen; I, load cell.

tion of the ice holder. The winch, wheel, ice holder and ram are mounted in a stiff aluminum frame approximately 2.3 m in height. The arms of the ice holder are fitted with strain gages (E) so that tension and bending forces are both measured. The base of the friction apparatus is bolted to the concrete floor of the laboratory to minimize vibration during experiments. A portable coldroom houses the apparatus and can maintain temperatures as low as $-28 \pm 1°C$. There is a window in the wall of the coldroom at the level of the ice holder to facilitate careful positioning of the ice holder prior to experiments and to allow for photographic and video tape recording of tests.

3. Data Acquisition System

Experimental data were collected using a digital acquisition system consisting of a Tatung 7000 AT computer with an RTI-815 (Analog Devices) analog-to-digital (12 bit resolution) acquisition board installed in one of the expansion slots. The software package ASYST (ASYST Software Technologies Inc.) was used to control the acquisition board, and analyze and display the data. Four input channels measured tension and bending in both sensing arms of the ice holder and two other channels monitored the pulling force on the holder and its position. Data were acquired at the same sampling rate and duration for all tests, 300 Hz over a period of 6 s. This sampling rate was chosen because preliminary tests carried out at sampling rates up to 2 kHz revealed that the resonant frequencies for tension and bending in the sensing arms of the ice holder were approximately 77 Hz and 212 Hz respectively. The data from bending indicated that the behavior of the ice during tests was predominantly confined to frequencies lower than the resonant frequency for tension.

The relatively low resonant frequency in tension introduced a considerable amount of noise in the data which, for graphic display purposes only, was effectively filtered out by applying a 4 point smooth operation. All of the computational analysis was carried out on unsmoothed data.

4. Sample Preparation

Three types of freshwater ice were used for the experiments reported here; small ($\sim$ .7 cm diam.) and large ($>$ 3 cm diam.) columnar-grained bubble-free ice and bubbly randomly-oriented fine-grain ice, hereinafter referred to as Type 1, 2 and 3 respectively. The columnar ice was grown from filtered, deionized, de-aerated water. Open containers with insulated bottoms and walls were filled with water near $0°C$ and left to freeze in the coldroom with the temperature set at approximately $-11°C$. Some of the water-filled containers were seeded on the surface with snow so that randomly-oriented small columnar-grained ice grew while large platelets of ice were let form on the surfaces of other containers, resulting in large irregular-shaped columnar-grained ice with some preferred c-axis orientation, as indicated by thin section analysis. The bubbly ice was made by flooding containers of snow with tap water and letting them freeze. The resulting ice was opaque, due to the bubble content. Blocks 10 cm wide, 11 cm long and 8 cm in height were cut with a bandsaw. Columnar grains were oriented in the vertical for ice Types 1 and 2. The tops of the blocks were cut to shallow truncated peaks so that during tests the initial loads would be localized in the center of the specimens.

5. Test Parameters

The test parameters which were controlled in this series of experiments were sliding speed, temperature, and ice characteristics. The ice samples were crushed into the wheel at the same rate (.55 cm/s) for all tests. The face of the mild steel wheel was basically an unprepared friction surface with a roughness factor which varied between .5 and 7 μm. The four sliding speeds were 5.95, 17.49, 39.94 and 82.35 cm/s. Four tests were usually carried out at each speed. Three temperatures were investigated; -5, -10 and $-19°C$. Three types of freshwater ice were used; small-grain (Type 1) and large-grain (Type 2) bubble-free columnar ice to study the influence of grain size, and bubbly ice (Type 3) to investigate the influence of bubbles.

6. Analysis

Because the experiments reported here involved a curved friction surface, the analysis was more complicated than in standard experiments where friction force and normal load are easily determined quantities. In the present work, the effective point of application of normal load between the ice and the steel wheel moved around on the wheel surface as an ice specimen was crushed (Figure 2), so that a component of the normal load

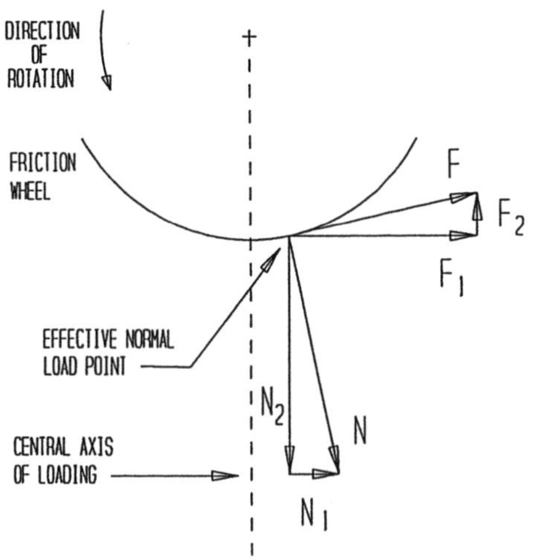

Figure 2. Components of normal load (N) and friction force (F). Vertical components register as bending in the sensor arms and horizontal components register as both tension and bending. The magnitude of the friction force has been greatly exaggerated.

registered as tension in the sensor arms of the ice holder along with the friction force. To separate these forces, it was necessary to also measure the amount of bending stress in the arms. From the knowledge of tension, bending, pulling force, ice holder position and the geometry of the situation, a computer program was written which could resolve all the forces. The expression for the friction force was,

$$\text{Friction Force} = (T - N\sin\theta - L\sin\phi)/\cos\theta, \tag{1}$$

where T was the measured tension in the sensor arms, L was the load measured by the load cell, ϕ was the angle by which the sensor arms deviated from the normal to the central loading axis of the apparatus, N was the normal load and θ was the angle between the friction force and the sensor arms. Both N and θ were functions of L, T, ϕ and the bending force.

An interesting and valuable artifact of the computation was that the effective application point of normal load on the ice by the curved wheel surface was calculated and could be traced throughout the duration of each test.

Friction coefficients (Table 1) were determined from the data by first estimating where in each test crushing friction began (see following discussion) and then performing a first order least-squares fit of friction

Table 1. Ice Friction Data

Ice Type	Temp. (°C)	Sliding Speed (cm/s)	Friction Coeff.	Peak Normal Load (kN)	Mean Normal Load (kN)
3	−10	5.95	0.079	9.02	4.08
3	−10	5.95	0.079	9.62	4.74
3	−10	5.95	0.047	9.99	3.52
3	−10	5.95	0.080	8.93	3.74
3	−10	17.49	0.047	8.71	4.02
3	−10	17.49	0.050	12.62	6.36
3	−10	17.49	0.045	11.41	4.26
3	−10	17.49	0.042	14.55	6.59
3	−10	17.49	0.047	10.57	3.57
3	−10	39.94	0.028	13.56	2.70
3	−10	39.94	0.032	11.02	4.07
3	−10	39.94	0.027	11.51	4.33
3	−10	39.94	0.023	15.72	4.31
3	−10	39.94	0.032	13.27	3.53
3	−10	39.94	0.026	13.86	3.34
3	−10	82.35	0.025	7.11	2.29
3	−10	82.35	0.027	8.38	4.02
3	−10	82.35	0.026	12.09	3.48
3	−10	82.35	0.020	11.60	5.20
3	−10	82.35	0.020	7.65	3.61
1	−10	5.95	0.074	6.90	2.50
1	−10	5.95	0.069	5.84	1.65
1	−10	5.95	0.072	6.77	2.76
1	−10	5.95	0.076	6.44	2.60
1	−10	17.49	0.050	6.18	2.35
1	−10	17.49	0.039	5.87	1.61
1	−10	17.49	0.052	6.40	2.11
1	−10	17.49	0.046	6.03	2.92

Table 1. Ice Friction Data (cont'd)

Ice Type	Temp. (°C)	Sliding Speed (cm/s)	Friction Coeff.	Peak Normal Load (kN)	Mean Normal Load (kN)
1	−10	39.94	0.029	7.92	1.91
1	−10	39.94	0.038	6.70	2.67
1	−10	39.94	0.036	6.78	2.72
1	−10	39.94	0.033	6.31	1.92
1	−10	82.35	0.030	6.05	1.36
1	−10	82.35	0.024	7.50	1.74
1	−10	82.35	0.024	8.42	2.68
1	−10	82.35	0.024	9.35	1.67
2	−5	5.95	0.041	7.57	1.75
2	−5	5.95	0.036	10.83	3.60
2	−5	5.95	0.041	8.84	1.61
2	−5	17.49	0.040	11.95	1.83
2	−5	17.49	0.032	11.19	1.80
2	−5	17.49	0.014	12.02	3.43
2	−5	17.49	0.029	8.58	1.84
2	−5	39.94	0.014	5.50	2.18
2	−5	39.94	0.019	10.67	2.47
2	−5	39.94	0.023	9.80	1.47
2	−5	82.35	0.018	8.04	2.24
2	−5	82.35	0.013	11.20	2.38
2	−5	82.35	0.012	10.68	1.90
2	−10	5.95	0.063	10.99	4.10
2	−10	5.95	0.072	11.15	3.32
2	−10	5.95	0.077	7.94	3.98
2	−10	5.95	0.061	10.36	4.95

Table 1. Ice Friction Data (cont'd)

Ice Type	Temp. (°C)	Sliding Speed (cm/s)	Friction Coeff.	Peak Normal Load (kN)	Mean Normal Load (kN)
2	−10	17.49	0.039	7.31	2.89
2	−10	17.49	0.032	8.49	3.24
2	−10	17.49	0.052	11.45	2.96
2	−10	17.49	0.044	6.94	3.44
2	−10	39.94	0.040	7.17	2.54
2	−10	39.94	0.033	8.90	3.08
2	−10	39.94	0.031	7.38	3.22
2	−10	39.94	0.037	8.49	2.94
2	−10	82.35	0.025	6.05	2.85
2	−10	82.35	0.025	6.42	2.20
2	−19	5.95	0.134	12.45	1.64
2	−19	5.95	0.104	7.57	2.48
2	−19	5.95	0.094	9.22	2.78
2	−19	5.95	0.076	7.89	2.98
2	−19	17.49	0.062	8.13	2.55
2	−19	17.49	0.067	6.83	2.86
2	−19	17.49	0.058	6.95	2.33
2	−19	17.49	0.075	7.50	2.01
2	−19	39.94	0.043	8.38	2.96
2	−19	39.94	0.044	11.15	4.06
2	−19	39.94	0.036	10.07	1.80
2	−19	39.94	0.044	8.37	2.82
2	−19	82.35	0.028	8.05	1.72
2	−19	82.35	0.041	10.84	2.86
2	−19	82.35	0.039	9.95	2.21

Figure 3. Results from a test run at $-19°C$ with a sliding speed of 17.49 cm/s. Time series plots are shown for normal load (A), friction force (B) and point of normal load application (C). Friction force versus normal load (D) is also shown. The normal load time series has been left unsmoothed so that the periodic crushing is more obvious. The arrow in (D) points to a small segment of the data, at the start of the test prior to the onset of crushing, where the friction coefficient is considerably lower.

force versus normal load for all the data beyond that point. The slope was taken as the friction coefficient.

7. Typical Test

Figure 3 shows the results of a typical test, in which 3 (D) shows friction force versus normal load. The majority of the data fall near the line whose slope yields the friction coefficient. At the very beginning of the test,

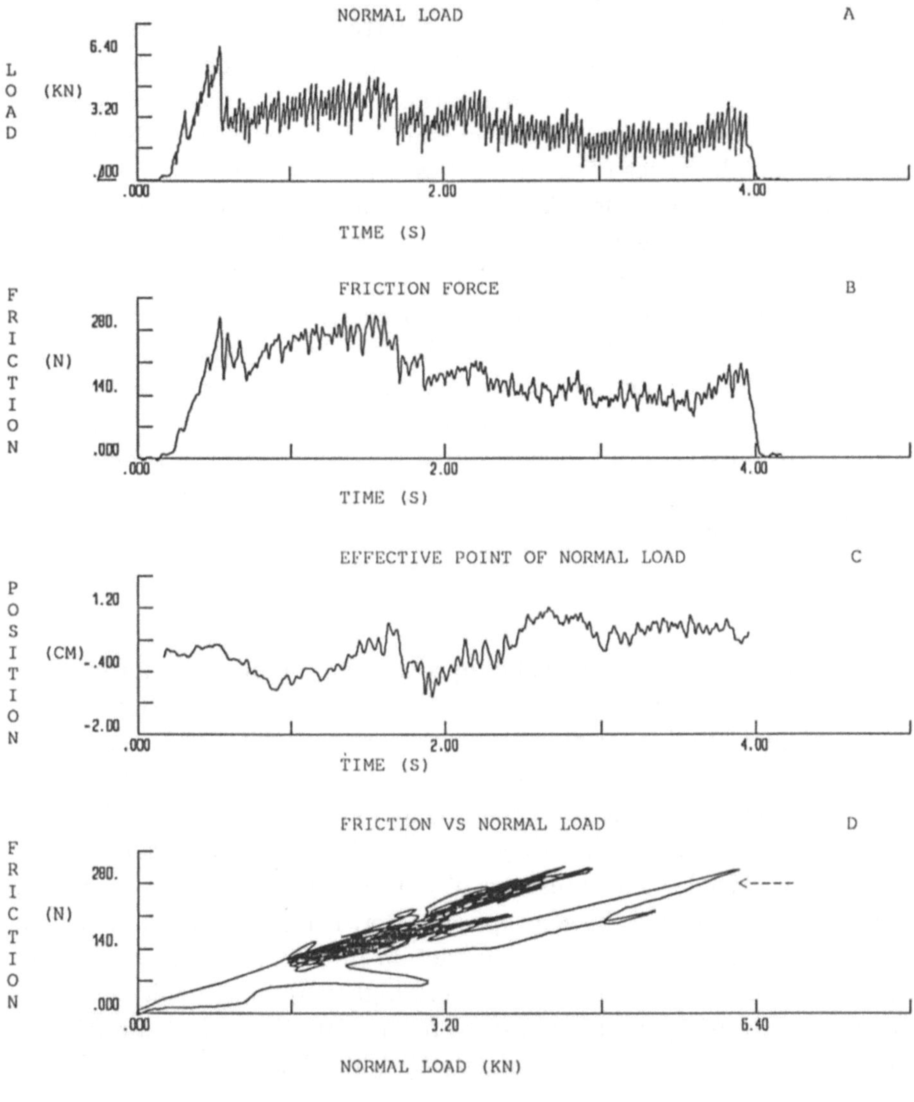

however, a small section of data falls on a significantly lower slope (arrow). There appeared to be two distinct friction regimes. To try and understand why this was so, video records (30 frames/s at 1 ms exposures) were taken during several tests. These revealed that the initial contact between the wheel and the ice specimen usually did not cause immediate cracking or crushing in the area of contact. During this time of initial stress build-up, prior to the onset of crushing, friction behavior was the same as in standard non-crushing friction tests. The transparency of the ice samples made the presence of a film of water, produced by frictional melting of ice and lack of cracks in the area of contact early in the test, clearly evident in some of the videos. High normal loads would often occur during this segment of the tests. Stress would eventually reach a point where cracking began around the contact area and crushing of the ice into small particles commenced. The onset of crushing was evident in some of the plots of effective point of normal load application. The layer of pulverized ice was quite thin, at most a few millimeters in thickness. A number of possible mechanisms to explain why the production of pulverized ice results in a higher friction coefficient are discussed in Section 9.

Figure 3 (C) shows the deviation, in the plane of the wheel, of the effective point of application of normal load away from the central axis of loading of the test apparatus. The periodic nature of the crushing and extrusion of pulverized ice, which varied between 20-50 Hz, is evident in Figure 3 (A). This was also quite clear in the video and high-speed film records of various tests.

8. Results

Figure 4 shows the influence of sliding speed and temperature for bubble-free large-grained columnar ice (Type 2). There was a reduction in friction coefficient at all three temperatures as the speed increased, seeming to asymptotically approach some constant value. This is consistent with some non-crushing ice friction experiments (Akkok et al., 1986; Evans et al., 1976; Tatinclaux and Murdey, 1985; Forland and Tatinclaux, 1985; Saeki et al., 1984). The presence of a thin lubricating film of water, created by friction, has been proposed (Bowden and Hughes, 1939) as the governing factor in the kinetic friction of ice. As temperature decreased less meltwater was produced and higher friction coefficients were observed. In raising the temperature from $-19°C$ to $-5°C$ the friction coefficient decreased by approximately 60% at all sliding speeds.

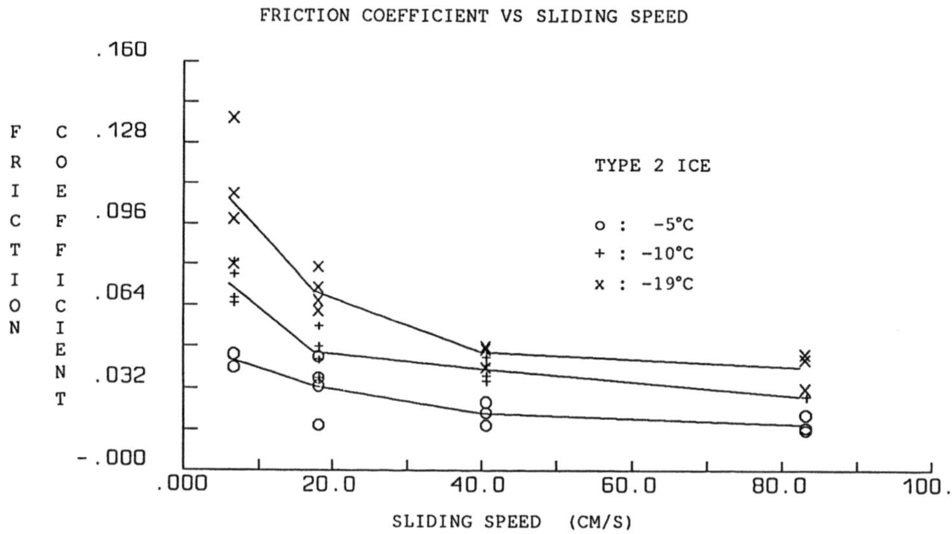

Figure 4. Friction coefficient versus sliding speed for Type 2 ice at three temperatures.

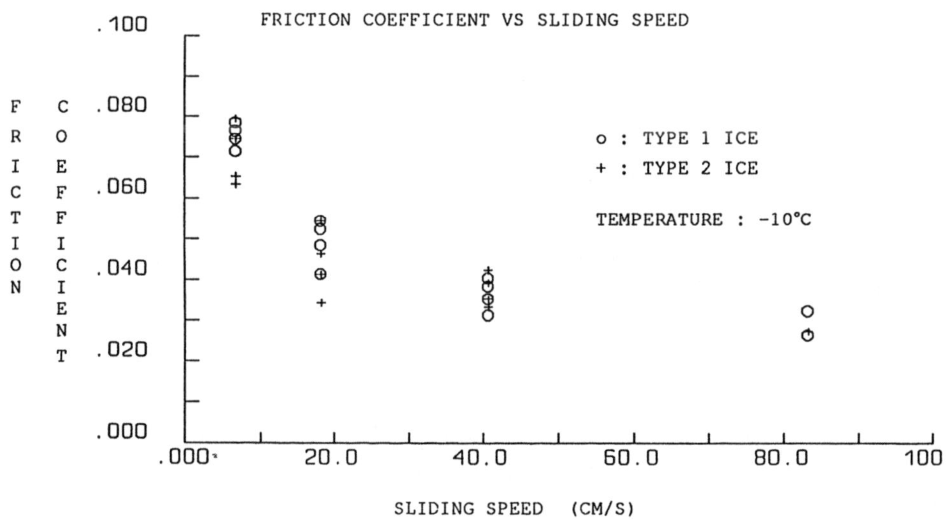

Figure 5. Friction coefficient versus sliding speed for bubble-free ice, Types 1 and 2, at $-10°C$.

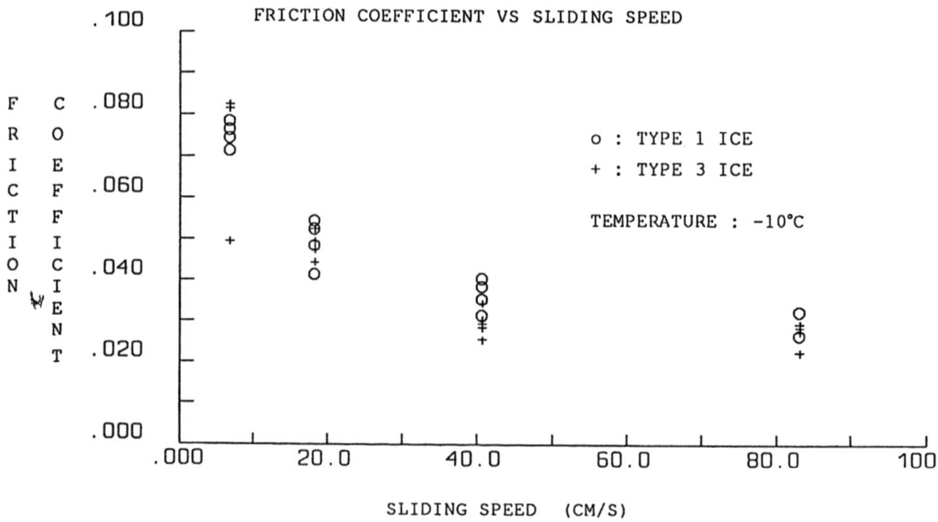

Figure 6. Friction coefficient versus sliding speed for bubbly ice (Type 3) and bubble-free ice (Type 1) at $-10°C$:

Figure 5 compares friction coefficients of Type 1 and Type 2 ice samples at $-10°C$. Considering the degree of scatter in the data it is clear that there was essentially no difference between the two ice types as far as friction coefficients were concerned. Similarly, Figure 6 compares ice with a large number of bubbles (Type 3) to bubble-free ice (Type 1) and again there appears to be no difference. Apparently neither fabric, texture or the presence of bubbles influences crushing friction coefficients.

During individual tests in general or in the comparison of different tests, there did not appear to be any unambiguous correlation of friction coefficient with normal load. Figure 7 illustrates this point for peak and mean normal load versus friction coefficient at various sliding speeds, where the peak normal load was the highest registered load for the entire test and the mean normal load was the average of the data points in the crushing segment of the test. This result disagrees with ship-based experiments, which probably involved crushing, where the friction coefficient appeared to decrease as normal loads increased (Hoffman, 1985; Enkvist and Mustamaeki, 1986).

The lack of correlation was maintained regardless of whether the peak in the normal load occurred before or after the onset of crushing.

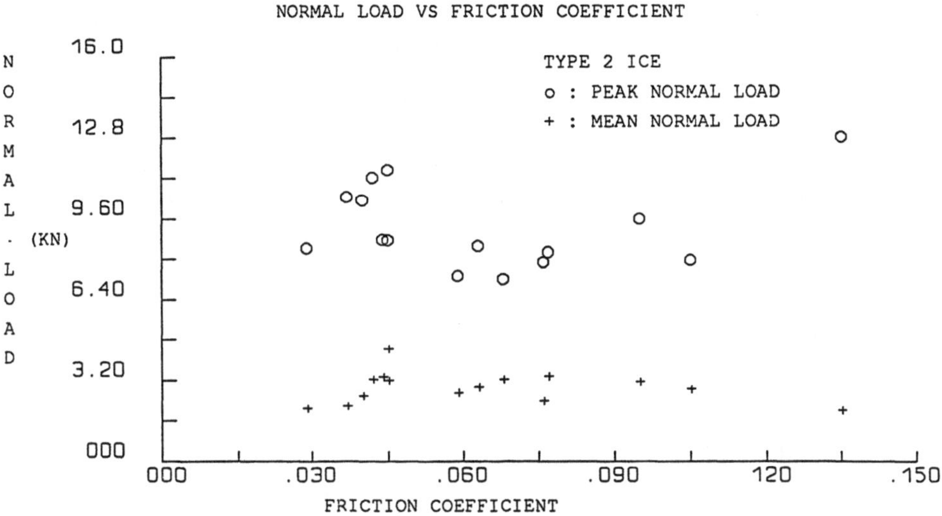

Figure 7. Peak and mean normal load versus friction coefficient for Type 2 ice at various sliding speeds at $-19°C$.

9. Discussions and Conclusions

This dynamic simulation of field conditions is unusual for a laboratory study, but has proven to be a technique from which useful information can be obtained, notwithstanding that the friction regime changed during a test, as is evident from both visual observation of the ice and the records of friction versus normal load.

The sliding velocity and ambient temperature were constant during the few seconds of each test, while the normal load initially increased and then varied at the maximum sustainable by the ice as it crushed.

The friction coefficient in these tests decreased with increasing sliding speed and temperature. It was unaffected by fabric, texture and presence of bubbles.

The coefficient of friction was generally higher during the crushing regime than during contact prior to crushing.

There was no unambiguous correlation of normal load with friction coefficient.

The factors which contribute to the increase in coefficient of friction during crushing may be the heat removed with the extruded crushed ice, the work expended in crushing and the work expended in friction at contacts throughout the crushed layer where sliding and rolling occurs, all of which reduce the production of lubricating meltwater. There could also be temporary adhesion between ice particles and the metal surface so that the bulk of the frictional dissipation is within the crushed layer rather than at the metal surface. Furthermore, a drastic reduction in the thickness of any frictional meltwater layer by being spread over all the particle surfaces, may lead to essentially solid-solid contact at a large number of points where particles penetrate any melt film present.

The work is continuing with a more detailed study expected of the crushing friction behavior.

Acknowledgements

The development of the equipment was supported by the Natural Sciences and Engineering Research Council under a strategic grant to one of the authors (J.M.) and Dr. P.N. Smith, who were assisted in the design by Dr. L.C. Wong and undergraduate engineering students on work terms. The current work is conducted under contracts from the National Research Council of Canada to Memorial University of Newfoundland, and the authors wish to acknowledge the work of Mr. Frank O'Dea and Mr. Douglas Maloney, senior engineering work term students, in the conduct of the experiments.

References

Akkok, M., Ettles, C. M. McM. and Calabrese, S.J. 1986. Parameters affecting the kinetic friction of ice. ASME Journal of Tribology, October, pp. 1-8.

Bowden, F.P. and Hughes, T.P. 1939. The mechanism of sliding on ice and snow. Proc. Royal Society London, Ser. A, Vol. 172, pp. 280-298.

Enkvist, E. and Mustamaeki, E. 1986. Model and full-scale tests with an innovative icebreaking bow. SNAME Transactions. Vol. 94, pp. 325-351.

Evans, D.C.B., Nye, J.F. and Cheeseman, K.J. 1976. The kinetic friction of ice. Proc. Royal Society London, Ser. A, Vol. 347, pp. 493-512.

Forland, K.A. and Tatinclaux, J.-C. 1985. Kinetic friction coefficient of ice. CRREL Report 85-6, pp. 1-40.

Hoffman, L. 1985. Impact forces and friction coefficient on the forebody of the German polar research vessel "POLARSTERN". POAC 85, Vol. 3, pp. 1189-1202.

Saeki, H., Ono, T., Nakazawa, N., Sakai, M. and Tanaka, S. 1984. The coefficient of friction between sea ice and various materials used in offshore structure. 16th OTC, Houston, Texas. pp. 375-382.

Tatinclaux, J.-C. and Murdey, D. 1985. Field tests of the kinetic friction coefficient of sea ice. CRREL Report 85-17, pp. 1-20.

Discussion

Q: R.A. Schapery, Texas A and M University

1. Do you know if the friction coefficient μ is affected by the amount of confinement of the sample? For example, if the rate of friction-surface dimensions to thickness is increased does μ change?

2. Can you say if the friction coefficient obeys a power-law in sliding speed? I believe it would be interesting to see a graph of log(μ) vs. log(speed) for different temperatures; for example, even if the lines are not straight, they may differ from one another only through their position on the log(speed) axis - this implying a master curve of μ vs log(speed $\times$ α_T) can be constructed.

A: R.E. Gagnon

1. We haven't looked into the questions of confinement yet but we definitely plan to.

2. No single power-law seems to fit all the data sets. We have only taken preliminary steps in the kind of useful analysis that you have suggested.

Q: R. Weiss, Exxon Production Research

1. Do you plan to conduct tests with snow?

2. Do you plan to conduct tests with sea ice?

A: R.E. Gagnon

1. We have already conducted a few preliminary tests with a thin snow cover on the specimens and plan more for the future. The presence

of the snow cover seems to eliminate the low friction often observed at the very beginning of tests before the onset of crushing.

2. Yes, we do plan to test sea ice.

Q: M. Metge, CANATEC

In the test shown, the normal load varied from a region of high average value to a lower value. I suggest that it might be worthwhile to isolate these two areas, and look at the results. This would give another check of the lack of influence of normal pressure on friction.

A: R.E. Gagnon

I think that is a good idea when you have a very uniform friction surface. In future experiments where we will use bands of various materials with uniform roughness we certainly will look more closely at small segments of the data for each test.

Q: R. Frederking, NRC

What sort of average contact pressures could you determine and how did these compare with uniaxial compression strength of similar ice?

A: R.E. Gagnon

We haven't determined any average contact pressures as yet though we are confident that accurate measurements of contact pressure can be made by using the visual records of tests and through analysis of the topography of the crushed specimens. We plan to observe the crushing-friction phenomenon at the contact interface by video taping through the bottoms of the specimens where little damage occurs.

Q: W.A. Nixon, IIMR

Do you plan to explore your hypothesis that crushing friction is controlled by sliding of crushed particles by using low thermal conductivity materials on the wheel?

A: R.E. Gagnon

We plan to use a wide array of materials in future experiments, metals with various thermal conductivities and non-metallic materials with low conductivities. We will also be looking at different paints.

Q: D.S. Sodhi, USA CRREL, Hanover, NH

1. Perhaps, the normal load can be reduced after ice crushing has taken place to determine the effect of normal load on the friction coefficient.

2. A transparent container around the ice may prevent the crushed ice from falling off the area of contact.

A: R.E. Gagnon

1. We have conducted experiments where the specimen holder was stopped during the crushing portion of the test. Crushing immediately stopped and the normal load decreased as ice wore away from the contact area. Sufficient analysis hasn't been performed yet to definitively say what happens to the friction coefficient in this situation.

2. We plan to look at the effect of confinement in future experiments.

Do Cracks Reduce Thermal Ice Stresses?

L. Å. Fransson

Department of Civil Engineering, Luleå University of Technology
S-951 87 Luleå, Sweden

Abstract Theoretical predictions of thermal ice loads often result in overestimations. The most important discrepancy between theory and practice is the idea of an ice sheet without cracks. Field measurements of in situ stresses have shown that the ability to build up thermal stresses may vary from 130 to 290 kPa/°C in the same effectively confined ice sheet.

Large volumes of natural ice contain thousands of fine dry cracks and also systems of wider cracks. This affects thermal ice loads in two ways:

- The expansion needed to close dry cracks delays the pressure start.
- The lateral restraint of uncracked ice decreases because of concentrated creep processes around the cracks.

A nonlinear continuum model is used to analyse the stress measurements. This model treats the ice as an uncracked continuum on average.

In this paper also a simplified stress-strain relationship for cracked ice is proposed. In this simplified model the ability to build up compressive stresses increases linearly with stress level and the creep rate is proportional to the square of the stress level.

1. Introduction

During thermal expansion of an ice sheet the typical strain rate is 10^{-8} $\sec^{-1}$. Extensive research on ice behaviour at low strain rates has been published but most of this deals with glacial ice (see Paterson, 1981). Very close to the melting point, long time creep experiments are extremely sensitive to heat flow and therefore difficult to perform in a laboratory having only limited space. Stresses can be predicted fairly accurately from strain rates but the process of building up thermal stresses is strongly dependent on ice temperature during the stage of maximum stress. At

S. Jones · R. F. McKenna · J. Tillotson · I. Jordaan (Eds.)
Ice-Structure Interaction
IUTAM-IAHR Symposium St. John's, Newfoundland, Canada

this stage the temperature is usually higher than -5°C for ice sheets in Swedish lakes.

Before thermal expansion the ice sheet must cool and hence contract. These contractions generate cracks, ranging from a few large wet cracks to thousands of fine dry cracks, depending on the ice thickness and temperature variations (Metge, 1976).

In this paper, different effects of cracks on the stress-strain relationship are discussed. This discussion is based on in-situ measurements of ice stresses in the Lule River (Fransson, 1988). It is possible that the gradual accumulation of large horizontal ice loads depends on the cementation of cracks formed during periods of cold temperatures. In most cases cold ice appears to be cracked. These cracks might explain why most reservoir dams and bridge piers in Nordic countries remain undamaged.

2. The Continuum Approach

The unrestricted thermal expansion of ice can be written

$$d\varepsilon = \alpha \cdot d\Theta, \tag{1}$$

where $d\varepsilon$ = expansion per unit length caused by a change in temperature $d\Theta$,
$d\Theta$ = change in temperature and
α = linear coefficient of thermal expansion.

The strain rate associated with this thermal expansion is therefore given by

$$\dot{\varepsilon} = \alpha\dot{\Theta}. \tag{2}$$

It is assumed that the viscous creep in deviatoric stress may be described with the flow law

$$\dot{\varepsilon} = A\left(\frac{S_I}{S_*}\right)^n, \tag{3}$$

where A and n are constants,
S_I is the principal stress-deviator having the largest magnitude and
S_* is the unit stress-deviator (1 Pa).

The flow law in Equation (3) is used because this has been found to be very satisfactory in glaciology (Glen, 1953) where ice movements have been predicted by using this equation (first used in metallurgical research). The principal stress deviator S_l may be replaced by an effective shear stress as first proposed by Odqvist (1934) if the third invariant is omitted. By following this procedure an expression for one stress component in a confined ice sheet may be calculation from the equation:

$$\dot{\sigma}_\varphi = \frac{f_1 E}{1-\beta\nu}[\dot{\varepsilon} - \frac{f_2}{(f_1)^n}(\frac{\sigma_\varphi}{\sigma_*})^n], \tag{4}$$

where $\dot{\sigma}_\varphi$ = stress rate measured at an angle φ from the principal stress direction,
f_1 = $\cos^2\varphi + \beta\sin^2\varphi$,
β = σ_2/σ_1,
ν = Poisson's ratio,
E = Young's modulus,
$\dot{\varepsilon}$ = expansion rate $(= \dot{\varepsilon}_x = \dot{\varepsilon}_y)$,
f_2 = $\lambda/B_0(\frac{\Theta_*}{\Theta_1-\Theta})^m$, temperature-dependent creep rate,
λ = B/B_0, ratio of the creep rate B in the actual stress field to the creep rate in a uniaxial test, B_0,
Θ_* = $\Theta_1 = +1°C$, $\sigma_* = 1Pa$ and
n, m = constants dependent on ice type.

In a case of pure thermal expansion, the biaxial strain rate $\dot{\varepsilon}$ can be calculated from Equation (2) if the rate of temperature change is known.

In Figures 1 and 2 the practical use of the continuum model for a confined ice sheet is shown. Unfortunately this nice fit was an exception during 5 years of field measurement.

3. Fine Cracks

Each time the temperature of a laterally confined ice sheet decreases, the top surface will contract more than the bottom surface of the sheet, and cracks may form if the rate of temperature change is sufficiently high. Metge (1976) claimed that most cracks stay dry when the ice is cold. If so, an expansion following a temperature rise is unlikely to cause much higher lateral pressure than the preceding expansion.

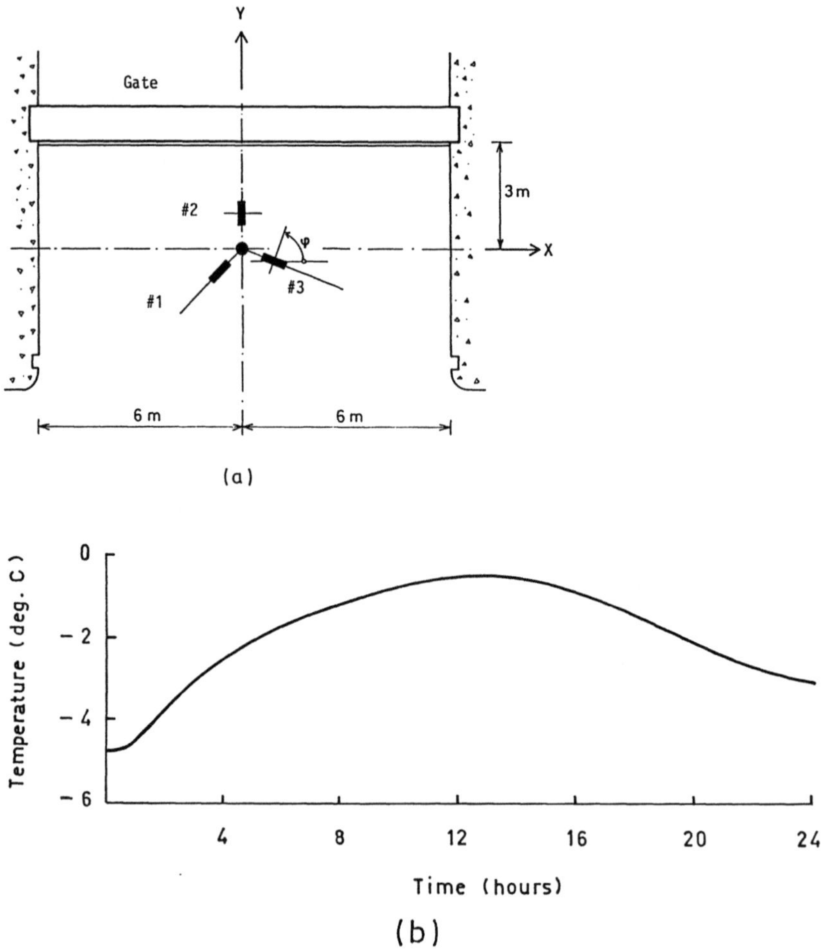

Figure 1. (a) Plan of the confined ice sheet in Vittjärv reservoir dam with three flat stress sensors installed. $\varphi_1 = 42^o, \varphi_2 = 0, \varphi_3 = 74^o$ (b) Measured ice temperature at the centres of the stress sensors.

It seems natural to assume that the temperature can rise up to a certain point earlier corresponding to zero pressure without creating pressure (if no cracks have been cemented with ice in the mean time). This "free temperature rise" can be correlated with the crack density in a confined ice sheet.

An idealized ice sheet with fine cracks is shown in Figure 3. All fine cracks at the confined length must be closed before the pressure can start to increase within the upper portion of the ice.

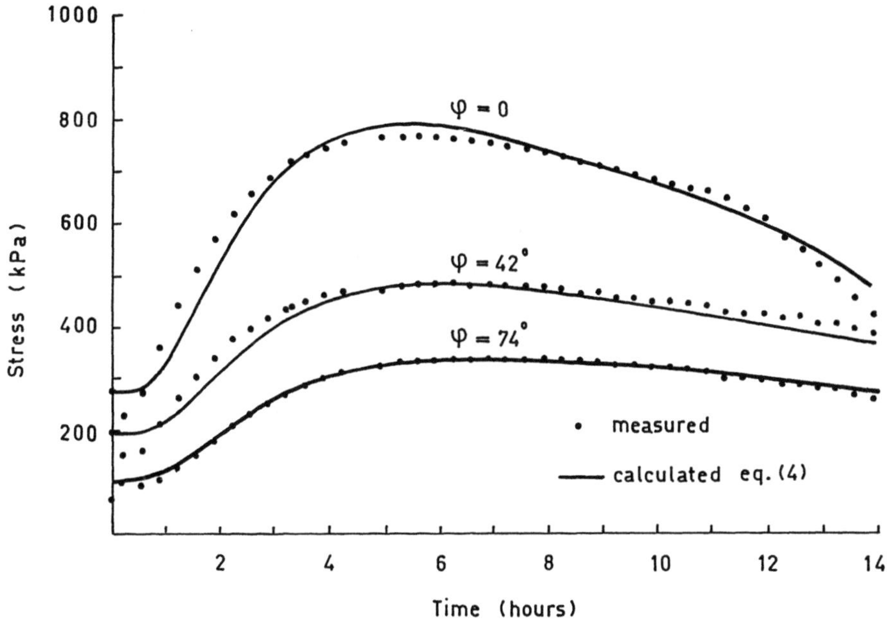

Figure 2. Adjustment with Equation (4) to measured stress in three directions at Vittjärv. The parameters used have the following values:

$E = 5 \times 10^9$ Pa	$\alpha = 50 \times 10^{-6}$ ° C^{-1}*	$\sigma = 1$ Pa	$n = 2$
$\nu = 1/3$	$\beta = 0.4$ *	$\Theta = 1$°C	$m = 0.2$
$B_0 = 2 \times 10^{-20}$ s^{-1}	$\lambda = 0.6$	$\Theta_1 = 1$°C	

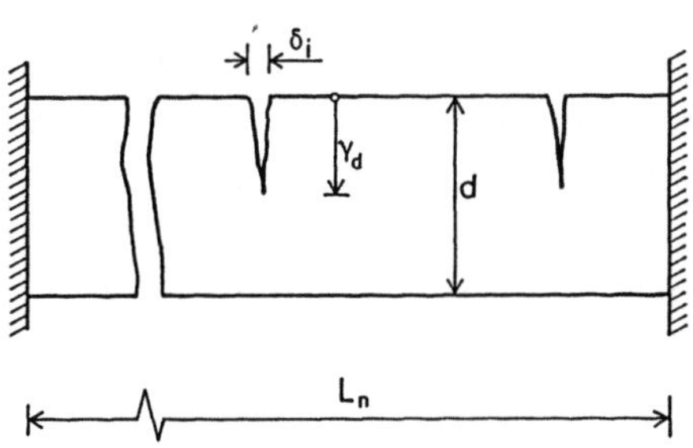

Figure 3. Fine cracks in a laterally confined ice sheet.

The crack density is difficult to measure physically. Thousands of fine cracks may contribute more than a few wider cracks. The case of dry cracks in river ice is shown in Figure 4.

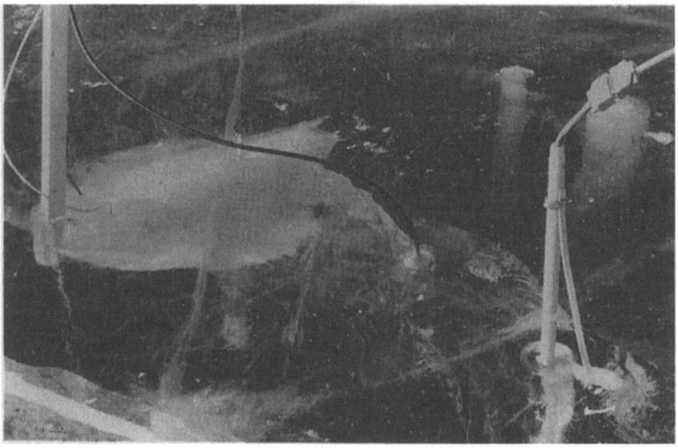

Figure 4. Crack pattern of snow-free and unflooded river ice, approximately 1 metre thick.

The average crack width per confinement length is

$$\Delta\varepsilon = \frac{\Sigma\delta_i}{L_n}, \tag{5}$$

and the free temperature rise at the surface is

$$\Delta\Theta = \Delta\varepsilon/\alpha. \tag{6}$$

By measuring the actual free temperature rise, the crack density $\Delta\varepsilon$ can be estimated from Equation (6). The value of $\Delta\varepsilon$ may be an important parameter in all applications where the mechanical behaviour of the cracked ice volume is involved.

The measured stress, shown in Figure 5, did not change much during the first 6 days of the temperature rise. The free temperature rise was about 7°C which corresponds to

$$\Delta\varepsilon = 7 \times 50 \times 10^{-6} = 350 \text{ microstrain}. \tag{7}$$

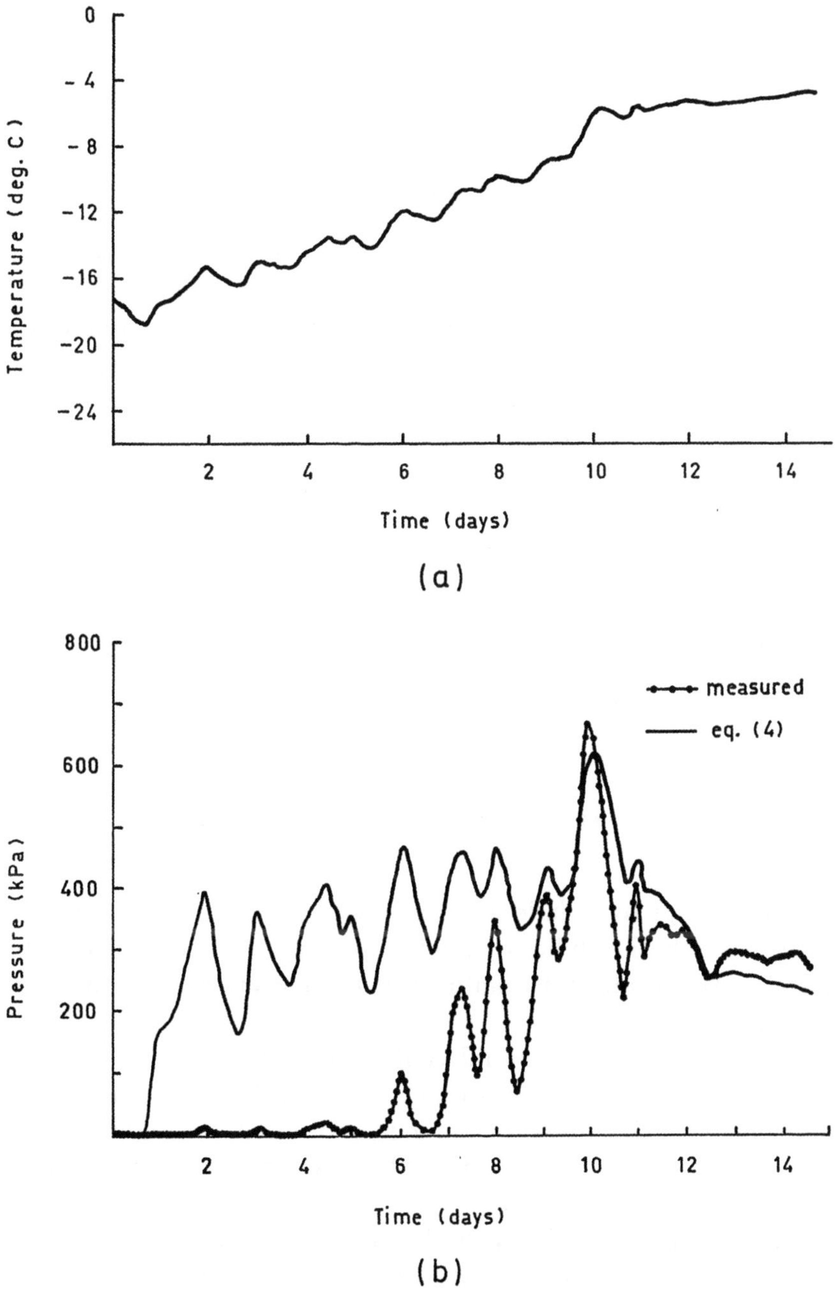

Figure 5. a) Temperature rise at the centre of stress sensor number 2 measured from a minimum temperature of −19°C. (b) Stress response of sensor number 2 for the temperature rise. Geometry shown in Figure 1.

This strain, if purely elastic, would have caused a stress several times higher than the yield strength. It is therefore assumed that the ice cover was cracked even though only a few cracks were visible on the surface.

4. Stress-Strain Relationship for Cracked Ice

Most visible cracks are dry cracks that remain open. These cracks have usually formed early by thermal contraction. When the ice sheet is subjected to lateral compression, the deformation of the highly stressed ice under the crack will be larger than its average value for the entire ice mass. This will affect confinement of the uncracked ice in a complex manner. A typical dry crack is shown in Figure 6.

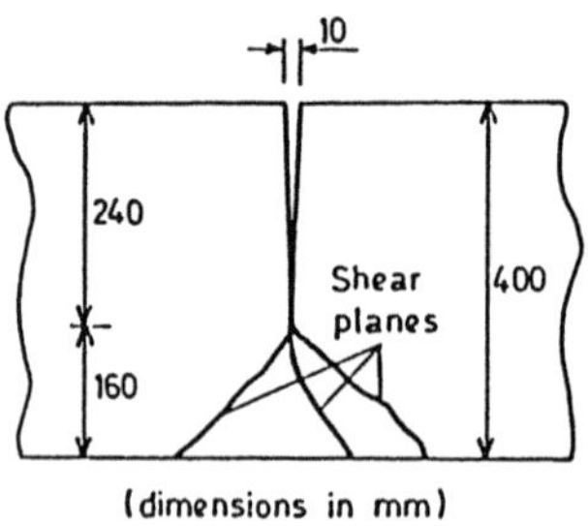

Figure 6. Section of an ice sheet with a dry crack (after Metge, 1976).

As a first attempt to include wider cracks in the continuum model it is assumed that only the ability to build up pressure is affected. The rate of stress relaxation evaluated from field measurements was found to be less sensitive to the ice quality.

The average reduction of lateral confinement of cracked ice can be expressed as an apparent modulus which is assumed to be a function of the reduction in contact area. In the idealized ice sheet (Figure 6) the average reduced modulus is a function of crack depth and the distance between two neighbouring cracks.

The contact area perpendicular to the most compressed direction may increase during compression because of crack closure. This will involve shear deformation under the crack tip and also crushing of soft ice and snow in the wider cracks. It is not known how this process is related to stress and time. Figure 7 illustrates one possibility.

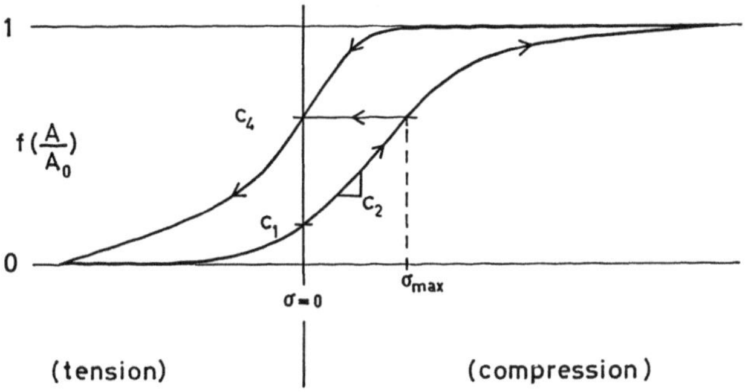

Figure 7. Proposed stress-hardening of a cracked ice sheet.

For large volumes of biaxially confined, cracked ice, a simplified strain-stress relationship is proposed. The resulting stress rate is given by the equation

$$\frac{\dot{\sigma}}{E} = (C_1 + C_2\sigma)(\dot{\varepsilon} - C_3\sigma^2) \quad \text{for } \dot{\sigma} > 0, \tag{7a}$$

$$\frac{\dot{\sigma}}{E} = C_4(\dot{\varepsilon} - C_3\sigma^2) \quad \text{for } \dot{\sigma} \leq 0, \tag{7b}$$

where
$$\begin{aligned} C_1 &= f \text{ (contact area when } \sigma = 0) \\ C_2 &= f \text{ (rate of change of contact area)} \\ C_3 &= f \text{ (creep rate for unit stress)} \\ C_4 &= f \text{ (contact area when } \sigma = \sigma_{max}) \end{aligned}$$

It is of course possible to include temperature dependence on the creep rate but in some applications other variables may be more important.

It can be seen from Figure 6 that the response to a lateral expansion of a confined ice sheet varies with depth. When the cracks are no longer perpendicular to the direction of principal stress, it is likely that Equation (7) does not apply.

The measured stress response at different relative depths is compared in Figure 8. These few recordings together with three similar comparisons in the Lule River indicate that the confinement increases with depth. However, the results are scattered and there might be good reasons for that. The presence of wet cracks and shore-cracks as later described in this paper is probably one reason why the confinement at the upper portion varies from one case to another.

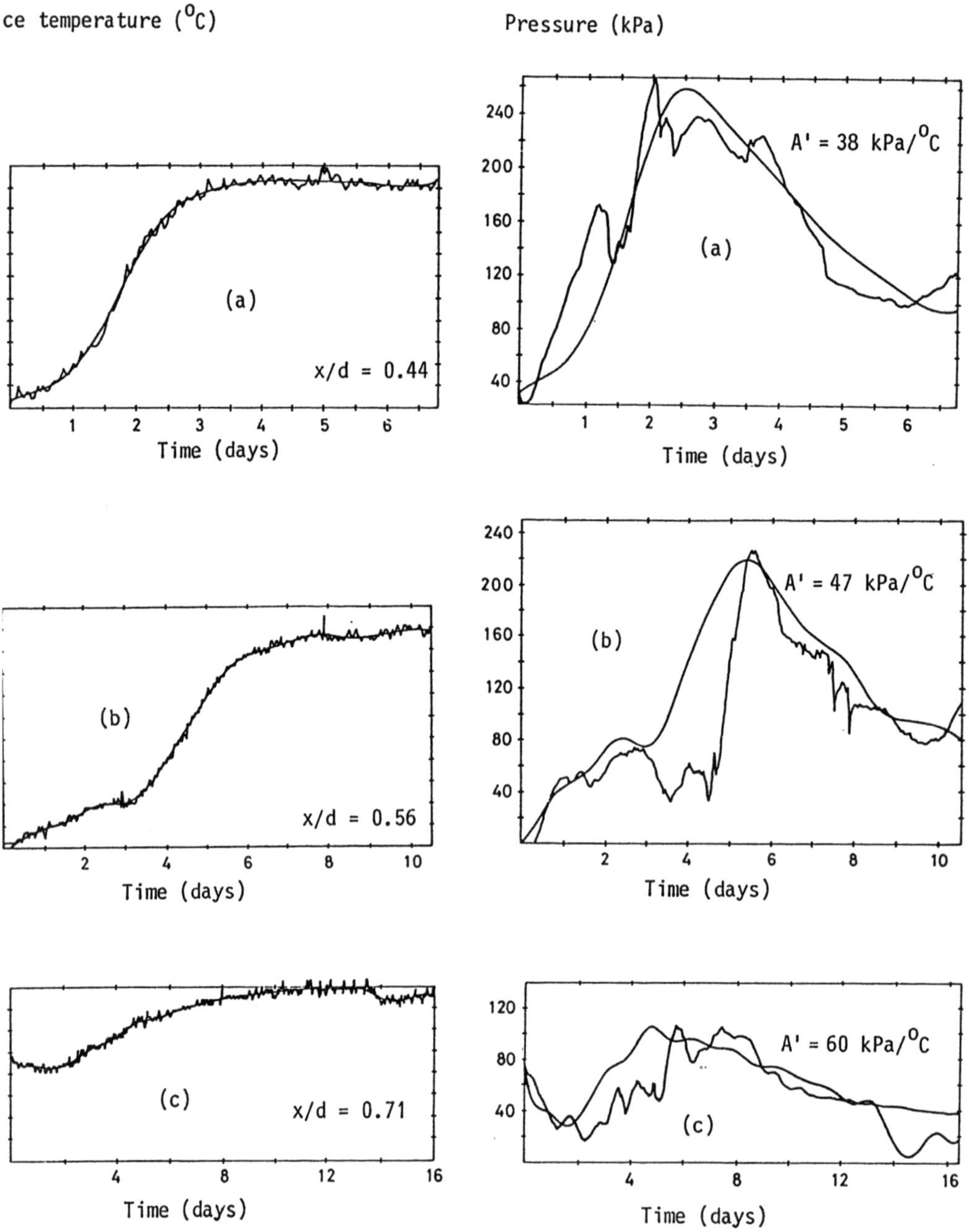

Figure 8. Adjustment with Equation (4) to measured ice pressure events in the Lule River. x/d is the ratio of sensing depth to ice thickness. A' is the elastic response due to measured temperature rise.

5. Wet Cracks

When a crack is formed early during a contraction, further contraction will open the crack bottom and water will penetrate into the crack up to the free water line, see Figure 9. These cracks refreeze, but usually they remain weaker than the surrounding ice. The next crack is therefore likely to form at a refrozen crack. This process can be repeated several times and this explains why compressive stresses usually dominate in an ice sheet.

Water level variation generates wet cracks with a predictable pattern along the shore-line and close to fixed structures. A typical shore-crack at different stages of development is shown in Figure 10.

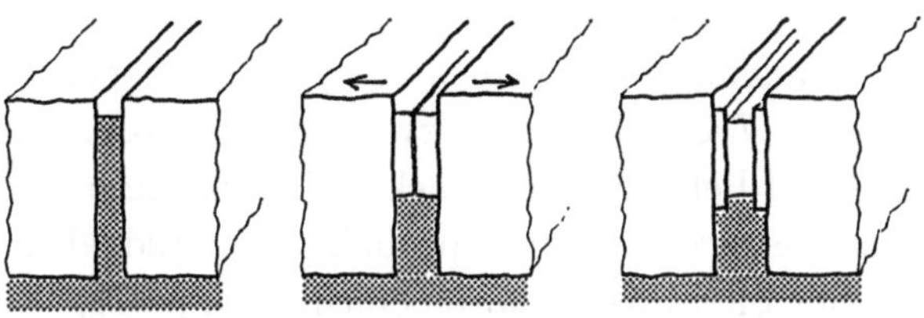

Figure 9. Ice growth in a wet crack.

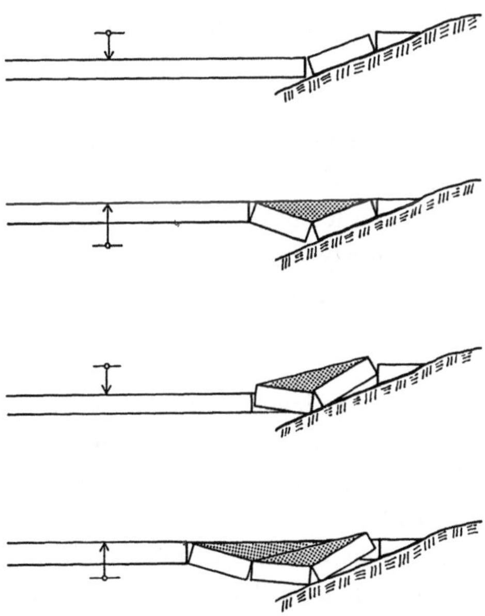

Figure 10. The development of bending cracks and growth of flooded ice area around the shore.

Cracks close to a vertical wall are created in a similar manner, but there the superimposed ice layer is limited only by the amplitude of the water level variation. Wet cracks in most shapes will probably be weak zones, but in connection with thermal expansion, wet cracks may increase the risk of high stresses. Flooding of cold ice creates a rapid expansion and the uplift of an expanded and curved ice sheet may also generate lateral forces, (Löfquist, 1989). The temperature rise, shown earlier in Figure 1b, was actually caused by flooding.

6. Discussion

The buildup of thermal stresses may vary from 130 to 290 kPa/°C measured in the same effectively confined ice sheet (see Figures 5 and 2). The response to a thermal expansion is sometimes delayed, corresponding to a free temperature rise of 7°C. These observations are probably connected with the existence and behaviour of cracks. The assumption that large volumes of ice behave as a continuum is probably violated when cracks are present. The presence of cracks could be considered partly by introducing a variable confinement of uncracked ice.

It is tempting to try to model large volumes of cracked ice by using discrete linear elastic elements. All time dependent deformations could simply be assumed to take place at the nodal point modelled with linkage elements.

However, such models cannot be applied without a better understanding of the shear deformation under cracks and the break-up of cracks. On a larger scale, bonds between large broken ice floes could be modelled by the same technique. This type of "boundary ice", refrozen snow and water mixed with crushed ice, must be studied separately with respect to mechanical properties.

In connection with in-situ tests on the flexural strength, Frederking and Svec (1981) have observed that in-situ cantilever beams often break at some distance from the root. When the ice is reinforced with a thin flooded layer, this rarely happens. With such experiments, the reduced flexural capacity of cracked ice can be studied.

During break-up of lakes and rivers, most cracks become weak and the ice sheet breaks up along these pre-existing cracks. Loads on bridge piers and harbours might be related to the maximum size of an unbroken ice floe when the ice is movable. Cracks do reduce the thermal loads on structures in a predictable manner. The reduction of dynamic ice loads from cracked ice might be better understood if it can be modelled numerically.

Acknowledgement

The research described in this paper was financially supported by the Swedish Council for Building Research.

References

Fransson, L. 1988. Thermal ice pressure on structures in ice covers. Doctoral Thesis, Dept. of Civil Eng., Luleå University of Technology, Luleå, Sweden. 161 p.

Frederking, R., and Svec, O. 1981. Cantilever beam tests in an ice cover: influence of plate effects at the root. Cold Reg. Sci. and Tech., No. 4, pp. 93-101.

Glen, J.W. 1953. The creep of polycrystalline ice. Proceedings of the Royal Society, Ser. A, Vol. 228, No. 1175, pp. 519-538.

Löfquist, B. 1989. Transverse static ice pressure on bridge piers. Proc. from 10th. Int. Conf. on Port and Ocean Eng. under Arctic Conditions (POAC'89), Luleå, Sweden, Vol. 1, pp. 568-574.

Metge, M. 1976. Thermal cracks in lake ice. Doctoral Thesis, Dept. of Civil Eng., Queen's University, Kingston, Ontario, Canada.

Odqvist, F.K.G. 1934. Plasticitetsteori och Tillämpningar (Theory of Plasticity with Applications), IVA, Stockholm.

Paterson, W.S.B. 1981. The Physics of Glaciers. 2nd. Ed. Pergamon Press, 380 p.

SESSION V:

Design Methodologies

Statistical Analysis of Ice Forces

T. J. O. Sanderson

Venture Research, BP International
London EC2Y 9BU, UK

Abstract Statistics offer a method for quantifying unmanageable complexity. During impact of large ice masses with structures, a vast number of local failure events take place. These local failure events may themselves be straightforward to understand, at least in principle, but understanding of their gross effect is an exercise in statistics. Simple statistical models based on non-simultaneous failure of discrete independent zones provide a realistic picture of observed force histories at, for instance, Hans Island (1980, 1981 and 1983). This is illustrated by an example numerical simulation using constant sized failure zones and stochastic occurrence of events. The treatment can be further generalised using fractal geometries. These models of non-simultaneous failure provide a plausible explanation of the exceptionally low forces measured at Hans Island. Observations since 1986 on the Molikpaq structure, however, show strikingly regular periodic force records, indicating that local zone failure may be coherent rather than stochastic and independent. The regularity of these records appears to be associated with structural response characteristics, but it is likely that pulverised ice at the contact zone plays an essential role in enabling apparently simultaneous failure to be observed over a wide front. The forces observed during such failure at Molikpaq are, nonetheless, low, and conform to the broad trend evident from the "pressure-area" diagram. Non-simultaneous failure may not be the appropriate explanation of the Molikpaq data, but the trend still apparently provides a basis for statistically based design load calculation.

1. Introduction

Suppose that a floating mass of sea ice, 3 km in diameter, 5 m in thickness and travelling at 1 m s^{-1}, impacts with a large fixed structure. Now, it is well known from countless laboratory studies that sea ice samples generally have an indentation "strength" of some 5-10 MPa, and it is therefore possible to calculate, in naive and simple fashion, what we might expect the total force during impact to be. As the floe impacts, ice crushes at the contact zone (with a failure pressure, let us say, of 5 MPa), and

S. Jones · R. F. McKenna · J. Tillotson · I. Jordaan (Eds.)
Ice-Structure Interaction
IUTAM-IAHR Symposium St. John's, Newfoundland, Canada

the contact zone grows rapidly as the floe progressively penetrates. The moving ice loses energy as the process continues, until eventually it comes to rest. A simulated total force history using this simplistic model is shown in Figure 1(a): the impact lasts just 4 s, and the total force rises monotonically to a maximum of 2250 MN.

But now look at Figure 1(b). This, of course, is the real thing: the trace of Figure 1(b) represents the total force history as a multiyear floe 3.2 km in diameter, 5.2 m in thickness and travelling at 0.88 m s^{-1} impacted with Hans Island in Kennedy Channel on 10th August 1983 (Danielewicz and Cornett, 1984).

Several features of these data alert us immediately to the suspicion that we are viewing a statistical process. Firstly, the trace is complex and irregular, not smooth, with force peaks occurring at apparently random

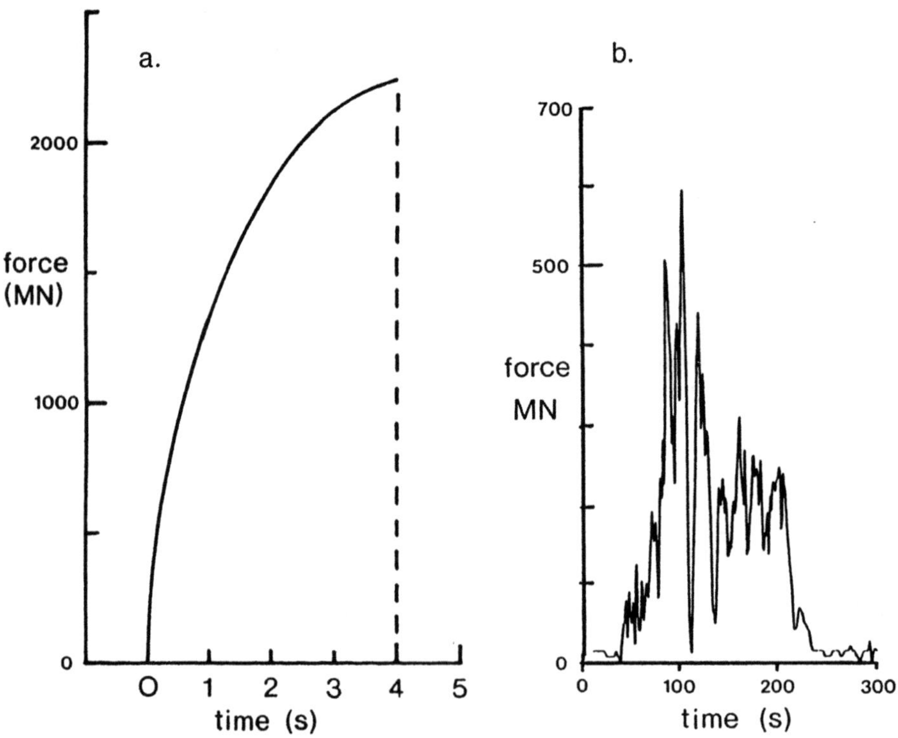

Figure 1. Force exerted during impact of a floe with a rigid obstacle, (a) simplistic theory based on crushing "strength" 5 MPa, (b) real data, from Event 10, Hans Island 1983 (after Danielewicz and Cornett, 1984). Floe is approximately 3 km in diameter, 5 m in thickness, travelling at 1 m s^{-1}. Island is 1 km diameter.

intervals. Secondly, the maximum force (600 MN) is very significantly less than expected by theory, and the impact process lasts much longer (several minutes). Thirdly, though not apparent from a single event, we should find that successive subsequent impacts with nominally identical starting conditions would give widely dissimilar force histories.

The importance of this observation — that full-scale ice impact cannot straightforwardly be understood by reference to laboratory tests — has not been lost on the ice community. Indeed, the story of the attempt to resolve the apparent inconsistency has essentially been the story of applied ice mechanics during the last decade.

2. Statistical Models

The central idea behind statistical models of ice failure is as follows. Ice is brittle, and on a local scale it does indeed fail at 5-10 MPa; however, when a large area of ice is in contact with a structure, the failure process is sporadic, local and non-simultaneous. The total force seen by the structure then becomes a statistical sum of a large number of localised failure events which are not correlated in time or space. This appealing picture accounts, in principle, for the two main observations cited above: firstly, the force trace is irregular, since from moment to moment a highly variable number of localised "zones" will actively be undergoing failure; and secondly, the total force is much reduced, because there is a very small probability that all zones will fail at once. It turns out that such statistical models are in some cases inappropriate. But before seeing how they may be wrong, let us see the many ways in which they are right.

There have been several formulations of the statistics of non-simultaneous failure, most notably Kry (1980), Slomski and Vivatrat (1983), Ashby et al. (1986) and Bhat (in press). Conceptually, that by Ashby et al. is the simplest to understand and the easiest to illustrate. The advancing front of impacting ice is represented as a set of discrete independent zones of dimensions $L_i \times L_i$, as shown in Figure 2. As each individual zone reaches the face of the structure it remains in contact while the ice advances a distance Δ_L and then fails cleanly at a peak stress $2P_L$ (P_L is then the average force during the contact process). As described elsewhere (Sanderson, 1988) the peak total load then shows realistic dependence on the total area of contact. Figure 3 shows a compilation of raw peak indentation pressures as a function of gross contact area for a wide range of scales of interaction, and the solid line shows

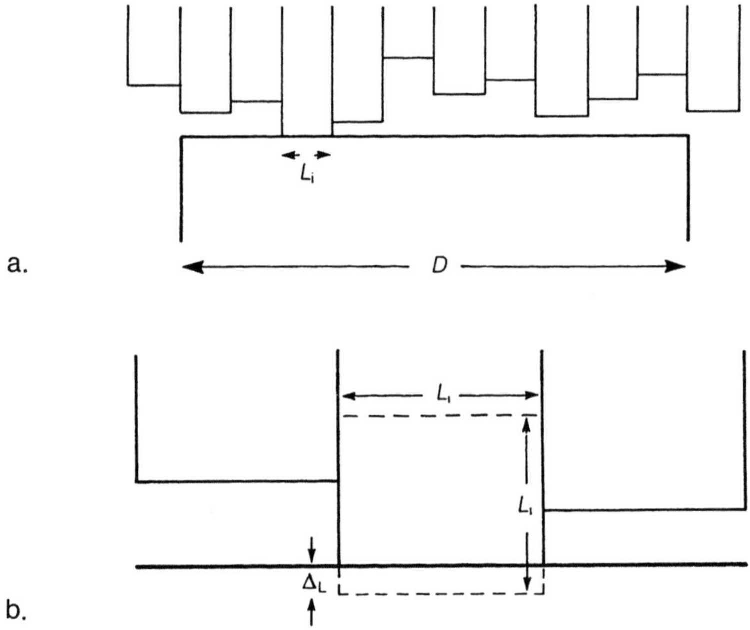

Figure 2. Conceptual model for analysis of non-simultaneous failure, (a) plan view, (b) detailed geometrical assumptions.

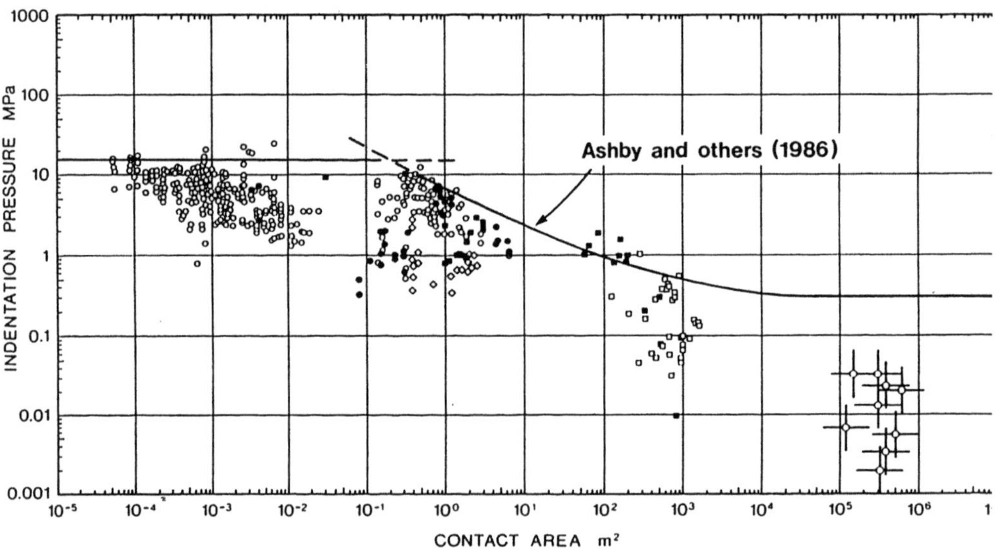

Figure 3. Theoretical curve of Ashby et al. (1986) constrained to the observed data of the pressure-area curve.

a calculation based on the statistics of Ashby et al. (1986) using the parameters L_i=1 m, P_L= 15 MN and Δ_L= 0.02 m, and assuming that the peak indentation pressure corresponds to the 3 standard deviation limit of a normal distribution. The agreement is far from perfect, but is plausible in form, showing an initial dependence of failure stress on the inverse square root of contact area.

What about the details of the impact process? Does the statistical approach account satisfactorily for the spikey characteristics of observed interaction histories (Figure 1(b)) as well as for the overall reduction in peak load? Apparently it does. Consider, for example, a model circular ice floe conceptually divided into independent zones of the form shown in Figure 4(a). Assume now that each zone has a local force history, as a function of penetration, as shown in Figure 4(b): ice fails at a stress σ_c, and once failed, is cleared from the failure zone with a "clearing stress" σ_d. The stress peaks in this idealization are assumed to be a stochastic process, with an equal chance of lying anywhere in the range 0 to $2L$. A concrete statistical realisation of such a model process is straightforward to carry out, using a random number generating programme, and the

a.

b.

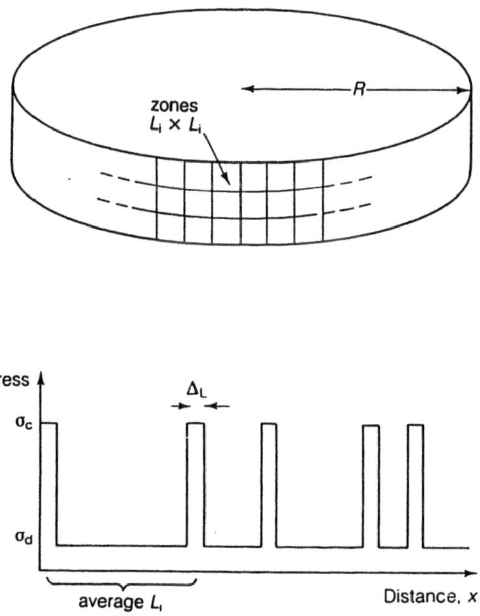

Figure 4. Modelling an impact event, (a) conceptual division of floe into discrete zones, (b) model stress history of each zone during impact.

result of such a simulation is shown in Figure 5. This simulation uses a floe of radius $R = 1000$ m, thickness $h = 10$ m, initial velocity $U = 0.5$ m s^{-1}, zonal failure strength $\sigma_c = 3$ MPa and clearing stress nominally 0.05 MPa. Further details are given in Sanderson (1988). The purpose here is simply qualitative and illustrative.

The model statistical stress history of Figure 5 has resemblance in form to that of Figure 1(b). Certainly the result is far superior to the simplistic model of Figure 1(a). The model is, however, dissatisfyingly arbitrary in many respects, and it sacrifices realism to conceptual neatness. Ice is not, of course, nicely divided into cubic zones of beautifully sculpted regularity. And why have we chosen a zone size of 1 m, or 3 m (or indeed any other size)? The ice at the failure zone of a real impact zone ranges in particulate size from sheer dust of pulverisation all the way

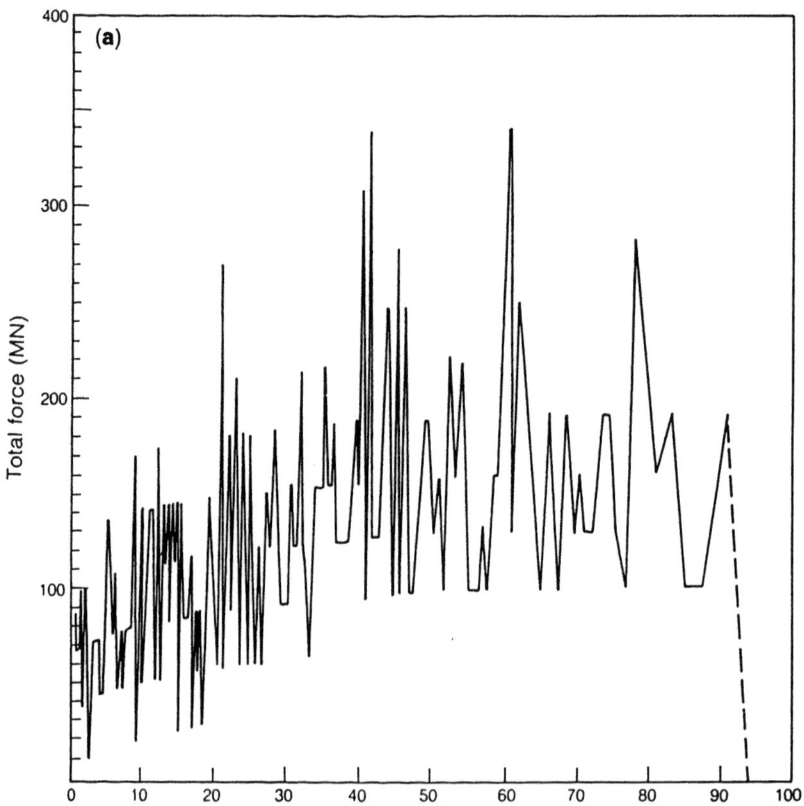

Figure 5. Force trace generated by simple numerical simulations.

to massive blocks of full through-thickness. Recently, Bhat (in press) has presented an elegant formulation of statistical failure which uses the powerful technique of fractal geometry, and which thereby removes at least one arbitrary dimension of the model so far described.

Fractal geometry is of particular use in describing surfaces and forms which have statistical irregularity, or roughness, on all scales. The classic example (Mandelbrot, 1982) is the cartographer's nightmare of how to map a coastline — however large the scale and however fine the detail, there is always a further level of detail which is too fine to map. A good example from the field of ice morphology is that of ice floe size distribution: Rothrock and Thorndike (1984) gave compelling illustrations that the statistical characteristics of floe size distributions are independent of the spatial scale at which they are viewed.

The same may be done for the irregular perimeter of an impacting floe. Figure 6 shows a two-dimensional fractal model of uneven contact geometry (Bhat, in press). The geometry consists of a series of generations of successively subdivided zones: a zone of width w in fact consists of N sub-zones of width $\lambda w/N$ (in Figure 6, we have $N = 2$ for simplicity). The sub-zone is itself divided into a further N zones of width $\frac{\lambda}{N}(\lambda/N)$, and so on. The parameter λ is a number between 0 and 1, characterising the openness of the roughness structure. In its infinite extension, such successive subdivision leads to what is termed a "fractal dust" with infinitesimal point-like contacts.

The importance of fractal theory to modelling the failure process is that it allows us to write a simple recursive relationship between the failure strengths of successive generations of subdivision. The zone of order $r + 1$ in Figure 6 can be be considered to have failed when both subdivisions of order r are failed.

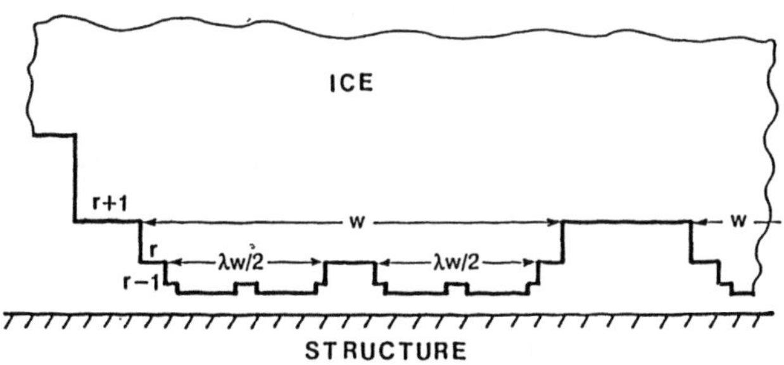

Figure 6. A fractal model of uneven contact geometry (after Bhat, in press).

Formally, if we denote the failure strength of the rth zone by σ_r, we can write, following Bhat:

$$\sigma_{(r+1)} = \lambda W/N \cdot \sigma_r \cdot N/W = \lambda \sigma_r. \quad (1)$$

We can also express the factor λ in terms of contact area A_r by the relationship:

$$\lambda = NA_r/A_{(r+1)}. \quad (2)$$

This means that we can proceed rapidly to a power-law expression of average failure stress in terms of gross contact area as:

$$\sigma_r \propto (A_r)^{-\alpha}. \quad (3)$$

where the exponent α is given by

$$\alpha = \frac{ln\lambda}{ln\lambda - lnN} \quad (4)$$

In Figure 7, Bhat has fitted such a relationship to the mean trend of the data of the pressure-area curve, using values of, for instance, $N = 2$

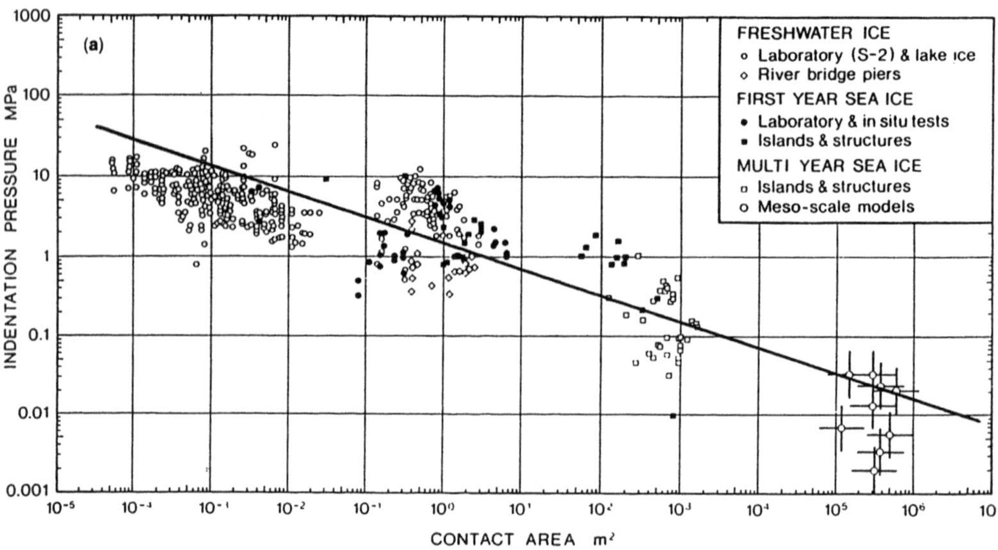

Figure 7. A fractal model fitted to the pressure-area data, using $\alpha = \frac{1}{3}$.

and $\lambda = 0.7$, which yield a value of $\alpha = \frac{1}{3}$. A slope of $A^{-1/2}$, that is $\alpha = \frac{1}{2}$, might be achieved by, for instance, $N = 2$ and $\lambda = 0.5$. These values are all plausible, but of course highly non-unique: all we are looking at here is a mechanical idealisation which explains the form of the data by a satisfactory mechanism, and we are not at this stage looking for a deductive tool. Notice that the introduction of self-similar fractal geometry, as opposed to just one generation of finite-sized roughness, leads to a single straight-lined power law (Figure 7) which accords better with the data than does the asymptotic behaviour of the curve shown in Figure 3.

3. Simultaneous Failure

Plausibility is not necessarily truth. The fact that statistical models account plausibly for a number of features of observed interactions does not necessarily mean that they are true or universal.

Evidence collected over the last 3 years from Gulf's Molikpaq structure (Jefferies and Wright, 1988) suggests that we cannot always rely on the non-simultaneous independent failure events which lie at the heart of the statistical failure model I have discussed here. A sample set of data is shown in Figure 8, representing loads monitored at three separate points of one caisson face of the structure. Two features are immediately apparent. Firstly the loads show a striking cyclicity: the load experienced by a given zone goes through a regular cycle of loading and (partial) unloading, with a period which in this case is approximately $3\frac{1}{2}$ s. This contrasts with the stochastic time series postulated by the model illustrated in Figure 4(b). Secondly (and this is a distinct additional feature) all three zones monitored appear to go through cycles of loading which are coherent and simultaneous. Their failure is, in the words of Jefferies and Wright, phase-locked.

I do not propose to comment here on the regular cyclicity of loading observed. Few data on load magnitudes and platform response are available in the public domain, and the subject has already been well-addressed by Jefferies and Wright (1988), Hallam and Pickering (1988), and Jordaan and Timco (1988). A feature that is of great interest, however, is the apparent simultaneity of failure.

Now, the ice failure mechanism observed at Molikpaq during such interaction events was almost invariably pure crushing: ice at the contact zone was pulverized to a fine powder and was then extruded upwards and (presumably) downwards. It seems that the layer of finely pulverized ice

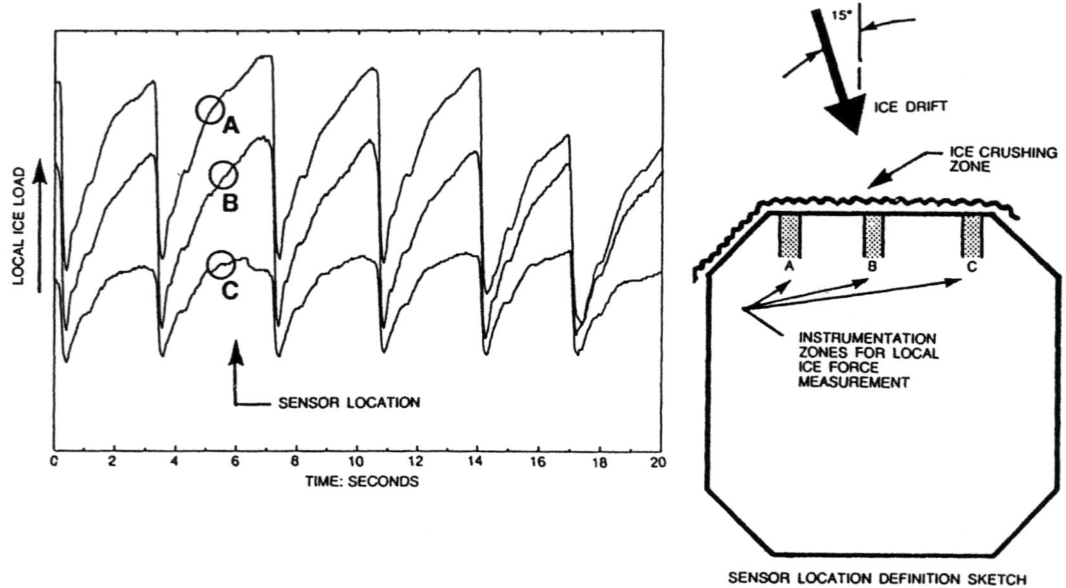

Figure 8. Phase locked loading at Molikpaq, showing simultaneous cyclic pressures at 3 separate sensor locations A, B, C (after Jefferies and Wright, 1988).

which exists at the interface between the structure and the advancing undamaged ice may itself be playing a vital role in enabling simultaneous failure to occur. It appears to be acting as a kind of cushion which redistributes contact stress and avoids the local stress concentrations which are essential to the non-simultaneous failure models described above in Section 2.

An effect of this sort has already been observed elsewhere during the course of some elegant experiments carried out by Ashby et al. (1986) on the non-simultaneous failure of zirconia foam. Zirconia foam is an open-celled brittle foam with a structure illustrated schematically in Figure 9. When it is indented by a rigid indentor, individual cell edges successively contact the indentor, snap off, and then clear efficiently by simply falling through the interstices of the foam. This corresponds closely to the simple mechanical model discussed in Section 2, and zirconia foam does indeed show a simple inverse square-root pressure-area dependence when indented by indentors of varying areas. A plot of peak indentation stresses

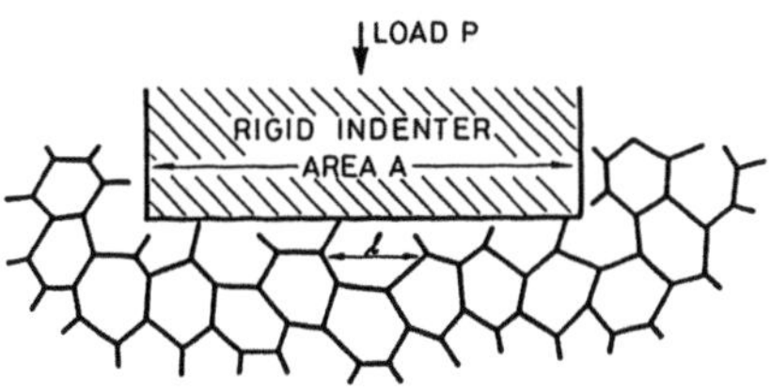

Figure 9. Schematic diagram of indentation of brittle zirconia foam (after Gibson and Ashby, 1988).

as a function of indentor area is shown in Figure 10. An interesting further experiment was, however, also carried out (Gibson and Ashby, 1988), but is not so widely reported: compliant facings, such as sheet rubber, were inserted between the rigid indentor and the zirconia foam (Figure 11(a)). The effect was dramatic: no longer was non-simultaneous failure observed, but, instead, the full maximum theoretical crushing strength was observed, since all cell walls which were even approximately in contact became more evenly stressed through the medium of the compliant facing.

This may very well be an important process occurring also in ice interactions under certain scenarios. Crushed, extruding ice may behave as an almost Newtonian fluid, (Jordaan and Timco, 1988) and under conditions

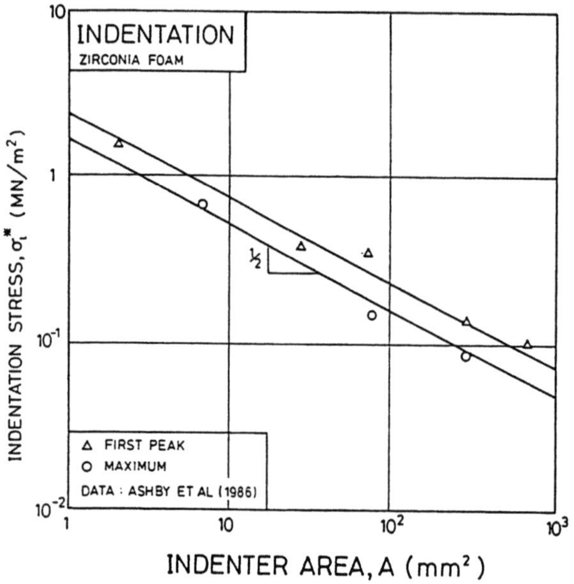

Figure 10. Data for the dependence of indentation stress on indentor area for brittle zirconia foam.

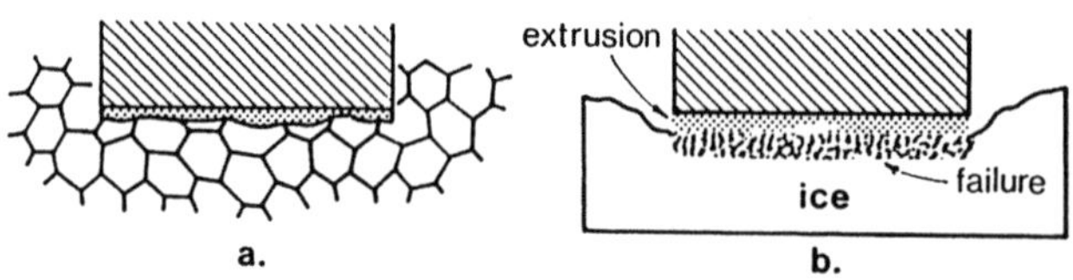

Figure 11. The effect of a cushioning compliant layer, (a) for zirconia foam, (b) the possible role of crushed ice during the ice-structure interaction process.

of sufficiently great ice thickness and sufficiently high rate of extension, the contact layer of pulverized ice may perhaps efficiently transmit an evenly distributed load high enough to fracture the advancing ice before true direct contact is made.

In practice, it is doubtful whether 100% simultaneous failure is ever actually achieved by such a mechanism, for if it were, we should surely expect to see global average pressures exceeding 5-10 MPa on those structures which have undergone such simultaneous failure. This is not the case: global pressures measured on Molikpaq at Tarsiut P-45 and Amauligak I-65 were generally less than 1 MPa, seldom exceeded 1.5 MPa, and never exceeded 2 MPa (see Sanderson, 1988, for data compilation). So, although some degree of simultaneity does seem to be introduced by the medium of the "compliant" crushed ice (and perhaps by the compliant structure as well), the simultaneity is partial, not total, and does not lead to the synchronized failure at full crushing strength which all ice engineers dread.

4. The Pressure-Area Diagram

Here and elsewhere (Sanderson, 1986, 1988) I have made much of the statistical trends evident in the pressure-area diagram for ice, and I have attributed this largely to the phenomenon of non-simultaneous failure. This still appears to be true under most circumstances, but it does now appear that non-simultaneous failure cannot be relied upon as the correct mechanistic explanation of the trend displayed by the Molikpaq data. However, it is worth making two points clear:

1. The pressure-area diagram is nothing more than a raw compilation of plotted data, and is quite independent of any explanations which may be advanced for its form.

2. The pressure-area diagram as shown here and elsewhere already <u>includes</u> the data publicly available for failure events which are described as "simultaneous" (that is, not non-simultaneous). Design methodologies based on the statistics of the curve continue to be valid.

It is sometimes suggested that the pressure-area diagram may be "wrong". A simple plot of data cannot be wrong (unless, of course, the data are faulty), though the plot can always be added to. Scientists who have new data should add them to the plot, and observe any changes they may make to the trend. The data may also be analysed further. It is particularly appropriate to ask whether there may be a family of pressure-area curves for different conditions (or structures of different compliance) which in aggregate give rise to the wide scatter of the pressure-area diagram.

Acknowledgement

I thank The British Petroleum Company plc for permission to publish this paper.

References

Ashby, M.F., Palmer, A.C., Thouless, M., Goodman, D.J., Howard, M.W., Hallam, S.D., Murrell, S.A.F., Jones, N., Sanderson, T.J.O., and Ponter, A.R.S. 1986. Non-simultaneous failure and ice loads on arctic structures. In Offshore Technology Conference 1986, Houston, Paper No. OTC 5127, pp. 399-404.

Bhat, S.U. (in press). Modelling of size effect in ice mechanics using fractal concepts (submitted to Journal of Offshore Mechanics and Arctic Engineering).

Danielewicz, B.W. and Cornett, S. 1984. Ice forces on Hans Island, 1983. APOA Project No. 202.

Gibson, L.J. and Ashby, M.F. 1988. Cellular solids: structure and properties. Oxford, Pergamon Press, 357 p.

Hallam, S.D. and Pickering, J.G. 1988. Modelling of dynamic ice loading of offshore arctic structures. In Polar Tech 88, Proceedings from the International Conference on Technology for Polar Areas, Trondheim, June 1988, Vol. 1, pp. 235-248.

Jefferies, M.G. and Wright, W.H. 1988. Dynamic response of "Molikpaq" to ice-structure interaction. In OMAE 1988, Proceedings of the 7th International Conference on Offshore Mechanics and Arctic Engineering, Houston, Vol. 4, pp. 201-220.

Jordaan, I.J. and Timco, G.W. 1988. Dynamics of the ice-crushing process. Journal of Glaciology, Vol. 34, No. 118, pp. 318-326.

Kry, P.R. 1980. A statistical prediction of effective ice crushing stresses on wide structures. In Proceedings of the ASCE Convention and Exposition, Portland, 14th-18th April 1980, Preprint No. 80-084.

Mandelbrot, B. 1982. The fractal geometry of nature. San Francisco, California, W.H. Freeman.

Rothrock, D.A. and Thorndike, A.S. 1984. Measuring the floe size distribution. Journal of Geophysical Research, Vol. 89 (C4), pp. 6477-6486.

Sanderson, T.J.O. 1986. A pressure-area curve for ice. In IAHR 86, Proceedings of the 8th International Association for Hydraulic Research Symposium on Ice, Iowa City, Vol.2, pp. 361-384.

Sanderson, T.J.O. 1988. Ice mechanics: risks to offshore structures. London, Graham and Trotman, 253 p.

Slomski, S. and Vivatrat, V. 1983. Selection of design ice pressures and application to impact load prediction. In POAC 83. Proceedings of the 7th International Conference on Port and Ocean Engineering under Arctic Conditions, 1983, Helsinki, Vol. 2, pp. 909-919.

Discussion

Q: K. Croasdale, Esso Resources Canada Ltd.

The pressure area curve developed by the authors, is a very useful tool for the designer because it displays real data. It also demonstrates the large differences between small-scale strength and real-world ice pressure. I agree however that changes in failure mode with area (width) need to be appreciated (and ideally understood). It is possible that different pressure-area (or pressure-width) relationships apply across regions of similar failure models. Would the author comment on this issue?

A: T.J.O. Sanderson

I thoroughly agree with this view. The spread of data in the pressure-area diagram is very wide, and it would be most disappointing if that was as far as we could go. I would very much like to see the rational development of a family of curves relating to different loading scenarios. This would substantially reduce the current great (though quantified) uncertainty.

Q: R. Weiss, Exxon Production Research

1. In your book, you admit the possibility that the pressure-area data which you have plotted could also be shown as demonstrating a relationship between pressure and structure diameter divided by thickness. Have you now rejected this as a possibility?

2. It would seem to me that the possibility could not be disregarded that the apparent decline of nominal ice pressure with increasing area indicated by the data which have been plotted are solely due to the selection and availability of data. At small areas, only perfect ice has been tested and only peak events have been shown. These indicate high pressures. At large areas only random unselected samples have provided data and those samples also have high diameter to thickness ratios. Thus due to the absence of selectivity possible with the small area samples and to the lack of confinement inherent in the high D/T ratios, the large area data show much lower pressures.

A: T.J.O. Sanderson

1. No, I have not rejected this possibility: the data do not currently allow a distinction, since we have no large area, low aspect ratio tests. (Full-scale iceberg tests might fill this gap). The distinction

depends, from a theoretical standpoint, on what determines the size of a local failure zone - if the zones are of order thickness h, we would expect pressure to be a function of D/h; if however they are of invariant size, we would expect pressure to be a function of area Dh.

2. I do not think that this is the explanation, since even sets of data acquired under consistent selection conditions display the effect very strongly. Consider, for example, icebreaker data from the USCGS Polar Sea (St. John and Daley, 1984; and p.140 of my book). Here, impact pressures measured over areas between $0.16m^2$ and $8m^2$ showed a systematic decline by a factor of about 20. No systematic selection procedure changes were involved; furthermore the tests are essentially all low aspect ratio.

Q: I.J. Jordaan, Memorial University of Newfoundland

The pressure-area relationship in the crushing mode is not necessarily entirely related to crushing. For instance, in spherical indentation there is considerable spalling around the edges and a similar effect can occur in ice sheets. The crushing could then occur at quite high stress, although both crushing and spalling would include non-simulta-neous aspects. Could you please comment?

A: T.J.O. Sanderson

I agree that crushing is by no means the only important failure mode. What is remarkable is that the pressure-area relationship seems to hold true even if a change in failure mode occurs. Any local failure mode (crushing, spalling, flexural failure) is likely to be subject to non-simultaneous effects, though global modes of failure (creep and perhaps gross radial cracking) behave differently.

Q: M. Metge, CANATEC

I was very interested in the compliant platen work of Ashby. In ice interaction with the Molikpaq, the compliance can be in the structure, the ice and/or the crushed ice, i.e. there could be a lot of compliance.

The "simultaneous" failure observed on the Molikpaq is not limited to the Molikpaq, similar behaviour has been observed on the SSDC and on the CIDS.

A: T.J.O. Sanderson

Thank-you; I was not aware of simultaneous failure observations at SSDC and CIDS. Certainly I do not think there are data available on simultaneous measured stresses at these structures.

Q: D.S. Sodhi, USA CRREL

The role of failure modes can, to some extent, explain the decrease of effective pressure with increase in contact area. For instance, the pressure for geophysical scales (10 km) used in the drift models of arctic ice cover is derived on the basis of potential energy required to pile broken ice in the sails and keels of pressure ridges, resulting in very low pressure estimates.

A: T.J.O. Sanderson

The work on geophysical scale ice strength is not intimately tied to a particular failure mode. The "strength" quoted is simply a parameter which has to be inserted into meso-scale models of Arctic ice circulation, convergence and dilatation in order to make the models fit observations. I agree that failure modes may help explain the statistical variability we see in the pressure-area diagram, but this has not yet been satisfactorily achieved.

Q: S. Shyam Sunder, MIT

Three specific comments:

1. There is an uncertainty of about two orders of magnitude in the indentation pressure at any given contact area. A better understanding of physical mechanisms such as failure mode, indentor velocity, ice thickness and confinement will lead to reduced uncertainties and less conservative design.

2. Winkler presented a paper at POAC '89 where, for a given contact area, he showed data for two platforms one with high pressures (corresponding to lower velocities) and the other with lower pressures (corresponding to higher velocities). This seems to suggest that these might be a strong dependence of indentation pressure on velocity at a given area.

3. Finally, with regard to the Dorris plasticity model presented yesterday. In the case of initial contact, where the entire ice sheet is in

virgin form it is possible for the ice to go through the peak stress before softening to the residual stress. What are your thoughts on these issues?

A: T.J.O. Sanderson

1. I agree that a narrow statistical distribution is highly desirable, and that more detailed understanding of failure mechanisms might allow subdivision of the data-set into smaller groups for specific scenarios or conditions. In particular, we might imagine a family of pressure-area curves (for multi-year, first-year ice, dynamic interaction, compliant structures, etc.). At present there are insufficient data to permit this - also, I think, insufficient understanding.

2. The pressure-area diagram is a statistical plot, and it is easy to find apparent counterexamples to its generality. But if I assert, for instance, that men are generally taller than women, I cannot be disproven by the discovery that one particular woman is taller than one particular man. Having said that, I fully accept that the pressure-area diagram may conceal many hidden explainable deviations of this type.

3. With perfect initial contact conditions, yes, you are right. In reality one might expect initial poor contact leading to non-simultaneous failure, followed by full-width coordinated crushing of strain-damaged ice.

Q: J. Lever, USA CRREL

Do you think that decreasing average ice pressure versus contact area holds when only one failure mode is present? The Pond Inlet indentation tests for iceberg ice in crushing show a plateau at larger contact areas. How important an effect is change in failure mode on pressure-area curve?

A: T.J.O. Sanderson

Yes, if the failure mode is one which admits non-simultaneous failure, then certainly I would expect to see a decreasing average ice pressure with contact area. I believe the apparent plateau in the Pond Inlet data to be illusory: at large areas the indentor has penetrated deeply and is subject to increasing triaxial confinement. I would not expect such behaviour if tests were conducted with a series of flat indentors of increasing size. In answer to your second question - yes, failure mode changes may well explain parts of the pressure-area diagram. This diagram does not depend on explanations, however, it is simply a compilation of raw data. The influence of failure mode changes is not at this stage very clear.

Q: J. Dorris, Shell Development

There seems to be some confusion about the term "size" effect in ice mechanics. Traditionally size effects have been related to a decrease in strength with an increase in material volume. This phenomenon is attributed to an intrinsic flaw distribution associated with the material. In ice mechanics plots of contact and indentation pressure versus contact area show a large decrease in pressure with increasing area. This decrease is sometimes referred to as a size effect. However, I believe this decrease should really be associated with a change in the state of stress from plane strain to plane stress. Would you please comment?

A: T.J.O. Sanderson

The term "size effect" is indeed potentially misleading. The pressure-area diagram is no more than a raw plot of data, and a true "size effect" is only one possible explanation. Alternatively, it may be an aspect ratio effect, a stress-state effect, a strain history effect, etc. These are all explanations — and the pressure-area diagram does not depend on these for its validity. I do not believe it is simply a question of plane strain vs. plane stress — the Hans Island data, at aspect ratios about 10-50, show very much lower average pressures than all comparable small-scale tests at the same aspect ratio. Incidentally, I do not agree with comments made elsewhere during this discussion that "size" and "scale" are two distinct things. To me they are the same — though "shape" is indeed a distinct concept.

Probabilistic Determination of Iceberg Collision Design Loads for Floating Production Vessels

M. K. Fuglem and D. Duthinh

Centre for Cold Ocean Resources Engineering,
Memorial University of Newfoundland
St. John's, Newfoundland, Canada A1B 3X5

J. H. Lever

U.S. Army Cold Regions Research and Engineering Laboratory
Hanover, New Hampshire, USA

I. Jordaan

Faculty of Engineering and Applied Science
Memorial University of Newfoundland
St. John's, Newfoundland, Canada A1B 3X5

Abstract A probabilistic method is presented for estimating global iceberg collision design loads for floating production vessels. Preliminary design loads are estimated for a concrete semi-submersible at the Terra-Nova site on the north-east part of the Grand Banks. A large number of simulated iceberg collision loads are determined using Monte Carlo techniques to choose the required input parameters (such as iceberg size and collision velocity) from estimated distributions. Using the probability distribution of these loads and the estimated frequency of iceberg collisions, design loads corresponding to probabilities of exceedance of 10^{-3} and 10^{-4} per year are estimated. The results are presented as curves giving the design load as a function of the success of the operators in detecting and avoiding icebergs.

1. Introduction

The possibility of icebergs colliding with offshore vessels or platforms must be considered when designing facilities to produce oil on the Grand

S. Jones · R. F. McKenna · J. Tillotson · I. Jordaan (Eds.)
Ice-Structure Interaction
IUTAM-IAHR Symposium St. John's, Newfoundland, Canada

Banks off Canada's east coast. The use of floating semi-submersible systems designed to move off location if a threatening iceberg cannot be successfully deflected away is one solution being considered. Since it is not always possible to detect small icebergs in storm conditions, adequate strengthening for collisions is required. The iceberg collision design loads used for different parts of the vessel should reflect the probability of the loads being exceeded, the consequences if they are exceeded, and the degree of risk which society is willing to accept. It was felt that a probabilistic method was the most rational approach to use given the variety and random nature of the iceberg types and the environmental conditions that could be encountered.

In this paper, preliminary global design loads are determined for the NEKTON 8000 FPV, an ice strengthened concrete semi-submersible proposed by Bouygues Offshore (Figure 1). The first step is to estimate the frequency of collisions and the probability distributions for the loads acting on the particular members being considered. The factors which determine the load distributions include: 1) the areal density of icebergs of different sizes and shapes; 2) the influence of waves, currents and other environmental factors on the velocities of the icebergs and vessel and on the operator's ability to detect the icebergs and avoid a collision; 3) the failure mode and strength of the ice; and 4) the structure's response.

Unlike wave loading events, iceberg collisions are relatively rare. For example, we estimate that for the area of study there would be a collision on average only every 5 years if no ice management system were used. With ice management the return period would be significantly reduced. In terms of avoiding collisions and the magnitude of the collision load, the most critical situation occurs when an iceberg approaches the platform during high seas. This is because detection of the iceberg may become difficult and the iceberg drift and wave induced velocities are increased. Also, the ability of the operator to tow the iceberg or move off location may be impeded.

While the probability of damaging collisions increases with the sea state severity, the probability of occurrence of high sea states decreases rapidly as the severity increases. For example, while sea states of greater than 15 meters significant height can occur in the area of study, only 4 percent of the time is the sea state greater than 5 meters. Therefore in determining iceberg collision return periods and load distributions, the decreased probability of detection and the increased motions of the

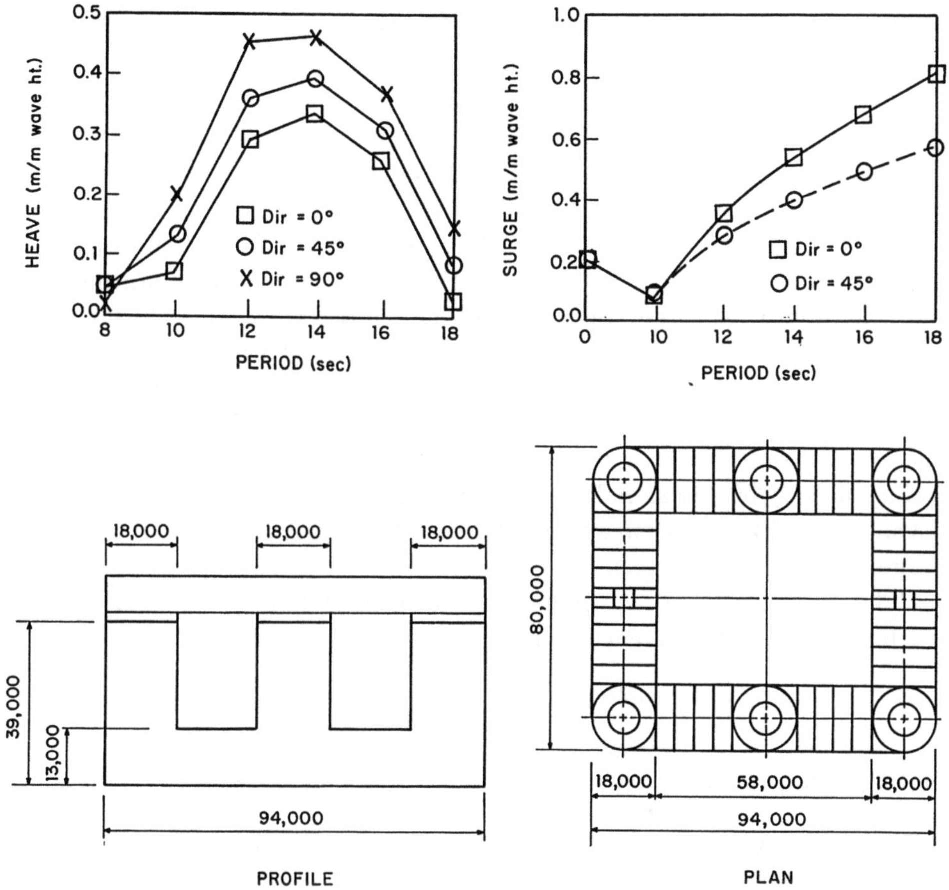

Figure 1. Characteristics of the Nekton 8000 FPV.

icebergs in sea states of increased severity must be weighed against the reduced probability of occurrence of these sea states.

The second step is to choose appropiate values from the load distributions so that the probability of failure is acceptably low. The CSA Standard S471 (CSA, 1989) recommends that the design loads be determined based on specified maximum allowed probabilities of exceedance per year. Different probability limits are used for different failure events depending on the consequences of failure. For instance, if the failure could result in severe consequences, the maximum allowed probability of exceedance of load is set very low, for example 10^{-4} per year. Failure events that result in damage without danger to lives or the environment could have higher maximum allowed probabilities of exceedance.

The use of limiting probabilities of exceedance is more appropriate for iceberg collision loads than traditional methods used for wave loads such as the 50 or 100 year load, even when these methods are combined with a load factor such as 1.3-1.4. This is because iceberg collisions are relatively rare and there is a large variance in the collision load when a collision does occur. For example, in the Hibernia Environmental Impact Statement, the maximum expected wave heights associated with 50 and 100 year return periods were estimated to be 28.0 m and 29.6 m respectively. If the structure is designed for the 50 year wave with a load factor and a 100 year wave is encountered, the structure will most likely survive. For iceberg collisions, with successful ice management there may not even be a collision every 50 years, making the concept of a 50 year load meaningless. Even the difference between the 500 and 1000 year load would be very large.

The overall method as applied to the NEKTON 8000 FPV is briefly outlined here with reference to Figure 2:

1. the iceberg and sea conditions are determined,
2. an iceberg management model is applied to determine how many collisions are avoided by towing or moving off location,
3. the frequency of collisions and the size distribution of colliding icebergs for each sea state are determined,
4. the probability distribution for the maximum load during a collision is determined using Monte Carlo simulation, and
5. the collision load probability distribution and collision probability are combined to get the probability distribution for the maximum load per year, from which the design load can be chosen.

Monte Carlo simulations are run separately for each sea state, then the resulting load distributions are weighted by the sea state probabilities and summed. Thus, fewer simulations are used without loss of accuracy since the number of simulated collisions in lower sea states, which occur frequently but do not result in high loads, is reduced.

Example global design loads are given corresponding to maximum allowable probabilities of exceedance per year of one in one thousand and

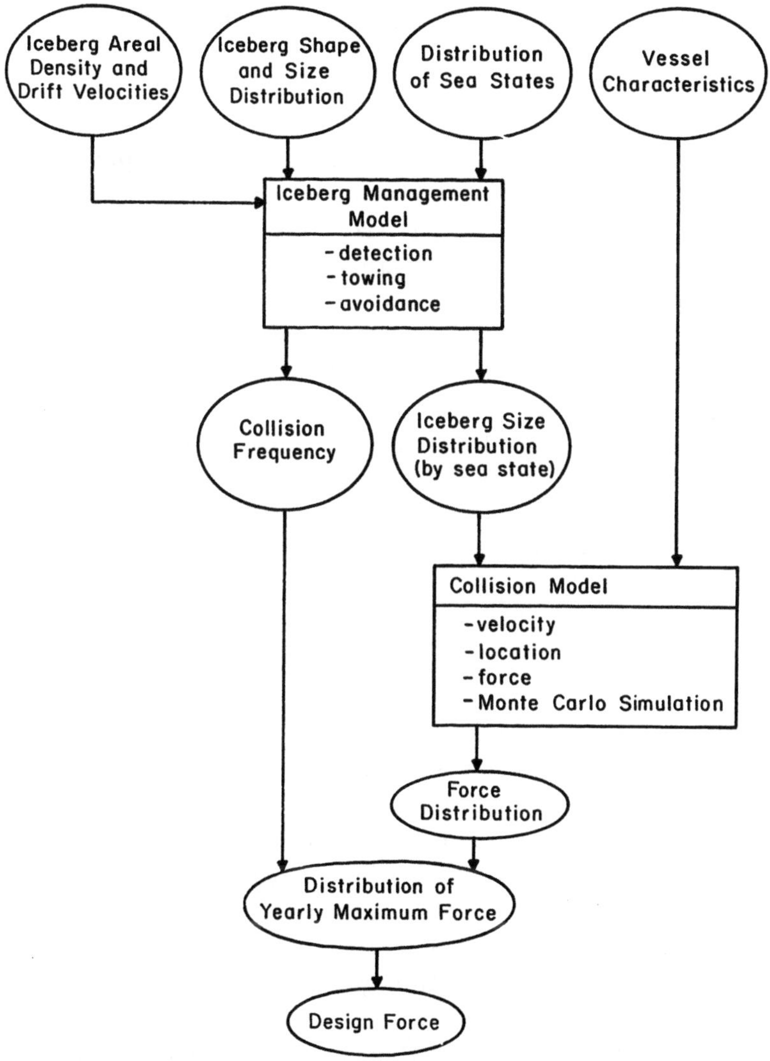

Figure 2. Steps for determining iceberg collision loads.

one in ten thousand. The results are only preliminary because further definition of the probability of detecting icebergs and the location of impact is required. Emphasis is placed on determining the environmental inputs; the modes of failure of the structure and the consequences are left for the designer to consider. For example, forces during collisions fluctuate rapidly as spalling and clearing of crushed ice occur, causing the highest loads to be of short duration. The degree of damage the vessel can withstand without significant danger must be considered. Because the design of the structure affects the modes of failure, several iterations may be needed in the design process.

A number of related studies are of interest because they use different approaches. Lindberg and Anderson (1987) estimate return periods of various damage scenarios to different steel semi-submersible designs based on estimated encounter rates for 4 iceberg size classes, 3 sea states, and 3 ice strengths. Iceberg and vessel motions are determined analytically and combined using superposition. The response of the impacted structural members are considered. Christensen (1989) shows how iceberg collision forces for fixed structures can be determined by explicitly integrating over the probability distributions for the input parameters. Isaacson and McTaggart (1989) determine design iceberg collision kinetic energies, for Hibernia, using a second order reliability method. They model the icebergs as vertical cylinders and determine their motions analytically. Putot et al. (1988) deal with various aspects of damage to concrete and steel structures under iceberg impact.

2. Iceberg Data

Icebergs reaching the Grand Banks are initially calved off ice shelves in Greenland and northern Canada and then drift southward with the Labrador current, typically taking 2 to 3 years to reach the Banks. North of the Grand Banks, the main part of the Labrador current swings east, taking most of the icebergs with it. Some icebergs reach the Banks due to current eddies and storm driven currents and winds. Because the Banks are relatively shallow (95 meters versus depths further north greater than 200 meters) many of the larger icebergs ground near the edge of the Banks.

Gathering iceberg data on the Grand Banks is both difficult and expensive, and there are still uncertainties in some parameters. One problem is the large variations in the number of icebergs from year to year and in different locations. This requires long term records; however the means of detecting and measuring icebergs have changed over time. Another problem is the inability to detect small icebergs in waves, even using modern techniques such as side looking airborne radar. Two main sources of data on icebergs are available. First, the International Ice Patrol (IIP) records the sighting of icebergs from a number of sources (including dedicated IIP overflights) with the objective of warning marine shipping of the presence of icebergs. Second, oil companies have conducted a number of studies in support of drilling operations, including measuring above water dimensions of icebergs and recording iceberg drift motions.

Iceberg shape and size influence (i) the frequency of collisions (because larger icebergs are more likely to collide with the vessel than smaller ones), (ii) the probability of detection (through iceberg projected areas), (iii) the collision velocities (smaller bergs are more wave-driven than larger bergs), and (iv) the collision forces (because of iceberg shape at the location of impact and the overall shape which affects rotation). We assume a spherical shape when determining iceberg wave induced motions and impact contact areas because 1) iceberg shapes are complex and some simplification is necessary 2) very few underwater profiles of icebergs are publicly available; and 3) motion characteristics of icebergs have been determined only for idealized shapes. The required iceberg dimensions are calculated from records provided by Husky-Bow Valley of iceberg water-line length L, above water height H, water-line width W, and shape class (Table 1). In the Monte Carlo simulation, the value for water-line length is chosen from the probability distribution shown in Figure 3 and other iceberg dimensions are determined from this value. Iceberg mass M is determined using the empirical relation $M = .959 \times L^{2.689}$ which we determine using a least squares fit on the logarithms of estimated mass and length. Estimated mass was determined using the relation $M = L \times W \times H \times 8 \times .89 \times SF$ where the factor 8 accounts for 1/8th of the iceberg being above water, .89 is the ice density (tonnes/m^3), and SF is a shape factor for deriving above water iceberg volumes (Table 1).

Table 1 - Iceberg Shape Factors SF (M. El-Tahan, unpublished)

SHAPE	SHAPE FACTOR
Tabular or Blocky	0.50
Dome	0.41
Pinnacle	0.25
Drydock	0.15
Wedge	0.33

The radii of the spherical icebergs modelled are based on equivalent masses. For the detection model, the iceberg above water height, H, is determined from the water-line length, L, using the empirical equation, $H = .389L^{.895}$, which we also determine using a least squares fit.

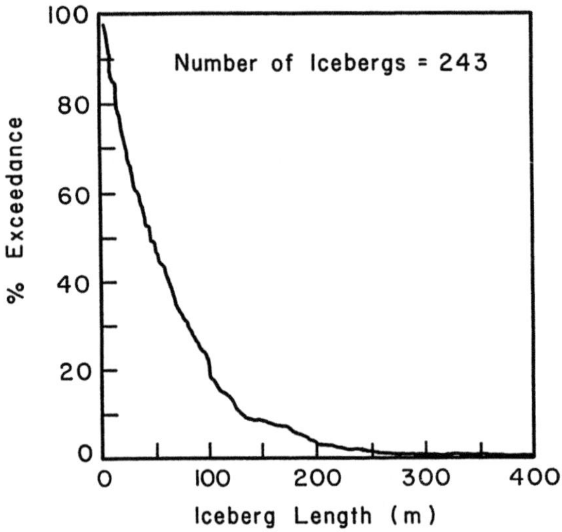

Figure 3. Percent exceedance of iceberg length (El-Tahan and Simms, 1988).

We estimate the number of collisions per year, R, that would occur with no iceberg management, from the areal density ρ of icebergs in the vicinity of the vessel, the mean effective iceberg width $\overline{W}_i$, the mean iceberg drift velocity $\overline{V}_i$, and the effective width W_s of the vessel using the equation (Nessim et al., 1984) $R = \rho \times (W_s + \overline{W}_i) \times \overline{V}_i$. The areal density ($1.02 \times 10^{-4}$ icebergs per km^2) is estimated from IIP maps supplemented with data from drill sites. We estimate the average iceberg width as the average iceberg water-line length, 63 m, times 1.26 (the ratio of the diameter to water-line length of a spherical iceberg). An average iceberg drift velocity of .35 m/s is used as determined from Husky-Bow Valley drill site records (El-Tahan and Simms, 1988). The result is .208 collisions/year.

3. Analysis of Sea State

In the study we characterize the sea state in terms of a single variable, the significant wave height H_s (the average height of the highest one third of the waves). Because the Monte Carlo simulations are run separately by sea state, it is necessary to determine the proportion of collisions that will occur in each sea state. Both the flux of icebergs and the distribution of sea states change through the year (Figure 4), therefore

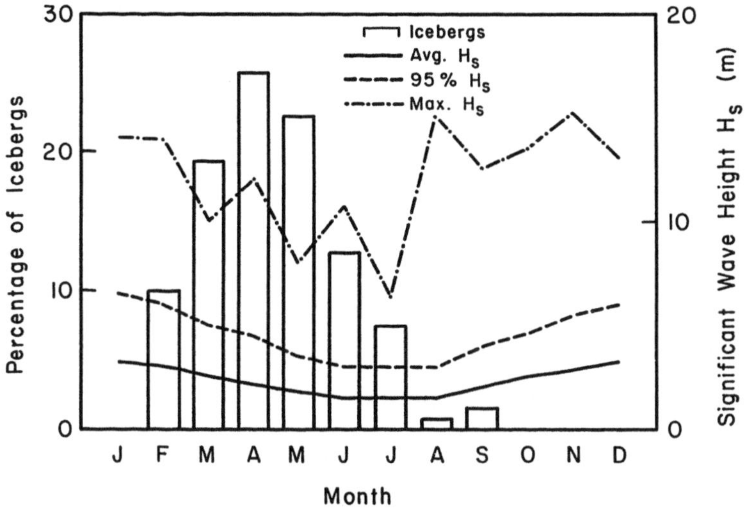

Figure 4. Monthly distributions of icebergs and significant wave height.

the proportion of collisions in a given sea state is set as the proportion of the sea state in each month times the proportion of collisions occuring in that month (Figure 5). Monthly sea state distributions from Atmospheric Environment Service Canada and monthly iceberg frequencies derived by El-Tahan and Simms (1989) from IIP data are used.

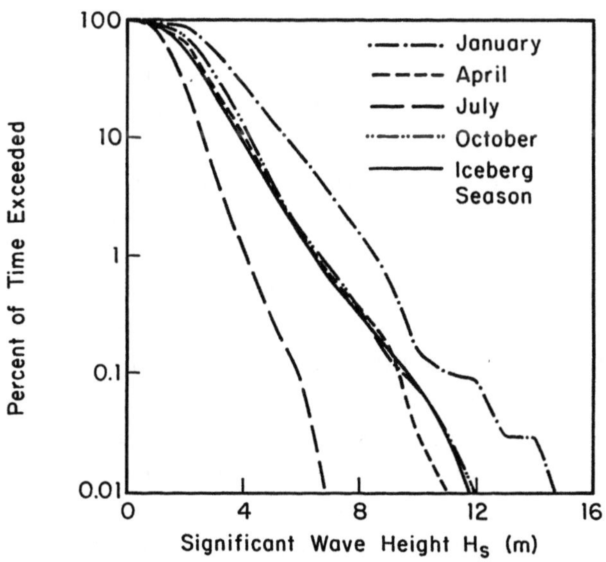

Figure 5. Distribution of significant wave height.

The iceberg velocities are based on model tests (Lever et al., 1988a, b) which were conducted for JONSWAP spectra of the form:

$$S_w(f) = \frac{A}{f^5} \exp\left[\frac{-5}{4}\left(\frac{f_0}{f}\right)^4\right] \gamma^a, \tag{1}$$

where

$$\begin{aligned} a &= \exp[-(f-f_0)^2/(2\sigma^2 f_0^2)] \\ \sigma &= 0.07 \text{ for } f < f_0 \\ \sigma &= 0.09 \text{ for } f > f_0 \\ A &= 5H_s^2 f_0^4/(16\gamma^{\frac{1}{3}}) \text{ for } 1 < \gamma < 4 \\ f_0 &= 1/T_p \text{ where } T_p \text{ is the peak period} \end{aligned}$$

These spectra closely resemble the sea state on the Grand Banks in storm conditions as determined by LeBlond et al. (1982) based on empirical fits. To derive the JONSWAP spectra from H_s alone, we use the approximation, $T_p = 4.43H_s^{\frac{1}{2}}$ and set γ to 2.2 (also based on LeBlond et al.).

4. Iceberg Management Model

Iceberg management involves detecting the icebergs, deflecting them if possible, and finally moving the vessel off location if necessary. Icebergs may be detected from the production or support vessels visually or using marine radar, and from overflights using systems such as synthetic aperture radar or side looking airborne radar. Good detection using rig mounted marine radar systems is important because spray, precipitation, fog, and nightfall may limit visual detection; detection from support vessels is not good; and the use of aircraft and helicopters may not be feasible in bad weather.

Radar detection is a complicated process which depends upon the system used, the sea state, the shape and size of the iceberg, noise, precipitation, and other factors such as multi-pathing. The probability of detection increases as the ratio of signal power return from the iceberg to return from other sources increases. The return from the iceberg depends on the size and shape of the iceberg and on the reflective properties of the ice. The iceberg may also be partially hidden by waves.

Sea clutter is the most important limiting factor on detection. The return from the sea state increases with the number of wind created capillary waves. Sea clutter varies with direction, being worst from upwind. Both the iceberg return and the sea clutter return decrease with range, while noise remains relatively constant and losses due to effects such as precipitation increase. Using a more powerful system can help increase range limitations due to noise and precipitation. The return from the iceberg, sea clutter, precipitation, and fog all vary with the radar frequency and the antenna height, making detection predictions difficult. To complecate things further there is only limited data on iceberg cross sections and on clutter in high sea states.

Because of the difficulty in verifying a detection system based on marine radar, visual detection, and overflights, a very simple detection model is used for this study and an analysis of the sensitivity of the design loads to the success of the radar system is performed. In the model, detection is determined based on the ratio of the height of the iceberg to the significant wave height of the sea state. Four different limiting ratios are tested, i.e. detection is limited to icebergs with sail heights above 1, 1.5, 2, and 2.5 times the significant wave height, respectively. It is recommended that for a final design analysis, the particular radar system to be used on a vessel be carefully modelled and verified to determine the design forces. If necessary, additional iceberg radar cross section and sea clutter data should be obtained.

For the iceberg management model, we assume that if an iceberg can be detected, it can be avoided by towing or moving off location. It should be noted, however, that even if disconnection and moving off site can be done as quickly as in 5 minutes, it would be necessary to detect an iceberg drifting very quickly (1.3 m/s) 400 meters away. In the model we apply the iceberg detection model for different sea states to determine the proportion of collisions that can be avoided and to determine the distribution of sizes of colliding icebergs in each sea state. In Figure 6, the limiting detectable iceberg masses are shown for the different detection limits. The collision return period, with iceberg management, is determined as the encounter period (4.8 years) divided by the fraction F of icebergs undetected. F is calculated as the integral over all sea states of the frequency of the sea state in the iceberg season $p(H_s)$, times the fraction of icebergs with length L less than the limiting detectable length $L_{DET}(H_s)$:

$$F = \int_0^{H_s MAX} p(H_s) \times f[L < L_{DET}(H_s)] dH_s. \tag{2}$$

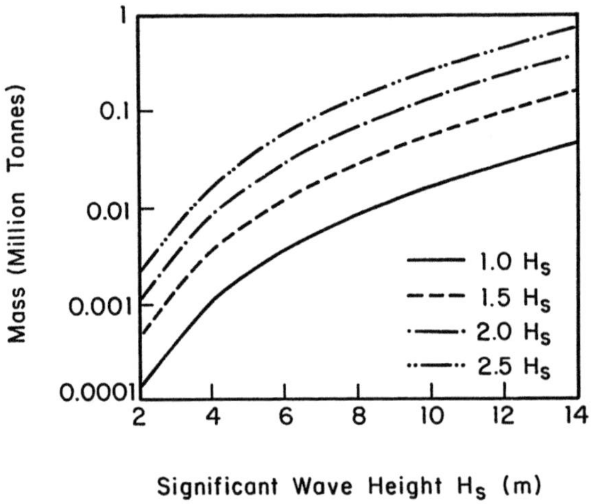

Figure 6. Iceberg detectable masses.

The return periods for the detection limits used are:

MODEL	FRACTION OF ICEBERGS COLLISIONS NOT AVOIDED	COLLISION RETURN PERIOD (YEARS)
1.0 Hs	.09	53
1.5 Hs	.15	32
2.0 Hs	.22	22
2.5 Hs	.27	17

5. Collision Velocity

The collision velocities are determined following the method presented by Lever et al. (1989). This method is based on Bayesian updating of the open water wave-induced velocities of icebergs in random seas to account for the increased probability of collision at higher velocities. The collision velocities are based solely on the wave-induced motions of the iceberg, i.e. we neglect hydrodynamic interaction effects and the wave-induced velocity of the vessel. Only iceberg surge (forward) and heave (upward) impact velocities are used. Lever et al. found the surge and heave impact velocities to be uncorrelated in random seas, so we treat them as independent for the purposes of the present investigation. For simplicity, drift is neglected in the present study.

First, we use linear superposition to estimate the iceberg response spectra and significant velocity components from their response amplitude operators (R.A.O.'s), as modelled in regular sea states, and from the energy spectra of the random sea state. The collision velocity distributions are then derived as follows. As with the sea surface elevation itself, we may assume that the probability distributions for iceberg open-water instantaneous velocities are Gaussian. The impact velocity probability distributions are different, however. Here, we assume that the probability of collision in surge (or heave) at any instant is proportional to the surge (or heave) velocity of the iceberg (i.e. to the volume swept out by the iceberg per unit time). Bayes' theorem then yields Rayleigh probability distributions for impact velocities, u.

$$P(u) = \frac{u}{\sigma_u\sqrt{2\pi}}e^{-u^2/2\sigma_u}, \tag{3}$$

where σ_u^2 is the variance of the iceberg heave or surge velocity in open water and m_0 is the zero moment of the open water velocity spectrum. For surge, u is equal to or greater than zero. Note that we have completely defined these Rayleigh distributions with the open water significant component velocities alone. Lever et al. showed these distributions approximate the measured impact velocity statistics for a semi-submersible composed of fairly slender members. Thus for the NEKTON 8000 FPV, which diffracts significantly more water, these wave-induced impact velocity distributions should be conservative.

In the Monte Carlo simulation we 1) determine the significant surge and heave velocities of the icebergs based on their size and the sea states using dimensionless significant velocity curves for spherical icebergs (Lever et al., 1988a, b) 2) independently choose the collision surge velocity, V_s, and heave velocity, V_h, from the corresponding Rayleigh velocity distributions and 3) determine the total collision velocity as $V = \sqrt{V_s^2 + V_h^2}$.

6. Global Collision Loads

To determine the maximum load in a collision we developed a time domain simulation that tracks the ice crushing force and the accelerations of the iceberg and semi-submersible. Spherical icebergs impacting directly (i.e. no eccentricity) on a vertical column are considered. The effect of ignoring the eccentricity of collision is not explictly addressed, though Duthinh and Marsden (1986) estimate that 75 percent of colliding icebergs would lose 70 percent or more of their kinetic energy during collision based on a model of a spherical iceberg hitting a fixed structure and rotating around the initial contact point with no slip. This indicates that the assumption of direct impacts is not extremely conservative.

The failure mechanism of the ice is a major factor in determining the collision forces. Unfortunately there is very limited experimental data on ice strength at the scale and rates which would occur in iceberg collisions with a structure. Perhaps the most realistic iceberg collision data to date comes from a pilot experiment in Antarctica (Duthinh et al., 1990). In this experiment, a small iceberg of 1700 tonnes was towed at a velocity of 0.89 m/s into a rocky cliff, colliding in 2.25 seconds with a force between 4.5 and 6.5 MN. An average pressure between 1.3 and 3.6 MPa was calculated over the contact area which was estimated to be between 1.8 and 3.6 m^2. While small scale laboratory investigations of the compressive strength of ice are available, the large scale forces are not simply the product of the 'nominal' area of contact times the observed small scale strengths. The "scale effect" is due to the distribution of natural flaws in the ice, non-simultaneous failure zones, etc.

Johnson and Benoit (1987) report on a series of medium scale ice strength tests on an iceberg in Pond Inlet. We applied the results of these tests to this study, though care was required in doing so. The Pond Inlet tests covered up to 3 square meters contact area and lasted approximately 3 seconds versus estimated design load level collisions of approximately 25 square meters and 0.5 seconds. In the experimental program a hydraulic ram was used to push spherical indentors at pre-programmed rates into the walls of tunnels cut into an iceberg. As the nominal area increased during the tests, the total force also increased. We portray this in Figure 7. The fluctuating component of the total force is due to spalling of ice fragments and clearing of the crushed ice. The variation due to clearing of the ice may result from buildups of pressure in the crushed and undamaged ice until pulverization and flow of the crushed ice are initiated and the undamaged ice springs back.

To estimate the maximum force during a collision, we first determine the maximum collision penetration and contact area based on the average ice crushing pressure, then multiply this area by a peak pressure chosen randomly from a distribution derived from the Pond Inlet data. The rationale for this method is that, on average the collision area is increasing fast enough that the largest impact force can be attributed to the last pressure peak. To determine the distribution of peak loads from the Pond Inlet data, we had to estimate the small scale, uniaxial compressive

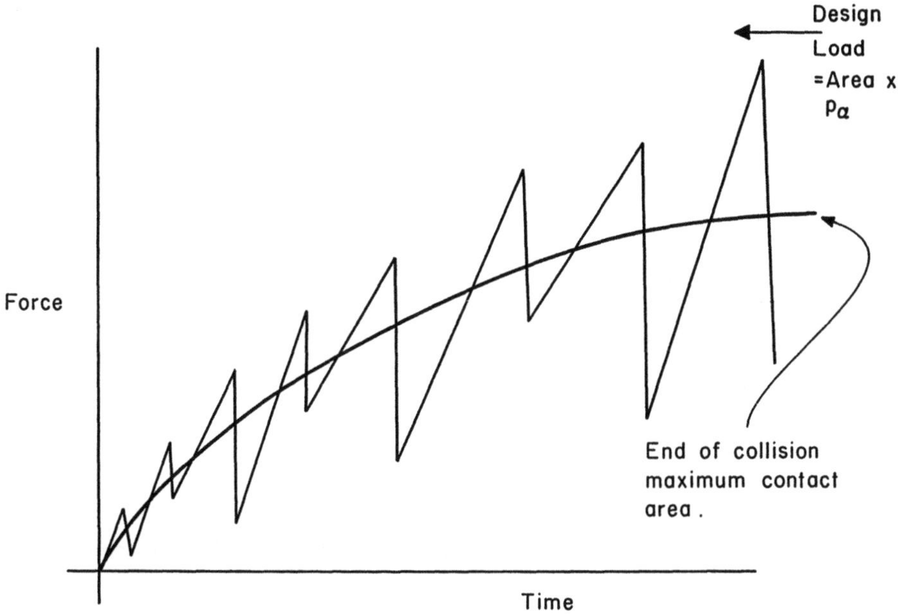

Figure 7. Typical progression of total ice crushing force.

ice strength because Johnson and Benoit normalized their results by this value. We chose a conservative value of 10 MPa based on Nadreau (1986). Johnson and Benoit also did not report the average ice crushing pressure and for this we assumed a value of one half of the mean peak pressure.

7. Calculation of Design Load

To determine the design load, first we determine the probability distribution of the peak load in a collision for each sea state. This is done by Monte Carlo simulations, randomly choosing the sizes of icebergs and using the collision velocity and force models already described. The resulting probability distributions are combined, weighted by the frequencies of the sea states. Figure 8 shows the resulting load probability distribution as well as the component probabilities of the load occurring in different sea states for the 2.0 H_s detection limit. Figure 9 shows the corresponding load exceedance probability distribution.

Next, we determine the exceedance probability distribution for the peak load in a year. For rare events such as iceberg collisions, the proba-

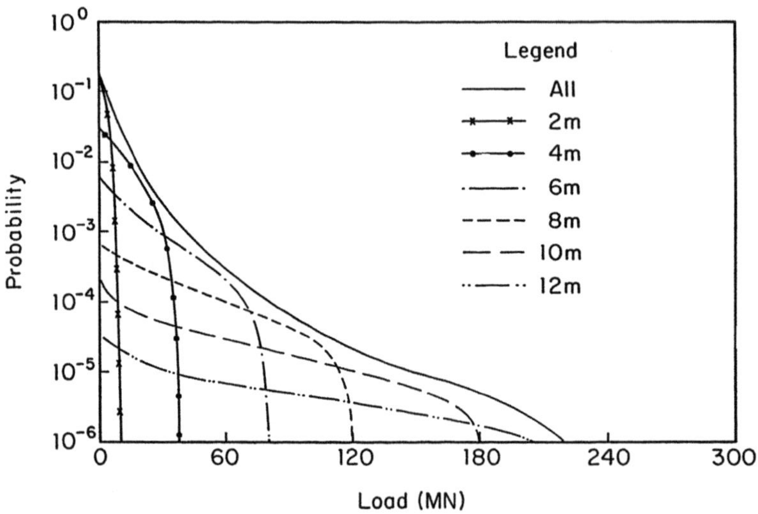

Figure 8. Collision load distributions by significant wave height for 2.0 Hs detection mode (final avg. load)

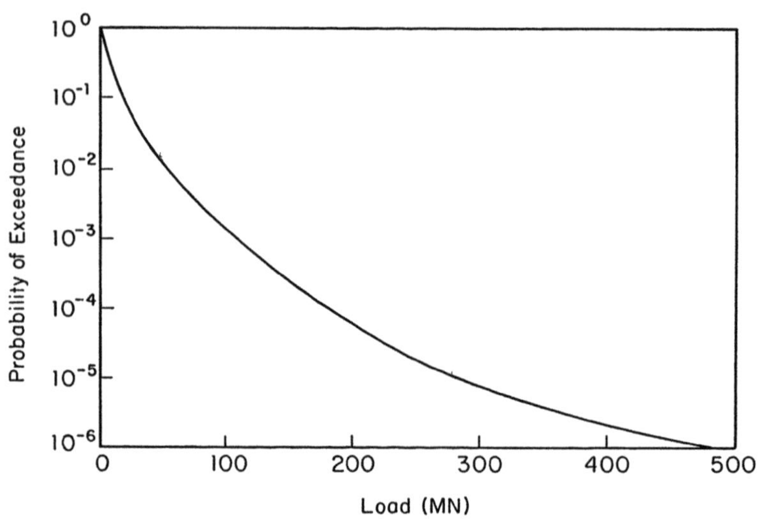

Figure 9. Collision load exceedance distribution for 2.0 Hs detection model (final avg. load).

bility of more than one collision per year is small. Thus, the probability, $P(L)$, of a load, L, being exceeded in a given year is approximately the probability of a collision in that year, ν, times the probability of that load being exceeded given a collision, $P(L|C)$, i.e. $P(L) \simeq \nu P(L|C)$. Figure 10 shows an example 'parent' distribution, $p(L|C)$, and resulting extreme distribution, $p(L)$. The probability of no collision (zero load) in a year is $1 - \nu$. The design load is determined similarly by scaling Figure 9 by the collision frequency $\nu = .21$ collisions/year and choosing the load associated with the appropriate probability of exceedance. The 10^{-3} and 10^{-4} probability of exceedance per year kinetic energies, final average loads and final peak loads are shown in Figure 11.

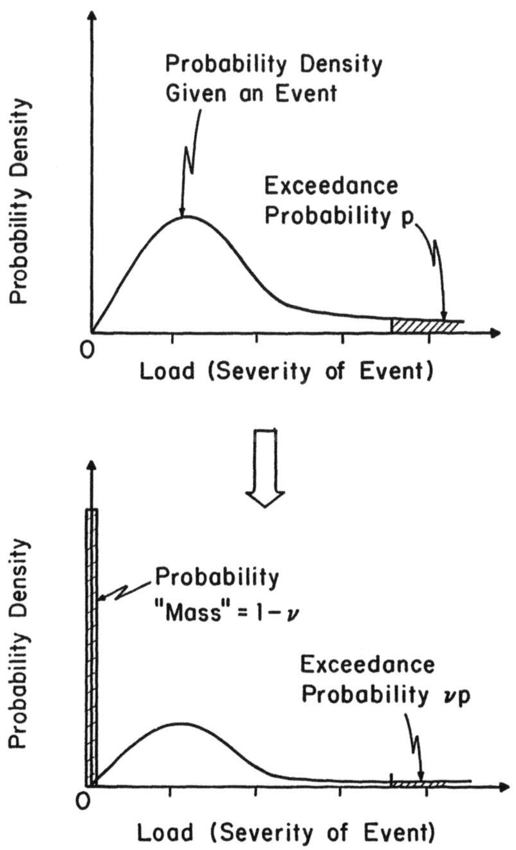

Figure 10. Method to choose design load.

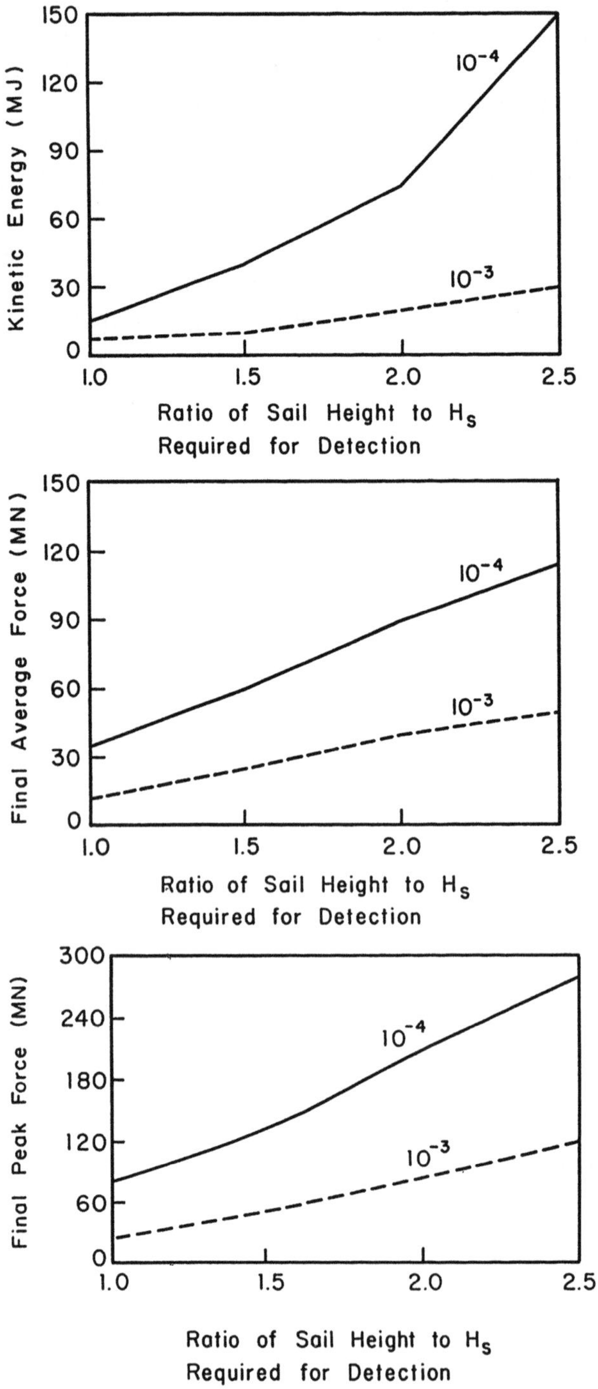

Figure 11. Design kinetic energy, final average load, and final peak load.

8. Conclusions and Recommendations

A probabilistic method has been developed and applied to the determination of design loads on floating production systems. The method explicitly models iceberg detection, the random nature of ice crushing pressures (as seen in the Pond Inlet data), and the increased wave induced collision velocities due to Bayesian updating. While improvements are required in a number of areas, the general method presented can easily be expanded as improved component models are developed.

By conducting separate Monte Carlo simulations for each range of sea states, the number of simulations is reduced and the probability of different collision loads occurring in different sea states can be determined. It was found that for a probability of exceedance of 10^{-4} per year, the design load is most likely to occur in sea states with significant wave heights in the range 5 to 11 meters for the area of study. This result held independently of the detection model used.

The design loads are very sensitive to the success of the detection system and it is very important to quantify how many and what sizes of icebergs will go undetected in given sea states. The load estimates can be improved by accounting for the motion of the vessel and the effect of waves diffracted from the vessel on the iceberg. Model tests by Lever et al. (1989) for random seas have shown that in the higher sea states the iceberg and structure will move in phase more often than not, reducing the collision loads. These results, and the effect of diffracted wave energy still need to be quantified for different sea states and iceberg sizes. The drift velocity components of the icebergs need to be added in for larger icebergs.

The collision model should be enhanced to handle eccentric collisions and to model the shape of the iceberg and the location of contact on the NEKTON 8000 FPV more accurately. Because of the short duration of the fluctuating loads, both the maximum average load and the maximum peak loads during the collision are presented. The short duration is an important factor in determining structural response. There are still uncertainties as to the failure mechanism of iceberg ice at large contact areas and varying impact rates (i.e. the Pond Inlet experiments were conducted for rates of .1 m/s whereas collisions of over 4 m/s can occur). The assumption of a fixed average ice strength should also be examined.

Acknowledgement

The support of Bouygues Offshore Canada and the Newfoundland Department of Development is gratefully acknowledged.

References

CSA Preliminary Standard S471-M1989. General requirements, design criteria, the environment, and load.

Christensen, F. 1989. Efficiency of detachable platforms in the Arctic. Journal of Cold Regions Engineering, Vol. 3, pp. 37-54.

Dawe, B.R. 1985. A radar cross section model for icebergs. IEEE 1985 International Radar Conference, Arlington, Virginia. pp. 259-264.

Duthinh, D., Marsden, S. 1986. Iceberg impact load on a gravity based structure. Cold Regions Eng., Proc. of the Fourth Int. Conf., Anchorage, AK, ASCE. pp. 82-92.

Duthinh D., Klein, K., Rerettier, J.F., Guichard, A. and Engler, M. 1990. Full scale iceberg impact: a pilot experiment in Antarctica. IAHR Symposium on Ice, Espoo, Finland. pp. 890-901.

El Tahan, M., Simms, A. 1989. The interaction of the NEKTON 8000 FPV with sea ice and bergy bits, Phase I: Ice environmental data. Final report. C-CORE Contract Report 88-C9.

Fuglem, M., Jordaan, I. 1989. The interaction of the NEKTON 8000 FPV with sea ice and bergy bits, Phase IV: Risk analysis. Final report. C-CORE Contract Report 89-C7.

Isaacson, M., McTaggart, K. 1989. An iceberg collision reliability model incorporating hydrodynamic effects. Proc. of the Eighths Int. Conf. on Offshore Mech. and Arctic Eng., The Hague, The Netherlands, Vol. 4, pp. 293-308.

Johnson, R.C., Benoit, J.R. 1987. Iceberg impact strength. 19th Annual Offshore Technology Conference, Houston, TX, Vol. 4, pp. 417-423.

Jordaan, I.J. 1983. Risk analysis with application to fixed structures in ice. Seminar/Workshop on Sea Ice Management, Memorial University of Newfoundland, St. John's, Nfld., 19p.

Jordaan, I.J. 1987. Probabilistic analysis of environmental data for design of fixed and mobile arctic offshore structures. Fifth Int. Conf. on Applications of Statistics and Probability in Soil and Structural Eng. (ICASP). Vol. 2, pp. 1130-1138.

Leblond, P.H., Calisal, S.M. and Isaacson, M. 1982. Wave spectra in Canadian waters. Canadian Contractor Report of Hydrography and Ocean Sciences, No. 6, Dept. of Fisheries and Oceans, Ottawa.

Lever, J.H., Attwood, D. and Sen, D. 1988a. Factors affecting the prediction of wave induced iceberg motion. Cold Regions Science and Technology, 15, pp. 177-190.

Lever, J.H., Sen, D. and Attwood, D. 1988b. The influence of shape on iceberg wave induced velocity statistics. Seventh Int. Conf. on Offshore Mechanics and Arctic Engineering, Houston, TX, Vol. IV, pp. 125-132.

Lever, J.H., Colbourne, B. and Mak, L. 1989. A model study of the wave-driven impact of bergy bits with a semi-submersible platform. Proc. of Eighth Int. Conf. on Offshore Mech. and Arctic Eng., The Hague, The Netherlands, Vol. IV, pp. 393-403.

Lindberg, K., Anderson, L. 1987. Ice impacts on semi-submersibles. Sixth Int. Offshore Mech. and Arctic Eng. Sym., Houston, TX, Vol. 4, pp. 313-320.

Nadreau, J.P. 1986. Survey of physical and mechanical properties of icebergs. Proc. of Workshop on Extreme Ice Features, Banff, AB, NRC Technical Memorandum No. 41, pp. 155-192.

Nessim, M.A., Murray, A.M., Maes, M.A., and Jordaan, I.J. 1984. Risk analysis of mobile offshore units operating in ice-infested waters. Ice Tech '84, Calgary, AB, Paper H.

Putot, C., Molin, B. and Valechon, C. 1988. Impact of ice bergy bits on a concrete semi-submersible. Polartech '88, Trondheim, Norway, Vol. 1, pp. 191-207.

Sittrop, M. 1977. On the sea-clutter dependancy on wind speed. Radar-77, International Conference, London, England.

Skolnik, M.I. 1970. Radar handbook. McGraw-Hill, New York.

Discussion

Q: M. Nessim, DnV Canada, Ltd

Have you considered the correlation between iceberg velocity and structure velocity since they are responding to the same wave?

A: M.K. Fuglem

It is true that in the larger sea states the two bodies will most often be responding to the same wave. At present we consider the vessel to be initially fixed. This will be conservative for the higher sea states where the two bodies are likely to move in phase, but may be overly optimistic in smaller waves where they move out of phase. Accounting explicitly for the phase distribution is one area for improvement to the model.

Q: C. Daley, Fleet Technology

Was the ice force model deterministic? Did you do a sensitivity analysis on the governing parameters? What is the most crucial parameter?

A: M.K. Fuglem

The ice force model is deterministic in as far as it determines the maximum contact area during the collision based on the average ice strength of 1.5 MPa. The model then also calculates the peak short term final force based on a peak pressure chosen randomly from a distribution of values derived from Johnson and Benoit's paper.

We did check the sensitivity of the model to various parameters such as ice strength and collision velocity. The most significant unknown at this time is the ability to detect icebergs of given sizes in different sea states. Because of the complex nature of radar detection and the lack of data, it is difficult to verify existing radar models for all conditions.

Q: Hussein El-Tahan, Crescent Consulting

Your model is sensitive to iceberg detectability and since the IIP data base has a gap in (lack of) bergy bit and growler data, how did you fill that gap? Could you please also comment on your assumption that iceberg draft velocity is independent of mass. What would be the effect on the design load of velocity dependence on mass.

A: M.K. Fuglem

To determine the areal density of icebergs at the site, IIP maps showing the locations of icebergs sighted on the Banks on given dates were used. Because bergy bits and growlers were often missed, only records of small, medium and large icebergs were considered. The distribution of iceberg lengths in the study was solely based on Husky-Bow Valley records. The areal density estimated using IIP data was increased according to the percentage of bergy bits and growlers in the Husky-Bow Valley length distribution.

The assumption is often made that the speeds of icebergs are independent of their mass, based on scatter diagrams of iceberg mass versus speed. Intuitively, it is felt that small icebergs in storms would move faster because more of their mass is in the upper currents which will move faster in winds. It may be that the apparent high velocities of larger icebergs are due to wave diffraction. Also, the recorded data of iceberg speed versus mass could be biased because of the increased probability of detecting large icebergs in high sea states. We are presently evaluating these possibilities. The effect of a drift velocity dependence on mass would depend on the magnitude of the changes in estimated drift velocities of the smaller and larger icebergs.

Q: A. Tunik, ABS

Have you distinguished which part of the vessel is subjected to the collision, whether it is the bow or side?

A: M.K. Fuglem

An analysis of general iceberg drift patterns in the area of concern shows that, though on average icebergs travel from north to south, there is a significant probability of the instantaneous velocity being in other directions. The direction of approach of a particular iceberg could be determined from the current, wave, and wind conditions and the shape of the iceberg, although these parameters, especially iceberg shape are often difficult to measure. The side on which a vessel is impacted will depend on the aspect of the vessel and on how it is positioned relative to the predominant currents and waves (for example, whether it is moored or dynamically positioned).

For this study we were looking at a semi-submersible moored by 12 lines and did not address the most likely side of impact. We did do some rough estimates of the probability of impacts on the pontoons versus different columns and determined that the corner columns were more likely to be impacted. We hope to carry this work further and determine the load distributions on the different members. For example, larger icebergs are more likely to impact the pontoons.

Q: T.H. Jacka, Australian Antarctic Division

Your considerations of the probability of an iceberg structure collision assumed a random distribution of icebergs and random velocities and movement directions of icebergs. Iceberg distributions and movement rates are governed however, primarily by currents, which are not random. Can you not include current data in your model?

A: M.K. Fuglem

The main emphasis in the study is on the long term distribution of collision forces acting on the structure. For this it is necessary to estimate the frequency of collisions by icebergs of different shapes and sizes. Relatively good historical data on the motions of icebergs are available from drill rig locations (though the data may be biased in some respects, for example small icebergs are hard to detect in higher sea states and may be underrepresented). To use current information, the variations in current with respect to location and depth would be required along with wind and wave information and the shape of the iceberg. Because of uncertainties in these values and in forecasting techniques, it was determined that the method used was more appropriate.

The Selection of Design Ice Loads and Design Ice Features For Fixed Structures in the Beaufort Sea

M. Nessim

Centre for Frontier Engineering Research
C-FER Building, University of Alberta Campus
Edmonton, Alberta, Canada T6G 2E2

I.J. Jordaan

Faculty of Engineering and Applied Science
Memorial University of Newfoundland
St. John's, Newfoundland, Canada, A1B 3X5

Abstract As is the case for other environmental conditions, design ice loads for offshore structures are characterized as the extremal loads corresponding to a certain probability level (e.g., annual probability of exceedance or associated return period). In the case of ice loads, the load level can be dependent on a large number of uncertain parameters such as the ice velocity, thickness, geometry, and mechanical properties. The estimation of design ice loads therefore requires deriving the load probability distribution from the probability distributions of the input parameters and an ice-structure interaction model, and then using the result in appropriate extremal analysis models based on the characteristics of the ice loading process.

This paper describes an integrated approach for selecting ice loads based on the above procedure for different loading scenarios encountered in the Beaufort Sea. The calculation of ice load probability distributions is based on a probabilistic approach developed in connection with reliability theory. This approximate approach gives good results, in a small fraction of the computer time required for the commonly used Monte Carlo approach. The method also results in an estimate of the most likely set of input parameters which could give a certain load level. This is useful because it can be used as a basis for selecting a "design ice feature" associated with the design load value. The calculations also result in a set of sensitivity factors which define the relative influence of different input parameters on the load. The approach is illustrated by examples dealing with design ice loads for multiyear floes and first year ice interaction with a fixed structure in the Beaufort Sea.

S. Jones · R. F. McKenna · J. Tillotson · I. Jordaan (Eds.)
Ice-Structure Interaction
IUTAM-IAHR Symposium St. John's, Newfoundland, Canada

1. Introduction

The concept of reliability, is now widely accepted as a consistent measure of structural safety and performance. Reliability of a structure in relation to a given mode of failure (or limit state) is defined as the probability that the structure will survive that failure mode for a given period of time. Uncertainty regarding the survival of the structure usually arises from more basic uncertainties relating to the parameters influencing the load (e.g., environmental ice parameters for ice load) and the resistance (e.g., structural dimensions and material properties) for the limit state considered. The estimation of structural reliability for complex limit state functions depending on a large number of parameters has been the subject of extensive research and efficient solutions are available for many practical cases (e.g., Thoft-Christensen and Baker, 1982 and Madsen et al., 1986).

A design problem is the reverse of the reliability estimation problem; the designer's task is to select a structure that achieves a target reliability level for a given (uncertain) load. For practical situations, codes of practice simplify the designers' problem by defining deterministic and design values of the load and resistance which are selected and calibrated such that the target reliability is, on average, achieved. The design load value S_d is usually defined by multiplying a certain specified (or characteristic) load value S_s by a partial factor f_s. Selection of S_s and f_s reflect the characteristics of the load probability of distribution. For environmental loads, characterized by large variability and uncertainties, a stable value of S_s can be obtained by selecting a value on the distribution tail. The CSA-S471 (1988) standard, for example, specifies the load value with an annual probability of exceedance of 10^{-2} for frequent environmental loads, while Det norske Veritas (1977) uses the most probable largest value in 100 years.

The foregoing discussion shows that the selection of design ice loads requires a probabilistic description of the maximum ice load occurring in a given period of time. Unlike other environmental loads such as wind and waves, ice loads cannot be described by a small number of environmental parameters (e.g., wind speed for wind load and wave height and period for wave loading). Ice loads can depend on a large number of environmental parameters, and are determined from those parameters using models that can be complex. This has the following implications:

1. A model for selecting design ice loads must incorporate ice load calculation models and probabilistic description of the influencing environmental parameters (see Figure 1). Lists of the environmental parameters affecting ice loads, and discussions of load calculation approaches are given in API-RP2N (1987) and the CSA-S471 (1988).

2. There is an infinite number of potential input parameter combinations that could lead to the specified load value. If a design ice feature is required, a criterion for selecting one out of these combinations is required.

The objective of this paper is to describe a model which addresses (1) and (2) above with emphasis on ice load scenarios encountered in the Beaufort Sea. The model has been described in previous publications (e.g., Nessim et al., 1986, 1988).

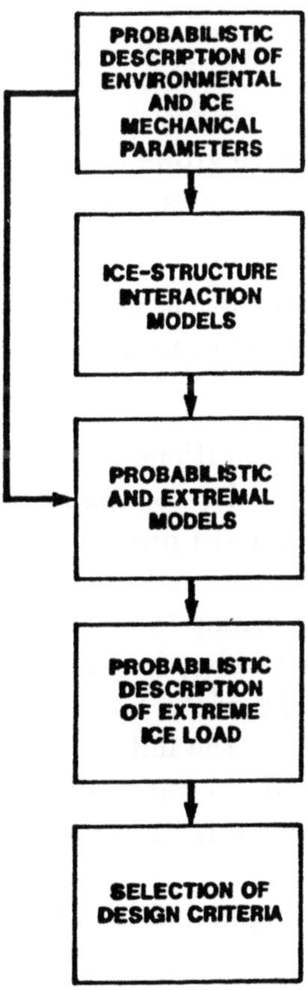

Figure 1. Overall methodology for ice load selection model.

2. Hazard Characterization

Ice features encountered in the Beaufort Sea include: (i) level first year ice, (ii) deformed first year ice features including rafted ice and first year ridges, (iii) multiyear ice floes and hummock fields, and (iv) glacial ice features including ice islands and ice island fragments. Details of the mechanisms involved in the formation of these types of ice and their mechanical properties are given in the literature (e.g., Pounder, 1965; Weeks and Mellor, 1983 and Lainey and Tinawi, 1984). The ice regime describing the cycle of ice cover formation and decay in the Beaufort Sea is given, for example, in Markham (1981). The two most significant events for ice loads are freeze-up which occurs around late October and break-up which takes place around late June. Maximum thickness of the first year ice cover just before break-up can be up to 2 m. This cycle results in three distinct types of ice conditions that are addressed in the model:

1. Open water (summer) conditions characterized by free movement of individual ice features.

2. Full ice cover (winter) conditions where first year ice grows between any multiyear or glacial ice features, resulting in close to 100% ice cover.

3. Partial ice cover (break-up) conditions where the sea is not completely frozen but the total ice concentration is too high to represent open water conditions.

The highest movement speeds occur in open water conditions whereas the lowest occur in full ice cover conditions. Conversely, confinement effects of the ice during interaction with a structure are most pronounced in full cover conditions and are not present in open water conditions.

The Beaufort Sea is divided into two ice zones according to the freedom of movement of the ice, the landfast zone containing ice that is mostly grounded and attached to the shoreline (in the winter) and the pack (active) ice zone which is constantly moving under the influence of environmental driving forces. The landfast zone extends to water depths of approximately 20 m in the Beaufort Sea, beyond which is the pack ice zone.

Based on the discussion in this section, the following global load scenarios can be envisaged (see Figure 2):

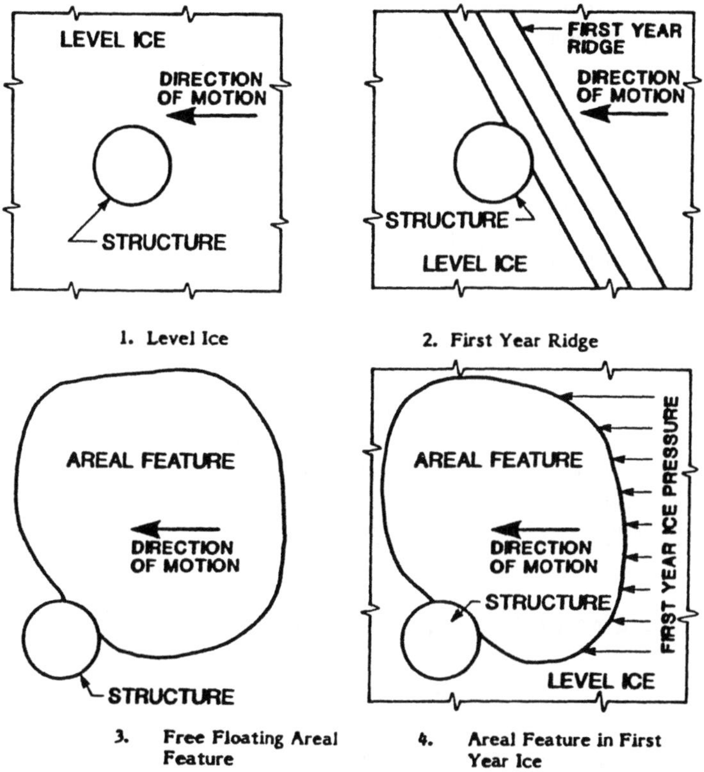

Figure 2. Illustration of the basic global ice load scenarios in the Beaufort Sea.

1. Interaction with a level ice sheet
2. Interaction with a first year ridge
3. Interaction with a free floating areal ice feature, including first year floe, multiyear floe, ice island or ice island fragment
4. Interaction with an areal ice feature (as in (3) above) which is surrounded by first year ice cover.

The applicability of these scenarios to different ice condition (season) and ice zone combinations in the Beaufort Sea is summarized in Table 1. It is noted that in the landfast ice zone, interaction with discrete features in full ice cover conditions is not likely due to the extremely slow ice movement rates.

Table I
Ice Load Scenarios for Different Ice Condition and Ice Zone Combinations

Ice Condition	Landfast Ice Zone	Pack Ice Zone
Partial Ice Cover	- first year ridges - level ice - multiyear floes - ice islands - ice island fragments	- first year ridges - level ice - multiyear floes - ice islands - ice island fragments
Full Ice Cover	- level ice	- first year ridges - level ice - multiyear floes - ice islands - ice island fragments
Open Water	- first year floes - multiyear floes - ice island fragments - ice islands	- first year floes - multiyear floes - ice island fragments - ice islands

3. Stochastic Process Models for Ice Loading

Ice loads are random quantities which vary with time and can thus be modelled as stochastic processes. In this context, a stochastic process models the magnitude of ice load due to a given scenario at a given point in time and can also be used as a basis for estimating the probability distribution of the maximum (extremal) ice load occurring in a given period of time. The latter provides the basis for selecting ice loads.

Three types of stochastic processes are used in the model (see Figure 3):

1. Discrete process: this is suitable for multiyear floe, ice island and first year ridge interactions, in which the load assumes actual values at localized points in time, between which the random load has a value of zero. These are modelled as a compound Poisson process which models the number of loading events N by:

$$P_N(n) = \frac{e^{-\lambda t}(\lambda t)^n}{n!}, \tag{1}$$

where λ is the average rate of the process and t is the time interval considered. The compound Poisson process also models an auxiliary random quantity X (here denoting the load magnitude) representing the height of each spike in Figure 3(a).

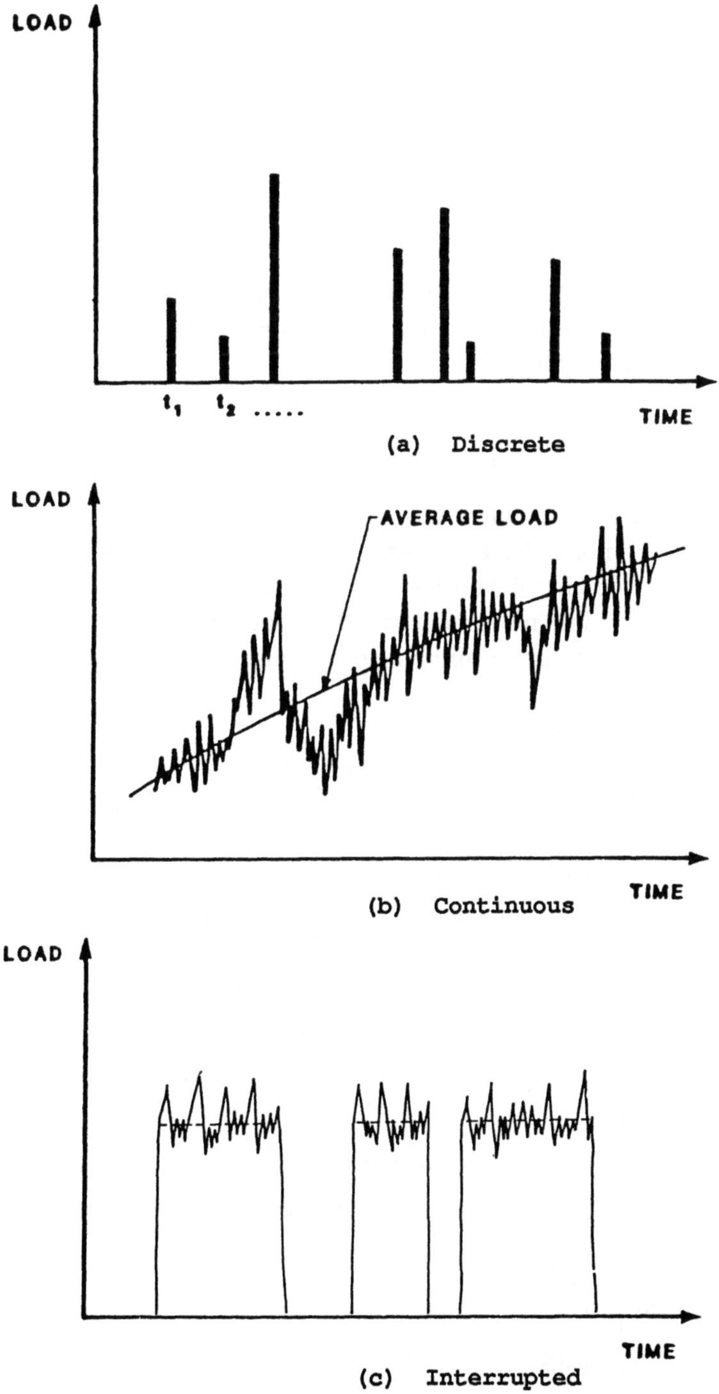

Figure 3. Illustration of ice load processes: (a) discrete (b) continuous (c) interrupted.

2. Continuous process: this is suitable for example for level ice during the full ice cover (winter) conditions in the pack zone, in which case the load is continuously defined at any point in time. Note the gradual increase in the load magnitude being caused by the gradual growth of the ice cover and implying that the process is non-stationary.

3. Interrupted process: for level ice load in partial ice cover (break-up) conditions a continuous process interrupted by open water periods is used. The process is stationary since ice thickness does not change significantly during the short break-up season.

Table 2 summarizes the random process models describing different load scenarios. The following comments are made on Table 2:

1. Level ice in the landfast zone winter conditions is modelled by a discrete process at normal ice speeds. The reasoning behind this choice is that movement of landfast ice is characterized by sharp peaks (of relatively high movement rates) separated by periods of low movement rates. It is reasonable to assume that load maxima

Table 2
Types of Stochastic Processes for Each Ice Load Scenario

Scenario	**Ice Conditions**	**Ice Zone**	**Movement Speed**	**Stochastic Process**
Level Ice	Full Cover	Landfast	Normal	Discrete (at velocity peaks)
			Break-off	Continuous - Non-Stationary
		Pack	Creep range	Discrete (at fixed intervals)
			Crushing range	Continuous - Non-Stationary
	Partial Cover	All Zones	Creep range	Discrete (at fixed intervals)
			Crushing range	Interrupted - Stationary
First Year Ridges	All Conditions	Pack	All Speeds	Discrete
		Landfast		N/A
Areal Ice Feature	All Conditions	Pack	All Speeds	Discrete
		Landfast		N/A

will occur at the velocity peaks and the process is quantified at those peaks.

2. Break-off is an event where a large area of the landfast ice breaks off (usually during a storm) and starts moving at speeds higher than those typical of landfast ice movement. This is modelled by a continuous process characteristic of ice crushing at high movement speeds.

For modelling purposes, all continuous and interrupted processes can be dealt with using a basic continuous stationary process. For interrupted processes this can be done by eliminating periods of zero load (open water). For a non-stationary continuous process for winter level ice conditions, reasonable time intervals (e.g, one month) are defined during which the ice thickness can be treated as constant. The process is then modelled as a sequence of stationary processes in consecutive one-month periods. For level ice scenarios where ice movement speed alternates between the creep range and the crushing range, the proportion of time during which the speed is in each of these ranges is calculated. The appropriate process for each range (see Table 2) is then used for the corresponding time period and the two processes combined to model the scenario. Based on the foregoing, the characterization of design loads requires extremal models for a discrete process and a continuous stationary process, and a method to find the maximum of two or more processes which do not overlap in time.

4. Extremal Analysis

For design, the engineer is interested in the maximum or extremal load Y occurring in a series of load values $\{X_1, X_2, ..., X_n\}$ generated during the loading process. The probability distribution of the individual load values X_i is called the "parent" distribution and denoted $F_{X_i}(x_i)$. The required tool for selecting design criteria is the probability distribution $F_Y(y)$ where Y is given by:

$$Y = \max\{X_1, X_2, ..., X_n\}. \quad (2)$$

For the Poisson process with rate λ in time t this can be shown to be:

$$F_Y(y) = \exp[-\lambda t\{1 - F_x(y)\}], \quad (3)$$

which assumes that X_i's are independent and have the same parent distribution $F_X(x)$.

For a continuous process the double exponential distribution is used. This is given by:

$$F_Y(y) = \exp[-\exp - (\frac{y-a}{b})], \tag{4}$$

where a is the distribution mode and b is a measure of dispersion. These parameters can be estimated directly from time series data defining the spectral characteristics of the process. This distribution is asymptotically valid as the number of measurements in the time series $n \rightarrow \infty$. It can be derived on the basis of the assumption that the individual measurements of the process are independent and identically distributed and that the parent probability density function (p.d.f.) has an exponentially decreasing tail (see Benjamin and Cornell, 1970 for simplified proof). The double exponential distribution is also valid as an asymptotic distribution for high values of the process (of interest for design) under more general conditions (see Cramer and Leadbetter, 1967). The distribution parameter a in Equation (4) depends on time in the following manner:

$$a = c_1 + b ln(c_2 t), \tag{5}$$

where c_1 and c_2 are constants. Equation (5) shows that the mode of the distribution is an increasing function of the time interval during which the maximum is sought. Thus, the distribution shifts toward higher values as t increases.

If different independent scenarios are denoted 1, 2, ..., m and if the associated extremal distribution are $F_{Y_1}(y_1)$, $F_{Y_2}(y_2)$, ..., $F_{Y_m}(y_m)$, then the final extremal can be calculated from:

$$F_Y(y) = \prod_{i=1}^{m} F_{Y_i}(y_i), \tag{6}$$

where $\prod$ denotes the multiplication of the functions $F_{Y_i}(y_i)$ for all i. Equation (6) follows from the fact that $Y \leq y$ is the union of $Y_i \leq y$ for all i.

The foregoing indicates that the extremal load probability distribution depends on:

1. The shape and parameters of the parent distribution which represents the <u>severity</u> of the interaction scenario.

2. The frequency and time period of interaction, or more generally the level of exposure of the structure to a particular scenario.

Models for the exposure and severity (magnitude) of different interaction scenarios are discussed in Sections 5 and 6.

5. Models for Exposure to Ice Loading

The exposure of the structure to ice loads is a function of the season length during which a certain scenario is applicable, and in the case of discrete scenarios, the rate of interaction. Some of the aspects that need to be addressed in estimating these quantities are addressed briefly in this section.

Season lengths in the Beaufort Sea exhibit considerable variability and should thus be treated as random quantities. It was found that season boundary dates can be modelled by a normal probability density function (p.d.f). The boundaries can be defined by reasonable criteria regarding ice concentration and thickness (e.g., the beginning of the open water season is the first day on which the concentration of first year ice is less than $\frac{3}{10}$). The standard deviation of such dates in the Canadian Beaufort Sea is in the order of 20 days. If season dates are assumed independent, the length of the season can be modelled by a truncated normal distribution with a mean and standard deviation which can be calculated from the mean and standard deviation of the season start and end dates. Details of this model can be found in Nessim et al. (1986).

For discrete features, impact rates can be estimated from the frequency of the feature type and its movement pattern and speed. These models are either based on the flux of ice features for cases that are dominated by drift in a given direction, or the density of ice features for cases where the movement is random. Such models are discussed in detail in Nessim et al. (1986).

6. Probability Distribution of the Interaction Load

The ice load X in a given interaction scenario can be represented as:

$$X = g(Z_1, Z_2, ..., Z_k), \tag{7}$$

where Z_i; i = 1, ..., k are parameters representing the environment (ice thickness, velocity and geometry) and the ice mechanical properties. In

Equation (7), g represents the ice-structure interaction model which depends on the structural geometry, the type of ice feature and the movement speed. Detailed reviews of such models are available in the literature (e.g., Croasdale, 1980; Nessim et al., 1987 and Sanderson, 1988) and is outside of the scope of this paper. In the work representing the basis for this paper (Nessim et al., 1986 and Nessim and Cormeau, 1988) the ice-structure interaction models were based on the following principles:

1. For areal ice feature collisions energy dissipation models can be used to account for the limits imposed on the force exerted on the structure by global considerations such as loss of contact and clearing of the feature due to an initial eccentricity of the collision. These models can use time integration approaches for a detailed solution or simplified solutions based on an upfront split of energy into local deformation and rotational components (Nevel, 1986). For vertical structures, the local contact force can be calculated on the basis of contact pressures observed in large scale experiments, whereas for conical structures, a series of crushing and ride-up events controlled by out-of-plane bending of the ice feature can be used. The same approach can be applied to winter collisions by considering the effect of the forces exerted by the ice sheet on the feature. Multiyear floes were treated within this overall model as thickened parts of the floe, but can be considered separately with minor modifications to the model.

2. For level ice movement scenarios a creep solution is applicable for low strain rates (e.g., based on the reference stress method, Ponter et al., 1983), while a crushing model should be used for higher strain rates.

3. For first year ridges, a Coulomb-Mohr material model can be used (Prodanovic, 1981). The ridge fails either by developing a velocity field to clear the rubble around the structure or by shear along two vertical planes parallel to the direction of ice motion. The interaction mode is governed by the lower of the two forces.

For multiyear floes and ice islands the p.d.f.'s used for the velocity and diameter of the feature are usually based on data obtained from randomly selected floes. In order to obtain correct estimates of the load severity it must be taken into account that the velocity and diameter probability distributions of a discrete ice feature which collides with a structure are different from those of a randomly selected ice feature (Jordaan, 1983). It can be seen that faster and larger floes travel over a larger water area in a given period of time and are therefore more likely to impact a randomly positioned structure. It follows that those floes which collide with a structure are more likely to be large and fast. Bayes' theorem can be used to derive the p.d.f. for colliding floe velocity and diameter from those of randomly selected or generic ones. Closed form and numerical solutions result, depending on the distribution type (see Nessim and Cormeau, 1988).

Once the input parameter distributions $f_{Z_i}(Z_i)$ are obtained, the remaining problem is to calculate the load probability distribution $F_X(x)$ from the $f_{Z_i}(Z_i)$ and the ice-structure interaction model $g(Z_1, ...Z_k)$. One possible approach is to use the Monte Carlo simulation method in which input values are generated randomly from $f_{Z_i}(Z_i); Z_i = 1, ...k$, and used in the model g to generate a set of load data. This set is used to generate a histogram and a distribution of X. This method has the advantage of being simple, but could be time consuming especially if g is a complex model and if a large number of simulations is required to obtain a good definition of the distribution tail. On the other hand, a more efficient approximate approach using algorithms developed for structural reliability theory can be used. In this approach the cumulative probability function at any given value of the ice force x is by definition:

$$F_X(x) = p(X \leq x) = p\left[(x - X) \geq 0\right]. \tag{8}$$

Using Equation (7) in Equation (8), we can write:

$$F_X(x) = p\left[g'(x, Z_1, Z_2, ..., Z_K) \geq 0\right]. \tag{9}$$

Equation (9) resembles the problem of calculating the probability of failure in structural reliability theory from the basic random quantities $Z_i; i = 1, ...k$ defining the limit state function g' (e.g., see Thoft-Christensen and Baker, 1982).

In order to visualize the problem, we consider a case with only two basic variables, Z_1 and Z_2. In this case the function g can be plotted in the (Z_1 and Z_2) plane (see Figure 4). According to Equation (9), g' divides the (Z_1, Z_2) plane into two regions: a (failure) region where $g' \geq 0$ and a (safe) region where $g' < 0$. If one envisages a perpendicular axis to the (Z_1, Z_2) plane on which a bivariate probability density function of

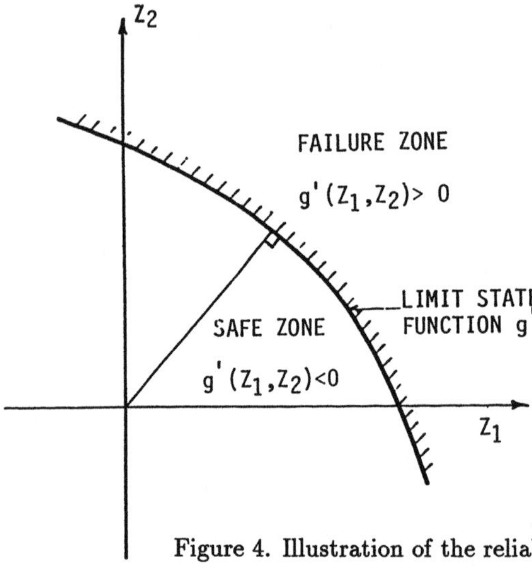

Figure 4. Illustration of the reliability problem in two-dimensional space.

Z_1 and Z_2 is defined, the probability given in Equation (9) is represented by the volume under the density function and over the failure region.

Efficient solutions and software programs to evaluate Equation (9) for an arbitrary value of k and an arbitrary format of g' are available (e.g., Moghtaderi-Zadeh and Madsen 1987 and Tvedt, L., 1986). The solution is based on the special case where all $f_{Z_i}(Z_i)$ are normal and independent and where the function g' is linear, for which a simple solution exists. This solution is used for the general case by transforming all Z_i into equivalent normal quantities (i.e., creating an equivalent function in a transformed normal space) and replacing g' by a linear function at the point closest to the origin (point d in Figure 4). The equivalent solution has been found to give a close approximation for a wide range of practical cases. The calculation can be repeated at different load levels (x) to calculate the corresponding cumulative probability and thus develop the cumulative probability distribution of X. In addition to large improvements in solution efficiency over the Monte Carlo approach, the method has the following advantages:

1. The solution can be used at large (design) load levels to calculate directly the corresponding probability, thus eliminating the problems associated with defining a smooth tail when using Monte Carlo simulations.

2. The design point is defined as the closest point to the origin (see Figure 4) and is therefore the point with the highest probability density. It defines the most likely combination of input parameters leading to a given load level, and is thus a good estimate of a design ice feature or design ice conditions if needed by the designer.

3. The solution calculates the direction cosines of the function g' in the transformed normal space which serve as sensitivity factors defining the relative importance of different input parameters in influencing the probability of exceedance of each load level. This would provide guidance regarding efficient data collection programs and sensitivity analyses.

7. An Overall Model For Selecting Ice Loads

An overall approach for calculating the annual extremal ice load probability distribution is given in Figure 5. This figure is a simplified flow chart of a software developed by Det norske Veritas (Nessim et al., 1986 and Nessim and Cormeau, 1988) for the development of ice design criteria for global loads on fixed offshore structures. The model contains built-in probability distributions of the relevant environmental parameters in the Beaufort Sea, and offers the option of redefining these distributions and their parameters if better or more site specific data are available. The approaches used for different components of the model are as described in this paper. The program deals with each season and scenario combination separately, and allows the user to select any number of combinations to be considered. The extremal load distributions for each scenario are calculated and can later be combined to obtain seasonal or annual extremes. The structural geometries considered are upward and downward breaking cones, and vertical sided structures with circular or polygonal plan shapes. The geometric parameters related to size and slopes (of conical structures) are user defined.

An example result for a vertical sided circular structure with a diameter of 100 m, using typical environmental parameter values for the shear zone of the Canadian Beaufort Sea is given in Figures 6 and 7. Typical application of these results in selecting design and operational criteria are as follows:

1. An overall extremal load distribution can be used in selecting a design load according to the criteria discussed in Section 1. For this

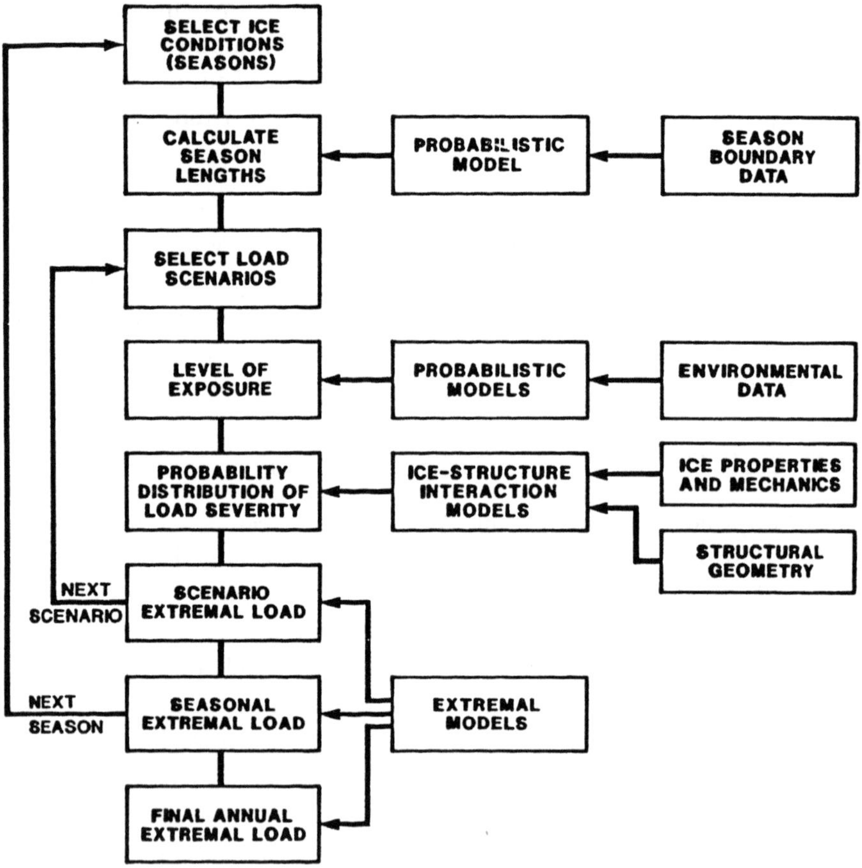

Figure 5. Overall approach for probabilistic ice load calculations.

example, the extreme load corresponding to 100 year return period is approximately 400 MN.

2. Figure 6 gives the cumulative density functions and return period relationships of the monthly extremal loads and the overall maximum load resulting from first year ice loading. The monthly extremals reflect higher loads as the winter progresses due to the increase of ice thickness with time. Such a result can be used in making decisions regarding month to month operations during the winter. For example, by reading the loads corresponding to a certain return period in different months from the figure, decisions can be made regarding how long into the winter a certain operation can be continued.

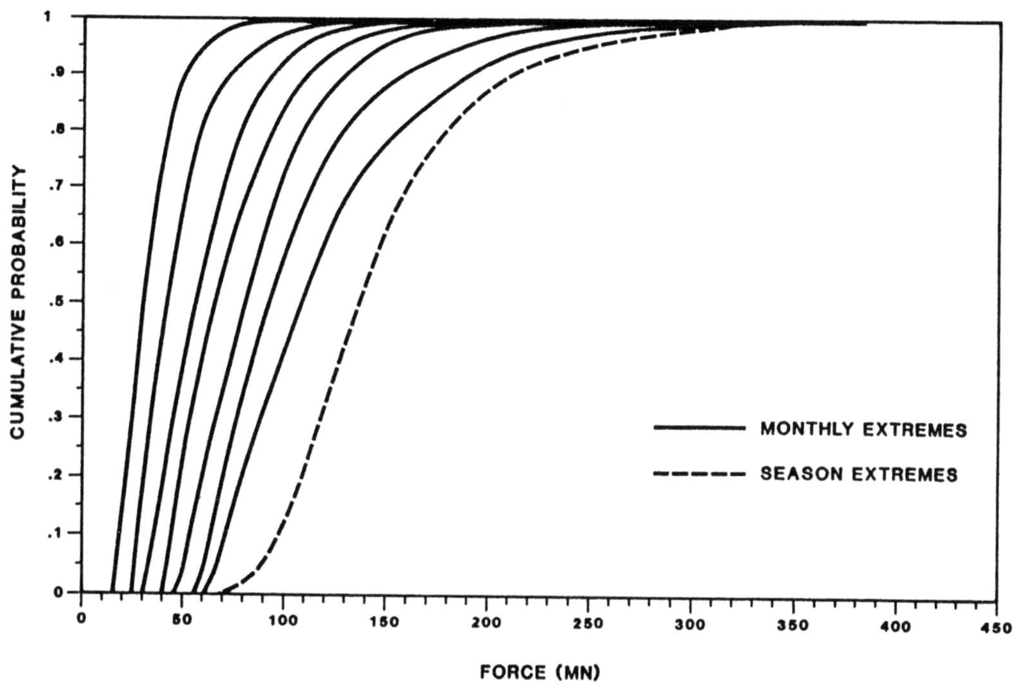

(a) Cumulative Density Function

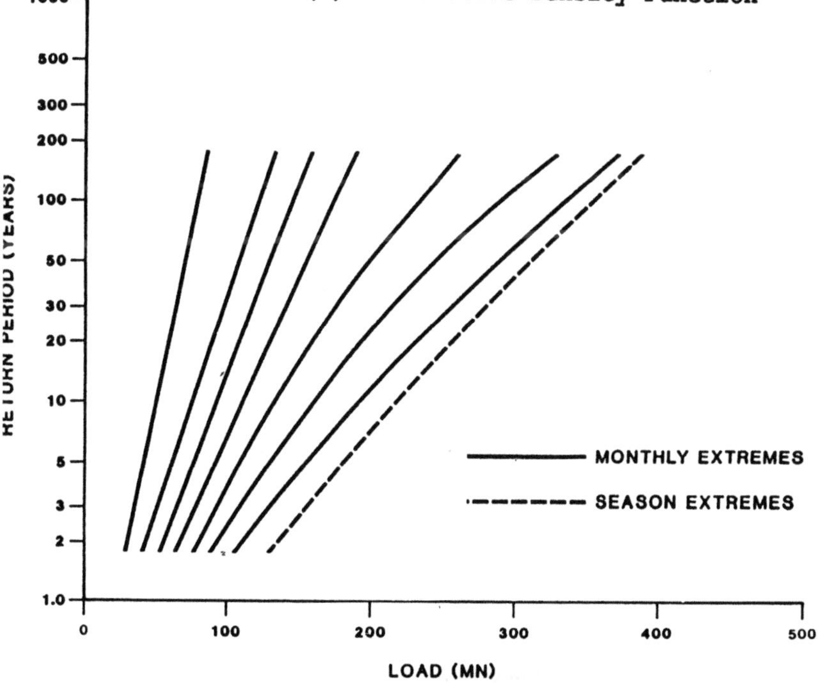

(b) Return Period Relationships

Figure 6. Illustration of the probabilistic description of winter first year ice loads (a) cumulative density function (b) return period relationships.

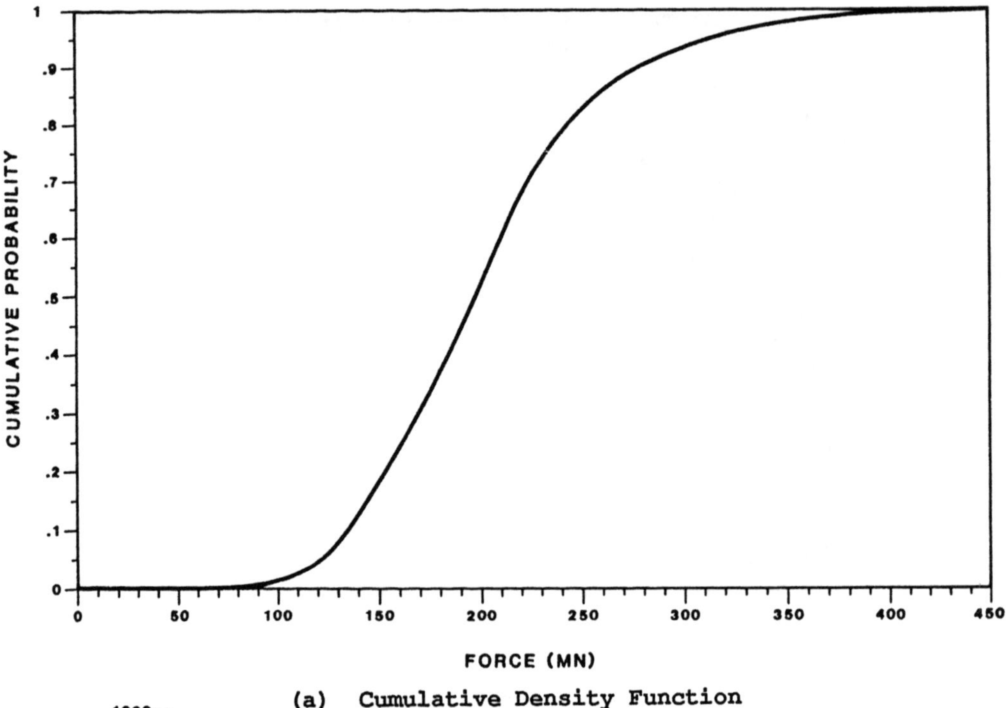

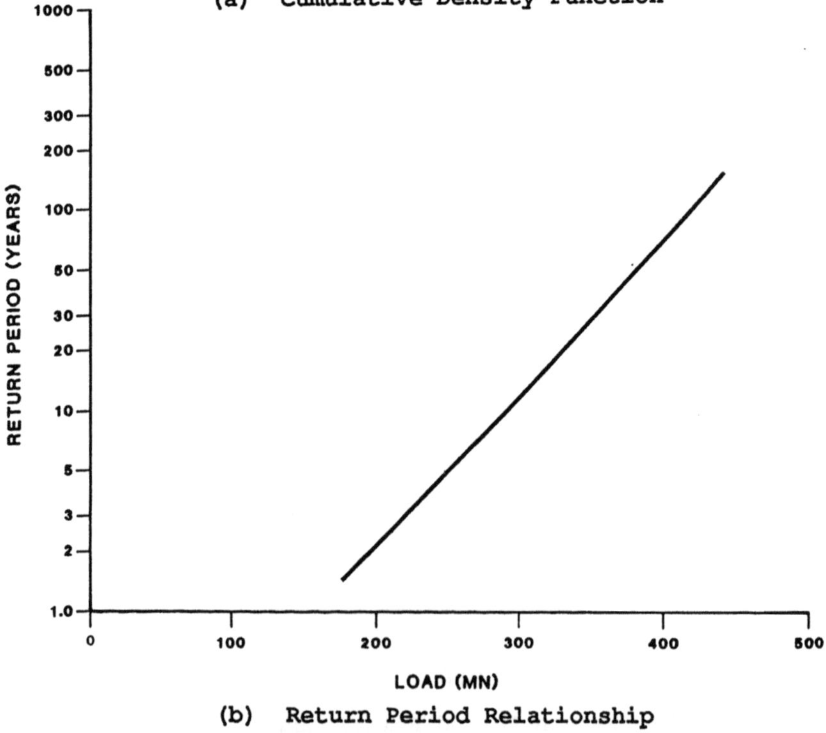

Figure 7. Illustration of the probabilistic description of multiyear floe loads (a) cumulative density function (b) return period relationship.

3. Figure 7 shows the extremal load resulting from multiyear interactions in all seasons. This result can be used to make decisions specific to the protection of the structure against multiyear floes, e.g., ice management or operation interruption for larger floes. The load

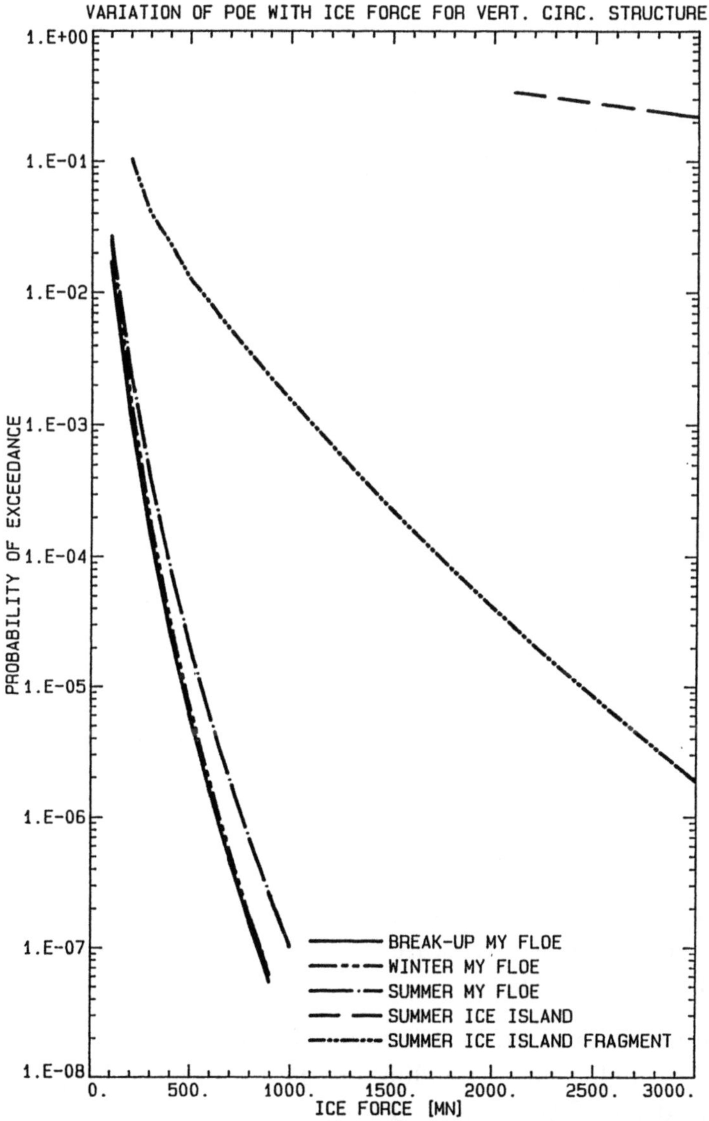

Figure 8. Probability of exceedance versus load for different loading scenarios on a vertical circular structure.

distribution due to multiyear floes can then be modified accordingly before combining it with loads due to other scenarios.

For the same structure and input data, the parent load probability of exceedance relationships (i.e., for one impact) are shown in Figure 8. This was developed using a reliability-based approach as discussed in Section 6, and illustrates the power of the approach in obtaining extremely low values of the probability of exceedance. To illustrate the selection of a design ice feature and sensitivity factors (see Section 6), assume that a design load for winter multiyear floes of 400 MN was selected based on the extremal distributions as before. One can examine the parent distribution and find the most likely input parameters "design winter floe" and the sensitivity factors at that load level. These are given in Table 3 for selected input parameters.

The table shows that the load at this level is most sensitive to multiyear pressure but that none of the four parameters are dominant.

Table 3
Design Winter Ice Floe and Sensitivity Factors at 400 MN

Parameter (units)	**Thickness (m)**	**Velocity (m/s)**	**Multiyear Diameter (km)**	**Multiyear Ice Pressure (MPa)**
Design Value	7.27	0.38	0.753	2.128
Sensitivity Factor	0.47	0.45	0.47	0.58

8. Conclusions

Experience with the probabilistic models described in the paper leads to the following conclusions:

1. Proper characterization of the behaviour and mechanical properties of ice during large scale ice-structure interaction is essential for the evaluation of reliable ice loads.

2. Ice loads are not dominated by one scenario, but different scenarios may lead to high loads depending on the location and type of structure considered. There is a need to give some attention to load calculation models for scenarios that have been viewed as secondary such as first year ridge and first year level ice interactions.

3. It would be useful to develop appropriate interfaces between a probabilistic model such as the one described here and different environmental and ice-structure interaction data bases. This would facilitate application of the model especially for sensitivity analyses.

Acknowledgements

The models described in this paper were developed at Det norske Veritas Canada Ltd. in a Joint Industry Study sponsored by ARCO Alaska Inc.; Conoco Inc.; Sohio/BP; Gulf Canada; I.H.I. Ltd.; Kajima Corporation; Minerals Management Service (US Department of Interior); Mobil Research and Development Corp.; Norwegian Contractors; Taisei Corp.; Union Oil; Tecnomare S.P.A. and Elf Aquitaine Inc., and a subsequent study for Public Works Canada using PERD funding. The authors are grateful to Mr. Andre Cormeau for carrying out the computer work required to combine the reliability software with the ice load software.

References

American Petroleum Institute. 1987. API recommended practice for planning, designing and constructing fixed offshore platforms. Revised API Bul. 2N, draft.

Benjamin, J.A. and Cornell, C.A. 1970. Probability, statistics, and decision for civil engineers. McGraw Hill.

Canadian Standards Association. 1988. General requirements, design criteria, environment and loads: Part I - CSA Code for the Design, Construction and Installation of Fixed Offshore Production Structures. CSA-S471, Eighth Draft.

Cramer, H. and Leadbetter, M.R. 1955. Stationary and related stochastic processes. John Wiley and Sons.

Croasdale, K.R. 1980. Forces on fixed rigid structures. Working Group on Ice Forces on Structures - A State-of-the-Art Report. U.S. Army Cold Regions Research and Engineering Laboratory, Hanover, New Hampshire. Special Report No. 80-26.

Det norske Veritas. 1977. Rules for the design, construction and inspection of offshore structures. Det norske Veritas, Høvik, Norway.

Jordaan, I.J. 1983. Risk analysis with application to fixed structures in ice. Seminar/Workshop on Sea Ice Management. Memorial University of Newfoundland, St. John's, Newfoundland.

Lainey, L., and Tinawi, R. 1984. Mechanical properties of sea ice - a compilation of available data. Can. J. of Civil Engineering, 11, pp.884-923.

Madsen, H.O. Krenk, R. and Lind, N.C. 1986. Methods of structural safety. Prentice-Hall.

Markham, W.E. 1981. Ice atlas: Canadian arctic waterways. AES Environment Canada.

Moghtaderi-Zadeh, M. and Madsen. H.O. 1987. Theoretical manual for PROBAN: The probabilistic analysis program, A.S. Veritas Research, Høvik, Norway, Report No. 868-2036.

Nessim, M.A., Jordan, I.J., Lantos, S. and Cormeau A. 1986. Probability-based design criteria for ice loads on fixed structures in the Beaufort Sea. Det norske Veritas (Canada), Calgary, Joint Industry Report No. 86-CGY-43, 2 volumes.

Nessim, M.A., Cheung, M.S. and Jordaan, I.J. 1987. Ice action on fixed offshore structures - A state-of-the-art review. Can. J. of Civil Engineering, 14, pp.381-407.

Nessim, M.A. and Cormeau, A. 1988. Probabilistic framework for ice loads on fixed marine structures. Det norske Veritas (Canada) Report No. 88-CGY-005, for Public Works Canada (PERD) funding.

Nevel, D.E. 1986. Iceberg impact forces. IAHR Symposium Iowa City, Iowa, Vol.3, pp. 345-369.

Ponter, A.R.S., Palmer, A.C., Goodman, D.J., Ashby, M.F., Evans, A.G. and Hutchinson, J.W. 1983. The force exerted by a moving ice sheet on an offshore structure. Cold Regions Science and Technology, pp.109-118.

Pounder, E.R. 1965. The physics of ice. Pergamon Press, Oxford, England.

Prodanovic, A. 1981. Upper bounds of ridge pressure on structures. Proceedings, POAC, Quebec, Canada, Vol.3, pp.1288-1302.

Sanderson, T.J.O. 1988. Ice mechanics, risks to offshore structures. Graham and Trotman, London, England.

Thoft-Christensen, P. and Baker, M.J. 1982. Structural reliability theory and its applications. Springer-Verlag, Berlin.

Tvedt, L. 1986. PROBAN users manual, Det norske Veritas Research Division, Høvik, Norway, Report No. 86-2037.

Weeks, W.F., and Mellor, M. 1983. Mechanical properties of ice in the arctic seas. Arctic Technology and Policy. Hemisphere Publishing Corp., New York, N.Y., pp.235-259.

Discussion

Q: S. Shyam Sunder, MIT

My comments are in three parts:

1. Can you treat the nonstationarity of the loads due to thickness changes in terms of a reduced variable with a moving average?

2. During breakup, is the random process really stationary or are there discrete sudden events due to storms which may be better described via a Poisson model?

3. Instead of using Monte Carlo simulations which are time-consuming, have you considered the use of response surface analysis techniques?

A: M. Nessim

1. The same models can be applied if the process can be transformed into a stationary process by applying a transformation which accounts for the changing average.

2. Breakup refers to the season between winter and summer where the sea is partially covered by the disintegrating ice cover. This is modelled by an interrupted process which can be visualized as a stationary process by eliminating open water periods. We have also treated break-off events where an area of the first-year ice would break off from the landfast zone under the influence of storms. A Poisson model would be appropriate for this, although we have used only a fixed rate of occurrence.

3. We have used first order reliability methods which approximate the response surface. The approach is described in the paper and has proven to be much more efficient than Monte Carlo simulations.

Q: O. Mahrenholtz, Technical University Hamburg - Harburg, Germany

What is the consequence of the return period relationship (Figure 7b) for the design of the structure concerned?

A: M. Nessim

The development of a return period-load relationship allows the designer to select a load which corresponds to a certain return period (e.g. 100 year). When the structure is designed for this load, with the application of appropriately calibrated partial factors, the target safety level would, on average, be achieved.

Q: T.J.O. Sanderson, BP International

This is a very good paper - risk analysis is the essential key to design load specification. It is worth emphasizing that a "100 year" design load does not imply that failure of the structure occurs on average once every hundred years. In practice, the design load is multiplied by a safety factor in such a way that the overlap of the load occurrence p.d.f. and the structural resistance p.d.f. give a failure probability of 10^{-4} to 10^{-6} per annum. Safety factors are a traditional short cut to this but the failure probability could alternately be developed on the basis of a full risk analysis without safety factors.

A: M. Nessim

I agree.

Structural Risk Model of Arctic Shipping

C. G. Daley

Fleet Technology Limited
311 Legget Drive, Kanata, Ontario, K2K 1Z8

C. Ferregut

Dept. of Civil Engineering, University of Texas at El Paso
El Paso, Texas, 79968-0516, USA
(Formerly at Fleet Technology Limited, Kanata, Ontario)

R. Brown

Fleet Technology Limited
311 Legget Drive, Kanata, Ontario, K2K 1Z8

Abstract This paper discusses Arctic shipping risk analysis. A program of research was started in 1982 to develop a risk analysis model. A computer program, ASPEN (Arctic Shipping Probability Evaluation Network) has been the focal point of the research. The elements of the program mirror the elements of the problem. These include: environmental definition, the navigation and ice avoidance process (including human behaviour), ship/ice collision mechanics and structural limit states (failure mechanics). All aspects of the problem are variable and must be treated statistically. Related studies to obtain the statistical parameters are discussed. The numerical methods employed by the program are described. Many fundamental issues remain unsolved, particularly those related to correlated variables. The paper concludes with a discussion of required developments.

1. Introduction

Arctic sea ice is a complex natural system which is largely dynamic, random and transient. The source of the randomness cannot be stated, but is doubtless connected with the random natural forces which cause and modify the ice cover.

S. Jones · R. F. McKenna · J. Tillotson · I. Jordaan (Eds.)
Ice-Structure Interaction
IUTAM-IAHR Symposium St. John's, Newfoundland, Canada

The following premise underlies the work presented here:

> "All ships navigating in the Arctic face a finite risk of being damaged. This is a result of the random nature of ice coverage and strength as well as uncertain human behaviour."

This statement cannot be proven. Empirical evidence will support this assumption.

The past two decades have seen considerable increase in the level of ship traffic in Arctic waters. There has been a corresponding increase in the scientific and technical investigations of ships in ice. The work presented here has its direct origins during the years 1980-83. Many technical studies of ice loads and Arctic ship design were underway (Edwards et al., 1981; Voelker et al., 1981). In Canada the existing Arctic Shipping Polution Prevention Regulations (Government of Canada, 1972) were thought to be in need of updating. The Canadian Coast Guard (CCG), responsible for ensuring safe shipping in the Arctic, was sponsoring much of the research. CCG felt that a methodology was neeeded to evaluate proposed rule changes, and to identify the crucial design factors of concern for safe shipping. Consequently, CCG sponsored the development of an Arctic shipping risk model. The model was to be capable of determining ice damage risk on the bow and shoulders of any vessel transiting any part of the Canadian Arctic (waters north of 60°N — west of Greenland[1] and east of 141°W) during any month of the year.

When the project was initiated, almost none of the necessary data (ice coverage, ship navigation, ice loads and ship strength) existed in the required form. That situation was not realized at the time, but became apparent during the development of the model. The original assumptions as to the nature and complexity of the problem seem overly simplistic from today's perspective. They were:

(1) The primary uncertainty in ice loads was thought to be due to impact velocity variations which result from manoeuvering variability and variability in ice floe size.

(2) The random ship manoeuvres were thought to result [primarily] from random floe positions. A simple (i.e. "dumb") manoeuvering al-

[1]Note: the Greenland coast waters are outside Canadian jurisdiction, but were included in the risk analysis domain for completeness.

gorithm, coupled with random floe placement would have allowed impact statistics to be gathered with a quite simple Monte Carlo simulation.

(3) The random impact values (ship speed, impact coordinates on the ship, ship and ice floe rigid body motions) could have been coupled with random ice edge shape values to produce (by a deterministic ice extrusion model) a random ice pressure, (Noble and Menon, 1981; Kheisin et al., 1973; Tam, 1980).

(4) Ship strength was thought to be definable deterministically with damage probabilities calculated directly from the random ice pressures.

All four of these assumptions were later found to be inadequate.

2. Background to Arctic Shipping Risk Analysis

Risk analysis is the designation that has become common within the marine industry to indicate the reliability studies that account for all possible failure modes. In general, risk is a quantitative measure that combines the probability of occurrence of an undesirable event and its consequences. Risk analysis can be used as a framework for accepting a system, for improving the system's design or for implementing a risk control program.

Probabilistic treatments of problems associated with ice-going vessels are relatively scarce. The earliest works appear to be those of the Russian scientists, Kheisin (1972), and Kudishkin (1972), who described ice loads in homogeneous level ice as being a stationary random ergodic process. Records of random ice loads can be found in Varsta (1977), Vuorio et al. (1979), Daley et al. (1984), Hakala and Riska (1983), and Ghoneim (1983).

There have been probabilistic analyses of the ice environment including attempts to build probabilistic models of ice failure (i.e. Bercha, 1979).

The objective of this short review is to describe the situation at the time the ASPEN project was initiated. While the fundamental theory of statistics and risk analysis was available, there was very little practical information to guide the development of an Arctic shipping risk model.

3. Overall Problem Formulation

For a ship to be damaged in ice, a particular series of events is assumed to occur. Let us begin at the end and work through all the necessary and/or possible preconditions for damage indicated by the arrows in Figure 1.

On the path to damage, the events E_1 to E_6 may form a simple sequence. Complicating factors are the chances of not detecting the ice, or

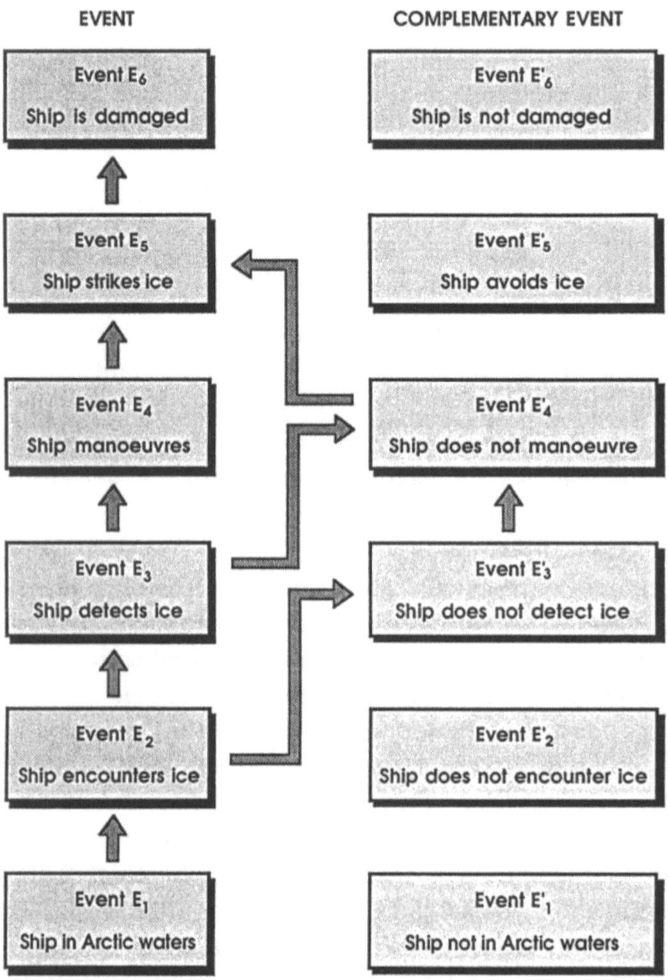

Figure 1. Events and complementary events leading to ice damage.

if detected, choosing not to avoid it. To calculate the probability of event E_6 it is necessary to calculate the probabilities of all preconditions. In general, the event probabilities are calculated as

$$P(E_i) = P(E_i|E_{i-1})P(E_{i-1}) + P(E_i|E'_{i-1})P(E'_{i-1}), \tag{1}$$

[although many of the conditional probabilities for complementary events are zero (i.e. $P(E_6 =$ ship is damaged $|E'_5 =$ ship avoids ice$) = 0$)].

The above description is the essence of the ASPEN model. Consider the situation of one ship with one limit state, travelling through one cell which contains one floe of one type of ice (randomly positioned). Assume single values for encounter, detection, avoidance and damage probabilities. Clearly in this case the calculation of risk would be trivial.

In the case of the ASPEN risk model, many more variables are added. But the essential risk calculation is the same. The overall flowchart of the risk calculations is shown as an event tree in Figure 2. The logic of the event tree can easily be implemented in a computer program. Figure 3 shows the structure of the ASPEN program, in a kind of pseudo code which most programmers should readily follow.

Let us now get into the details. The following sections will address the main aspects of the ASPEN program:

- probabilistic ice database,
- calcuation of collision probabilities,
- calculation of damage probabilities and
- the ASPEN workstation.

The first two topics have been covered quite extensively in previous papers (Perchanok et al., 1986b; Perchanok et al., 1989; Ferregut et al., 1987). A short review of these topics will be followed by a more comprehensive treatment of the last two topics.

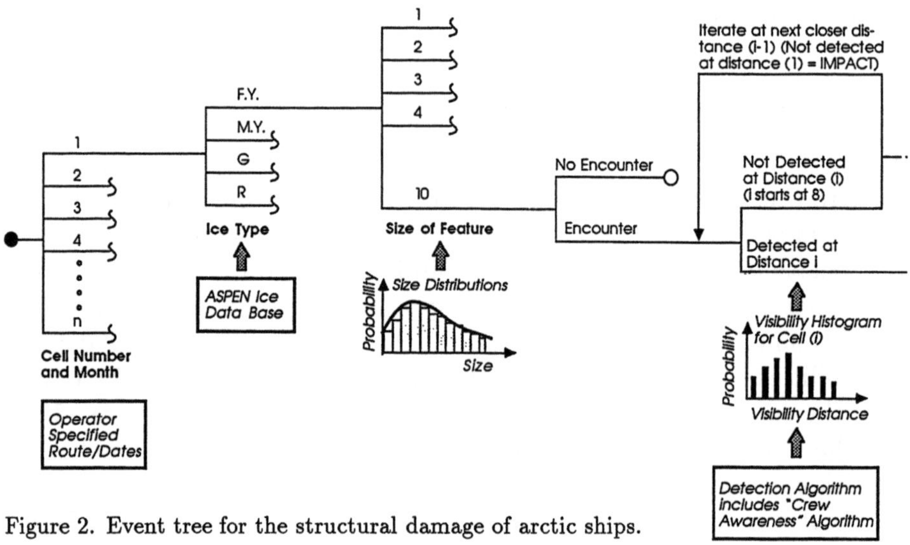

Figure 2. Event tree for the structural damage of arctic ships.

```
Routine     ASPEN

Title       Main Calculation for Arctic Shipping Risk Model

Variables   Cell:        counter for cell/month of route
            Maxcell:     last cell in route
            Ice Type:    ice type counter (first year, glacial, ridged and multi-year ice)
            Ice Size:    ice feature size  (sizes depend on ice type)
            Structure:   types of ship structure  (shell, main frame, stringer, hull girder)
            Limit:       limit states (number depends on type of structure)

Begin       Init Variables
            Get Route Data
            Get Ship Data
            Get Ice Data
            Cell = 1
            WHILE (cell ≤ maxcell) DO
                    For Ice Type = 1 to 4
                      For Ice Size = 1 to M (Ice Type)
                        Calc _ Prob _ Encounter
                        Calc _ Prob _ Detection
                        Calc _ Prob _ Manoeuvre
                        Calc _ Prob _ Collision
                        For Structure = 1 to 4
                            For Limit = 1 to N (Structure)
                               Calc _ Prob _ Damage
                            Next Limit
                          Next Structure
                        Next Ice Size
                      Next Ice Type
                    Cell = Cell + 1
                  END WHILE
                  Output Results
End
```

Figure 3. ASPEN program (main routine in pseudo code).

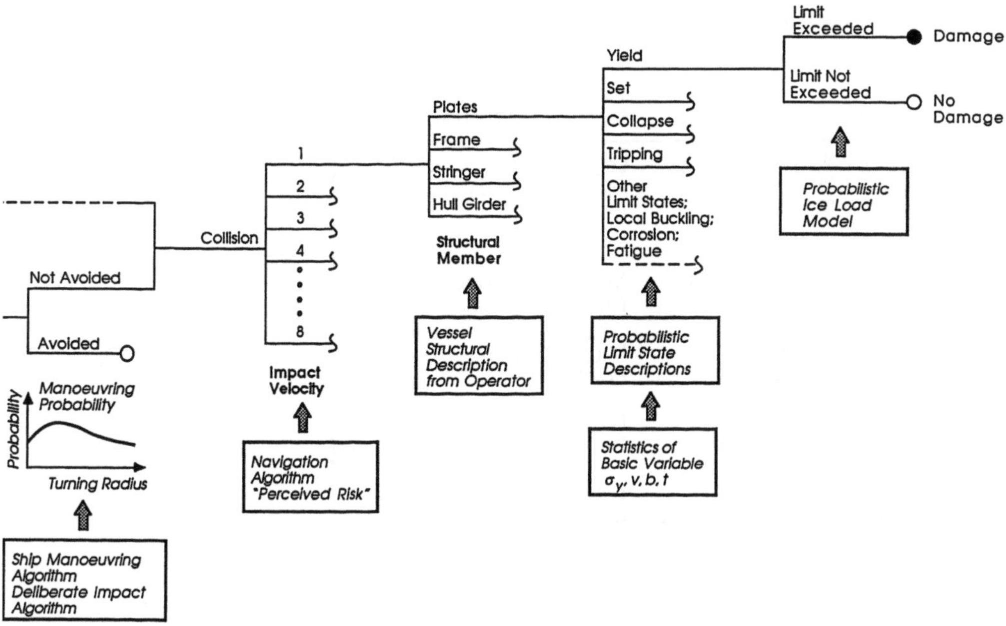

4. Probabilistic Ice Database

The ASPEN program requires a probabilistic description of several environmental variables:

- sea ice concentration,
- sea ice thickness,
- sea ice floe size,
- iceberg frequency,
- iceberg size,
- first year ridge heights,
- first year ridge frequency and
- visibility.

For each variable, a mean value and coefficient of variation are stored. Alternative distribution functions were examined for each variable. Chi-square goodness-of-fit tests were used to select the best functions for each variable type. Table 1 lists the environmental variables and the types of distributions chosen. Figure 4 shows a typical frequency diagram for raw data on ice thickness, together with the fitted lognormal distribution.

Table 1. Statistical distribution used for ASPEN variables.

ASPEN Variable	Type of Statistical Distribution
F.Y. Ice Thickness	Lognormal
F.Y. Ice Concentration	Mean and CoV
F.Y. Floe Size	Exponential
M.Y. Ice Thickness	Lognormal
M.Y. Ice Concentration	Mean Value and CoV
M.Y. Floe Size	Exponential
F.Y. Ridge Height	Lognormal
F.Y. Ridge Frequency	Poisson
Glacial Ice Diameter	Lognormal
Glacial Ice Frequency	Poisson
Visibility	Discrete Histogram Values

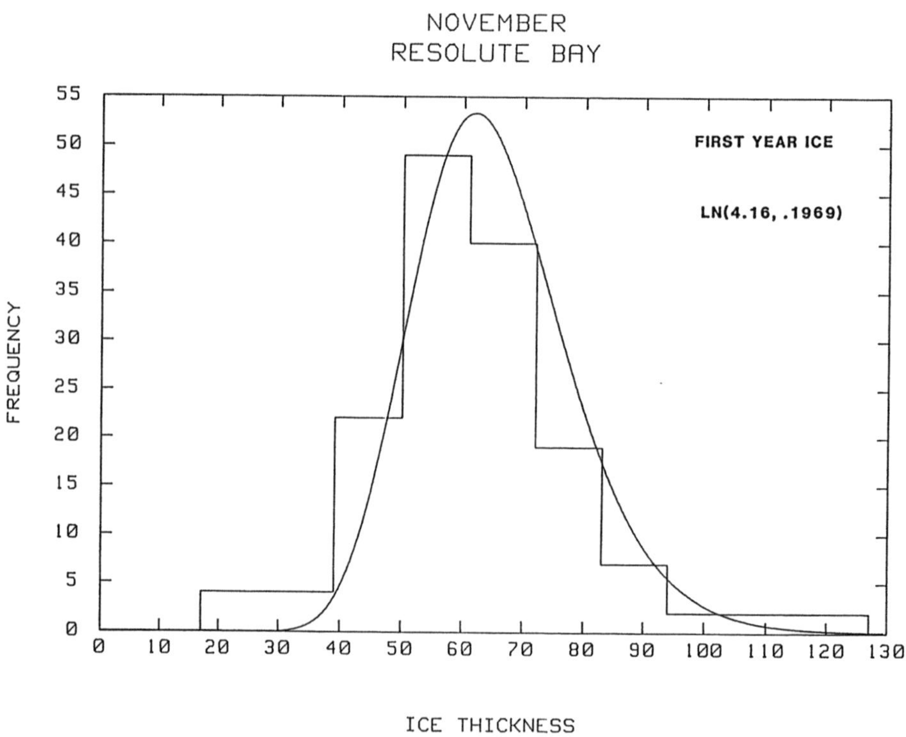

Figure 4. Frequency distribution for first year ice thickness in the vicinity of Resolute Bay, NWT, for November.

Further details can be found in Perchanok et al. (1989) and in Section 7.1.

5. Calculation of Ice Collision Probabilities

The risk model is based on the premise that the amount of damage is a function of the number and intensity of collisions with floating ice along a given route. Consequently, the calculation of collision probabilities is a necessary step prior to calculating the damage probabilities.

The conceptual model of the encounter/detection/manoeuvre/collision process is illustrated in Figure 5. While it is realized that ice floes are clustered and often avoided in groups, the mathematical treatment of the problem assumes that hazards are encountered and avoided individually (Ferregut et al., 1987). Figure 6 illustrates the geometry involved in the encounter calculations. The probability of encounter of a floe in a cell is

$$P(A) = \frac{b+d}{W}, \tag{2}$$

where

$P(A)$	= probability of event A (encounter),
b	= beam of ship (in meters),
d	= diameter of floe (in meters) and
W	= width of cell (in ASPEN, $W = 15$ miles).

The number of encounters is assumed to occur according to a Poisson distribution of the type

$$P(M = m) = \frac{\nu^m e^{-\nu}}{m!}, \tag{3}$$

where

$P(M = m)$	= probability that the number of ice encounters, M, takes on the specific value m and
ν	= mean rate of encounters (mean number of floes per cell times the probability of encountering a floe).

Clearly the encounter probabilities are very easy to calculate. Calculation of detection and avoidance probabilities is somewhat more difficult. Detection (i.e. knowledge of encounter) probabilities are dependent on

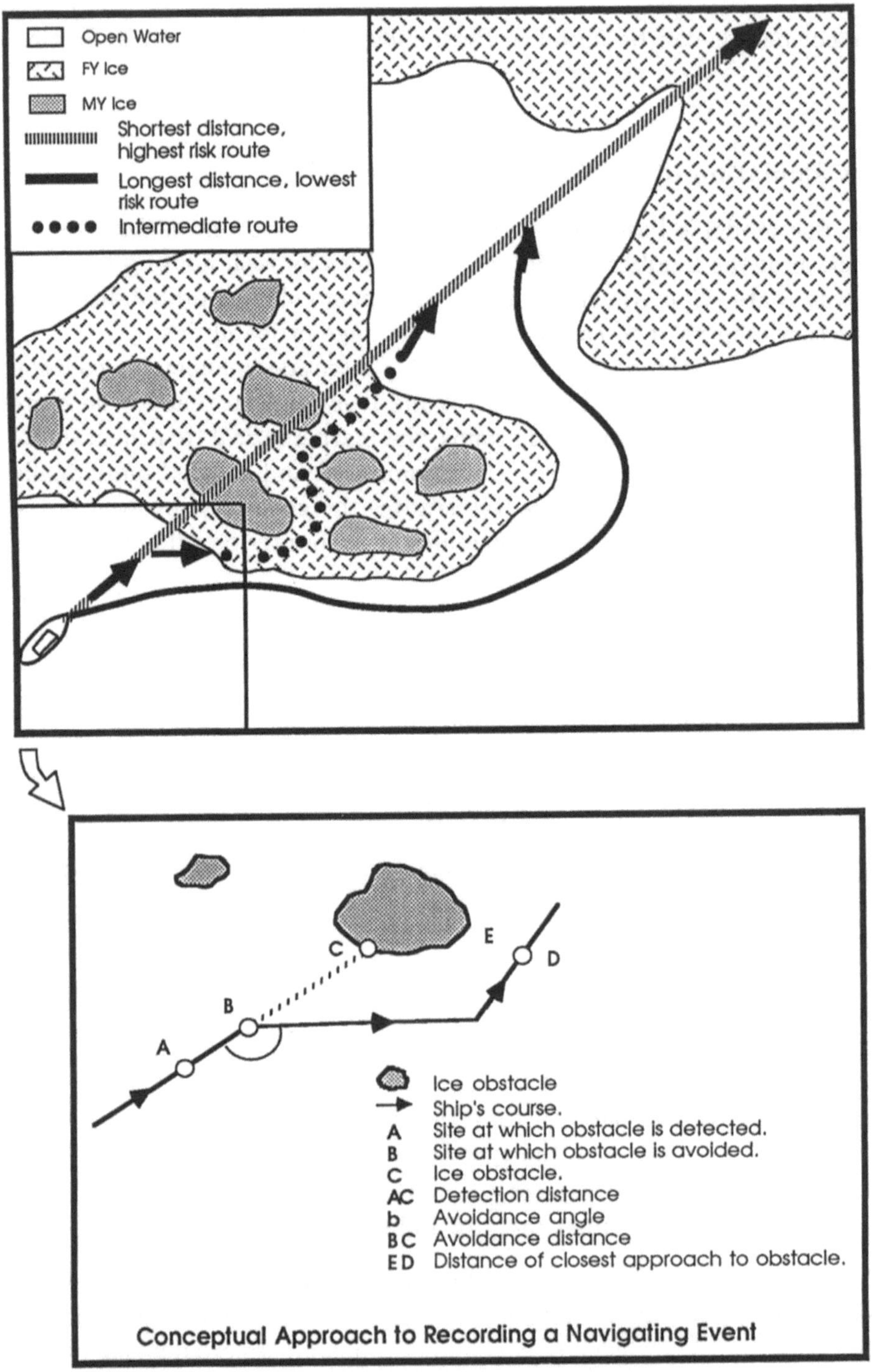

Figure 5. Conceptual model of encounter, detection, manoeuvre, and collision with ice (after Ferregut et al., 1987).

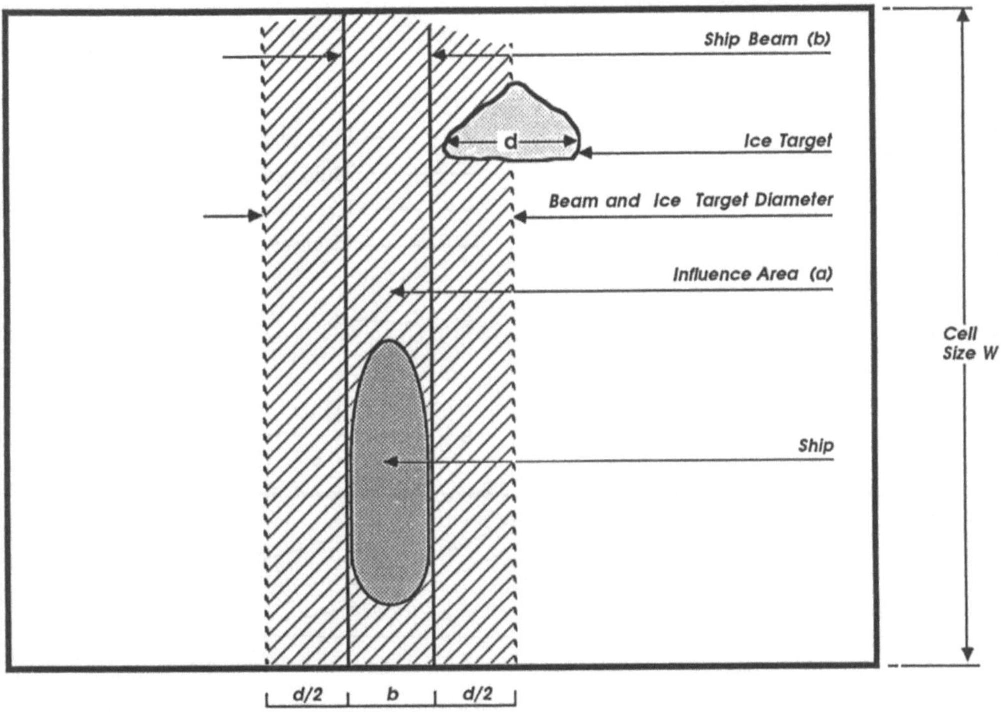

Figure 6. Geometry of encounter probability calculations (after Ferregut et al., 1987).

- atmospheric clarity (termed **visibility**) and
- capacity of an ice feature to be recognized and distinguished from the background scene (termed **detectability**).

Beta distribution models have been selected for the detectability functions as shown in Figure 7. When the visibility is less than the distance to the object, the detection probabilities are equal to the detectability probabilities, truncated by the visibility distances. The outcome of detection calculations is the distribution of distances at which floes will be detected. Different floe sizes and ice types result in different distributions.

Conditional avoidance probabilities are based on the following assumptions:

- Upon detecting the floe, the ship will decide to attempt avoidance or not.

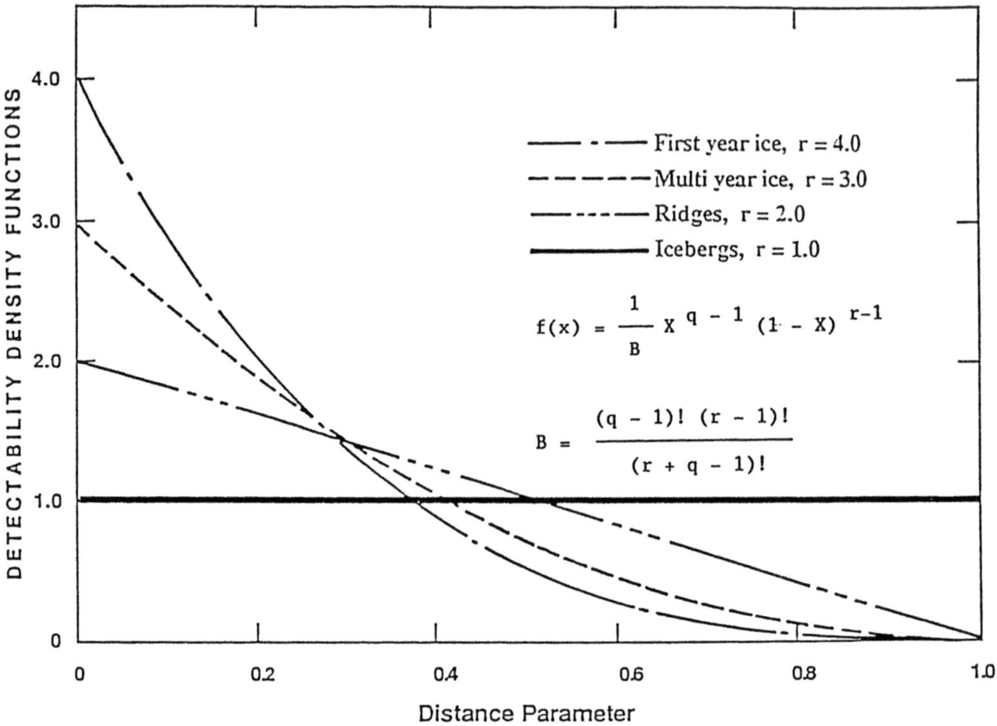

Figure 7. Standard beta density function, $q = 1.0$.

- If avoidance is attempted, the ship will follow an arc defined by the minimum turning circle.
- The radius of the minimum turning circle is dependent on ice conditions, and is a random variable described by a normal distribution.

For ice floes of a given size and type, the detection/avoidance probabilities follow the same distribution for all such floes. The result of this is that collision probabilities will also follow a Poisson distribution (with parameters dependent on ice type and size). Again the reader is referred to an earlier paper for a more detailed explanation of the collision process (Ferregut et al., 1987).

6. Calculation of Damage Probabilities

The probability of damage is the probability that load exceeds strength. For this definition strength must be defined as the load which causes

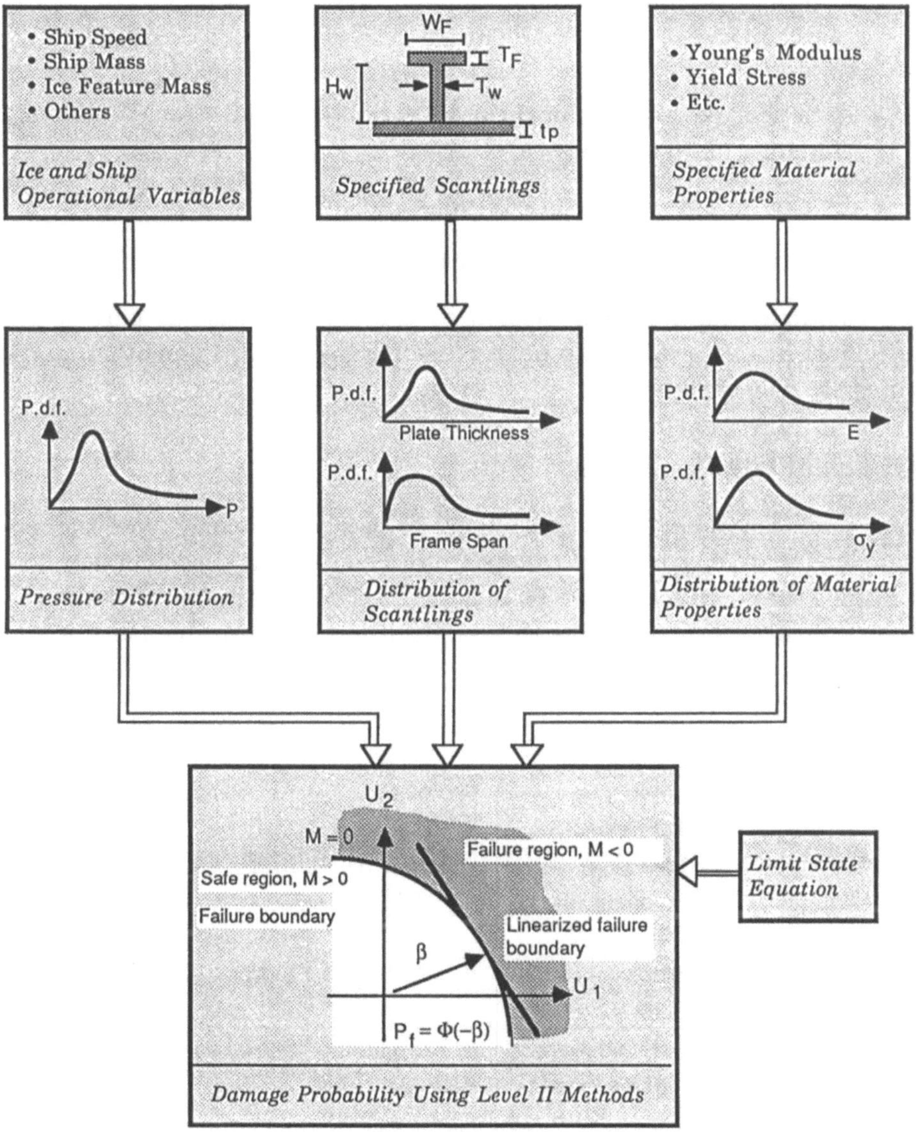

Figure 8. Schematic flow diagram for the computation of the probability of damage, given a ship/ice collision.

the structure to experience a given limit state. Figure 8 illustrates the development of load and strength probability distributions with all the calculation of the probability of damage (for a single limit state). The calculations require five initial quantities:

- scantlings (nominal),
- limit state descriptions,
- statistical distribution for the basic variables (yield, strength, Young's Modulus, plate thicknesses, etc.),
- distribution of number of collisions and
- statistical distribution of ice pressure for a single collision.

The nominal scantlings are defined by the ASPEN user, as part of the specification of the vessel (to be discussed in Section 8). The other quantities are discussed below.

6.1 Limit State Descriptions

The condition of a ship can vary more or less continuously from the undamaged state to a state of collapse. As a convenient simplification it is assumed that a structure can be in a finite set of limit states. Usually the limit states correspond to the thresholds of some undesirable behaviours.

Table 2. Limit state descriptions.

Structure	Limit State	Limit State Equation	Load Area
Shell Plating	Initial Yield	$P_y = (2\,\sigma_y/(1\text{-}v^2))\,(t/b)^2$	$A = b^2$
	Edge Hinge	$P_{eh} = (3\,\sigma_y/(1\text{-}v^2))\,(t/b)^2$	$A = b^2$
	Permanent Set *(10% of Span)*	$P_{set} = (.74\,\sigma_y/(1\text{-}v_p^2))\,t/b$	$A = b^2$
	Collapse *(Ultimate Strength)*	$P_{ult} = (1.11\sigma_y/(1\text{-}v_p^2))\,t/b$	$A = b^2$
Frames & Stringers	Initial Yield	$P_y = 12\,\sigma_y S/dL^2$	$A = L \times d$
	Set *(Edge Hinge)*	$P_{eh} = 12\sigma_y Z/dL^2$	$A = L \times d$
	Collapse *(3 Hinges)*	$P_{ult} = 16\sigma_y Z/dL^2$	$A = L \times d$
	Tripping	$P_{trip} = 58\,S/dL^2\,((\pi^2 E/L^2) \times ((a^2 I_y + \Gamma)/I_{pc}) + GJ/I_{pc})$	$A = L \times d$
	Local Buckling	$P_{LB} = ((48\,KE)/\,1-v^2))\,(t_w/h_w)^2$	$A = L \times d$
Hull Girder	Initial Yield	$P_y = D_s\sigma_y(AF/2 + Aw/6)/.2L_s\,A_L$	$A = 1000$
	Ultimate Strength	$P_{ult} = Z_H\sigma_y/.2L_s\,A_L$	$A = 1000$

The set of limit states which ASPEN considers is given in Table 2. Each limit state is a deterministic equation of strength, expressed as the load capacity (pressure for a given load area) in terms of basic dimensions and properties. These equations are taken from Ferregut and Daley (1988). The plating equations are quite standard, in this case taken from Ratzlaff and Kennedy (1986). The frame strength equations are standard. The frame stability equations were derived in Daley and Ferregut (1988). The hull girder strengths were taken from Daidola and Basar (1980) and Caldwell (1965). Typical statistical parameters of the basic variables on the limit state equations used by the ASPEN program are given in Table 3.

Table 3. Statistics for material and geometric variables.

Variable Name	Symbol	Mean/ Specified	Coefficient of Variation	Reference
Area of Flange	A_f	1.02	0.012	Kennedy and Gadaly, 1980
Area of Web	A_w	1.02	0.012	Kennedy and Gadaly, 1980
Young's Modulus	E	1.02	0.012	Galambos and Ravindra, 1978
Shear Modulus	G	1.02	0.012	Galambos and Ravindra, 1978
Height of Web	H_w	1.0	0.001	ISSC, 1985
Moment of Inertia with Respect to Toe	I_{pc}	1.03	0.02	Kennedy and Gadaly, 1980
Moment of Inertia About y	I_y	1.03	0.02	Kennedy and Gadaly, 1980
Plate Constant	K	1.1	0.1	Assumed
Unsupported Length	L	1.0	0.03	Allen, 1975
Elastic Section Modulus	S	0.99	0.021	Kennedy and Gadaly, 1980
Thickness of Flange	T_f	1.0	0.016	ISSC, 1985
Thickness of Web	T_w	0.99	0.01	ISSC, 1985
Thickness of Plating	t	1.0	0.05	Allen, 1975
Width of Flange	W_f	1.0	0.003	ISSC, 1985
Plastic Section Modulus	Z	0.99	0.038	Kennedy and Gadaly, 1980
Poisson's Ratio	ν	1.0	0.03	Galambos and Ravindra, 1978
Yield Stress	σ_y	1.1	0.09	Kennedy and Baker, 1984

6.2 Statistical Distribution of Ice Pressure

Ice forces and pressures are best described as statistical quantities. ASPEN requires a general statistical model of ice pressure in order to perform the risk calculation. A complete understanding of the ship/ice contact phenomena, and the statistical processes involved, does not yet exist. Much work is continuing to understand this subject. The general model that will be described below is presented only as a workable procedure. It was developed in 1984 (Daley, 1984) and has not been significantly updated since.

In 1983 an extensive set of local pressure measurements were made on the USCGC *POLAR SEA* (Daley et al., 1984). The local ice pressures in multi-year ice were analyzed statistically as a function of area. Figure 9 shows cumulative distribution functions for ice pressures over a range of areas. The data represent 368 multi-year ice impacts that were recorded in April of that year in the northeastern Chukchi Sea. Pressures reported for more than one measuring panel represent contiguous areas. The form of the Type I extreme value equation (Benjamin and Cornell, 1970) fitted to the data is

$$F_P(P) = e^{-e^{-\alpha(A)(P-U(A))}}, \tag{4}$$

where

$F_P(P)$ = probability the ice pressure will exceed the value P,
$\alpha(A)$ = dispersion parameter (function of area) = $1.563A + .35$ (in this case),
$U(A)$ = characteristic value (function of area) = $-.046A + .5$ (in this case) and
A = area (m^2).

At a probability of .99 (1% probability of exceedance) the equation for pressure as a function of area is

$$P^*(A) = .37\frac{F}{A}\left[1 - \frac{1}{1.24\left[\left(\frac{139A}{F}\right)^{.85}\right]}\right]. \tag{5}$$

where

$P^*(A)$ = ice pressure (probability .99),
F = total impact force (MN) and
A = area (m^2).

To generalize this equation for all ships and ice types, the following assumptions were made:

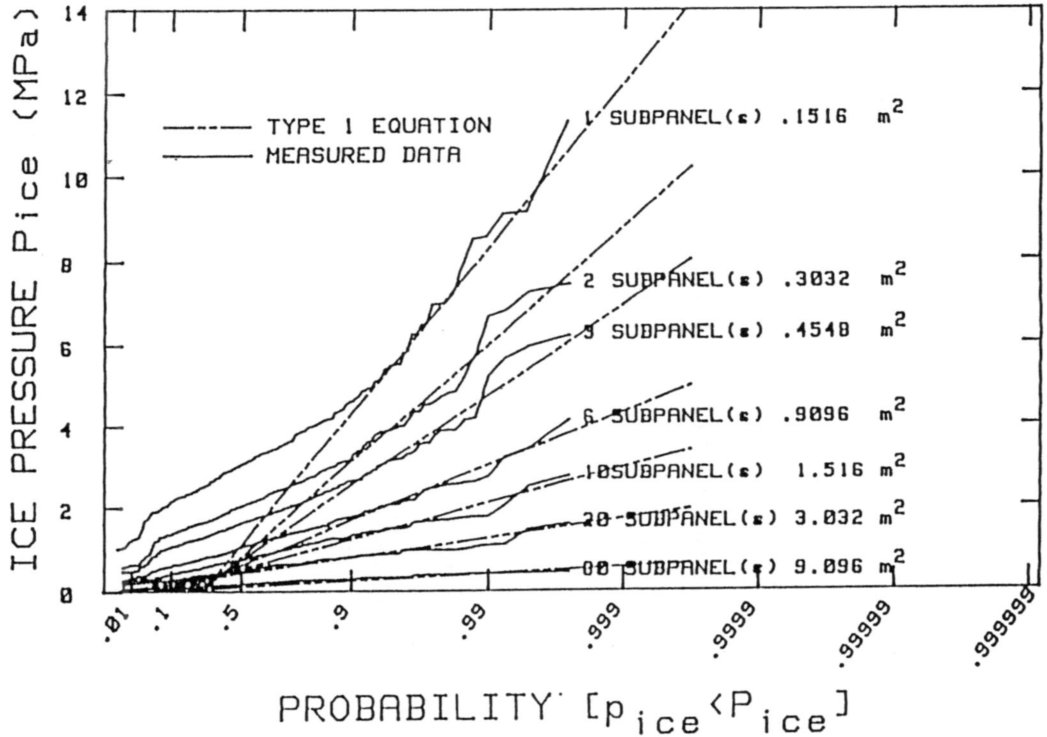

Figure 9. Cumulative distributions of multi-year ice pressure on the USCGC *POLAR SEA* (after Daley et al., 1984).

- the term $U(A)$ $(= -.046A + .5)$ is constant for all ships,
- total force, F, is a function of ship and ice parameters, and
- equation (5) above is valid for all situations of multi-year ice impacts.

Equation (5) can be inverted and combined with equation (4) to allow the calculation of $\alpha(A)$ for all multi-year impacts as follows:

$$\alpha(A) = \frac{4.6}{P^*(A) - U(A)} \,. \tag{6}$$

With F and $\alpha(A)$ determined, equation (4) defines the distribution of pressure (versus area) for all multi-year impacts.

A simple equation for the force F when ramming infinite ice is

$$F = .68V\Delta^{.9}, \tag{7}$$

where

F = total ice force (MN),
V = ship speed (m/s) and
Δ = vessel displacement (kT).

When the ice mass is non-infinite, the term Δ in equation (7) should be replaced by

$$\Delta_e = \frac{1}{\frac{1}{\Delta} + \frac{1}{\Delta_i}}, \tag{8}$$

where

Δ_e = effective displacement (kT) and
Δ_i = mass of ice (kT).

In this case of level ice of finite thickness (where flexural failure will dominate) the equation for force F is (Daley, 1985)

$$F = 1.05\sigma_f h^{1.875}(V/\tan\beta)^{.25}, \tag{9}$$

where

σ_f = flexural strength of ice (MPa),
h = ice thickness (m) and
β = shear angle of ship (from vertical).

Ice forces and pressures in other ice types can be modified using quasi-empirical factors (Daley, 1984).

6.3 Probability of Damage

The performance function that expresses the relationship between the strength and load effect of a structural member according to a specified limit state (or failure mode) is usually given by

$$M = R(x_1, x_2, \ldots, x_m) - L(x_{m+1}, \ldots, x_n), \tag{10}$$

in which $R(x_1, x_2, \ldots, x_m)$ is the strength of the structural components according to a specified limit state, $x_1, \ldots, x_m$ are the basic strength random variables, $L(x_{m+1}, \ldots, x_n)$ is the load effect on the specified failure mode and $x_{m+1}, \ldots, x_n$ are the basic load variables. The failure event is defined as the event $M < 0$ and the survival event as the event $M > 0$. Thus the probability of failure or damage can be evaluated by the following integral:

$$P_f = \int \ldots \int f_x(x_1, x_2, \ldots, x_n) dx_1, dx_2, \ldots, dx_n, \tag{11}$$

where f_x is the joint density function of the x_i's and the integration is performed over the region where $M < 0$. Because each of the basic random variables has a unique distribution and they interact, the integral in equation (11) cannot be easily evaluated. An approximate estimate to the probability of damage can be obtained using the following procedure (Rackwitz and Fiessler, 1978). The original basic variables space (or x-space) is transformed into a standard normal variable space or (U-space) within which the variables are all mutually independent. In general, this is achieved by means of the Rosenblatt transformation. In the ASPEN program, because the basic variables of the original x-space are assumed all mutually exclusive, the required transformation is

$$x_i = F_{x_i}^{-1}(\Phi(u_i)) \text{ for } i = 1 \text{ to } n, \tag{12}$$

where $\Phi(\ )$ is the standard normal cumulative distribution function, and $F_{x_i}^{-1}(\ )$ is the inverse of the cumulative distribution function of the ith basic variable.

The probability of failure is approximated by the probability content of the failure region obtained by linearizing the transformed failure boundary at the point of its closest approach to the origin of the U-space using a truncated Taylor's series. The distance from the origin of the U-space to the linearization point is defined by Hasofer and Lind (1974) as the reliability index β, see Figure 8. The probability of failure is defined in terms of the reliability index by

$$P_f = \Phi(-\beta). \tag{13}$$

7. The ASPEN Workstation

All aspects influencing the structural risk of Arctic shipping discussed in previous sections have been integrated into a computer model. The software was written using the *C* programming language. The program layout is shown in Figure 10. It can be seen that the program consists of three main modules: the environmental database, the set-up run preprocessor and the examine results post-processor.

7.1 Environmental Database

The environmental database was implemented together with a Geographical Information System (GIS). Figure 11 illustrates the screen display of the GIS with its 80×120 Arctic data cells. The database contains all the statistical information on the Arctic environmental variables used by the risk analysis algorithms. It stores a permanent database of 26 parameters representing eight environmental variables for each of the 4340 grid cells over the ocean according to month (Figure 12). This database provides default input to the risk analysis algorithms. In addition to the permanent database, the user can enter project specific data by defining a region on the screen with rubber band line and keying values into a dialogue box. The data in the database can be displayed by listing all 26 parameters for a single cell and month as shown in Figure 12, or by selecting a single variable and displaying a map on which different values are shown using different shadings. They can also be extracted on a shipping route basis which uses all parameters along a linked series of grid cells defined by the user.

7.2 Set-Up Run Preprocessor

This module gives the user a similar display to that shown in Figure 11. When in this module the user can define new ships, specify the path of a route and perform the risk analysis calculations. The main inputs required to define the ship are the overall dimensions, the operational characteristics, structural scantlings and the material properties as shown in Figure 13a or 13b. The values used to define the ship are nominal values (e.g. design values). The second order statistical descriptions for those variables that are considered basic random variables are estimated using the specified nominal values and the relationships given in Table 3.

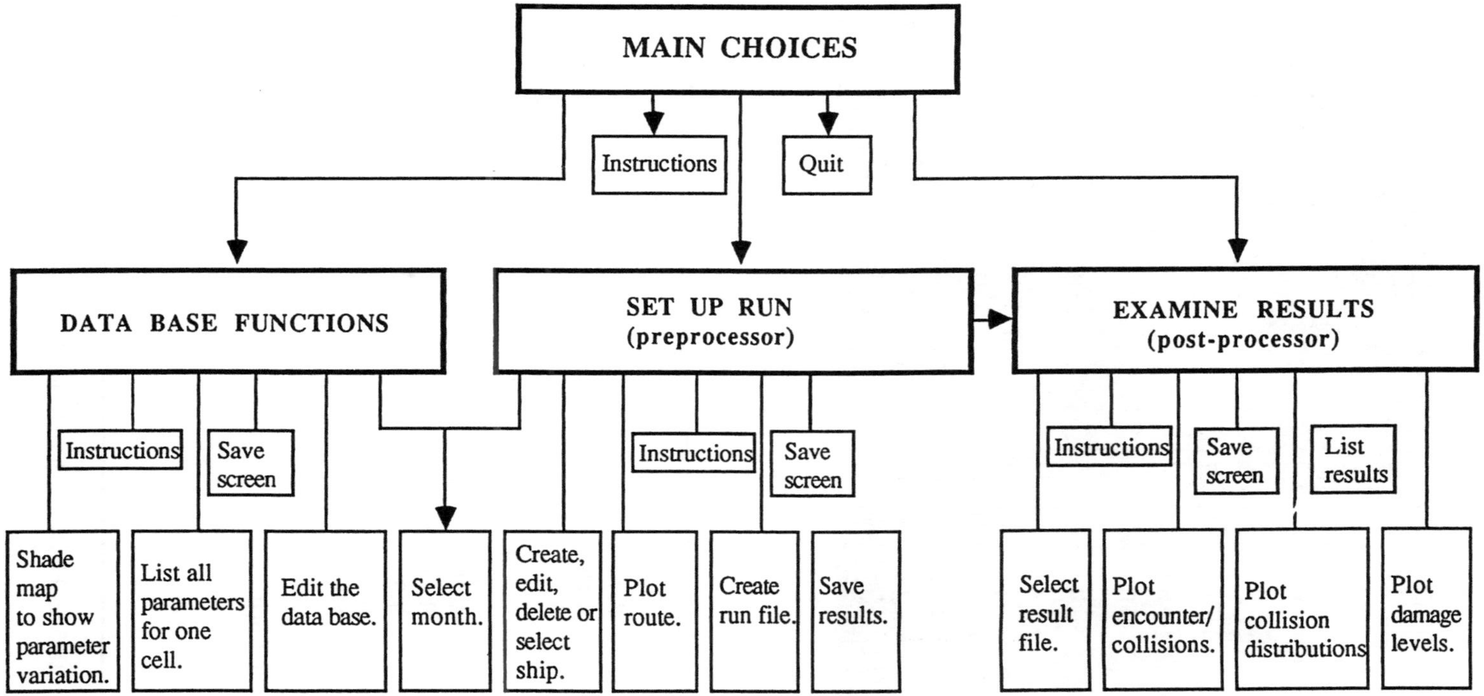

Figure 10. ASPEN flowchart.

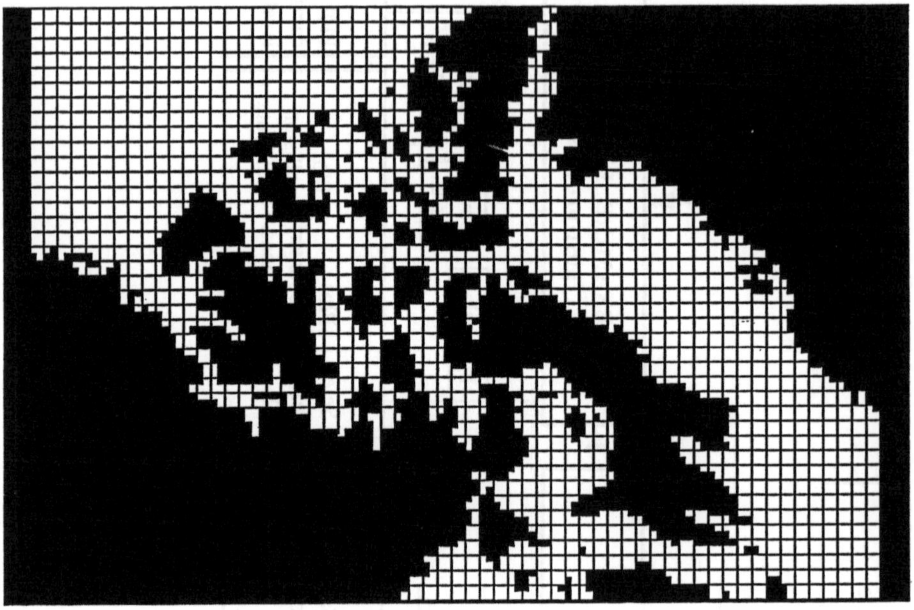

Figure 11. ASPEN map.

	Ridges	Glacial
Mean Height/Dia,m	0.73	119.9
Height/Dia CoV	0.35	0.98
Average Number	6	1

	F. Yr	M. Yr
Mean Thickness,m	1.55	2.6
Thickness CoV	0.17725	0.17725
Mean Conc, Tenths	9.9	0.1
Concentration CoV	0.5	2.5
Mean Floe Dia,m	60	60
Floe Dia. CoV	1	1

Visibility	Prob
Dist:0.025 km	0.008
0.125	0
0.35	0.021
0.75	0.015
1.5	0.045
3.0	0.055
7.0	0.161
15.0	0.695

Cell Number 4393 in April

Okay

Figure 12. Environmental variables for each cell/month.

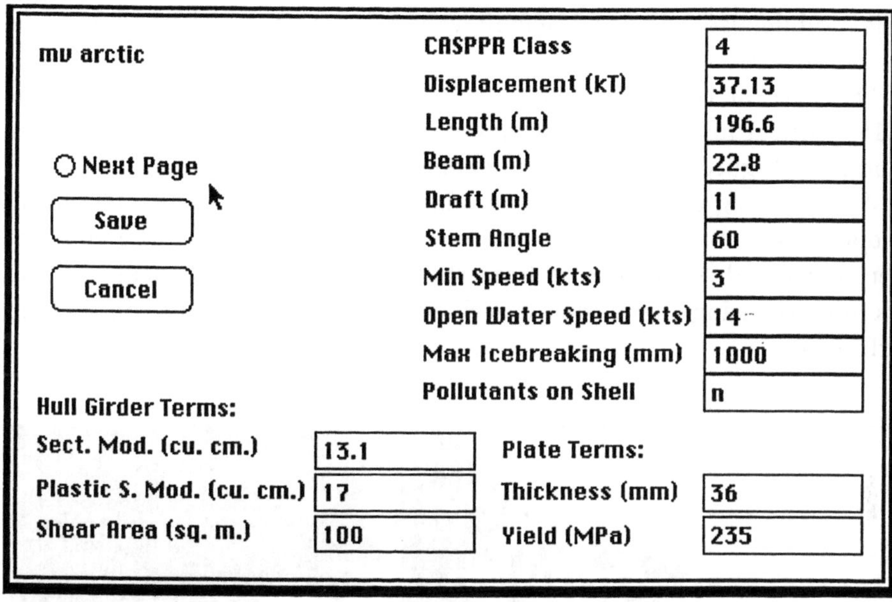

Figure 13a: Ship parameters.

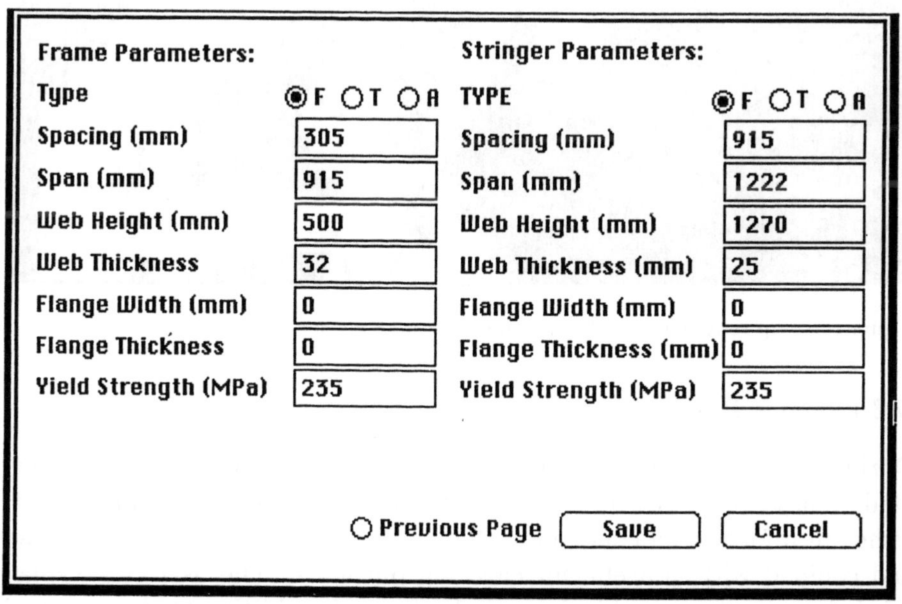

Figure 13b: Ship parameters.

This module also calculates default values for some variables that might not be known.

The path of the route (which consists of a sequence of square cells of 15 nautical mile per side) may be drawn interactively using a rubber band line (Figure 14). The route defined can be stored for further use.

When setting up a run, the user also is required to specify the month of operation. Based on this, the environmental data for the route is then assembled by reading it from the environmental database and the risk analysis using the models described in previous sections can then be performed.

Figure 14: Ship route on the ASPEN map.

7.3 Examine Results Post-Processor

During the course of the analysis a greal deal of information is generated. Usually only the final probability of damage results are of interest, although information concerning the number of ice features encountered and collided with is useful for checking purposes. The user can display this information in graphical or tabular form. In addition to the expected number of encounters and collisions, the user can display the distribution of collisions which gives an idea of the range for the expected number of collisions.

The final step of any risk analysis is to compare the results with criteria of acceptability determined by the original objectives. Depending on the purpose of the analysis the acceptability criteria are most often established by industry, government or a classification society. Thus the conclusions of a risk analysis are usually taken in the form of ranking of risks for various design and operation alternatives and recommended options, such as go-no-go evaluation, or a recommended framework for management and regulation to reduce the risks. To facilitate these processes of interpretation, analysis, and comparison, the final probabilities of damage on each limit state are displayed in graphical form grouped by element as shown in Figure 15. The interesting problem of acceptable risk levels is still a matter of on-going research and debate and it is out of the scope of this paper.

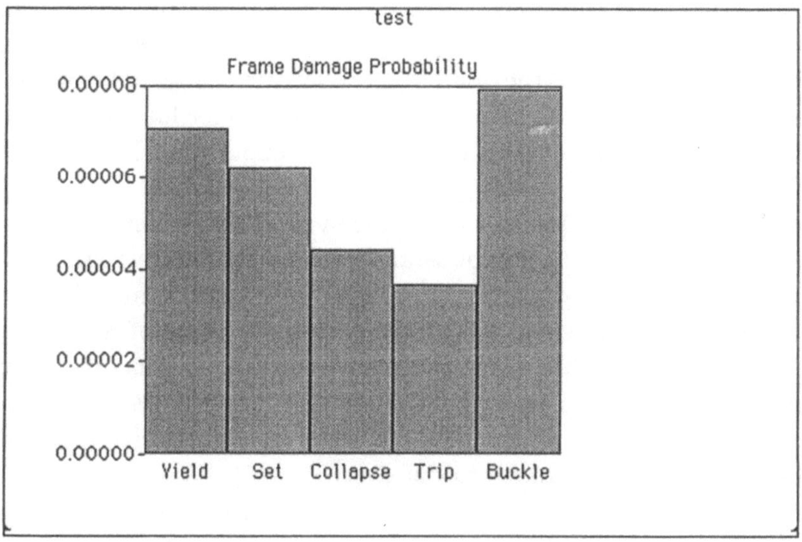

Figure 15: Probabilities of damage of main frames.

8. Conclusions

An operational model for predicting risk of ice damage on Arctic ships has been presented. The model is in principle quite simple. Complexities arise in the assembly of all the data required to execute the calculations, and in the numerical methods needed to obtain a result efficiently.

The primary strength of the model lies in the fact that it is complete in the sense that it models the environment, navigation, ice loads and structural failure. As such within the limits of its data/assumptions it can show the effect of changes in various parameters (route, time of year, hull strength).

Where should the model be improved? There are endless possiblities for further improvements. These include further refinements of the present scope (i.e. refined avoidance algorithm) and extensions of the current scope (i.e. inclusion of risk to bottom plating). The authors suggest two areas for improvement, one pragmatic and one very basic.

ASPEN has seen the improvement of several algorithms since it was first written. One that is still in its original form is the ice load algorithm. This could readily be improved by enhancing the load shape description and by improving the statistical models used for the load. Evidence suggests that the Type I extreme value model might not be best for all ice types and areas of coverage. As well, no account is presently taken of a preferred direction for the load (i.e. transverse and longitudinal structures are currently treated equally). An improved ice load algorithm would be an asset in keeping with the improved strength models that have been recently added.

The second suggested improvement is much more fundamental. Ships may avoid hazardous objects in groups and with exceptionally high quality information. The asumptions illustrated in Figure 4 may be too simplistic (i.e. independence of ice position, independence of avoidance events). To tackle this problem it is necessary to establish the correlation among the basic variables. Aside from improving ASPEN itself this exercise may serve to identify what one might call independent ice regimes. These would be valuable features to discover.

The principal value of risk models, such as the one described in this paper, is in its use as a relative measure for comparison between alternative systems (e.g. routes, ships, seasons). In this vein, the ASPEN model can be used by different organizations involved in Arctic shipping to assist ship designers in selecting among design alternatives; demonstrate the required safety level of a ship to governmental authorities in order to obtain approval to proceed with an intended voyage; aid shipbuilders and operators in making non-mandated decisions concerning allocation of resources and introduction of measures to reduce risk; provide a basis upon which the governmental authorities and classification societies can formulate more rational codes, standards of practice and regulations; and aid operators and governmental and certifying agencies in investigating the causes and effects of accidents.

Acknowledgements

The funding for the development of ASPEN has come from Coast Guard Northern. The authors are grateful for the opportunity to pursue this course of work. In particular, Mr. John McCallum and Mr. Andrew Hart of that organization are to be thanked for their direction, assistance and support over many years. The authors would also like to acknowledge the collaboration of Mr. Max Perchanok who compiled most of the data stored in the ASPEN database. Finally, the enthusiastic support provided by Mrs. Rosemary Gardner and Mr. Arthur Karton at different stages of the project is also gratefully acknowledged.

References

Benjamin, J.R. and Cornell, C.A. 1970. Probability, Statistics and Decision for Civil Engineers. McGraw-Hill.

Bercha, F.G. 1979. The development and application of multi-model ice failure theory. IUTAM Symposium on Physics and Mechanics of Ice. Ed. P. Tryde, Copenhagen, pp. 16-27.

Caldwell, J.B. 1965. Ultimate longitudinal strength. Transactions RINA, Vol. 107, pp. 411-430.

Daidola, J.C. and Basar, N.S. 1980. Probability design for ship hull structural strength. Spring Meeting/STAR Symposium, SNAME, California. pp. 105-124.

Daley, C.G. 1984. ASPEN, Arctic shipping probability evaluation network — group III studies. Report by Arctec Canada Limited 1305C to Coast Guard Northern.

Daley, C. G. 1985. Studies in Connection with updating the Arctic shipping pollution prevention regulations group IV — further development of ship/ice collision probability model — ASPEN. Reported by Arctec Canada Limited to Coast Guard Northern, Transport Canada Report TP 6274E.

Daley, C.G. and Ferregut, C. 1988. Strength and stability of framing for ice strengthening. Report by Fleet Technology Limited to Coast Guard Northern, Transport Canada, Report No. TP 9157E.

Daley, C.G., St. John, J.W. and Seibold, F. 1984. Analysis of the extreme ice loads measured on the USCGC *POLAR SEA*. Transactions of SNAME, Vol. 92, pp. 241-252.

Edwards, R.Y. et al. 1981. Results of full scale trials of the CCGS *PIERRE RADISSON*. STAR Symposium, Ottawa. pp. 291-310.

Ferregut, C. and Daley, C.G. 1988. ASPEN — Structural modelling using limit states. Fleet Technology Report to Coast Guard Northern, Transport Canada Report TP 9472E.

Ferregut, C., Perchanok, M. and Daley, C.G. 1987. Ship/ice collision probabilities in Arctic shipping. Port and Ocean Engineering Under Arctic Conditions, Vol. I, W.M. Sackinger and M.D. Jeffries (Eds.) University of Alaska, Fairbanks, pp. 631-643.

Galambos, T.V. and Ravindra, M.F. 1978. Properties of steel for use in LRPD. Journal of the Structural Division, ASCE, Vol. 104, No. 519, pp. 1459-1468.

Ghoneim, G.A. 1983. *KIGORIAK* and *ROBERT LEMEUR* full scale impact tests for 1983. Final Report by Canadian Marine Drilling Limited to Coast Guard Northern.

Government of Canada. 1972. Arctic shipping pollution prevention regulations. Canada Gazette, Part II, Vol. 106, No. 20.

Hakala, M. and Riska, K. 1983. Results from statistical measuring system onboard MV *ARCTIC* 1982. Report by VTT to Canadian Coast Guard Northern.

Hasofer, A.M. and Lind, N.C. 1974. An exact and invariant first-order reliability format. J. Engineering, Mech. Div., ASCE, Vol. 100, pp. 111-121.

ISSC. 1985. Applied design. Report of Committee V. 2, Proceedings of the Ninth International Ship Structures Congress, Vol. 2, Genova, pp. 405-471.

Kennedy, D.J.L. and Baker, K.A. 1984. Resistance factors for steel highway bridges. Canadian Journal of Civil Engineering, Vol. 11, No. 2, pp. 324-334.

Kennedy, D.J.L. and Gadaly, M.M.A. 1980. Limit states design of steel structures — performance factors. Canadian Journal of Civil Engineering, Vol. 7, pp. 45-77.

Kheisin, D.E. 1972. Use of probability methods in estimating the manoeuvering qualities of ships in ice. Publ. in Ice Navigation Qualitites of Ships, Ed. by Kheisin and Popov, CRREL Translation 417, pp. 40-58.

Kheisin, D.E., Likhomanov, Y.A. and Kurdyumov, V.A. 1973. Determination of specific breaking energy and contact pressures produced by the impact of a solid against ice. Symposium on Physical Methods of Studying Snow and Ice, Leningrad, Translation by CRREL No. TL539, pp. 210-219.

Kudishkin, Y.S. 1972. On the distribution of probabilities of the number of impacts of ship's hull against the ice. Publ. in Ice Navigation Qualities of Ships, Ed. by Kheisin and Popov, CRREL Translation 417. pp. 136-141.

Noble, P. and Menon, B. 1981. Laboratory investigation of ice impact pressure. Report 450 by Arctec Canada to Canadian Coast Guard Northern.

Perchanok, M.S., Ferregut, C. and Brow, R. 1989. A probabilistic ice information system for Canadian waterways. Marine Technology Society Journal, Vol. 23, No. 2, pp. 28-32.

Perchanok, M.S., Wells, D.G. Lowings, M.G. 1986. Merchant ship ice navigation studies in the Canadian Arctic. Intl. Polar Transportation Conference, Vancouver, pp. 859-877.

Rackwitz, R. and Fiessler, B. 1978. Structural reliability under combined random load sequences. Computers and Structures, Pergamon Press, Vol. 9, pp. 489-494.

Ratzlaff, K.P. and Kennedy, D.J.L. 1986. Behaviour and ultimate strength of continuous steel plates subjected to uniform transverse loads. Can. Journal of Civil Engineering, Vol. 13, No. 1, pp. 76-85.

Tam, G. 1980. Development of an ice extrusion model — interim report in task 4 of group I studies for update of ASPPR. Report IR 586A by Arctec Canada Limited to Canadian Coast Guard.

Varsta, P. 1977. Measurement and analysis of ice induced stresses in the shell of an icebreaker. Helsinki Winter Navigation Research Board, Report 21.

Voelker, R.P. et al. 1981. Evaluating commercial Arctic marine transportation with polar class icebreakers. STAR Symposium Ottawa, pp. 145-158.

Vuorio, J. et al. 1979. Long term measurements of ice pressure and ice-induced stresses on the icebreaker *SISU* in winter 1978. Helsinki Winter Navigation Research Board, Report 28.

Notation

A	Loaded area
A_f	Area of flanges of hull girder cross section
A_w	Area of webs of hull girder cross section
a	Distance from frame (stringer) toe to shear center
b	Diameter of floe; frame (stringer) spacing
D_s	Depth of hull cross section
d	Floe diameter; frame (stringer) spacing
E	Young's modulus
E_i	Event i
F	Total impact force
G	Shear modulus
H_w	Height of web plate
h	Ice thickness
I_{Pc}	Moment of inertia with respect to toe
I_y	Moment of inertia about y axis
J	Saint Venant's torsional constant

K	Plate constant
L	Unsupported length
$L(\,)$	Load function
L_s	Ship length
M	Number of ice encounters
$P(A)$	Probability of event A
P_f	Probability of failure
$P^*(A)$	Pressure with 1% probability of exceedance
P_i	Ice pressure that causes limit state i
q	Parameter of the Beta density function
$R(\,)$	Strength function
r	Parameter of the Beta density function that depends on the type of ice
s	Elastic section modulus
t	Thickness of plating
T_w	Thickness of web plate
$U(A)$	Parameter of Type I extreme value distribution
V	Ship speed
W	Width of cell
x_i	Basic variable i
Z	Plastic section modulus of web (frame)
Z_H	Plastic section modulus of hull girder
$\alpha(A)$	Parameter of Type I extreme value distribution
β	Shear angle of ship; safety index
Γ	Warping constant
$\triangle$	Ship displacement
$\triangle_e$	Effective displacement
$\triangle_i$	Mass of ice
ν	Rate of ice encounters; Poisson's Ratio
ν_p	Poisson's Ratio in the plastic range
π	3.141596...
σ_f	Flexural strength of ice
σ_y	Yield stress
Φ	Standard normal cumulative distribution function

Discussion

Q: A.F. Aboulazm, Marine Institute

1. In the probability of a ship-ice feature encounter, would your model handle an ice feature consisting of a number of ice floes?
2. Would your model account for the response of the ship to initial impact with the ice feature such as yawing or speed reduction?

A: C.G. Daley

1. The model does not yet deal with a cluster of ice floes. In such a case ASPEN would treat the cluster as a series of independent encounters. Treatment of floe clusters is one of our recommended enhancements, as discussed at the end of the paper.
2. The model treats speed reduction by linking speed to ice and visibility conditions. However, for a given ice/visibility combination, the speed of impact is considered fixed. The model does not consider reflected impacts (since all floes are treated as single events). It should be noted that ASPEN is not a transit simulation of a Monte Carlo model. ASPEN performs a direct closed form risk calculation on a cell by cell basis.

Q: M. Nessim, DnV Canada

1. Could you explain how your limit states were arrived at?
2. Design equations may not be appropriate as damage limit states because when damage occurs the behaviour of the structure is much more complex than is normally assumed in developing design equation for individual members.

A: C.G. Daley

1. The details of the limit state derivations are given in reference [Fergut and Daley, 1988] in the paper. The equations represent the authors' opinions of the best available equations to describe strength on an element by element basis.
2. This question is the most difficult to answer, because it concerns the core philosophy of the ASPEN model. To begin with, it is true that

in cases of extreme damage, the limit state equations in ASPEN are (far too) simple. It is also true that in this case, advanced methods (finite element methods) could be used to better describe the structural behaviour. Further, I admit that in the case of a particular damage on a particular vessel such methods can help improve a design and reduce risk.

However, there are many reasons why in ASPEN's case, the current approach is the best.

- ASPEN considers individual element limit states which are good descriptions of small damages (visible damage up to small fractures). These small damages are frequent enough to allow calibration of the model.

- Extreme levels of damage are not treated for the same reasons that extreme consequences (environmental, political, social, economic) are not. These factors pose problems which are beyond the scope of engineering risk analysis, primarily because one must answer the question "How will the people react in the crisis?"

- The ice load description was developed from ice loads on elastic (nearly rigid) structures. It is suspected that the loads might be very inappropriate if applied to highly deforming structures. It is my opinion that development of elaborate structural failure limit states is out of keeping without relatively poor understanding of ice loads.

- ASPEN is meant as a general purpose risk model for use as a decision aid. It is best used to make risk comparisons between different structural configurations, routes, ice conditions etc. While admitting that the limit states represent relatively light damage, they are nevertheless good indicators of the risk involved.

- A final point - as ASPEN has developed we have attempted to keep the sophistication of various parts of the model in balance. While the structural strength descriptions could be improved (even greatly) with current finite element technology, it should be realized that even as is, the strength part of ASPEN is the most precise. Further work on this part is not warranted at this time. Understanding the complex interdependencies in the ice environment and the navigation process are the real challenges.

Q: J. Lever, CRREL

Is the probability of encounter adjusted for the design or class of vessel?

A: C.G. Daley

Vessel parameters including strength (class) affect the avoidance (not encounter) probabilities. We performed a number of navigation studies (see paper references) to determine the navigation parameters. We found that the perceived level of risk (by the master and bridge personnel) affected the avoidance probabilities. Weaker (lower class) vessels were much more careful when entering an ice field. Also ship length and power were used to determine speed and turning circle (used in avoidance manoeuver algorithm).

Q: D. Duthinh, C-CORE

Does your model account for dependence of ship velocity on ice presence?

A: C.G. Daley

Yes, the ship velocity is dependent on the presence of ice. In an ice cover the base velocity is determined from a simple resistance formulation. Velocity is also dependent on the presence of glacial ice in the area (connected to perceived risk). Our studies found that the master would slow by 1 knot when any glacial ice was detected. Once the density of glacial ice exceeded a critical level, the master would slow to half speed. In extreme conditions the master would slow to "dead slow". These step changes in speed were easy to implement in ASPEN.

Q: A. Tunik, ABS

1. How is the mode of ice navigation considered in the model (whether the ship breaks ice continuously in level or ridged ice, or sails in broken ice)?
2. Distance, not time is selected as the basic variable. How is this compatible with the fact that the navigation in particular Arctic Zones is seasonally restricted depending on ice class?

A: C.G. Daley

1. The usual (baseline) mode of navigation considered is the ship in open pack. In this way each floe is considered individually. When

the coverage becomes high enough (9+/10 to 10/10 unconsolidated to 10/10 consolidated), the rate of collisions increases because the avoidance algorithm results in no avoidances. As such the mode of navigation is not actually changed - it must be stressed again that ASPEN is not a transit model, but a risk model. The emphasis is on the correct treatment of hazardous ice. A ship will not (cannot) transit continuously in ice which will seriously damage it. In a 10/10 ice cover with 1/10 multiyear (hazardous) ice, the important thing is the treatment of the multiyear floes. In this situation the presence of the first year ice cover affects the detection algorithms (harder to see multiyear ice against a background of first year ice) and the avoidance algorithms (harder to turn but easier to stop in first year ice compared to open water).

2. The basic (potentially) hazardous event is a transit impact with ice. The number of encounters (potential impacts) is related to the transit distance, not time spent in an area. Seasonal restrictions by Arctic Zone is the way in which the present ASPP Regulations attempt to limit the type and amount of hazardous ice which a vessel will experience. These are separate issues. ASPEN does permit examination of risk to vessels following the current ASPP Regulations. However, one must specify what routes (month and route) are transitted.

Probabilistic Ice Forces on Offshore Structures

Donald E. Nevel

Conoco Inc.
Houston, Texas, USA

Abstract This paper presents an easy to understand method for predicting probabilistic ice forces on offshore structures. It discusses the development of probabilistic distributions from environmental data, and the formulas which are used to compute the ice forces. For one type of ice feature, many forces are calculated and a probability is assigned to each force to obtain the probability distribution. A method is developed for combining ice force distributions for more than one type of ice feature. The return period of a force is defined and the selection of its design value is discussed. The probability of exceeding the force in a time period, such as the life of the structure, is considered. Future developments are discussed with emphasis on obtaining better ice data.

1. Introduction

Return periods have been used to select design conditions for floods and waves. In these cases, there has been one statistical variable considered, i.e. the discharge rate or the wave height. Return periods have also been used to select design conditions for one type of an ice feature, such as an iceberg. In this case, more than one statistical variable has been considered. In this paper, a method for combining the ice force statistical distributions for more than one kind of an ice feature is developed. The purpose of this paper is to provide the structural designer the understanding, confidence and limitations that are presently available for predicting the probabilistic ice loads and the associated return periods.

2. Probabilistic Forces Measured from an Ideal Experiment

Consider an offshore structure that is instrumented to measure the global forces from the movement of ice. The forces are recorded over many

S. Jones · R. F. McKenna · J. Tillotson · I. Jordaan (Eds.)
Ice-Structure Interaction
IUTAM-IAHR Symposium St. John's, Newfoundland, Canada

years. Different types of ice features move against the structure, such as first year floes, multiyear floes, first year ridges, multiyear ridges, rubble fields, icebergs and ice islands. A video camera documents the type of ice features for correlation to the measured forces. The ice forces vary because of different:

1. types of ice features,
2. size and shape of ice features,
3. properties of ice features, and
4. ice failure modes.

The size and shape of the structure also influences the measured ice forces. The maximum ice force that occurs during a single interaction with an ice feature is used to represent that interaction. The maximum force is chosen because the designer wishes to be conservative.

The variation of the peak ice force during a single interaction is not considered. This variation may be caused by non-uniform geometry or properties of the ice as well as the failure process during bending or crushing failures. The variation during a single interaction is important for analysis of dynamic response of structures, but is not considered here for global static analysis.

Over the time period of measurement, the average rate at which each type of ice feature interacts with the structure is defined by

$$\text{Average Rate} = \frac{\text{Number of Interactions}}{\text{Time Period}} \tag{1}$$

The above ideal experiment can not be performed for a number of reasons. However it should be kept in mind when using other methods to simulate the ice force statistics.

3. Ice Force Formulas

An alternative to measuring the ice forces is to calculate the ice force from mathematical formulas. A mathematical formula is required for each kind of failure mode occurring for each type of ice feature against each structure. The force obtained from each formula depends upon environmental variables, such as ice thickness, ice strength, etc. Each formula should predict the maximum ice force accurately or conservatively for design.

An example formula is the simple crushing force F of

$$F = S\ W\ h, \tag{2}$$

where S is the effective crushing strength of the ice, W is the width of the structure and h is the ice thickness. The ice strength and thickness are considered to be environmental variables with each having a probability distribution.

Other force formulas are also available, of which the more recognized ones are discussed in the American Petroleum Institute RP 2N, Recommended Practice for Planning, Designing, and Constructing Fixed Offshore Structures in Ice Environments (API, 1988).

There is Korzhavin's formula for estimating the crushing strength from small samples (API, 1988). For slow crushing, there is a reference stress formula (Ponter, 1983), and a reference stress formula with an upper cut off (Walden, 1987). There is a formula for impact crushing with limited energy (Nevel, 1986). There is a shear failure formula for first year ridges (Prodanovic, 1981).

There are formulas for the elastic bending and buckling of beams (Hetenyi, 1946) and wedges (Nevel, 1968, 1979). A general review of buckling is available (Sodhi and Nevel, 1980). There is a review of the forces required to create floating and grounded pressure ridges (Kovacs and Sodhi, 1980). There is a formula for ice breaking in bending and ice riding up a pressure ridge (Nevel, 1983).

For ice floes against conical structures, there is a plastic limit formula (Ralston 1977, 1979). Elastic beam and wedge formulas have also been used. For multiyear ridges against conical structures, there is a plastic limit formula (Wang, 1984), and a beam bending formula (Kim and Kotras, 1973) which includes the attached ice floes. There is a formula for short ridges sliding up a cone (Winkler and Nordgren, 1986).

The above is not a complete list of available force formulas, but represents the most quoted ones. More sophisticated force methods are available using finite element techniques for boundary value problems or time stepping techniques for impact problems. Although these techniques can produce accurate results, they require more computation time than the simpler ones quoted above. It is recommended that these more sophisticated techniques be used to select and calibrate the simpler force formulas because the statistical procedure requires the force formula to be calculated thousands of times.

4. Probability Distribution of Environmental Variables

Statistical variables are described by either a density, a cumulative or an exceedance probability distribution. Each is uniquely determined from any of the others. A distribution is usually determined from experimental data. Traditionally, data has been grouped into bins and the density distribution determined. Grouping reduces the amount of data to be handled, but it also loses information. Today, large amounts of data can be handled easily by a digital computer which reduces the need for grouping. The statistical procedure to be used requires either a cumulative or an exceedance distribution. Either distribution can be obtained directly from the data without determining the density distribution.

Consider the example of the force formula in Equation 2 which depends upon the ice thickness. Ten values of multiyear ice thickness in meters have been measured in a random manner such that it is believed that each value had an equally likely chance of being observed. A probability of 1/10 is assigned to each measurement. This probability is a segment on the exceedance probability scale. The thicknesses are ordered according to magnitude. Each successive thickness is assigned to the mid point of each successive segment since this is least bias assignment. A conservative, but biased, approach is to assign each thickness to the bottom of the segment for a cumulative distribution and to the top of the segment for an exceedance distribution.

Nature may provide a lower or upper limit for each statistical variable. Both lower and upper limits must be considered realistically since both influence the probability distribution of the force. For this example, it is believed that the lower limit is two meters since that is the maximum first year thickness. It is believed that the upper limit is 14 meters since this is the water depth at the site. The following shows the least bias assignments p and the conservative assignments p_c for the exceedance probabilities of the ice thickness h.

h	2.0	3.0	3.5	4.0	5.0	5.5	6.0	6.3	7.2	8.0	10.0	14
p	1.0	.95	.85	.75	.65	.55	.45	.35	.25	.15	.05	0
p_c		1.0	.90	.80	.70	.60	.50	.40	.30	.20	.10	0

The difference between the two methods is small when there are many values. The least bias method is preferred since it represents the data better. Conservatism can be introduced when representing the distributions mathematically. In practice many more than 10 values are required for a valid distribution.

In the past, a great effort has been placed on determining a single mathematical representation for an experimental distribution. This allows the distribution to be described by a few parameters, usually the mean and standard deviation. The main deficiency of a single mathematical distribution is that it does not represent the experimental data well over the entire range. Piece-wise mathematical representations are recommended since they are easily handled by the computer. The simplest piece-wise curve is a straight line, but the points which define the various sections must be reasonably close together.

Care must be taken when projecting the distribution beyond the data at the tail. For flood and wave distributions, great care has been taken since only one variable was considered to be statistical. Usually a type of exponential function is used.

For more than one statistical variable, the resulting force distribution depends less on a single tail than that for only one statistical variable. In any case, conservative estimates should be made. A simple approach is to select a conservative upper limit whose probability of exceedance is zero, and use a straight line from the last data point to the upper limit.

The statistical variables may have lower and upper limits which will determine the lower and upper limits of the force distribution. The lower limit of the force is determined from all the lower limits. The upper limit of the force is determined from all the upper limits. All upper and lower limits for statistical variables should be determined from nature.

If an arbitrary lower limit is placed on one statistical variable, the calculated forces may be missing some of the lower forces. The missing forces will be below the force obtained using the arbitrary lower limit for that variable and the upper limit for all other variables. It may be prudent to neglect forces below this value. The rate of interaction must reflect any arbitrary lower limits. It is not a good idea to use an arbitrary lower limit of a statistical variable because it complicates the analysis. When there is not a great deal of confidence in a probability distribution for an environmental variable, especially in the region of the tail, sensitivity studies should be performed to see how changing of the distribution affects the force distribution and the design force.

5. Monte Carlo Method

A technique called Monte Carlo Distribution Sampling is used to obtain the force distribution when the force depends upon more than one statistical variable. For each of the statistical variables, a random number between 0 and 1 is generated which represents either the cumulative or the exceedance probability. This is justified because the probability distribution was based on equally likely occurrences. From the probability distribution, the value of the variable is determined. All the statistical variables are determined in this way, and a force is calculated from the force formula. Repeating this process many times gives many forces. The forces can be arranged according to magnitude and probabilities assigned as will be explained below. Either a cumulative or exceedance distribution can be produced.

The probability of interaction is proportional to the rate of interaction with the structure. For discrete ice features the rate of interaction is determined from a "rain drop" method. That is the center of the ice feature is assumed to be a rain drop which is uniformly distributed and is moving past the structure. The rate R of interaction for an ice feature of diameter D_i against a structure of width W is

$$R = \text{Flux}\ (W + D_i), \tag{3}$$

where the Flux is the number of ice features per unit of time per unit of swath length. The Flux is determined from

$$\text{Flux} = \frac{\text{(Concentration) (Velocity)}}{\text{Area}}, \tag{4}$$

where the Concentration is expressed as a fraction, and the average Velocity and average Area are used. The average rate R is obtained when the average diameter D_i is used.

The probability of obtaining the force F_i is proportional to the rate which provides

$$P(F_i) = \frac{(W + D_i)}{\Sigma(W + D_i)}, \tag{5}$$

where the sum Σ is on i from 1 to N which is the number of simulations. Each force is assigned a probability as given in Equation 5 which is its segment on the probability of exceedance scale. Then the force is assigned to the center or end of the interval as explained earlier.

Some authors have assigned equal probability to the forces generated by the Monte Carlo procedure which is not correct. Others have modified the diameter distribution using Bayes' theorem and assigned equal probabilities to the forces. It has not been clearly shown that this method produces the same distribution as that obtained from using Equation 5.

Still others have assumed that the probability depends upon the velocity of the ice feature. This implies that an ice feature has a velocity permanently attached to it, and that different ice features have different velocities. In this case, the probability assignment would be

$$P(F_i) = \frac{(W + D_i)\ V_i}{\Sigma(W + D_i)\ V_i}. \tag{6}$$

In the real world, the velocity varies accordingly to that of the regional ice velocity which depends on the wind and current. Hence, the premise of Equation 6 is not correct.

The forces at the center of the distribution are determined with fewer simulations than at the tail where the large, but more rare, forces occur. A sufficient number of simulations must be performed to obtain a stable distribution of the force in the region of interest on the tail. Typically the number of simulations in the order of 1,000 to 5,000 is usually adequate. Since only the forces above a reasonable value are needed for design, the forces generated on the lower part of the distribution need not be saved.

6. Force Distribution for more than one Type of Ice Feature

The Monte Carlo simulation could be performed for long time T during which each type of ice feature is simulated $R\ T$ times. The period of simulation should be long enough to obtain a stable distribution for the most infrequent occurring type. This means that the frequently occurring types are simulated many more times than necessary. To make the simulation more efficient, the Monte Carlo simulation can be performed for each type of feature until a stable distribution is obtained. The resulting force distributions must be combined together in a way that obtains the same distribution as that obtained for simulating a long time T.

The number of simulations N that would have occurred in a desired period T for each ice feature is

$$N = R\ T. \tag{7}$$

The number of simulations that would have occurred above a force is

$$n = p\,N = p\,R\,T, \tag{8}$$

where p is probability of exceedance of the force. If two ice features are considered, the combined exceedance probability for the force is

$$p = \frac{n_1 + n_2}{N_1 + N_2} = \frac{p_1\,N_1 + p_2\,N_2}{N_1 + N_2} = \frac{p_1 R_1 + p_2 R_2}{R_1 + R_2}. \tag{9}$$

The rate of occurrences for the combined distribution is

$$R = \frac{R_1\,T + R_2\,T}{T} = R_1 + R_2. \tag{10}$$

These formulas can be extended to include more than two ice features. Equations 9 and 10 can be used to assess the significance of the force distributions. If a rate is sufficiently small, the corresponding terms in these equations may not affect the combined probability or rate significantly. Similarly, if the probability of exceedance goes to zero, the distribution should not be considered.

7. Return Period

The time period in Equation 1 is called the return period if the number of interactions is the expected number of interactions to occur before a specified force is exceeded. The probability of obtaining a force F greater than F_0 on the N-th observation is

$$P(F > F_0|N) = (1-p)^{N-1}\,p, \tag{11}$$

where p is the exceedance probability for F_0. The expected number of interactions N_e to occur before the force F_0 is exceeded is

$$N_e = \sum_{n=1}^{\infty} N\,(1-p)^{N-1}\quad p = 1/p. \tag{12}$$

On the average, the force F_0 will occur once during the time of the return period which is

$$\text{Return Period} = \frac{1}{p}\;\frac{1}{R}. \tag{13}$$

8. Probability of Exceedance in a Time Period

The probability of a force F exceeding a force F_0 in N interactions is

$$P(F > F_0) = 1 - (1 - p)^N. \tag{14}$$

Equation 14 is also the exceedance probability in a time period T if N equals $R\,T$, where R is the average rate. The rate can vary during time T which makes N vary. This variation of N can be accounted for by considering the number N to be Poisson distributed as

$$p(N) = \frac{(R\,T)^N\, e^{-R\,T}}{N\,!}. \tag{15}$$

This is the least bias distribution for N when only the average number $R\,T$ is known. Then the expected probability for the force F to exceed the force F_0 in a time period T becomes

$$P = 1 - \sum_{N=0}^{\infty} (1 - p)^N \frac{(R\,T)^N\, e^{-R\,T}}{N!} = 1 - e^{-p\,R\,T}, \tag{16}$$

Equations 14 and 16 give similar answers when p is small. It depends how close the value of (1 - p) from Equation 14 is to the value of (e^{-p}) from Equation 16. Comparison of these expressions for various values of p are:

p	.2	.1	.05	.01
(1 - p)	.80000	.90000	.95000	.99000
(e^{-p})	.81873	.90484	.95123	.99005

These numbers must be raised to the power N or $R\,T$ to obtain the expressions in Equations 14 and 16. A distinction should be made between the probabilities of Equations 14 and 16 when p is not small. It is believed that Equation 16 is a better probability estimate for a shorter time period. However, probability estimates for short times are not generally reliable.

The exceedance probabilities for a time should not be confused with the force exceedance probability p. The return period has been defined in terms of the force exceedance probability p. The force exceedance probability p can be substituted from Equation 13 into Equation 14 or 16 to obtain the exceedance probability in a time, as a function of the return period. For Equation 16, this is

$$P = 1 - e^{-(T/RP)}, \tag{17}$$

which is independent of the rate. From Equation 17, the return period RP is 99.5 years when P is 0.01 and T is 1 year.

9. Design Criteria

For design, a return period is usually selected several times the life of the structure. The selected return period and the known rate determine the exceedance probability of the force in Equation 13. The design force is determined from its exceedance distribution. Typical return periods that have been previously selected for design are 10, 25, 100, and 2000 years. The confidence one places on the prediction may well influence the selection of the return period.

The Canadian Standard Association (CSA, 1989) is developing a code for the design, construction, and installation of fixed offshore structures which will consider probabilistic ice loadings. This is the first code to consider probabilistic ice loadings.

10. Conclusions and Future Directions

The statistical procedures for predicting the return period of ice forces are presented in this report. To use these procedures, a computer program is required along with mathematical formulas and probability distributions of the environmental data. A computer program is not yet generally available, but this should not be difficult to produce.

Many force models have already been developed. Little attention has been given to force models for first year ridges and unconsolidated rubble fields. There is a need to continue improving force models. Field observation will be our best guide in determining the failure processes and ice strengths.

The gathering of the required environmental data is expensive. There is a need to reduce these costs and to increase the amount and quality of data that is availabe. A Synthetic Aperture Radar system is planned for the European Space Agency satellite due to be launched in late 1990. This system may provide information on the area, size, shape, and movement of ice features in the Arctic. Possibly information on the thickness and instantaneous velocity could be obtained from submarines or sensors on the seabed.

It is hoped that this paper clarifies the statistical procedures for predicting the probabilistic ice loads against offshore structures. With this information and dedicated effort, probabilistic ice forces will become as commonplace as predicting the one hundred year wave on offshore structures.

References

API. 1988. Recommended practice for planning, designing and constructing fixed offshore structures in ice environments. API Recommended Practice 2N, Washington, DC: American Petroleum Institute

CSA. 1989. Canadian Standards Association Preliminary Standard S471. Part I - Code for the Design, Construction and Installation of Fixed Offshore Structures.

Hetenyi, M. 1946. Beams on elastic foundation. University of Michigan Press, Ann Arbor, MI.

Kim, J.K. and Kotras, T.V. 1973. Mathematical model to describe the behavior of a moving ice field encountering a conical structure. APOA Project 57, Vol.1.

Kovacs, A. and Sodhi, D.S. 1980. Shore ice pile-up and ride-up. Cold Regions Science and Technology, 2, pp.209-298.

Nevel, D.E. 1968. The general solution of a wedge on an elastic foundation. CRREL Research Report 247, Hanover, NH.

Nevel, D.E. 1979. Bending and buckling of a wedge on an elastic foundation. IUTAM Symposium on Physics and Mechanics of Ice, Copenhagen, Denmark, pp.278-288.

Nevel, D.E. 1983. Pressure ridge forces. Proc. POAC '83, Helsinki, Finland, Vol.1, pp.212-220.

Nevel, D.E. 1986. Iceberg impact forces. Proc. IAHR Ice Symposium, Iowa City, Iowa, Vol.3, pp. 345-369.

Ponter, A.R.S. et al. 1983. The force exerted by a moving ice sheet on an offshore structure, Cold Regions Science and Technology, Vol.8, pp.109-118.

Prodanovic, A. 1981. Upper bounds of ridge crushing pressure on structures. Proc. POAC '81, Quebec, Canada, Vol.3, pp. 1288-1302.

Ralston, T.D. 1977. Ice force design considerations for conical offshore structures. Proc. POAC '77, St. John's, Newfoundland, pp. 741-752.

Ralston, T.D. 1979. Plastic limit analysis of sheet ice loads on conical structures. Proc. IUTAM Symposium on Physics and Mechanics of Ice, Copenhagen, Denmark, pp. 289-308.

Sodhi, D.S. and Nevel, D.E. 1980. A review of buckling analysis of ice sheets. Working Group on Ice Forces on Structures, CRREL Special Report 80-86, Hanover, NH, pp. 131-146.

Walden, J.T., Hallam, S.D. and Baldwin, J.T. 1987. An explicit technique for calculating first year ice loads on structures. Proc. Offshore Mechanics and Arctic Engineering Symposium, Houston, Texas, Vol.4, pp.267-272.

Wang, Y.S. 1984. Analysis and model tests of pressure ridges failing against conical structures. With discussions. Proc. IAHR Ice Symposium, Hamburg, Germany, Vol.2, pp.67-76.

Winkler, M.M. and Nordgren, R.P. 1986. Ice ridge ride-up forces on conical structures. Proc. IAHR Ice Symposium, Iowa City, IA, Vol.1, pp.171-183.

Discussion

Q: T.H. Jacka, Australian Antarctic Division

Your considerations of the probability of an iceberg - structure collision assumed a random distribution of icebergs and random velocities and movement directions of icebergs. Iceberg distributions and movement rates are governed primarily by currents, which are not random. Can you not include current data in your model?

A: D.E. Nevel

In this study, the number of iceberg collisions was estimated using the equation documented by Nessim et al., (1984). This method requires the areal density of icebergs, the average drift velocity, and the average iceberg length. While improvements could be made to the model, such as relating the drift velocity of icebergs to their mass or to the sea state, the method is a good first approach. The above inputs were determined directly from data on icebergs and their motions. While there are some problems with the data used (for example, the velocity data could be slightly biased because small icebergs are hard to detect in high sea states) it is the best available.

While data on currents, winds, and waves are necessary for forecasting iceberg motions, they are not as useful for hindcasting. To estimate iceberg motions, the shapes of icebergs, the waves and wind conditions and the spatial variations in the current (especially w.r.t. depth) would be required. Because of the uncertainties with these inputs, it is more efficient and accurate to use directly measured iceberg drift data.

Q: A. Tunik, American Bureau of Shipping

Have you distinguished which part of the vessel is subjected to the collision, whether it is the bow or the side?

A: D.E. Nevel

An analysis of iceberg drift patterns in the area of concern shows that the predominant (southward) drift component of the icebergs is small compared to the short term motions due to tides, winds and eddy currents. There is therefore a reasonable probability of an iceberg approaching from any direction.

If the vessel were dynamically positioned it would be possible to orient it towards the perceived most likely direction of approach of icebergs based on the current, wave and wind conditions at that instant.

We have developed a model to determine the location of impact (on different columns and pontoons of a semi-submersible) for different sea states and iceberg sizes. The model assumes that icebergs are equally likely to approach from any direction but ignores the wave energy diffracted from the vessel. The model is presently being updated to account for the probability distributions of the different input parameters and to account for the phase between the iceberg and the vessel.

Q: Hussein El-Tahan, Crescent Consulting

Your model is sensitive to iceberg detectability and since the IIP data base has a gap in (lack of) bergy bit and growler data, how did you fill that gap? Could you please also comment on your assumption that bergy bit speed is independent of its mass. What would be the impact of accounting of the velocity dependence on mass.

A: D.E. Nevel

The distribution of iceberg lengths in the study was solely based on Husky-Bow Valley records. To determine the areal density of icebergs at the site, IIP maps showing the locations of icebergs sighted on the Banks on given dates were used. Because bergy bits and growlers were often missed, only records of small, medium and large icebergs were considered. The areal density determined was then increased by the proportion of bergy bits and growlers calculated from the Husky-Bow Valley data. It should be noted that even in the Husky-Bow Valley data set, small icebergs may be detected less often than large icebergs. Also, because the IIP could possibly miss small, medium and large icebergs due to cloud cover, high sea states or non-uniform flight coverage, we are reviewing their reconnaissance procedures.

The assumption is often made that the speeds of icebergs are independent of their mass, based on scatter diagrams of iceberg mass versus speed. Intuitively, it is felt that small icebergs in storms would move faster because more of their mass is in the upper currents which will move faster in winds. It may be that the apparent high velocities of larger icebergs are due to wave diffraction. Also, the recorded data of iceberg speed versus mass could be biased because large icebergs are more likely to be detected than small icebergs in high sea states. We are presently evaluating these possibilities. The effect of higher velocities for smaller icebergs would be to increase the collision probabilities, velocities and forces due to these icebergs.

Q: Claude Daley, Fleet Technology Ltd.

Was the ice force model deterministic? Did you do a sensitivity analysis of the governing parameters? What is the most crucial parameter?

A: D.E. Nevel

The ice force model is deterministic in as far as it determines the maximum contact area during the collision based on the average ice strength of 1.5 MPa. The model then also calculates the peak short term final force based on a peak pressure chosen randomly from a distribution of values derived from Johnson and Benoit's (1987) paper.

We did check the sensitivity of the model to various parameters such as ice strength, collision velocity, and detectability, etc. The most significant unknown at this time is the ability to detect and avoid icebergs of given sizes in different sea states. Because of the complex nature of radar detection and the lack of data, it is difficult to verify existing radar models for all conditions.

Q: Maher Nessim, DnV Canada

Have you considered the interaction between iceberg velocity and structure velocity since they are responding to the same wave?

A: D.E. Nevel

It is true that in the larger sea states the two bodies will most often be responding to the same wave. At present we consider the vessel to be initially fixed. This will be conservative for the higher sea states where the two bodies are likely to move in phase, but may be overly optimistic in smaller waves where they move out of phase. Accounting explicitly for the vessel velocity and phase distribution is one area for improvement to the model.

Q: Peter Wadhams, Scott Polar Research Institute, England

In using the Monte Carlo method to calculate the force probabilities from the exceedance probabilities of strength, width and thickness, does not a complication occur because the thickness and strength are not entirely independent variables?

A: D.E. Nevel

If variables are not independent, a procedure must be used to insure that the resulting force reflects the dependence in some way. Various ways have been used to solve this problem and a few examples are:

1. Sometimes the variables can be treated independently if the resulting force distribution is conservative. This has been applied for icebergs between mass and velocity. It has been postulated that the larger icebergs may move with a slower velocity, but this result has not been substantiated with observations. Considering the mass and velocity as independent parameters produces a conservative ice force estimation.

2. Sometimes the variables can be made dimensionless. This has been applied to icebergs where the Moment of Inertia was expressed dimensionless in order that it be independent of the mass. In this case, the radius of gyration divided by the maximum radius of the iceberg produced a distribution that was independent of the mass.

3. Sometimes a correlation method has to be used between dependent variables. This has been applied for icebergs between the mass and the draft.

Q: Ian J. Jordaan, Memorial University of Newfoundland

As a matter of information, there is a computer program on probabilistic analysis of loads on such structures in the public domain. This program, "Probabilistic Framework for Ice Loads on Fixed Marine Structures, 1989", was developed by Det norske Veritas (Canada) Ltd. for Public Works Canada, Ottawa. This might well prove useful as a basis for others to use as well. Have you considered using fixed time periods, such as a year, to tie return periods and exceedance probability together?

A: D.E. Nevel

I am aware of the Public Works Canada program as well as its predecessor, the joint industry program AOGA 297, "Probability based Design

Criteria for Ice Loads on Fixed Structures in the Beaufort Sea" which was also developed by DnV (Canada). Both of these programs suffer from certain deficiencies of which the major ones are:

1. Incorrect assignment of probability to forces from the Monte Carlo simulation.

2. Incorrect adjustment of the rate of interaction due to a lower bound of a mathematical distribution.

3. Loss of accuracy of the force distribution due to grouping of the forces into bins.

4. Not considering multiyear pressure ridges as a separate ice feature.

Because some of these corrections require a major reorganization of the above programs, I have found it easier to write a new program for an IBM compatible PC rather than correct the above mentioned programs.

In my paper, I have discussed the probability in a fixed time period. This probability is defined from the exceedance probability, the rate of occurrence of the data in the distribution and a selected fixed time period. The return period does not depend upon a fixed time period. For a selected fixed time period, there is a one to one relation between the return period and this probability. The probability in a fixed time period may seem to be a better criterion for design than the return period because the life of the structure can be selected for the time period. However, the design return period is usually selected several times the life of the structure, and therefore has a time period associated with its selection.

If the rate of occurrence of the data is Poisson distributed, the average probability in a given time period is a better estimate for shorter time periods than the probability obtained from using the average rate.

Q: T.J. O. Sanderson, British Petroleum International, England

Thank you for an excellent summary. It is essential that we all work towards ensuring that risk and reliability principles are accepted by management and incorporated in codes. One question, do you not agree that the exceedance functions for floe diameter and velocity have to be modified according to Bayes' Theorem in order to give the diameter and velocity for impacting floes (as opposed to all floes)?

A: D.E. Nevel

I do not agree. I will respond to your question by separating the discussion between that for velocity and that for diameter. The probability of a floe hitting a structure depends upon the floe's diameter, but not upon its velocity.

The velocity of floes within an ice field are the same as that for the ice field. The velocity of ice field will change with time, but all floes with the same diameter in this field will have the same probability of hitting the structure. I believe other authors have assumed, but not stated, that the velocity attached to a floe in the Monte Carlo simulation remains permanently with that floe in time. If this premise is accepted, then floes with higher velocities sweep out more area in a given time, and the probability of hitting the structures depends upon the velocity. However, this premise is not correct for the real world.

The probability for a force generated from the Monte Carlo simulation is proportional to the probability of hitting the structure. My paper describes what I call the "rain drop" method which shows how the probability of hitting the structure depends on the diameter of the floe. Hence, each force from the Monte Carlo simulation is assigned a different probability. Other authors have assigned equal probability to the forces from the Monte Carlo simulation, but have computed the forces by selecting the diameter from a modified diameter distribution which was determined from Bayes' Theorem. This approach should give the same numerical result, but is more indirect and not as simple to understand.

CHAIRMAN'S REMARKS

R.W. Marcellus, C.M.E.L. Enterprises Ltd.

In this session a broad range of topics and geographical areas have been addressed. Advances to the science of calculating ice forces have taken more of the form of organizational advances as opposed to real advances in the theoretical basis for calculating ice forces over the past five years. The alternative design methodologies presented reflect the degree of uncertainty that currently prevails for certain failure modes.

From the information that was presented during the session a size effect for ice failure pressure is evident, yet a proper explanation of size effect is not available. This leads us to a partial recognition of the high reliance that the design methodologies have on the available empirical information (e.g. see Sanderson (1988) and Marcellus et.al. (1989)). Probabilistic models are stochastic to the point of calculating the probabilities of certain ice areas which may interact with the structure, but the designer generally opts for a conservative or what he feels is an upper bound on the actual failure pressure for the ice. What is lacking are adequate theoretical and stochastic bases for calculating the ice failure pressures for alternative failure modes. Past work for example on the flexural strength of the ice have assumed that the material follows classic elastic theory. For what we know about ice as a material today, we know, for example, from Sinha's work as presented in Marcellus et.al. (1989), that the initiation of microcracking in the material loaded in tension is very much dependent on the load path that the sample undergoes prior to failure. The current practice of specifying flexural strength for design applications however, does not currently consider the load path that ice undergoes prior to failure. This is indeed one area where future research could be directed.

The situation for compressive failure strength is somewhat better, but adequate relationships for describing the degradation of the strength of the material for all loading situations are not available, and the current work (the topic of other sessions at this conference) in the area of constitutive modelling promises to provide a better understanding of the failure processes in ice within the next few years.

From an overall perspective, the design methodologies currently in use are firmly based upon the available empirical data. Development of firm theoretical bases for ice failure in combination with the development of accurate phenomenological models for different failure modes are required in order to make advances design methodologies.

References

Marcellus, R.W., M. Sander, I.J. Jordaan, A.B. Cammaert, N.K. Sinha, and M.A. Maes. 1989. "Ice forces on marine structures review of deterministic models and analysis of large scale data", Report # 1051. Prepared for Public Works Canada. March, Ottawa.

Sanderson, T.J.O. 1988. "Ice mechanics: The risks to offshore structures", Graham and Trotman, London, 253p.

SESSION VI:

Ice-Structure Interaction

Ice Interaction with Structures

M. P. Määttänen

Faculty of Mechanical Engineering, Helsinki University of Technology
Espoo, Finland

Abstract Different mechanisms in various ice-structure interaction scenarios are described. Only bottom-founded structures are considered, and the aim is to give insight for ice forces design starting from the physical background. The effect of structural shape and size at the water-line, type of foundation, structural rigidity, mass and damping are related to ice type, properties, thickness, integrity and velocity, as well as to other environmental effects to result in typical ice-structure interaction modes. Ice failure can be visco-plastic yielding, crushing, shearing, bending or splitting. Structural response can be static or dynamic. Also the ice can exhibit dynamic response. A coupling of the dynamic response of the structure with the ice failure process may develop. After ice failure has occurred, the clearing mechanisms of broken ice are again dependent both on the structure and the ice.

1. Introduction

Ice interaction with structures covers most of the engineering applications in the field of ice mechanics: from thermal ice thrust to iceberg impact. Interaction occurs whenever contact is established and any relative movement takes pláce. The interaction process can be quiet, as in the case of thermal ice expansion, or dynamic, as in the case of a floe hitting a bridge pier. Either ice or structure, or both can be moving in relation to the other. The properties of the ice and the structure have an effect on ice failure and clearing mechanisms. There is energy exchange between the interacting bodies. The structure can exhibit significant dynamic response. Thorough knowledge of interaction mechanisms is required in order to predict reliable values for ice pressures and forces which usually are the limiting design criteria for structures.

S. Jones · R. F. McKenna · J. Tillotson · I. Jordaan (Eds.)
Ice-Structure Interaction
IUTAM-IAHR Symposium St. John's, Newfoundland, Canada

The first encounter of ice interaction with man made structures was bridge piers and marinas which have been prone to ice action for centuries. Trial and error, in addition to creative thinking, has been the main tool to develop these structures to withstand ice effects both during the winter and ice break up in the spring. Winter navigation is relatively new, just over a hundred years old, and its early development was also based on in-field experiences in a trial and error process. Engineering and scientific approaches in the research of ice- structure interaction are less than fifty years old. The first pioneers in the 1930s and the 1940s were Japanese, (Kubo, 1980) with ice bearing capacity, and Russians, (Gamayunov, 1947) with ice forces against bridge piers. The promises of arctic oil resources in the 1970s were a strong catalyst to accelerate the research on ice interaction with structures. The vibration incident of the Molikpaq in 1986 widened interest in dynamic interactions.

The field of ice interaction with structures is so wide that it can be covered only partly in this context. Only bottom founded structures are covered, ships and floating structures are omitted. Thermal and uplift cases are excluded even though otherwise ductile ice behaviour is considered. Ice pile-up on beaches and artificial islands, as well as wear and abrasion of walls under ice action, are not treated either.

The objectives are to describe the physical reasoning behind the mathematical models, which form a backbone for analyzing ice interaction with structures. This involves physical models from experimental work, general mathematical models, simplifications and relevant parameters of both the structure and ice.

2. Interaction Model

2.1 Process model

A schematic model of ice interaction with structures has three elements: structure, ice failure process, and ice sheet, Figure 1. Each can be described by a mathematical model. For the structure and ice there are well known governing differential equations, the dynamic equations of equilibrium. The two interacting bodies are connected by the center element, the ice failure process, which includes clearing mechanisms of the broken ice masses. The conditions of continuity also have to be taken care of in the center element.

2.2 Physical model

Structures with inclined walls at the water-line include sloping faced bridge piers, conical structures and ships, that are intended to break moving ice in bending mode even though crushing and shearing modes are simultaneously present to some extent. Both horizontal and vertical interaction forces will develop. The mode of ice failure depends in addition to the shape and size of the structure, also on ice properties, the velocity of the ice and friction between the structure and ice. In a loading cycle local crushing and shearing will initially occur, smoothing a large enough contact area, making higher ice loads possible. Increasing vertical forces first cause radial cracks and later a circumferential crack, which yields to final bending failure. Thereafter a new cycle can start. The ice load is at its maximum just before the formation of the circumferential crack. The clearing mechanism for upwards breaking structures involves pushing broken ice pieces upwards and to the side of the structure. This mechanism is the opposite with downwards breaking structures.

With vertical structures, as the pressure is increasing at the contact zone ice behaviour depends on the loading rate. At low loading rates, ice behaves like a viscous fluid and passes smoothly by the structure. Ductile deformation may include cracking effects. With increasing loading rate plastic yielding decreases and cracking activity increases. A damaged ice zone is then formed. Ice starts to break into floes and small fragments. At high loading rates ice failure is totally brittle. In certain conditions, ice is practically being pulverized. The clearing mechanism of broken ice masses with ice floes or fragments is different to that of pulverized ice: the former pops out "explosively" while the latter is extruded.

Another question is how the actual ice failure occurs at high loading rates. Recent studies using high speed photographical techniques and transparent structure walls (Joensuu and Riska, 1988) indicate that there is only a narrow "contact line", in which high interaction pressure acts and crushes the ice, while the rest of the contact zone is involved only in the clearing mechanism.

2.3 Mathematical model

Mathematical models for inclined walled structures are well established and verified by scale model tests, e.g. elastic ice bending failure by Bercha and Danys (1975) and plastic limit analysis by Ralston (1977). Full scale testing, (Määttänen, 1986) indicated rubble pile formation, which was not observed in scale model tests or in mathematical models. The effect of the rubble pile can be observed with minor adjustments in these mathematical models. For vertical structures the mathematical model of ice failure for the contact zone is complicated: even the physical model is not yet well understood. Due to the Molikpaq incident (Jefferies and Wright, 1988), research activity was aroused to learn what is going on between the ice edge and the structure. Lately, Hallam and Pickering (1988) and Jordaan et al. (1988) have developed new mathematical models to describe ice failure and pulverized ice extrusion. They used assumed viscous flow and cohesive material models in the clearing phase. The measured contact line model (Joensuu and Riska, 1988) awaits the formulation of a mathematical model.

A schematic mathematical model of ice interaction with structures, Figure 1, can be formulated with two coupled equations of equilibrium, Equation 1 and 2,

$$[k]\{\delta\} + [d]\{\dot{\delta}\} + [m]\{\ddot{\delta}\} = \{f(\delta, \dot{\delta}, \ddot{\delta}, \Delta, \dot{\Delta}, \ddot{\Delta}, t, V)\}, \tag{1}$$

$$[K]\{\Delta\} + D\{\dot{\Delta}\} + [M]\{\ddot{\Delta}\} = \{F(\delta, \dot{\delta}, \ddot{\delta}, \Delta, \dot{\Delta}, \ddot{\Delta}, t, V)\}, \tag{2}$$

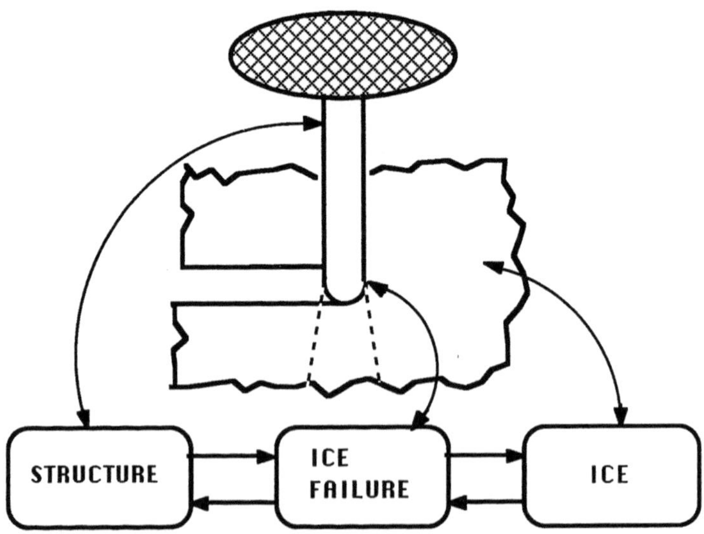

Figure 1. Elements in ice interaction with structures.

which represent the governing differential equations of the structure and ice in a discretized form. The symbols $\{\delta\}$ and $\{\Delta\}$ stand for the displacement vectors of the structure and ice, dots above mean the derivative with respect to time, t, giving velocities and accelerations; $[k]$, $[K]$, $[d]$, $[D]$, $[m]$, and $[M]$ are stiffness, damping and mass matrices respectively, V is the initial ice velocity. The coupling is through nonlinear functions $\{f\}$ and $\{F\}$ that describe interaction forces from the structure to the ice failure process, and from the ice failure process to the ice sheet.

If all elements in Figure 1 could be replaced by a mathematical model correctly, such as Equation 1 and 2, the problem of ice interaction with structures could be solved. Then most problems could be analyzed by using numerical ice mechanics. The main difficulty is in the ice constitutive model: elastic, viscous, and plastic behaviour including cracking should be modeled in three dimensions. It is only recently, that visco-elasto-plastic continuum damage constitutive models have been presented for ice (Sjölind, 1988; Pulkkinen, 1988; Santaoja, 1989; Choi and Karr, 1989). It should be noted that a lot of experimental and numerical work is needed to calibrate and verify these models.

2.4 Simplifications

Equations 1 and 2 are in themselves an approximation from the reality, because the continua - ice and structure - with infinite number of degrees of freedom, is replaced by a discretized model with a limited number of degrees of freedom. However, if the modelling is otherwise accurate, a sufficient accuracy can be maintained by a proper choice of the number of degrees of freedom in the discretization.

The interaction forces in Figure 1 are distributed along the whole contact area giving many nonzero terms in the coupling forces $\{f\}$ and $\{F\}$ in the discretized model. An evident simplification is to use only a single resulting interaction force. From the force balance in the center element it follows that the resulting $f = F$.

In Equation 1 and 2 interaction forces are not known beforehand, they depend on the state of the system. The most common approach is to assume an interactive forcing function, which enables us to separate the two systems. The physical interpretation of Equation 1 is then ice with no response, and of Equation 2, an infinitely rigid structure. Still simpler is to exclude all dynamics, or to use only a single degree of freedom dynamic model.

There are many simplifications in modelling ice behaviour and failure. The nonlinear visco-elasto-plastic constitutive equations of ice are linearized and/or replaced by a simple constitutive law which considers only one effect to be significant. This way it is possible to have a linear elastic material model for fast loading cases, reference stress method for slow loading cases, ideal plastic behaviour for limit load analysis, etc.

3. Static Interactions

Static ice interaction with structures is characterized by the absence of energy interchange between the structure and ice. Other interaction mechanisms are still present. The shape of the structure at the water-line controls the ice failure mode: crushing, splitting, shearing, bending and also controls the broken ice clearing mechanisms. The structure is only passively absorbing the ice failure reaction forces.

Usually the main interest is the design of the structure for ice forces. If the structure is very stiff, its response will be static even though the interaction ice force has dynamic fluctuations.

Another form of static interaction is with multilegged structures. When the spacing of legs at the water-line is less than about five times the leg diameter, ice failure against one leg promotes ice failure against the other. This leads to lower resulting ice loads than if all the legs were alone without interference from the others. Full scale data is scarce, but scale model evidence exists (Mizikos, 1986; Saeki et al., 1986; Timco, 1986, etc.). These have led to experimental design methods. However, it should be noted that close spaced legs may lead to ice jamming and adfreeze.

4. Dynamic Interaction

Ice-structure interaction implies that both the ice and the structure are active partners. In the case of a rigid structure, the only contribution of the structure to the ice failure process comes from the shape of the structure at the contact zone. In terms of vibrations, the stiffness, mass and damping properties are passive in relation to the ice forces.

In the case of a flexible structure, the dynamic response of the structure plays an active role with the force originating from the ice failure process.

Hence all structural properties, shape at the contact zone, stiffness, mass and damping will have their effect on the interactive forces between the ice and the structure.

The distinction between a rigid or flexible structure is arbitrary. Even a solid rock undergoes elastic deformations when loaded by ice action. Also the ice field itself undergoes elastic deformations. From the practical point of view a structure can be considered flexible when its displacements are significant when compared to the elastic deformations of the ice, e.g. of the same order as the ice grain diameter.

In dynamic ice-structure interaction, energy from the moving ice is being transferred and stored as elastic energy in the structure. After ice failure and at the structural spring back phase, stored energy is released. It is converted into breaking ice and into kinetic energy. Energy exchange implies also that dynamic response occurs, either in the structure, in the ice, or in both. As ice strength is strongly dependent on loading rate, dynamic response will alter the average loading rate, and hence the interaction forces. Ice failure will occur at high loading rates with low ice strength, and load build up at low loading rates with high ice strength. This leads to the synchronization of ice failure and structural response.

Of course there are other parameters to control sychronization. Load build-up, possible gap between the structure and ice edge, and broken ice clearing are controlled by the relative displacement between the interacting bodies. It is also possible to postulate the effect of acceleration as in the case of wave loads.

Theoretical explanations for dynamic ice-structure interaction are based on forced or self-excited vibration models. In the former, ice is fractured into floes of certain size, which is only related to ice thickness (Neill, 1976; Sodhi and Morris, 1986; and Sodhi, 1988). Hence ice force build-up and failure frequency is directly proportional to ice velocity. The interaction ice force depends only on the advancement of ice sheet. Ice force history is then known *a priori* when the ice velocity is known. If the dynamic response of the structure is insignificant there is no dynamic interaction, only dynamic reaction forces. A mechanical model for this kind of an interaction using breaking brittle cantilevers was presented by Matlock et al. (1969), which also observes the dynamic response of the structure.

In the self-excited ice-induced vibration model (Blenkarn, 1970; Määttänen, 1978, 1988), the interaction ice force is dependent on the dynamic response of the structure at the ice action point with no *a priori* known ice force time history. The model observes the varying ice failure strength with varying loading rate. Hence the feedback, or coupling, variable is the dynamic response velocity. The system is autonomous and the only chance for vibrations to occur is when the system becomes dynamically unstable, which leads to limit cycles in self-excited vibrations. An autonomous vibration system is one in which dynamic response and the interaction force originate from the system itself without need to know beforehand any history of this force.

The ice failure process is a nonlinear event from load build-up to ice failure and clearing. Neither the forced nor self-excited models observe the clearing phase. A new self-excited vibration model by Hallam and Pickering (1988) is based on load build-up and a clearing mechanism. According to measurements and observations by Jordaan et al. (1988), maximum interaction force occurs at the end of the clearing phase when the gap between the structure and solid ice edge is narrowest just before ice failure in the next cycle. By observing the load build-up, ice failure and the dependence of broken ice mass extrusion pressure on the gap size, a self-excited model is derived. For that model the dynamic response displacement is the coupling feed-back variable. Another displacement controlled self-excited model is by Kärnä (1988), which is based on the high frequency and subsequent short distance ice edge fracture and clearing mechanism.

5. Ice Impact

The impact of ice against a structure has different scenarios: transients during ice sheet edge hitting, splitting of floes, pushing aside floes, growler or iceberg impacts.

If a sufficiently large ice sheet moves, and its edge hits a structure, there will be a short duration transient after which continuous ice structure interaction will be established according to the previous models. During the transient phase, initial load build-up may induce higher dynamic response in the structure than in the continuous phase. If the laws for ice load build-up can be assumed, the calculation of dynamic response is straightforward. Typically, dynamic amplification of the transient is less than twice that of the static loading. If a resonant loading in dynamic ice structure interaction is possible, its dynamic amplification is much higher due to the inherently low damping of the structure.

Splitting of a small floe during an impact occurs easily. The interaction is just to initiate splitting. Theoretical models based on force equilibrium (Michel, 1978) and on fracture mechanics (Bhat, 1988; Sanderson, 1988) explain the low load levels needed for splitting the ice. The latter models also explain the splitting of floes with a diameter of several kilometres.

A floe rarely hits symmetrically. Asymmetric contact causes the floe to rotate and to be pushed aside from the structure. Interaction forces can be calculated according to laws of conserving energy and momentum (Nevel, 1986).

The hitting of a growler or iceberg can be treated like the hitting of a floe. However, more emphasis is now in the load build-up history. It is dependent on the velocity and the size of contact area. The former can be calculated from the energy balance and the latter from the geometry of impacting bodies (Nevel, 1986). An important factor is ice strength dependence on the contact area (Sanderson, 1986; Nadreau et al., 1987).

6. Structural Parameters

The first distinction is due to the angle from horizontal level of the leading edge of the structure. Angles below 75 degrees promote ice bending failure, which usually results in lower total ice forces. Provided that water depth is deep enough to ensure no jamming against the bottom, the inclination angle can be negative to produce ice bending failure downwards with still lower total forces. This is a result of the clearing mechanism. If the ice is broken upwards the broken floes have to rise above the water, which needs about nine times more energy than to submerge them below the water with a downward breaking structure (Ralston, 1979). Surface roughness also plays an important role both in ice failure and clearing against an inclined face structure. High friction can totally prevent the vertical sliding, changing the ice bending failure into crushing failure with much higher loads.

The planform shape of the structure at the water-line can be circular, square, or wedged. With the same width the total load for a circular shape is a little lower, about 10%, than with a square in the crushing mode (Korzhavin, 1962). A wedge shaped leading edge in a bridge pier promotes floe splitting, which reduces total ice load. The most efficient shape to reduce ice forces is, according to scale model tests (Haynes et al., 1983), a wedge with a progressively steepening leading edge.

With vertical structures crushing is the dominating ice failure mode even though spalling, splitting or buckling can also occur, (Blanchet et al., 1988). Aspect ratio B/h, the width of the structure at water-line, B, divided by ice thickness, h, is one parameter which determines the failure mode. At low aspect ratios the state of stress in the ice at the contact zone is three dimensional resulting in higher average ice pressures than at high aspect ratios. From many experimental equations for the ratio of average ice pressure to uniaxial compressive strength σ_c, the one by Afanas'yev et al. (1971), is the most often quoted:

$$p_e/\sigma_c = (1 + 5h/B)^{0.5}. \tag{3}$$

For inclined face structures, the aspect ratio effect is more pronounced in clearing mechanisms. As an example, a low aspect ratio conical test structure in the Okhotsk Sea, did not induce a rubble pile formation which was present in the Gulf of Bothnia at a higher aspect ratio (Määttänen, 1986).

The number of legs in the case of a multilegged structure is related to interference factors. If the spacing of the legs is more than five leg diameters at the water-line, no reduction is to be expected in the total ice force (Saeki et al., 1986). The closer the legs are, the lower the resulting total ice force is. However, the danger of ice jamming in between the legs increases with closer spaced legs. If jamming occurs, the total ice load will increase significantly. Another risk is ice adfreezing simultaneously to all legs making all the leg-bounded area to behave as one single block (Mizikos, 1986).

For dynamic ice-structure interaction, the natural modes and frequencies of the structure are important (Määttänen, 1978, 1984). They are controlled by the mass and stiffness distribution in the structure, and by the boundary conditions, namely how stiff is the bottom foundation attachment. Also structural damping is important in suppressing ice-induced vibrations. In scale model, and full scale tests, it has been measured, that with a sufficiently high internal damping in the structure, resonant type self-excited vibrations can be totally avoided (Määttänen, 1987).

Another point in the dynamic ice-structure interaction analysis is the number of degrees of freedom in the model. As ice failure occurs the reduction of ice load level is almost instantaneous. For a mathematical model to be capable of following these abrupt changes, a multi-degrees of freedom model is required. It is the higher natural modes that can follow faster sudden load changes. Hence, results by using a single degree of freedom model in dynamic ice-structure interaction analysis cannot be reliable.

7. Ice Parameters

All the normal mechanical properties of ice have their effect in ice interaction with structures, and they are not discussed here. Just to mention the most important ones: crystal structure, temperature, salinity. These properties determine ice strength and rigidity for different failure modes.

Ice thickness appears first in the aspect ratio dependence, which has its effect on average crushing pressure, etc., as explained in Section 6. A general trend is increasing ice forces with increasing ice thickness in all modes of ice-structure interaction. Also the mode of ice failure can change, e.g. from buckling to crushing or from bending to crushing or shearing, with increasing ice thickness.

In dynamic ice-structure interaction, the energy input from the ice to the structure - negative damping - is increased with increasing ice thickness. With thin ice the internal damping of the structure can overcome the effect of ice induced negative damping and suppress resonant vibrations, but with thicker ice the situation can become the opposite and resonant type vibrations can occur because the structural damping is constant.

Ice load fluctuations against rigid structures are caused by the properties of the ice alone. Subsequent ice edge flaking failures is one explanation, tendency to fail into floes of certain size another, but also the dynamic response in the ice field together with the ice strength sensitivity to strain rate can contribute to resulting interaction ice force fluctuations.

Internal friction of ice depends on ice type, salinity and temperature, and its effect appears in ice piling. A direct consequence is in all clearing mechanisms of the broken ice mass in front of the structure. Snow above the ice increases the friction considerably. In the Kemi-I test cone project it was learnt in full scale measurements with snow on ice, that a rubble pile will form in front of the cone (Määttänen, 1986), a phenomenon which was not postulated theoretically or observed in scale model tests beforehand.

Ice velocity determines the average strain-rate level, which is an important factor in ice behaviour from ductile to brittle and in ice-induced vibrations. If the strain-rate is very low, the ice behaves like a viscous fluid and there is no source for dynamic excitation. At high strain-rates, ice behaves in a brittle manner causing random ice force variations in crushing failure mode without any other strain-rate effects. It is around the transitional strain-rate from ductile to brittle, 10^{-3} - $10^{-2}s^{-1}$ or stress rate around 0.2 - 0.6 MPa/s, where self-excited vibrations are most pronounced.

In general, the lower the ice integrity, the lower are the interaction ice forces. Cracks and leads reduce integrity. Maximum integrity is with a solid, even ice cover, unless a spheroidal iceberg is considered. If ice is formed of loosely packed floes, it is easy to understand that floes can be pushed aside easily without higher ice breaking forces. This was evident in Cook Inlet measurement data (Blenkarn, 1970). Broken ice masses and pack ice can, however, be under pressure, which on the contrary increase ice forces. There is no published comparison of ice loads against a bottom-founded structure in the same ice mass with and without pressure as with ships, where there is a significant increase. One factor in ice integrity is the boundary conditions. If the ice movement is bounded on one or both sides, its integrity could increase.

8. Conclusions

The knowledge of ice interaction with structures has made major strides during the last two decades. The catalyst for ice mechanics research has been the need to utilize arctic oil and gas resources. Another factor has been increasing winter navigation.

The physical model of ice-structure interaction can be divided into three elements: the structure, the ice, and the ice fracturing process including clearing mechanisms.

Theoretical models have been based on experimental observations, both in scale model laboratory and in-field full scale tests. Full scale tests have indicated interaction forms that have not been observed in laboratory, such as rubble pile formation in front of a conical structure and resonant type vibrations in a caisson retained island.

Mathematical modelling of the structure is straightforward, the modelling of ice and ice failure process more complicated. It is only lately that three dimensional constitutive models for ice visco-elasto-plastic behaviour with cracking have been developed. The validation of these models still remains to be proven.

Many simple models and equations to calculate ice-structure interaction forces have been developed by simplifying the complicated reality. Most often used are the Korzhavin equation for ice crushing, the Ralston limit analysis for conical structures, and the reference stress method by Ponter et al., for crushing in slow ice movement.

In static ice-structure interaction, no energy interchange from ice to structure, or vice versa, occurs. The ice failure mode is controlled by the shape of the structure, and the ice force fluctuations by the properties and velocity of ice.

In dynamic interaction the energy interchange from ice to the structure and vice versa, has its effect on interactive ice forces. A synchronization of the dynamic response of the structure and ice fracture process may develop. Important are both the properties of the structure and ice: natural modes and frequencies, ice strength thickness and velocity.

Dynamic ice force fluctuations are a result of the tendency of ice to fracture into floes of a certain size or to be self-excited by nature. The latter model is based on ice strength dependence on loading rate, and hence on ice velocity and structural response rate. Lately other self-excited models have been developed to observe the structural displacement response in ice fracture and clearing phases.

The impact of ice against structures can be modeled by considering the energy and momentum balance. Interaction is dynamic and the transient phase is decisive in ice forces design. Fracture mechanics models explain splitting phenomena.

The most important structural parameters are the shape of the structure at the water-line in controlling the ice failure mode, and its stiffness, mass and damping in dynamic interaction. The most important ice parameters are its strength, thickness, integrity and velocity.

To understand ice-structure interaction better, and to be able to design safe and economical offshore structures in future, more effort is needed in full-scale testing, developing ice constitutive models, and verifying computational ice mechanics software.

References

Afanas'yev, V.P., Dologopolov, I.U. and Shvaysteyn, Z.I. 1971. Ice pressure on separate supporting structures in the sea. CRREL Draft Translation, No.346 (1972), Hanover, NH, U.S.A.

Bercha, F.G. and Danys, J.V. 1975. Prediction of ice forces on conical offshore structures. Marine Science Communications, 1975, pp.365-380.

Bhat, S.U. 1988. Analysis for splitting of ice floes during summer impact. Cold Regions Science and Technology, Vol.15, pp.53-63.

Blanchet, D., Churcher, A., Fitzpatrick, J. and Badra-Blanchet, P. 1988. An analysis of observed failure mechanisms for laboratory, first-year and multi-year ice. IAHR 1988, University of Hokkaido, Sapporo, Japan, Vol.3, pp.89-136.

Blenkarn K.A. 1970. Measurement and analysis of ice forces on Cook Inlet structures. Proc. 2nd Offshore Technology Conference, Houston, TX, OTC 1261, Vol.II, pp.365-378.

Cammaert, A.B. and Muggeridge, D.B. 1988. Ice interaction with offshore structures. Van Nostrand Reinhold, New York.

Choi, K. and Karr, D.G. 1989. A damage mechanics model for uniaxial creep and cyclic loading of polycrystalline ice. OMAE 1989, The Hague, Netherlands, Vol.IV, pp.67-74.

Gamayunov, A.I. 1947. Opredelenie davleniya l'da na opory mostov v naturnykh usloviyakh, (Determination of ice pressure on the supports of bridges under natural conditions). Teckhnika Zheleznykh Dorog, Nos.4 and 12.

Hallam, S.D. and Pickering J.G. 1988. Modelling of dynamic ice loading of offshore arctic structures. Proc. Polartech'88, Tapir Publishers, Trondheim, Norway, pp.235-248.

Haynes, F.D. and Sodhi, D.S. 1983. Ice forces on model marine structures. POAC 1983, Helsinki, Finland, Vol.II, pp.778-787.

Jefferies, M.G. and Wright, W.H. 1988. Dynamic response of "Molikpaq" to ice-structure interaction. OMAE 1988, Houston, TX, U.S.A., Vol.IV, pp.201-220.

Joensuu, A. and Riska, K. 1988. Jään ja rakenteen välinen kosketus. (Contact between ice and structure, in Finnish only). Report M-88. Helsinki University of Technology, Laboratory of Naval Architecture and Marine Engineering, Espoo, Finland.

Jordaan, I.J., Maes, M.A, and Nadreau, J-P. 1988. The crushing and clearing of ice in fast spherical indentation tests. OMAE 1988, Houston, TX, U.S.A., Vol.IV, pp.111-116.

Kärnä, T. and Turunen, R. 1988. Dynamic response of narrow structures to ice crushing. Cold Regions Science and Technology, Vol.17, pp.173-187.

Korzhavin, K.N. 1962. Action of ice on engineering structures. USSR Academy of Science, Siberian Branch, CRREL Draft Translation No. 260.

Kubo, Y. 1980. Introductory remarks on iceological engineering, (in Japan). Publisher Y Kubo, Tokyo, Japan.

Määttänen, M. 1978. On conditions for the rise of self-excited ice-induced autonomous oscillations in slender marine structures. Finnish-Swedish Winter Navigation Board, Finland, Research Report 25, 98p.

Määttänen, M. 1984. The effect of structural properties on ice-induced self-excited vibrations, IAHR 1984, Hamburg, West Germany, Vol.II, pp.11-20.

Määttänen, M. 1986. Test cone project. Proc. Polartech '86, Technical Research Centre of Finland, Symposium 71, Espoo, Finland, Vol.II, pp.749- 761.

Määttänen, M. 1987. Ten years of ice-induced vibration isolation in lighthouses. OMAE 1987, Houston, TX, U.S.A., Vol.IV, pp.261-266.

Määttänen, M. 1988. Ice-induced vibrations in structures -self-excitation. IAHR 1988, University of Hokkaido, Sapporo, Japan, Vol. 2, pp.658-665.

Matlock, H., Dawkins, W. and Panak, J. 1969. A model for the prediction of ice-structure interaction. Journal of Engineering Mechanics, ASCE, EM4,pp.1083-1092.

Michel, B. 1978. Ice mechanics. Les Presses de l'Université Laval, Quebec, Canada.

Michel, B. and Toussaint, N. 1977. Mechanisms and theory of indentation of ice plates. Journal of Glaciology, Vol. 19, No. 81, pp.285-300.

Mizikos, J-P. 1986. Ice design of multi-legged structures. Proc. Polartech '86, Technical Research Centre of Finland, Symposium 71, Espoo, Finland, Vol.II, pp.735-748.

Montgomery, C.J, Gerard, R. and Lipsett, A.W. 1980. Dynamic response of bridge piers to ice forces. Canadian Journal of Civil Engineering, Vol.7, pp.345-356.

Nadreau, J-P., Christian, F. and Stone, B. 1987. Spherical indenter tests at Resolute Bay. (In preparation).

Neill, C. 1976. Dynamic ice forces on piers and piles: an assessment of design guidelines in the light of recent research. Canadian Journal of Civil Engineering, Vol.3, pp.305-341.

Nevel, D.E. 1986. Iceberg impact forces. IAHR 1986, University of Iowa, Iowa City, IO, U.S.A., Vol.III, pp.345-369.

Peyton, H. 1968. Sea ice forces. Ice Pressure Against Structures, National Research Council of Canada, Ottawa, Canada, Technical Memorandum 92, pp.117-123.

Ponter, A.R., Palmer A.C., Goodman D.J., Evans A.G. and Hutchinson, J W. 1983. The force exerted by a moving ice sheet on an offshore structure. Part 1. The creep mode. Cold Regions Science and Technology 8, pp.109-118.

Pulkkinen, E. 1988. Numerical modelling of ice behaviour. Acta Universitas Ouluensis, Ser.C, Technica 46. 122p.

Ralston, T. 1977. Yield and plastic deformation in ice crushing failure. Proc. Symposium on Sea Ice Processes and Models, Seattle University, Washington, Vol. 2, pp.147-156.

Ralston, T. 1979. Plastic limit analysis of sheet ice loads on conical structures. Proc. IUTAM Symposium on the Physics and Mechanics of Ice. Technical University of Denmark, Copenhagen, Denmark, pp.289-308.

Saeki, H., Ono, T., Takeuchi, T., Suenaga, T.L. and Sakai, M. 1986. Total ice forces on clusters of cylindrical piles. OMAE 1986 Symposium, Tokyo, Japan, Vol.IV, pp.461-466.

Sanderson, T. 1986. A pressure-area curve for ice. IAHR 1986, University of Iowa, Iowa City, IO, U.S.A., Vol.II, pp.361-384.

Sanderson, T. 1988. Ice mechanics, risks to offshore structures. Graham and Trotman, London, UK.

Santaoja, K. 1989. Continuum damage mechanics approach to describe the multidirectional microcracking of ice. OMAE 1989, The Hague, Netherlands: The American Society of Mechanical Engineers, Vol.IV, pp.55-66.

Sjölind, S-G. 1987. A constitutive model for ice as damaging visco-elastic material. Cold Regions Science and Technology, Vol.14, pp.247-262.

Sodhi, D. 1988. Ice-induced vibration of structures. IAHR 1988, Sapporo, Japan, Vol.2. pp.625-657.

Sodhi, D. and Morris, C. 1986. Characteristic frequency of force variations in continuous crushing of sheet ice against rigid cylindrical structures. Cold Regions Science and Technology, Vol. 12, pp.1-12.

Timco, G. 1986. Ice forces on multi-legged structures. IAHR 1986, University of Iowa, Iowa City, IO, U.S.A., Vol.II, pp.321-338.

Toyama, Y., Sensu, T., Minami, M. and Yashima, N. 1983. Model tests on ice-induced self-excited vibration of cylindrical structures. POAC 1983, Helsinki, Finland, Vol.II, pp.834-844.

Xu, J. and Wang, L. 1988. Ice force oscillator model and its numerical solutions. OMAE 1988, Houston, TX, U.S.A., Vol.IV, pp.171-176.

Discussion

Q: John Dempsey, Clarkson University

Concerning the various dynamic ice-structure interaction events, has anyone quantitatively categorized and correlated the size of the broken ice pieces that occur?

A: M.P. Määttänen

I think that the only assessment of broken ice pieces is by Charles Neill in his bridge pier ice force research. In field data, the characteristic floe size is about 0.2 to 0.5 times the ice thickness.

Q: M. Metge, CANATEC

In one of your pictures, you showed a cone which with fast ice movement was covered with rubble. What was the angle of the cone, and what were the failure modes observed? In particular, was flexure still effective in lowering the loads?

A: M.P. Määttänen

The cone angle was 55° from the horizon. During ice piling in front of the cone all forms of ice failure were present, including crushing, shearing, and bending. It was not possible to see what was actually going on under the ice rubble and the snow covered ice sheet. There were indications that circumferential bending failure cracks did occur.

Adjusting Local Stiffness for Improving Dynamic Ice-Structure Interaction of Bohai Jackets[1]

Xu Jizu

Ocean Engineering and Naval Architecture Department,
Tianjin University, Tianjin, China

Wang Lingyu

Civil Engineering Department,
Tianjin University, Tianjin, China

Abstract Flexible structures built in sea ice infested waters require countermeasures against ice-induced vibrations. A design strategy to improve the dynamic ice-structure interaction is proposed, considering the behaviour of sea ice forces. According to the controlled displacements of the structure, the mode shapes are optimized by adjusting the local structural stiffness. The approach and its efficiency are illustrated by a Bohai jacket.

1. Introduction

In some sub-Arctic areas, flexible structures such as jackets or other piled structures are used for petroleum production and aid-to-navigation purposes. Practical experiences indicate the large dynamic amplification as the ice sheet is crushing through the structure. In this case, the ice forces and their effects on the structure will evidently be influenced by the response and hence the dynamic behaviour of the structure. This is called dynamic ice-structure interaction, and is also regarded as self-excited vibration. Therefore, the engineers always pay much attention to a reasonable structural design.

[1]This research project is supported by the National Natural Sciences Foundation of China.

S. Jones · R. F. McKenna · J. Tillotson · I. Jordaan (Eds.)
Ice-Structure Interaction
IUTAM-IAHR Symposium St. John's, Newfoundland, Canada

In the Finnish waters of the Gulf of Bothnia, after a ten year engineering test and improvement, vibration isolation systems have been popularized in the flexible lighthouse constructions in this area (Määttänen, 1987).

In the Bohai Gulf, offshore of Northern China, new types of ice resistant platforms for oil production are being investigated with the aim of replacing the traditional jackets, which are regarded as being appropriate to conditions of waves but not to those of ice, partially due to their sensitivity to ice-induced vibrations (Reference [3], 1988).

Among the various possible ways to reduce the dynamic amplification for an offshore structure, it seems more reasonable and economical to adjust the mode shapes according to the characteristic of environmental loads than to change its natural frequencies. For improving an existing structural design, it is feasible to keep the control displacements on some specified degrees of freedom within a certain limit by only properly adjusting the local structural stiffness, and leaving the rest of the structure unchanged. This comes to an optimization of the mode shapes. In addition to the sea ice forces, the wind loads applied on the upper part of the platform should also be considered in the optimization.

2. Expression of Dynamic Displacements on specified Degrees of Freedom

A structure in motion can be discretely described as:

$$[M]\{\ddot{X}\} + [C]\{\dot{X}\} + [K]\{X\} = \{F\}, \tag{1}$$

where

$[M]$, $[C]$, $[K]$ – mass, damping and stiffness matrices of the structure.

$\{\ddot{X}\}$, $\{\dot{X}\}$, $\{X\}$ – structural acceleration, velocity and displacement vectors.

$\{F\}$ – load vector.

Using the following coordinate transformation

$$\{X\} = \Phi_1 Y_1 + \Phi_2 Y_2 + \cdots + \Phi_N Y_N = \sum_{n=1}^{N} \Phi_n Y_n, \tag{2}$$

or

$$\{X\} = [\Phi]\{Y\}, \tag{3}$$

where $\{Y\}$ is the modal coordinate vector, $[\Phi]$ is the modal matrix with its columns $\Phi_1, \Phi_2, \cdots \Phi_n$ indicating the mode shapes, N and n are respectively the total and ordinal numbers of modes; Equation (1) can be changed into a new set of equations consisting of the following single equations for each mode:

$$M_n\ddot{Y}_n + 2\xi_n\omega_n M_n\dot{Y}_n + \omega_n^2 M_n Y_n = F_n \,. \tag{4}$$

All the symbols with suffix n in the equation refer to th nth mode. If the external forces are applied on $j1$ degrees of freedom, ıe modal load Fn then can be written as

$$F_n(t) = \Phi^T_{nj1}F_{j1}(t), \tag{5}$$

where Φ_{nj1} and F_{j1} are $j1 \times 1$ vectors.

Assuming a single harmonic excitation $F_{j1}(t) = F_{oj1} = F_{oj1}$ sinΩt with frequency Ω, the stationary solution of Equation (4) can be written as:

$$Y_n = \frac{\Phi^T_{nj1}F_{oj1}}{M_n\omega_n^2}[(1-\beta_n^2)^2 + (2\xi_n\beta_n)^2]^{-1/2}sin(\Omega t - \Theta_n), \tag{6}$$

where

$$\beta_n = \Omega/\omega_n, \qquad \Theta_n = tan^{-1}\left(\frac{2\xi_n\beta_n}{1-\beta_n^2}\right). \tag{7}$$

The requirement for operation and safety of the structure in vibration can be defined by the "control displacements" $\{X_{j2}\}$ on some specified degrees of freedom j_2 (for instance, the horizontal displacements at the deck level), and from Equation (2) the control displacements can be expressed as

$$\{X_{j2}\} = \sum_{n-1}^{N} \frac{\Phi_{nj2}\Phi^T_{nj1}F_{oj1}(t)}{M_n\omega_n^2}[(1-\beta_n^2)^2+(2\xi_n\beta_n)^2]^{-1/2}sin(\Omega t-\Theta_n). \tag{8}$$

3. Optimization of Mode shapes

Equation (8) indicates that the coefficient matrix $C^n_{j2\times j1} = \Phi_{nj2}\cdot\Phi^T_{nj1}$ plays a big role in determining the control displacements $\{X_{j2}\}$. Generally, the maximum structural response always happens under a single mode resonance or locking in, implying the major contribution of a single mode to control displacements (Määttänen, 1981). Experience has also identified that a change of local stiffness of the structure will have an evident effect on the mode shapes or distributions of modal displacements along the height of a structure but not on the order of modes. This knowledge confirms that the coefficient matrix $C^n_{j2\times j1}$ can be regarded as a characteristic index in restraint of the control displacements by adjusting the local structural stiffness K:

$$C^n_{j2\times j1} = \Phi_{nj2}(K) \cdot \Phi^T_{nj1}(K), \tag{9}$$

and the following equation can be employed to denote the optimized coefficient matrix $PC_{j2\times j1}$ with the associated local stiffness K^*:

$$PC_{j2\times j1} = min_{K=K^*}\ [MC_{j2\times j1}(K)], \tag{10}$$

where $MC_{j2\times j1}(K)$ indicates the coefficient matrix resulting in the maximum modal control displacements among all modes considered here under a given local stiffness K. Assuming the number of degrees of freedom j_3 to which the stiffness is adjusted, the local stiffness matrix becomes

$$K = K^* = K_{j3\times j3}, \tag{11}$$

4. Calculation of Local Stiffness

There are different ways to change the local stiffness of the structure. A device shown in Fig. 1 could be used. In the following discussion, the local stiffness of a column section between the loading point and the upper node (see Figs. 2 and 3) is assumed to be adjustable.

Having the assumption of lateral motion of nodes without rotation, the force-deflection relation for the beam element can be set up in the following cases:

(1) Case 1 (Fig. 2)

From the "force method" equation

$$\delta_{22}M_2 + \Delta_{2F} = 0 \tag{12}$$

$$M_2 = \frac{-\Delta_{2F}}{\delta_{22}} = (\frac{F_1 l_1^2}{2EI_1})/(\frac{l_1}{EI_1} + \frac{l_2}{EI_2}). \tag{13}$$

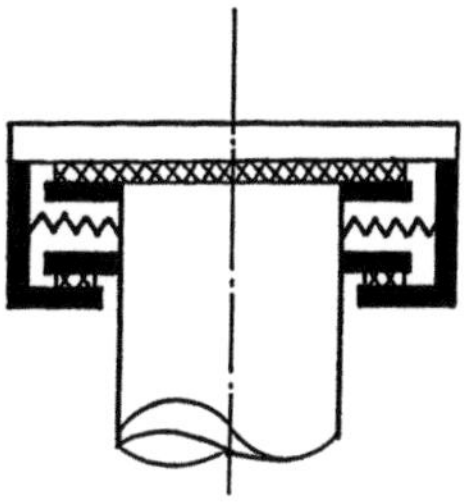

Figure 1. A device to change the local stiffness of the structure.

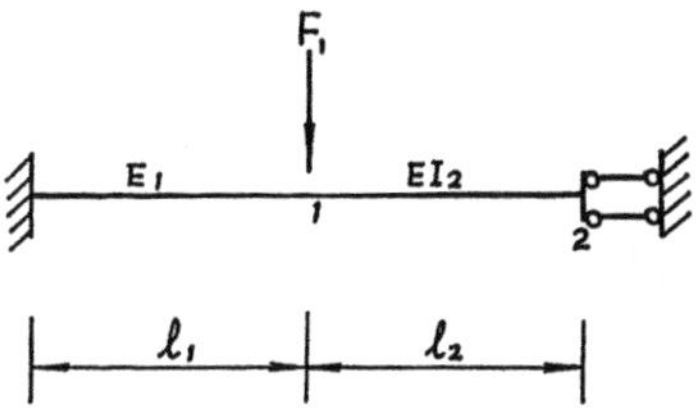

Figure 2. Loading case I

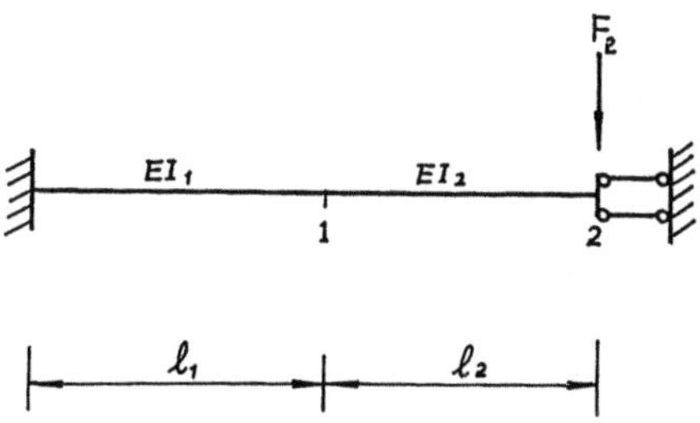

Figure 3. Loading case II

The lateral displacement Δ_x at x can then be expressed as follows:

(a)$x \leq l_1$:

$$\Delta_x = \frac{F_1}{k_1}(\frac{x}{l_1})^2[(6 - \frac{2x}{l_1}) - 3/(1 - \frac{k_1}{k_2}.\frac{l_1^2}{l_2^2})], \tag{14}$$

where $k_1 = \frac{12EI_1}{l_1^3}$ and $k_2 = \frac{12EI_2}{l_2^3}$, denote the stiffness of the two sections l_1 and l_2 respectively.

(b) $l_1 < x < l_1 + l_2$:

$$\begin{aligned}\Delta_x &= \frac{2F_1}{k_1}(\frac{3x}{l_1} - 1) - \frac{3F_1}{k_1}(\frac{2x}{l_1} - 1)/(1 + \frac{k_1}{k_2}.\frac{l_1^2}{l_2^2}) \\ &-\frac{3F_1}{k_2}.\frac{l_1}{l_2}(\frac{x}{l_2} - \frac{l_1}{l_2})^2/(1 + \frac{k_1}{k_2}.\frac{l_1^2}{l_2^2}),\end{aligned} \tag{15}$$

(c) $x = l_1$:

$$\Delta_1 = \frac{F_1}{k_1}[4 - 3/(1 + \frac{k_1}{k_2} \cdot \frac{l_1^2}{l_2^2})], \tag{16}$$

(d) $x = l_1 + l_2$:

$$\begin{aligned}\Delta_2 &= \frac{2F_1}{k_1}(2 - \frac{3l_2}{l_1}) - \frac{3F_1}{k_2}(1 - \frac{2l_2}{l_1})/(1 + \frac{k_1}{k_2} \cdot \frac{l_1^2}{l_2^2}) \\ &- (\frac{3F_1}{k_2} \cdot \frac{l_1}{l_2})/(1 + \frac{k_1}{k_2}\frac{l_1^2}{l_2^2}),\end{aligned} \tag{17}$$

(2) Case 2 (Fig. 3)

The right side end moment M_2 will become

$$M_2 = \frac{F_2(1_1 + 2l_2)}{2}/(1 + \frac{k_1}{k_2} \cdot \frac{l_1^2}{l_2^2}) + (\frac{F_2 l_2}{2})/(1 + \frac{k_1}{k_2} \cdot \frac{l_1^2}{l_2^2}), \tag{18}$$

and the lateral displacement Δ_x can be calculated by:

$$\Delta_x = \frac{F_2 x^2}{2EI_1}(l_2 + \frac{2x}{3}) - \frac{M_2}{2EI_1}x^2 \qquad (x \leq l_1), \tag{19}$$

and

$$\begin{aligned}\Delta_x &= \frac{F_2 x^2}{2EI_1}(l_1 + l_2 - \frac{x}{3}) - \frac{F_2}{2EI_1}(x - l_1)^2[l_2 - \frac{1}{3}(x - l_1)] \\ &- \frac{F_2}{2EI_2}(x - l_1)^2[l_2 - \frac{1}{3}(x - l_1)] - \frac{M_2 l_1}{EI_1}(x - \frac{l_1}{2}) \\ &- \frac{M_2}{2EI_2}(x - l_1)^2; \quad (l_1 \leq x \leq l_1 + l_2).\end{aligned} \tag{20}$$

It is noted that the local stiffness K in Equation (9) or Equation (10) can be obtained by $K = \frac{F_2}{\Delta_2}$ from the above equation.

(3) Case 3:

In case of $EI_1 = EI_2 = EI$, the lateral displacement Δ_x comes to be

$$\Delta_x = \frac{Fx^2}{12EI}(3l - 2x), \qquad (x \leq l), \tag{21}$$

where $l = l_1 + l_2$.

5. Illustration of the Approach

A typical Bohai wellhead jacket with 4 columns (Fig.4) is taken to be an example to illustrate the approach described above, and it is simplified, for convenience of illustration, into a plane frame structure (shown in Fig. 5) with rigid girders moving horizontally without rotation. The

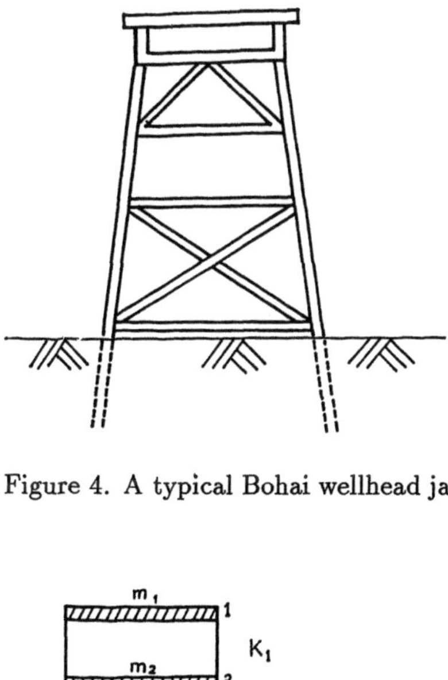

Figure 4. A typical Bohai wellhead jacket with four columns.

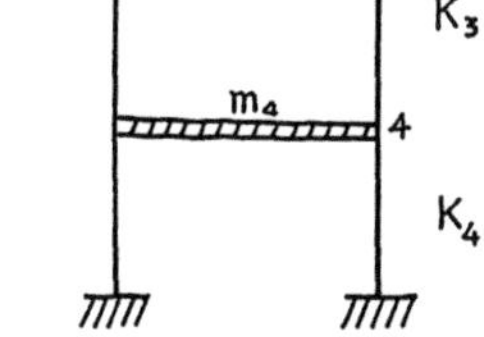

Figure 5. A plane frame structure.

horizontal deflection at deck level (at node 1) is taken to be the control displacement, and the local stiffness k_2 of the column section between loading point p and node 2 is being adjusted.

Assuming single harmonic processes, both the ice force (applied on point p) and wind force (applied on node 1) are considered in the mode shape optimization. The wind force is taken as having an amplitude only $\frac{1}{10}$ or $\frac{1}{20}$ that of the ice force.

Under various representative load conditions, the coefficient matrix MC in Equation (10) and associated control modal displacement Xc are calculated for different specified local stiffness K (Figs. 6 and 7).

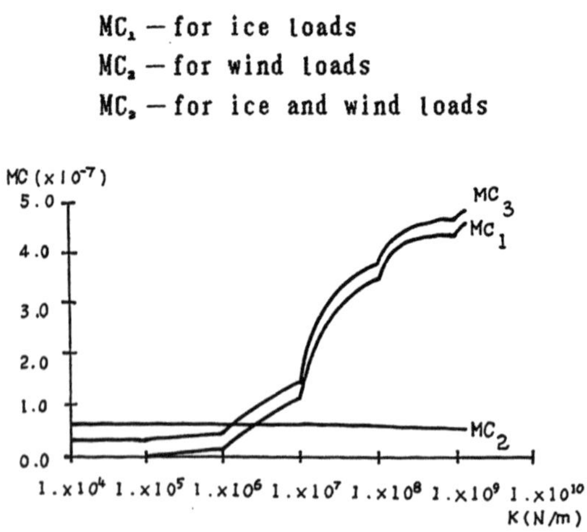

Figure 6. Coefficient matrix MC for different specified local stiffness K.

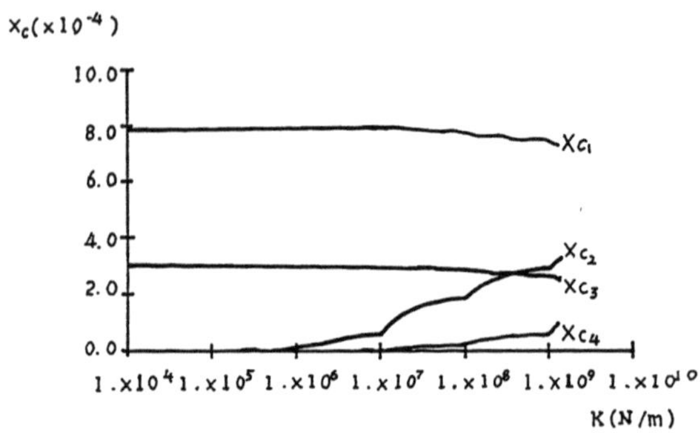

Figure 7. Control modal displacement Xc for different specified local stiffness K.

An optimized interval of local stiffness values can thus be obtained for structural design. As a reference value, the modal displacement at the loading point, Xp, is also calculated for different local stiffness K (Fig. 8).

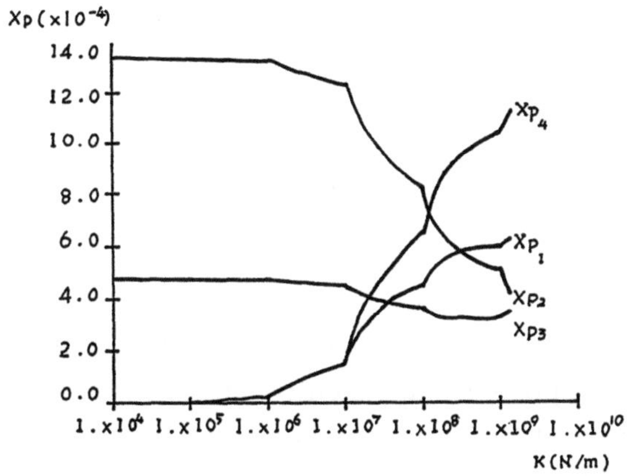

Figure 8. Modal displacement at the loading point Xp for different local stiffness K.

Taking a local stiffness value $K = 1.0 \times 10^7$ N/m from the optimized interval, the first four modes and associated frequencies of the "new" structure with changed local stiffness can be obtained as shown in Fig. 9, and can be compared with those of the original structure (Fig. 10).

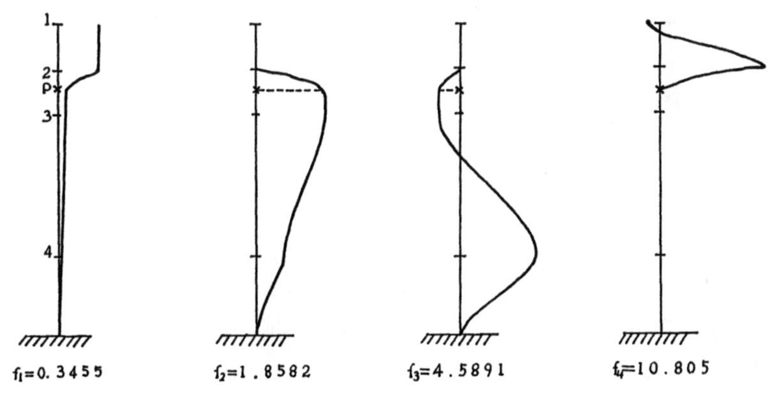

Figure 9. First four modes of the new structure.

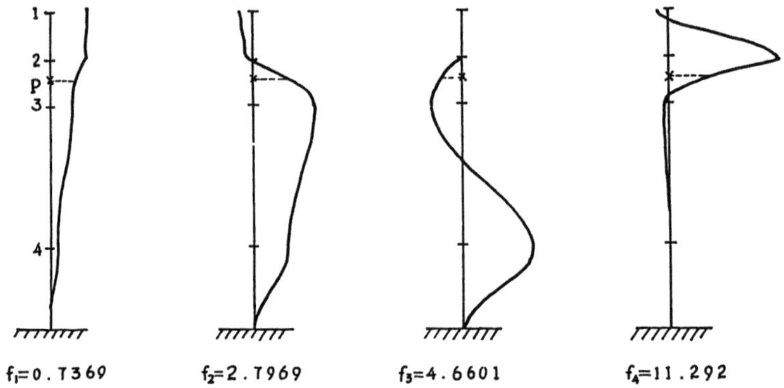

Figure 10. First four modes of the original structure.

It is indicated that the modal displacements at the deck level as well as the point on which the ice force applied are apparently reduced, and this would most improve the situation in ice-induced vibrations.

An ice-induced vibration analysis is carried out for both the original and improved structures under the same design condition of ice in the Bohai Gulf. The results, which are given in Figs. 11 and 12 respectively,

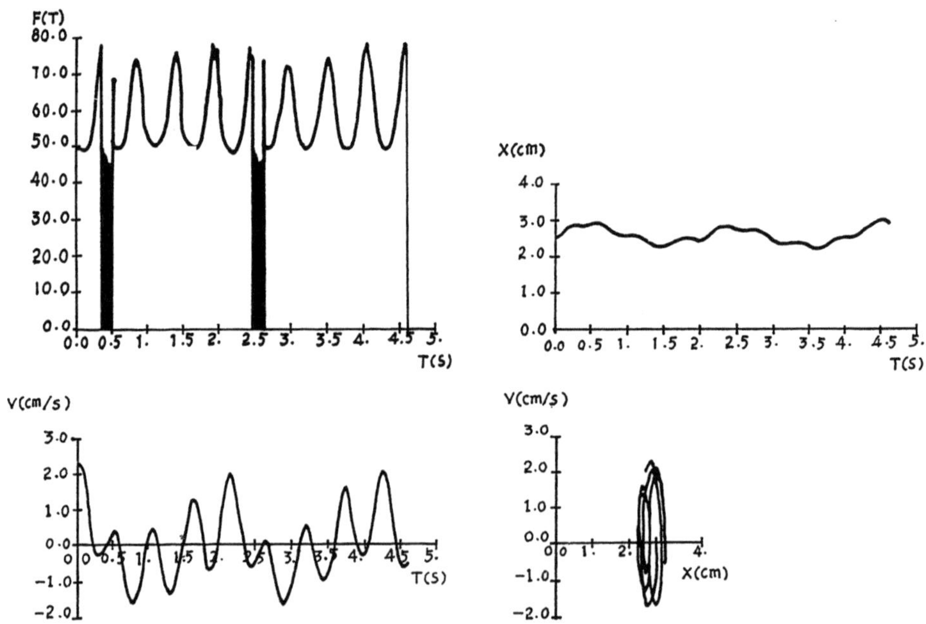

Figure 11. Ice-induced vibration analysis for the original structure.

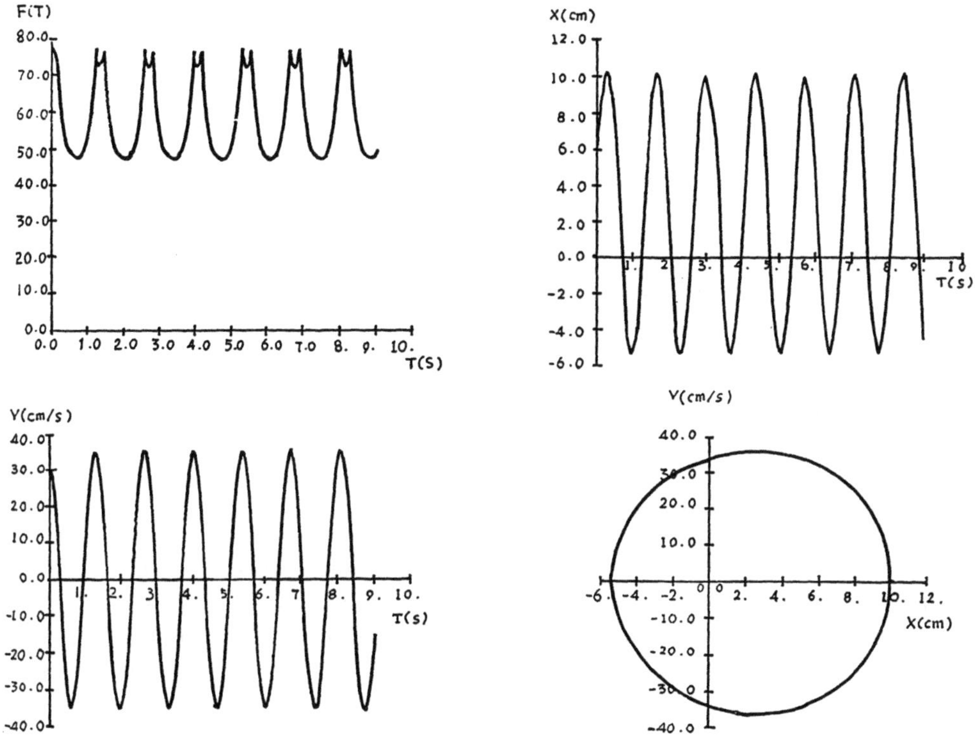

Figure 12. Ice-induced vibration analysis for the improved structure.

demonstrate the effect of the approach. The dynamic deflection at the deck level is substantially reduced in the new structure compared to that in the original one. The response spectra (Figs. 13 and 14) also indicate the much bigger contribution of the second mode in the new structure as contrasted with that in the original structure.

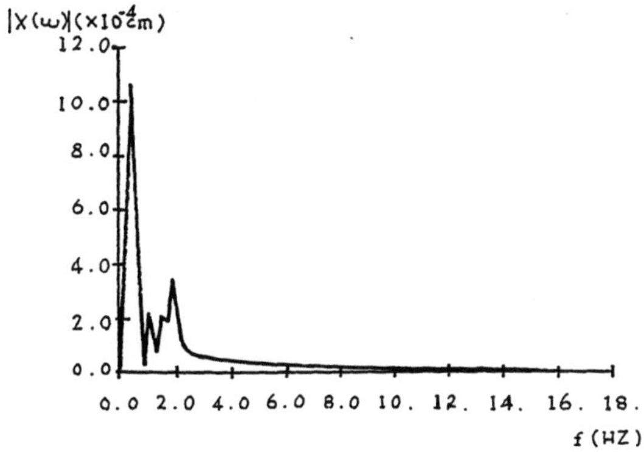

Figure 13. Response spectra, new structure.

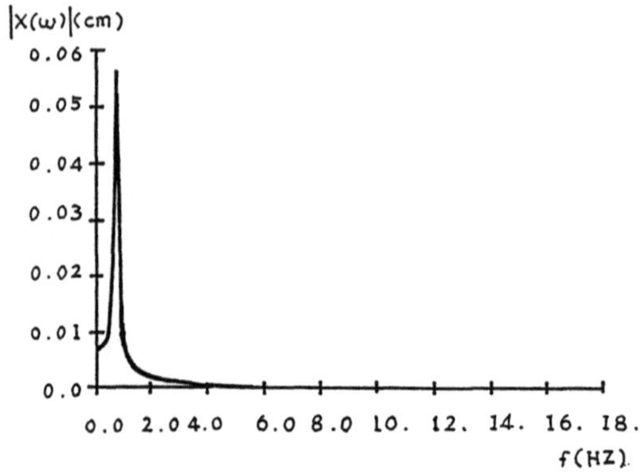

Figure 14. Response spectra, original structure.

6. Concluding Remarks

Flexible structures can be used in the sub-Arctic areas, when the ice-induced vibration has been considered and carefully treated in the structural design, i.e., the dynamic ice-structure interaction must be restrained (by adding damping), avoided (by tuning the natural frequency) or improved (by adjusting the mode shapes).

It seems more reasonable and economical for an ice resistant structure to improve the dynamic ice structure interaction by adjusting the mode shapes. This can be a thoroughly considered selection of mode shapes in a new design concept, or a local change of stiffness in improving an existing structural design. In the latter case, it will be more efficient when the location of stiffness changing is closer to the ice load (i.e., closer to water surface) due to the feature of ice-induced vibration.

The optimization of the coefficient matrix MC discussed in the paper is based on a single harmonic load and a shearing structure model. However, it would not be very difficult to introduce a load spectrum and a more realistic structural model with rotation of nodes in the analysis.

The approach requiers a delicate design of structure and construction for adjustment of the local stiffness.

References

Määttänen, M. 1981. Ice-structure dynamic interaction, ice forces versus velocity, ice-induced damping, International Symposium on Ice, IAHR, Quebec City, Canada, Vol. 2, pp. 783-796.

Määttänen, M. 1987. Ten years of ice-induced vibration isolation in lighthouses. Proc. Sixth International Offshore Mechanics and Arctic Engineering (OMAE) Symposium, Houston, Texas, V. 4, pp.261-266.

Report on feasibility study of the deeply embedded platform in the Liaodong Bay. Dept. of Ocean Naval Architecture, Tianjin University, 1988.

Discussion

Q: M. Määttänen, Helsinki University of Technology

The proposed method to adjust stiffness looks promising when considering the experience with vibration isolated steel lighthouses in the Baltic. The first and second natural modes of the Bohai jacket platform are similar to the first and third natural modes of steel lighthouses. The former is insensitive (dynamically stable) to ice induced vibration and the latter exhibits no significant superstructure movement.

A: J. Xu

Thank you for your comment. The concept of and experience with vibration isolation systems used on lighthouses in the Bothnia Gulf are helpful in solving the ice-induced vibration problem of the Bohai jackets. In spite of some differences in the latter case (multiple columns, bigger deck weight and so on) compared to the former one, there are general principles in the design of flexible offshore structures in the sub-Arctic areas.

Theoretical Modelling of Ice-Structure Interaction

K. Riska

Helsinki University of Technology, Laboratory of Naval Architecture and Marine Engineering
Otakaari 4, 02150 Espoo, Finland

Abstract The principles of analysing theoretically the structure/ice interaction are described. A general framework and a solution method to perform a simulation of the interaction is presented. The interaction is divided into three constituents: structure motions, ice motions and penetration into ice. All these are described but especially the penetration is analysed in detail. Finally an example of the procedure described is given by analysing a ship ramming into a finite ice floe.

1. Introduction

Ice loads on structures are encountered due to contact between an ice feature and a structure. The maximum forces may be attributed to relative motion between the ice and the structure in which case the ice, or the structure, or both, fail making the motion possible. Historically the assessment of structural reliability has been based on a description of maximum forces and the assessment of ship performance on the description of time or space averaging of forces. These descriptions have strong empirical background.

There are some occasions when the structural response influences the contact force thus creating an interaction between ice deformation, ice failure and structural response. The modification of ship hull vibratory response in ramming large ice floes and local ice pressure distribution on ship shell plating are examples of interaction between ice force or pressure and structural deformation. The vibration of offshore structures under ice loads may prove to be interaction in this sense. A description of interaction requires analysis of the whole loading event i.e. simulation

S. Jones · R. F. McKenna · J. Tillotson · I. Jordaan (Eds.)
Ice-Structure Interaction
IUTAM-IAHR Symposium St. John's, Newfoundland, Canada

of the contact between the structure and ice. The purpose of this paper is to try to set a framework for this simulation and review some factors important to it. This paper may be seen as a continuation of a paper published six years ago (Varsta, 1983). Much work has been done since to clarify the interaction. However, this paper is not to be seen as a state-of-the-art review but rather as a part of the author's research plan.

2. Principles of Modelling Ice/Structure Interaction

In mechanical systems, the quantities used for modelling are the time dependency of forces and displacements. Thus for simulation it is sufficient to solve the time histories of displacements and forces from the relevant equations of motion. The selection of the relevant equations of motion requires some analysis of the case under consideration. First the structure/ice interaction is split into two subsystems: ice, and structure, with equations of motion for each. The next step is to consider what forces are significant in the behaviour of the subsystems. Finally the exciting forces must be described. These forces contain the interaction because it is considered to be caused solely by contact between the two systems. Some other ways of interaction are possible e.g. through motion excited waves.

The quantities describing the contact are pressure and surface traction distribution on the contact area. It is possible to define an average unit normal to the contact area, e_n, and subsequently a tangent plane. These unit vectors are shown in Fig. 1. Further it is possible to define a relative average displacement between the structure and ice for the contact area. This can be divided into components in the normal and tangential direction. The normal component u_p is thus

$$u_p = u_s - u_i, \tag{1}$$

where u_s and u_i are displacements of structure and ice in normal direction which is, in general, changing during the contact. Fig. 2 describes Eq. (1) schematically. The contact force, F, is a vector having magnitude and direction. Usual assumption is to separate these two aspects. This is done by assuming the force component in the tangent plane to oppose the sliding motion. Thus the contact force is

$$F(u_p) = F_n(u_p)(e_n - \mu e_t), \tag{2}$$

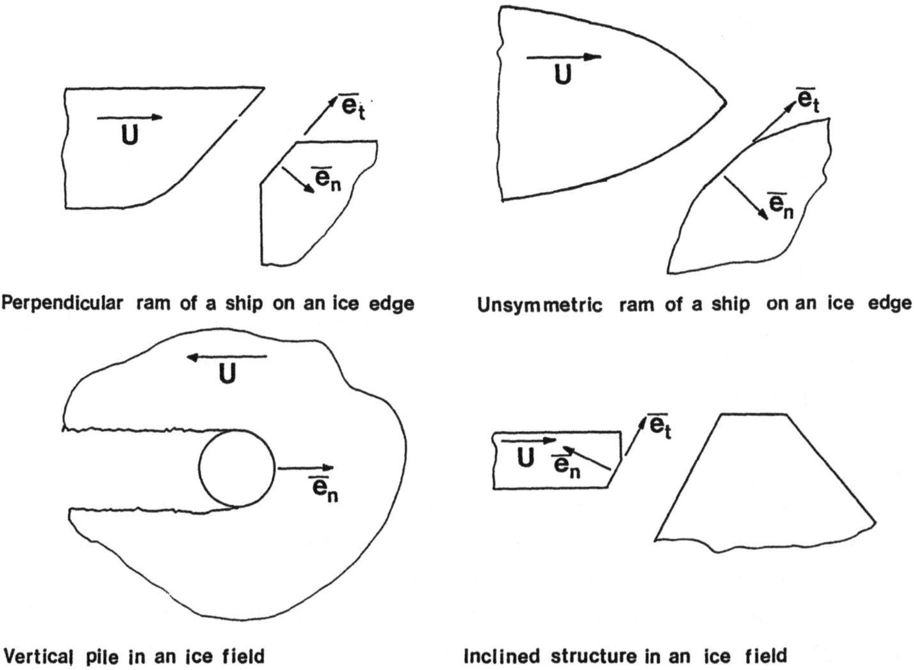

Figure 1. Four different ice-structure interaction scenarios. The unit vectors normal and tangential to the contact are shown.

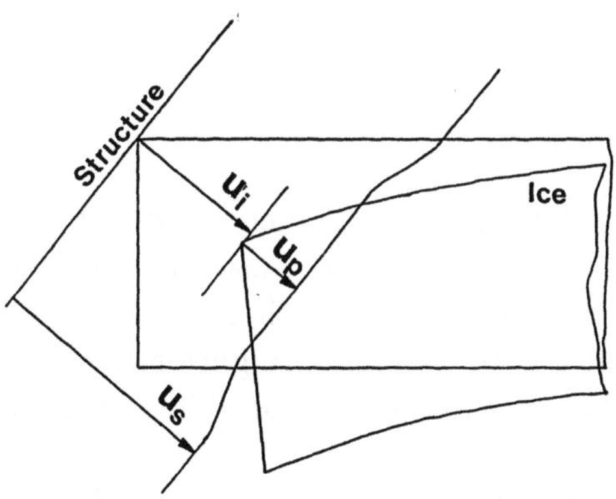

Figure 2. A sketch defining the displacements of the structure and ice u_p and u_i and the penetration into ice u_p.

where the unit vector e_t is in the direction of tangential relative motion, and μ is the coefficient of friction. The emphasis on the dependency between the normal force F_n and the penetration does not exclude the influence of other parameters on the force. The equations of motion to be solved in ice/structure interaction may now formally be stated as

$$\begin{aligned} L[r_s] &= F(u_p), \\ L[r_i] &= -F(u_p), \end{aligned} \tag{3}$$

where r_s and r_i contain the relevant displacement components of the structure and the ice feature and $L[\cdot]$ stands for the appropriate equations of motion. The penetration into ice can be expressed as a function of structure and ice displacements r_s and r_i through geometric relations $u_s(r_s)$ and $u_i(r_i)$. Formally, Eq. (3) is a coupled set of equations, nonlinear even if $L[\cdot]$ were linear because the contact force depends in a nonlinear manner on the penetration. Moreover the equations of motion may contain different equations for subsystems, e.g., due to rigid body motions and elastic deformations. Thus the solution of Eq. (3), i.e., carrying out the mathematical simulation, is a large task. One way to facilitate the solution is to assume the time history of force $F_n(t)$ to be an independent variable. The equations of motion may then be formally solved to give $u_p(F_n)$ and $u_i(F_n)$. The dependency between the normal force and penetration gives $u_p(F_n)$. When these three displacement components are inserted in Eq. (1), an equation for F_n emerges. This equation is still formal.

The solution is made possible using a time stepping algorithm in which the contact force is assumed to be linear between two instants in time t_n and t_{n+1}, and of the form

$$F_n(t) = F_n(t_n) + b(t - t_n), \quad t_n \leq t \leq t_{n+1}, \tag{4}$$

where the constant b is unknown and $F_n(0) = 0$. The equations of motion may now be integrated using b as a parameter. Inserting the displacements into Eq. (1) gives an equation for b which may be solved numerically. The procedure outlined above seems straightforward but the problem is hidden in creating the equations of motion. This problem is discussed in the subsequent sections.

The simulation procedure where the force is assumed piecewise linear, seems to contain strong assumptions. The full set of Equations (3) must, however, also be integrated stepwise. Thus an algorithm where force is considered unknown and Eq. (1) is considered as an extra equation gives the possibility to split the problem into parts: ice feature motion, structure motion and penetration into ice. Standard solution techniques such as modal superposition may be used for the first two. This way the attention is focused in one problem at a time and coupling is concentrated in Eq. (1), the so-called kinematic condition.

3. Analysis of the Components of Displacement

3.1 Structure

The motion of a structure may be divided into two parts; rigid body motions and elastic deformation. These could be analyzed together (see e.g. Bishop and Price, 1979) but usually the division facilitates the analysis. The rigid body motions are relevant only for floating structures, see e.g. Fig. 3. The general motion has six degrees of freedom which are commonly written as a vector U_{rb}. Then the general equations of motion are (Ogilvie, 1964; Cummins, 1957; Liapis, 1986)

$$M\ddot{U}_{rb} + B\dot{U}_{rb} + CU_{rb} + \int_0^t K(t-\tau)\dot{U}_{rb}(\tau)d\tau = F, \tag{5}$$

where F is the excitation force matrix due to the contact with ice. Matrix M is related to added mass but is not speed or frequency dependent. The

Figure 3. Canmar Kigoriak ramming a large ice ridge presents a case where the ship rigid body motions are significant.

matrices B and C are related to the conventional damping and buoyancy matrices, see (Ogilvie, 1964) for definitions. All the memory effects due to e.g. radiating waves are contained in the matrix function $K(t)$. The matrices are difficult to evaluate even in a zero speed case (van Oortmerssen, 1976) and further simplification is needed. The present status is that the speed dependent terms are disregarded and results obtained by assuming harmonic motion are used. Once the vector U_{rb} is known the displacement component along the contact normal u_{rb} can be determined from geometry.

The structural hierarchy allows usually the division of elastic deformation into parts. The basic division is division into two parts according to St. Venant's principle: one due to deformation of the local impact zone and the other due to the response of the whole structure. An example of this division is the separation of ship shell structure deflection and ship hull bending deflection. One of the reasons for the division is that further simplifications may be introduced. The local deformation may be considered static and linear because the lowest natural frequency of local structure is usually many hundred Hz. Thus the relationship between contact force and the average displacement on the contact area along the normal requires determination of a spring constant.

The lowest natural frequencies of the whole structure are about 1 Hz. Thus the elastic deformation of the whole structure should generally be considered dynamic. The structural behaviour in the elastic range is linear. Thus the modal superposition technique is applicable. This facilitates the solution as the modes are solved only once for each structure. Usually only a few lowest modes are necessary in the summation (Riska, 1987; Räty, 1986). If the displacement component along the contact normal due to global deformation is u_q and due to local deformation u_e then the displacement of the structure is

$$u_s = u_{rb} + u_g + u_e. \tag{6}$$

This division is not perhaps well defined in some cases as the components influence each other. It can be argued e.g. that ship vertical motions and hull bending should not be separated. Each case should be analysed separately and the level of accuracy possible for each component must guide the decision.

3.2 Ice Feature

As in the case of the structure, the motion of an ice edge may be divided into two parts: motion of the ice feature as a rigid body and motion due to elastic deformation. The analysis of rigid body motions of the ice feature poses problems similar to that of the structure. If the ice feature moves noticeably as a rigid body it must be relatively small in size. Then for practical purposes the uncertainties in modelling hydrodynamic forces are overshadowed by uncertainties in exact dimensions of the ice feature. This allows the use of simplest forms of hydrodynamic forces. Mass of the ice feature is the dominating factor in modelling the motion. The displacement component due to rigid body motion is termed u_{rb} as in the case of structure. The displacement of the ice edge could be divided into local and global parts but it is more rational to include the local part in the penetration into the ice. The analysis of elastic global deformation of ice (u_g) is very dependent on the thickness of the ice feature. If ice is relatively thin, it is bent to failure as Fig. 4 shows. When the ice thickness increases, the pure bending changes to a case in which it is

Figure 4. Icebreaker Urho in level ice of thickness about 80 cm. The breaking pattern formed by bending is evident.

difficult to separate local and global effects. Fig. 5 shows failure of thick ice under loads inclined to vertical.

The examples above suggest that in thin level ice the global ice displacement is due to bending of ice as a thin plate. The inertial forces due to water under the plate are also important. These could be about 75% of all inertial forces (Sörensen, 1978), suggesting that the solution of hydrodynamics for the transient case is necessary. The use of added mass, however, gives results close to measured ones (Varsta, 1983). To the knowledge of the author, the case of thicker ice has not been analyzed.

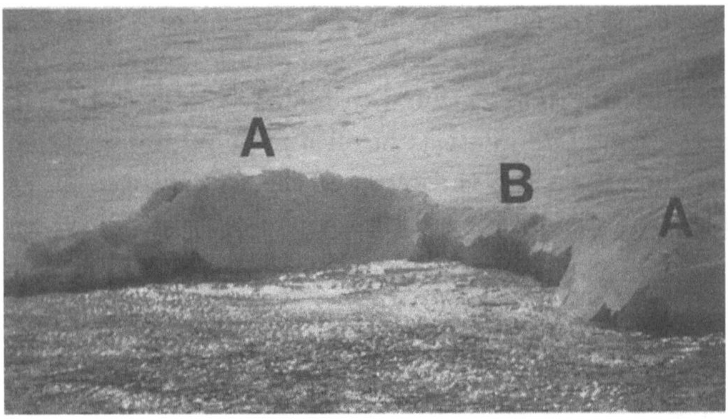

Figure 5. Bow imprint left on ice edge after a ship ram. The ice was relatively thick, about 3 m, so ice was predominantly crushed in some locations (marked A) and at some locations about vertical cracks occurred breaking of big lumps of ice (location marked B).

To summarize, the displacement of ice features is divided into two components, one due to rigid body motions and other due to bending

$$u_i = u_{rb} + u_b. \tag{7}$$

The displacement due to local deformation is not included in Eq. (7) but is analyzed together with penetration into ice.

3.3 Penetration Into Ice

Lately the penetration of structures into ice has received much attention. Despite many measurements made and calculations performed, the constituents of penetration into ice are still obscure. It is easy to observe that when a structure, e.g. ship bow, penetrates into ice, crushed ice flows out from the contact zone but otherwise the ice remains intact. An example of crushing penetration is shown in Fig. 6. The conclusion that the failure is localized to the contact area is also corroborated by investigation of crushing test specimens after a test. They show no major cracking inside the specimen, Fig. 7. Full scale observations in relatively thin ice under inclined loads show major cracking inside the ice as shown in Fig. 8. Cracking inside the ice has also been observed in laboratory tests (Timco, 1986). In these, the ice was thin and cracks occurred when the loaded area was wider than the ice thickness.

In view of the interaction simulation, the analysis of penetration should provide the displacement quantities. At present, the major questions are where and how the ice is crushed and consequently whether there exists a boundary between crushed and intact ice.

Crushing tests where high local pressures, up to 50 MPa, have been measured (Glen and Comfort, 1983; Joensuu and Riska, 1989) indicate direct contact between structure and ice. This has lead to the so called "hot spot" model of contact shown schematically in Fig. 9. The contact in this model is made of two types: direct ice/structure contact and contact through a layer of crushed ice. That the crushed ice participates in the contact is supported by the observation that injection of water into the contact zone changes the average contact pressure (Varsta, 1983). Recent

Figure 6. Ship ramming the edge of a multi-year ice floe showing only crushing of the edge.

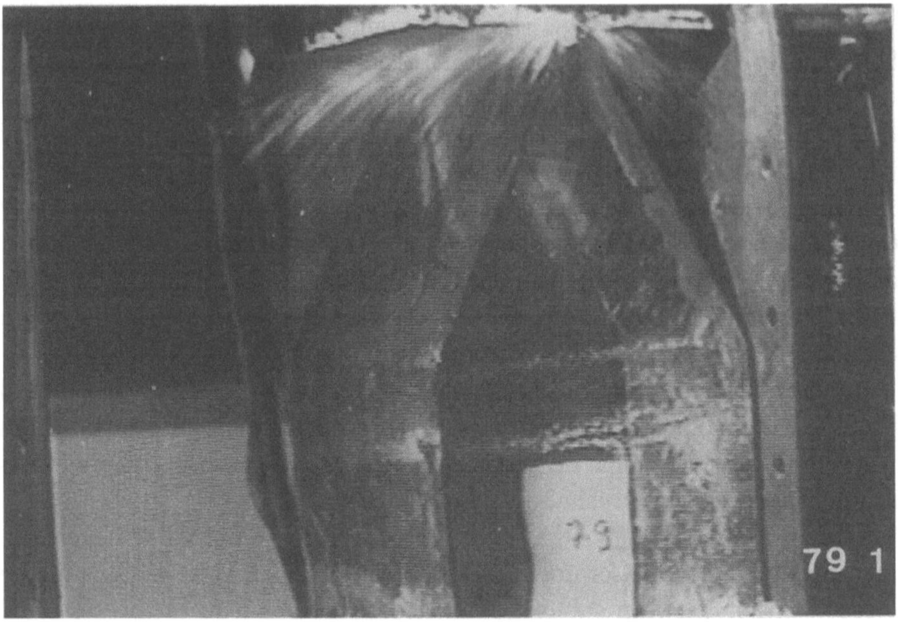

Figure 7. A snapshot from a crushing test showing crushed ice coming from the contact. Note that the test specimen is intact apart from the crushing.

Figure 8. A photograph taken at the bow of a vessel. The bow is at the left edge of the picture and the vessel is travelling towards bottom left corner. The ice thickness is about 20 cm. The car tyres set a rough scale. Note the multitude of cracks.

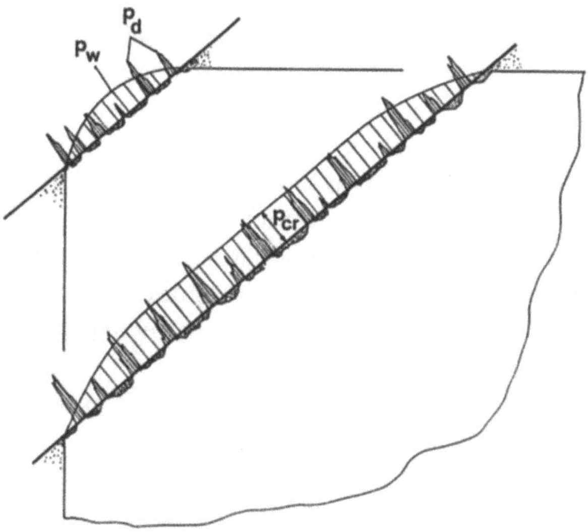

Figure 9. A sketch showing the direct contact and layer of crushed ice. The pressure in crushed ice p_w is depicted assuming it to be viscous fluid. Further ice is assumed to crush under the layer in pressure p_{cr}. The pressure from direct contact is p_d.

measurements seem to endorse the existence of two contact types but point out that the direct contact is not made up from points but is a line as presented in Fig. 10. It must be noted here that the agreement on the existence of direct contact is not unanimous. Contact only through the

Figure 10. Ice edge after crushing penetration of about 1 cm. There is crushed ice on top of the piece. The line-like contact is to be noted.

layer of crushed ice (Popov et al., 1969; Kurdyumov and Kheisin, 1976) was the first model presented. This has been augmented with a finite crushing depth where, instead of continuous crushing, ice is periodically crushed to a finite depth.

The cracking inside the ice suggests that perhaps no clear boundary exists between the crushed ice and relatively intact ice. If this is the case, the penetration cannot be analyzed by dividing it into parts, and consequently methods given by damage theories may be applied (Jordaan, 1986). A weaker form of this reasoning is to include the boundary but to consider ice as a damaged continuum.

The brief discussion above outlines one of the fundamental questions which is still open. The question is where and how crushed ice is created. If the direct contact is in all cases one dimensional, part of the crushing then takes place at the line. Finite crushing depth and thus a rapidly changing location of the line would account for all the crushing. No crushing would take place under the layer of crushed ice. The uncertainty of available knowledge makes it difficult to describe the penetration. It may be assumed, however, that the penetration into ice includes three factors. One is due to crushing of ice, one due to squeezing of crushed ice from the contact area and one due to ice deformation. If the displacements related to these factors are u_{cr}, u_{μ}, and u_{ϵ} respectively then the penetration into ice may be separated into two parts:

$$u_p = u_c(u_{cr}, u_{\mu}) + u_{\epsilon}, \tag{8}$$

where u_c is the displacement of the boundary between crushed and intact ice averaged for the contact area and u_{μ} the average thickness of the layer of crushed ice. The displacement u_c is a direct sum of the components in the case of no direct contact, or a more complicated function of these. In Fig. 11, a sketch of these components is presented.

The displacement due to ice deformation may be solved assuming ice to be an elastic material. The static solution for different contact geometries, assuming the ice to have three dimensions, is similar to a Boussinesq solution

$$u_{\epsilon} = C\frac{F_n}{ED}, \tag{9}$$

where C is a constant dependent on the geometry, D characterises the contact size and E is Young's modulus of ice. The value of constant C for the case of uniform pressure on diameter D on an elastic half space is

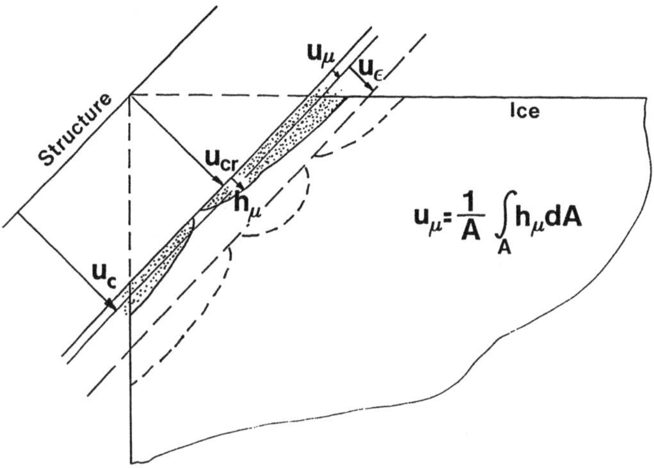

Figure 11. The components of displacement constituting the penetration into ice u_p.

about 0.9. As the elastic displacement is small, usually a few millimeters, and loading rates are not large, the static solution is adequate. The static solution of displacement for an elastic half plane, i.e., a two dimensional body loaded in its plane contains a singularity. This question has some bearing for the vertical structures applying in-plane loads in level ice. The use of ice floes of finite size or dynamic solutions has been considered.

The displacement of the boundary of intact ice is related to the question how ice fails, i.e., how crushed ice is formed. The ice failure under direct contact can occur continuously by creating fragments of finite size. Visual observations where a line-like contact was seen wavering about an average position indicate this. Fragment size created is small, about 1 mm diameter (Joensuu and Riska, 1989). As a first estimate, the pressure to fail the ice on the line may be assumed to be only a function of ice strength, and there is a line load q describing the direct contact.

Ice must fail also under the crushed ice layer if the direct contact is one dimensional and its location does not move relative to the contact surface during the penetration. The mechanism for this failure is obscure at present. The speed of penetration u_p determines the mass flow of crushed ice created. The penetration speed is roughly equal to the speed of crushing, $u_p \approx u_{cr}$. Thus it is plausible that the pressure in the crushed ice layer is mainly a function of the local layer thickness h_μ and of u_{cr} i.e. $p_{cr} = p_{cr}(h_\mu, u_{cr})$. This relationship is obtained once the nature of crushed ice is known. Its behaviour as viscous or Mohr-Coulomb material has been suggested. The above assumptions lead to the contact force being the sum of two components as

$$F_n = qL + p_{cr}A, \tag{10}$$

where A is the contact area and L is the length of direct contact which is related to the contact area. The proportionality between area and length of direct contact is obscure. Different assumptions give A proportional L^n where $n \in [1,2]$. Thus dividing Eq. (10) by contact area gives the average pressure on the contact area as

$$p_{av} = KqA^{1/n-1} + p_{cr}(h_\mu, u_{cr}), \tag{11}$$

where K is a constant dependent on contact shape. Equation (10) or (11), seen as equations for u_μ and u_{cr}, need to be supplemented by a relationship describing ice failure under the crushed layer. At present this is not available. Consequently most of the simulation routines assume that $u_c = u_{cr}$, or that no direct contact exists, and $u_c = u_{cr} + u_\mu$. In the first case an empirical force-penetration relationship is used (Riska and Frederking, 1987). In the second case, the formulation assuming crushed ice to be a viscous fluid is used (Kurdyumov and Kheisin, 1976).

3.4 Analysis of Interaction Cases

All the displacement components are present in most of the interaction scenarios. However, practical considerations limit the number of components to be included in the simulation. Criteria to restrict the number of components are difficult to set. An adequate criterion cannot be formed by comparing orders of magnitude of the components nor comparing the mechanical energies involved with each component. An example of the difficulty is provided by the longitudinal ramming of a ship into a massive ice floe in which the elastic displacement of the ice influences the nature of interaction despite the fact that it is small, about 1 mm, compared to ship motions which are about 10 m. Also the energy involved in elastic deformation of ice is vanishingly small compared to energies involved in ship rigid body motions (Riska, 1987). A large simplification is gained if it is known that some component involves constant speed. Usually this happens for rigid body motions. When a large ice floe impacts a light house with velocity v then the rigid body motion component of ice is simply $u_{rb} = vt$ (assuming a suitable time origin). A similar case is presented by a ship navigating in relatively thin level ice; then for the structure $u_{rb} = v_n t$ where v_n is ship speed along the contact normal.

Another possibility to simplify calculations is to ignore some displacement components. Some judgement must then be exercised in selection. A comparison with the exact solution or with some test results can justify the selection but, as these rarely exist, the judgement is necessarily subjective. A sketch of this kind of selection for three interaction cases is shown in Table 1. The table is based partly on the author's own calculations and partly on cases presented in literature.

Table 1. The components of normal displacement at the contact which are more important in modelling different interaction scenarios. Relatively more important components are marked with "1".

	Structure			Ice		Penetration	
	u_{rb}	u_g	u_e	u_b	u_{rb}	u_c	u_ϵ
Longitudinal ship ram on							
massive ice	1	1	0	0	0	1	1
small floe	1	1	0	1	1	1	1
thin level ice	0	0	1	1	0	1	1
Nonsymmetric ship ram on							
massive floe	1	0	0	0	0	1	0
thin level ice	1	0	0	1	0	1	0
Bottom founded vertical pile in							
level ice	0	1	0	0	0	1	1
ice floe	0	1	0	0	1	1	1

4. Sample Calculations

The above procedure in simulating interaction is demonstrated by the case of a ship longitudinally ramming a finite ice floe. The floe is assumed to be thick enough not to bend significantly. The ship used in calculations is M.V. Arctic before the bow change (Arctic, 1978). The ship rigid body motions (heave and pitch) were coupled together with the Timoshenko beam behaviour of the hull (Bishop and Price, 1979; Riska, 1987). Surge motion was assumed to be undamped.

Rigid body motions i.e. heave, pitch and surge of the ice floe were calculated assuming the ram being along the horizontal symmetry axis of the floe. The floe is of constant thickness, of rectangular horizontal area and also of rectangular cross section. Thickness is H and side length is D. Only the crushing displacement and the elastic displacement were included in the ice penetration.

The simulation procedure contains three phases: crushing, sliding and free phase. During the crushing phase, u_{cr} was increasing and during the sliding phase, it stayed constant while the ship rode on the ice. The free phase, when the bow lost contact with ice, was encountered only in the end of the ram. The first two phases alternated during the ram and, in the case of a smaller ship, the bow could have lost the contact with the ice after the first crushing phase and free motion would have ensued. The calculation procedure is shown in Fig. 12.

The results from the simulation are given in Fig. 13. Two rams are shown, both with ramming speed of 3 m/s, one on an infinite ice floe and the other on floe of thickness 5 m and side length 100 m. The vertical motion of ship bow is w_y. It becomes clear from the figure that the floe size has large influence on the ram. This is further elaborated in Fig. 14 where the vertical bow ice force and the maximum ship hull bending moment are shown using floe mass as a parameter and assuming constant thickness of 10 m. The vertical force may be converted to normal force knowing the stem angle 30° and coefficient of friction $\mu = 0.12$. The ice floe masses used were 22.5, 90, 360, 1440 and 5760×10^6 kg. The mass of M.V. Arctic in these units is 38 so when the ice mass is about 10 times ship mass or more, the floe size does not influence the forces.

The maximum vertical bow ice force and hull bending moment are drawn versus the floe thickness H in Fig. 15 using the side length as a parameter. The figure shows that the floe mass is the major parameter influencing the forces. The dimensions of the floe have, however, some influence on the force and response. This conclusion is valid as long as the ice does not bend significantly.

The ramming interaction may also be investigated using the mechanical energies involved. When the ship rams an infinite ice floe, the ramming process is such that basically the initial kinetic energy in longitu-

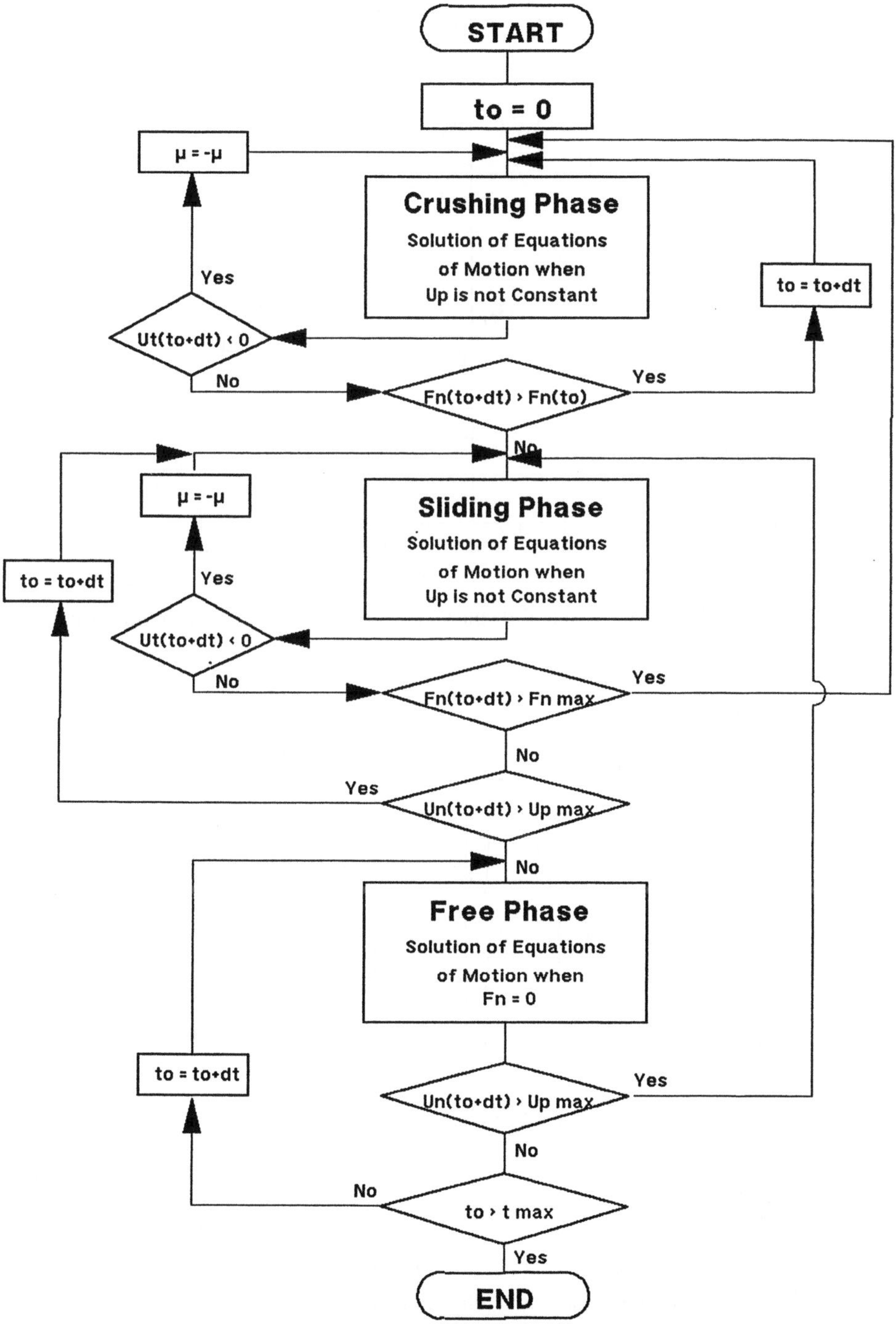

Figure 12. Block diagram of the simulation procedure for longitudinal ship ram on thick ice floe.

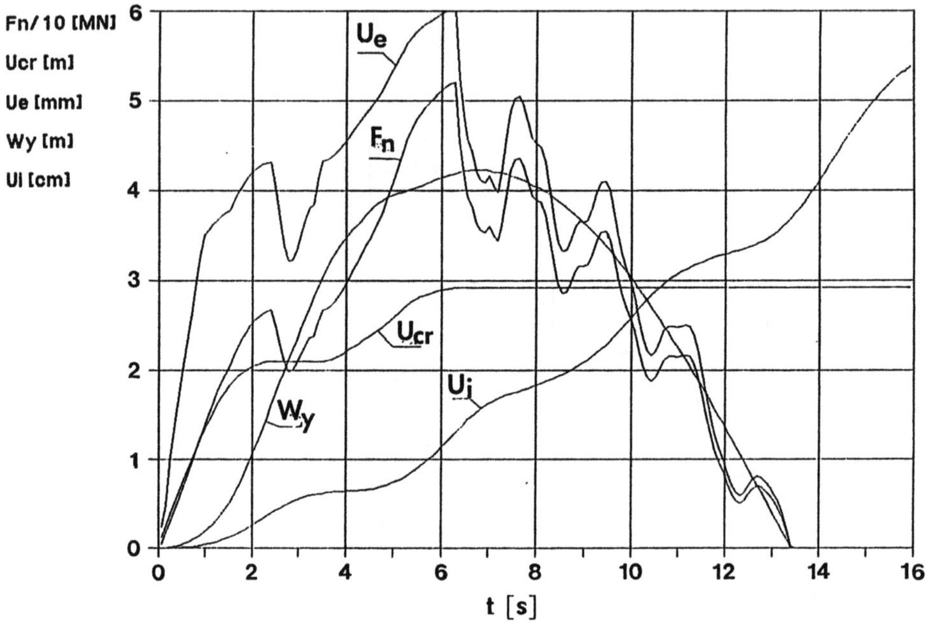

Figure 13a. Normal bow ice force and displacements during a ram on an infinite ice floe. Ramming speed is 3 m/s.

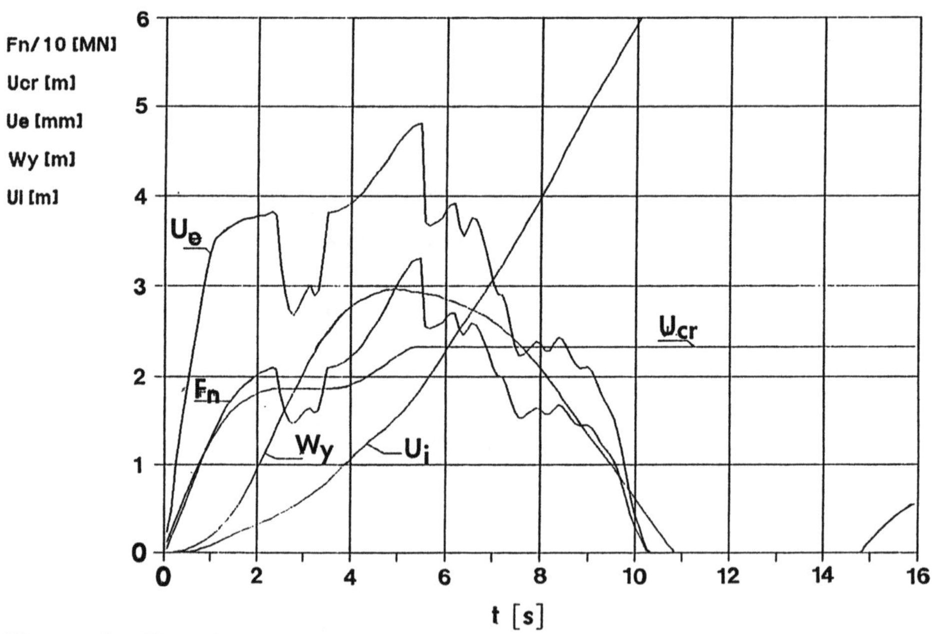

Figure 13b. Normal bow ice force and displacements during a ram on an ice floe of thickness 5 mm and side 100 m. Ramming speed is 3 m/s.

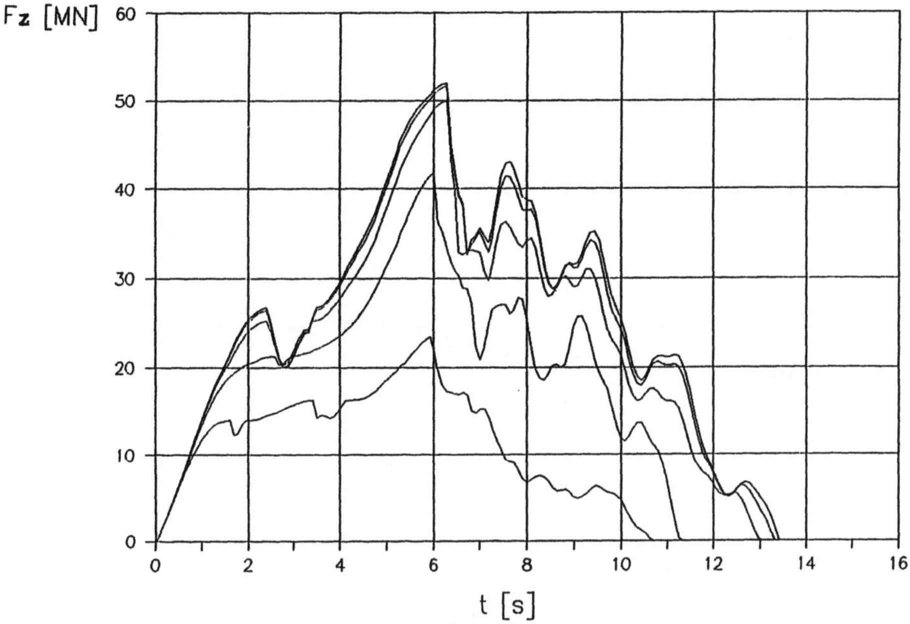

Figure 14a. The vertical bow ice force from a 3 m/s ram on an ice floe of thickness 10 m and side of 50, 100, 200, 400 and 800 m.

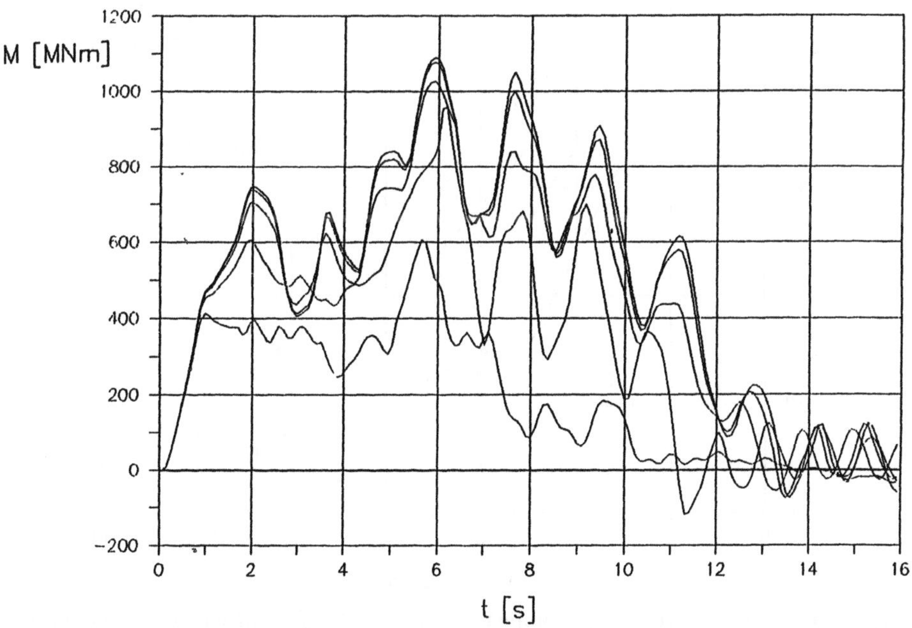

Figure 14b. The maximum hull bending moment from a 3 m/s ram on an ice floe of thickness 10 m and side of 50, 100, 200, 400 and 800 m.

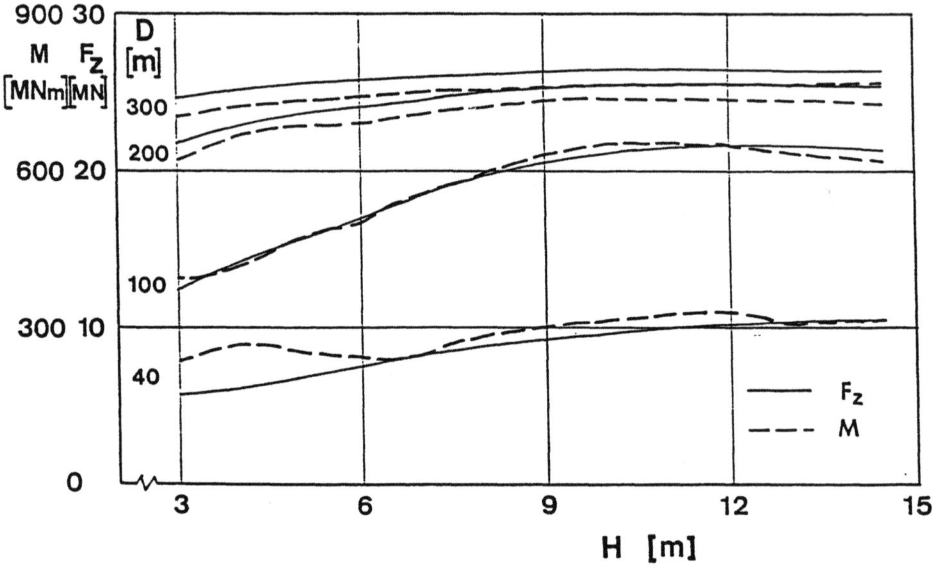

Figure 15. Maximum vertical bow ice force and hull bending moment depicted versus the thickness of the ice floe. Side length is used as a parameter.

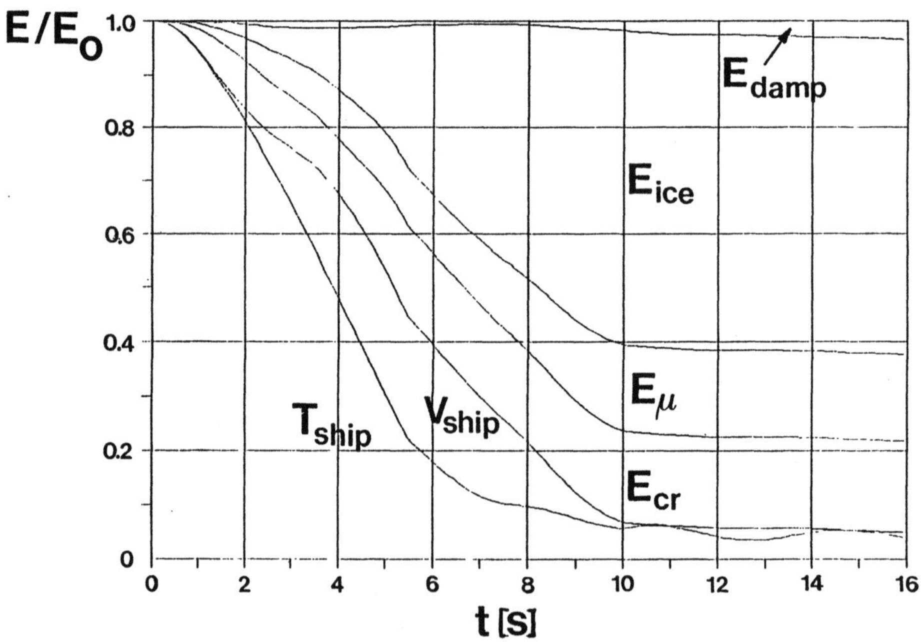

Figure 16. The relative energy components from a 3 m/s ram on an ice floe of size $5 \times 100 \times 100\text{m}^3$. Ship kinetic and potential energies consumed in crushing ice, sliding on ice and structural and hydrodynamic damping are E_{cr}, E_{μ} and E_{damp}. E_{ice} is mechanical energy in ice and E_o is the initial kinetic energy.

dinal direction is exhausted in crushing the ice and in sliding on the ice. Only some energy is recovered in rebound velocity. If the ice floe is finite, the exchange occurs between ship and ice kinetic energies as Fig. 16 shows.

5. Concluding Remarks

The ice/structure interaction has been analyzed theoretically in some idealized cases such as ship proceeding in thin level ice, ship longitudinally ramming thick ice floes and vertical piles in level ice field. From these different cases, it is possible to extract similarities based on which a more general framework containing these cases may be formulated. The value of a more general framework can be questioned because it does not advance the analysis of specific cases which are encountered in reality.

The value of a general framework lies, in the author's opinion, in making it possible to see the influence of the constituents in the interaction. In this way, research can be focused on factors which are relatively more obscure. Much work has been done in clarifying the structural behaviour and many efficient methods are already in the toolbox of the ice/structure interaction scientist. The present analysis, albeit subjective, shows that today the greatest lack of knowledge lies in describing the penetration into ice. This fact is manifested in the variety of theories explaining the origin of vibration of offshore structures in ice.

Once it is possible, in a reliable way, to model single interaction cases, the research can move to cases which combine different single cases. An important case for naval architects is the ship proceeding in level ice. Here the ice is first bent, then turned and finally it slides under the ship hull. The breaking pattern, the trajectories of broken ice floes and interaction between ice floes are some of the areas needing clarification. Usually the appearance of theoretical analysis in some field shows that the empirical work moves further to a more complicated case. This can already be foreseen in structure/ice interaction.

Acknowledgements

The subject for this paper formed the background for the author when working in a research program initiated in the Laboratory of Naval Architecture and Marine Engineering at Helsinki University of Technology.

The program is funded be Technology Development Centre in Finland and Winter Navigation Research Board. The funding and the opportunity to work at the university are gratefully acknowledged.

References

Arctic. 1978. The Motor Ship, 10, pp. 35-41.

Bishop, R. and Price, W. 1979. Hydroelasticity of ships. Cambridge, Cambridge University Press, 423 p.

Cummins, W.E. 1957. The force and moment on a body in a time varying potential flow. Journal of Ship Research, Vol. 1, pp. 7-18.

Glen, I. and Comfort, G. 1983. Ice impact pressure and load: investigation by laboratory experiments and ship trials. Proceedings of the 7th International Conference on Port and Ocean Engineering under Arctic Conditions, Espoo, Finland, Vol. 1, pp. 516-533.

Joensuu, A. and Riska, K. 1989. Contact between a structure and ice. Results from joint measurements with Wärtsilä Arctic Research Centre, spring 1988. Laboratory of Naval Architecture and Marine Engineering, Helsinki University of Technology, Report M-88, Otaniemi, 57 p. + app. 154 p.

Jordaan, I.J. 1986. Numerical and finite element techniques in calculation of ice-structure interaction. Proceedings of 8th Symposium on Ice Problems, International Association for Hydraulic Research, Iowa City, pp. 405-440.

Kurdjumov, V. and Kheisin, D. 1976. Hydrodynamic model of the impact of a solid on ice [in Russian]. Prikladnaja Mekhanika, Vol. 12, No. 10, pp. 103-109.

Liapis, S.J. 1986. Time-domain analysis of ship motions. Department of Naval Architecture and Marine Engineering, The University of Michigan, Report No. 302, 125 p.

Ogilvie, T. F. 1964. Recent progress toward the understanding and prediction of ship motions. Proceedings of 5th Symposium on Naval Hydrodynamics, Washington D.C., pp. 3-128.

Oortmerssen, van G. 1976. The motions of a moored ship in waves. Netherlands Ship Model Basin, Wageningen, The Netherlands, 141 p.

Popov, Yu. et al. 1969. Strength of ships sailing in ice (translation). US Army Foreign Science and Technology Center, Washington, DC, USA, FSTC-HT-23-96-68, 228 p.

Riska, K. 1987. On the mechanics of the ramming interaction between a ship and a massive ice floe. Technical Research Centre of Finland, Espoo, Finland, Publication 43, 86 p.

Riska, K. and Frederking, R. 1987. Ice load penetration modelling. Proceedings of the 9th International Conference on Port and Ocean Engineering Under Arctic Conditions, Fairbanks, Alaska, Vol. 1, pp. 317-328.

Räty, R. 1986. On the dynamics of three classic nonlinear oscillators (in Finnish). Thesis submitted to the Helsinki University of Technology, Espoo, Finland, 87 p.

Sörensen, C. 1978. Interaction between floating ice sheets and sloping structures. Institute of Hydrodynamics and Hydraulic Engineering, Technical University of Denmark, Lyngby, Denmark, Series Paper No. 19, 175 p.

Timco, G.W. 1986. Indentation and penetration of edge-loaded freshwater ice sheets in the brittle range. Proceedings of 5th International Symposium on Offshore Mechanics and Arctic Engineering, Tokyo, Japan, Vol. 4, pp. 444-452.

Varsta, P. 1983. On the mechanics of ice load on ships in level ice in the Baltic Sea. Technical Research Centre of Finland, Espoo, Finland, Publication 11, 91 p.

Varsta, P. 1983. Modelling of impact between ship hull and ice. Proceedings of the 7th International Conference on Port and Ocean Engineering under Arctic Conditions, Espoo, Finland, Vol. 2, pp. 760-777.

Discussion

Q: J. Dempsey, Clarkson University

Could you comment on the 'Hot Spot' theory?

A: K. Riska

The 'Hot Spot' theory for ice-structure contact stems from the idea that the contact may be divided into two types: direct ice/structure contact and indirect contact through a layer of crushed ice. If the direct contact is geometrically localized in small randomly distributed points, "hot spots", then this theory would explain the high contact pressures. The present knowledge supports hot spots in some cases, in other cases however, the contact is again one dimensional. The matter waits for an explanation.

Q: A.F. Aboulazm, Marine Institute

1. What are the speeds of the model (and full scale) used in the ramming test?
2. In the ramming test, what type of model constraints were used? Was the model constrained during the ram, after the initial contact?
3. What is the coefficient of friction between the ship model and ice?

A: K. Riska

These questions refer to a video which was shown as the author presented his paper. This paper did not address the details of the model test program.

1. Model speeds up to about 0.8 m/s were used. The geometric scale is 1:40 so this speed range corresponds in full scale up to 5 m/s.

2. The model was ramming freely i.e. not constrained.

3. The coefficient of friction was about 0.12.

Ice - Structure Interaction During Indentation Tests

Devinder S. Sodhi

U.S. Army Cold Regions Research and
Engineering Laboratory
72 Lyme Road, Hanover, NH
USA 03755

Abstract To study dynamic ice-structure interaction during crushing failure of ice, indentation tests were conducted by pushing a vertical, flat indentor into the edges of floating ice sheets. The indentor was supported on three load cells to measure interaction forces at the interface. The displacements of the carriage and the indentor were measured separately. The indentor displacement relative to the carriage and the indentor acceleration were also measured. These measurements provided comprehensive data on the dynamic ice-structure interaction during crushing failure of an ice sheet. Three basic modes of ice behavior were observed: creep deformation at low velocities, intermittent crushing at intermediate velocities, and continuous crushing at high velocities. Based on these measurements, a theoretical model is proposed which produces results similar to those of the experiments.

1. Introduction

Under certain conditions, an interaction between a moving ice sheet and a structure results in structural vibrations as a result of the time-varying forces that are generated by repetitive crushing of ice in the contact zone. At some locations, these vibrations and the high ice forces have caused severe damage to structures, such as lighthouses and bridge piers. It was believed that only slender, vertical structures vibrate as a result of dynamic ice-structure interaction. In 1986, however, there were severe vibrations of Molikpaq, a 110 m wide structure deployed at Amauligak in the southern Beaufort Sea. The vibrations, which were caused by crushing of a multi-year ice floe, induced fatigue of the sand core, the principal element by which Molikpaq resists ice loads (Jefferies and Wright, 1988).

S. Jones · R. F. McKenna · J. Tillotson · I. Jordaan (Eds.)
Ice-Structure Interaction
IUTAM-IAHR Symposium St. John's, Newfoundland, Canada

Some of the earliest experiences with the ice-induced vibrations of structures were reported by Peyton (1968) and Blenkarn (1970), who measured the ice forces and vibrations by instrumenting the vertical, cylindrical legs of a number of platforms in Cook Inlet, Alaska, over a number of years (1963-1969). Since then, many field programs have been undertaken to measure ice forces. Neill (1976) presented a review of the results obtained in the early seventies. Small-scale experiments and theoretical models of dynamic ice-structure interaction have been attempted in different ways, and, for a detailed discussion of these, readers are referred to the review paper by Sodhi (1988). Previous small-scale experiments on indentation of flat and round structures into freshwater, floating ice sheets were performed by Hirayama et al. (1973), Zabilansky et al. (1975), Michel and Blanchet (1983), Timco (1986) and Sodhi and Nakazawa (1988).

In most of the previous small-scale experimental work, the interaction forces, commonly referred to as ice forces, have been measured by instrumenting the support of the structure. This methodology has the drawback that the response of the structure and the interaction force in a dynamic situation are not necessarily related. A transfer function approach was proposed (Montgomery and Lipsett, 1981; Määttänen, 1982) to obtain the interaction forces between an ice sheet and a structure. Though this approach produced successful results in some situations (Montgomery and Lipsett, 1981), it does not produce satisfactory results in all situations because the structural response is a narrow band filtering of the interaction force signal. The results of tests conducted by Timco et al. (1989) indicate that this approach is useful for determining the ice forces on flexible structures only if the frequency of the loading events is not close to the natural frequency of the structure. At or near resonance conditions, this technique cannot be used because of reasons stated above. Evers and Wessels (1986) and Frederking and Timco (1987) obtained similar results during their testing programs.

In this experimental study, indentor plates of different widths were mounted on three load cells so that the true interaction forces could be measured. This methodology was developed by Sodhi and Nakazawa (1988) to determine the position of the resultant interaction force. The adoption of this methodology in this study played a key role in measuring the forces during dynamic ice-structure interaction independently of the structural motion. Besides measurements of the interaction forces, the displacements of the carriage and the indentor plate were measured separately with respect to a fixed datum. Further, the displacement of the indentor was measured with respect to the carriage. The acceleration of the structure, on which the indentor was stiffly mounted, was also measured during the interaction. These measurements provided a complete set of data on the interaction forces and the response of the structure. Based on the results obtained in this study, a model was developed to describe the dynamic ice-structure interaction. Suggestions are made on the methodologies to conduct model experiments and to interpret acceleration data taken in the field during ice-structure interaction involving repeated ice crushing.

2. Experimental Set-up and Procedure

To simulate a vibrating structure, a single-degree-of-freedom system was chosen to avoid the more complex vibrations of a multi-degree-of-freedom system. The vibrating system consisted of a mass and a spring, as shown in Figure 1. It consisted of two horizontal plates attached to each other with four hinged links. The top plate was mounted under the carriage of a high-force module, which in turn could be installed under the main

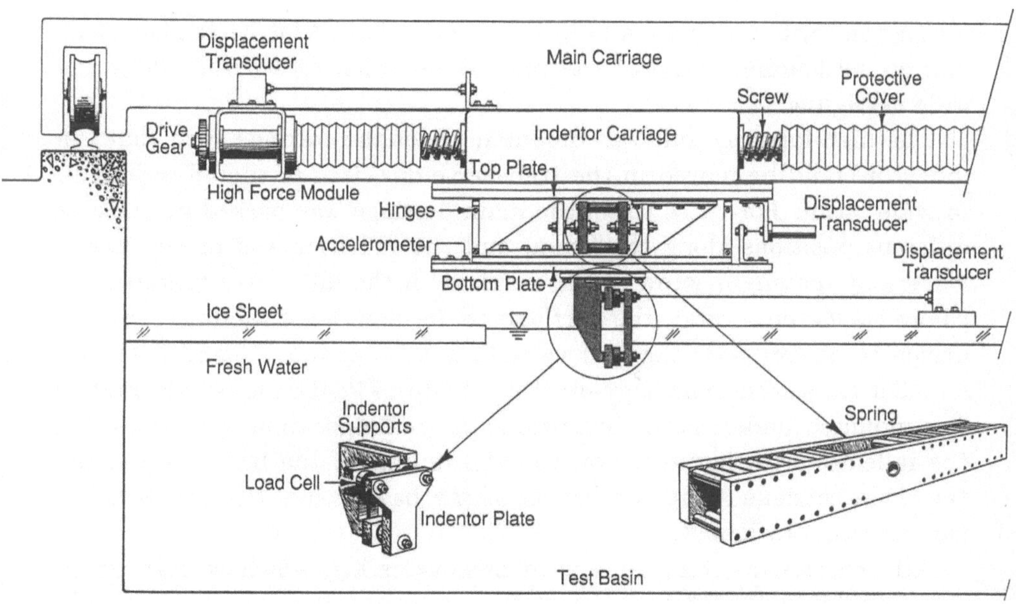

Figure 1. Schematic sketch of the experimental set-up.

carriage that spanned the test basin of the Ice Engineering Facility of the U.S. Army Cold Regions Research and Engineering Laboratory, Hanover, NH. On the bottom plate, the supports for the indentor plate and about a 400-kg mass of lead weights were mounted. Between the two plates, two brackets were attached, one to each plate, and a spring was mounted between the two brackets, as shown in Figure 1. The stiffness of the spring could be varied as described later. The total mass of the lower plate and other things installed on it was about 600 kg. In the rest of this paper, reference to the displacement of the structure or the indentor are synonymous because the indentor was mounted rigidly on the lower plate. Also, reference to velocity means the velocity of the carriage, which was held constant. The indentor velocity was highly variable as per the response of the structure to interaction forces.

The spring consisted of two rectangular aluminum bars that were 25 mm (1 in.) by 152 mm (6 in.) in cross section and 1.8 m (6 ft.) long. The two bars were bolted to many 25 mm diameter, 100 mm long aluminum cylinders at a spacing of 50 mm, as shown in Figure 1. The compliance of the spring came mostly from the flexure of the aluminum bars. By removing or installing these cylinders, we could change the span in the middle of the two bars to vary the stiffness of the spring from about 0.9 to 3.2 MN m^{-1}. The stiffness of the spring was determined for a test by taking the ratio of incremental load to incremental relative displacement during the loading phase of the interaction when structural vibrations were negligible.

The high-force module was mounted under the carriage such that the indentor could be moved in the transverse direction of the 36 m long, 9 m wide basin. For these tests, the main carriage was parked in place at different positions along the basin, and the indentor was moved transversely by moving a screw-driven carriage in the high-force module. By changing the gear ratio, the carriage of the high-force module could be driven to obtain two ranges of velocities: 0.5-7.5 mm s^{-1} and 3-41 mm s^{-1}. For tests at velocities greater than 40 mm s^{-1}, the high-force module was mounted under the main carriage in the longitudinal direction, and the indentor was moved forward in the middle of the basin by moving the main carriage along the length of the basin while the screw-driven carriage was stationary.

All experiments were done with freshwater ice, which was grown in the basin by first bubbling air into the water to mix the water to obtain a uniform temperature close to 0°C, spraying a fine mist of air and water into the -10°C ambient air above the basin to seed the ice, and letting the ice grow at an ambient temperature of -20°C to the desired thickness. The resulting ice sheet had the columnar features of the seeded ice. A discussion on the thin-section and the crystallographic structure of similarly grown S2 ice sheets is given by Gow et al. (1988) and Bentley et al. (1988).

Ice sheets of different thicknesses (15-57 mm) were grown during this program. To determine an effective modulus of elasticity, the characteristic length of each ice sheet was measured by placing a dead weight on the floating ice sheet and recording the resulting deflection. On each ice sheet, experiments were conducted by moving the indentor at three discrete velocities, and this was repeated for two or three different values of spring stiffness. While performing experiments later in the program with the main carriage in the middle of the basin, the velocity of the carriage was changed in discrete steps without stopping an experiment. After each test, the ice thickness was measured, and notes were made on cracks observed in the ice sheets.

All the transducers were connected to a signal conditioner, which also provided the excitation signal to the transducers. The conditioned signals were amplified, filtered and sampled at a constant rate. Depending on the indentation velocity, the sampling rate was between 100 and 500 Hz, the faster rate for the higher indentation velocity. The data were stored in the computer memory during a test, transferred to floppy disks after the test, and processed later for permanent storage. All the force signals were filtered with low-pass, 100 Hz filters, whereas the displacement and acceleration signals were filtered with low-pass, 10 Hz filters because these signals were found to have a frequency spectrum below 10 Hz.

During the course of the experiments, the following parameters were changed: ice thickness, velocity, spring stiffness and indentor width. Most experiments were conducted with a 50 mm wide indentor, but 100 mm and 150 mm widths were also used.

3. Results and Observations

During a low-velocity test, microcracks developed in front of the indentor in a zone where the ice was deformed by the pressure generated during the interaction. As more and more microcracks developed, the transparent, freshwater ice became opaque with a white, milky appearance. From visual observations, the density of microcracks was greatest near the indentor and decreased with increasing distance from the indentor. Soon after the first contact but before much microcracking and ice failure, a macrocrack would run from the contact zone to a few meters ahead of the indentor.

Beside these observations, the deformation of the ice at the interface was recorded by using a 50 mm wide, acrylic, transparent indentor and installing an underwater video camera behind the indentor. When the indentor was moved at low speeds ($\approx$ 1-10 mm/s), there was an initial, non-uniform contact between the ice and the indentor. As the ice deformed, the contact between the ice and the indentor became full and uniform across the whole area. With the progression of time (<5 s) as the interaction pressure increased, many microcracks were observed to form both in the vertical and horizontal directions. Just before the crushing failure of the ice, the microcrack density was great enough to make individual cracks indiscernible; the ice was crushed or pulverized. From these images, it is evident that at a low loading rate the ice is crushed or pulverized by a continuous microfracturing process. Though there may be some deformation taking place in the pulverized or crushed ice, it was not visible. Our attempts to record images during subsequent crushing events were not successful because the crushed ice is not transparent. However, the crushed ice was observed to move up and down at the interface during subsequent crushing events. The recorded images during high-speed ice-structure interaction were not useful at all, because the vibrations caused by the interaction produced images that were full of noise.

It was possible to establish from the force records the times when the indentor and the ice were in contact with each other and when there was no contact between them. It is important to note that the interaction force can develop only in one direction to compress the ice during the duration of contact and that the force is zero when the contact is lost. This aspect of determining the contact between ice and indentor plate has been helpful in developing a theoretical model for this interaction. The inability to generate tension forces during loss of contact introduces a non-linearity which must be taken into account in numerical modeling of the interaction.

Figure 2 shows typical results of experiments conducted at different velocities. In these figures and others to follow, the record of the following measured quantities with respect to time are presented: (a) total interaction force (sum of three transducer signals), (b) displacement of the structure with respect to the carriage, i.e. the spring deflection, (c) acceleration of the structure, (d) carriage displacement, and (e) indentor displacement. The force-time histories shown in Figure 2 shows distinct differences in the interaction between an ice sheet and a structure at different speeds: (a) creep deformation of ice, resulting in gradual loading and unloading of the structure and subsequently a steady load, (b) repetitive cycles of gradual loading and sudden unloading due to failure of the ice, resulting in intermittent crushing of ice and a stick- slip type of structure motion (also see data shown in Figure 4), (c) continuous crushing failure of the ice, resulting in an almost constant interaction force and practically no vibration of the structure.

Figure 3, along with Figure 2b, shows the data for tests in which all the parameters were the same except the stiffness of the spring. For the records shown in Figures 3a, 2b and 3b, the values of spring stiffness were 3.23, 1.71 and 0.89 MN m^{-1}, respectively, and the estimated natural frequencies were 11.68, 8.50 and 6.13 Hz, respectively. The data clearly show a definite dependence of the ice failure frequency on the stiffness of the spring, because the number of failures per unit time was proportional to the stiffness of the spring, even though the average ice failure force was approximately the same.

For the case of intermittent ice crushing, the force and the displacement records are plotted on an expanded time scale in Figure 4 to show a typical ice-structure interaction during a few cycles of loading and ice failure. In these detailed plots, it can be seen that there is a time lag in the measured indentor displacement in comparison to the measured forces. This is believed to be caused by the delayed response of the displacement transducer because of the inertia of the spring-loaded pulley system. The corresponding points (A to E) are marked on the displacement and force records in Figure 4. From the displacement records of the carriage and the indentor, it is evident that these two were close to each other before the start of the loading (point A in Figure 4). As the interaction force increased, the indentor lagged behind the carriage, resulting in a displacement of the indentor with respect to the carriage. It should be noted that the superposition of carriage displacement and indentor displacement with respect to the carriage gives the total displacement of the indentor. The acceleration record shows almost no acceleration of the structure during the time of loading. When the ice failed, the loading part of the cycle ended with a sudden decrease in the interaction force to a much lower level (point B in Figure 4). At the instant of ice failure,

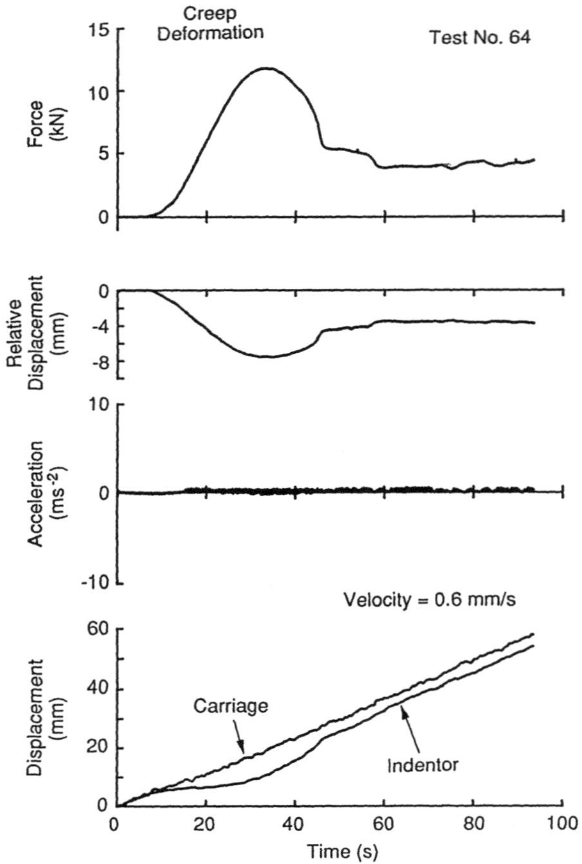
Creep
Deformation
Test No. 64
Force
(kN)
15
10
5
0
Relative
Displacement
(mm)
0
-4
-8
Acceleration
(ms-2)
10
0
-10
Velocity = 0.6 mm/s
Displacement
(mm)
60
40
20
0
Carriage
Indentor
0
20
40
60
80
100
Time (s)

Fig. 2a.

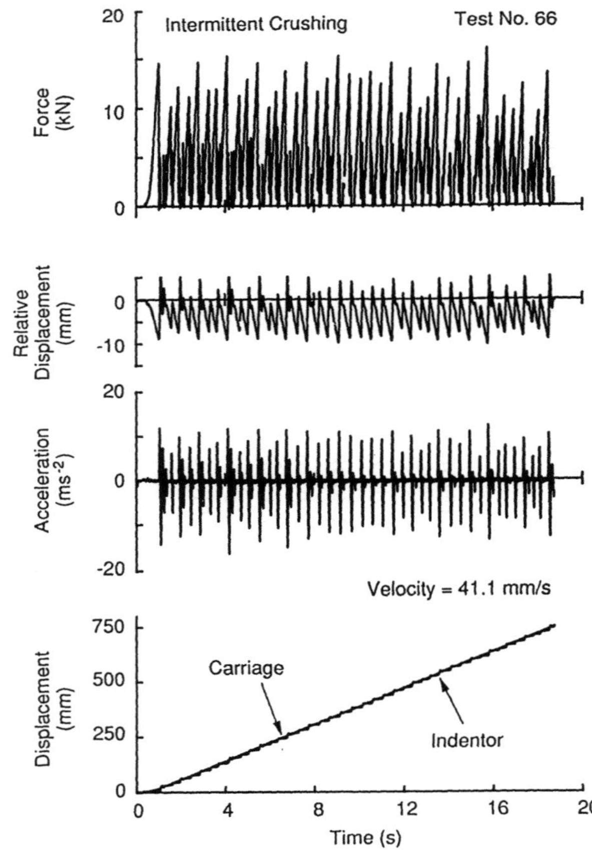
Intermittent Crushing
Test No. 66
Force
(kN)
20
10
0
Relative
Displacement
(mm)
0
-10
Acceleration
(ms-2)
20
0
-20
Velocity = 41.1 mm/s
Displacement
(mm)
750
500
250
0
Carriage
Indentor
0
4
8
12
16
20
Time (s)

Fig. 2b.

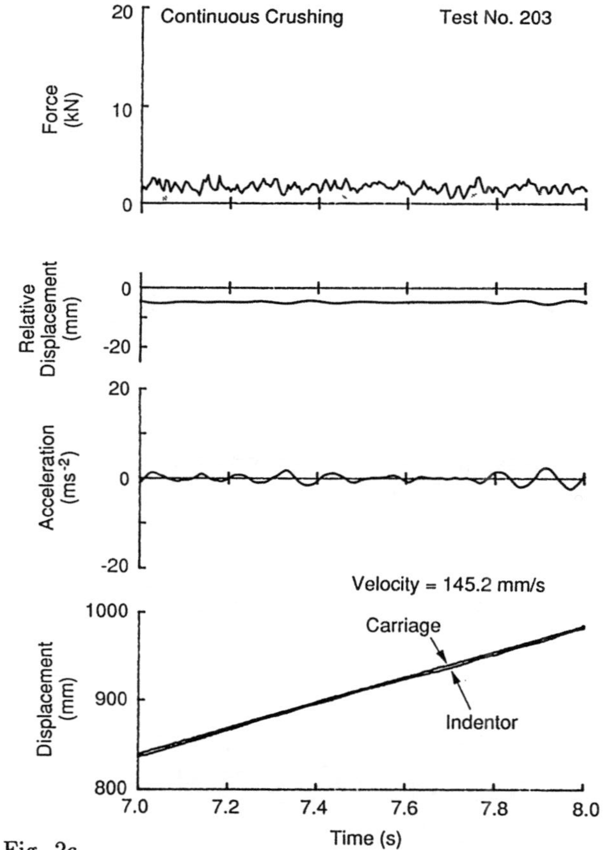

Fig. 2c.

Figure 2. Typical experimental results of indentation tests at different velocities: (a) 0.6 mm s^{-1}, (b) 41.1 mm s^{-1} and (c) 145.2 mm s^{-1}. The indentor width was 50 mm. For the respective records, the ice thickness and the structural stiffness, respectively, were (a) 27.1 mm, 182 MN m^{-1}, (b) 27.5 mm, 1.71 MN m^{-1}, and (c) 24.0 mm, 1.13 MN m^{-1}. The velocity indicated on each set of records is obtained by the method of least squares from the data of the carriage displacement during the time intervals of the record.

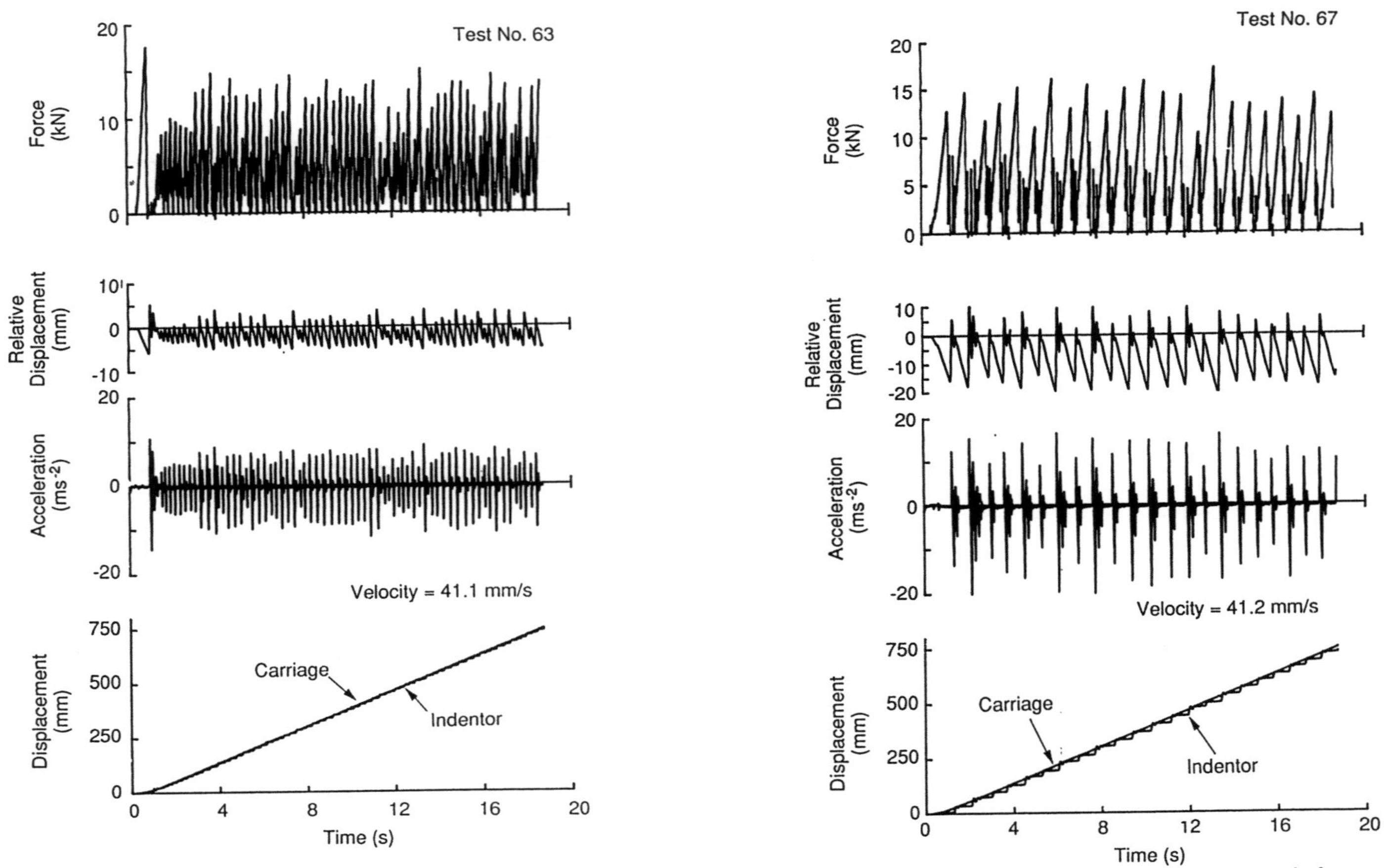

Figure 3. Experimental results of indentation tests on an ice sheet in which all parameters (e.g. ice thickness = 27 mm, indentor speed = 41.1 mm s^{-1} and indentor width = 50 mm) were the same except the spring stiffness which had the following values: (a) 3.23 MN m^{-1} and (b) 0.89 MN m^{-1}. The data in Figure 2(b) refer to a similar test conducted at a stiffness value of 1.71 MN m^{-1}.

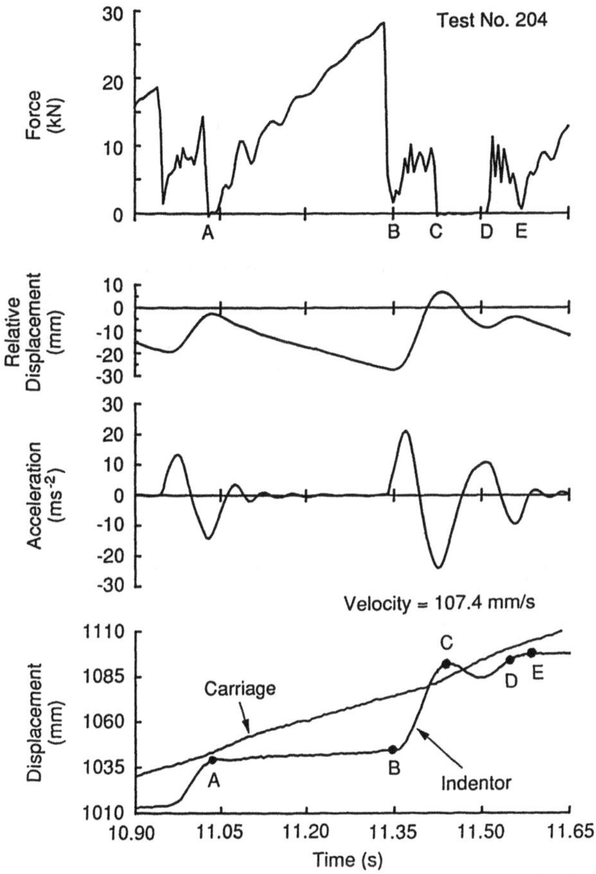

Figure 4. Detailed plots of force, relative displacement, acceleration and displacements of the carriage and the indentor with respect to time. During the test, the ice thickness was 24 mm, and the indentor width 100 mm.

the spring was deflected considerably, and the stored energy in the spring was then expended to move the structure forward at an estimated velocity of 0.85 m s^{-1}, during which time (between points B to C in Figure 4) the crushed ice in front of the indentor was extruded by moving it up and down. For the event depicted in Figure 4, the structure moved past its equilibrium position. After its maximum excursion relative to the carriage, the structure moved back, and during that time it lost contact (between points C and D in Figure 4) with the ice when its absolute velocity with respect to the ice was in the negative (backward) direction, as indicated by the negative slope of the displacement-time record. The structure oscillated a few times relative to the carriage as it was being driven forward by the carriage. During these oscillations, the indentor may have contacted the ice as indicated from the force records. Eventually, the carriage and the structure displacements were nearly equal when the vibrations decayed. A new cycle of loading (point E in Figure 4) began when the ice started to resist the indentor and to hold it back while the carriage kept moving forward.

Figure 5 shows the force, acceleration and displacement records for a test when the structure vibrated at a frequency close to its natural frequency. The force record shows an almost periodic "spike" that occurred

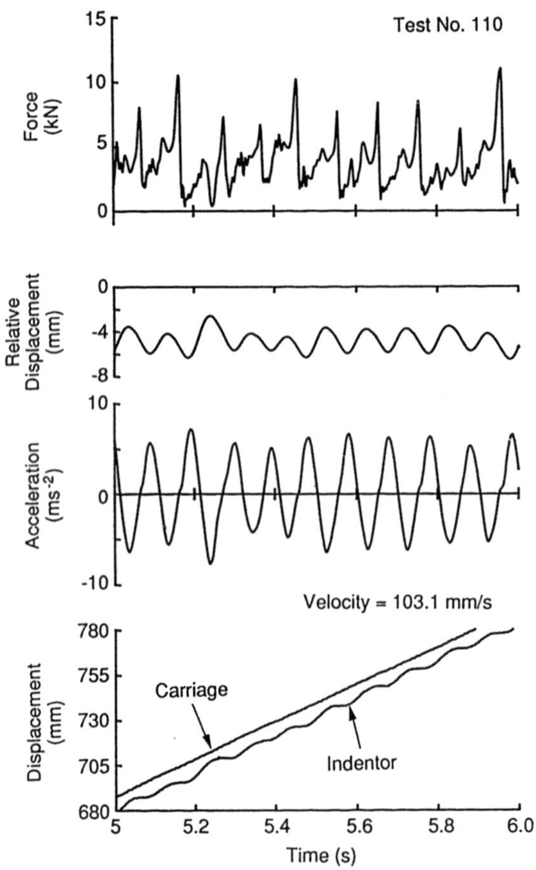

Figure 5. Experimental results from a test when the structure vibrated at a frequency close to its natural frequency. During this test, the ice thickness was 30 mm, and the indentor width 50 mm.

in phase with the backward relative displacement of the structure. The records of the carriage and the structure displacements show that the carriage had a steady rate of displacement and that the motion of the indentor was a superposition of a ramp and an oscillatory displacement. A detailed analysis of the structure motion reveals that the absolute velocity of the indentor never reversed (see Figure 5); instead it merely slowed down when the interaction "spike" load built up, supplying energy to the carriage-transported structure periodically at its natural frequency. Contact between the indentor and the ice during this test always existed because the interaction forces were never zero. Because the indentor pushed the ice in the same direction it travelled, its rate of work done on the ice was always positive, i.e. the energy was always dissipated in crushing and removing the ice.

The data presented here are not a complete listing of the results obtained from all of the tests conducted in this program. The author plans to publish them in a later paper or a CRREL Report.

4. Discussion

As discussed earlier, we observed three modes of ice-structure interaction: creep of ice at low velocities, intermittent ice crushing at intermediate velocities, and continuous ice crushing at high velocities. One of the intriguing results from these experiments is the effect of velocity on the energy supplied by the carriage and dissipated by the indentor and on the effective pressure (p) that develops between the indentor and the ice during these modes of interaction.

Because the displacements of the carriage and the indentor were measured separately, we could compute the amount of work done by each in removing a unit volume of ice between any two instants in time. This was done by summing the area under their respective plots of force versus displacement and dividing the result by the volume of ice removed (= contact area x distance travelled). Units of work done per unit volume are the same as that for pressure, i.e. N m^{-2} or Pa. In the case of creep deformation of ice and continuous crushing, the ratio of work done per unit volume by the indentor to that by the carriage is almost equal to 1. During intermittent crushing, this ratio varies from 0.67 (for data shown in Figure 4) to 0.96 (for data shown in Figure 5). This means that part of the energy supplied by the carriage is dissipated by the indentor in crushing and removing the ice; the remainder is dissipated by the vibrations of the structure. The amount of work done by the carriage in removing a unit volume of ice was in the range of 1.4 MPa for continuous crushing (Figure 2c) to 4.4 MPa for intermittent crushing (Figure 3b). Similar values for creep deformation of ice were about 2.8 - 2.9 MPa for data shown in Figure 2a.

At low carriage velocity (1-2 mm s^{-1}), the ice deformed in creep mode. After an initial peak force (e.g. $p = 8.8$ MPa from Figure 2a), the resistance offered by the ice was low (e.g. $p \approx 3.0$ MPa from Figure 2a) and almost constant. During this process, the ice may be thought of as being microfractured and extruded continuously. It should be noted that contact between the ice and the indentor existed all the time.

At higher carriage velocities, the mode of interaction (or crushing) was either intermittent or continuous. Depending on the spring stiffness and the indentor width, we observed that the crushing mode changed at a certain velocity, hereafter defined as the transition velocity.

For tests conducted below transition velocity (Figure 2b or 4), the force record shows that the effective failure pressure was high ($p_f \approx$ 8-13 MPa) at the instant of ice failure. After ice failure the indentor surged forward at velocities between 0.4 and 0.8 m s^{-1}, estimated from the slope of the indentor displacement records (e.g. Figure 4). During the forward motion, the indentor encountered an almost constant extrusion force ($p_e \approx$ 2-4 MPa) as long as the contact between the indentor and the ice existed. As indicated by force records, there was loss of contact between the indentor and the ice for short durations of time. When the indentor vibrations decayed, a new cycle of loading began.

For tests conducted above the transition velocity (e.g. Figure 2c), the force record shows an almost constant level of force with small random variations superimposed on it. The indentor moved at a constant velocity and almost negligible acceleration. There was a constant relative displacement between the indentor and the carriage as would be expected with a constant force. A steady process of microcracking, pulverization and extrusion was active ahead of the indentor, similar to the low-velocity experience. The measured effective pressure was in the range of 1.4 to 1.7 MPa. Contact between the ice and the indentor existed at all times.

The reason for different effective pressures during different types of crushing events in this study may be attributed to the differences in the actual indentor velocities relative to the ice because ice is a rate-sensitive material. During intermittent crushing, the indentor velocity during the time of loading was very low (e.g. 10-15 mm s^{-1} in Figure 4). At this rate, the ice underwent microcracking and pulverization resulting in crushed ice. After an ice failure event, the forward indentor velocity was about two orders of magnitude higher than that mentioned above. During the continuous crushing process, the indentation velocities were also an order of magnitude higher.

The exact mechanism for the ice crushing at a high indentation rate is still poorly understood. But arguments can be made by comparing the stick-slip motion of the indentor during intermittent crushing to the motion of an object sliding on a frictional surface and being pushed or pulled through a flexible spring. The difference in the static and kinetic coefficients of friction between the object and the surface results in a stick-slip motion. Similarly, the loss of contact between the indentor and the ice results in unloading of the ice sheet, and subsequent loading may have to be higher to overcome something like static friction in the ice sheet. This feature is incorporated into a theoretical model described in the next section.

5. Theoretical Model

Based on the experimental results, a model of the ice-structure interaction is shown in Figure 6. It is assumed that the carriage moves forward at a constant velocity v and that its displacement $x(t)$ is obtained as vt, where t is the elapsed time. Though this model is developed for a carriage-transported structure, the case of a stationary structure and a moving ice sheet can easily be implemented by superimposing a constant

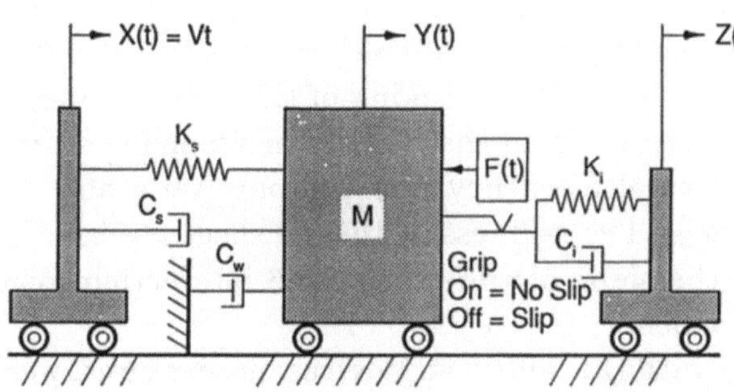

Figure 6. Theoretical model for the ice-structure interaction.

velocity in the opposite direction to that of the carriage. Other things in the model remain the same as described below.

The structure is assumed to be a single-degree-of-freedom system with a mass M, a spring of stiffness K_s and a damping element C_s. The damping between water and the indentor is represented by another damper C_w. The position of the mass M is denoted by $y(t)$.

The mass M is assumed to interact with the ice, which is initially assumed to be a spring K_i and a damper C_i during the loading phase of a cycle. A stick-slip grip is assumed to exist between the ice and the face of the structure. The other side of the ice elements (K_i and C_i) are attached to a distant point in the ice sheet. The position of this point is denoted by a non-decreasing function $z(t)$, which is assumed to be a constant during the loading phase. When the effective pressure generated between the indentor and the ice as a result of compression of the spring K_i reaches a critical value (i.e. failure pressure = 8-13 MPa), the grip between the structure and the ice starts to slip, and the structure begins to move forward rapidly, encountering a resistance $F(t)$, equivalent to an extrusion effective pressure ($\approx$ 2-4 MPa). During this forward motion, the base of the ice spring and damper is assumed to move with the mass M, i.e. $z = y$. If at any time mass M moves in the reverse direction (i.e. $dy/dt < 0$), the variable z is assumed to be constant at its last value because it is a non-decreasing function. This means that the contact between the ice and the structure does not exist as long as $y < z$. When the contact is made (i.e. $y = z$), a new loading cycle begins by turning on the grip between the ice elements and the mass and keeping the value of z a constant.

Some of the parameters used in the simulation model were determined by performing a dynamic calibration of the structural system. After the testing program, the structure was loaded with a wire rope attached between a fixed point and the indentor. A breakable bolt was incorporated in the connections of the wire rope. The breaking of the bolt provided a sudden unloading of the structure. From the measurements made during the calibration tests, the natural frequency, the damping ratio and the spring stiffness were obtained. The stiffness of the ice sheet used in the model was obtained from the slope of the plot of the interaction force versus the indentor displacement.

These parameters, along with the failure and extrusion pressures, were input into the model described above, and the results obtained are shown in Figure 7. The model results were found to be highly sensitive to

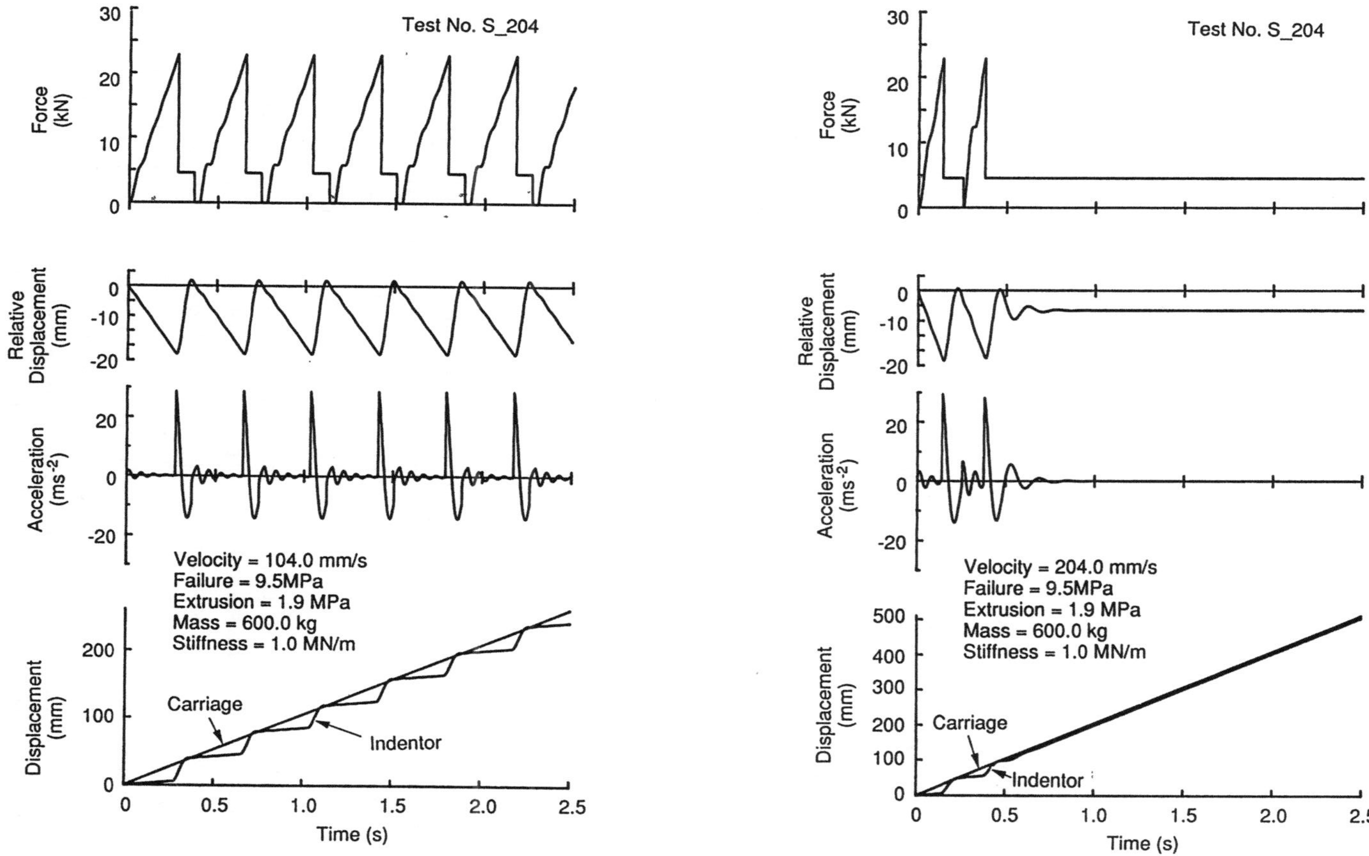

Figure 7. Results of the theoretical model for the dynamic ice-structure interaction at (a) below transition velocity and (b) above transition velocity.

the values of failure and extrusion pressures and slightly sensitive to the coefficients of damping C_s, C_i and C_w. When the model was run for high carriage velocities, a transition from intermittent to continuous crushing was predicted by the model at a transition velocity, which depended on the failure and extrusion pressures. That the model captures the essential features of the experimental results, particularly the prediction of the intermittent and continuous crushing below and above a transition velocity, is very encouraging.

6. Recommendations

On the basis of these experimental results, the following recommendations are made for conducting model tests and for measuring ice forces in the field.

1. The interaction forces should be measured by installing a rigid, instrumented panel in the area where ice contacts the structure. Such measurements will provide the true interaction force that is not influenced by the response of the model structure. The same recommendation is made for field measurements of the forces on slender, light-weight, flexible structures.

2. The stiffness and the mass of the model structure should be carefully scaled to simulate the prototype structure, because these influence the ice-structure interaction.

3. In many field measurements it is convenient to record the acceleration at a point on the structure, but the interpretation of acceleration data is difficult. During the loading period of an intermittent ice crushing event, the acceleration is nearly zero, and the relative displacement of the structure increases with the increase in interaction force (e.g., Figure 4). Immediately after the ice failure, the structure moves forward due to the spring force, and the acceleration record will show positive and negative acceleration. The acceleration signal after an ice failure event can be integrated twice to obtain the structural deformation prior to ice failure, from which an estimate of the maximum force deflecting the structure can be made. Obviously, such an estimation will have some errors associated with it, and an estimate of the error bound should also be provided.

7. Conclusion

To obtain detailed information on dynamic ice-strucuture interaction, indentation tests were conducted at different velocities on freshwater ice. Interaction forces as well as response of the structure in terms of displacements and acceleration were measured.

Depending on the indentation velocity, the interaction resulted in three modes of ice failure: creeping mode at low velocities, intermittent crushing at intermediate velocities, and continuous crushing at high velocities. A theoretical model is proposed that simulates intermittent and continuous crushing of the ice-structure interaction, and the simulated results compare well with the experimental results of this study.

One of the important results of this study is that the effective pressures during continuous crushing were found to be an order of magnitude lower than the peak pressures during intermittent crushing. This is similar to the field measurement of effective pressure during impact situations where the effective pressures have also been found to be an order of magnitude lower those measured in small-scale experiments. The exact reasons for this decrease are, at present, not known, but the ice velocity and the structural parameters play an important role in determining the mode of ice-structure interaction. The implication of this finding is important from the point of view of designing structures for interactions at high ice velocities.

Acknowledgement

The author is indebted to Carl Martinson, Randy McGilvary and Naoki Nakazawa for their assistance in conducting the experiments.

References

Bentley, D.L., Dempsey, J.P., Sodhi, D.S. and Wei, Y. 1988. Fracture of S2 columnar freshwater ice: floating double-cantilever beam tests. Proceedings, IAHR Symposium on Ice, Sapporo, Japan, Vol.I, pp.152-161.

Blenkarn, K.A. 1970. Measurement and analysis of ice forces on Cook Inlet structures. Proceedings, 2nd Offshore Technology Conference, Houston, TX, USA, OTC 1261, Vol.II, pp.365-378.

Evers, K.-U. and Wessels, E. 1986. Model study of level ice forces on cylindrical multi-legged structures. Proceedings, POLARTECH-86, VTT, Helsinki, Finland, Vol.3, pp.251-275.

Frederking, R. and Timco, G.W. 1987. Ice loads on a rigid structure with a compliant foundation. Proceedings, 9th International Conference on Port and Ocean Engineering under Arctic Conditions (POAC), Fairbanks, AK, USA, Vol.III, pp.409-418.

Gow, A.J., Ueda, H.T., Govoni, J.W. and Kalafut, J. 1988. Temperature and structure dependence on the flexural strength and modulus of freshwater model ice. U.S. Army Cold Regions Research and Engineering Laboratory, Hanover, NH, CRREL Report 88-6.

Hirayama, K.-I., Schwarz, J. and Wu, H.-C. 1973. Model technique for the investigation of ice forces on structures. Proceedings, 2nd International Conference on Port and Ocean Engineering under Arctic Conditions (POAC), Reykjavik, Iceland, pp.332-344.

Jefferies, M.G. and Wright, W.H. 1988. Dynamic response of "Molikpaq" to ice-structure interaction. Proceedings, 7th International Conference on Offshore Mechanics and Arctic Engineering (OMAE), Houston, TX, USA, Vol.IV, pp.201-220.

Määttänen, M. 1982. True ice force by deconvolution. Proceedings, 1st International Model Analysis Conference (IMAC), Orlando, FL, USA, (Union College, Schenectady, NY), pp.586-590.

Michel, B. and Blanchet, D. 1983. Indentation of an S2 floating ice sheet in the brittle range. Annals of Glaciology, Vol.IV, pp.180-187.

Montgomery, C.J. and Lipsett, A.W. 1981. Estimation of ice forces from dynamic response. Proceedings, IAHR International Symposium on Ice, Quebec, P.Q., Canada, Vol.II, pp.771-782.

Neill, C.R. 1976. Dynamic ice forces on piers and piles: An assessment of design guidelines in the light of recent research. Canadian Journal of Civil Engineering, 3:305-341.

Peyton, H.R. 1968. Sea ice forces. Ice pressures against structures. National Research Ccouncil of Canada, Ottawa, Canada, Technical Memorandum 92, pp.117-123.

Sodhi, D.S. 1988. Ice-induced vibrations of structures. Proceedings, IAHR International Symposium on Ice, Sapporo, Japan, Vol.II, pp.625-657.

Sodhi, D.S. and Nakazawa, N. 1988. Results from indentation tests on freshwater ice. Proceedings, IAHR International Symposium on Ice, Sapporo, Japan, Vol.I, pp.341-350.

Timco, G.W. 1986. Indentation and penetration of edge-loaded freshwater ice sheets in the brittle range. Proceedings, 5th International Conference on Offshore Mechanics and Arctic Engineering (OMAE), Tokyo, Japan, Vol.IV, pp. 444-452.

Timco, G.W., Frederking, R. and Singh, S.K. 1989. The transfer function approach for a structure subjected to ice crushing. Proceedings, 10th International Conference on Port and Ocean Engineering under Arctic Conditions, Lulea, Sweden, Vol.I, pp.420-430.

Zabilansky, L., Nevel, D.E. and Haynes, F.D. 1975. Ice forces on model structures. Canadian Journal of Civil Engineering, 2(4):400-417.

Discussion

Q: S.A.F. Murrell

What was the temperature of the ice in your tests? If the temperature was close to 0°C, then perhaps melting (due to friction) may have occurred at the highest indentor velocity (e.g. in Figure 2c this was 145.2 mm s^{-1}) when unusually low forces occurred during continuous crushing. These low force values are otherwise puzzling.

A: D.S. Sodhi

The temperature of freshwater ice was measured a few times during the experimental program and it was found to be in the range of -1°C to 0°C. As suggested above, let us consider the possibility of ice melting by assuming that all the work done by the indentor is transformed into heat by internal friction. In the case of continuous crushing (Figure 2c), the amount of work done by the carriage (or the indentor) in crushing and removing a unit volume of ice was 1.4 MPa or 1.4×10^6 J/m^3 (see Section 5 entitled "Discussion"). Assuming the latent heat of ice fusion to be 3.34×10^5 J/kg and ice density equal to 920 kg/m^3, we obtain a value of 3.0728×10^8 J/m^3 for complete melting of ice. Taking a ratio of the work done by the indentor to the energy needed for complete melting, we get a value of 0.0046, i.e. about 0.5% of the volume of ice would melt in this process.

The suggestion of friction melting does offer an explanation for low interaction forces during continuous crushing, but similar arguments for intermittent crushing lead to a higher volume fraction of melted ice and also a higher interaction force. There is a need for further research to explain the difference between forces during intermittent and continuous crushing. There are indications that it is related to ice-structure interaction and the resulting structural vibrations.

Q: I.J. Jordaan

I would like to compliment you on your experimental techniques and interesting test series. Your conclusion regarding when extrusion takes place may result from the flexibility of your loading device; when the load drops there is a large amount of energy released and this causes rapid clearing, almost "flicking" the ice particles away. This is manifested by the load dropping to zero. In the case of more rigid structures, this would not be the case, and I consider that extrusion can occur during the drop or increase of load, and in general during both. The question of how accurate load measurements are in indentation tests would depend on the particular structure and stiffness of the loading device as well as location of load cell. It may be difficult to generalize. Could you please comment on these points?

A: D.S. Sodhi

Thank you very much for your kind words regarding the experimental techniques used in this study.

My conclusions regarding the ice extrusion not taking place during the loading phase of a dynamic ice-structure interaction was based on visual observations and experimental data on interaction force and displacements. Your statement that "in the case of more rigid structures" ice "extrusion can occur during the drop or increase of load" needs experimental verification, particularly during the increase of load. All structures have some amount of flexibility, or a finite amount of rigidity, and there is, at present, no established criterion to classify a structure as rigid or flexible in terms of other parameters related to structure and ice. That the structural stiffness plays an important role in the ice-structure interaction has clearly been demonstrated in this study (see results presented in Figure 2b, 3a and b), but no difference in the extrusion process was observed even when the structural stiffness was increased by a factor of 3.63.

The main purpose of this experimental study was to understand what goes on in a dynamic ice-structure interaction during indentation tests, hence the title of the paper. Whether the results of this study can be generalized to other situations remains to be seen by comparing the results of this study with future small-scale and full-scale studies.

Structure Geometry and Ice Interaction

K. R. Croasdale

Research Department, Esso Resources Canada Ltd.
Calgary, Alberta

M. Metge

Canatec Consultants Ltd.
Calgary, Alberta

Abstract The presence of ice usually requires that structures be of different geometry than those in ice free areas. In fact structural form is usually chosen to minimize ice loads. However the engineer often has to balance the sometimes conflicting requirements of ice loads, wave and current effects, foundation stability, structure use, installation, and costs. In order for such an optimization to be conducted, a reasonable understanding of the relationship between structural form and ice interaction is necessary. This paper reviews the current state of knowledge of how ice interaction is influenced by structural shape.

Ice failure modes, ice clearing processes, global ice loads (and their dynamics) are reviewed and contrasted for a variety of structural forms including sheet ice, ridges, isolated floes, ice islands and icebergs. The paper comments on the level of confidence associated with these comparisons and highlights future research with special emphasis on the comparison between sloping and vertical structures. These are compared in the context of three types of application; narrow sub-arctic structures, wide arctic structures and structures for iceberg infested waters.

1. Introduction

Practitioners of ice engineering are well aware of the dilemma relating to the choice of geometry for ice-resistant structures. Not only must the engineer achieve a safe design against ice loads, but he also has to consider the effect of geometry on other factors such as function, strength, installation, costs and other types of environmental loads. Furthermore,

S. Jones · R. F. McKenna · J. Tillotson · I. Jordaan (Eds.)
Ice-Structure Interaction
IUTAM-IAHR Symposium St. John's, Newfoundland, Canada

these deliberations are often undertaken in a framework of uncertainty, primarily the uncertainty relating to ice loads for the structural shapes being considered. As a result, it is not clear that we have yet achieved a rational basis for selecting structure geometry. This paper discusses some of the alternatives for ice resistant structures, and compares ice interaction phenomena. A major focus of the paper is the comparison between sloping and vertical structures in sub-arctic, arctic, and iceberg infested regions.

2. Overview

There are three well-known forms of structure geometry, each is associated with distinct failure modes which in turn lead to different magnitudes of ice loads. These structural forms are:

- Narrow vertical-faced structures: against which (in most cases) the ice will fail in crushing.
- Narrow sloping-faced structures: against which (in most cases) the ice will fail in bending.
- Wide structures around which an ice-rubble field can form. The predominant ice failure mode is then rubble-building, which is manifested by random, non-simultaneous out-of-plane predominantly bending failures of the ice.

These three distinct structure geometries and failure modes are illustrated in Figure 1. Also in Figure 1 are typical ice loads associated with each geometric form and failure mode (for 1m ice thickness). Typically the narrow vertical structure can be subject to ice loads of 2 to 6 MNm^{-1} compared to the lower unit forces on the sloping structure (of about 0.5 to 1.0 MNm^{-1}) and wide structure (of about 0.2 to 1.0 MNm^{-1}).

This simple comparison would suggest the superiority of a sloping structure over a vertical structure. This also seems to be intuitively correct in the light of the experience of ice breaking vessels, which are essentially sloping structures, and which probably would not be feasible if they had to progress through the ice by creating crushing failures! So why is it that many actual and proposed fixed platforms in ice have vertical sides (and is this reasonable)? That is a key question which this paper will review and deliberate upon. Another point to recognize in

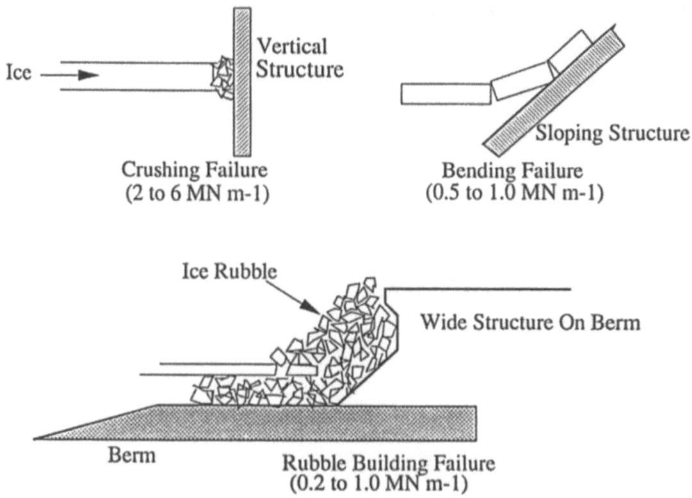

Figure 1. Ice failure modes and approximate range of loads for ice 1 m thick ice.

this general overview is the issue of how ice failure modes (and hence unit ice loads) can vary as the width of a structure (or contact area) is increased. This is a classic problem in ice mechanics and has been the subject of much speculation, theoretical analyses and experiments, as well as examination of full-scale data. This topic will not be examined in any detail in this paper, but has been recently reviewed in several key publications. For example Blanchet et al. (1988) examined extensive laboratory and full-scale data against vertical structures. They proposed that three major failure modes be categorized as buckling, crushing and bending (out-of-plane rubble building). They also declared that, from a theoretical standpoint, there are three distinct regimes defined by ice thickness and aspect ratio (structure diameter divided by ice thickness). For ice thickness less than 0.75m and aspect ratios greater than 20, a buckling failure mode will occur. For aspect ratios less than 5, crushing failures will occur independent of thickness. For aspect ratios greater than 10, all three failure modes are possible depending on ice thickness and the presence of eccentricities in the ice (e.g. thickness and modulus variations).

Sanderson (1988) addressed the issue of size effect as a function of contact area. By lumping together data which spanned several orders of magnitude in size, he demonstrated that an upper bound to all the data had a slope of $A^{-0.5}$ (A being the contact area). Also by examining large scale data obtained from ice action on Beaufort Sea Structures he proposed an empirical relationship for ice action on vertical structures (at strain rates high enough to cause brittle failures). It is as follows:

$$\sigma = \sigma_o \left(\frac{A_o}{A}\right)^{0.5}, \tag{1}$$

where σ is the ice pressure or failure stress at contact area A (width times thickness), σ_o is the reference ice pressure at reference area A_o derived from large scale data. For the 95% confidence limit, Equation (1) becomes for arctic first year ice: (A in m^2),

$$\sigma = 1.77 \left(\frac{100}{A}\right)^{0.5} \text{ MPa} \tag{2}$$

and, for arctic multi-year ice,

$$\sigma = 2.04 \left(\frac{100}{A}\right)^{0.5} \text{ MPa.} \tag{3}$$

A criticism of Sanderson's empirical approach might be that it doesn't definitively recognize the influence of changing failure modes as width or contact area change. A related concern would be whether the $A^{-0.5}$ relationship holds across each failure mode. Another factor controlling failure modes is ice velocity. At low velocities ice creeps and appears ductile whereas at higher velocities fracture processes dominate the ice loads. This topic is beyond the scope of this paper and readers are referred to key references in Sanderson (1988) and Cammaert and Muggeridge (1988).

The remainder of this paper will focus on a comparison between vertical and sloping structures as this is one of the most difficult choices facing the engineer. First we will consider narrow sub-arctic structures; typically piers are 5 to 10m wide subject to ice up to 2.0m thick.

3. Narrow Structures: Sloping Versus Vertical

3.1 Conical Structure — Methods

As already discussed, ice loads on sloping structures are controlled by bending failures of the ice and subsequent clearing of the broken ice. The prediction of ice loads on sloping structures has been addressed by several investigators (e.g. Bercha and Danys, 1975; Semeniuk, 1975; Edwards and Croasdale, 1976; Ralston, 1977; Croasdale, 1980; Frederking and Timco, 1985). A comparison of horizontal ice forces calculated by methods developed by several of the above investigators is given in Figure 2. Readers are referred to Marcellus et al. (1988, 1989), for a critical discussion of these various methods, including comparisons with physical data. All the methods include both ice breaking and ice clearing forces.

Compared to physical data, Ralston's method overpredicts the ice loads (Semeniuk, 1975; Marcellus et al., 1989). However some conservatism is usually preferred, so Ralston's method will be used in this paper for the purpose of comparisons with vertical structures. The method expresses ice loads on conical structures in the following way

$$H = A_4[A_1\sigma_f t^2 + A_2\rho_w g t D^2 + A_3\rho_w g t(D^2 - D_T^2)], \tag{4}$$

$$V = B_1 H + B_2\rho_w g t(D^2 - D_T^2), \tag{5}$$

where H and V are the horizontal and vertical ice loads; σ_f is the bending strength of the ice; t is the ice thickness; ρ_w is water density; g is the acceleration due to gravity; D is the structure diameter at the water line; D_T is the structure diameter at the top of the slope; A_1 and A_2 are coefficients dependent on $\frac{\rho_w g D^2}{\sigma_f t}$; A_3, A_4, B_1 and B_2 depend on cone angle and ice to structure friction. In Equation (4) it will be noted that the first two terms depend on σ_f and t and represent the ice breaking terms; the third term depends on structure geometry and t, and represents the ice ride-up term. For a narrow structure (e.g. 3m) the ice breaking terms

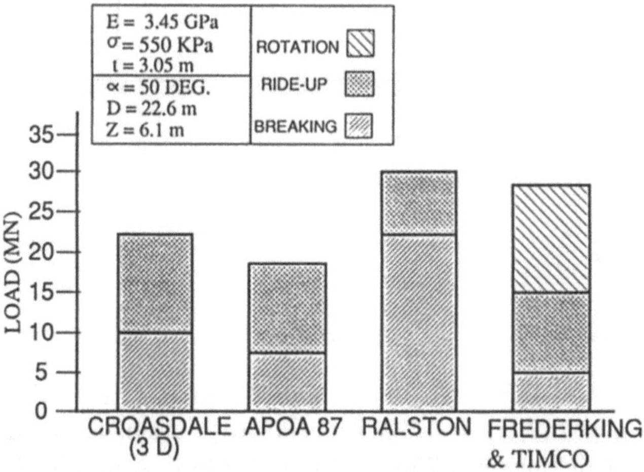

Figure 2. Comparison of sheet ice ice loads on conical structures.

typically account for about 98% of the total load; whereas for a wider structure (e.g. 60m) the ice breaking component reduces to about 50% or less, and the ice clearing force becomes significant (see Croasdale, 1980 and 1987).

In this part of the paper, a comparison will be made between the two structures illustrated in Figure 3. One is a cylindrical pier 7m diameter at the water line, the other is a conical equivalent being 11m diameter at the water line and 5m diameter at the top of the slope.

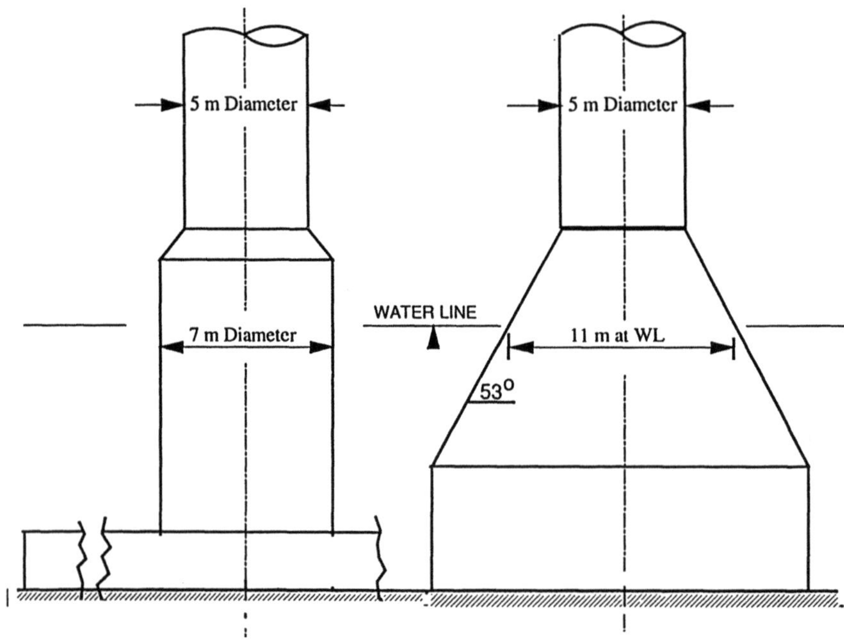

Figure 3. Comparative narrow and vertical structures.

3.2 Conical Structure — Loads

For Ralston's equations the structure being considered has the values $D = 11\text{m}$; $D_T = 5\text{m}$; $\mu = 0.2; \sigma_f = 0.7$ MPa (100 psi). Also, loads are calculated for three ice thicknesses 0.75m, 1.0m and 2.0m. A sensitivity is performed for half the ice strength. The results are shown in Table 1.

Table 1. Comparison of Loads for Narrow Structures.

	Horizontal Ice Force (MN)		
	$t = 0.75$m	$t = 1.0$m	$t = 2.0$m
A Conical Structure (11m Dia. at WL; Angle= 53°; 5m Dia. at top; $\mu = 0.2; \sigma = 0.7$ MPa)	5.5	7.1	15.5
B Conical Structure (As above but half the ice strength i.e. $\sigma = 0.35$ MPa)	4.0	5.1	10.1
C Adfreeze Loads On Conical Structure			
(a) High Strain Rate	19.1	25.4	50.1
(b) Creep	3.8	5.1	10.2
D Vertical Structure Dia. = 7m			
(a) Indentation Equation ($\sigma_c = 6.0$ MPa; $k = 0.4$)	34.1	45.5	91.0
(b) Using Size Effect Curve	40.6	47.0	66.2
(c) Using an Effective ice pressure of 2.0 MPa	10.5	14.0	28.0

3.3 Adfreeze Loads

A potential problem with a conical or sloping structure relates to the issue of adfreeze. This is the term used to describe the condition when the surrounding ice sheet is frozen to the structure. This could happen if ice movement is intermittent. If adfreeze does occur then the adfreeze bond has to be broken before ice can move against the cone and fail in flexure at the loads previously calculated. The adfreeze issue has been addressed theoretically (e.g. Croasdale, 1978; Cammaert et al., 1986). Also several laboratory experiments have been conducted to measure adfreeze strengths (on a small scale) (e.g. Arctec Canada Ltd., 1975; Sackinger and Sackinger, 1977; Oksanen, 1983).

The adfreeze force is of the form (Cammaert et al., 1986)

$$H_A = \frac{C_a C_s \pi D t q}{\sin\alpha \cos\alpha}. \tag{6}$$

This assumes ice frozen to the structure all around. C_a and C_s are factors to allow for uneven bonding and uneven stress distribution. In this example we will assume $C_a C_s = 0.5$. Also it is assumed that $q =$ 0.7 MPa (100 psi); $\alpha = 53°; D = 11$m. The adfreeze loads are then as shown in Table 1. It will be noted that they are greater than those previously calculated due to ice breaking, and would control the design. However they are unduly conservative because they do not account for the effect of strain rate.

Generally adfreeze strengths quoted in the literature are obtained from high strain-rate or rapid loading tests. Yet in the real world, an adfreeze bond will likely be broken at low strain rates (or loading rates) as a build-up in wind stress causes a slowly increasing strain in the ice sheet. This effect should be accounted for in our assessment of adfreeze loads on sloping structures. For the purpose of this paper, in-lieu of any experimental data, it is proposed that the adfreeze bond is governed by the same creep laws as ice itself. For ice, the dominating creep regime is secondary creep for which Glen's law applies, i.e.

$$\dot{\varepsilon} \propto \sigma^n, \tag{7}$$

where $\dot{\varepsilon}$ is strain rate, σ is stress and n is a power which is approximately equal to 3.0. Thus, if we know the strength at high strain rates, we can calculate the "strength" corresponding to a lower strain rate according to:

$$\sigma = \sigma_o \left(\frac{\dot{\varepsilon}}{\dot{\varepsilon}_o}\right)^{0.33}. \tag{8}$$

For adfreeze strength (q), the same kind of relationship would be

$$q = q_o \left(\frac{\dot{\varepsilon}}{\dot{\varepsilon}_o} \right)^{0.33} , \tag{9}$$

where q_o and $\dot{\varepsilon}_o$ represent the peak adfreeze strength and corresponding strain rate.

In this case, let us use $q_o = 0.7$ MPa and assume this is associated with a strain rate of 10^{-3} sec^{-1} (which is the strain rate usually associated with peak compressive strength of ice). Then;

$$q = 0.7 \left(\frac{\dot{\varepsilon}}{10^{-3}} \right)^{0.33} \quad \text{MPa}, \tag{10}$$

from which the following values can be derived

$\dot{\varepsilon}$	(sec^{-1})	10^{-3}	10^{-4}	10^{-5}	10^{-6}	10^{-7}	10^{-8}
q	MPa	0.7	0.33	0.15	0.072	0.034	0.016

Using a strain rate of 10^{-6} as typical for the beginning of an ice movement event leads to a design adfreeze strength of 0.07 MPa. In the example previously presented, this lower value of adfreeze strength leads to adfreeze loads on the conical structure which are now low enough not to control the design, see Table 1. It should also be noted that adfreeze bond strength can also be lowered by heating or by using special coatings (Arctec Canada, 1975).

3.4 Vertical Structure — Methods and Loads

Three methods will be used. The first is based on the contact area (size effects) approach as proposed by Sanderson (1988) and to which reference has already been made. The empirical values developed for first-year sea ice will be used, i.e. Equation (2). Loads are developed below for a 7m dia. pier.

	$t = 0.75$m	$t = 1.0$m	$t = 2.0$m
Area (m)^2	5.25	7.0	14.0
σ (MPa)	7.7	6.7	4.7
Load (MN) ($= \sigma D t$)	40.6	47.0	66.2

The second method uses the indentation equation as originally developed by Korzhavin, i.e.

$$\text{Load } (Hc) = Dt\sigma_c Ikm \tag{11}$$

where σ_c is the uniaxial compressive strength,
I is an indentation factor accounting for confinement,
k is a contact factor and
m is a shape factor.

Inserting typical values of $\sigma_c = 6.0$ MPa; $I = 3.0$ (for columnar ice); $k = 0.4$; $m = 0.9$, yields the loads as a function of thickness as shown in Table 1.

The third method is based entirely on measurements and experience with the Cook Inlet platforms (Blenkarn, 1970). Load measurements on a platform with leg diameters of 4.3m and ice 0.4m thick indicate typical effective ice pressure of about 0.8 MPa in the crushing mode of ice failure. An effective ice pressure of 1 MPa bounds all the steady ice loads. Dynamic peak loads were in some cases about twice the steady peak loads. On the basis of this experience, we will use a design ice pressure of 2.0 MPa. The ice loads are then as shown in Table 1 and are the lowest of the three methods. The concern with this approach relates to matching ice conditions, ice properties, driving forces etc. between Cook Inlet and other proposed sites for a structure.

3.5 Overall Comparison — Narrow Structures

Table 1 summarizes the horizontal ice loads which have been developed for narrow sloping and vertical structures. Figure 4 plots the loads as a function of ice thickness. The results clearly show lower ice loads on the conical structure. The concern with the conical structure surrounds the adfreeze situation. If conservative adfreeze strengths are used, the conical structure becomes less attractive especially in relation to the lowest load estimated for the vertical structure. On the other hand if the lower adfreeze loads are used (and it is believed they should be) then the cone has a clear advantage over the vertical structure. In addition it is perceived that the dynamics associated with the ice loads will be less severe on the conical versus the vertical structure.

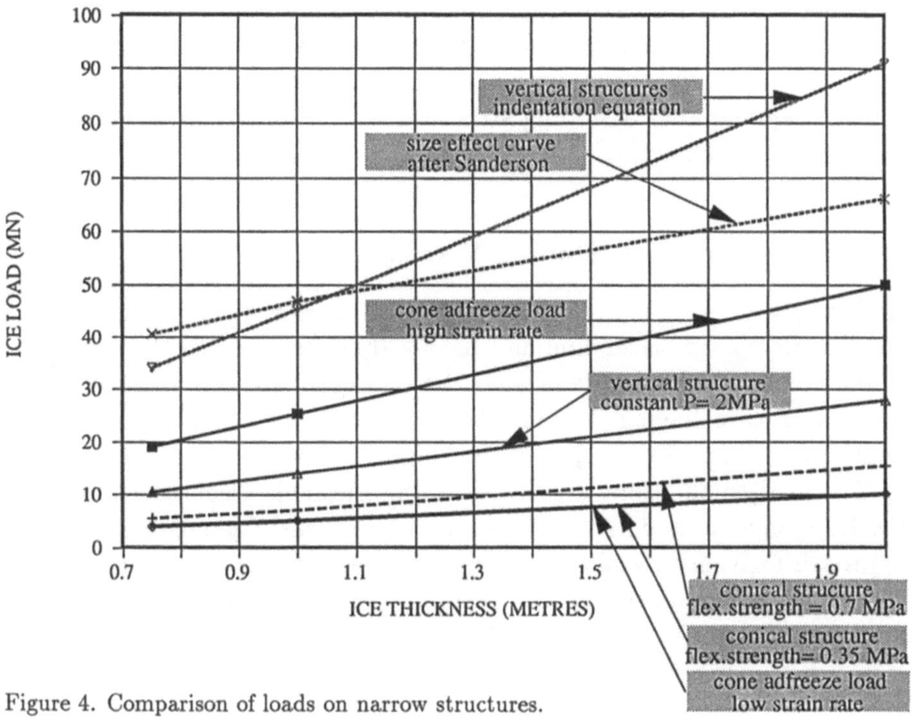

Figure 4. Comparison of loads on narrow structures.

3.6 Ice Load Dynamics

Moving ice acting on a fixed platform creates fluctuating ice loads. The fundamental issue is whether these cyclic loads create adverse vibrations of the platform. Recent experience has demonstrated that ice crushing against a platform can create unacceptable vibration (Jefferies and Wright, 1988). The nature of the vibration in ice crushing is very complex (Jordaan and Timco, 1988). The load build-up and decay involves not only ice failure, but extrusion of the crushed ice. Also it has been speculated that movement of the structure itself has an influence on the mechanics of ice failure. Experience and intuition indicates that load fluctuations are more severe and recurrence intervals between loads are shorter in the crushing mode than in the bending mode (Blanchet et al., 1988). This is because the distance between peak loads as the ice advances is much greater in flexural failure than for crushing failure. For flexural failure the distance might typically be five times the ice thickness, for crushing failure the distance is less than the thickness. For a 2m thick ice feature moving at $0.5ms^{-1}$, a typical frequency of peak loads would be about 0.05 Hz for flexural failure compared to about 1 Hz for crushing failure. Typical platforms and piers have natural frequencies closer to 1 Hz than to 0.05 Hz. Thus, from the point of view of dynamic excitation, it would appear that a sloping structure is generally more advantageous than a vertical structure.

4. Wide Structures: Sloping Versus Vertical

Let us now consider a typical arctic conical or sloping structure which is 150m wide at the water line and has a 20m freeboard to the top of the slope which is at an angle of 45°. We will compare this with a vertical faced structure having a diameter of 110m (which corresponds to the width at the top of the slope of the sloping structure - see Figure 5).

The load calculations are based on the methods discussed previously, i.e. Ralston's method for the sloping structure, and Sanderson's empirical approach for the vertical structure. However, a further method based on aspect ratio is also used for the vertical structure. This is the EIS curve for ice pressure (Marcellus, 1982). Because this is an arctic structure,

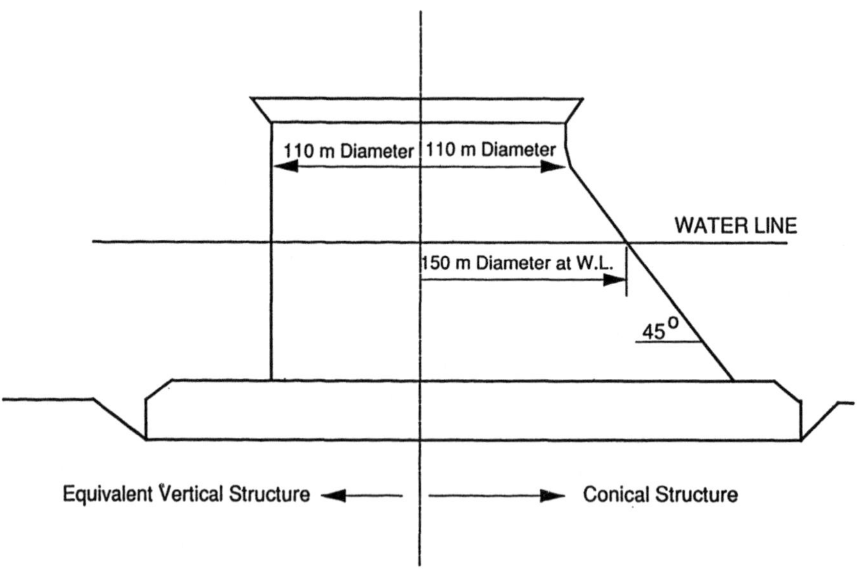

Figure 5. Comparative Arctic vertical and sloping structures.

Table 2. Comparison of Loads for Wide Structures.

	Horizontal Ice Force (MN)				
	Sheet				Ridge
	$t = 5$m	$t = 10$m	$t = 15$m	$t = 20$m	$t = 20$m
A Conical Structure (150m Dia. at WL; 110m Dia. at top; Angle= 45°; $\mu = 0.2; \sigma = 0.7$ MPa)	400	964	1650	2134	657
B Conical Structure (As above but half the ice strength)	328	783	1296	1726	≈ 480
C Vertical Structure 110m Dia. at WL					
(a) Using Side Effect Curve	480	622	825	946	946
(b) Using Aspect Ratio Curve	385	880	1650	3300	3300
(c) Constant 1.0 MPa ice pressure	550	1100	1650	2200	2200

ice thicknesses up to 20m will be considered. The load comparisons are given in Table 2 and plotted in Figure 6.

The results present the engineer with some difficult choices. In terms of the global loads, it appears that the conical structure offers some advantages but not significantly, especially at the extreme ice thickness of 20m. (For example; 2134 MN for the conical structure versus 2200 MN for the vertical structure; based on 1.0 MPa ice crushing pressure). In this situation the engineer has to weigh other factors in making a choice. Some of these are as follows:

- The conical structure will be more costly than the vertical structure.
- Severe vibrations may occur with ice crushing directly against the vertical structure; they are less likely with the conical structure.

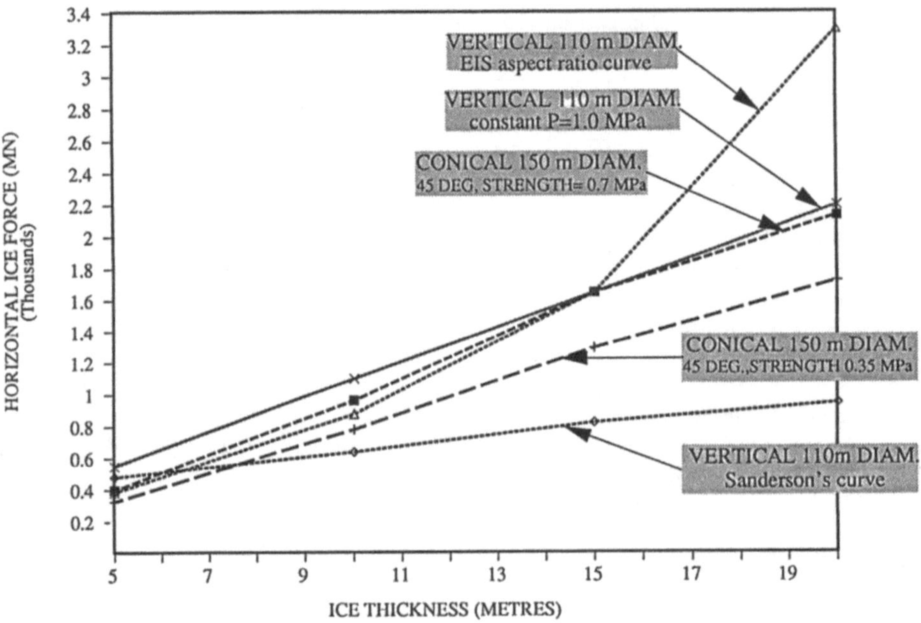

Figure 6. Comparison of loads on wide structures.

- If ice doesn't properly clear around the conical structure then the loads may be higher than those calculated.
- Set down of a conical structure is more difficult than a vertical structure.
- Wave loads should be lower on the conical structure.
- For the conical structure, a vertical force acts downwards in association with the horizontal ice load which could help foundation stability.

It should also be noted that, depending on water depth, an ice rubble field may form around a wide structure in the winter months.

4.1 Ridges

The above comparison should also be further qualified by two additional perspectives. First, if a narrow-necked cone is used instead of a wide-necked cone, the ride-up forces which are implicit in Table 2 will be lower. Second, if the extreme design ice feature is in the form of a linear ridge-like feature rather than a thick floe, then the advantages of a cone are enhanced. For example, typical ridge loads as derived by Winkler and Nordgren (1986) are compared with the equivalent crushing loads in the last column of Table 2. In this comparison, the advantages of the cone are more apparent. A significant question then arises which relates to the issue of what are the most critical design ice features for arctic conical structures. Much emphasis in the past has been on linear ridge-like features of extreme thickness. It may be necessary to look more closely at the loads which can arise from less thick but a really more extensive multi-year hummock fields.

5. Limiting Situations for Ice Loads and the Effects of Structure Geometry

So far in this paper it has been assumed that the ice loads are controlled by local interaction between the ice and the structure. This condition is known as limit-stress because local stresses in the ice control the load. However other limiting situations for ice loads can occur (Croasdale, 1984) and may control. These include limit-momentum (or energy) and limit-force. It is appropriate to consider how structure geometry influences these other limits to ice loads.

6. Effects of Structure Geometry on Iceberg Impact Loads on a Vertical Structure

This type of loading is an example of the limit-momentum (or energy) condition. In this situation the initial momentum or kinetic energy of the ice feature creates a limit to the ice load. In fact the work done by the ice load on the ice feature and the structure is equated to the initial kinetic energy of the ice in order to calculate the resulting load (Cammaert and Tsinker, 1981; Johnson and Nevel, 1985).

For a vertical structure in addition to the initial kinetic energy, the ice load generated by an impacting iceberg is also a function of the ice crushing strength, and the local interface geometry between the iceberg and the structure (Croasdale, 1983; Cammaert and Muggeridge, 1988). These parameters control the force/penetration curves as the ice interacts with the structure. The smaller the slope of the force penetration curve, the lower the ice load. This is because for a given work done (the area under the curve) a lower peak force is required. Schemes to flatten the force penetration curve are based on scalloped or wedge shaped geometries (Figure 7). A comparison of the force penetration curves for a typical example presented by Cammaert and Muggeridge is shown in Figure 8. In the example, the overall structure diameter is 106m, the ice-

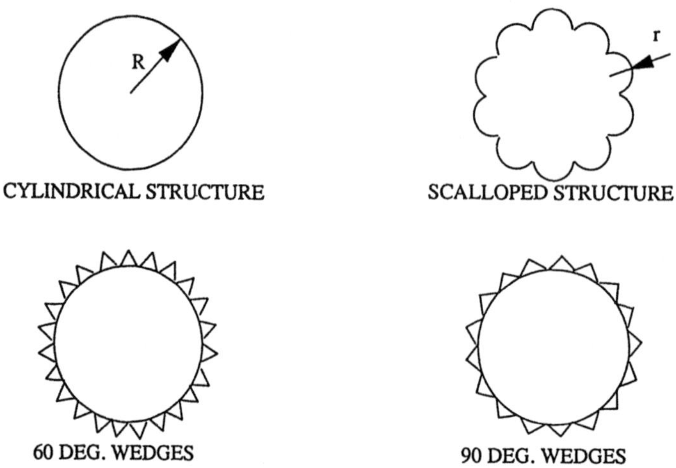

Figure 7. Local geometries, iceberg resistant structures.

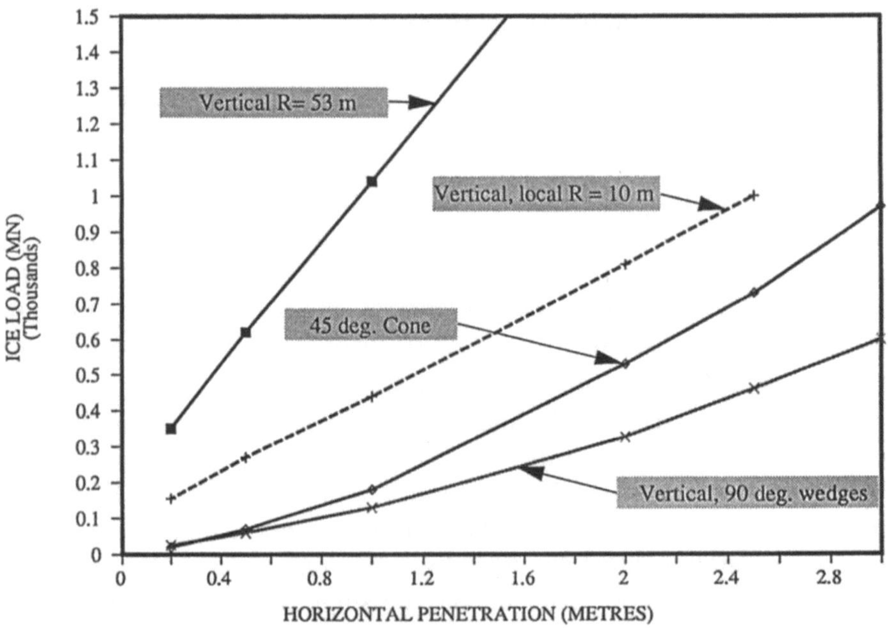

Figure 8. Force penetration curves for different shapes (after Cammaert and Muggeridge, 1988).

berg mass is 6 million tonnes, added mass coefficient is 0.5, speed 0.575 ms^{-1}, resulting in a kinetic energy of 1490 MJ. The calculation assumes a crushing strength of 7 MPa (constant) and initial vertical contact length of 5m increasing at a 60° angle top and bottom. Loads are calculated for three interface geometries, a plain radius of 53m, a locally scalloped radius of 10m, and vertical wedges with 60° points.

A comparison of global ice loads is given below:

	$R = 53$m	Local $R = 10$m	60° Wedges
Ice Load (MN)	1570	1050	810
Maximum penetration (m)	1.6	2.6	4.5

It can be seen that, theoretically for a head-on collision, the wedge geometry interface is advantageous. It should however be emphasized that the loads are also quite sensitive to the iceberg geometry which cannot be controlled. Also it is not clear how the wedges will perform when collisions are not head-on. It is also a fair comment to note that, if the average ice crushing strength was much lower than 7 MPa, the need for the ice wedges would be lessened. Other geometries which might be designed to resist iceberg impacts include a tiered caisson (Gerwick et al., 1984; Croasdale, 1985), a caisson protected by a rock berm, and a sloping structure (Croasdale, 1983).

In the context of a sloping structure, it is worth noting that the kinetic energy of the iceberg used in the previous example can be completely dissipated by raising the iceberg only 1.9m. To the authors' knowledge, to date, no formal analysis has been presented for the impact of an iceberg against a sloping structure. The remainder of this paper addresses this deficiency and compares the loads on a conical structure with those just reviewed for the vertical structure.

7. Iceberg Impact With a Sloping Structure

It is assumed for this analysis that the iceberg is rectangular and that the structure is conical. Variations on these geometries will be discussed later. The general arrangement and the interaction forces are shown in Figure 9.

The assumed sequence of events is as follows. As the iceberg crushes against the structure, a normal force N is generated; accompanying N is a friction force μN as the iceberg tries to ride up the structure. N and

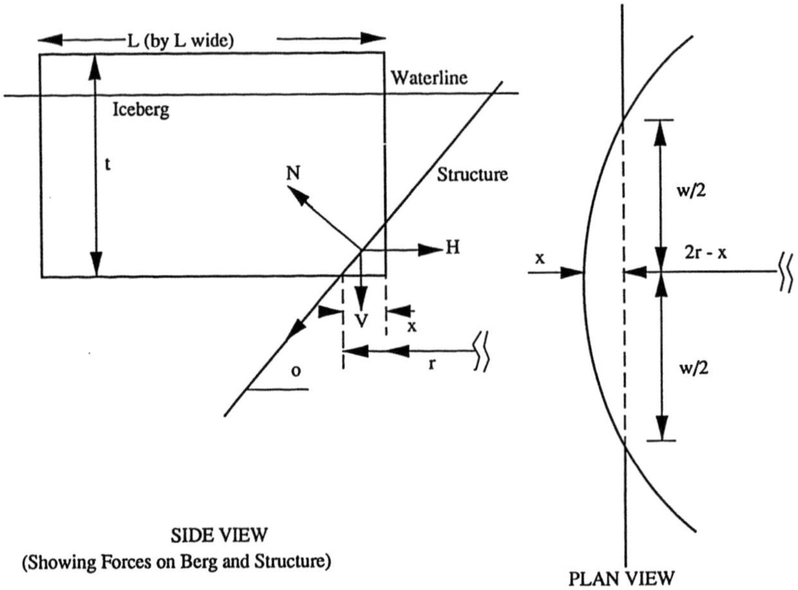

Figure 9. Forces and geometry for iceberg impact on sloping structure.

μN are the only interface forces generated on the iceberg and structure by the interaction. These can be resolved into horizontal and vertical forces (H and V). These forces will tend to lift and rotate the iceberg.

As H and V develop, work is done which dissipates the initial kinetic energy of the iceberg. In relation to the forces applied to the iceberg as depicted in Figure 10, the energy balance is written as

$$(1 + C_m)\frac{Mv^2}{2} = \int_0^{x_m} H dx + \int_0^{y_m} V dy + \int_0^{\theta} M d\theta, \tag{12}$$

where C_m is the added mass coefficient, M is the iceberg mass, V is velocity, x is horizontal penetration, y is uplift and θ is rotation. The forces and mass are given by

$$H = N \sin\alpha + \mu N \cos\alpha, \tag{13}$$

$$V = N \cos\alpha - \mu N \sin\alpha, \tag{14}$$

$$M = 0.5VL - 0.5Ht, \tag{15}$$

where α is the slope angle, μ is friction, L is iceberg length and width, and t is iceberg thickness.

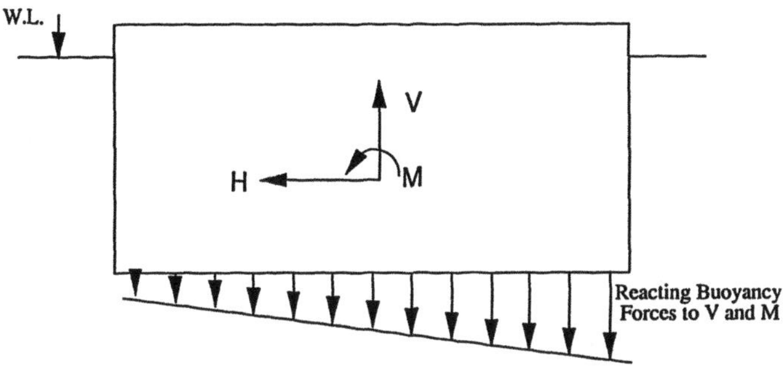

Figure 10. Forces acting on iceberg.

Obviously the development of N as the ice crushes against the structure is very dependent on the interface geometry. Initially this will be calculated for the geometry assumed here, variations on this will then be examined.

For the interaction between a rectangle and a cone, the relationship between contact area (A) and penetration (x) can be expressed as:

$$A = \frac{0.67xw\sigma}{\cos\alpha}, \tag{16}$$

where

$$w = \sqrt{8rx} \qquad (\text{ if } x \text{ is small compared to } r). \tag{17}$$

The normal force is then

$$N = \frac{x^{1.5}(3.56r)^{0.5}\sigma}{\cos\alpha}. \tag{18}$$

For $\sigma = 7$ MPa; $\alpha = 45°$; $r = 100$m, the above equation yields the following force penetration:

x	m	0.2	0.5	1.0	1.5	2.0	2.5	3.0
N	MN	16.7	66	187	343	528	739	971

This is also plotted on the force penetration curves previously derived for vertical structures (Figure 8); it lies between the two best curves for the vertical structure. Recognizing that this is an artifact of the assumptions for the iceberg shape, a factor (k) will be added to Equation (18) so that the effect of less advantageous force/penetration curves can be assessed.

Equation (18) becomes:

$$N = \frac{kx^{1.5}(3.56r)^{0.5}\sigma}{\cos\alpha}, \tag{19}$$

where k is a factor which is 1.0 for the base case. Now combining Equations (13), (14) and (19) yields

$$H = kx^{1.5}\sigma(3.56r)^{0.5}(\tan\alpha + \mu), \tag{20}$$

and

$$V = kx^{1.5}\sigma(3.56r)^{0.5}(1 - \mu\tan\alpha). \tag{21}$$

Work done by H (horizontally) (W_H)

$$W_H = \int_0^{x_m} kx^{1.5}\sigma(3.56r)^{0.5}(\tan\alpha + \mu)dx \tag{22}$$

$$W_H = 0.4kx_m^{2.5}\sigma(3.56r)^{0.5}(\tan\alpha + \mu) \tag{23}$$

Work done in uplift by V (W_v)

$$W_v = \int_0^{y_m} Vdy = \int_0^{x_m} V\frac{dy}{dx}dx. \tag{24}$$

There is a relationship between V and y which depends on buoyancy, i.e.

$$V = yL^2\rho_w. \tag{25}$$

Combining Equations (14), (24) and (25), and integrating, gives

$$W_v = \frac{1.78k^2x_m^3\sigma^2r(1 - \mu\tan\alpha)^2}{L^2\rho_w}. \tag{26}$$

Work done by M (W_m)

Assume rotation about the centre of mass and that the edges of the rectangular berg stay above the water line. Then the relationship between M and θ can be defined as

$$M = \frac{\rho_w\theta L^4}{12} \tag{27}$$

and

$$W_m = \int_0^{\theta_m} Md\theta = \int_0^{x_m} M\frac{d\theta}{dx}dx. \tag{28}$$

Combining Equations (15), (27), (28) and integrating gives

$$W_m = \frac{5.3k^2x_m^3\theta^2r}{\rho_wL^4}[L - L\mu\tan\alpha - t\tan\alpha - \mu t]. \tag{29}$$

Equation (12) can be rewritten

$$(1 + C_m)\frac{MV^2}{2} = W_H + W_v + W_m. \tag{30}$$

For any iceberg impact we can now solve Equations (23), (26), (29) and (30) to obtain x_m when the iceberg is brought to rest. Using this value of x, the peak values of V and H can be calculated using Equations (20) and (21).

This will now be done for the previous iceberg impact example but with a sloping structure having the characteristics $\alpha = 45°$; $\mu = 0.1$; $r = 100$m (at point of contact). Also the 6 million tonne iceberg is represented by a rectangle 80m thick and 290m by 290m in plan; $\sigma = 7$ MPa; the initial kinetic energy is 1490 MJ. Inserting the above values yields

$$1490 = 58kx_m^{5/2} + 19.6kx_m^3. \tag{31}$$

For $k = 1.0$, $x_m = 3.04$m, then $H = 773$ MN and $V = 632$ MN. The horizontal iceberg impact load calculated above is less than those previously calculated for the vertical structures (which were in the range 810 to 1570 MN). The above comparison is based on a 45° angle at the point of contact with thick icebergs. One advantage of the sloping structure for iceberg impact is that some optimization is possible with respect to this angle according to the depths and sizes of the expected extreme icebergs.

The base-case sloping structure with a constant 45° angle is illustrated conceptually in Figure 11. A possible optimization with shallow angles at deeper depths is also shown in Figure 11. Using the previously discussed methodology, a 30° angle at the −80 to −60m water depth yields $H = 504$ MN and $V = 701$ MN. This value of H is 38% lower than the best vertical faced design. The effects of other parametric variations are shown in Figure 12.

It has to be recognized however, as discussed previously, that the relative geometry between an impacting iceberg and structure will never be easily specified for a design situation. To examine the sensitivity to this issue, the factor k has been incorporated into this methodology. A value of $k = 1$ gives the idealized geometry between a rectangular iceberg and a cone. By increasing k, we can examine the sensitivity of the loads to an interface geometry which is more severe; i.e. the load/penetration curve has a steeper slope.

A sensitivity with $k = 1.5$ and 2.0 as well as other parametric variations are presented in Table 3. In addition to the issue of interface

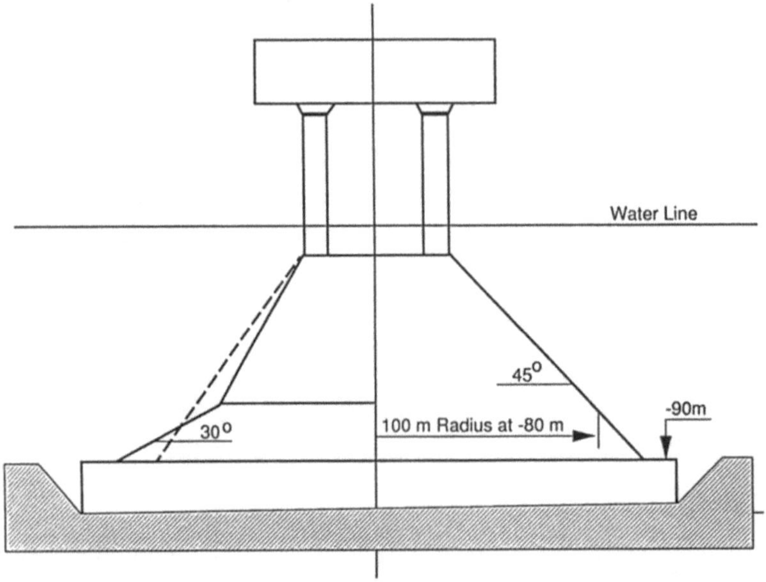

Figure 11. Possible conical structure arrangements for iceberg infested areas.

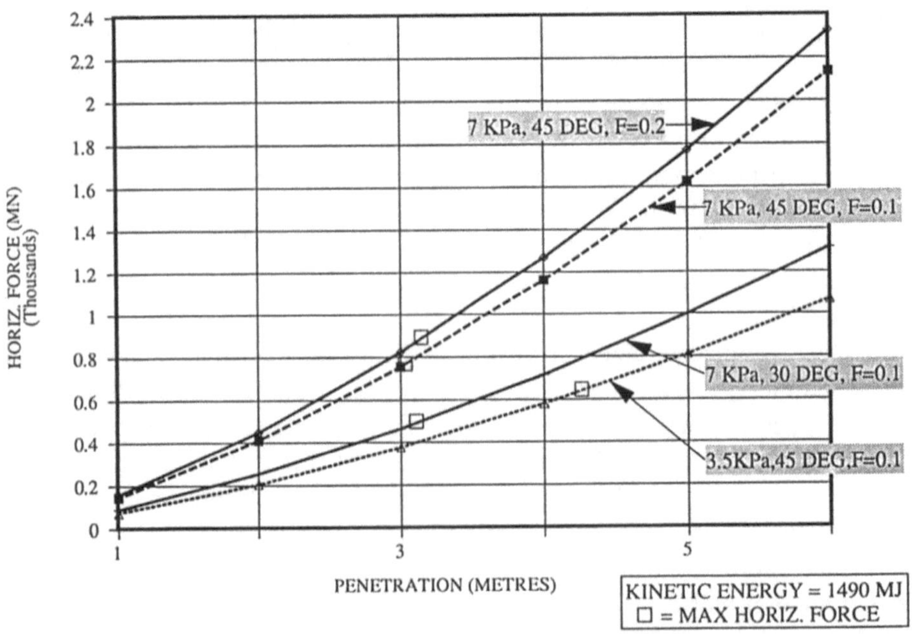

Figure 12. Horizontal force versus penetration conical structures.

Table 3. Icebergs on Sloping Structures: Parametric Variations

			Horizontal Load (MN)		
σ MPa	α ^{o}	μ	$k = 1.0$	$k = 1.5$	$k = 2.0$
7	45	0.1	773	845	894
7	30	0.1	504		
7	30	0.2	591		
7	45	0.2	871		
3.5	45	0.2	709		
3.5	45	0.1	645	720	773
3.5	30	0.1	445		
3.5	30	0.2	511		

geometry between iceberg and structure, other simplifying assumptions have been made in the method presented. For example inertia effects have been neglected (this is implicit in the energy method). Other simplifications relating to the effect of uplift and rotation on penetration, and the rotational resistance of non-rectangular bergs, are considered to be conservative and are areas for future refinement of the method which should yield lower forces.

We can conclude this section on structures to withstand iceberg impacts by noting that sloping structures can be designed which appear to result in lower global ice loads than for vertical structures. Intuitively we might expect that a sloping structure will be less sensitive to the specific interface geometry between the iceberg and the structure. This is because independent of any energy dissipation by ice crushing, substantial energy sinks exist in relation to raising and rotating the iceberg.

8. Limiting Driving Force

This limit to ice forces has been proposed and reviewed by one of the authors (e.g. Croasdale, 1981, 1984). In this situation, with thick ice and large structures, there is insufficient driving force on the thick ice feature (from the surrounding thin ice) to cause local ice failure. Clearly, the structure geometry has no influence on this limit to ice loads.

9. Conclusions and Recommendations

This paper has considered how structure shape affects ice failure modes, clearing processes and ice loads. It has also attempted to provide the engineer with specific guidance on how to compare vertical and sloping structures (from the perspective of ice interaction). For narrow structures, the conical shape appears to offer the advantages of lower ice forces and less severe dynamic excitation than the equivalent vertical structure. An often expressed concern for sloping structures relates to adfreeze loads. This paper has examined the creep failure of the adfreeze bond. This is considered a more realistic loading scenario than the high strain rate case which previous investigators have addressed, and results in low adfreeze forces. It should be noted that adfreeze forces can also be minimized by special coatings. It is recommended that further work on adfreeze be undertaken (including large scale basin tests) in order to confirm that adfreeze loads can be minimized.

For wider arctic structures, the advantage of a conical structure over a wide vertical structure is not as clear-cut. On the one hand, load levels are generally lower, and dynamic fluctuations less severe for conical structures. On the other hand, depending on the surrounding water depth, the presence of a grounded rubble field in the winter could diminish these advantages. Also the issue of thick ice riding up and striking the deck and superstructure of a conical structure needs to be considered.

In the context of iceberg impacts, a conical structure appears to offer the advantage of lower ice loads which are also less sensitive to the actual geometry of the ice/structure interface. This is because the energy sinks of uplift and rotation of the iceberg are also provided. Further work on this topic is recommended and would include an investigation of hydrodynamic and inertia effects. Model testing at as large a scale as possible is recommended. It should also be noted that a sloping structure can be configured to give smaller wave loads than a massive cylindrical structure. On the other hand the conical structure may be more costly to build and the installation can be more difficult.

In conclusion, it appears that sloping structures do offer many advantages over vertical structures; research to better understand ice-structure interaction on sloping structures and specifically the issues identified in this paper should continue.

References

Arctec Canada Limited. 1975. Preliminary assessment of the adhesion shear strength of ice-steel and ice-frozen sand bonds. APOA Project 85, Montreal.

Bercha, F.G. and Danys, J.V. 1975. Prediction of ice forces on conical offshore structures. Proc. IAHR Ice Symposium, Hanover, NH. pp. 447-458.

Blanchet, D., Churcher, A., Fitzpatrick, J. and Badra-Blanchet, P. 1988. An analysis of observed failure mechanisms for laboratory, first-year and multi-year ice. Proc. IAHR Ice Symposium, Sapporo, Japan. Vol. 3. pp. 89-136.

Blenkarn, K.A. 1970. Measurement and analysis of ice forces on Cook Inlet structures. Proc. OTC., Houston, TX, Vol. 2, Paper 1261. pp. 365-378.

Cammaert, A.B. and Tsinker, G.B. 1981. Impact of large ice floes and icebergs on marine structures. Proceedings of the Sixth International Conference on Port and Ocean Engineering under Arctic Conditions. Quebec City, Canada. Vol. 2. pp. 653-667.

Cammaert, A.B. and Muggeridge, D.B. 1988. Ice interactions with offshore structures. Van Nostrand Reinhold, New York.

Cammaert, A.B., Kimura, T., Noma, N., Yashima, N., Yano, S., and Matsushima, Y. 1986. Adfreeze forces on offshore platforms. Proc. Int. OMAE Symp., Tokyo, Japan, Vol. 4. pp.541-548.

Croasdale, K.R. 1980. Ice forces on fixed rigid structures. In Working Group on Ice Forces on Structures, A State-of-the-Art Report, CRREL Special Report 80-26. Hanover, N.H.

Croasdale, K.R. 1983. The present state and future development of arctic offshore structures. Seventh POAC Conference, VTT, Helsinki, Finland. pp. 489-518.

Croasdale, K.R. 1985. Arctic fixed platforms for water depths greater than 50 meters. Proceedings of the First Spilhaus Symposium, Arctic Engineering for the 21st Century, B.C. Gerwick (ed.), The Marine Technology Society, Williamsburg, Virginia. pp. 169-180.

Croasdale, K.R. 1987. Ice loads: experience, methods and remaining issues. Proc. Symp. on Design of Offshore Struct. for Canadian Frontiers, London, Ontario. pp. 63-72.

Croasdale, K.R., and Person, A. 1984. A logical approach to ice loads. Presented at the Second Global Marine Arctic Offshore Symposium, Houston, Texas, 1984.

Croasdale, K.R., Metge, M. and Verity, P.H. 1978. Factors governing ice ride-up on sloping beaches. Proc. IAHR Ice Symp., Pt. 1, Lulea, Sweden. pp. 405-420.

Edwards, R.Y. and Croasdale, K.R. 1976. Model experiments to determine ice forces on conical structures. Symp. on Applied Glaciology, Int. Glaciological Society, Cambridge, U.K.

Frederking, R.M. and Timco, G.W. 1985. Quantitative analysis of ice sheet failure against an inclined plane. Proc. Int. OMAE Symp., Vol. 2, Dallas. pp. 160-169.

Gerwick, B.C., Potter, R.E. and Rojansky, M. 1984. Development of a structural concept to resist impacts from multi-year floes, ridges and icebergs. Offshore Technology Conference, Houston, TX, Vol. 3, Paper 4799. pp. 193-201.

Jefferies, M. and Wright, B. 1988. Dynamic response of Molikpaq to ice structure interaction. Seventh OMAE Int. Conference, Houston, Texas, Vol. IV., pp. 201-220.

Jordaan, I.J., and Timco, G.W. 1988. Dynamics of the ice crushing process. J. of Glaciology, Vol. 34, No 118, pp. 318-326.

Marcellus, R.W., Morrison, T.B., Allyn, N.F., Croasdale, K.R., Iyen, H.S. and Tseng, J. 1988, 1989. Ice forces on marine structures. Volumes I and II, for Public Works Canada.

Marcellus, R.W. and Morrison, T.B. 1982. Ice design statistics for the Canadian Beaufort Sea. CMEL Report 1015 prepared for the EIS partners.

Oksanen, P. 1983. Adhesion strength of ice. Proc. Int. Conf. POAC, Vol. 2, Helsinki. pp. 710-719.

Ralston, T.D. 1979. Plastic limit analysis of sheet ice loads on conical structures. Proc. IUTAM Symp., Copenhagen, Denmark. pp. 289-308.

Sackinger, W.M., and Sackinger, P.A. 1977. Shear strength of the adfreeze bond of sea ice to structures. Proc. Int. Conf. POAC, Vol. 2, St. John's, Newfoundland, Vol. 2. pp. 607-614.

Sanderson, T.J.O. 1988. Ice mechanics, risks to offshore structures. Graham and Trotman, London.

Semeniuk, A. 1975. Computer program to evaluate the forces generated by a moving ice field encountering a conical structure. APOA Project No. 87, Calgary.

Discussion

Q: A. Prodanovic, Mobil R & D Corp.

Comments on cone structures against icebergs.

1. By providing a slope/ramp we increase the possibility of iceberg flowing up onto deck, which must be avoided.

2. Vessels are a risk, can't see a submerged cone.

3. Tow-out hydrostatic consideration (decreased waterplane area).

A: K. Croasdale

1. As mentioned in the paper, an iceberg can be stopped by raising it only 1 - 2m. Also rolling of an iceberg can presumably occur even

during impact with a vertical structure (due to imbalance caused by damage and failure).

2. and 3. I agree that these are disadvantages which have to be considered in balancing the pros and cons of a sloping versus a vertical structure.

Q: M. Määttänen, Helsinki University of Technology

As the author proposed, adhesion is not a decisive design criteria in conical structures due to strain rate effects. Our in-field data of Kemi-1 test cone, with a 10m diameter at waterline, confirms the point. Ice, including a significant pile-up, was stationary for a week at temperatures well below freezing point, after which a storm made the ice mass move. The initial adhesion break up did not cause the highest loads.

Q: I. Jordaan, Memorial University of Newfoundland

In view of experience in the Beaufort Sea on actual structures, do you feel that it might be justified to reduce load requirements for the adfreeze condition?

A: K. Croasdale

Intuitively, we agree, however we don't know of any documented codes of the adfreeze bond being broken against sloping structures in the Beaufort Sea. For vertical structures it would seem that high break-out loads (analogous to the adfreeze condition) have not been experienced; which perhaps support the hypothesis that ice movement events are preceded by low-strain rate movements.

Q: A.C. Palmer (comment on Määttänen's discussion)

Tide must be a factor that helps to break the adfreeze bond. Professor Määttänen observes that there is little adfreeze effect in the Baltic, which has almost no tide, and so there ought to be still less effect elsewhere.

A: K. Croasdale

We agree.

Q: R.P. Nordgren, Shell Development Co.

Consideration of multi-year ridges would be of interest in the comparison of structures for the Beaufort Sea. Foundation considerations also are relevant.

A: K. Croasdale

Certainly, thick multi-year ridge interaction would likely favour a sloping structure (even a wide one) over a vertical structure, and we agree that the paper should have examined this loading condition.

Foundation conditions are certainly one of the ingredients in the decision relating to choice of structural shape. Factors such as ice load dynamics, and the vertical components of ice loads need to be included in foundation considerations. For poor foundation areas and extremely thick ice ridges, the sloping structure would probably be advantageous.

Dynamics of Continuous Icebreaking by an Icebreaker Hull

Robert Ettema and Frederick Stern

Iowa Institute of Hydraulic Research and College of Engineering
The University of Iowa
Iowa City, Ia, USA

Abstract An overview is presented of the results from an ice towing tank study conducted in order to document the relationships between icebreaking pattern, resistance, and hull motions for a Polar-Class icebreaker model hull moving continuously through sheets of level ice. Two sets of experiments were conducted: for one, the hull was free to pitch, heave, and undergo limited roll (free hull); for the other, the hull was restrained from motions (fixed hull). For each condition, the hull was towed at the same range of speeds through urea-ice sheets of similar properties. Measurements were made of resistance as well as motions and restraining forces and moments for the free- and fixed-hull conditions, respectively. The temporal records were analyzed to yield values of their mean and standard deviation and spectral densities. A series of experiments was conducted to determine values of zero-speed natural frequencies and added-mass and damping coefficients for ice-covered conditions. The presence of an ice sheet significantly reduces the natural frequencies and increases the added-mass and damping coefficients. It was found that hull motions and icebreaking pattern determined the dominant cycles of resistance for the free hull. The fixed hull responded at higher frequencies than the free hull. The mean resistance was larger for the free hull than the fixed hull. The relationship between mean resistance and hull speed was linear for the fixed hull, but not for the free hull, which was characterized by a larger and similar slope at low and high speeds, respectively, than for the fixed hull. The speed range at which the slope changes coincides with that at which the motions are maximum. This correlation indicates that the ice-induced hull motions of the free hull are responsible for its increased resistance over that of the fixed hull.

1. Introduction

A ship hull, continuously breaking its way through a level sheet of ice, comprises a complex nonlinear dynamic system with six degrees of freedom that as yet is insufficiently understood. Especially ill-defined are the relationships between resistance, icebreaking pattern, and hull motions.

S. Jones · R. F. McKenna · J. Tillotson · I. Jordaan (Eds.)
Ice-Structure Interaction
IUTAM-IAHR Symposium St. John's, Newfoundland, Canada

In order for these relationships to be documented, an extensive program of experiments was conducted using an ice towing tank, urea model ice, and a scale-model Polar-Class hull (Figure 1), the largest class of icebreaker operated by the U.S. Coast Guard (USCG). The program consisted of two sets of experiments. For the free-hull condition, the hull was instrumented so that its pitch, heave, and roll motions (both displacements and accelerations) could be measured, together with resistance. For the fixed-hull condition, the hull was rigidly restrained from motions, and the restraining forces and moments were measured, together with resistance. The latter set of experiments was conducted for the purpose of relating icebreaking and resistance without the influence of hull motions. Comparison of the two sets of experiments would aid in determining the influences on resistance and icebreaking pattern of hull motions. For each condition, the hull was towed at the same range of speeds through tank-long ice sheets of similar properties (thickness, strength, etc.). Temporal records of resistance, hull motions, and motion-restraining forces and moments were analyzed to yield values of their mean and standard deviation and spectral densities. Herein, an overview of some of the results is presented and discussed. The complete results and further discussions are provided by Ettema et al. (1987, 1989) and Stern et al. (1989).

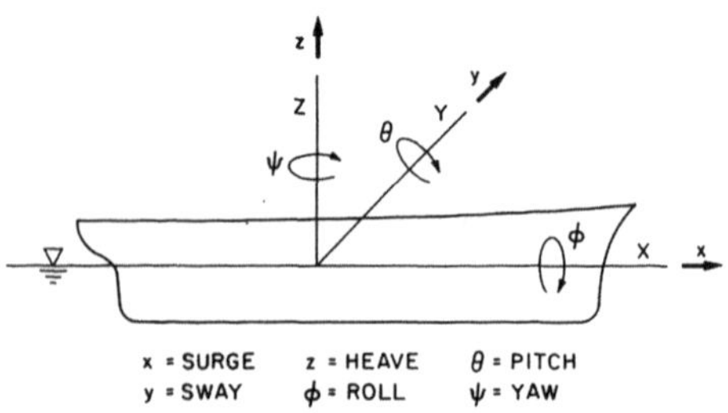

Figure 1. Abbreviated lines of the USCGC Polar-Class hull.

2. Equations of Motion

Consider an icebreaker hull transiting an ice sheet in continuous-mode icebreaking. Hull speed as well as the ice-sheet properties are assumed constant. Under certain conditions, motions are of sufficient amplitude that nonlinear effects are important. However, as a first approximation and to provide a framework for discussion, hull motions are assumed to be linear such that the linear superposition principle is applicable; i.e., the response at a given frequency depends only on the excitation at that frequency.

2.1 General Equations of Motion.

Let (X, Y, Z) be a right-handed coordinate system fixed with respect to the mean position of the ship, with Z vertically upward, X in the direction of forward motion, and the origin at the hull's center of gravity. The coordinate system and translatory and angular displacements are shown in Figure 2. Under the assumptions that the responses are linear, six linear coupled differential equations of motion can be written in the form

$$(M_{mn} + A_{mn})\ddot{q}_m + B_{mn}\dot{q}_m + C_{mn}q_m = f_m, \tag{1}$$

where $q_m = (x(t), y(t), z(t), \phi(t), \theta(\tau), \psi(t))$ are the responses corresponding to surge, sway, heave, roll, pitch, and yaw, respectively; M_{mn} is the generalized-mass matrix; A_{mn} and B_{mn} are the added-mass and damping coefficients, respectively; C_{mn} are the hydrostatic-restoring coefficients; and $f_m = (f_x, f_y, f_z, m_\phi, m_\theta, m_\psi)$ are the excitation forces and moments.

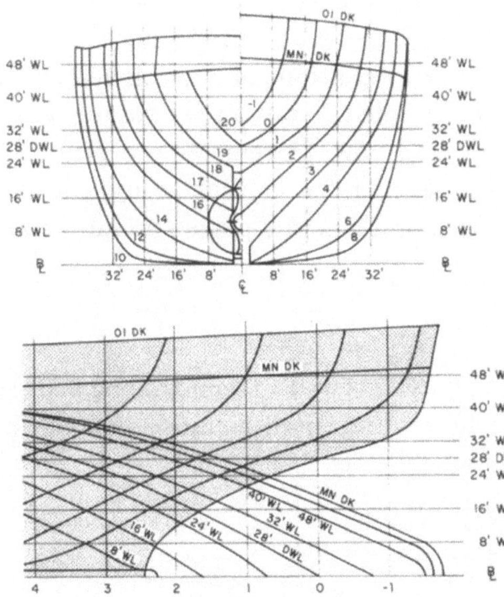

Figure 2. Definition sketch.

If both the hull geometry and loading are laterally symmetric, the coefficient matrices $(M_{mn}, A_{mn}, B_{mn}, C_{mn})$ are of such a form that the equations of motion for surge, heave, and pitch are uncoupled from those of sway, roll, and yaw. For a hull in open water, the hydrodynamic loading is, in fact, laterally symmetric. For a hull moving through an ice sheet, the distributions of ice loads at any instant are not necessarily laterally symmetric. This means that A_{mn} and B_{mn} are full matrices resulting in complete coupling between all six modes of motion. As an approximation, it is assumed that loading is symmetric for a hull moving through an ice sheet.

The only nonzero hydrostatic-restoring coefficients C_{mn} are

$$\begin{aligned} C_{33} &= \rho g A_{wp} \\ C_{44} &= \rho \mathcal{V} \overline{GM} \phi \\ C_{55} &= \rho g I_{wp} \\ C_{35} &= C_{53} = -\rho g M_{wp}, \end{aligned} \tag{2}$$

in which A_{wp}, M_{wp}, and I_{wp} are the area, moment, and moment of inertia of the waterplane, respectively; $\overline{GM}$ is the metacentric height; and $\mathcal{V}$ is the displaced volume. The values of the coefficients C_{33}, C_{55}, and C_{35} were experimentally determined for the hull to be 7950 N/m, 2450 Nm and -7.18 N, respectively.

If the appropriate generalized-mass matrix, the damping coefficients, and the hydrostatic-resorting coefficients are substituted into the equations of motion (1) then the six coupled equations of motion reduce to two sets of equations; one set of three coupled equations for surge, heave, and pitch, and another for sway, roll, and yaw. Herein, we are primarily concerned with the former. However, during the present study, roll was found, under certain conditions, to be significant and played an important role in the dynamics of icebreaking.

2.2 Pitch and Heave Motions.

The experiments involved a hull that was towed at constant speed, so that $\ddot{x} = \dot{x} = 0$. Consequently, the heave and pitch equations of motion have the form

$$\begin{bmatrix} M + A_{33} & A_{35} \\ A_{53} & I + A_{55} \end{bmatrix} \begin{bmatrix} \ddot{z} \\ \ddot{\theta} \end{bmatrix} + \begin{bmatrix} B_{33} & B_{35} \\ B_{53} & B_{55} \end{bmatrix} \begin{bmatrix} \dot{z} \\ \dot{\theta} \end{bmatrix}$$

$$+ \begin{bmatrix} C_{33} & C_{35} \\ C_{53} & C_{55} \end{bmatrix} \begin{bmatrix} z \\ \theta \end{bmatrix} = \begin{bmatrix} f_z(t) \\ m_y(t) \end{bmatrix}. \tag{3}$$

The solution to (3) for harmonic motions and excitations, i.e.

$$q_m = \tilde{q}_m(\omega)e^{i\omega t}, \qquad f_m = \tilde{f}_m(\omega)e^{i\omega t}, \tag{4}$$

is well known (e.g., Meirovitch, 1975) and is generally written as

$$\tilde{q}_m(\omega) = H_{mn}(\omega)\tilde{f}_m(\omega), \tag{5}$$

where $\tilde{q}_m$ and $\tilde{f}_m$ are the complex motion and excitation amplitudes, respectively, and H_{mn} is the transfer function defined as

$$H_{mn} = \begin{bmatrix} -\omega^2(M + A_{33}) + B_{33}\omega i + C_{33} & -\omega^2 A_{35} + B_{35}\omega i + C_{35} \\ -\omega^2 A_{53} + B_{53}\omega i + C_{53} & -\omega^2(I + A_{55}) + B_{55}\omega i + C_{55} \end{bmatrix}^{-1}. \tag{6}$$

When the hull is free to heave and pitch, $\tilde{q}_m(\omega)$ is known because the hull's heave and pitch were measured. In order to determine the forces causing the motions, it is necessary to determine the excitations. This is accomplished by inverting (5); i.e.,

$$\tilde{f}_m(\omega) = H_{mn}^{-1}(\omega)\tilde{q}_m(\omega), \tag{7}$$

which can be solved when the transfer function H_{mn} (6) is known. The terms in H_{mn} depend on ω, A_{mn}, B_{mn}, and C_{mn}. Generally, ω and C_{mn} are known, but the added mass A_{mn} and damping B_{mn} must be determined.

For a hull in open water, potential-flow methods are available for predicting A_{mn} and B_{mn}. Little is known, and no theory is available, for predicting the influences of ice-sheet presence on A_{mn} and B_{mn}. The influences are probably large because the surrounding ice sheet and distribution of ice pieces near the hull strongly influence the flow field, including the wave pattern. In the present study, the added-mass and damping coefficients were determined experimentally.

2.3 Natural Frequencies and Added-Mass and Damping Coefficients.

A series of experiments was conducted in order to determine values of natural frequencies and added-mass and damping coefficients associated with hull motions during icebreaking. Generally, the Froude number F_n $(= V/\sqrt{gL})$ for an icebreaker hull is small, typically less than about 0.19, and for the open-water condition, zero-speed values are a good approximation. Herein, zero-speed values are used, although it should be recognized that the influence of forward speed, even for small F_n, for the ice-covered condition may be significant. Ice-sheet thickness is another factor influencing natural frequencies and added-mass and damping coefficients.

The usual experimental procedure for determining natural frequencies and added-mass and damping coefficients involves forced-oscillation tests. However, the equipment required for this was not available. Instead, three free-vibration experiments were performed: (i) the hull was restrained from heaving and the bow was displaced downwards and then released such that the hull oscillated in pitch only; (ii) the hull was restrained from pitching and displaced downwards such that the hull oscillated in heave only; and (iii) the hull was unrestrained and displaced such that it both heaved and pitched. The tests were conducted both for open water and for two thicknesses of ice, $h = 30$ and 38mm, when the hull was amid, and overrode, broken ice, in a track opened after a resistance experiment. For each experiment, the pertinent hull motions were measured and expressed as

$$q_m(t) = \tilde{q}_{mn} e^{s_n t}, \tag{8}$$

in which $\tilde{q}_{mn}$ is the amplitude and $s_n = \xi_n + i\omega_n$; with $\xi_n =$ damping rate, and $\omega_n =$ frequency of oscillation. For experiments (i) and (ii), $n = m$ in (8).

The procedure is straight-forward for experiments (i) and (ii); i.e., (8) is first used to evaluate s_n and then substituted into (3) to determine A_{33}, B_{33} and A_{55}, B_{55}, respectively. However, experiment (iii), which determines the s_n for coupled motion and $A_{35}(= A_{53})$ and $B_{35}(= B_{53})$, requires some further discussion. In general, as indicated by (8), the coupled heave and pitch responses include contributions from both natural frequencies. Thus, the evaluation of the s_n, although possible, is more complicated. Fortunately, however, the initial displacements used for experiment (iii) resulted in synchronous harmonic oscillations in both heave and pitch (i.e., for experiment (iii), $n = m$ in (8) also). Therefore, the procedure for experiment (iii) was similar to that described above for experiments (i) and (ii). No doubt, these experiments involved some error in the determination of the natural frequencies and added-mass and damping coefficients; however, for the present purposes, they are of sufficient accuracy.

The values of ξ_n and ω_n and A_{mn} and B_{mn} determined from these experiments are summarized in Tables 1 and 2, respectively. Also shown for comparison in Table 2 are open-water values predicted using Lewis forms. As might be expected, the presence of an ice sheet significantly reduces the natural frequencies of heave and pitch. Consistent with this, the added-mass and damping coefficients increase substantially. As shown in Table 1, the values of the heave and pitch natural frequencies for coupled motion are quite close; therefore, in the following discussions the natural frequencies for coupled heave and pitch are simply referred to as ω_n which designates the average value of 0.75 Hz.

Table 1
Measured values of damping rates and frequencies of oscillation

Ice-Sheet thickness (mm)	Pure Pitch frequency (ω) (Hz)	Pure Pitch damping (ξ)	Pure Heave frequency (ω) (Hz)	Pure Heave damping (ξ)	Pitch-heave frequency (ω) (Hz)	Pitch-heave frequency (ω) (Hz)
*0	1.09	0.47	0.91	0.22	1.00	1.00
30	0.83	1.3	0.66	1.21	0.76	0.74
38	0.83	1.3	0.66	1.21	0.74	0.72

*Open Water

Table 2
Added-mass and damping coefficients

	"Ship amid ice" (experimental)	"Open water" (Lewis forms)
A_{33}	250.80	80.31
A_{55}	72.23	16.75
A_{53}	38.20	2.57
B_{33}	714.07	
B_{55}	265.99	
B_{53}	128.22	

2.4 Roll Motion

Ice loads are not always laterally symmetric because icebreaking may not always occur simultaneously around the bow. This induces asymmetric responses, i.e., sway, roll, and yaw. The equations governing these motions can be readily obtained in a similar manner as for coupled surge, heave, and pitch. The hull used in the experiments was instrumented such that sway was restrained completely and yaw was confined within small margins, but not measured, and only roll was measured in several tests to determine its influence. Initially, measurements of roll were not planned.

However, during the course of the experiments, under certain conditions, large-amplitude roll motions were observed. As roll undoubtedly plays an important role in resistance, it was decided that some limited measurements of it should be made. As an aid in identifying frequencies related to roll, the zero-speed open-water natural frequency of roll was experimentally determined to be 0.45 Hz. The model was heeled to an angle and released such that the hull oscillated in roll only. The influences on roll natural frequency of ice and yaw motion were not determined.

3. Experiments

The free hull was instrumented using a dynamometer for measurement of resistance, and with two linear-voltage-displacement potentiometers located on plates spaced 1.9 m apart for measurement of pitch and heave. Pitch and heave accelerations were determined using accelerometers mounted 1.8 m apart. Yaw and roll were limited by a guide located near the stern. The limited roll measurements were made using a helipotentiometer and a cable attached to one side of the hull's midsection. The fixed hull was rigidly connected to a multicomponent load cell that enabled measurement of surge (resistance) and heave restraining forces and pitch and yaw restraining moments. The model was towed at a range of constant speeds (0.05 to 0.9 m/s) through sheets of urea (model) ice grown to thicknesses varying from 10 to 40 mm, and with flexural strengths of 20 to 25 kPa. Transits of the hull were recorded on videotape. Photographs were used to record icebreaking patterns, and the dimensions of ice cusps in the channel were measured with calipers.

4. Icebreaking Patterns

For the free hull, continuous-mode icebreaking began with the bow depressing and nosing onto the ice sheet, which increased the hull's pitch and heave. Simultaneously, the resistance increased, as ice crushed and sheared at the bow stem, and as the bow slid against the ice sheet. Sequentially, ice was flexurally broken around both sides of the bow. A pattern of icebreaking was initiated when a circumferential, or arced crack developed and propagated toward the bow's stem and shoulder, reaching the latter first because of the forward motion of the hull, thus forming an ice cusp. A major crack would develop either side of the bow, but not always simultaneously. Generally, the crack length was equal to the average waterline distance from the stem to the shoulder, for the condition of hull trim variation during which the crack formed. However, it varied slightly with ice-sheet thickness and strength, and with speed. Radial cracks developed through each cusp as the bow commenced over-riding it. After the first flexural icebreaking event, resistance, as well as pitch and heave momentarily decreased.

The hull's pitch may not diminish completely due both to the influence of the dynamics of hull motions and the fact that ice cusps freshly broken have to be submerged and displaced beneath the bow. Consequently, during a subsequent flexural-breaking event, the bow may impact the ice sheet with altered contact perimeter or geometry. For a Polar-Class hull, the contact perimeter diminishes (Figure 1) with increasing pitch and heave, such that a subsequent cusp formed with a lesser length, but of the same width as the first cusp. Usually, the failure of this second, smaller cusp, together with the submergence and displacement of the first cusp, was accompanied by the hull's pitch abruptly decreasing. A third cusp also formed, as shown in the general ice breaking pattern presented as Figure 3. The icebreaking patterns for the fixed hull were similar, but more uniform. If either the free or fixed hull, were moving with creeping speed, $V = 0.05$ m/s, the depressed ice sheet would be flooded prior to flexural failure. With increasing speed, the time available for flooding would diminish such that, for moderate and higher speeds, say $V = 0.5$ m/s, there was insufficient time for the depressed ice sheet to flood. For these speeds, the hull depressed both the sheet and the water level around the bow.

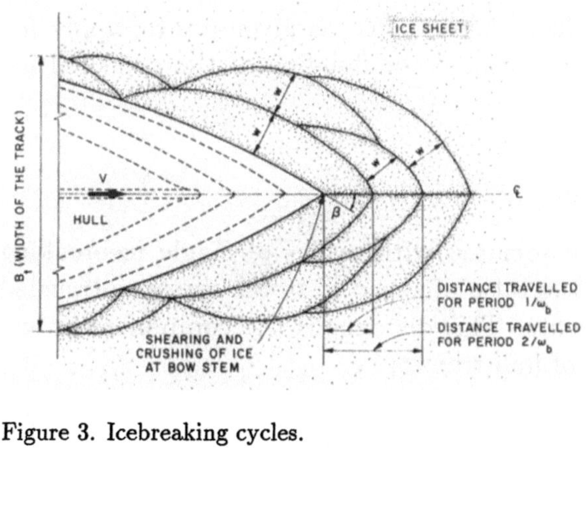

Figure 3. Icebreaking cycles.

Measured values of cusp width W for both the free and fixed-hull experiments are given by Ettema et al. (1987, 1989). It was found that the ratio W/h ranged from about 3 to 5, for speeds of 0.5 m/s and above, to about 7 for creeping speeds of 0.05 m/s. The variation of W/h with V can be attributed both to the hydrodynamics of water displacement during ice-sheet deflection and to altered geometry of loading associated with the simultaneous shearing, crushing, and flexing of the ice sheet. Cusp widths produced by the fixed hull did not vary significantly from those of the free hull. However, the track contained a more uniform pattern of broken ice. This is attributed to the steady trim maintained by the fixed hull. Icebreaking occurred in a more regular pattern at the bow of the fixed hull, although some irregularity was inevitable due to slight nonuniformities in ice-sheet properties along the length of the tank.

It is common practice to relate W to a characteristic length of ice sheet, $\ell_c = \{Eh^3/[12\rho g(1-\nu^2)]\}^{0.25}$. Generally, values of W/ℓ_c were less than one, ranging from about 0.8 to 0.4, decreasing with increasing speed V. Values of W/ℓ_c did not vary with hull condition. The effects on W and, thereby, on the icebreaking pattern, of h and V are explainable in terms of the flexure of a semi-infinite plate or, more simply, a semi-infinite beam on an elastic foundation (water). Through consideration of the locus of the maximum stress in an edge-loaded plate (e.g., loaded by an icebreaker bow), it can be shown that W increases with h. The relationship between W and h becomes more complicated if submergence of the sheet's loaded end is taken into account. W diminishes with increasing speed because increased impact speed results in increased foundation modulus as the hydrodynamic forces associated with displacing water and ice increase. Correspondingly, characteristic length should be modified to account for increasing foundation modulus. Values of ℓ_c, as given above, apply for slow loading rates and do not account for flooding of the deflected ice sheet.

5. Important Frequencies

The simplest frequency is associated with cycles of single icebreaking events ω_b and is defined as the ratio of speed V to the distance traveled between two icebreaking events. This distance is the component of ice-cusp width in the direction of hull transit; i.e.

$$\omega_b = V\cos\beta/W, \tag{9}$$

in which β is the angle between the average normal to the waterline in contact with the ice sheet and the hull's centerplane (Figure 3). Values of ω_b were determined with $\beta = 44°$. However, β varied somewhat so that there was some error in the determination of ω_b. When icebreaking did not occur simultaneously along both sides of the hull, a frequency higher than ω_b – notably $2\omega_b$ – was evident, as icebreaking events occurred successively on both sides of the bow.

A more complex frequency is associated with multiple icebreaking events. This frequency arises as a consequence of the cyclic variation of contact geometry between bow and ice sheet during hull motions. It is V divided by the distance traveled by the hull as it completes an icebreaking pattern; i.e.,

$$\omega_b/n = V \cos\beta/(nW). \tag{10}$$

For most cases, $n = 2$, when $\omega_b < \omega_n$, and increases to $n = 3$, when $\omega_b > \omega_n$. Figures 3 and 4 schematically illustrate icebreaking pattern and resistance cycles, respectively.

Other important frequencies are the natural frequencies of coupled heave and pitch motions. As discussed above, these have quite close values and are simply referred to as ω_n which designates the average value $\omega = 0.75$ Hz (see Table 1). These frequencies are very important when hull motions are excited at frequencies close to, and exceeding ω_n. Also, frequencies associated with antisymmetric motions, in particular roll and yaw, are important under certain conditions.

For constant ice-sheet properties identifiable cycles in resistance, hull motions, and restraining forces and moments occurred at one or more of the frequencies $\omega_b, \omega_b/2$, and ω_n. For the free hull, when ω_b was less than ω_n the major cycle of resistance and hull motions occurred at $\omega_b/2$ and a minor cycle at ω_b. When ω_b equaled or exceeded ω_n, the major cycle

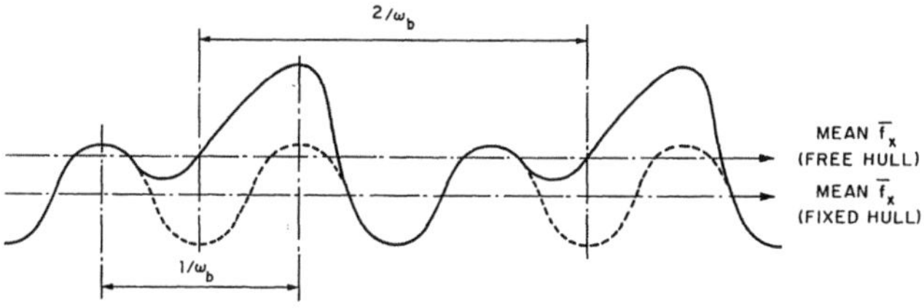

Figure 4. Resistance cycles and variation of mean resistance $\bar{f}_x$ associated with single (ω_b) and double ($\omega_b/2$) icebreaking events.

occurred at ω_n. In general, the fixed hull responded at higher frequencies than the free hull. When ω_b was less than ω_n, the major cycle of resistance and restraining forces and moments occurred at ω_b and a minor cycle at $\omega_b/2$. When ω_b equaled or exceeded ω_n, the major cycle occurred at $\omega_b/2$. The fixed hull response is more broad banded than that of the free hull.

A short segment of time history presented in Figure 5 shows that cycles of resistance for the free hull coincide with $\omega_b/2$ when $V = 0.1\text{m/s}$ and $h = 30$ mm. For the same values of V and h, the major cycles of resistance for the fixed hull occurred at frequencies close to ω_b (Stern et al., 1989). The spectral analysis procedures and power spectra associated with resistance, pitch, and heave, together with the motion excitations are presented by Stern et al. (1989). Herein, space dictates that only the power spectra for the above mentioned experiment be presented. This is done in Figure 6. The dominant frequency of resistance is 0.2 Hz, again coinciding with $\omega_b/2$. Frequency peaks corresponding to ω_b are also evident in the band 0.4 to 0.5 Hz. The power spectra of heave and pitch and their excitations show peaks at both $\omega_b/2$ and ω_b. Note that in the power spectra of heave and pitch, relatively small peaks occur at ω_n. The hull was observed to oscillate after each icebreaking impulse.

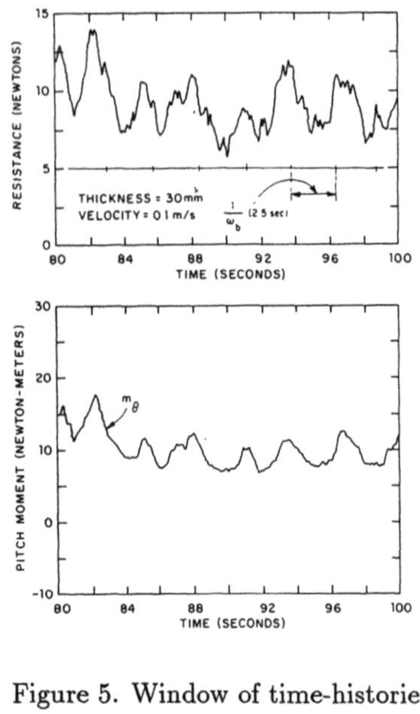

Figure 5. Window of time-histories of f_x, θ, and $\ddot{\theta}$ for free hull.

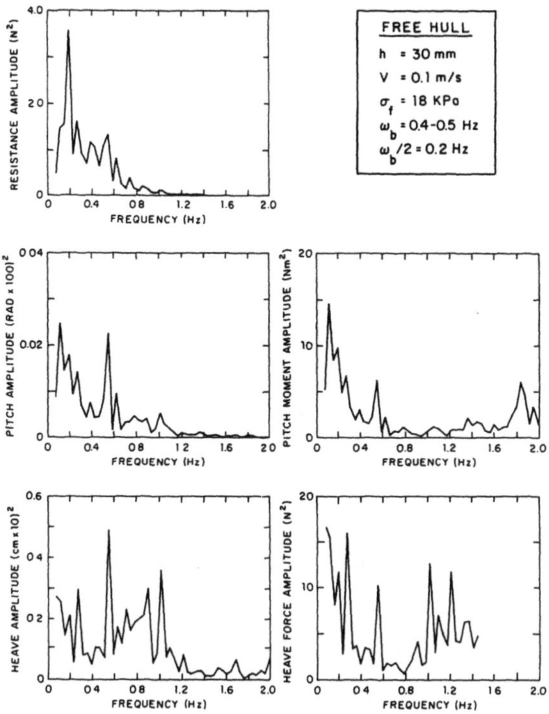

Figure 6. Free-hull power spectra: $V = 0.1$ m/s, $h = 30$ mm, $\sigma_f = 18$ kPa.

6. Resistance

The variation of mean resistance $\bar{f}_x$ with ice-sheet thickness h and speed V is now considered. Reference is also made to the mean plus two standard deviations: $\bar{f}_x + 2\sigma_{fx}$, where σ_{fx} is the standard deviation of f_x. For clarity of presentation , the subscript on σ_{fx} has been dropped; however, in cases where confusion might arise, it is retained. Note that σ_f refers to flexural strength of the ice sheet. Values of $\bar{f}_x + 2\sigma$ are used to indicate peak values. Values of f_x higher than $\bar{f}_x + 2\sigma$ may result from ice-sheet irregularities and, therefore, be unrepresentative.

6.1 Comparison of Free- and Fixed-Hull Conditions.

For the free hull, $\bar{f}_x$ increased with both h and V as shown in Figure 7. Note that for the lowest ($V = 0.05$, 0.1 m/s) and highest ($V = 0.5$, 0.9 m/s) speeds, the resistance curves are in close proximity compared to those for the mid-range speeds ($V = 0.1$, 0.25, 0.5 m/s); admittedly, only

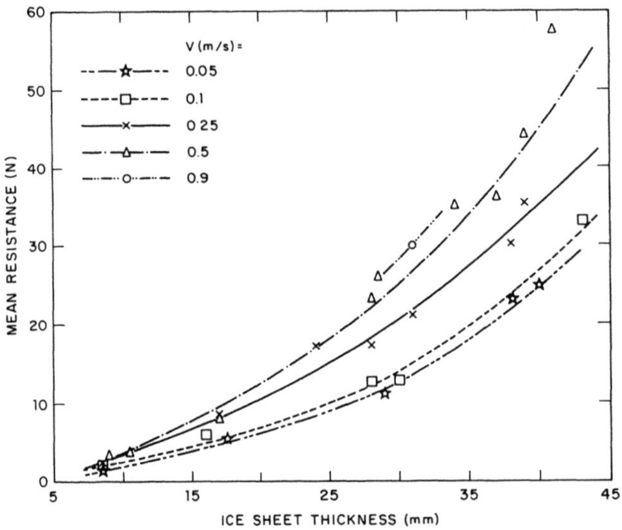

Figure 7. Variation of mean resistance $\bar{f}_x$ with ice-sheet thickness and hull speed: free hull.

a single data point was obtained for $V = 0.9$ m/s. Significantly increased values of $\bar{f}_x$ occurred as speed increased from 0.1 to 0.25 and 0.5 m/s. For the fixed hull, $\bar{f}_x$ also increased with h and V, as shown in Figure 8. However, $\bar{f}_x$ is larger for the free than for the fixed hull, at least for

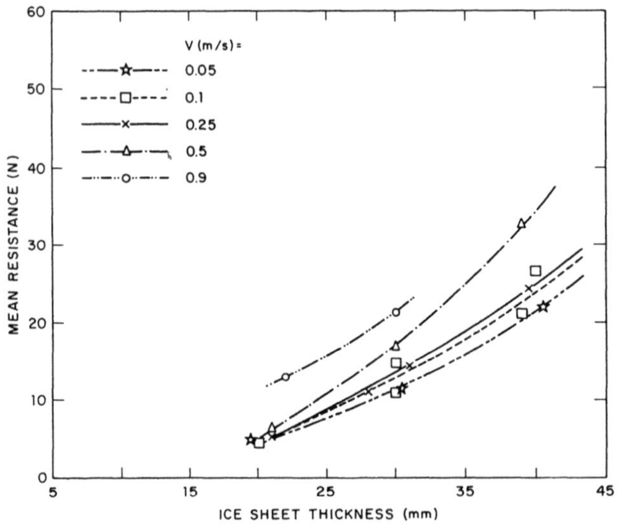

Figure 8. Variation of mean resistance $\bar{f}_x$ with ice-sheet thickness and hull speed: fixed hull.

increasing values of h and V. The distance between the resistance curves is more uniform for the fixed than for the free hull. To display further the nature of the differences in $\bar{f}_x$ for the free- and fixed-hull conditions, Figure 9 shows a comparison of $\bar{f}_x$ versus V for $h = 30$ and 20 mm. Also shown are values of $\bar{f}_x + 2\sigma$. Values $\bar{f}_x$ and $\bar{f}_x + 2\sigma$ vary linearly with V for the fixed hull, but not for the free hull. For the two slower speeds ($V = 0.05$, 0.1 m/s), there is little difference in the values of $\bar{f}_x$ and $\bar{f}_x + 2\sigma$ for the free-and fixed-hull conditions. With increasing speed ($V = 0.25$, 0.5 m/s), the values of $\bar{f}_x$ and $\bar{f}_x + 2\sigma$ are larger for the free than the fixed hull (i.e., in this speed range, the slope of the free-hull resistance is larger than that of the fixed hull). The free-hull resistance is 60 to 70% (when $V = 0.5$ m/s) larger than that of the fixed hull. At the greatest speed tested ($V = 0.9$ m/s), the curves of $\bar{f}_x$ and $\bar{f}_x + 2\sigma$ tend to converge for the free and fixed hull (i.e., in this speed range the slope of the free-hull resistance is similar to that of the fixed hull). Additionally, it can be observed that the difference in $\bar{f}_x$ for the free- and fixed-hull conditions is smaller for thinner ice. Practical implications of Figure 9 are that, for Polar-Class hulls, $\bar{f}_x$ does not vary linearly with V, and simple regression analysis cannot be used to relate them.

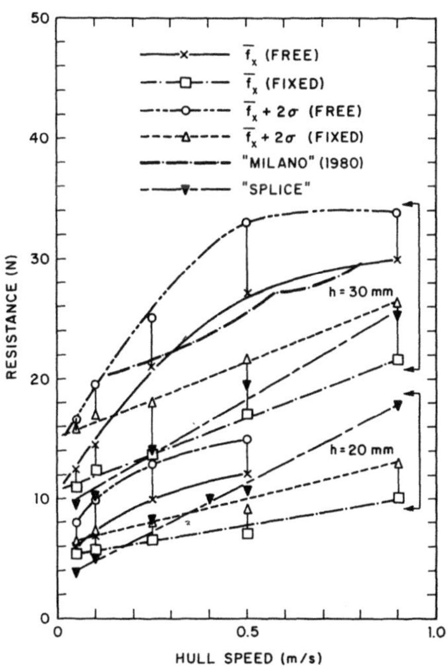

Figure 9. Comparison of $\bar{f}_x$ and $f_x + 2\sigma$ for the free- and fixed-hull conditions: $h = 30$ mm and 20 mm.

7. Hull Motions and Restraining Forces and Moments

Considered now are variation with h and V of the mean and standard deviation of hull motions and restraining forces and moments. For the free hull, the mean values of pitch and heave, $\bar{\theta}$ and $\bar{z}$, for each V increased with h as shown in Figures 10 (a) and (b). Their dependency on h is

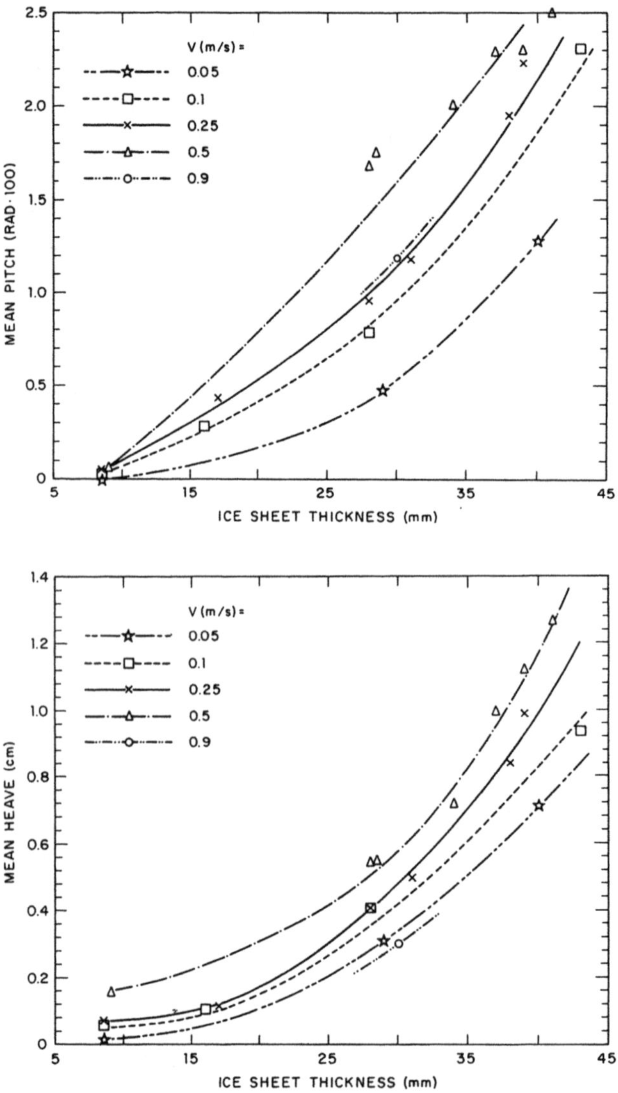

Figure 10. Variation with ice-sheet thickness and hull speed of (a) $\bar{\theta}$ and (b) $\bar{z}$.

similar to that for $\bar{f}_x$, although the distances between the $\bar{z}$ curves are both smaller and more uniform than those for either $\bar{\theta}$ or $\bar{f}_x$. For all but the largest V (= 0.9 m/s), the dependency on V is also similar to that of $\bar{f}_x$; i.e., $\bar{\theta}$ and $\bar{z}$ increase with increasing V. However, for $V = 0.9$ m/s, both $\bar{\theta}$ and $\bar{z}$ decrease substantially. In fact, $\bar{z}$ has a smaller value than that obtained for all other speeds. In order to elucidate the dependency of motions on speed, values of $\bar{\theta}$ and $\bar{z}$ versus V for $h = 30$mm are shown in Figure 11. Note that $\bar{\theta}$ and $\bar{z}$ attained maximum values for $V \simeq 0.5$ m/s, which coincides with the location of the change in slope in the free-hull resistance curve for $h = 30$ mm (Figure 9). This is discussed further below.

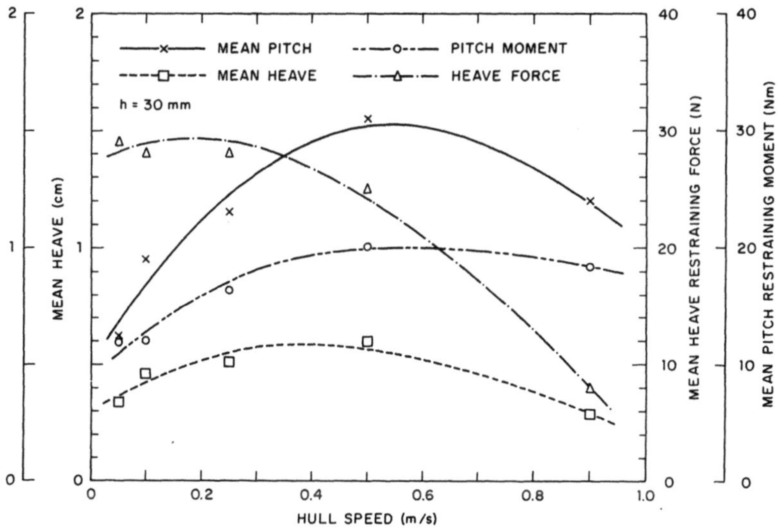

Figure 11. Variation of mean pitch $\bar{\theta}$ and heave $\bar{z}$ and pitch- and heave-restraining moment/force $\bar{m}_\theta$ and $\bar{f}_x$ with hull speed: $h = 30$ mm.

Though the mean values of pitch and heave accelerations, $\ddot{\theta}$ and $\ddot{z}$, were near zero for all experiments, because of the oscillatory nature of hull motions, their standard deviations, $\sigma_{\ddot{\theta}}$ and $\sigma_{\ddot{z}}$, increased with h and V, as shown in Figures 12 (a) and (b). For the two lowest speeds, $V = 0.05$ and 0.1 m/s, $\sigma_{\ddot{\theta}}$ and $\sigma_{\ddot{z}}$ were small, and only varied slightly with h. An increase in V from 0.5 to 0.9 m/s did not result in as large an increase in $\sigma_{\ddot{\theta}}$ and $\sigma_{\ddot{z}}$ as occurred from 0.25 to 0.5 m/s. As mentioned above, $\bar{\theta}$ and $\bar{z}$ were diminished for the former speed range compared

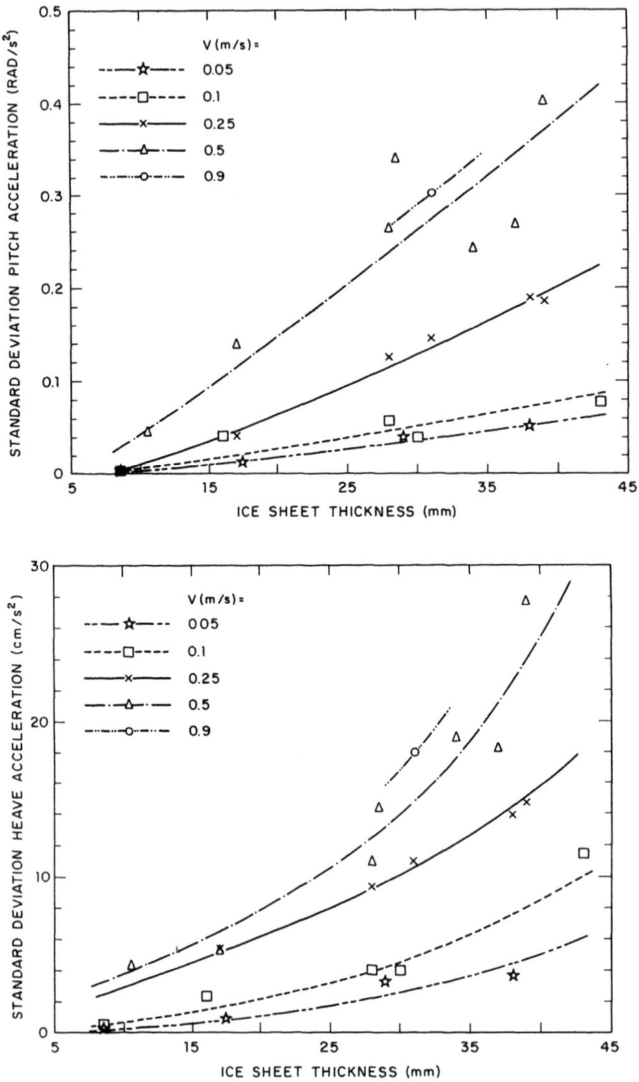

Figure 12. Variation with ice-sheet thickness and hull speed of (a) σ_{θ} and (b) $\sigma_{\bar{z}}$.

to the latter. The variations of $\bar{\theta}$, $\sigma_{\ddot{\theta}}$, and $\bar{z}$, and $\sigma_{\ddot{z}}$ with h and V can be interpreted, using the equations of motion for an icebreaker hull, to assess the relative importance of inertia and hydrostatic-restoring forces in balancing resistance, or excitation, forces. The equations of motion (1) show that, excluding damping forces, resistance/excitation forces are balanced by both inertia and hydrostatic-restoring forces. For large values

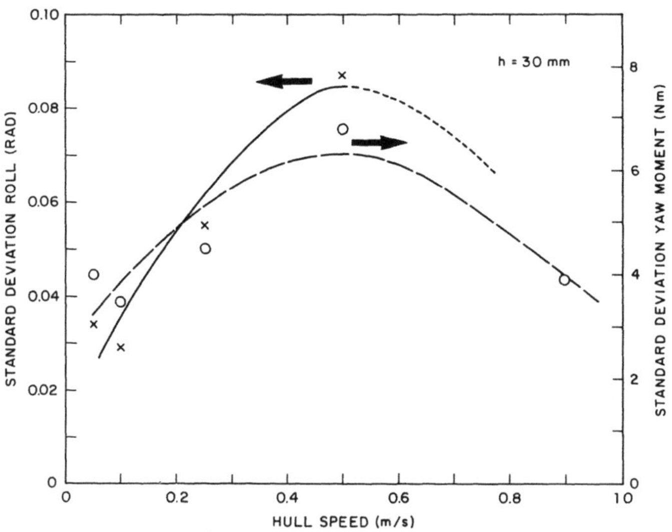

Figure 13. Variation of standard deviation of roll σ_ϕ and yaw restraining moment $\sigma_{m\phi}$ with hull speed: $h = 30$ mm.

of h and V, the increasing importance of inertia terms is reflected by increasing values of $\sigma_{\ddot{\theta}}$ and $\sigma_{\ddot{z}}$, and by the waning of $\bar{\theta}$ and $\bar{z}$ after having attained maximum values.

As the free hull rolled almost equally to either side, mean values of roll ϕ were near zero. However, σ_ϕ increased dramatically for V up to 0.5 m/s, and diminished thereafter. Figure 13 shows values of standard deviation of roll σ_ϕ versus speed for $h = 30$ mm. Although roll was not measured for $V = 0.9$ m/s, the hull was observed to undergo considerably less roll than at $V = 0.5$ m/s.

For each V, the mean values of pitch-restraining moment $\bar{m}_\theta$, heave-restraining force $\bar{f}_z$, and standard deviation of yaw-restraining moment $\sigma_{m\psi}$ for the fixed hull increased with h as shown in Figures 14 (a)-(c). For $\bar{m}_\theta$ and $\bar{f}_z$, the dependency on h is similar to that for $\bar{\theta}$, $\bar{z}$, and $\bar{f}_x$, and is explainable in terms of the hull having to exert increased force, through increased pitch and heave, on thicker ice sheets in order to break them. For $\sigma_{m\psi}$, the dependency on h appears to be linear. The dependencies on V of $\bar{m}_\theta$ and $\sigma_{m\psi}$ are similar to those exhibited by $\bar{\theta}$ and σ_ϕ, as shown in Figures 12 and 13 and can be explained as follows. At slow speeds, flexure of ice sheets was accompanied by flooding of the depressed sheet which assisted in ice-sheet loading. However, with increasing V, the hull flexurally depressed and broke ice with diminished flooding such that $\bar{f}_x$, $\bar{\theta}$, and $\bar{m}_\theta$ increased. At high speeds, $\bar{\theta}$, and $\bar{m}_\theta$ cease increasing with V. Observations, with the underwater video camera, indicate that with further increase of V the hull maintained a steadier trim and planed through the ice sheet with considerable crushing and shearing at the bow stem, more than at lesser speeds. Interestingly, $\bar{f}_z$ is less dependent on V than $\bar{m}_\theta$ and $\sigma_{m\psi}$, especially for thinner ice. For thick ice, the trends of $\bar{f}_z$

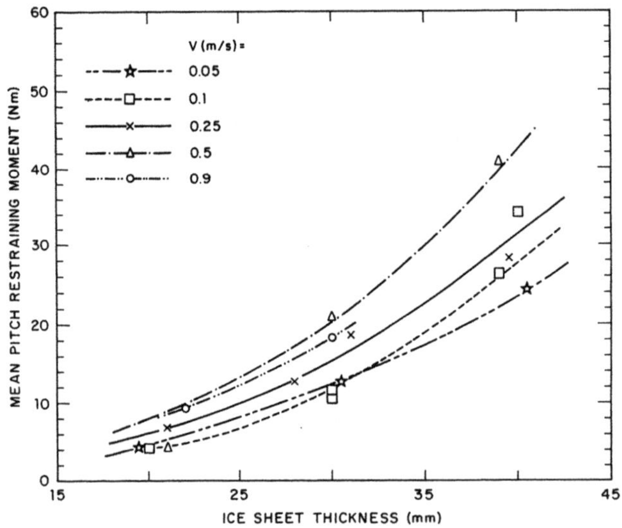

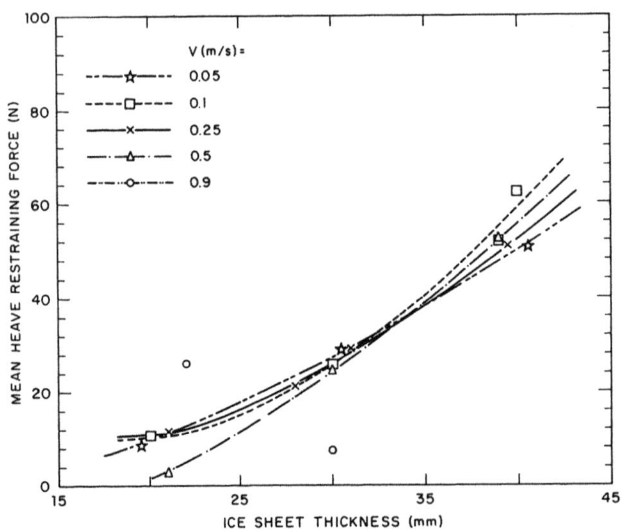

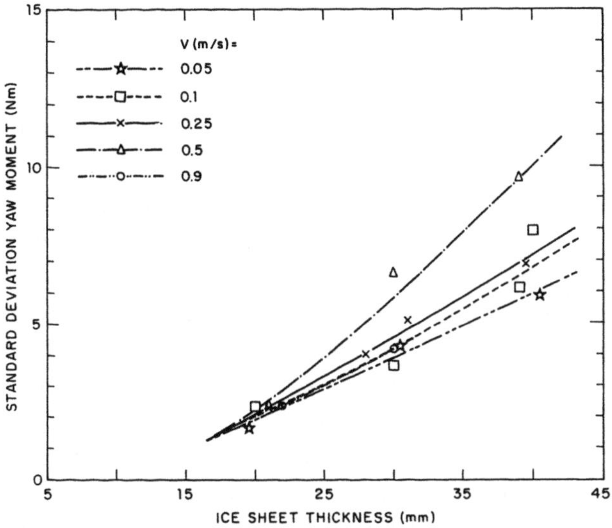

Figure 14. Variation with ice-sheet thickness and hull speed of: (a) $\bar{m}_\theta$, (b) $\bar{f}_z$ and (c) $\sigma_{m\phi}$.

with V are similar to those of $\bar{m}_\theta$ and $\sigma_{m\psi}$; however, the maximum value occurs for a lower speed ($V = 0.1$ m/s). For $V = 0.9$ m/s, the reasons for the quite large and small values for small and medium thickness, respectively, are uncertain. However, one explanation for the latter case is that, when the hull planes through the ice sheet, $\bar{f}_z$ is small.

In summary, both hull motions and restraining forces and moments increased with thickness in a manner similar to that of the resistance. The dependency on speed is such that, in most cases, maximum values are attained for $V = 0.5$ m/s; but this differs from the dependency on speed of resistance and, as is discussed next, explains the differences in resistance for the free- and fixed-hull conditions. There are two reasons for the decrease in hull motions and restraining forces and moments at high speed: (i) as is discussed earlier, the dominant excitation frequencies exceed motion natural frequencies and (ii) the icebreaking mechanism appeared to involve considerably more shearing and crushing for the hull in both the free and restrained conditions.

8. Conclusions

It has been shown that the major cycles of resistance, hull motions, and restraining forces and moments occur at one or more of the frequencies

ω_b (icebreaking), ω_b/n (pattern), and ω_n (natural frequencies). For low speeds ($V = 0.05$, 0.1 m/s), the free hull responded primarily at the pattern frequency, whereas for medium and high speeds ($V = 0.25$, 0.5, 0.9 m/s), it responded at the natural frequencies. In general, the fixed hull responded at higher frequencies associated with icebreaking. Thus, it is concluded that the freedom of the free hull to undergo motions resulted in its interacting with the ice sheet such that icebreaking pattern and natural frequencies dominate the resistance cycles.

The mean resistance of the free hull was larger than that of the fixed hull for all speeds, whereas the fixed-hull resistance varied linearly with speed. The difference in resistance between the two conditions is largest for speeds $V = 0.25$ and 0.5 m/s. It is for these speeds that the motions are maximum. This correlation indicates that the ice-induced hull motions of the free hull are responsible for its increased resistance over that of the fixed hull. This unexpected result is attributed to the following: (1) hull motions consume energy and, thereby, lead to increased resistance; (2) roll motion affects the symmetry of icebreaking, resulting in a broadened pattern and increased resistance; and (3) for increased pitch, the bow stem of a Polar-Class hull impacts the ice sheet with an increased angle causing increased shearing and crushing and resistance. The results from spectral analysis indicate that at speeds $V = 0.25$ and 0.5 m/s the icebreaking and pattern frequencies are close to the hull natural frequencies of coupled heave and pitch, and roll, respectively, such that resonance conditions are evident. Consistent with this is the fact that the motions are quite large under these conditions. It might have been expected that, for ω_b near ω_n, motions would be large. However, the identification of the important role of the pattern frequency ω_b/n (with $n = 2$, usually) is a new and, again, an unexpected result.

The present work has several implications with regard to the future development of prediction methods for continuous-mode icebreaking. It is quite evident that hull motions play a significant role and that their influence should be included.

The influence is speed dependent such that consideration must be given not only to the icebreaking frequency but also to pattern frequency and the natural frequencies of hull motions. The presence of the ice sheet must be considered in determining the natural frequencies of hull motions.

Lastly, with regard to the direction of future work, further experiments are needed to determine the influence of bow form and to aid in the development of improved methods for evaluating ice forces, including the influence of hull motions. Such an experimental program is required in support of the future development of prediction methods as discussed above.

Acknowledgements

This research was sponsored by the United States Department of Transportation, Maritime Administration (MARAD) under Contract No. DTMA91-85-C-50115.

References

Ettema, R., Stern, F. and Lazaro, J. 1987. Dynamics of continuous mode icebreaking by a Polar-Class icebreaker hull. Iowa, Institute of Hydraulic Research, The University of Iowa. IIHR Report No.314.

Ettema, R., Stern, F. and Lazaro, J. 1989. Dynamics of continuous-mode icebreaking by a Polar-Class hull - part 1: Mean response. Journal of Ship Research, Vol.33, No.2, pp.115-134.

Meirovitch, L. 1975. Elements of Vibration Analysis. McGraw-Hill, New York.

Price, W.G. and Bishop, R.E.D. 1974. Probabilistic theory of ship dynamics. Chapman and Hall, London.

Stern, F., Ettema, R. and Lazaro, J. 1989. Dynamics of continuous-mode icebreaking by a Polar-Class hull - part 2: Spectral analysis. Journal of Ship Research, Vol.33, No.3, pp.236-251.

Discussion

Q: Scott Newbury, Institute for Marine Dynamics

How was the icebreaking pattern quantified?

A: R. Ettema

The ice pattern beneath and around the hull was measured using vernier callipers, tape measure and still photography, when the hull was raised from the ice tank. In addition, widths of ice pieces were determined by means of underwater video camera and scaled video monitor.

Impact Ice Pressure: More Questions than Answers

Alfred L. Tunik

American Bureau of Shipping
Paramus, NJ, USA

1. Introduction

Impact interaction between ice and engineering structures is always accompanied by ice failure. The failure modes can include crushing, cracking, bending, spalling, etc. Which mode is predominant depends mainly on the geometry of the colliding bodies, while the ice-induced impact loads depend both on the ice failure mode, and on the impact conditions. The earlier authors (Runeberg, 1888; Makarov, 1901) hypothesized that the ice-induced loads cannot exceed the strength of ice under corresponding types of loading. This presumption, which is now called the limiting ice strength approach, has been widely used as the basic criterion in specifying the design ice loads due to static (or quasi-static) and dynamic interactions. Here, the term "static" implies that the ice-structure interaction is not affected by the inertial forces of the interacting bodies. The "dynamic" or "impact" interaction has been thought of as the interaction which is either governed mainly by the inertial forces or when the latter may not be ignored. The wind-driven ice fields advancing on an offshore structure, a ship continuously progressing in consolidated ice, both can be treated as static or quasi-static interactions regardless of the advancement rate and the cyclicity of the induced load. A collision of a ship with an ice floe, or a wave-driven bergy bit hitting an offshore structure are the typical "impact" (dynamic) scenarios addressed in this work.

Impact ice loads are the prevalent loads for ship hulls and can, under certain conditions, be high for offshore structures as well. However, the bulk of information on the mechanical properties of ice acquired to date is related to the static load. This discussion is focussed at ice-induced impact pressure which is an important design factor.

S. Jones · R. F. McKenna · J. Tillotson · I. Jordaan (Eds.)
Ice-Structure Interaction
IUTAM-IAHR Symposium St. John's, Newfoundland, Canada

2. Ice Induced Impact Load Formulation

Earlier laboratory and in-situ experiments of dropping steel spheres on ice cover have revealed (Kheisin and Cherepanov, 1970) that an interface layer of crushed ice is being formed within the contact zone and extruded from there. This fact, which has been verified later by many other studies (Glen and Comfort, 1983; Johnson and Benoit, 1987; Jordaan and McKenna, 1988), has been the keystone for developing an analytical model of ice-solid body impact (Kheisin et al., 1975) idealizing the crushed ice mass as a viscous liquid being pressed out between two rigid surfaces (Figure 1). The strict analytical solution yielded the ice-induced impact pressure and force to be dependent on the impact mass (M) and velocity (V), as well as on two parameters μ and k characterizing the mechanical properties of the crushed ice mass. The maximum (in time) pressure at the center of the contact zone (P_{max}) and the maximum (in time) global force (F_{max}) are:

$$P_{max} = K_p(MD)^{1/9}V^{4/9}(3\mu k^3)^{2/9}, \tag{1a}$$

$$F_{max} = K_f(MD)^{5/9}V^{11/9}(3\mu k^3)^{1/9}, \tag{1b}$$

where D - sphere's diameter; μ and k - viscosity and pressure gradient factor of the viscous medium idealizing the crushed ice mass; K_p and K_f - numerical coefficients. Since μ and k appear in the equations in a single combination of μk^3, they can be combined into a single parameter, taken

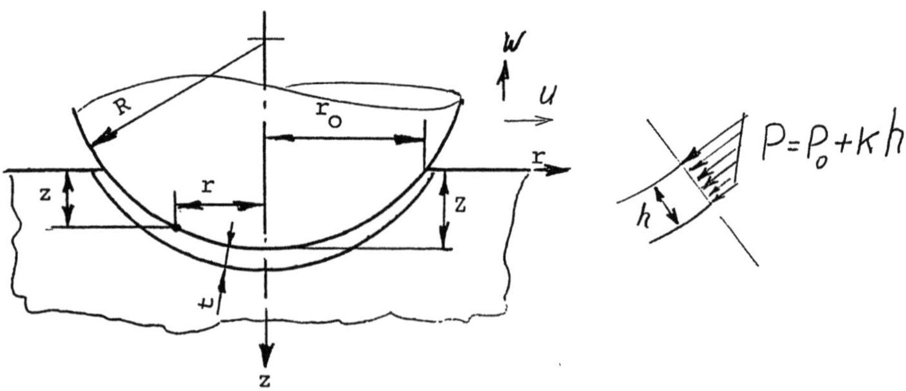

Figure 1. Schematic of impact indentation.

by the authors to be:

$$a_p = (3\mu k^3)^{5/24} \tag{2a}$$

and by Tunik (1984) to be:

$$A = (3\mu k^3)^{1/4}. \tag{2b}$$

Assuming that both μ and k, as well as their product μk^3, are material constants of ice, one may conclude from (1) that both pressure and force are proportional to the mass and velocity at the appropriate fractional powers. However, the quantities μ and k have never been measured directly, while their products a_p and A have only been estimated by recalculating back the ship structural response data. Moreover, a direct measurement of these parameters is hardly feasible. Therefore, the ice loads should better be expressed using another parameter, which like μk^3, characterizes the resistance of ice to impact crushing, and unlike μ and k, can be measured directly from reproducible experiments. The specific impact crushing energy of ice satisfies both requirements. It was introduced first by Kheisin and Likhomanov (1973) and defined as the crushing energy per mass of the crushed ice. However, in their following development of the model of ice-rigid body impact (Kheisin et al., 1975; Kurdyumov and Kheisin, 1976) the authors have not used this characteristic for describing the impact loads. The specific energy, re-defined as the kinetic energy dissipated for ice crushing per volume of the crushed ice, W, was introduced into Kheisin and Kurdyumov's model by Nevel (1986) and Tunik (1989).

$$\varepsilon = MV^2/2W. \tag{3}$$

The definition (3) yields the dimension of J/m^3 which is identical to N/m^2, i.e. the specific crushing impact energy has the dimension of stress (Pa). Since it characterizes the resistance of ice to crushing failure during impact, it can be called the impact crushing strength of ice - the name used hereinafter. For the drop-ball tests, this definition yields

$$\varepsilon = 0.64(MD)^{1/9}V^{4/9}(3\mu k^3)^{2/9}, \tag{4}$$

while the maximum pressure and force can be expressed via ε, as follows:

$$P_{max} = 1.20\varepsilon, \tag{5a}$$

$$F_{max} = 1.60(\varepsilon MD)^{1/2}V \tag{5b}$$

3. Ice Testing Data

The impact crushing strength of ice can be obtained directly from the drop-weight and pendulum impact tests. The available experimental data related to the impact crushing strength of ice were analyzed by Tunik (1989). The analysis reveals that the amount of the experimental data on the impact crushing strength, summarized in Table 1, is barely sufficient for estimating the range of variation. Considerable spread has been reported by the authors who performed numerous tests with or without parametric variations of impact velocity, mass or energy (Yen et al., 1970; Likhomanov and Kheisin 1971, 1973; Garcia et al., 1985; Sodhi and Morris, 1986). For example, the root square mean of ε for all 180 tests reported by Kheisin and Likhomanov (1973) is around 0.23 of the mean, and the magnitudes of ε have the normal (Gaussian) distribution. Comparable data spread can be found in Yen et al., Garcia et al., and other publications. However, this data spread does not exceed typical data

Table 1
Experimental Data on Impact Crushing Strength of Ice "ε"

Ice type and conditions reported	ε, MPa mean (st.dev.) [range]	Test type and data source
Fresh-water ice cover, Ladoga Lake		
a) January, t_{air}=-30 to -20°C	9.8 (3.4)[4.7-26.9]	d/b, Kheisin & Likhomanov
b) February, t_{air} =-10 to -5°C	10.0 (1.9)[6.1-13.9]	d/b, Kheisin & Likhomanov
c) February, t_{air}=-22 to -5°C, c-axis horizontal	13.7 (2.7)	d/b, Kheisin & Likhomanov
d) March, $t_{air} \approx$ +2 to +4°C	3.3 (0.8) [2-5]	d/b, Kheisin & Likhomanov
Natural iceberg ice		
in-situ, t_{ice} = -6 to -9°C	16.4 (9.9) [2.5-34]	d/b, Johnson & Benoit
laboratory samples, t_{ice} = -5°C	9.4 (2.7) [6.9-15.6]	d/w, El-Tahan et al.
Natural sea ice cover		
Can. Arctic, Fox Channel, July, m. -y. ice	[4.5-6.2]	d/b, Pounder & Little
Kara & Laptev Seas	[4.4-26.3]	d/c, Dementyev
North Caspian Sea	[0.4-4.1]	d/w, Tsurikov et al.
Antarctic Ice	[1.2-3.7]	d/w, Tsurikov et al.
unspecified, $t_{ice} \approx$ -1 to -14°C	[2.2-19.4]	d/b, Khrapaty et al.
Laboratory ice		
Artifical snow ice, t_{ice} = -5°C	15.4 (3.3) [9.8-22.1]	d/w, El-Tahan et al.
Laboratory polycrystalline ice,-5°C	[1 to 35]	d/w, Garcia et al.
Lab. saline ice, -2 to -25°C; S=6; 1 ppt	[0.4 to 6.0]	p., Glen & Comfort
Laboratory urea ice	[0.12 to 3.3]	p., Sodhi & Morris
Refrozen river ice, -1°C	14.0 (5.2)[8.3-20.3]	p., Fransson & Sandkvist

Notations: d/b = drop-ball, d/c = drop-cone, d/w = other drop-weight, p = pendulum spread of standard laboratory tests of compressive and tensile strengths, and is noticeably smaller than that for in-situ bending strength tests. Thus, the drop-ball test can provide stable and reproducible information on the ice impact resistance.

It is important to know how the testing procedure affects the measured and calculated values of the kinetic energy of crushing, the volume of the crushed mass, and the resulting impact crushing strength of ice. In particular, what are the mass and velocity effects? The effects of impact velocity and mass on the impact crushing strength is still unclear. On one hand, a number of authors have presented data implying that the effects are minimal, if any. Within the range from 1.5 to 6 m/s, no velocity effect can be traced from the data by Kheisin and Likhomanov (1973) - see Figure 2. Two different masses (300 and 150 kg) were used in these tests - no mass effect was distinguished in three series of field tests, while the fourth series showed a noticeable difference - see the table in Figure 2. The data reported by Sodhi and Morris can be used for estimating the velocity effect (the mass has been constant). The data in Figure 3, both plotted separately for different bending strength of the ice and plotted regardless of the bending strength, do not imply a velocity effect. Data reported by Glen and Comfort provide only a few points for each of three testing temperatures - again, no velocity effect can be distinguished (Figure 4). On the other hand, the numerous data points reported by Garcia et al., (1985) clearly demonstrate a reciprocal effect of the kinetic energy on specific crushing energy (Figure 5). Given the energy in Garcia's tests varied within two orders of magnitude, while the velocity varied only from 1.4 to 3.1 m/s, the energy dependence of ε should be attributed, at least partly, to a mass effect. It should also be noted that, unlike all other drop-ball tests, in Garcia's tests the volume was directly measured including all cracks. A few data points published recently by Fransson and Sandkvist (1989) also imply a reciprocal velocity effect (Figure 6), but the results are preliminary.

The theory suggests in equation (4), the impact crushing strength does depend on both mass and velocity, provided that μk^3 is constant. The effect of mass is rather small ($\approx M^{1/9}$). Changing the mass by a factor of 2 (as in Kheisin's tests) can cause only 8 percent change in ε which is too small to be detected for sure in such experiments, especially in in-situ tests. The mass variation should span at least one order of magnitude to be surely detectable in carefully controlled laboratory experiments, or two orders of magnitude for in-situ tests. This range of mass variation

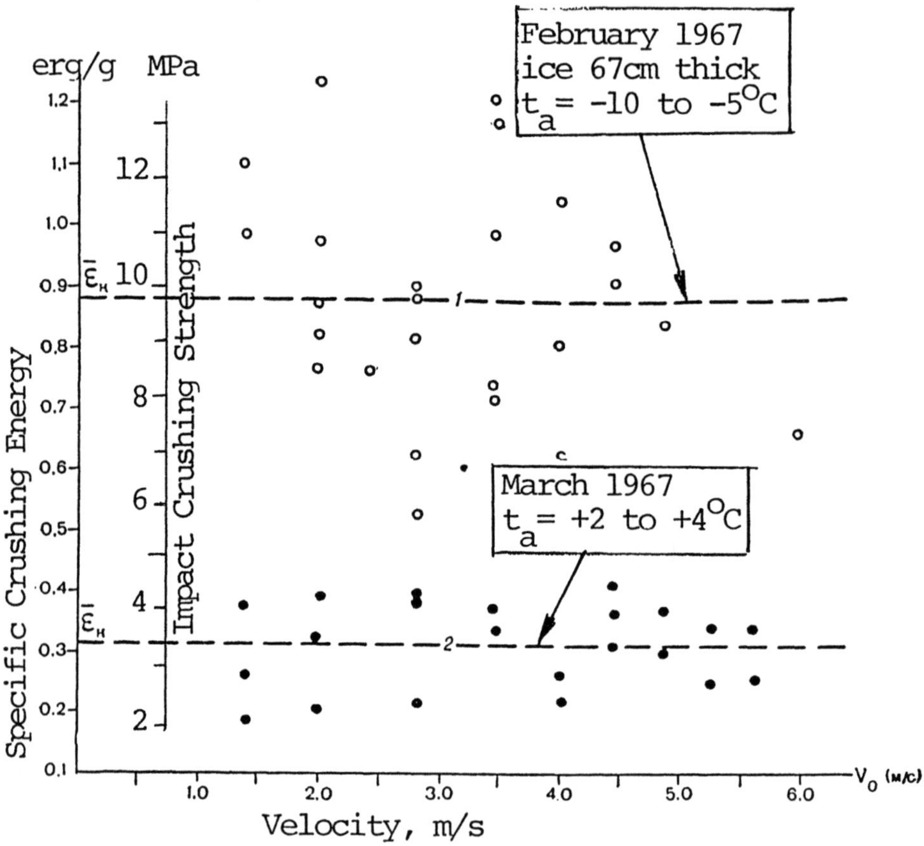

Mean Values (ε) and standard deviations (σ_ε) of ε versus mass M

Test Series in Table 1	ε (σ_ε), MPa	
	M=300 kg	M=156 kg
a	9.8 (3.7)	10.1 (2.2)
b	9.8 (2.0)	10.3 (1.8)
c	15.6 (3.1)	10.3 (1.9)
d	3.4 (0.8)	3.2 (0.8)

Figure 2. Impact crushing strength, ε, versus velocity, V, for the Ladoga Lake ice for two test series, and table of the mass effect for four test series, after Kheisin and Likhomanov (1973).

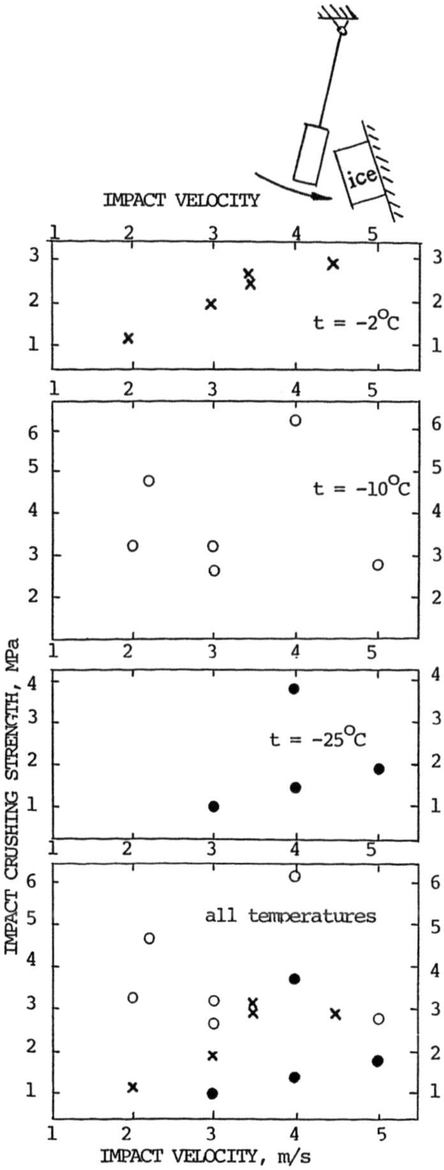

Figure 3. Impact crushing strength, ε, versus velocity, V, for urea ice, plotted for different ranges of the bending strength σ_b, obtained from the data by Sodhi and Morris.

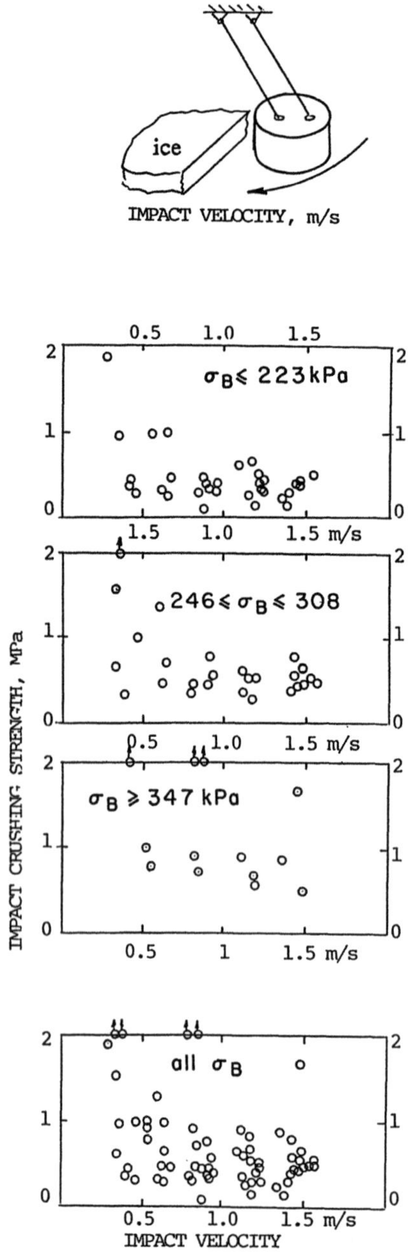

Figure 4. Impact crushing strength, ε, versus velocity, V, for laboratory saline ice, at different temperatures, obtained from the data by Glen and Comfort.

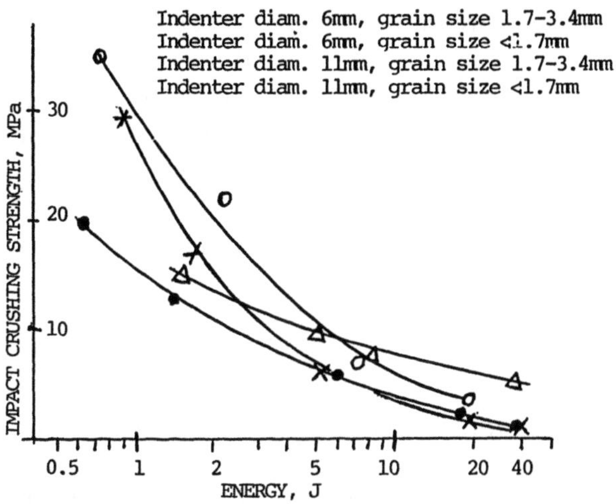

Figure 5. Impact crushing strength, ε, versus the impact energy, obtained from the data by Garcia et al.

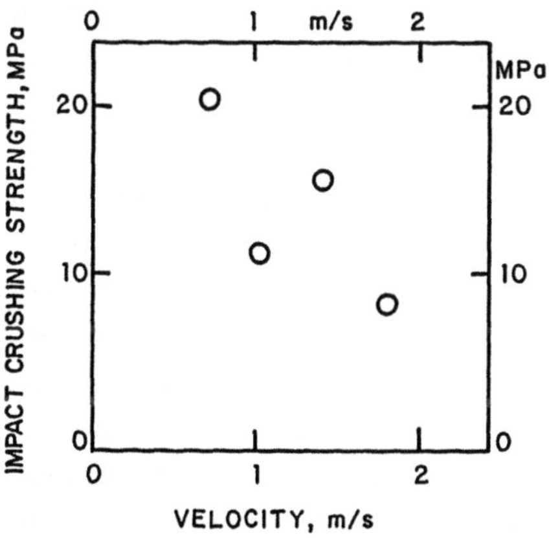

Figure 6. Impact crushing strength of refrozen river ice ε, versus pendulum's velocity, V, obtained from the data by Fransson and Sandkvist.

took place in Yen's tests but the authors studied only restitution, not measuring the ice crushing characteristics. The impact velocity effect is, theoretically, more pronounced than the mass effect ($\approx V^{4/9}$) and supposedly it could be easily detectable for the velocities ranging by a factor of 2-4, provided (again) that μk^3 is constant. However, if μk^3 is either velocity or mass-dependent, then it could balance or reverse the velocity or mass influence in equation (4). This assumption is physically plausible since both the viscosity, μ, and pressure gradient, k, are expected to be inversely proportional to the fineness of the crushed ice mass, which, in turn, is supposed to be proportional to the mass and velocity of the hitting indenter. Thus, further experiments are necessary to clarify the sensitivity of the impact crushing strength of ice to the principal testing procedure parameters: impact velocity and mass. After this, a standard procedure can be developed for testing ice at varying conditions (structure, temperature, salinity, etc.).

4. Full-Scale Test Data

The importance of studying the impact crushing strength of ice and its dependence on the velocity and mass is in the fact that the local pressures (both maximum and average) due to an ice impact are specified as the impact crushing strength factored. The impact ice pressures measured on ships and fixed structures subjected to the impacts can also provide information both on the very phenomenon of the impact crushing of ice and on the applicability or adjustment of the analytical model. Although hundreds of ships have been tested, the available information on the pressures measured is rather scanty with some important details missing.

Earlier data acquisition systems (Ryvlin and Kheisin, 1980) included an array of strain gauges measuring the response (bending strains, mainly) to ice actions at predetermined points spread along hull structural elements (frames, sometimes stringers and shell plating). The parameters of actual ice loads were then reconstructed comparing the response measured and calculated assuming a certain ice load distribution. The obtained ice pressures were, hence, indirect estimates. In another approach, a relatively small (several frame spacings) area of hull framing was heavily instrumented with strain gauges measuring compressive or shear strains to be used as the primary ice pressure measuring tool onboard a ship. Such panels have been instrumented at bow side sections of MV *Arctic* (Glen et al., 1985), MV *Kashira* (Kivimaa and Kujala, 1988), and *Polarstern* (Muller et al., 1988, 1989), at bow side sections and the stem of USCGC *Polar Sea* (St. John and Daley, 1984; Daley et al., 1984). Specially designed pressure gauges have also been installed onboard the icebreaker *Canmar Kigoriak* (Ghomeim et al., 1983), the icebreaker *Sisu* (Riska et al., 1983) and CCGS *Louis S. St. Laurent* (Glen and Blount, 1984), as well as a special ice force measuring panel onboard the *Po-*

larstern (Hoffman, 1985) and MV *Kashira* (Kivimaa and Kujala, 1988). Not all of these systems have been equipped with a synchronized speed recording device.

The acquired information on the local ice pressure and its relationship with the ship speed can be summarized as follows:

- The local ice pressures reported in all sources, except one, do not exceed ≈ 9MPa, with a majority of data points not exceeding 6-7 MPa. Only Glen and Blount (1984) reported the peak ice pressures reaching values as high as 24-36 MPa, with mean peaks of 19 MPa for continuous level icebreaking and 23 and 30 MPa for ramming icebreaking. It should be noted, however, that membrane diameter of the transducers used was 8mm, which is comparable with the grain size of ice. At such a small diameter of the diaphragm, the gauge could measure the strength of single ice crystals rather than the strength of an ice macrostructure, thus exaggerating the ice-induced pressures. This hypothesis is supported indirectly by the authors' conclusion that the ice forces derived from the pressure transducers were greater than the strain gauge derived forces. Pressures of the same order of magnitude and higher were measured recently by a similarly small (area 0.7 cm^2) piezoelectric force transducer during in-situ drop-sphere tests (Chin and Williams, 1989). The results are, however, preliminary.
- Based on the statistical data acquired onboard many ships, Ryvlin and Kheisin (1980) showed the frame bending stresses measured in similar conditions increase statistically with an increase in ship speed. The velocity measuring technique and strain gauge arrangement used in the Soviet tests of ships in ice did not allow the speed effect to be attributed to the impact pressure or force. The experimental data on the highest ice pressures on a single instrumented panel, acquired onboard the USCGC *Polar Sea* and plotted versus her horizontal speed (St. John et al., 1984) reveal virtually no speed effect both in first-year and in multi-year ice conditions - see Figure 7. (The speeds were read manually from the standard shipborne log for specific events during earlier deployments and recorded synchronously from a Doppler radar during later tests.) Muller and Payer (1988) have also found a weak, if any, speed effect on the maximum local ice loads derived from the structural response measured onboard the *Polarstern*. It should be noted that the strain

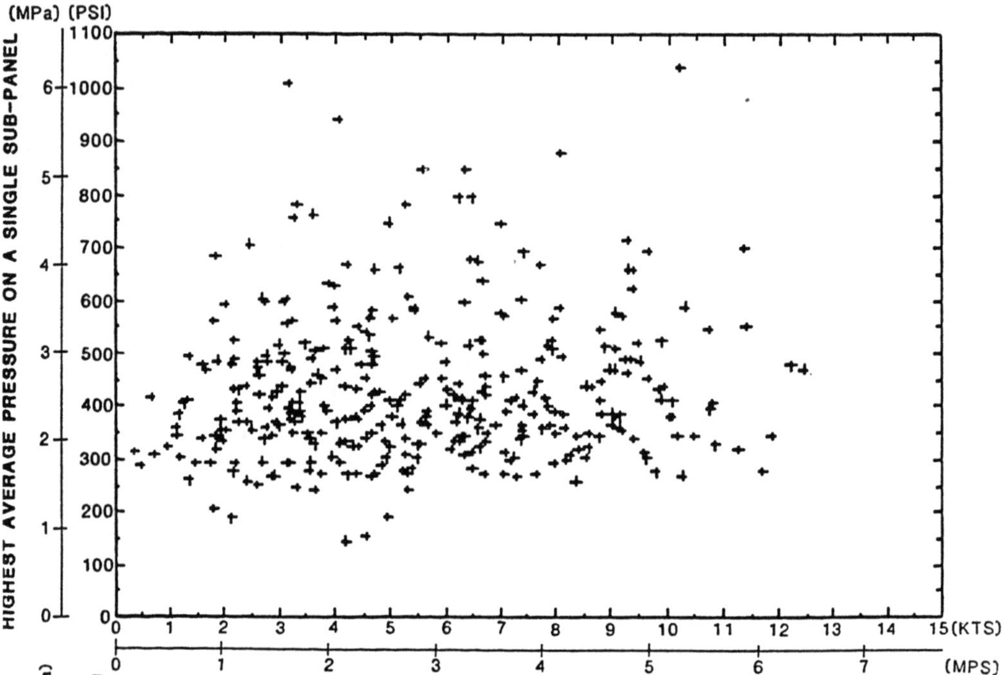

SHIP SPEED FOR FIRST-YEAR EVENTS

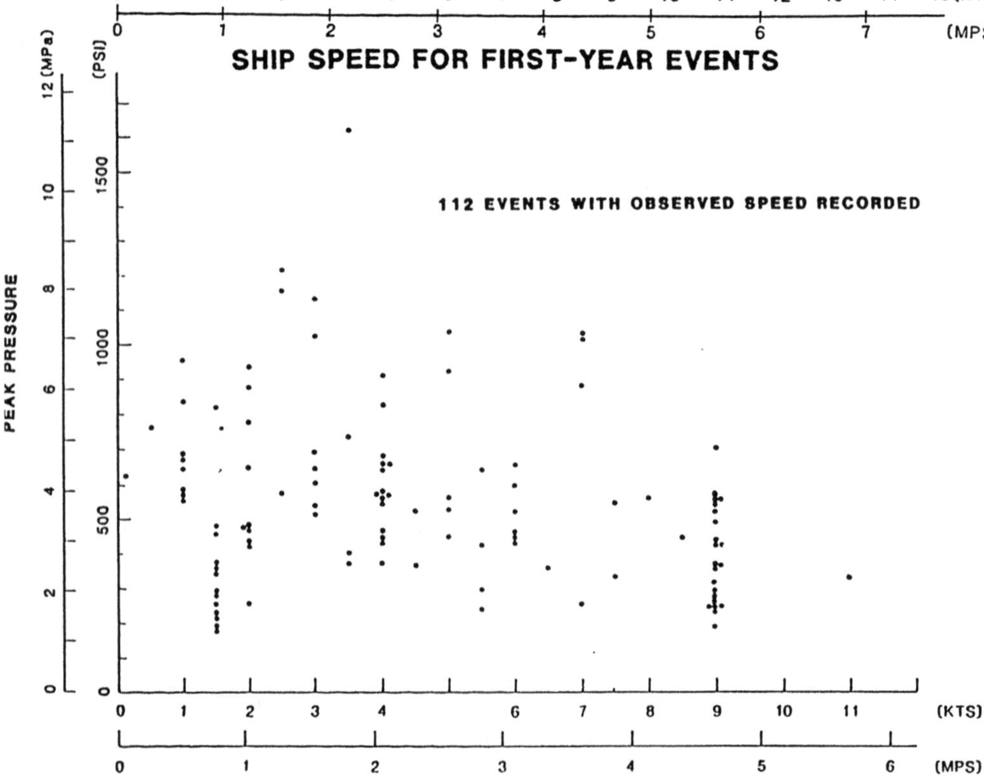

HIGHEST AVERAGE PRESSURE ON A SINGLE SUB-PANEL vs IMPACT SPEED FOR THE BEAUFORT SEA DATA (SUMMER)

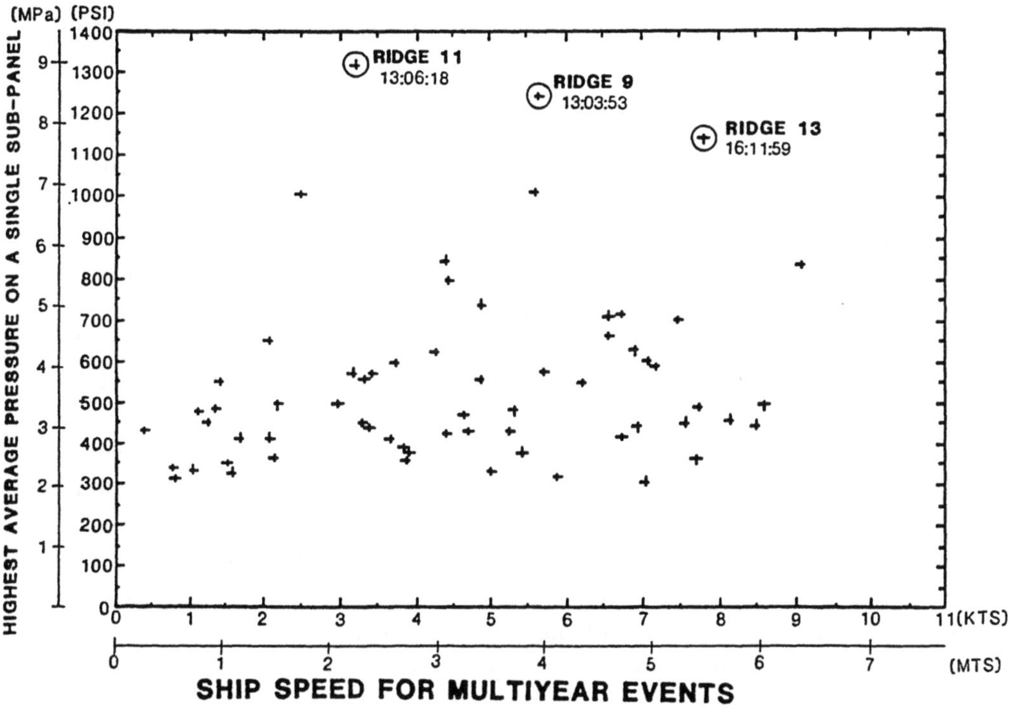

Figure 7. Pressures measured on the instrumented side panel of the USCGC POLAR SEA plotted versus ship speed in first-year and multi-year ice, after St. John et al., (see report).

gauges were arranged on side longitudinals at the waterline to measure the fibre bending stresses in the span and the shear stresses at the supporting frames. The ice load was then calculated as a fictitious uniform pressure which would result in the bending stress at the support to be as high as that measured at the midspan - the technique which is virtually identical to the Russian one. Thus, the "equivalent" pressures obtained by analyzing the strain-gauge measurements are indirect estimates, as well as those reported by Ryvlin and Kheisin (1980). No details are given on the velocity measurements onboard the *Polarstern.* Based on the *Louis S. St. Laurent* tests, Glen and Blount (1984) also came to a conclusion that the impact pressure has no apparent relationship to the ship speed, while the impact force is certainly speed-dependent. The local ice pressures measured by the bow pressure gauges aboard the *Canmar Kigoriak* did not show any dependence on the ramming velocity.

Thus, the data obtained to date from the full-scale trials of ships do not provide conclusive information on the impact ice pressure. For the sake of completeness, the Hans and Tarsiut Islands experiments should be included in the review (APOA 1982, 1984). In experiments of this kind when freely floating ice hits a rock, the impact velocity measurements are as simple as in the drop-ball tests, while the ice mass determination is rather vague. Only a few data points on impact pressures have been obtained. The pressures were well below 2MPa, and the number of data points did not allow the plotting of a pressure-velocity relationship.

It should be emphasized that, unlike the drop-ball tests where the impact velocity can easily be (and has usually been) measured accurately before, during and at the end of the impact, the ship-ice impact velocity is not easy to measure, especially at side impacts. The horizontal speed of ship motion, often measured during the trials, is not the impact velocity. The latter is the difference between the initial horizontal ship speed, factored by the directional cosines taking into account both the ship hull geometry and the instantaneous angles of ship rolling-pitching-heaving motion, and that at the end of the impact. For a majority of trials where the average horizontal speed readings were reported in accompanying notes, the impact velocities could be estimated rather vaguely. When the horizontal ship speed is recorded synchronously with the gauge records, the impact velocity can be calculated more or less accurately for ramming icebreaking mode (the technique described by Tunik et al. (1988) in analyzing ice ramming forces on the *Polar Sea*). For side impacts, at least two velocity (or acceleration) components should be synchronously recorded. None of the publications referred to contains velocity records of this kind.

It should also be noted that the prevailing ice failure modes should be reported for each impact event to distinguish between the impact crushing and bending failures. The absence of ice failure mode records can lead to a confusion as illustrated in Figure 8. When moving in thinner ice at higher speeds, the ship breaks the ice continuously by bending with low ice loads experienced: as a result, the data points would fill in area A of Figure 8a. In thicker ice at slower speeds, both bending and crushing failures can occur, with the crushing failures causing much higher loads - the corresponding data points would fill in the areas B and C (Figures 8b and 8c) for the bending and crushing ice failure modes, respectively. In the thickest ice which can be broken continuously by the ship, the load-vs-speed plot would fill in area D (Figure 8d). When navigating in a channel

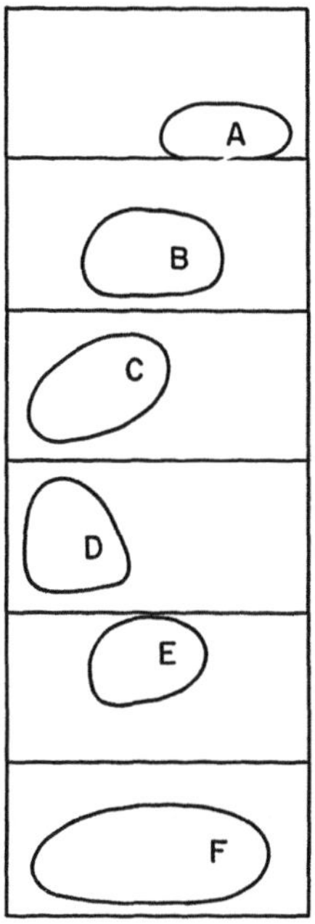

Figure 8. Hypothetical plot areas for local ice impact pressures versus ship speed for testing in various ice conditions: (a) continuous icebreaking in relatively thin ice; (b) same in moderate ice; (c) motion in continuous and broken ice of moderate and high thickness;(d) continuous icebreaking in the thickest ice breakable by bending by the ship in question; (e) motion in a channel broken in thick ice; (f) motion in ridged ice.

in thick ice which cannot be broken by bending by this ship, the loads due to ice crushing failure would fill in the area E (Figure 8e). Somewhat similar but with wider data spread could be observed in ridged ice (Figure 8f). The thicker the ice, the more often the crushing failure occurs, and the slower is the apparent speed of ship motion. However, the actual impact velocity is not necessarily higher at a higher ship speed, since it is defined not by the average ship speed but by the difference between the initial and final impact velocities reduced to the impact point. Thus, plotting all acquired data points regardless of the ice failure mode and not reducing the velocities to the normal can result in a misinterpretation of the actual ice load-velocity relationship. This is illustrated in Figure 9 which is self-explanatory.

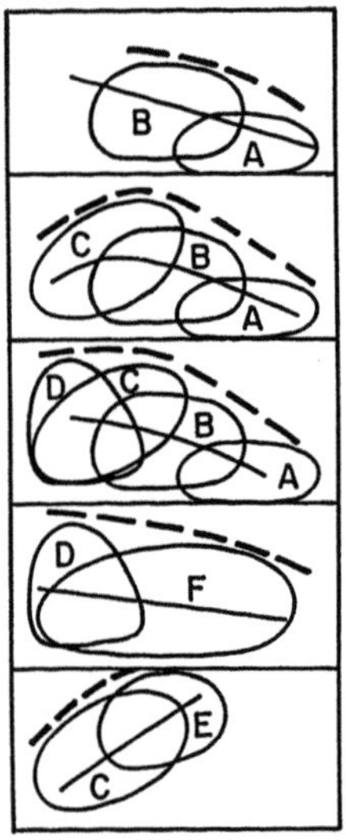

Figure 9. Data plots combined from Figure 8 and their mean (solid) and envelope (dashed) curves.

5. Conclusions and Hypotheses

1. The Kheisin-Kurdyumov hydrodynamic model of ice-rigid body interaction is able to predict comprehensively the parameters of the impact ice loads on ships and offshore structures. It has been developed in conjunction with drop-ball experiments performed to verify the basic assumptions and to quantify some values, and is generally in good agreement with the drop-ball test data. A number of undisputable experimental facts recorded in laboratory tests and full-scale trials are fairly predicted by the theory, at least qualitatively (for example, the time histories of impact pressure, global force, velocity and penetration depth). However, practical application of this theory for ship-ice and offshore structure-ice impacts has been limited by insufficient experimental data to show which of the properties of ice are responsible for its resistance to impact penetration and how to quantify them. These properties can not be measured directly; only indirect estimates of their values are available. An alternative representation of the ice resistance to impact penetration is necessary to make the theory more effective in predicting the impact loads.

2. The impact crushing strength of ice, also known as the specific impact crushing energy, and denoted here and in other references by ε, is the appropriate parameter characterizing the resistance of ice to impact penetration. It is defined as the kinetic energy dissipated for ice crushing per volume of the crushed ice, and has a dimension of stress (Pa). It can be measured directly from simple tests, yielding stable reproducible values with satisfactory data spread. It is incorporated into the theory being the only parameter characterizing the ice resistance to impact against a rigid body.
3. The experimental data on ε acquired to date are insufficient for clear understanding of the relationships between the parameters of the ice load and the impact conditions. Similar to other strength characteristics, the impact crushing strength of ice should be further tested to investigate its dependence on the impact velocity, mass and scale, as well as on ice grain/crystal structure, temperature, salinity, etc.
4. The ice impact pressures on the testing indenters, as well as on the ships and offshore structures, are specified by the theory as depending on the impact crushing strength only, i.e. velocity-and-mass independent provided that the impact crushing strength of ice is velocity-mass independent, as well. The question whether the impact pressure is indeed independent of the impact velocity or mass is not yet answered undisputably. There are experimental data sets of ice impact tests showing that the impact velocity effect on the impact crushing strength is weak, if any. By the same token, there are a few data sets, on the contrary, implying a strong velocity effect. Full-scale trials data available from publications provide inconclusive information regarding the ship speed effect on the impact pressure. It implies that the speed effect is rather small, if any. Its quantitative estimates are uncertain. The mass effect on the impact crushing strength of ice has not been detected in the drop-ball tests. Since the mass variation range has been rather narrow, the mass-independence is not a conclusion but rather a hypothesis.
5. The impact global force, as defined by the theory, depends strongly on the impact velocity and mass, as well as the impact crushing strength of ice. Virtually all experimental data sets (both laboratory and full-scale) support this conclusion but differ in quantifying the dependence.

6. An experimental program of carefully controlled drop-ball tests can clarify the problem of ice impact pressures and forces on engineering structures, without (or at least prior to) carrying out expensive full scale trials. The testing program should include parametric variations of mass within at least two, but preferably three or more, orders of magnitude, velocity within one order of magnitude, and diameter of the spherical indenter.
7. When measurements of ice impact loads are included in a ship trial program, the data acquisition system should include synchronous measurements of the velocities of ship motion. At least two velocity components should be recorded for ramming icebreaking and three or more for side impacts. Ice failure modes should also be reported, where feasible.

Acknowledgements

The author gratefully acknowledges the American Bureau of Shipping for the opportunity to publish this work.

References

APOA (Arctic Petroleum Operators Association) 1982. Ice forces on Hans Island. APOA Project No.181, 1982.

APOA (Arctic Petroleum Operators Association) 1984. Tarsiut Island data analysis. APOA Project No.213.

Chin, S.N., Williams, F.M. 1989. Measurement of impact force on sea ice. Proc. POAC '89, Lulea, Sweden, June 12-15, Vol.2, pp.638-647.

Daley, C.G., St. John., J.W., Seibold, F., Bayly, I. 1984. Analysis of extreme ice loads measured on USCGC *Polar Sea*. Trans. SNAME, Vol.92, pp. 241-252.

Dementyev, Kh.N. 1961. Determining the impact hardness of ice. Problems of Arctic and Antarctic, 7, pp.52-53, in Russian.

El-Tahan, H., Swamidas, A.S.J., Arockiasamy, M., Reddy, D.V. 1984. Strength of iceberg and artificial snow ice under high strain rates and impact loads. Proc. OMAE '84, New Orleans, LA, Vol.3, pp.158-165.

Fransson, L., Sandkvist, J. 1989. Pendulum indentation tests on manufactured ice. Proc. POAC '89, Lulea, Sweden, Vol.2, pp.669-678.

Garcia, N.B., Farrel, D., Mellor, M. 1985. Ice penetration tests. Cold Regions Science and Technology, 11, pp.223-236.

Ghomeim, G.A.M., Keinonen, A.J. 1983. Full-scale impact tests of *Canmar Kigoriak* in thick ice. Proc. POAC '83, Helsinki, Finland, Vol.3, pp.329-346.

Glen, I.F., Comfort, G. 1983. Ice impact pressure and load: investigation by laboratory experiments and sea trials. Proc. POAC '83, Helsinki, Finland, Vol. 1, pp.516-533.

Glen, I.F., Blount, H. 1984. Measurement of ice impact pressures and loads onboard CCGS *Louis S. St. Laurent*. Proc. OMAE '84, New Orleans, LA, Vol.3, pp.246-252.

Glen, I.F., Nawwar, A.M., Brown, R. 1985. MV *Arctic* structural performance. Summary report TP 5681E (Available from Transportation Development Centre, Montreal, Quebec, Canada).

Hoffman, L. 1985. Impact forces and friction coefficient on the forebody of the German polar research vessel *Polarstern*. Proc. POAC '85, Narssarssuaq, Greenland, Sept. Vol.3, pp.1189-1202.

Joensuu, A. 1988. Ice pressure measurements using PVDF Film. Proc. OMAE '88, Houston, Texas, Vol.4, pp.153-158.

Johnson, R.C., Benoit, J.R. 1987. Iceberg impact strength. Proc. OTC '87, Houston, TX, Paper OTC-5599, Vol.4, pp.417-423.

Jordaan, I.J., McKenna, R.F. 1988. Modelling of progressive damage in ice. Proc. IAHR Ice Symposium 1988, Sapporo, Japan, Vol.2, pp.585-624.

Kheisin, D.E., Cherepanov, N.V. 1970. Changing the structure of ice in the zone of a rigid body impact against ice cover surface. Problems of Arctic and Antarctic, 34, pp.79-84, in Russian.

Kheisin, D.E., Likhomanov, V.A. 1973. Experimental determination of specific energy of mechanical crushing of ice due to impact. Problems of Arctic and Antarctic, 41, pp.55-61, in Russian.

Kheisin, D.E., Kurdyumov, V.A., Likhomanov, V.A. 1975. Determining the specific crushing energy and contact pressures due to rigid body-ice impact. Trans. Arctic and Antarctic Research Institute, Leningrad, Vol.326, pp.210-218, in Russian.

Khrapaty, N.G., Tsuprick, V.G. 1976. Experimental study of rigid body-ice impact. Proc. Coordination Meeting on Hydraulic Engineering, Leningrad, Energia Publishing House, issue 111, pp.166-169, in Russian.

Kivimaa, S., Kujala, P. 1988. Measuring ice loads onboard the MV *Kashira*. Proc. Soviet-Finnish Seminar on Ice Strength of Ships, Leningrad in Russian.

Kurdyumov, V.A., Kheisin, D.E. 1976. A hydrodynamic model of ice-rigid body impact. Applied Mechanics, XII, 10, 1976, pp.103-109.

Likhomanov, V.A., Kheisin, D.E. 1971. Experimental study of rigid body-ice impact. Problems of Arctic and Antarctic, 38, pp. 105-111, in Russian.

Makarov, S.O. 1901. *The Ermak* in ice. St. Petersburg, 670p, in Russian.

Minnick, P., St. John, J., Cowper, B., Edgecombe, M. 1986. The 1985 global ice impact tests on the USCGC *Polar Sea*. Report No. SR-1313 for Ship Structure Committee.

Muller, L., Payer, H.G. 1988. Loads on research vessel *Polarstern* under arctic conditions, Proc. POAC '87, Fairbanks, Alaska, USA, Vol.1, pp.495-508.

Muller, L., Payer, H.G., Moore, C. 1989. Ice impact of ship hulls. Working Group on Ice Forces, 4th State-of-the-Art Report, Special Report 89-5, CRREL, Hanover, New Hampshire, pp.339-375.

Nevel, D.E. 1986. Iceberg impact forces. Proc. IAHR Ice Symposium 1986, Iowa City, IA, USA, Vol.3, pp.345-369.

Pounder, E.R., Little, E.M. 1959. Some physical properties of sea ice. Canadian J. of Physics, Vol.37, pp.345-369.

Riska, K., Kujala, P., Vuorio, J. 1983. Ice load and pressure measurements onboard IB *Sisu*. Proc. POAC '83, Helsinki, Finland, April 1983, Vol.2, pp.1055-1069.

Runeberg, R. 1888. On steamers for winter navigation and ice-breaking. Proc. Institution of Civil Engineers, Vol. 97, Pt. III, pp.277-301.

Ryvlin, A.Ya., Kheisin, D.E. 1980. Tests of ships in ice. Leningrad, Sudostroenie Publishing House, 208p.

St. John, J.W., Daley, C. 1984. Shipboard measurements of ice pressures in the Bering, Chukchi, and Beaufort Seas. Proc. OMAE '84, New Orleans, LA, Vol.3, pp.260-266. (see also St. John, Daley, Blount - Ice Loads and Ship Response to Ice. Final Report No. SR-1291, Dec. 1984, by Arctec Group Companies prepared for SSC and Transport Canada).

Sodhi, D.S. and Morris, C.E. 1986. Impact ice force and pressure: An experimental study with urea ice. Proc. OMAE '86, Tokyo, Japan, Vol.4, pp. 569-576.

Swearengen, J.C. 1988. Deformation and fracture of arctic sea ice during kinetic energy penetration. Proc. OMAE '88, Houston, Texas, Vol.4, pp.331-336.

Tsurikow, V.L., Veselova, L.E. 1973. Dynamic hardness of the Caspian Sea ice. Trans. "Studies of Ice of the USSR Southern Seas", Moscow, Nauka Publishing House, pp.60-68.

Tunik, A.L. 1984 Dynamic ice loads on a ship. Proc. IAHR Ice Symposium 1984, Hamburg, W. Germany, Vol.3, pp.297-313.

Tunik, A.L. 1989. Impact crushing strength of ice. Proc. POAC '89, Lulea, Sweden, 1989, Vol.2, pp.685-670.

Tunik, A.L., Minnick, P.V., St. John, J.W., Chen, Y.K., Chen, A.P.Y. 1988. Ramming forces on the USCGC *Polar Sea*. Proc. IAHR Symposium on Ice, Sapporo, Japan, 1988, Vol.1, pp.746-761.

Yen, Y., Odar, F. and Bracy, L. 1970. Impact of spheres on ice. Proc. ASCE, J. of the Engineering Mechanics Division, Vol.96, No. EM5, pp.641-652.

Discussion

Q: A.F. Aboulazm, Marine Institute

1. Do you think that part of the initial energy of bodies in impact is used in rebound e.g. not all energy is used in crushing?
2. Is there any coefficient of restitution between ice and other materials? Do you think that this coefficient of restitution should be established in the future (under different conditions)?

A: A.L. Tunik

1. Of course, a part of the kinetic energy is dissipated for rebound. The coefficient of restitution (CR), measured by Yen et al. (1970) and Likhomanov and Kheisin (1971), is shown in Figure 10. As seen, it is effectively below 0.3 for masses higher than a kilogram at speeds less than 1 m/s (which corresponds only to the 4cm-height of dropping the sphere). One should also bear in mind that the ratio of rebound-to-total kinetic energies $E_r/E = CR^2$, thus if CR = 0.3, then only 9% of the total kinetic energy is reserved for rebound.
2. The work by Yen et al. is the only source of experimental data on the coefficient of restitution for materials other than steel impacting ice. The materials tested (nylon, acrylic, teflon) have much higher coefficients of restitution than steel. For teflon, for example, CR > 0.8, i.e. the teflon/ice impacts were mainly elastic with no ice crushing at all. I agree that more studies are needed.

Q: Ian Jordaan, Memorial University of Newfoundland

This is an important area of research and I believe you are taking the best approach available. In determining the breakup energy, I feel more attention should be given to the division between viscous extrusion and the breakup process, with the first of these not necessarily being negligible. Would you please comment.

A: A.L. Tunik

Distinction between the viscous extrusion and breakup processes within a model of ice-structure interaction is essential for understanding such phenomena as load pulsation during continuous indentation.

Q: B.L. Parsons (IMD/NRCC)

I would like to draw the author's attention to recent work by Chin and Williams (POAC '89) showing time records of pressure velocity and force for the first few milliseconds, that shows pressures may reach 60 MPa.

A: A.L. Tunik

This is an interesting work. Pressures of the same order (up to 56 Mpa) were also measured by Glen and Blount (1984) onboard an icebreaker. The pressure transducers in both the Chin and Williams, and the Glen and Blount tests had different types of gauges (piezoelectric force transducer, and strain-gauge diaphragm, respectively) but were of almost equal diameters (9 and 8mm, respectively). At such small sizes, which are of the order of the grain size, the measured pressure peaks should perhaps be attributed to the strength of single ice grains rather than to the strength of ice macrostructure. A better understanding of the nature of such high peaks could probably be obtained if 2 or 3 piezoelectric force transducers are installed on the indenter near each other with synchronous recording of their readings.

A Method of Scaling for Ship-Ice Model Tests

D. B. Colbourne

Institute for Marine Dynamics, National Research Council Canada
St. John's, NF Canada

J. H. Lever

U.S. Army Cold Regions Research and Engineering Laboratory
Hanover, NH USA

Abstract A dimensional analysis of the problem of vessels transiting level continuous ice is presented. Resistance is divided into components and each analyzed to identify dimensionless terms relevant to the individual component. These expressions provide a framework for scaling ship-ice model test results. A workable subset of scaling requirements for a three component division of the problem is derived from the general case. Some experimental results are presented demonstrating the utility of the method. It is shown that it may be possible to compensate for poor scaling in ice properties by scaling each component based on its own law of similitude.

1. Introduction

Ship model testing in ice has developed over a relatively short period and is a commonly used method of predicting icebreaker performance. Although a number of ship-ice testing facilities exist throughout the world, a standard method of testing and scaling results has not yet been developed.

The major problem with modelling ships in ice is that many different mechanisms resisting the vessel forward motion act at the same time. These include hydrodynamic effects, friction, ice crushing, ice fracture, buoyancy and inertial effects. To date, no model ice-fluid combination correctly scales all these phenomena simultaneously.

S. Jones · R. F. McKenna · J. Tillotson · I. Jordaan (Eds.)
Ice-Structure Interaction
IUTAM-IAHR Symposium St. John's, Newfoundland, Canada

Most researchers address scaling difficulties by dividing icebreaker resistance into components. Over the years, non-dimensional expressions for such components have been derived by dimensional analysis. In most cases, coefficients for these expressions have been obtained by regression analysis of gross ice resistance model tests, not by separate measurement of each component. Experience has shown that full-scale predictions based on this approach are not very reliable. This lack of reliability indicates that there are different scaling requirements for different but equally important components of the total resistance. Furthermore, regression analysis on complete model test data does not confirm that the correct non-dimensional expressions have been formulated.

To facilitate separate scaling, we must separately measure the principal components of icebreaking resistance. This paper presents the scaling requirements for three icebreaking components (hydrodynamic, breaking and clearing resistances), the means to measure those components experimentally and a set of results which demonstrate the good collapse of the data achieved by the selected non-dimensional expressions. Taken together, the method provides a practical means of scaling icebreaking resistance obtained in model tests.

2. Background

Kashteljan et al. (1968) were the first to divide icebreaking resistance into components. Although their approach is not strictly non-dimensional, it paved the way for later developments. The final equation (see Notation) was of the form;

$$R_t = C_B B \sigma h \mu_0 + C_s \rho_i B h^2 \mu_0 + C_v B^x h V^y / n_2 + R_{OW}, \tag{1}$$

where μ_0 and n_2 are hull shape factors. This divides the resistance into four parts. The first term is associated with breaking ice while the second is resistance due to submerging broken ice pieces. The third is a velocity dependent resistance associated with the inertia of ice pieces pushed aside by the ship. R_{OW} is the open-water resistance of the vessel.

Edwards et al. (1972) present non-dimensional groups applicable to the icebreaking problem;

$$\frac{R}{\rho_w g B h^2} \quad \text{Total Resistance Coefficient}$$

$$\frac{V}{(gh)^{\frac{1}{2}}} \quad \text{Froude Number}$$

$$\frac{\sigma}{\rho_w g h} \quad \text{Ice Strength Ratio}$$

$$\frac{h_s}{h} \quad \text{Snow Cover Ratio}$$

These groups provide a component basis for scaling of model tests. However, Edwards et al. (1972) did not separately measure the contributing resistance components.

Enkvist (1972) gives a review of ship-ice model testing and introduces a non-dimensional resistance function. He explicitly identifies friction in his expression:

$$\frac{R}{BhT\rho_D g} = [1 + C_\mu(\mu)][C_\sigma(S_\sigma, G_T) + C_s + C_V(F_T, G_T)], \tag{2}$$

where

$$S_\sigma = \sigma/(T\rho_D g) \quad \text{Ice Strength Ratio}$$

$$F_T = V/(gT)^{\frac{1}{2}} \quad \text{Froude Number}$$

$$G_T = T/h \quad \text{Ice Thickness Ratio}$$

$$\mu = \quad \text{Friction Factor}$$

Vance (1974) also developed scaling requirements for testing ship models in ice, based on force similitude. His scaling requirements are similar in form to the expressions derived by Enkvist.

Enkvist (1983) later developed an experimental technique to separate breaking resistance from the total resistance. This approach involved presawing the ice cover to achieve zero ice strength. Enkvist, however, assumed that the ice breaking component of resistance was speed independent, and hence only conducted presawn tests at a single low speed.

Atkins and Caddell (1974) present a non-dimensional parameter involving the fracture toughness of ice. This parameter, named the Ice Number, is said to be a Cauchy number modified to account for the effect of flaws in the material. Its appearance represents the first recognition that vessel velocity is relevant to the ice breaking component.

$$I_n = \frac{V^2 \rho (L)^{\frac{1}{2}}}{K}, \tag{3}$$

where

$$K = (EG)^{\frac{1}{2}} \qquad \text{Critical Stress Intensity.}$$

Although it is becoming more common to apply fracture mechanics to the problem of ice failure, the issue has not been extensively addressed and Atkins' Ice Number has not gained wide usage.

On the tank testing side, much emphasis has been placed on developing ice formulations to satisfy simultaneously all scaling requirements. Although steady progress has been made in improving mechanical properties of model ice (Timco, 1984), this is similar to attempting to formulate a test fluid to satisfy both Reynolds and Froude scaling at the same time. By comparison, it is standard practice in naval architecture to separate the wave making and viscous components of open-water resistance. This is also likely to be the most fruitful approach for scaling the more complex processes involved in icebreaking.

In order to establish a practical approach, we must be able to identify the non-dimensional terms relevant to each resistance component and be able to isolate the components experimentally. The following sections present methods of achieving both these objectives.

3. Component Identification

3.1 Dimensional Analysis

For this exercise, we treat each icebreaking resistance component independently. The total ship-ice resistance is assumed to be the sum of;

1. Open Water Resistance	=	R_O	(4a)
2. Ice Breaking Resistance	=	R_B	(4b)
3. Floe Submergence Resistance	=	R_S	(4c)
4. Floe Inertial Resistance	=	R_I	(4d)
5. Frictional Resistance	=	R_F	(4e)
6. Snow Resistance	=	R_C	(4f)

At present we choose to neglect snow resistance as there is no standard way of modelling it.

Physical variables associated with the identified components are;

$$\begin{array}{lll} \text{Open Water:} & R_O = f_1(\mu, L, B, T, V, g, \rho_w, S) & \text{(5a)} \\ \text{Breaking:} & R_B = f_2(B, h, \sigma_F, \sigma_C, \sigma_S, G, a, E, S, V, \rho_i) & \text{(5b)} \\ \text{Submergence:} & R_S = f_3(\rho_d, T, B, S, h, g) & \text{(5c)} \\ \text{Inertial:} & R_I = f_4(V, \rho_i, h, B, S, g, \rho_w) & \text{(5d)} \\ \text{Frictional:} & R_F = f_5(L, f_i, P, h) & \text{(5e)} \end{array}$$

The application of dimensional analysis (Sharp, 1981) to the equations (5a)-(5e) leads to:

$$\frac{R_O}{\rho_w V^2 LT} = \Phi_1\left(\frac{VL}{v}, \frac{V^2}{gL}, \frac{L}{B}, \frac{B}{T}, S\right) \text{ where } v = \mu/\rho_w \tag{6a}$$

$$\frac{R_B}{Gh} = \Phi_2\left(\frac{G}{\sigma_F h}, \frac{\sigma_S}{\sigma_F}, \frac{\sigma_C}{\sigma_F}, \frac{E}{\sigma_F}, \frac{B}{h}, \frac{G}{Ea}, \frac{\rho_i V^2}{E}, S\right) \tag{6b}$$

$$\frac{R_s}{\rho_d g BhT} = \Phi_3\left(\frac{T}{h}, \frac{B}{h}, S\right) \tag{6c}$$

$$\frac{R_I}{\rho_i V^2 Bh} = \Phi_4\left(\frac{V^2}{gh}, \frac{\rho_i}{\rho_w}, \frac{B}{h}, S\right) \tag{6d}$$

$$\frac{R_F}{PLh} = \Phi_5\left(f_i, \frac{h}{L}\right) \tag{6e}$$

The normal requirements for testing models in open water are defined by item (a). For an icebreaker, this component is different from resistance in the absence of ice because the broken floes and the unbroken ice cover modify the flow and wave radiation fields. Although beyond the scope of this paper, the main influence of ice appears to be its ability to absorb wave energy, thereby increasing that resistance component (Colbourne, 1989). Thus, strictly speaking, model open-water resistance should be measured along a slot in the ice approximately the width of the broken channel. Viscous fluid drag may be computed using one or another standard formulation (Comstock, 1967). Because the open-water component of icebreaking resistance is relatively small in most instances, small errors in its calculation do not substantially affect estimates of total resistance.

In consideration of dimensionless groups associated with the breaking component, (item b), ice thickness is scaled to the same geometric ratio as the ship (λ) and the ratios of strengths and elastic modulus for the ice must be similar at full and model scale. Normally, Froude velocity scaling is used (velocity scaled by $\lambda^{\frac{1}{2}}$). It then follows that elastic modulus and consequently all ice strengths must be scaled by λ (assuming equal ice density at both scales). Fracture toughness on the other hand must be scaled by λ^2. However, if we can independently measure the breaking component we do not need to use Froude velocity scaling for this component. We could vary model velocity to maintain the relevant dimensionless ratio for varying flexural strength or fracture toughness in the model ice.

Inertial resistance (item d) is similar to the submergence term (item c) with the addition of velocity effects. Indeed, it is sensible to eliminate submergence as a separate component and combine all ice-clearing resistance into one term. The only advantage of separating these two components is that the inertial resistance is negligible at low speed while the submergence component remains. However as speed increases, the distinction between the two components is less definite. For our purposes, we will treat submergence resistance as the zero speed limiting value of inertial resistance. Note that dimensionless groups associated with inertial resistance dictate maintenance of the ice-water density ratio and geometric scaling of ice thickness. An ice thickness based Froude number imposes $\lambda^{\frac{1}{2}}$ velocity scaling for this term.

Frictional resistance (item e) is primarily a function of the coefficient of friction between the ice and the ship. Theoretically, this coefficient should be the same at full and model scale. However, there are problems measuring ice friction coefficients and evidence in the literature (Forland and Tatinclaux, 1985) indicates that the frictional coefficient for ice is a function of both pressure and sliding velocity. It is also unlikely that friction is an independent resistance component; it probably influences the other components to varying degrees, depending on local velocities and pressures. Thus, rather than treating it as a separate entity, we will view it as an element of the other components.

3.2 A Three Component System

Recall that our objective is a practical scaling methodology which covers the principal icebreaking resistance components. To this end, we may reduce the components under consideration to three: hydrodynamic, breaking and clearing.

The hydrodynamic component remains unchanged. As mentioned, to account for additional energy absorption of the ice cover, we should measure wavemaking resistance in a slotted channel and scale its value using Froude scaling. We correct for viscous effects using the ITTC correlation line. As discussed above we may combine the submergence and inertial terms into a single ice clearing resistance. Because frictional forces are influenced by the magnitude of ice-breaking and ice-clearing forces (and vice versa) we cannot treat ice-vessel friction as an independent term, choosing instead to show it as a parameter associated with ice-breaking and ice-clearing resistance components. The original ice-breaking term (equation 6b) includes some strength terms which are also deleted. This is due to the fact that flexural strength is viewed as the most important. In addition, other strengths (compressive, shear) in model ice cannot be independently controlled, and track the flexural strength as it decreases with tempering time. While fracture is clearly an important process, we must know both the fracture toughness and the initial flaw distribution to compute a failure stress. Because flaw distributions at either model or full scale are essentially unknown, the fracture parameters are of little practical modelling value. Instead most researchers characterize the failure strength of an ice sheet using cantilever beam tests to yield flexural strength. While this is not, strictly speaking, a material property, flexural failure of a cantilever beam incorporates ice fracture properties in a similar fashion to vessel icebreaking. Its use avoids the need to measure and control these properties.

Based on the foregoing arguments, we may write the following non-dimensional expressions for the three principal icebreaking resistance components:

$$\frac{R_0}{\rho_w V^2 LT} = \Phi_1\left(\frac{VL}{v}, \frac{V^2}{gL}, \frac{L}{B}, \frac{B}{T}, S\right) \tag{7a}$$

$$\frac{R_B}{\rho_i V^2 Bh} = \Phi_2\left(\frac{\rho_i V^2}{\sigma}, \frac{E}{\sigma_F}, \frac{B}{h}, S, f\right) \tag{7b}$$

$$\frac{R_C}{\rho_i V^2 Bh} = \Phi_3\left(\frac{V^2}{gh}, \frac{\rho_i}{\rho_w}, \frac{B}{h}, S, f\right) \tag{7c}$$

The clearing component (7c) is a combination of the previous inertial and submergence components and is a function primarily of ice-thickness Froude number. Effects due to the ice/water density ratio, which is not correctly modelled in most model ices, are assumed to be second order in nature. The breaking component (7b) is primarily a function of $(V^2\rho_i/\sigma)$

which we will call the Strength Number. This shows vessel velocity to be significant in scaling which is not in keeping with previous developments, where most have assumed the breaking component to be velocity independent (Kashteljan et al., 1968; Edwards et al., 1972; Enkvist, 1972; Vance, 1974). This term indicates that the ratio of inertial forces to flexural-strength forces in the ice sheet is important in the ice-breaking resistance. We believe this to be the case, and that traditionally, the dynamic role of the underlying fluid as an elastic foundation has been underestimated when considering the breaking component. The influence of the E/σ ratio is also assumed to be second order. Although it is usually lower in model ice than at full scale, our experience indicates that changes in E/σ ratio do not have any effect on vessel resistance as measured in model ice.

This three component system provides a workable basis for scaling model test results provided we can separately measure the breaking and ice-clearing components. The most promising means of doing this is Enkvist's presawn method (Colbourne, 1989). By sawing the ice in a pattern similar to the cracking pattern observed during a normal ice resistance test, we may effectively reduce the ice strength to zero. When the model is run through this sawn ice, the measured resistance consists only of the hydrodynamic and ice clearing components. By subtracting these resistance values from those recorded during the normal ice test, we obtain the breaking resistance. However, we conduct presawn tests for a range of model velocities in recognition of the scaling importance of the Strength Number. If we then conduct open-water resistance tests and measure the hydrodynamic resistance, the clearing component can be derived by subtracting the open water test values from the presawn results.

4. Verification

An extensive experimental program was carried out at the ice tank facility of the Institute for Marine Dynamics (IMD). The objectives of this program were to confirm that the presawing technique was a reliable and practical technique for separating the icebreaking process into components, and to verify that the functional relationships developed from the dimensional analysis do indeed collapse the data. Experiments consisted primarily of resistance measurements on a simplified hullform towed through level ice. Ice mechanical properties, presawn pattern and vessel speed were varied within predefined ranges. The experiments and results are described in detail in Colbourne (1989). Summarized below are the key results relating to the practicality and reliability of the three component method.

Presawing was found to be a practical method of eliminating strength effects from model ice and allowing isolation of the breaking component. Test results were relatively insensitive to minor variations in the sawing pattern. Indeed, presawing has become a regular ice test procedure at the IMD. Some coupling was observed between the hydrodynamic resistance and the ice-clearing resistance, making it more difficult to clearly separate these two components. This is probably due to entrainment of the broken ice pieces in the flow around the vessel hull. Further research is planned to investigate this phenomenon.

Use of the non-dimensional formulations presented in 3.2 achieved a good collapse of the model test data for both the breaking and clearing components. Under this system, all data points, regardless of ice strength, thickness or model velocity collapse to the same curves (one

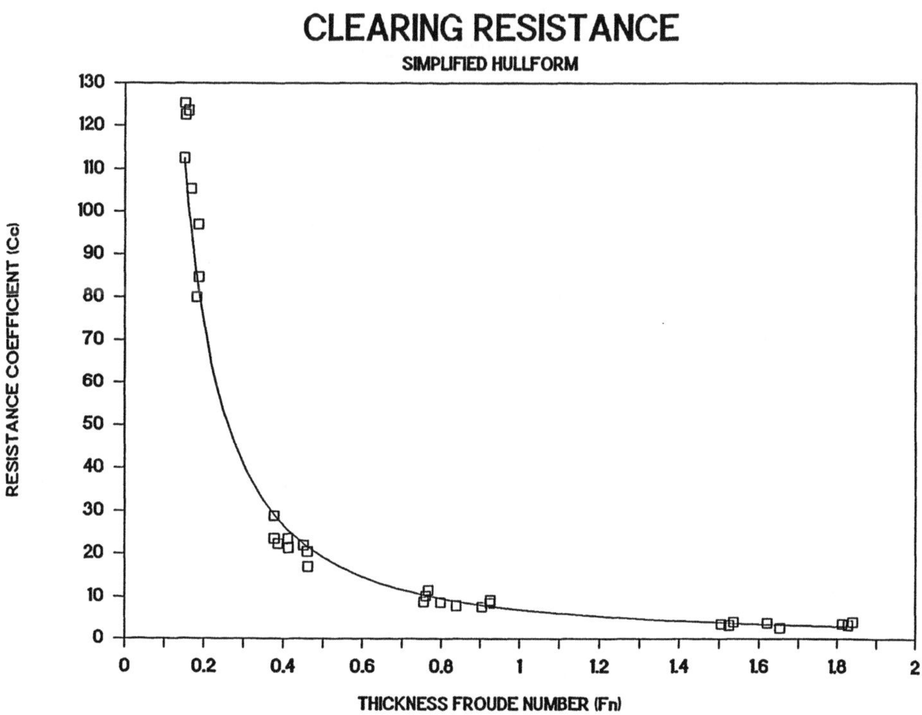

Figure 1. Clearing resistance: simplified hullform.

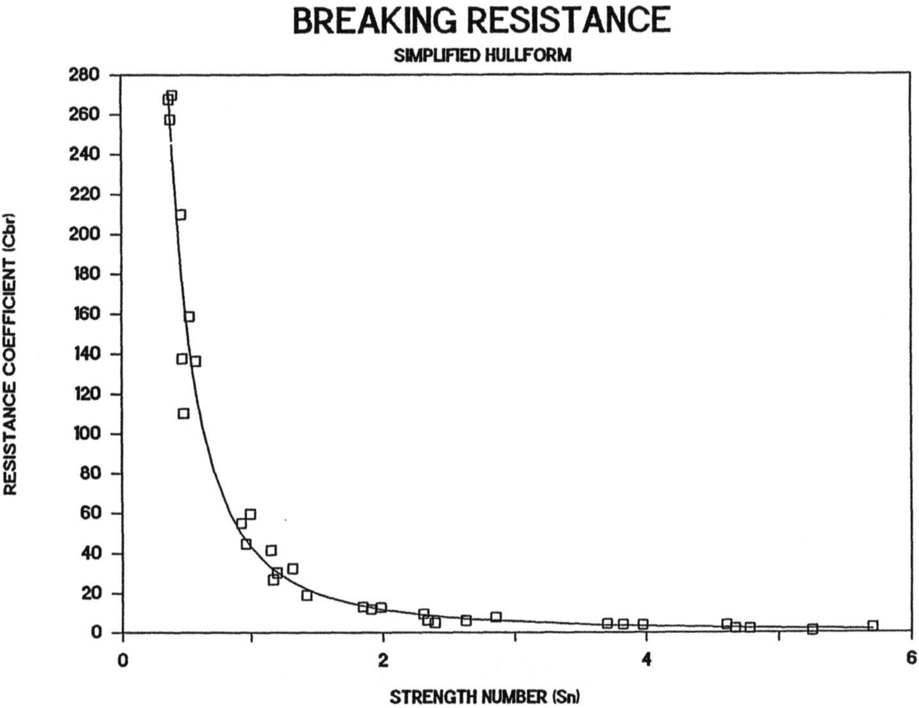

Figure 2. Breaking resistance: simplified hullform.

each for breaking and clearing). This was true for a highly simplified hull form and for a number of existing icebreaker forms. The effects of friction can also be more easily identified by obtaining one non-dimensional breaking curve and one clearing curve for each friction coefficient. Figures 1 and 2 show the non-dimensional results of resistance experiments on the simplified hullform. Very good collapse of the data is apparent. As expected from the dimensional analysis, ice clearing resistance is a definable function of the thickness Froude number and ice-breaking resistance is a similar function of the Strength Number. Points on both curves represent differing combinations of model velocity, ice strength and thickness. From these curves, we can derive relationships of the form:

$$\frac{R_C}{\rho_i V^2 Bh} = C_C = k_1(Fn)^{-a}, \qquad (8a)$$

$$\frac{R_B}{\rho_i V^2 Bh} = C_B = k_2(Sn)^{-b}, \qquad (8b)$$

where a, b, k_1, k_2 are regression coefficients.

The basic form of these equations have been confirmed for five conventional ship hullforms for which adequate test data were available, including a comprehensive test series of the Canadian R-class hull at 3 model scales (1:40, 1:20, 1:8).

5. Discussions and Conclusions

For a complex process such as vessel icebreaking, it is unrealistic to expect adequate scaling of all important parameters simultaneously. Instead, we have adopted the approach of separately identifying, measuring and scaling individual resistance components. Using dimensional analysis we have formulated relationships for each component of icebreaking resistance which for practical reasons are reduced to three principal resistance components: hydrodynamic, breaking and clearing. The key to overcoming competing scaling laws for these components lies in being able to separate them experimentally. Following a comprehensive series of tests we found the presawing technique to be both practical and reliable. The data for clearing resistance coefficient (C_C) collapses well as a function of ice thickness Froude number (Fn). Similarly the breaking resistance coefficient (C_B) collapses well on ice strength number (Sn). These results confirm the forms of the developed non-dimensional groups.

We believe this method offers improvement over scaling the total icebreaking resistance on the basis of a single scaling law. In the course of standard ice model tests, this procedure has been performed at minimal additional cost. Information gained from measurement of individual components provides a more rational means of analyzing test results and offers insight into the physics of the dominant ice breaking resistance phenomena. Separation of the breaking component may be of further advantage by allowing some problems with mechanical properties of model ice to be overcome. For example, tests could be run at lower speeds in higher strength model ice (constant Strength Number) which would have better E/σ ratios and not require long tempering times to yield lower strengths. This would improve test procedures and produce a longer test "window" when the model ice properties are acceptable. Also, because all data applies to the same non-dimensional curves, scatter in results and deviations from target values of ice strength and thickness are of less concern.

There is however room for more work in this area. We observed during the experiments that there is some coupling between the flow of water and the flow of ice pieces around a hull. We plan to investigate this phenomenon further, aiming toward better separation of the ice clearing and hydrodynamic resistance. Quantification of frictional effects would also be valuable. Tests conducted at three different friction values suggest that by breaking the problem into components, some insight can be gained into the role of friction in ship-ice interactions.

Although research into icebreaking is expensive, even at model scale, the ability to make reliable predictions for what are among the most expensive ships built, could offer substantial savings in design and construction. The presented system of scaling, based on division of the resistance into three components offers a good basis for further analysis of the icebreaking problem and an improved method of scaling results of model tests.

Notation

GROUP	PARAMETER	SYMBOL	DIMENSION
Ship	Length	L	L
	Beam	B	L
	Draft	T	L
	Velocity	V	L/T
	Shape Coefficients	S	-
Response	Resistance	R	ML/T^2
Ice	Flexural Strength	σ_F	M/LT^2
	Compressive Strength	σ_C	M/LT^2
	Shear Strength	σ_S	M/LT^2
	Elastic Modulus	E	M/LT^2
	Fracture Toughness	G	M/T^2
	Flaw Size	a	L
	Thickness	h	L
	Density	ρ_i	M/L^3
	Lateral Pressure	P	M/LT^2
	Friction Coefficient	f_i	-
	Stress Intensity	K	$M/L^{\frac{1}{2}}T^2$
Water	Density	ρ_w	M/L^3
	Viscosity	μ	M/LT
Other	Gravitational Acc.	g	L/T^2
	Dimensionless Coeff.	C	-
	Scale Factor	λ	-

References

Atkins, A.G., Caddell, R.M. 1974. The laws of similitude and crack propagation. International Journal of Mechanical Sciences, Vol. 16, No. 8, pp. 541-548.

Colbourne, D.B. 1989. A three component method of scaling ice resistance. Ph.D. Thesis, Memorial University of Newfoundland.

Comstock, J.P. 1967. Principles of naval architecture. Society of Naval Architects and Marine Engineers, New York.

Edwards, R.Y. et al. 1972. Full scale and model tests of a Great Lakes icebreaker. Transactions of The Society of Naval Architects and Marine Engineers, Vol. 80, pp. 170-207.

Enkvist, E. 1972. On the ice resistance encountered by ships operating in the continuous mode of icebreaking. The Swedish Academy of Engineering Sciences in Finland, Helsinki, Report No. 24.

Enkvist, E. 1983. A survey of experimental indications of the relation between the submersion and breaking components of level ice resistance to ships. POAC, Helsinki, Vol. 1, pp.484-493.

Forland, K.A., Tatinclaux, J.C. 1985. Experimental investigation of the kinetic friction coefficient of ice. CRREL Report No.85-6, Hanover, N.H.

Kashteljan, V.I., Poznjak, I.I., Ryvlin, A.J. 1968. Ice resistance to motion of a ship. Translation, Marine Computer Applications Corp.

Sharp, J.J. 1981. Hydraulic Modelling. Butterworths, Toronto.

Timco, G.W. 1984. Ice forces on structures: physical modelling techniques. IAHR Symposium on Ice, Hamburg, Vol.4, pp.117-150.

Vance, G.P. 1974. A modelling system for vessels in ice. Ph.D. Thesis, University of Rhode Island.

Discussion

Q: C. Daley, Fleet Technology Ltd.

Your justification for considering certain modelling parameters to be irrelevant (crushing, shear, friction etc.) seems to be based on a modelling medium in which these parameters are either irrelevant or uncontrollable. This seems to be circular reasoning. You don't show that these parameters are irrelevant in full scale. Please comment.

A: D.B. Colbourne, J.H. Lever

In no case did we state that any modelling parameters were irrelevant. We did however state that flexural strength is the most important index of ice strength of ship icebreaking. In addition we did not neglect friction but indicated it to be an important factor in each of the icebreaking components. Model ice does not allow independent control of all strength properties (ie. as flexural strength reduces, so do crushing and shear strengths) but we would suggest that this also happens at full scale.

In modelling any phenomena, one is not likely to be able to scale properly all things at one time. Choices must be made as to which parameters are most important. In this paper, we have attempted to break the problem into units for which a single dominant physical mechanism can be identified, thus singling out the most important parameters. This has nothing to do with the modelling medium and to us, seemed to be quite logical reasoning.

To date we have not been able to find any full scale data which would allow a conclusive confirmation or refutation of our assumptions. We have, however, tested our method over three model scales; 1:40, 1:20 and 1:8, a range over which other ice properties showed substantial variation, and found that our assumptions were valid.

Q: A.F. Aboulazm, Marine Institute

What is the width of the presawn channel and the size of broken ice in relation to model size?
Have you compared your results with earlier presawn model testing available in the literature?

A: D.B. Colbourne, J.H. Lever

For the purposes of our experiments we established a standard channel width equal to the model beam plus three times the ice thickness. Lateral cuts were made at 10 to 20 cm separation depending on ice thickness.
As far as possible we have compared our results with those published by Enkvist and Tatinclaux. In general we have found them to be similar.

Q: M. Wishahy, NORDCO Limited

How does the dynamic behaviour of the ship influence the breaking and clearing components of the ice, and how do you account for the difference in behaviour of the ship as it moves in presawn ice from that in a solid ice sheet?

A: D.B. Colbourne, J.H. Lever

We feel that the influence of ship dynamic motions on resistance in ice are probably relatively minor. Any augmentation of resistance due to greater motion amplitudes in solid ice as opposed to presawn ice would be attributed to the breaking resistance component.

Q: B. L. Parsons, IMD/NRCC

Could you comment on what you think the mechanism of interaction between the submergence and hydrodynamic resistance might be, and how one might separate the two phenomena in an experiment in the ice tank? Is this interaction relevant to the limit state of the icebreaker, or is it only important at the higher velocities?

A: D.B. Colbourne, J.H. Lever

We believe that the interaction mechanism is a combination of viscous drag and added mass effects, resulting in ice masses being "entrained" in the flow around the ship hull. In other words, at higher speeds, as the fluid is pushed aside the ice pieces are carried with it. We are not sure at this time how these phenomena might be separated by experiment, but plan further investigations as the follow on work to the present study.

We suspect that this interaction is not relevant for icebreakers operating at low speeds (< 3 kts) in heavy ice but may be evident at higher speeds (6 kts) in light ice. This is probably not the limiting case for most icebreakers.

CHAIRMAN'S REMARKS

G.W. Timco, National Research Council

There are five papers presented in the morning session on ice/structure interaction. There was good interest in all of the papers and significant discussion followed each paper.

The first paper, which was presented by Professor Mauri Määttänen of Helsinki University in Finland was a general lecture on "Ice Interaction on Structures". The theme of the paper centred around a review of the current problems in this area and the difficulties in understanding the complex nature of the interaction process. The emphasis of the paper was on bottom-founded structures. The paper described the process of combining the properties of both the structure and the ice in order to arrive at the failure processes in the ice. This task was found to be very difficult given the wide variety of factors which can affect the outcome. Määttänen described the role of several of these factors including the numerous failure processes in ice (visco-plastic yielding, crushing, shearing, bending and splitting) and the fact that the structural response can be either static or dynamic. This further complicates the situation since in certain instances a coupling of the structure to the ice failure process may occur. Finally, Määttänen presented a brief discussion of the clearing processes of ice after failure has occurred. In this case again both the properties of the structure and the ice are important.

This paper gave a good starting point for discussions in the area of ice/structure interaction. Following the presentation of this paper there were several questions, many of which were quite general in nature. The discussion centred around the problem areas in understanding the processed discussed by Määttänen. There was agreement that our understanding in this area has advanced to the state where many of the significant parameters and processes have been identified. However it was clear that our basic understanding of these processes is still very much in its infancy.

The second paper in the session was presented by Dr. D. Sodhi of the U.S. Army CRREL in Hanover, U.S.A. The paper described a detailed test series which had been performed by the author in an investigation of the ice indentation problem. The tests, which were performed using freshwater ice in the CRREL ice tank involved measurements of the loads, displacements, accelerations and failure processes of the ice when indented with a flat indentor bar in edge loading. Dr. Sodhi presented a comprehensive discussion of the test apparatus which had been developed for this test program. The system was designed to measure all of the important test parameters, and allowance was made to be able to change the relative stiffness of the indentor bar. During the presentation a number of the salient features of the test results were presented, but a time limitation precluded a full disclosure of all of the results. A short video was shown which gave the audience a good understanding of the test arrangement and the failure behaviour of the ice.

This piece of work was very well received by the audience and there was considerable discussion. One feature which was pursued in the question period centred about the general applicability of the range of the test results. In the video presentation it was clear that there was considerable microcracking in a large area surrounding the indentor in the ice. The microcracking extended for a distance several times the diameter of the indentor, and extended in a region along both sides of the indentor. The large extent of this zone and amount of microcracking contrasted significantly with the results of similar tests performed by Timco at NRC and discussed by Jordaan during his presentation. The tests by Timco, however, were performed at a higher rate than those shown by Sodhi in his video. This comparison clearly illustrated the importance of interaction rate in the failure processes of ice in experiments of this type. It was evident that general statements as to the processes involved will always have to be prefaced with the range in which the processes predominate.

The third paper in the session, which was presented by Professor Xu Jizu of Tianjin University, China, discussed aspects of designing a structure which may be excited by ice crushing through self- induced vibrations. Professor Xu gave an overview of some of the problems which a Design Engineer faces in designing a slender structure which will be used in ice-infested waters. He then related some of the experience which the Chinese have acquired in using jacket-type structures in the Bohai Bay region in China. In the past ten years they have experienced several examples of severe structural vibration of these structures due to ice loading. In a few instances the vibrations have been very severe leading to a collapse of the structure. Prof Xu and his colleagues have been investigating this problem for several years and an approach to help alleviate this problem was discussed in his paper. In this technique the local stiffness of the structure is adjusted to optimize the mode shape. This then helps to control the displacements of the structure. In the paper the techniques for doing this are described mathematically using a Bohai Bay jacket structure as an example.

The discussion on this paper was centred around a comment by Professor Määttänen who offered support for the approach described by Professor Xu. This support stems from experience which Professor Määttänen has had in trying to solve this problem for lighthouses in the Baltic Sea.

The fourth paper in the session was presented by Dr. K. Riska of the Helsinki University of Technology, Finland. The presentation gave an overview of a theoretical analysis of the ice/structure interaction with the salient processes considered in terms of the structure motions, the ice motions and the penetration of the structure into the ice. Although all three of these aspects were considered, the main emphasis was on the penetration process. The work was illustrated through an example of a ship hitting a finite ice floe. A short video was presented showing some model tests of ship impact which had been performed by Fleet Technology Ltd., Canada.

The theoretical approach presented by Dr. Riska was well received. The insight gained by breaking down the interaction process into the constituent components provides a useful means of treating this problem. In the presentation there was discussion of some experiments on ice crushing which had been performed by the author. In particular, questions were raised regarding the "hot spot" theory of localized crushing. It was felt that this concept is extremely interesting and should be explored in detail in future experimental work.

The final paper of the session was a very straightforward account of the influence of the geometry of the structure in influencing the ice loads. The paper, authored by Mr. K. Croasdale and Dr. M. Metge of Canada gave insight into the design requirements for an offshore structure. It discussed the need to consider the requirements of ice loads, wave loads, foundation stability, etc. which the designer must consider in arriving at the final design. Emphasis was placed on the role of the structural shape in influencing the ice loads, ice clearing processes, etc. on offshore structures. Three structural shapes were used in the illustration: narrow Arctic structures, wide Arctic structures and structures in iceberg infested waters.

This paper was very enlightening since it pointed out several of the problems which must be faced in designing an offshore structure for ice conditions. There was some surprise at the wide range of load values which could be obtained for the same nominal situation if calculated using different theories. The large uncertainties in estimating the influence of ice adfreeze was discussed in considerable detail. There appears to be little information for guidance in this area.

Overall the session presented five very interesting and informative papers. The audience participation was very good and some useful discussion ensued following each of the papers. It was quite clear however that the tone and scope of the papers presented in this session were very different to those presented in earlier sessions. In those sessions there was considerable detail given on the state-of-the-art of constitutive models for ice, as well as the experimental measurement of microcracking in laboratory samples. This session centred about the actual problems involved in determining and understanding ice/structure interactions. There is still a large gap between the laboratory studies of material properties and their application to solving the problems of determining the ice loads on an offshore structure. The papers and dialogue in this session should help in bridging this gap.

Contribution to General Discussions

D.M. Cole, USA CRREL

Re Jim Dorris' statements on the use of post peak or "residual" strength in his plasticity-based ice force model: At one point, the discussion turned to using the post peak strength for the calculations since by the time a particular element of ice enters the failure zone, it has been subjected to a gradually increasing loading rate for some time. Some of my experimental work has addressed this point (Cole, 1983). In creep experiments on freshwater granular ice, the expected sequence of decelerating primary creep followed by an inflection point and accelerating tertiary creep, was drastically altered if the load was ramped on slowly. In the extreme case, when the specimen was past its failure strain (0.01) by the time the full creep stress was attained, no decelerating creep was observed: the strain rate increased continuously. Additionally, the creep rates at higher strains were greater (significantly greater for tests performed at -5°C) than for the quickly-loaded specimens. These results indicate that variations in the loading path can cause significant deviations from the material response determined in "standard" laboratory experiments. The implication with regard to the ice-structure interaction problem is that damage accumulated by a unit of material approaching a compressive failure zone can have a significant effect on its ultimate failure characteristics.

Cole, D.M. 1983. The effect of stress application rate on the creep behaviour of polycrystalline ice. J. Energy Technol. Resources, Vol.105, pp.454-459.

T.J.O. Sanderson, BP International

J.F. Dorris raised the question of load history during slow indentation as potentially being the key to causing pre-failure damage - with consequently lower "residual stresses" being exerted. There may be something in this idea for certain loading scenarios: if ice approaches a structure of diameter 100 m at a rate of 0.1 ms^{-1} then during conditions of continuous indentation a piece of virgin undeformed ice may be expected to pass through a highly stressed "bulb" of extent of the order about 100 m in front of the structure. This would take about 1000 s, or 17 minutes, and if the ice were to reach a stress of 5 MPa at contact, this corresponds to a loading-rate of less than 5kPa s^{-1}. Referring to Figure 6.14 of my book (Sanderson, 1988), it is seen that Sinha's delayed elastic strain criterion would predict first crack formation at less than 2 MPa. This damage may be an explanation for the low stresses observed during such interactions.

Sanderson, T.J.O. 1988. "Ice mechanics: The risks to offshore structures." Graham and Trotman, London, 253 p.

Andrew Palmer, Andrew Palmer and Associates

I would like to join in the discussion about size effects.

The term "size effect" is a reflection of the present limits of our understanding. When we find that a large piece of wood has a greater mass than a small piece of the same wood, we do not call that a size effect: we know that there is a material property, density, and that mass is density times volume.

In ice mechanics, people have come to use the term size effect as a description of the observed fact that forces on large structures are smaller than forces predicted from experiments on small structures, on the basis of the present state of knowledge. It's an ignorance effect rather than a size effect.

In five/ten years, we shall understand what is happening, and then we shall not need to speak of size effects.

S.A.F. Murrell, University College London

I would like to comment on three aspects of the fundamental mechanics and micromechanics of ice indentation that we have been considering:

1. According to the modern concept of friction initiated by Bowden and Tabor, surfaces are in contact at asperities which deform in

compression so as to support the normal load. In the contact areas they adhere, and the shear strength of the adhering asperities determines the frictional forces. This leads to an expression for the coefficient of friction:

$$\mu_f = \frac{\text{shear strength of asperities}}{\text{compressive strength of asperities}}.$$

This ratio is much smaller for brittle solids (such as silicate rocks on ice) than it is for ductile metals, where strength is determined by crystal plasticity. Because ice readily shows both brittle and crystal plastic behaviour (with complications due to melting caused by pressure, or by frictional heat not easily dissipated due to low thermal conductivity), one might expect complex frictional behaviour in this case. Low friction values may be caused by brittle behaviour, but also by strongly anisotropic plasticity. On the other hand high friction values might occur under some conditions (perhaps at low sliding rates).

2. The formation of finely comminuted ice under some indentation conditions is interesting but puzzling. The phenomena brings to mind cataclastic flow, which is referred to in my paper. We envisage the process of brittle deformation in compression to involve the widespread nucleation, with limited subsequent growth, of microcracks throughout the solid volume. At low confining pressures, a process eventually occurs in which cracks interact in planar surfaces to form shear fractures (or axial splitting fractures in uniaxial compression). This process must have an important statistical aspect, which might perhaps be analysed by percolation theory. However, at sufficiently high confining pressures there is a transition to cataclastic flow (see Murrell et al., this volume) in which crack interaction occurs throughout the whole solid volume, leading to total fragmentation. In either case, however, the processes of crack interaction lead to comminution, both at sub-grain, grain, or super-grain scale, as described in Fonseka et al. (1985). I suspect the stresses (in particular the hydrostatic component) are too low for general cataclastic flow to have occurred in the ice experiments which were described. However, it could be that kinematic constraints due to the indenter and the semi-infinite ice mass cause shear deformation to be confined to a relatively thin layer in which strong crack interaction leading to ice comminution occurs. This process would be analogous to the

building up of a layer of highly comminuted fault gouge between shear fracture surfaces in faulted rocks. This process might also be assisted perhaps by frictional melting under some conditions.

3. I have been wondering why the plane wedge or flat indenter geometries have not been more widely used in ice mechanics studies. For a plane wedge the force (per unit thickness) on the wedge $F = 2p(\mu + \tan\theta)H$ where p is penetration depth, μ is friction coefficient, θ is wedge half-angle, and H is the indentation pressure (or hardness). For a plane flat indenter $F = wH$, where w is the indenter width. However, for a cylindrical (or spherical) indenter F is a non-linear function of $p(F = 2[p + \sqrt{2Rp - p^2}]H$ for a cylinder).

   The simple geometries would facilitate, I believe, the study of the indentation hardness or pressure (H) and its relationship to other measures of strength and dependence on physical variables (temperature, strain-rate, etc.).

   The use of a spherical indenter brings in additional considerations concerning contact stresses (the Hertzian Problem), which must complicate the study of H.

Fonseka, G.M., Murrell, S.A.F. and Parsons, P. 1985. "Scanning electron microscope and acoustic emission studies of crack development in rocks." Int. J. Rock Mechs. Min. Sci & Geomech. Abstr., Vol. 22, pp. 272-285.

Ian Jordaan, Memorial University of Newfoundland

Re damage and details of cracks: (Gordon Cox's question). The main idea behind damage mechanics is to average over a sufficient volume of material so that the details are "averaged out".

MIX
Papier aus verantwortungsvollen Quellen
Paper from responsible sources
FSC® C105338

If you have any concerns about our products,
you can contact us on
ProductSafety@springernature.com

In case Publisher is established outside the EU,
the EU authorized representative is:
Springer Nature Customer Service Center GmbH
Europaplatz 3, 69115 Heidelberg, Germany

Printed by Libri Plureos GmbH
in Hamburg, Germany